Biomathematical and Biomechanical Modeling of the Circulatory and Ventilatory Systems

Volume 6

For further volumes:
http://www.springer.com/series/10155

Marc Thiriet

Anatomy and Physiology of the Circulatory and Ventilatory Systems

Springer

Marc Thiriet
Project-team INRIA-UPMC-CNRS REO
Laboratoire Jacques-Louis Lions, CNRS UMR 7598
Université Pierre et Marie Curie
Place Jussieu 4
75252 Paris Cedex 05
France

ISSN 2193-1682 ISSN 2193-1690 (electronic)
ISBN 978-1-4614-9468-3 ISBN 978-1-4614-9469-0 (eBook)
DOI 10.1007/978-1-4614-9469-0
Springer New York Heidelberg Dordrecht London

Library of Congress Control Number: 2013953870

Printed on acid-free paper

Springer is part of Springer Science+Business Media (www.springer.com)

Preface

> So it is that being and non-being give birth each to the other; that difficulty and ease each produce the idea of the other; that the ideas of height and lowness arise from the contrast of the one with the other; that the musical notes and tones become harmonious through the relation of one with another; and that being before and behind give the idea of one following the other.
> (Attributed to Lao Tzu: Tao te Ching, [The classic of the way of virtue])

Living organisms react to environmental changes and maintain the stability of the internal medium by regulation systems (*homeostasis* introduced by C. Bernard). In addition, humans adapt the homeostasis level (*heterostasis*) to the environment (e.g., sea level vs. mountain and desert vs. icy region).

Breathing and blood circulation are anatomically and functionally coupled to deliver oxygen to the body's tissues and remove carbon dioxide that results from cell metabolism (Fig. 0.1). Two apparatus—the respiratory and cardiovascular systems—operate in series, interacting at the level of the alveolocapillary barrier. These physiological apparatus are strongly controlled by locoregional mechanisms and the nervous system to match the body's ventilation to blood circulation in each pulmonary compartment. This physiological coupling ensures proper ventilation-to-perfusion ratios and the body's homeostasis.

After reviewing main events and their associated signal transduction processes that occur in cells and biological tissues (nano-, micro-, and mesoscopic scales) in normal conditions in Vols. 1 to 5, Volumes 6 and 7 targets the organ scale (macroscopic scale) in normal and pathological conditions. Because of the amount of accumulated data on the physiology and pathophysiology of the circulatory and ventilatory apparatus, the initially scheduled one-volume book (referred as Vol. 6 "Circulatory and Ventilatory Conduits in Normal and Pathological Conditions" in previous volumes of the series "Biomathematical and Biomechanical Modeling of the Circulatory and Ventilatory Systems") has been split into two. Volume 6 focuses on anatomy and physiology as well as medical images and signals of the cardiovascular and ventilatory apparatus.

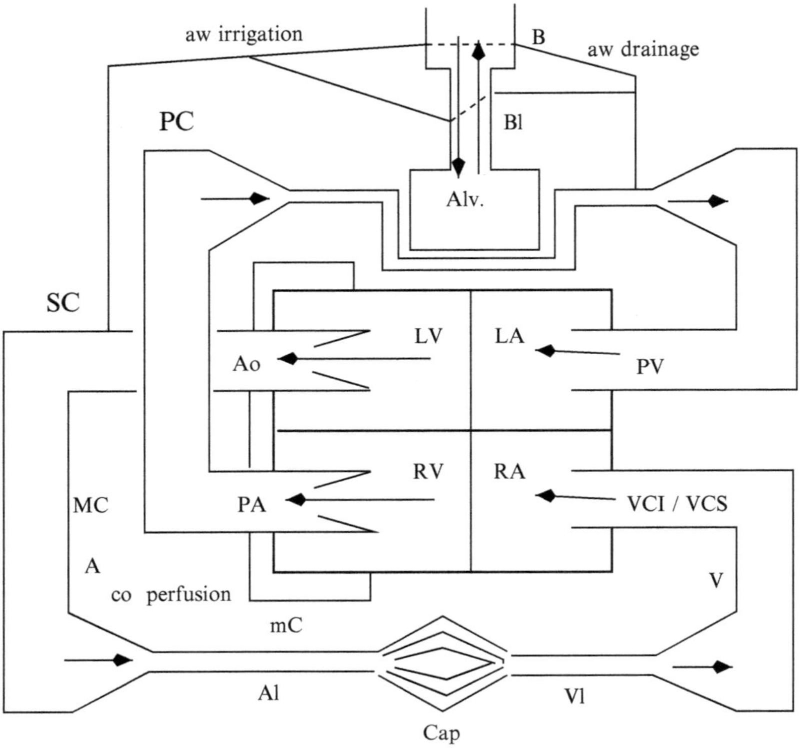

Fig. 0.1 Circulatory and ventilatory apparatus are coupled both anatomically and physiologically to ensure an optimal gas exchange (A: artery; Al: arteriole; Alv: alveolus; Ao: aorta; aw: airway; B: bronchus; Bl: bronchiole; Cap: capillary; co: coronary; I(S)VC: inferior (superior) vena cava; LA: left atrium; LV: left ventricle; MC: macrocirculation; mC: microcirculation; PA(V): pulmonary artery (vein); PC: pulmonary circulation; RA: right atrium; RV: right ventricle; SC: systemic circulation; V: vein; Vl: venule). The body's respiration is achieved by a back and forth (tidal) air motion (successive cycles of inhalation and exhalation) in the respiratory tract. The latter consists of a set of airways characterized by a single conduit outside the pulmonary parenchyma and a caudal compartment characterized by a high level of branchings inside the thoracic cage (tracheobronchial tree). Bronchi shorten and narrow progressively to generate bronchioles. In pulmonary acini, these pipes have walls endowed with alveoli and give rise to alveolar ducts, the walls of which are entirely constituted of alveoli. These air cavities are densely irrigated by pulmonary capillaries. Both physiological apparatus are coupled at the level of the alveolocapillary membrane and ensure an adequate ventilation–perfusion ratio in the pulmonary compartments. The body's ventilation is fulfilled using a circuit of airways open to the atmosphere by the thoracic pump, which is external to the intrathoracic tracheobronachial tree. On the other end, blood flows in a closed circuit, which includes 2 serial compartments, the systemic and pulmonary circulations, each endowed with its own intrinsic pump (left and right heart) that are apposed to work synchronously. Each blood circulatory circuit is composed of arteries (from heart, or efferents), arterioles, capillaries, venules, and veins (afferents ensuring blood drainage to the heart). The systemic circulation perfuses airways as well as the cardiac wall via coronary arteries. Therefore, lungs receive a double blood input (systemic and pulmonary)

The cardiovascular and respiratory systems, as any physiological system, are characterized by 4 main properties: (1) diversity, or namely the strong between-human subject variability in conduit network architecture; (2) variability associated with time-dependent adjustments in response to environmental constraints, any given description of the airway tract corresponding to a snapshot; (3) intricacy in shape and structure, as its curved deformable shape associated with a multilayered structure made of composite materials; and (4) organized complexity in behavior (self-organization, subdivision in many parts in intricate arrangement and correlated interaction, and nonlinear dynamics).[1] Its space- and time-dependent functioning is related to a huge number of interacting parameters that are difficult to measure. In any case, in the absence of complete knowledge of the system's behavior, the simple still needs to be solved before investigating the complex.

Air and blood flow behavior depend not only on ventilatory and cardiac pump quality and conduit architecture, but also on its structural constituents that are able to regulate locally the conduit lumen size, as they direct the broncho- and vasomotor tone. Moreover, both the rib cage and airways and both the heart and blood vessels are controlled by nervous and endocrine signals. In addition, cardiac output depends on blood circulation conditions—pre- and postload—and vice versa.

These complex physiological systems evolve between order and chaos, with sufficient order to auto-organize and freedom to adapt. They change their organization and possibly structure to adapt to environmental stimuli. They are constituted by many components that nonlinearly interact with feedback loops to contribute to the system bulk behavior. The bulk behavior and organization of these complex systems not only result from the simultaneous activities of their constituents, but also emerge from the sum of interactions between their constituents.

In chaos theory, chaos means that the final outcome of a process can be extremely sensitive upon initial conditions. Chaotic systems have the 3 main properties of sensitivity, mixing, and periodicity. Whereas the diseased heart has either a predictable constant or strong random frequency, the healthy heart is characterized by slight temporal variations of its frequency with a self-similar choatic pattern.

Fractals result from chaos theory. Fractals are associated with a fine structure determined by a recursive process, characterized by self-similarity at various length and/or time scales and fractal dimension. Fractals have been used to model, in particular, the structures of blood circulation and pulmonary airways.

---

[1]Organized complexity refers to nonrandom relationships between the elements of the system. Owing to the organized aspect of this type of complexity, the system "emerges", that is does not need external guidance. Complex adaptive systems are characterized by the following features: (1) the system has memory effects; (2) the system regulation includes feedback; (3) the system is influenced by and adapts itself to its environment; (4) the system is highly sensitive to initial conditions; and (5) the relations between the system components as well as between it and its environment are nonlinear. On the other hand, disorganized complexity results from the very large number of system components that mainly interact randomly such as a confined gas with gas molecules as its parts.

The microscopic unit, the cell, of these physiological apparatus is a complex system. Cells are characterized by self-organization and self-reproduction. Self-organization occurs when system components interact dynamically with each other to generate a system that acquires emergent properties. These properties cannot be directly predicted from the individual properties of their elements. Self-organization concerns dynamical cell shape and coherently associated functions that emerge from regulated molecular interactions within the cell, applied physical constraints, and collective behaviors of interacting cells.

The respiratory tract and vasculature constitute a pipe network equipped with a pump, which is intrinsic (heart) or extrinsic (musculoskeletal chest wall). The heart provides flow in the arterial bed via the internal upstream pressure. The thoracic cage operates via the intrathoracic pressure that is the external pressure applied on the intrathoracic respiratory tract to inflate and deflate air spaces and conduits, hence decreasing or increasing pressure with respect to atmospheric pressure in thoracic air space terminals, which is the end (alveolar) pressure, corresponding alternatively to the downstream and upstream pressure during inspiration and expiration, to generate flow.

Nothing is given. Everything is constructed.
(G. Bachelard) [1]

Modeling of any physiological apparatus can have multiple goals: (1) prediction; (2) development of pedagogical and medical tools; (3) computation of quantities inaccessible to measurements using inverse problems; (4) control; and (5) optimization.

Modeling entails several stages: (1) a definition stage during which the whole set of data related to the actual system is collected and the problem and its goals are identified (concrete level); (2) a representation stage of the system, in which the number of data is reduced (information filtering), as data relevant to the problem are extracted and kept to describe the model, a simplified version and falsification of the reality; (3) a validation stage that results from generality loss by comparisons with available observation data that yield a meaning to the model; and (4) possible model improvements. Two main varieties of models include representation and knowledge models. The former is associated with mathematical relations between input and output variables. The equation coefficients do not necessarily have physical or physiological meaning. The latter is aimed at analyzing the mechanisms that produce the explored phenomena.

Any model is developed in the framework of a theory, such as continuum mechanics, but the model is much more specific than the theory. The simplification degree of the model provides its application domain. Certain assumptions that are not strongly justified can be removed in iterative model refinement. Any model is characterized by (1) its potential to improve a system's knowledge as well as the underlying theory and to generate new concepts, and (2) its capacity to define new strategies.

Modeling of complex systems relies on a bottom-up approach, starting from the acquired basic knowledge of the system's parts. Models that integrate nano- and microscale processes target reacting adaptive systems. Furthermore, they can investigate responses of living tissues to administered substances, especially drugs delivered by nanotechnology-based methods or implanted medical devices. This type of modeling incorporates the dynamics of cell processes, that is spatial and temporal organization of biochemical reactions and major molecular interactions with positive and negative feedback that amplifies or limits the response, respectively.

Input data for integrative investigation of the complex dynamic cardiovascular and respiratory systems include knowledge accumulated at various length scales (multiscale modeling), from molecular biology (Vols. 3 and 4) to physiology (the present volume) and from histology (Vols. 1 and 2) to anatomy (also the present volume).

Tier architecture of living systems is characterized by its communication means and regulation procedures, thereby integrating environmental changes to adapt. Multiple molecules interact to create the adaptable activity of cells, tissues, organs, and the body. The huge quantity of these mediators forms a complex reaction set with feedback loops and a hierarchical organization. Studies from molecular cascades primed by chemical and physical stresses as well as mechanical loadings that exert at all scales, from organs and tissues to cells, need to be combined to study living systems with complex dynamics.

Any patient-specific model of a body's organ or apparatus integrates anatomy (shape, size, links to its neighborhood [boundary conditions]), histology (structure and constituents), physiology, and dynamics. In the case of the heart, a model combines anatomy, electrophysiology (generation, propagation, and disappearance of the command wave), structural dynamics (myocardial cycles of contraction and relaxation), and hemodynamics (blood storage and ejection). Models of the cardiac anatomy (two apposed pumps, each containing two thick-walled chambers communicating by valvular atrioventricular orifices, the ventricles expelling blood also through valvular ventriculoarterial orifices into the ascending aorta and pulmonary artery) are provided by three-dimensional reconstruction using data files acquired during three-dimensional imaging techniques.

An integrative model incorporates behavior at various time scales, including characteristic response times, from cardiac and breathing cycles (s) to diurnal periodicities (h), to efficiently describe the structure–function relations of explored physiological systems.

> All in the world know the beauty of the beautiful, and in doing so they have the idea of ugliness; they all know the good, and in doing so they have the idea of what is the bad. (Attributed to Lao Tzu: Tao te Ching [The classic of the way of virtue])

The present book includes 5 chapters that mainly refer to the macroscopic scale, although nano- and microscale pathophysiological mechanisms are described. Chapters 1 and 2 provide a brief introduction of the anatomy of the cardiovascular and ventilatory systems, respectively. Chapter 1 presents the anatomy of the cardiac pump and emphasizes peculiar segments of the vasculature. Chapter 2 is related to the architecture and structure of the respiratory tract as well as its vascularization and innervation. Chapters 3 and 4 deal with the physiology of blood circulation and ventilation. Chapter 5 focuses on some aspects of images and signals currently used to get the shape of the cardiovascular and ventilatory apparatus and to assess their functions. These signals and images can be used as input data and to build computational domains.

Acknowledgments

The author acknowledges the patience of his wife Anne, daughter Maud, sons Alrik and Damien, and their respective French (Julien, Jean, and Louis), American (Raphaëlle, Matthieu, and Alexandre), and Polish (Joanna and Frédéric) families.

Contents

1 Anatomy of the Cardiovascular Apparatus 1
1.1 Vasculature Architecture and Function 1
1.2 Systemic and Pulmonary Circulation 2
1.3 Fetal Blood Circulation 2
1.4 Stresses and Strains in the Vasculature 3
1.5 Space and Time Scales 4
1.6 Heart 5
1.6.1 Atria 7
1.6.2 Ventricles 11
1.6.3 Aortic Valve 12
1.6.4 Internal Anatomical Features of the Heart 16
1.6.5 Heart Vascularization 18
1.6.6 Heart Innervation 18
1.6.7 Fractal Geometry 25
1.6.8 Cardiogenesis 29
1.6.9 Early Stages of Cardiogenesis and Chemical Control 29
1.6.10 Cardiac Progenitor Cells and Repair 33
1.7 Vasculature 35
1.7.1 Systemic and Pulmonary Circulation 35
1.7.2 Blood Circuit 36
1.7.3 Blood Compartments 36
1.7.4 Local Geometry and Blood Flow 37
1.7.5 Particular Structures 39
1.7.6 Microcirculation and Macrocirculation 40
1.7.7 Adaptation 42
1.7.8 Fractal Vasculature 42
1.7.9 Circulatory Networks 45
1.8 Innervation of the Cardiovascular Apparatus 58
1.8.1 Parasympathetic and Sympathetic Control 58
1.8.2 Role of Reactive Oxygen Species 63
1.8.3 Inputs from Vascular Sensors 64

2 Anatomy of the Ventilatory Apparatus 73
2.1 Upper Airways 75
2.1.1 Nose 75
2.1.2 Mouth 83
2.1.3 Pharynx 83
2.1.4 Larynx 84
2.2 Thoracic Respiratory System 87
2.2.1 Chest Wall 87
2.2.2 Mediastinal Airways 91
2.2.3 Lungs 94
2.3 Tracheobronchial Tree 101
2.3.1 Bronchi 104
2.3.2 Bronchioles 108
2.3.3 Fractal Modeling 109
2.3.4 Bronchial Walls 113
2.3.5 Bronchiolar Walls 114
2.3.6 Pulmonary Alveolus 115
2.3.7 Between-Airway Communications 116
2.4 Development of Intrathoracic Airways and Airspaces 116
2.4.1 Developmental Phases 118
2.4.2 Development at the Macroscopic Scale 119
2.4.3 Development at the Microscopic Scale 120
2.4.4 Influence Factors 122
2.4.5 Computational Model of Early Intrathoracic Airway Morphogenesis 127
2.5 Pulmonary Vascularization 131
2.5.1 Communications Between Pulmonary and Bronchial Circulation 131
2.5.2 Development of Intrapulmonary Blood Vessels 132
2.5.3 Pulmonary Circulation 133
2.5.4 Pulmonary Arterial Circulation 134
2.6 Innervation of Ventilatory Organs 136
2.6.1 Neurons 140
2.6.2 Autonomic Nervous System 141
2.6.3 Parasympathetic and Sympathetic Control 141
2.6.4 Evolution of Innervation of the Ventilatory Apparatus 149
2.6.5 Nerve Subpopulations 150
2.6.6 Mechanosensors – Baroreceptors and Voloreceptors 151
2.6.7 Baroreflex 153
2.6.8 Chemosensors 153
2.6.9 Arterial Chemoreflex 154

3 Cardiovascular Physiology 157
3.1 Heart 159
3.1.1 Left Ventricle Cycle 161
3.1.2 Stroke Volume 165

3.1.3 Pressure–Volume Curve 167
3.1.4 Interferences Between Cardiac Cells in the Cardiac Function 171
3.1.5 Cardiomyocyte Organelles and Coupled Metabolism, Energetics, Signaling, and Contraction 172
3.1.6 Influence of Post-Translational Modifications of Regulators 185
3.1.7 Energetics 186
3.1.8 Ketone Bodies 219
3.1.9 Matching of ATP Supply to Demand 225
3.1.10 Myocardial Oxygen Consumption 226
3.1.11 Cardiac Electrophysiology 228
3.1.12 Calcium Handling During Exercise 244
3.2 Large Blood Vessels 245
3.2.1 Arterial Circulation 245
3.2.2 Venous Return 247
3.2.3 Metabolic Blood Flow Regulation 249
3.2.4 Coronary Circulation 250
3.2.5 Cerebral Circulation 255
3.2.6 Pulmonary Circulation 258
3.3 Microcirculation 259
3.3.1 Capillary Flow 260
3.3.2 Capillary Recruitment 261
3.3.3 Transcapillary Transfer of Materials 261
3.4 Rhythmicity 286
3.5 Convective Heat Transfer 289
3.6 Regulation of the Circulation 291
3.6.1 Autoregulation in the Arterial Bed 293
3.6.2 Neural Effects on Heart and Blood Vessels 312
3.6.3 Adrenergic and Cholinergic Receptors 329
3.6.4 Circulation Sensors 331
3.6.5 Short-Term Control of the Circulation 332
3.6.6 Response to Stress 333
3.6.7 Delayed Control of the Circulation 340

4 Physiology of Ventilation 353
4.1 Main Features and Input Data of the Body's Respiration 353
4.1.1 Oxygen Input and Cell Metabolism 353
4.1.2 Gas Mixture 355
4.1.3 Gas Transfer 357
4.1.4 Inhaled Air 360
4.1.5 Breathing Modes 366
4.1.6 Functions of the Body's Respiration 367
4.1.7 Body's Oxygen and Carbon Dioxide Content 368
4.1.8 Respiratory Epitheliocytes 369

4.1.9 Respiratory Cilia ... 370
4.1.10 Deformation of the Lung Parenchyma ... 371
4.2 Lung Volumes ... 371
4.2.1 Pressure–Volume–Temperature Relations of an Ideal Gas ... 372
4.2.2 Measurement of Pulmonary Volumes and Respiratory Flow Rates ... 375
4.2.3 Lung Volumes Measured by Spirometry ... 376
4.2.4 Dead Spaces ... 378
4.3 Mode of Ventilation ... 379
4.4 Mechanics of Ventilation ... 380
4.4.1 Monoalveolar Model ... 381
4.4.2 Quiet Breathing ... 383
4.4.3 Effort Breathing ... 386
4.4.4 Elastic Properties of the Lung and Chest Wall ... 387
4.4.5 Ventilatory Resistances ... 392
4.4.6 Ventilatory Work ... 392
4.5 Ventilation–Perfusion Ratio ... 393
4.6 Gas Transport ... 397
4.6.1 Inhaled Air Distribution ... 397
4.6.2 Gas Content in the Serial Compartments ... 398
4.6.3 Alveolar Ventilation ... 398
4.6.4 Gas Transport and Mixing ... 400
4.6.5 Alveolocapillary Transfer ... 401
4.6.6 Pulmonary Gas Transfer Capacity ... 403
4.6.7 Pulmonary Circulation ... 405
4.6.8 Blood O_2 Affinity Curve ... 407
4.6.9 CO_2 Blood Transport ... 410
4.7 Maintenance of pH Homeostasis—Hydrogen Ion Control ... 412
4.8 Control of the Respiration ... 414
4.8.1 Respiration Nervous Centers ... 417
4.8.2 Respiratory Clock, a Synchronous Bilateral Command ... 419
4.8.3 Respiratory Nerves ... 424
4.8.4 Ventilation Receptors ... 427
4.9 Oxygen Sensing ... 428
4.10 Immune Defense of the Ventilatory Apparatus ... 429
4.10.1 Organization of the Immune System of the Ventilatory Apparatus ... 431
4.11 Metabolic Functions of the Lung ... 436
4.11.1 Detoxificafion Enzymes ... 436
4.11.2 Circulating Substance Fate ... 438
4.11.3 Respiratory Cell Production ... 439

5 Medical Images and Physiological Signals 441
5.1 Imaging and Geometry Modeling 441
5.1.1 Imaging Techniques 442
5.1.2 3D Reconstruction 456
5.1.3 Meshing 460
5.2 Hemodynamics Signals 462
5.2.1 Volume and Pressure 462
5.2.2 Flow Rate and Velocity 466
5.3 Measurement of Heart Electric and Magnetic Properties 469
5.3.1 Electrocardiogram 470
5.3.2 Vectorelectrocardiography 474
5.3.3 Magnetocardiography 474
5.3.4 Impedance Plethysmography 474
5.4 Upper Airway Patency and Resistance 475
5.5 Lung Function Testing 476
5.5.1 Breathing Mechanics—Main Measuring Devices and Procedures 476
5.5.2 Breathing Function—Blood Gas Measurements 481
5.5.3 Carbon Monoxide Transfer Capacity 481
5.5.4 Clearance by Respiratory Epithelium 482
5.5.5 Respiratory Indices 483
5.5.6 Blood Gas 485

Conclusion 487

References 491

Notation Rules: Aliases and Symbols 525

List of Currently Used Prefixes and Suffixes 533

List of Aliases and Primary Symbols 537

Complementary Lists of Notations 567

Index 575

Chapter 1
Anatomy of the Cardiovascular Apparatus

The cardiovascular apparatus is a transport circuit that carries blood cells, oxygen and nutrients to cells of the body's organs, and wastes produced by working cells to their final destinations (mainly lungs, liver, and kidneys). It is involved in both mass and heat transfer.

1.1 Vasculature Architecture and Function

The blood circulation aims at perfusing and draining the body's organs and their constituting tissues. The blood circulation network is composed of serial compartments according to the blood vessel size and blood flow dynamics: (1) arteries (large, medium sized, and small); (2) arterioles (large and small); (3) capillaries; (4) venules (small and large); and (5) veins (small, medium sized, and large; Vol. 5 – Chap. 7. Vessel Wall).

The blood vessels are able to respond to local changes in stress exerted by the flowing blood on their walls. The arterial bed is especially characterized by its flow-dependent, myogenic, and metabolic[1] responses to limit cardiac afterload and adapt blood flow to tissue demand.

Tissue perfusion depends on: (1) state of irrigating arteries and draining veins; (2) flow input with its governing parameters (Vol. 8 – Chaps. 1. Hemodynamics and 2. Air Transport); (3) blood and vessel wall rheology (Vol. 8 – Chap. 4. Rheology), and (4) irrigation control (Chap. 3).

[1]Metabolic flow regulation is associated with local metabolite release.

M. Thiriet, *Anatomy and Physiology of the Circulatory and Ventilatory Systems*, BBMCVS 6, DOI 10.1007/978-1-4614-9469-0_1,

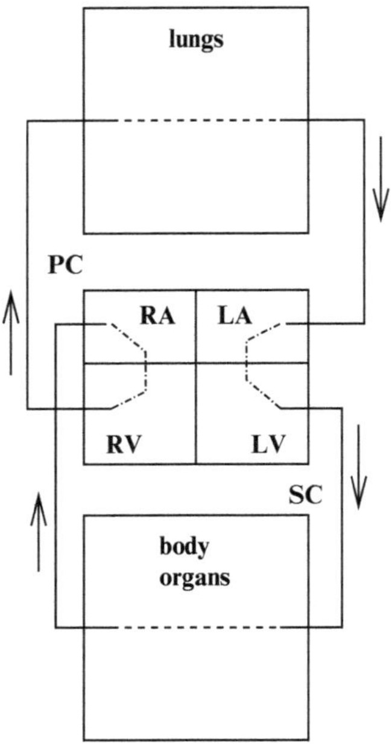

Fig. 1.1 Pulmonary (PC) and systemic (SC) circulation. The blood is conveyed from the drainage veins of the systemic circulation to the right atrium (RA). It is then convected into the right ventricle (RV) and expelled into the arteries of the pulmonary circulation for oxygenation into the lungs. The oxygenated blood is sent to the left atrium (LA) via the pulmonary veins. It then enters into the left ventricle (LV) to be propelled into the arteries of the systemic circulation, which distribute blood to every body organ (including heart and lungs) for energy and nutrient supply and waste removal

1.2 Systemic and Pulmonary Circulation

The cardiovascular apparatus is mainly composed of the cardiac pump and vasculature that constitute 2 circulatory networks: *systemic* and *pulmonary* circulation.

The heart is made of 2 synchronized apposed pumps in series, each being composed of 2 chambers. The left heart propels blood through the systemic circulation, the right heart through the pulmonary circulation (Fig. 1.1).

1.3 Fetal Blood Circulation

The heart is the first organ to develop. In the human embryo, the heart begins beating at about 21 days after conception.

The fetal circulatory system relies on a different anatomical network, because the fetus receives oxygen from the placenta rather than lungs. Before birth, the ductus arteriosus allows blood to bypass the nonfunctional fetal lungs, as it connects the pulmonary artery that supplies blood to lungs to the aorta that supplies blood to the rest of the body. The ductus arteriosus normally closes rapidly after birth. Platelets are recruited to the ductus arteriosus lumen within 20 mn after birth in mice to promote its closure [1].

The transition from fetal to neonatal circulatory circuit results from the suppression of the ductus arteriosus. The functional closure of the ductus arteriosus that occurs within minutes by vasoconstriction is followed by a structural closure that happens within days by a fibrogenic, proliferative mechanism.

Ductal smooth myocytes sense oxygen and contract, as oxygen increases mitochondrial-derived heme oxygenase that raises cytosolic calcium and activates RoCK kinase as well as provokes ductal smooth myocyte proliferation and mitochondrial fission via a cyclin-dependent kinase-mediated phosphorylation of dynamin-1-like protein Dnm1L (Ser616) [2].[2] Mitochondrial fission activates pyruvate dehydrogenase and increases mitochondrial production of heme oxygenase and subsequently ^{ETC}complex-I activity, thereby raising oxidative metabolism. Catalase overexpression stops cytosolic calcium influx and increased heme oxygenase action.

1.4 Stresses and Strains in the Vasculature

The heart and blood vessels are exposed to mechanical stresses and strains during their development as well as in their mature states.

Mechanical stress is a force per unit surface area. It can be decomposed into tangential (*shear*) and normal (*pressure* and *stretch*) components. The stress components depend on the selected coordinate system. Mechanical stresses can be computed using a local reference frame due to the complicated geometry (*Freynet coordinates*).

In distensible arteries, the endoluminal pressure generates parietal tensions with 2 major components, *axial* and *circumferential* (or *azimuthal*) *tension*. During inflation, as the structure is relatively uniform in the azimuthal direction and tissular surrounding is supposed to be homogeneous, the wall deforms homogeneously in the radial direction. Wall elasticity cannot be assumed to be uniform in the axial and radial directions. As the wall material is incompressible, the circumferential stretch is associated with a radial thinning.

Strains relate the deformed to the unstressed reference configuration. Strains result from change in length (*normal strain*) and angle (*shear strain*). Due to a nonlinear rheological behavior, the stiffness of the vascular wall increases progressively with the strain.

Stresses and strains are investigated at a given length scale yielded by the typical size of the mesh used for computational fluid (blood flow inside the deformable vascular lumen) and solid (viscoelastic vascular wall) dynamics. The material *elastic modulus* is the slope of the *stress–strain relation*.

In laminar regime, flow velocity distribution in a given vessel cross section display a given spatial velocity gradient, hence a *shear rate*, at a given radial position and phase of the cardiac cycle. Any viscous blood particle bear shear, but, as it is quasi-uniformly pressurized, it is neither stretched nor compressed. Moreover, although it experiences a *shearing torque* applied by apposed blood particles in the wall and flow-core direction that endow it with vorticity, it does not rotate because of the uniform pressure on all its faces.

---

[2] A.k.a. dynamin-related protein DRP1, dynamin-like protein-4, and dynamin family member proline-rich C-terminal domain less protein Dymple.

The velocity at the moving wetted surface of the vascular wall equals zero (no-slip conditions). The blood particle adjacent to the wall undergoes a shear by the apposed inward blood particle that flows more rapidly. A strong spatial velocity gradient is then created within the near-wall layer, the so-called *boundary layer*. The *wall shear stress* is the product of the wall shear rate by the fluid kinematic viscosity. This wall shear stress is sensed by components of the luminal surface of endothelial cells.

Hemodynamic stresses are transmitted to the intramural medium. Cells partly or entirely immersed in the vascular wall bear axial and azimuthal tensions that derive from blood pressure. Once stress components are sensed, cells adapts to the local stress field.

1.5 Space and Time Scales

Blood circulation is characterized by several space scales from mural cells of the cardiac and vascular walls to composite materials of wall layers and the continuous wall from the cardiac pump to capillaries and from these tiny vessels to the heart (blood return).

It also encompasses several time scales from the cardiac and circadian cycle to the entire set of time constants involved in the local regulation of blood flow rate (mechanotransduction from $\mathcal{O}[1\,\text{ms}]$ to $\mathcal{O}[1\,\text{h}]$; Vol. 5: *Tissue Functioning and Remodeling in the Circulatory and Ventilatory Systems*) and in the fast nervous and slower endocrine control, in addition to long-term adaptative and maladaptative remodeling (Table 1.1).

Table 1.1 Blood circulation length and time scales

Vasculature geometry	
0.1 μm	Endothelium cleft width
1 μm	Averaged capillary wall thickness
10 μm	Capillary lumen bore
2–25 μm	Blood cell size
10–80 μm	Endothelial cell size
1 mm	Large artery wall thickness
3–5 mm	Lumen bore of large artery and vein
1 cm	Ventricle wall thickness
1–3 cm	Lumen bore of aorta and vena cava
2–6 cm	Width of heart chambers
1–2 m	Body height
Blood circulation-related activities	
1 s	Cardiac cycle
10–100 s	Control
mn–h	Adaptation
d–wk	Remodeling

1.6 Heart

The heart is located within the *mediastinum*,[3] usually behind and slightly to the left of the sternum [3, 4]. About two-thirds of the heart is left of the midline, with its long axis oriented from the left hypochondrium to the right shoulder in the usual situation, with minor changes during respiration. The right border is given by the right atrium, the inferior border mostly by the right ventricle, and the lower and upper parts of the left border by the left ventricle and atrium, respectively. The cardiac position varies between subjects with possible mirror-image configuration.

The base of the heart is formed by vessels and atria and its apex by ventricles. The heart rests on the diaphragm. The heart has 4 cavities: upper left (LA) and right (RA) *atria*, and lower left (LV) and right (RV) *ventricles*. The left ventricle is located posteriorly and leftward from the right ventricle.

The heart chamber size varies during the ventricular cycle due to myocardium activity (Table 1.2). The left ventricle is the largest chamber with the thickest wall (Table 1.3).[4] The septum separates the left and right hearts.

The *pericardium* surrounds the heart and the roots of great blood vessels. It is attached by ligaments to the spine column, diaphragm, and other organs. The pericardium restricts excessive heart dilation, and thus limits ventricular filling.

Table 1.2 Echographic measurements of cardiac kinetics. Estimated width (w) range and mean (mm). The volume of the ventricle, assumed to have the configuration of an ellipsoid, is evaluated by $V = \pi/6\, Lw^2$ (L: base-to-apex length of the cavity). The shortening fraction (%) is equal to the ratio of the difference between the end-diastolic width and the end-systolic one to the end-diastolic width

	Range	Mean
LV width (diastole)	35–55	45
LV posterior thickness (diastole)	5–10	10
Interventricular septum thickness (diastole)	5–10	10
LA width	20–40	30
LV shortening fraction	35–45	36

Table 1.3 Estimates of heart wall thickness (mm)

Cavity	Wall thickness (mm)
RA, LA	1–3
RV	3–5
LV	10–15

[3]The mediastinum is the chest space between the sternum (front), spine column (rear), and lungs (sides). It contains the heart and its afferent and efferent vessels, thymus, trachea and main bronchi, esophagus, thoracic nerves and plexus, and lymph vessels and nodes.

[4]The right ventricle pushes blood into low-pressure pulmonary circulation; therefore, it has a moderately thick muscle layer. The left ventricle expels blood into high-pressure systemic circulation and thus has the thickest myocardium.

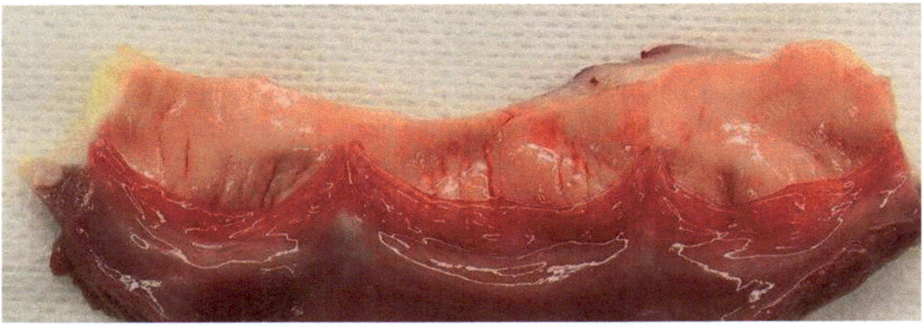

Fig. 1.2 Isolated pulmonary orifice axially cut at one commissure line to display the 3 retracted cusps with their curved insertion lines and the commissures. The ventriculoarterial valves are not connected to the cardiac wall by chordae tendineae

Epicardial adipose tissue can be observed on human hearts during surgery (e.g., grafting) or by imaging procedures (CT, MRI, and echocardiography) [5]. In the adult heart, adipose tissue lodges in the atrioventricular and interventricular grooves. Epicardial adipose tissue can increase to cover the entire epicardial surface.

Four valves at the exit of each heart cavity, between the atria and the ventricles, the *atrioventricular valves* (AVV), and between the ventricles and efferent arteries, the *ventriculoarterial valves* (VAV, Fig. 1.2), regulate blood flow through the heart and allow bulk unidirectional motion through the closed vascular circuit.[5]

The atrioventricular valves are inserted at the atrioventricular junctions, whereas the ventriculoarterial valves are hinged from semilunar insertions (without any ring attachment) at sinusal junctions.[6]

The *tricuspid valve* (TrV) regulates blood flow between the right atrium and the right ventricle. It is composed of antero-superior, postero-inferior, and septal leaflets. The tricuspid annulusis elliptical. The anterior leaflet is usually the largest. Three papillary muscle groups beneath each of the 3 commissures support the tricuspid valve leaflets.

The 3 tricuspid valve leaflets can be visualized simultaneously by real-time three-dimensional echocardiography with a parasternal (less degree of distortion), apical, or subcostal mode, but, unlike the aortic and mitral valves, only 2 leaflets can be imaged using transthoracic or transesophageal two-dimensional echocardiography [6].

---

[5]On a frontal chest radiograph, the pulmonary valve is the upper one, above the mitral valve. The aortic valve is positioned at the southwest corner of the mitral valve, above the tricuspid valve. The right-heart valves, widely separated from each other on the roof of the right ventricle, are located anteriorly from the adjacent left-heart valves on the roof of the left ventricle.

[6]Semilunar hingelines extend from the sinotubular junction to a virtual ring joining the basal valvar insertion, thus crossing the junction between the ventricular infundibulum and the arterial wall.

Table 1.4 Estimates of valvar orifice size (mm). Datum variability (Sources: [7,8])

Orifice	Perimeter	Caliber	Thickness
Tricuspid	110–130		
Mitral	90–110	20–25	
Pulmonary	75–85		0.4 ± 0.1
Aortic	70–80	32	0.6 ± 0.2
Coronary sinus		2–7	
Inferior vena cava		2–8	

The *pulmonary valve* (PuV) controls blood flow from the right ventricle into the pulmonary arteries, which carry blood to the lungs to pick up oxygen. The pulmonary valve is located at the end of the pulmonary infundibulum of the right ventricle.

The *mitral valve* (MiV) lets oxygen-rich blood from pulmonary veins pass from the left atrium into the left ventricle. It consists of 2 soft thin cups attached to the atrioventricular fibrous ring (oblique position): a large anterior (aortic) and a small posterior (mural).[7]

The *aortic valve* (AoV) guards the left ventricle exit; once opened, blood crosses it from the left ventricle into the aorta, where it is delivered to the body. Like the pulmonary valve, it consists of 3 semilunar cusps. Immediately downstream from the aortic orifice, the wall of the aorta root bulges to form the *Valsalva sinuses*. Orifice sizes are given in Table 1.4.

Trabeculae carnae, muscular columns, are observed in both ventricles, especially the left ventricle. Ventricular muscular pillars, *Papillary muscles*, protrude into both ventricular lumina and point toward the atrioventricular valves. They are connected to *chordae tendineae*, narrow tendinous cords that are attached to the leaflets of the respective atrioventricular valves (Fig. 1.3).

1.6.1 Atria

There is a remarkable remnant of primitive fibres persisting at the sino-auricular junction in all the mammalian hearts examined. These fibres are in close connection with the vagus and sympathetic nerves, and have a special arterial supply; in them the dominating rhythm of the heart is believed to normally arise.

(Keith A. and Flack M. [9])

[7]The large mural leaflet closes about two-thirds of the valvar orifice.

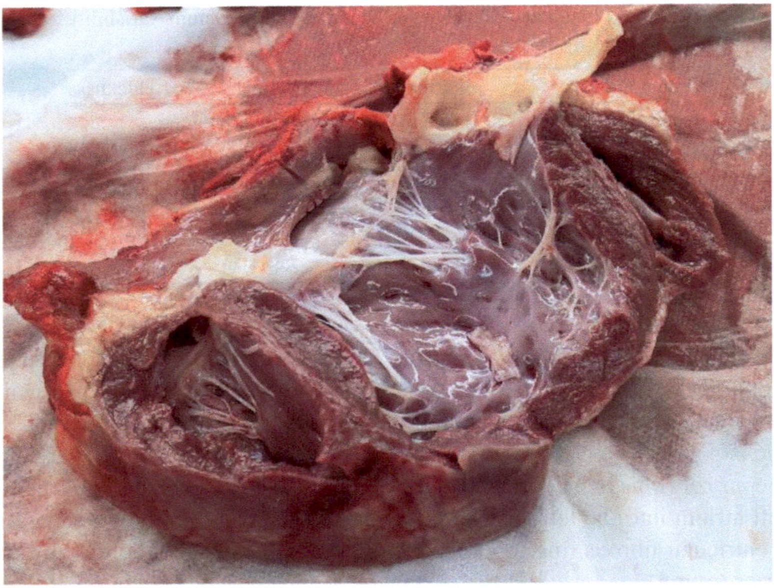

Fig. 1.3 Left (central open cavity) and right (open cavity with 2 parts located at bottom left and top right (small) photograph corners) ventricles of a calf. Papillary muscles and chordae tendineae, associated with the atrioventricular valves (with a focus here on the mitral valve), are larger in the left than in the right ventricle. The apparatus composed of papillary muscles and chordae tendineae allows the atrioventricular valves to remain closed during the systole, the sheet made by contact leaflets taking a parachute-like shape, hence avoiding leakage into the atrium. In the upper right region, the entrance segment of the ascending thoracic aorta is displayed with removed aortic valve and the 2 coronary ostia in their respective Valsalva sinuses. These sinuses are separated by the valvar commissure

The atria are composed of: (1) a body, quasi-absent in the adult right atrium, although clearly observed in the left atrium; (2) a venous chamber, with vein endings;[8] (3) a vestibule, leading to the atrioventricular valve; and (4) a muscular region, which consists of the atrial appendage and the pectinate muscle [10].[9]

[8]The venous chamber, or sinus venous, is the connection region of the terminal veins to the atria. Superior and inferior vena cavae and the coronary sinus are incorporated by the right atrium in the sinus venosus (venous chamber or systemic venous sinus of the right atrium), whereas the 4 pulmonary veins drain into the 4 corners of the left atrium posterior dome. These terminal veins can connect abnormally to the atria.

[9]The left atrial appendage constitutes a diverticulum containing pectinated muscles. The right atrial appendage, with its pectinated muscles, is interposed between the smooth-walled venous chamber and the vestibule. In the right atrium, the junction between the appendage and the venous chamber is internally defined by the prominent terminal crest and associated groove, the sulcus terminalis, on the external right atrium surface. The crista terminalis, a myocardium ridge, is formed from the superior part of the right venous valve. The sinoatrial node resides near the crista terminalis.

The atrial septum is composed of the septum secundum,[10] the inferior edge of which is the limbus and fossa ovalis on the right atrium side, and the septum primum, which is the flap valve of the oval foramen on the left atrium side.

The atrioventricular septum is connected to the atrioventricular valves. In the right atrium, the *Eustachian valve* (ridge), or valve of inferior vena cava (IVC), is an endocardial crescentic fold situated in front of the IVC orifice.[11] The *coronary valve* (ridge) is a semicircular fold that protects the orifice of the coronary sinus.

1.6.1.1 Right Atrium

The right atrium receives deoxygenated blood from the body's organs including the heart. The superior and inferior vena cava drain systemic venous blood from the body's organs into the posterior wall of the right atrium. The coronary sinus (CS) returns the coronary venous blood.

The coronary sinus is situated in the posterior grove at a level between the left atrium and left ventricle. The ostium of the coronary sinus is limited by the *Thebesian ridge*. The primary veins that empty into the coronary sinus comprise the *vein of Marshall* (VOM), which runs along the lateral wall of the left atrium, *great cardiac vein* (GCV; or middle cardiac vein), and *posterolateral vein* (PLV; or posterior left ventricular vein), the 2 latter being endowed with a valve at their confluence with the coronary sinus. The complete coronary sinus drainage network (CS, PLV, GCV, and VOM) can be visualized by venography after successful cannulation of the valve of Vieussens of the coronary sinus inlet. The *Vieussens valve* localizes between the coronary sinus and the great cardiac vein. This flimsy valve is composed of 1 to 3 leaflets.

The right atrium contains the natural cardiac pacemaker, the sinoatrial node. The latter localizes to the posterolateral wall near the orifice of the superior vena cava. Intra- and interatrial preferential conduction paths traverse the atrium anteriorly, superiorly, and posteriorly.

Among its anatomical structures, the *terminal crest*, also called crista terminalis, is a vertical ridge that extends from the lower posterior lateral wall to the region just below the sinoatrial node. It separates the trabeculated muscular and smooth right atrial walls.

Numerous pectinate muscles run from the anterior part of the terminal crest to the right free wall. The *right auricle*, or atrial appendage, is a pouch-like extension of the wall that consists of pectinate muscles. The right auricle is separated from the right atrium by the *terminal sulcus*, or sulcus terminalis, and, internally, by the terminal crest.

The flap valve of the fossa ovalis, a remnant of the oval foramen and its inferior rim are septal structures, whereas the other rims are infoldings enclosing fat [11].

[10]The septum secundum is an infolding of the atrial wall. The actual atrial wall is the flap valve of the foramen ovale.

[11]In the fetus, this ridge directs the blood from the IVC to the LA through the foramen ovale. In the adult, it occasionally persists.

The atrial septum contains the *foramen ovale*, or fossa ovalis, that corresponds to the opening between the right and left atrium during the intrauterine life and closes at the moment of birth. The *limbic band* is a ridge that surrounds the foramen ovale.

The *Eustachian ridge*, like the Thebesian ridge, runs along the anterior border of the orifice of the inferior vena cava and posteriorly toward the coronay sinus ostium, where it may connect with the Thebesian ridge.

The atrioventricular tricuspid valve is inserted to the annulus fibrosus. The ventricular side of its 3 leaflets is attached to the chordae tendineae.

The *triangle of Koch* is a region of the lower right atrium defined by: (1) the ostium of the coronary sinus, posteriorly; (2) the septal leaflet of the tricuspid valve annulus; and (3) the tendon of Todaro, a tendinous structure connecting the Eustachian ridge of the inferior vena cava ostium to the central fibrous body. The inferior right atrial isthmus between the Eustachian valve and the tricuspid annulus inferior to the coronary sinus constituted of interlacing muscular trabeculae is a zone of slow conduction. The atrioventricular node localizes to the center of the Koch's triangle.

1.6.1.2 Left Atrium

The left atrium is oriented leftward and posteriorly with respect to the right atrium. Its walls can be described as superior, inferoposterior, septal, left lateral, and anterior.

Its structure is characterized by a pulmonary venous territory situated posteriorly and superiorly with the orifices of 4 pulmonary veins at each corner, a lateral finger-like appendage with pectinate muscles, an inferior vestible that surrounds the mitral valve orifice, and a prominent body that shares the septum with the right atrium.

The walls are composed of overlapping layers of differently oriented myofibers. In addition to oblique myofibers, circular myofibers are more or less parallel to the atrioventricular valve plane and longitudinal myofibers run nearly perpendicularly. The wall thickness is varies markedly according to the atrial region.

Electrochemical impulse propagates through interatrial connections in the subepicardium. The Bachmann's bundle, the most prominent connection, runs anteriorly across the interatrial grove. Other connections include the coronary sinus musculature and posterior conducting bridges. Atrial activation arises from the lateral right atrial wall through the interatrial septum to the inferior, anterior, and lateral left atrial walls during sinus rhythm.

The left atrium ensures the transition from a continuous flow through the pulmonary veins to the intermittent filling of the left ventricle. The left atrium has 3 functions: (1) reservoir, storing pulmonary venous return during left ventricular systole and isovolumic relaxation (after the closure and before the opening of the mitral valve); (2) conduit for blood flow from pulmonary veins to the left ventricle in early diastole driven by a pressure gradient when the atrioventricular valve is open (passive ventricular filling); and (3) pump to increase left ventricle filling in late diastole when time is available (active ventricular filling), thereby determining the left ventricular end-diastolic volume.

The atrial contraction can be assessed by pulsed-wave tissue Doppler echocardiography. The contraction of the lateral wall can be delayed with respect to that of the anterior and inferior walls.

In each phase of the cardiac cycle, the left atrium and ventricle are closely related. In normal subjects, the reservoir, passive conduit, and pumping phases represent about 40, 35, and 25% of the atrial contribution to stroke volume, respectively.

Left atrial flow contributes to about 26% of the normal left ventricle filling volume [12]. When the left ventricle relaxation is altered, this contribution can rise to about 38% of the stroke volume. In restrictive left ventricle disease, the filling pressure progressively increasing, this contribution can decay to about 19%. The maximal left atrial volume-to-body surface area ratio and left atrial emptying fraction can serve as indices. The former increases and the latter decreases in hypertension and are correlated to left ventricle end-diastolic volume and ejection fraction as well as natriuretic peptide levels [13].

1.6.2 Ventricles

The right ventricle forms the larger part of the sternocostal surface and a small part of the diaphragmatic surface of the heart. Its upper left angle forms a conical pouch (conus arteriosus), which gives rise to the pulmonary artery. The anatomy of the right ventricle differs from that of the left ventricle. It has a more complicated shape; it is more trabeculated; it wraps the left ventricle. Its wall is thinner (thickness 3–5 mm).

The left ventricle is longer than the right. It forms a small part of the sternocostal surface and the larger part of the diaphragmatic surface of the heart. It corresponds to the apex of the heart.

Both ventricles have 3 types of muscular columns: (1) some are attached along their entire length on one side; (2) certain of them are fixed at their extremities, but free in the middle; (3) others—papillary muscles—are anchored by their bases to the ventricle wall and their apices serve as insertions for chordae tendineae.

Each ventricle possesses 2 papillary muscles connected to the anterior and posterior walls of the ventricle. In the right ventricle, chordae tendineae connect either the anterior papillary muscle to the anterior and posterior cusps or the posterior papillary muscle to the posterior and medial cusps of the tricupid valve. In the left ventricle, chordae tendineae from each papillary muscle are connected to both cusps of the bicuspid mitral valve.

The structural organization of the ventricular wall is characterized by a three-layered structure with a three-dimensional myofiber arrangement. The left ventricular wall contains epicardial myocytes oriented obliquely, midmyocardial myocytes more circumferentially directed (hence the circumferential shortening and radial thickening during contraction), and oblique endocardial myocytes. In the right ventricle, epicardial myofibers are oriented obliquely, the midwall circumferential layer is poorly developed, and endocardial myofibers are oriented longitudinally.

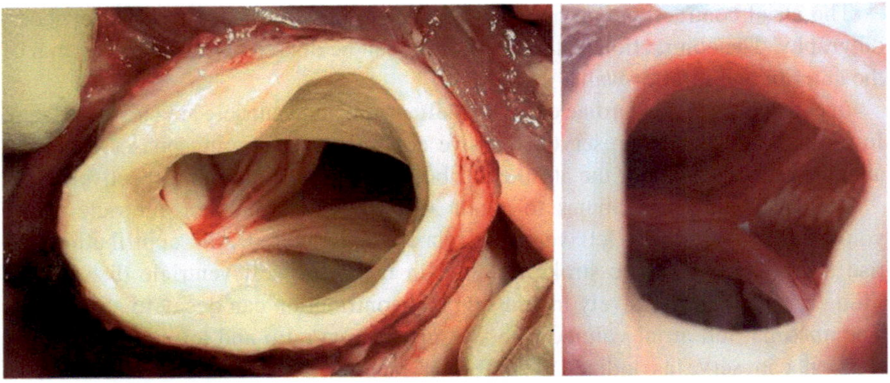

Fig. 1.4 View from the arterial lumen of the aortic orifice. (**Left**) Two displayed cusps with their free edges and the lunulae; the commissure is at the lower left corner. (**Right**) Three close cusps; the insertion line of the 3 semilunar leaflets, from the cusp nadir to the commissures are hidden by the aorta wall

The right ventricular ejection results by longitudinal shortening rather than by circumferential deformation. The longitudinal shortening progresses from the inflow to the outflow part of the right ventricle.

1.6.3 Aortic Valve

The aortic valve consists in 3 quasi-equal semilunar cusps: right, left, and noncoronary leaflets (thickness 0.2–0.4 mm;[12] Fig. 1.4) [14]. Dimensionless heights and bores are displayed in Fig. 1.5. Dimensions of the 3 cusps of the aortic valves are also given in Table 1.5 for porcine hearts, which provide orders of magnitude of cusp sizes [15].[13]

Each aortic valve leaflet is constituted of 3 layers (Vol. 5 – Chap. 6. Heart Wall): the *fibrosa* facing the aortic surface; the *ventricularis* in front of the left ventricle; and the *spongiosa* between these 2 layers. Whereas the fibrosa consists mainly of collagen fibers in circumferential direction, the ventricularis is mainly composed of elastin.

A *congenital bicuspid valve* is observed in approximately 0.5 to 2% of the general population. The valvular ring geometry is oval, whereas the tricuspid valve orifice is circular. In addition, the cusp opening angle is lowered with respect to a tricuspid valve [16]. Congenital bicuspid valve generates a different field of mechanical

---

[12]Higher thickness values up to 700 μm can be found.

[13]Between-species geometry variations may explain xenograft failure with porcine bioprostheses.

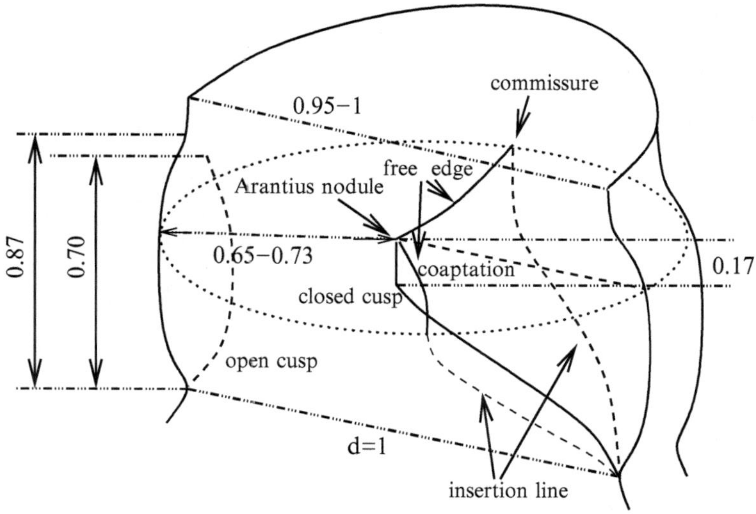

Fig. 1.5 Schematic drawing of the aortic valve and dimensionless lengths (Source: [14])

Table 1.5 Dimensions (mm; mean data of a set of 10 post-mortem porcine hearts) of right (RCC), left (LCC), and noncoronary (NCC) leaflets of porcine aortic valves [15]

	RCC	LCC	NCC
Width	13.3	13.9	13.7
Free edge length	33.0	31.5	32.7
Insertion line length	46.4	47.6	48.1
Perimeter	79.4	79.1	80.8

stresses both on the aortic valve and wall than that exerted by the flow expelled through a tricuspid valve.

The 3 *Valsalva sinuses* of the aorta root match the 3 valve leaflets (Figs. 1.6 and 1.7) [17]. The coronary arteries branch off from 2 sinuses (Fig. 1.8).

The intersection regions between the aorta wall and the cusp free edges are called the *commissures*. The *nodulus of Arantius* is a large collagenous mass in the central part of the coaptation region.[14] The *lunula* is the region on each side of the nodule. When the aortic valve is closed, the free margins of the cup-like leaflets seal each other, defining 2 planes, a vertical sealing (the aorta root axis being vertical) and a bottom oblique plane from the insertion line to the channel axis, with an angle of nearly 20 degrees.[15]

---

[14]The cusp coaptation region is the leaflet portion below the free edge, which comes into contact with the neighboring cusps.

[15]The more or less curved bottom cusp below the coaptation region, which is connected to the wall, is approximated by a straight line in a plane crossing the mid-cusp.

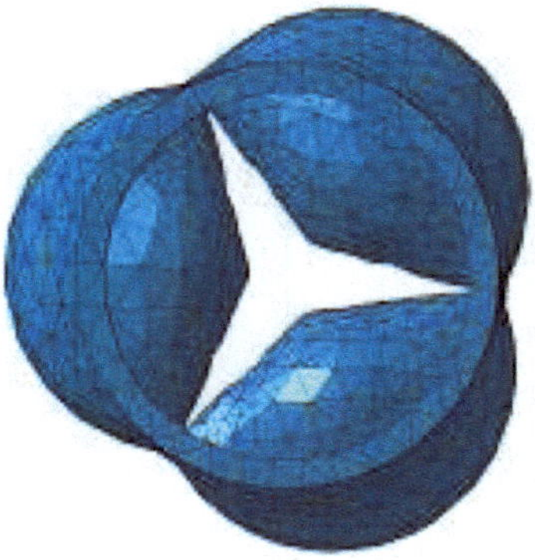

Fig. 1.6 Top view of a mesh of the aorta root with the 3 leaflets of the aortic valve and the Valsalva sinuses (from T. Cherigui)

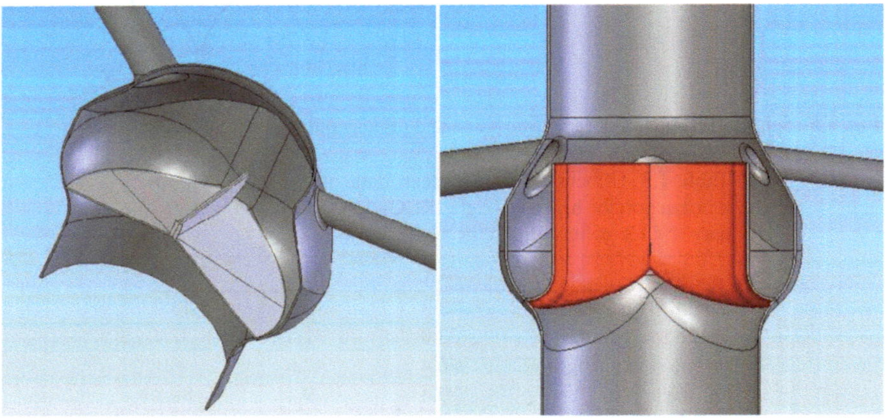

Fig. 1.7 Model of the aorta root with aortic valve cusps in open and closed configurations (from T. Tran)

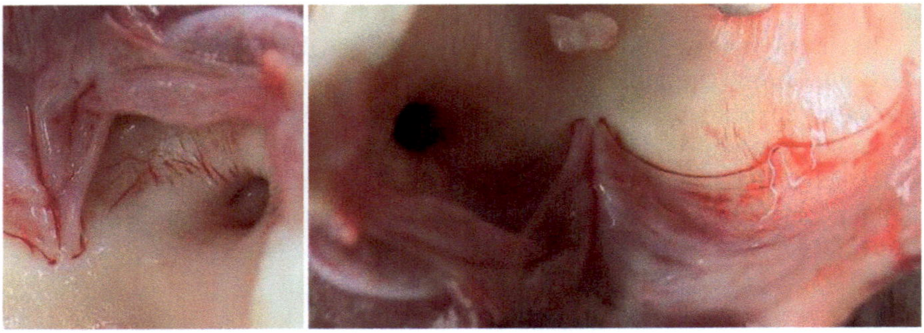

Fig. 1.8 Valsalva sinuses of the aortic root with the coronary ostium. In the calf heart, the commissure reaches the sinotubular junction, whereas it remains below in humans. (**Left**) Coronary ostium with its smooth convergent. (**Right**) View of the sinotubular ridge

Table 1.6 Mean dimensions of human aortic leaflets (mm; LCC: left coronary cusp; RCC: right coronary cusp; NCC: noncoronary cusp; Source: [19])

	LCC	RCC	NCC
Cusp height	15.2	15.0	14.9
Lunula width	4.6	4.4	4.3
Lunula length	30.6	30.1	30.0
Intercommissural distance	25.2	24.5	24.1
Intercommissural chord	19.8	19.2	20

When the reference length is the channel bore at the insertion line bottom, the sinus height is equal to 0.87, the length between the insertion line bottom and the commissure 0.71, the coaptation length 0.17, and the maximum sinus half width 0.65–0.73; the aortic bore downstream from the sinuses is almost equal to 1 [14]. The open leaflets form a triangular orifice when the cusps are not fully open. To limit the valve obstacle, the flow can flatten the cusp along the sinus cavity, the cusp end projecting slightly inside the cavity [18].

Certain dimensions of Valsalva sinuses and cusps of the aortic valve, left coronary cusp (LCC), right coronary cusp (RCC), and noncoronary cusp (NCC) have been measured [19]. The height of the cusps are assessed from the bottom of the Valsalva sinus to the cusp free margin at the middle point between the commissures. The mean values[16] of aortic valve dimensions (cusp height, lunula width and length, and intercommissural distance)[17] for the 3 cusps (LCC, RCC, and NCC) are given in Table 1.6. The position of the ostium can be determined by the distances between the ostium and the commissures at the left and right side of the ostium and the length between the ostium and bottom (nadir) of the corresponding Valsalva sinus. The authors have observed that both coronary ostia can be located in the left coronary sinus and be supracommissural. The left and right coronary ostia are located on average at a distance of 9.6 and 11.1 mm from the left commissure, 11.0 and 11.1 mm from the right commissure, and 13.3 and 14.8 mm from the Valsalva sinus bottom, respectively. The position of the aortic leaflets can also be defined with respect to the ventricular septum, using the distance between the septal extremity and RCC-LCC commissure, the septal end and the RCC-NCC commissure, and the septum and the NCC-LCC commissure. The mean values of these distances are 9.5 (± 5.3), 5.7 (± 4.2), and 19.2 (± 2.9) mm. The mean aortic diameter is equal to 21.8 (± 3.6) mm.

The aortic valve begins to close when the intraventricular starts to decay, i.e., during the decelerating phase of the systolic ejection. Blood pushing between aortic face of valve leaflets and Valsalva sinus walls can help to terminate valve closing. During diastole, when the valve is closed, the coaptation zone (between-

[16]One hundred healthy hearts of both sexes from people 9 to 86 years old have been studied.

[17]Semilunar leaflets have a thin free lunula with a dense nodule at the midpoint. The size of the lunulae is estimated by the width at the commissural level and the length of the free margins. The intercommissural distances, either straight (chord) or curved (arclength) along the aortic orifice circumference, have also been measured.

cusp contact region) is able to withstand backflow originating from deflating aorta. The valve leaflets are stretched in both circumferential and radial direction during diastole with respect to systole [16]. During the cardiac cycle, the aortic valve bears pressure on their wetted surface and parietal stretch as well as bending forces.

1.6.4 Internal Anatomical Features of the Heart

1.6.4.1 Nodal Tissue

Nodal tissue (Vol. 5 – Chap. 6. Heart Wall) is a specialized type of myocardium that generates (*heart automatism*) and propagates command electrochemical waves (*action potential*), which trigger myocardium contraction. Therefore, nodal tissue contracts like a striated muscle such as the myocardium and generates impulses such as nerves. When action potential depolarizes cardiomyocytes of left and right homologous chambers, these chambers contract (*excitation–contraction coupling*).

Action potential is generated by the *sinoatrial node* (*pacemaker*) located in the posterior wall of the right atrium near the superior vena cava. It spreads throughout the atria primarily by an intercellular conduction (speed ~0.5 m/s), although specialized atrial conducting routes—internodal tracts—exist. Action potential then reaches the *atrioventricular node* situated in the posteroinferior region of the interatrial septum. This impulse then enters the base of the left ventricle running along the *atrioventricular bundle*, or *bundle of His*, and the left and right *bundle branches* along the interventricular septum that divide into *Purkinje fibers* (Table 1.7).

1.6.4.2 Inner Anatomical Landmarks

The right atrium (volume ~60 ml [3]) is larger than the left, but its walls are thinner (~2 mm [3]). Its cavity consists of 2 parts, a main posterior cavity, or *sinus venarum* (sinus venosus), between the 2 venæ cavæ, and a smaller anterior region, the *auricula* (right auricular appendix or auricula dextra). The right atrium contains openings of (Fig. 1.9; Table 1.8): (1) the superior and inferior vena cava, the latter being guarded by the *ridge of the inferior vena cava*, a rudimentary semilunar valve, or (the *Eustachian ridge*; (2) coronary sinus protected by the *ridge of the coronary sinus*, or the (*Thebesian ridge*, a semicircular fold; (3) foramina venarum minimarum (foramina Thebesii), orifices of small veins; and (4) valved atrioventricular orifice. The *fossa ovalis* is an oval depression of the atrium that corresponds to the foramen ovale in the fetus, above the orifice of the inferior vena cava. The *annulus ovalis* is the prominent oval margin of the fossa ovalis.

Table 1.7 Nodal tissue, a specialized myocardium

Component (impulse time)	Features
Sinoatrial node ($t = 0$)	Located within the crista terminalis near the superior vena cava Cardiac pacemaker
Internodal tracts	Anterior, middle, and posterior path P wave on ECG trace
Atrioventricular node ($t \sim 70$ ms)	Located in the right atrium above the opening of the coronary sinus and the septal cusp of the tricuspid valve PQ segment on ECG trace
Atrioventricular bundle (or His bundle)	Passes through the right fibrous trigone into the muscular part of the interventricular septum
Bundle branches ($t \sim 160$ ms)	In the interventricular septum Right and left branches QRS segment on ECG
Left posterior fascicles	In walls of the left ventricle QRS segment on ECG
Purkinje fibers ($t \sim 180$ ms at apex) ($t \sim 220$ ms at base)	Conduct action potentials at high speed (~ 4 m/s) throughout the ventricles QRS segment on ECG

The upper left angle of the right ventricle (volume ~90 ml [3]) forms a conical pouch, the *conus arteriosus*, from which the pulmonary artery arises. A tendinous band (tendon of the conus arteriosus) extends upward from the right atrioventricular fibrous ring and connects the posterior surface of the conus arteriosus to the aorta. Trabeculæ carneæ (or columnœ carneœ) are muscular columns that project from the inner surface, except in the conus arteriosus.

The left atrium (volume ~35 ml [21]) is smaller than the right, but its walls are thicker (~3 mm). Like the right atrium, it consists of 2 regions, the principal cavity and (auricula). It contains: (1) openings of the 4 pulmonary veins; (2) mitral orifice surrounded by a dense fibrous ring, cusps of which are are larger, thicker, and stronger than those of the tricuspid valve; and (3) pectinate muscles (musculi pectinati) that are fewer and smaller than in the right atrium and confined to the auricula.

The left ventricle forms the cardiac apex (end-diastolic volume ~120 ml; end-systolic volume ~40 ml [22]). Its wall thickness is about 3 times larger than that of the right ventricle. It contains trabeculæ carneæ and chordæ tendineæ.

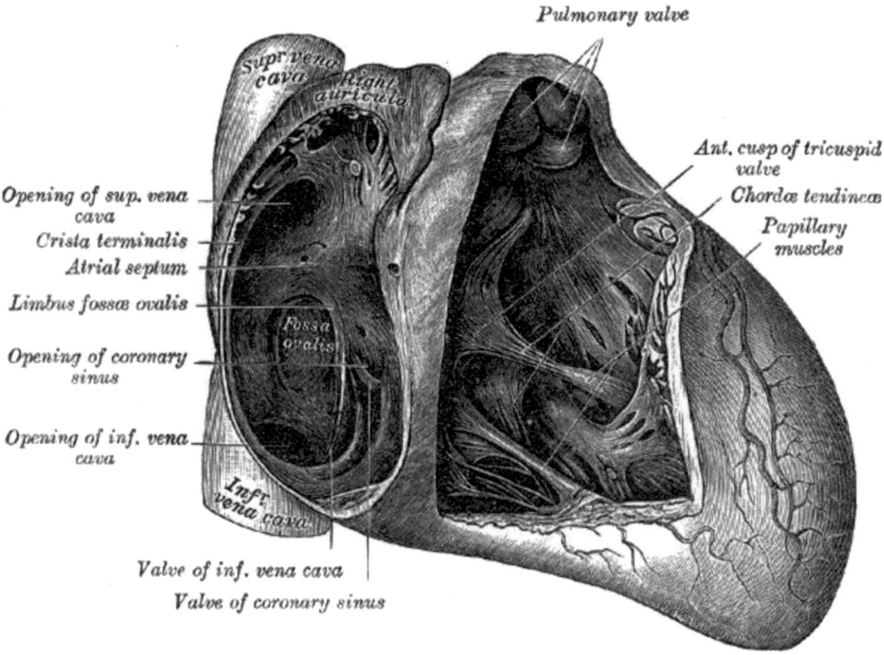

Fig. 1.9 View of interior right side of the heart (Source: [3])

1.6.5 Heart Vascularization

The heart is perfused by the right (RCA) and left (LCA) *coronary arteries*, originating from the aorta just above the aortic valve.[18] These distribution coronary arteries lie on the outer layer of the heart wall. These superficial arteries branch into smaller arteries that dive into the cardiac wall.

The right coronary artery branches successively into the conus, right ventricle, acute marginal, posterior descending, and posterior left ventricle arteries. It supplies blood to the right atrium and posterior part of the ventricles.

The left coronary artery divides into the left anterior descending and circumflex arteries, giving birth to obtuse marginal and diagonal branches, respectively. It supplies blood to the left atrium and the anterior part of the ventricles (Fig. 1.10).

1.6.6 Heart Innervation

The heart is capable of beating independently of any nervous or hormonal influences, as it is endowed with *intrinsic automaticity*. However, its activity is influenced by nervous signals as well as by regulatory chemicals.

[18]The lumens of the coronary arteries are not obstructed by open leaflets.

Table 1.8 Anatomical elements and landmarks of cardiac cavities (Source: [3,20])

Element	Description
Atrioventricular valves	
Tricuspid valve	Between the right atrium and right ventricle 3 cusps (anterior, posterior, and septal)
Mitral valve	Between the left atrium and left ventricle 2 cusps (anterior and smaller, posterior)
Chordae tendineae	Cords attaching atrioventricular valve cusps to papillary muscles Lodge only in the ventricles
Ventriculoarterial valves	
Aortic valve	3 semilunar valves (right, left, and posterior cusps)
Pulmonary valve	3 semilunar valves (anterior left and right and posterior cusps)
Right atrium	
Crista terminalis	Ridge of cardiac muscle separating the smooth-walled posterior sinus venarum from the roughened wall of the anterior atrium
Fossa ovalis	Shallow depression in the left wall Remnant of the foramen ovale
Limbus fossa ovalis	Ridge around the fossa ovalis in the left wall
Pectinate muscles	Prominent ridges of myocardium
Septomarginal trabecula (moderator band)	Ridge from the interventricular septum to the anterior papillary muscle
Left atrium	
Pectinate muscles	Prominent ridges of myocardium
Right ventricle	
Papillary muscles	Projections of myocardium 3 (anterior, posterior, and septal)
Pulmonary conus	Smooth region Below the opening into the pulmonary trunk
Trabeculae carnae	Ridges
Left ventricle	
Papillary muscles	Projections of myocardium 2 (anterior and posterior)
Trabeculae carnae	Ridges

As are other myocyte types and neurons, cardiomyocytes are excitable. They synchronously contract at a given rate when their plasma membrane depolarizes upon the arrival of electrochemical impulses—action potentials—that are initiated in and run through the nodal tissue.

The heart is innervated by both components of the autonomic nervous system (ANS; Sect. 1.8; Table 1.9; and Fig. 1.12). Normally, parasympathetic innervation represents the dominant neural influence on the heart. Maximal stimulation of

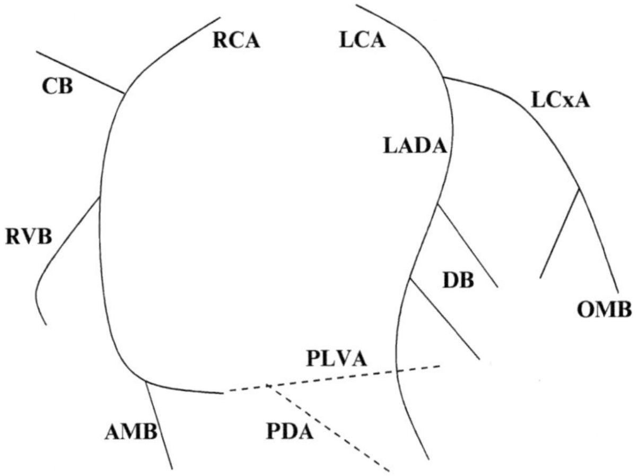

Fig. 1.10 Schematic drawing of the coronary artery network (LCA, RCA: left and right coronary arteries; CB: conus branch; RVB: right ventricle branch; AMB: acute marginal branch; PLVA: posterior left ventricle artery; PDA: posterior descending artery; LADA: left anterior descending artery; LCxA: left circumflex artery; OMB: obtuse marginal branch; DB: diagonal branches)

Table 1.9 Mediastinal nerves control blood circulation and respiration that affects the venous return

Nerve	Target
Sympathetic	Heart
	Blood vessels
	Tracheobronchial tree
Parasympathetic	Heart
	Blood vessels
	Bronchial smooth myocytes and mucous glands

vagal fibers can stop myocardium contractions. When stimulated, sympathetic fibers release noradrenaline (Sect. 3.6.2).

Parasympathetic innervation originates in the *cardiac inhibitory center* in the medulla oblongata of the brainstem and is conveyed to the heart via the vagus nerve (cranial nerve X; Table 1.10).

Sympathetic innervation comes from the *cardiac accelerating center* in the medulla and upper thoracic spinal cord. The cardiac sympathetic nerves extend from sympathetic neurons in stellate ganglia, which reside bilateral to vertebrae. Sympathetic nerve fibers project from the base of the heart into the myocardium. They are located predominantly in the subepicardium in the ventricle. The nodal tissue is abundantly innervated with respect to the myocardium.

Neurons involved in cardiac regulation localize to the insular cortex from the heart (Table 1.11) [23]. Intrathoracic ganglia are monosynaptic efferent relay

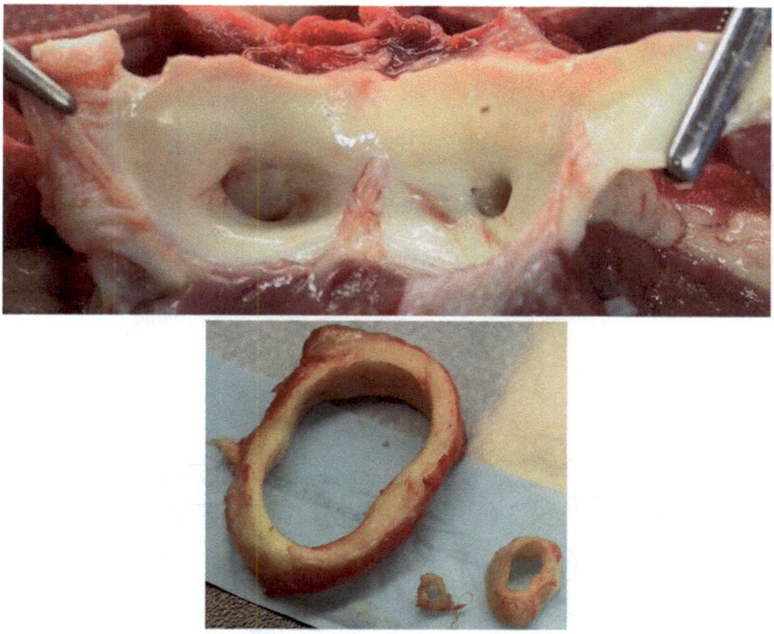

Fig. 1.11 Coronary artery calibers. (**Top**) Ostia in the Valsalva sinuses between the valve insertion lines and sinotubular ridge are separated by commissures of the right and left coronary cusps of the aortic valve. In this 3-month-old calf, the bore of the entrance segment of the right coronary artery is smaller than the left. (**Bottom**) Differences in the bore of the entrance segment slices of the aorta and right and left coronary arteries (from left to right) by artery slice sizes. Axial cut of vessel segments, like those displayed here, induces a circumferential lengthening of 10% to 20% (azimutal prestress in unstressed configuration)

stations that process efferent inputs to the heart. Sensory information transduced by cardiovascular afferent neurons in nodose and dorsal root ganglia is processed by the central nervous system to activate sympathetic postganglionic neurons in paravertebral ganglia via the spinal cord neurons. Parasympathetic postganglionic neurons in the target organ ganglia receive inputs from medullary neurons.

Cardiac sensory afferent neurons localize in intrathoracic extracardiac, intrinsic cardiac ganglia, nodose, and dorsal root ganglia from C7 to T4 levels of the spinal cord [23].

Activated cardiac motor neurons regulate cardiac frequency (chronotropy), nodal conduction (dromotropy), contractibility (inotropy), and diastolic relaxation (lusitropy). Activated sympathetic efferent neurons increase cardiac chronotropy, dromotropy, and inotropy and decrease left ventricular end-diastolic volume; activated parasympathetic efferent postganglionic neurons have opposite effects.

Fat pads on the surface of the heart contain neural ganglia. The interacting *sinoatrial ganglion* and *posterior atrial ganglion* reduce the cardiac frequency (negative chronotropic effect); the *atrioventricular ganglion* lowers the nodal conduction (negative dromotropic effect); and the *cranioventricular ganglion* attenuates the

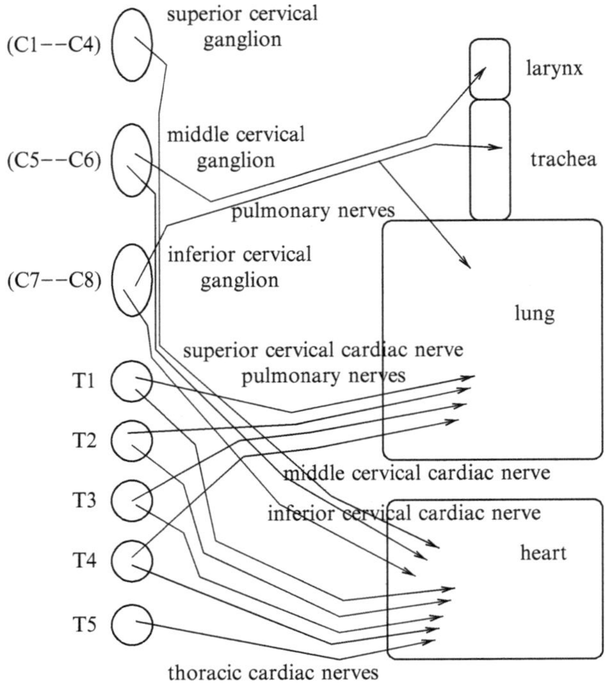

Fig. 1.12 Sympathetic innervation of heart and respiratory tract (C: cervical nervous root; T: thoracic (dorsal) nervous root/ganglion)

left ventricule contractility (negative inotropic effect) [24]. Intraganglionic neuronal circuits are found within the sinoatrial ganglion defining an intrinsic cardiac nervous system [24].

Cardiac parasympathetic efferent preganglionic neuronal soma reside primarily in the ventral lateral region of the nucleus ambiguus as well as in the dorsal motor nucleus and the intermediate zone between these 2 medullary nuclei [23]. These neurons project axons to parasympathetic efferent postganglionic neurons of atrial and ventricular ganglionated plexus (right [RAP], dorsal [DAP], left [LAP], and inferior vena cava–inferior [IVCIAP] atrial and right [RVP], ventroseptal [VSVP], and craniomedial [CMVP] ventricular ganglionated plexus).[19] Intrinsic cardiac

[19] A nervous plexus is a branching network of axons, or nerve fibers, which can include sympathetic and parasympathetic efferents and visceral afferents. It can be constituted by subdivisions of 2 or more nerves, which are interconnected with one another to form a network through which nerve fibers of the constituent nerves are regrouped (e.g., cervical, brachial, and sacral plexus). A ganglionated plexus contains axons as well as autonomic neuron soma. Cardiac ganglionated plexus lodge in epicardial fat pads. Atrial fibrillation is associated with an increased risk of stroke and heart failure. Antiarrhythmic drug-refractory atrial fibrillation is treated by minimally invasive, electrophysiologically guided, video-assisted surgery with ganglionated plexus ablation.

Table 1.10 Nerves of the thorax (Source: [3])

Nerve	Source and branches
Cardiac plexus	Formed by cardiac branches of the parasympathetic vagus nerve and cervical sympathetic trunk that has usually 3 cardiac branches (superior, middle and inferior) Supplies the coronary and pulmonary plexus
Pulmonary plexus	Formed by pulmonary branches of vagus nerve and the sympathetic trunk Supplies the bronchial tree and visceral pleura Parasympathetic: mainly bronchial smooth muscle and glands Sympathetic: mainly vascular smooth muscle
Vagus nerve (X)	Innervate heart, tracheobronchial tree, and lung parenchyma Branches in the jugular fossa: meningeal and auricular Branches in the neck: pharyngeal, superior laryngeal, recurrent, and superior cardiac Branches in the thorax: inferior cardiac, esophageal, anterior and posterior bronchial Branches in the abdomen: gastric, celiac, and hepatic Hence, 2 cervical cardiac (superior and inferior) and 1 or more thoracic cardiac branches Sensory fibers from cells of jugular and nodose ganglion (inferior ganglion: organs of head, neck, and thorax) Somatic motor fibers from cells of the nucleus ambiguus (intrinsic muscles of the larynx, pharynx except stylopharyngeus) Efferent fibers from the dorsal motor nucleus (airway smooth muscle, heart glands of the pharynx, larynx, and airways) Superior (jugular) ganglion communicates with the accessory nerve, the petrous ganglion of the glossopharyngeal, the facial nerve via auricular branch, and the sympathetic superior cervical ganglion Inferior (nodose) ganglion connects to the hypoglossal, the superior cervical ganglion of the sympathetic, and the loop between the first and second cervical nerves

cholinergic efferent postganglionic neurons receive direct inputs from medullary preganglionic neurons of these 2 medullary nuclei.

For example, the left atrial ganglionated plexus located in epicardial fat pads and ligament of Marshall[20] contain afferent neurons from the atrial myocardium,

[20]The ligament of Marshall is a vestigial structure on the epicardium between the left atrial appendage and left pulmonary veins. It contains muscle bundles (Marshall bundles) that directly connect to atrial myocardium and coronary sinus muscle sleeves. Ganglionated plexus lodge in and around the ligament of Marshall.

Table 1.11 Tiers of the nervous control of the heart (Source: [23]). Cardiac mechano- and chemosensory cues are transduced by receptor cells and transmitted by afferent neurons, soma of which localize in nodose, dorsal root, and intrathoracic extrinsic and intrinsic cardiac ganglia. This information engenders intrathoracic and central medullary and spinal cord reflexes transmitted by a massively parallel command. Intrinsic cardiac, mediastinal, middle cervical, stellate, and superior cervical ganglia are related to overlapping, spatially defined cardiac regions. Intrathoracic, dorsal root, and nodose ganglia signal directly or indirectly via interconnecting neurons to cardiac adrenergic and cholinergic motor neurons

Central nervous system	Higher centers Medulla Spinal cord
Intrathoracic ganglia (extracardiac)	Efferent neurons Afferent neurons Local circuit neurons
Intrinsic cardiac ganglia	Efferent sympathetic neurons Efferent parasympathetic neurons Afferent neurons Local circuit neurons
Nodose and dorsal root ganglia	Afferent neurons

efferent cholinergic and adrenergic neurons, and interconnecting neurons, which allow communication between ganglionated plexus. Stimulation of the ganglionated plexus produces both parasympathetic stimulation, which shortens action potential duration, and sympathetic stimulation, which increases calcium transients in the atrial and pulmonary vein myocardium.

Sympathetic efferent postganglionic neurons involved in cardiac regulation receive inputs from caudal cervical and cranial thoracic spinal cord preganglionic neurons [23]. Cardiac sympathetic efferent preganglionic neurons of the spinal cord project axons via the T1 to T5 rami that synapse with [23]: (1) cardiac sympathetic efferent postganglionic neurons of the superior and middle cervical ganglia; (2) cranial poles of stellate ganglia; (3) mediastinal ganglia adjacent to the heart; and (4) intrinsic cardiac ganglionated plexus. Sympathetic postganglionic neurons of intrinsic cardiac plexus project axons to widespread regions of the heart. Among them, some project 2 axons via different cardiopulmonary nerves to innervate distinct cardiac regions.

Many intrathoracic local circuit neurons receive inputs from additional extrathoracic sources. The intrathoracic local circuit comprises various sets of neurons. Many neurons of intrathoracic ganglia project axons only to other neurons within the same ganglion [23]. A second set of neurons projects axons to neurons in different intrathoracic ganglia. A third set of neurons projects axons to central neurons.

Intrinsic cardiac afferent neurons transmit signals from local mechano- and chemoreceptors as well as those of large intrathoracic vessels to other neurons of the same ganglion as well as to those in other intrinsic cardiac and intrathoracic extracardiac ganglia [23]. In other words, intrathoracic afferent neurons signal to both local circuit neurons in intrinsic cardiac and intrathoracic extracardiac ganglia and central neurons.

The cardiac innervation is characterized by [23]: (1) short-latency feedbacks relying on intrinsic cardiac ganglia that modulate cardiac function during each cardiac cycle; (2) medium-latency feedback communicated by intrathoracic ganglia that influence the cardiac behavior over a few cardiac cycles; and (3) relatively long-latency feedback signals from the spinal cord and brain responsible for sustained effects. Dorsal root and nodose ganglion cardiovascular afferent neurons initiate spinal cord- and brainstem-derived reflexes, respectively. Anatomical and functional connectivity among atrial and ventricular neurons supports long-term interactions.

Some intrathoracic extracardiac and intrinsic cardiac local circuit neurons have a respiratory-related activity: (1) right ventricular outflow tract mechanosensory neurons that react to changes in pulmonary arterial resistance and (2) pulmonary mechanosensory neurons that respond to regional pulmonary strains. Adaptation of cardiovascular and respiratory functions is thus ensured by a simultaneous coordinated control.

The intrinsic cardiac nervous system not only acts as a simple monosynaptic relay for central control of regional cardiac function, but also processes sensory information from various regions of the heart. On the one hand, sensory neurons in each major part of the intrinsic cardiac autonomic nervous system (ganglionated plexus) transmit signals received by relevant receptors throughout the heart. In the other, many atrial and ventricular neurons have stochastic behavior that may reflect the stochastic nature of sensory inputs [25].

Intrinsic atrial and ventricular neurons can be stimulated by numerous types of chemicals, including those liberated by the ischemic myocardium, such as adenosine and bradykinin [25]. Chemosensory signals may have a greater impact than mechanosensory inputs.

1.6.7 Fractal Geometry

At rest, i.e., in the absence of acute stimulation, physiological apparatus are in a steady state. The basic structural and functional unit—the cell—operates under optimal conditions of pH, temperature, osmolarity, and nutrient levels. However, the maintenance of equilibrium must cope with the circadian rhythm of the body driven by internal clocks (Vol. 2 – Chap. 5. Circadian Clock), among others.[21] Upon

[21] The circadian pacemaker in the suprachiasmatic nucleus with its transcriptional and translational feedbacks has a self-sustaining activity with a reset mechanism responsive to environmental signals

stimulation, biological entities switch to other states, continuing to ensure the body's homeostasis adapted to new environmental conditions.

Physiological apparatus are, in fact, characterized by complex behaviors. They experience 5 main types of behaviors [26]: (1) equilibrium (e.g., homeostasis maintenance); (2) periodicity produced by a single oscillator or synchronized multioscillator, a process used by biological clocks and anatomical pumps; the cycle is repeated with a given frequency; (3) quasi-periodicity created by multioscillators that can bifurcate, a mechanism involved when order is required; the ratio between involved cycle frequencies is not a rational number. (4) deterministic chaos generated by multioscillators when efficiency and flexibility is needed; the behavior is not confined to repeated cycles at a given rhythm; it is free to respond and adapt, although it is constrained by initial conditions; and (5) random behavior, usually a pathological state, which results from loss in coordination between involved components.

Equilibrium, periodicity, and quasi-periodicity are examples of linear dynamics. Causes produce effects in a magnitude-dependent manner. The behavior of the entire system can be deduced by adding the behaviors of each of its components. Equilibrium, periodicity, and quasi-periodicity can be studied using the reductionist approach, but not deterministic chaos, as a nonlinear system is more than the sum of its parts.

Nonlinear dynamics is used to study the controlled behavior of a deterministic system that experiences discontinuous sudden phenomena. In a nonlinear system, the system components cannot be added, as they participate in positive and negative feedback (amplification or reduction).[22] In a chaotic system, small changes in initial conditions are magnified (sensitivity to initial conditions); its behavior cannot be predicted over the long term.

The phase space enables the display of values of parameters that describe the state of a studied system. The system state, at any given moment, can indeed be represented as a point in the phase space. As the state of a dynamic system is continuously changing, all possible states of the system are graphically depicted.

The evolution of a dynamical system toward a given state results from a driving force that dampens effects of initial transient states, i.e., a given set of points in phase space, the attractor. The attractor of a deterministic nonchaotic system, hence a predictable behavior, can be a fixed point, a limit cycle (a periodic system), or a limit torus (a quasiperiodic system). Physiological apparatus can change their dynamics and move between order and chaos. In phase space, the attractor changes (bifurcation).

delivered by the retinohypothalamic tract. It regulates peripheral clocks using neurohormonal commands.

[22] A feedback loop is a regulatory loop that feeds the system, either negatively (an increasing output has a suppressing effect on the triggering signal) or positively (an increasing output amplifies the output), modulating the input by the output.

The attractor of a chaotic system—the strange attractor—is displayed as points loop endlessly within the boundaries set by the attractor toward a central point, but the trajectory traced by the point never repeats. The strange attractor contains an infinite number of possible trajectories through which the state of the system can cycle. Strange attractors are fractals [26].

Fractals[23] are mathematical contructs characterized by a never-ending cascade of similar structural details revealed upon magnification on all scales. These self-similar structures are governed by *power-law functions*[24] and *fractal dimensions*, i.e., fractional (noninteger dimensions). An empirical fractal dimension (D_f) is calculated between a property (P) and the resolution (r) of the general form [27]:

$$P = k\,r^{f(D_f)}, \tag{1.2}$$

where k is the prefactor for the power law and the exponent $f(D_f)$ is a simple function of the fractal dimension D_f. The power law enables us to correlate in a simple manner properties of a system to its structure and to the dynamics of its formation.

A fractal object, in the purely mathematical sense, requires infinitely multiple orders of magnitude of power-law scaling. Fractality at least requires many orders of magnitude, i.e., many iterations, or resolution levels. In particular, fractals are fragmented geometric figures that repeat themselves at progressively smaller scales, exhibiting progressively more complicated structure when observed at larger magnifications. Yet, when a small part of the overall self-similar structure is magnified, it exhibits nearly the same aspect as the entire object. In other words, split parts are, at least approximately, reduced-sized copies of the entire structure.

The smaller the fractal dimension, the higher the influence of determinism (the lower that of randomness). The fractal dimension of a self-similar pattern is typically larger than the topological dimension. However, nonlinear fractals are not self-similar.

The Julia and Mandelbrot sets are examples of fractal structures with perfect self-similarity. These sets can be obtained by iterative computations that relies on:

$$z_{n+1}^* = z_n^{*p} + c^*, \tag{1.3}$$

where z^* and c^* are complex numbers and p denotes power (a natural number [or positive integer] equal or greater than 2). If an object is subdivided into N elements to fill the original space using a scale factor r, then:

---

[23]Latin *fractus*: rupture; *fractus* (a, um [frango]): broken, divided up, fractured, interrupted, shattered, smashed.

[24]Kleiber's law is a metabolism scaling law, or power law, that predicts the metabolic rate (M) using a 3/4 power relation between the body mass (m) and energy consumption [28, 29]:

$$M \propto m^{3/4}. \tag{1.1}$$

$$N = k\, r^{-D_f}, \tag{1.4}$$

where k is a constant.

Diffusion-limited aggregation is a diffusion process whereby particles have a random walk behavior, but when a particle comes into contact with another particle it sticks to it and can no longer move, leading to an accretion. A spatial pattern is dynamically formed that has a fractal feature.

In a fractal system, the measurement of any parameter depends on the resolution at which the measurement is carried out. The value of the measured length generally increases when finer details are revealed. The scaling relates the measurement to the resolution.

Physiological apparatus are characterized by a *statistical self-similarity*, rather than a perfect self-similarity. Unlike mathematical examples, biological fractals are usually only self-similar over a limited range of orders of magnitude of length scales, i.e., are bounded by a maximal (global configuration) and a minimal (size of smallest component) scale. A limiting magnification exists, beyond which self-similarity is lost. Moreover, fractals are approximately self-similar at different scales.

In the cardiovascular apparatus, fractal geometry was used to model the architecture of venous (blood drainage from body's tissues) and arterial (blood distribution from the cardiac pump) circuits (e.g., from large feeding arteries to small arterioles, i.e., a range of 3 orders of magnitude [$\mathcal{O}$[1 cm], $\mathcal{O}$[1 mm], and $\mathcal{O}$[100 μm]).

Fractal analysis was also utilized in the heart, in addition to coronary vasculature (arteries and veins), especially to describe the structure of specialized tissues, such as the myocardium (from muscular bundles and myofibers to myofibrils and myofilaments), papillary muscles and chordae tendinae binding the auriculoventricular valves to the heart wall, aortic valve leaflets, and the His–Purkinje nodal circuit devoted to the propagation of electrochemical command waves.

The fractal structure may yield fault tolerance during development and growth. In addition, fractal branching provides a suitable surface area in the available volume. Moreover, fractal structure redundancy supports robustness against injury. Last, by not least, the fractal structure of the blood circulation enables tthe damping of eventual hammer blows generated by the heart, as fractal boundary conditions drastically alter wave excitations [30].

Moreover, fractal analysis was used in time-varying signal processing such as deterministic variability of the cardiac frequency. The beat-to-beat variability of the sinusal rhythm controlled by the parasympathetic nervous system was processed by a temporal fractal, as the self-similarity extends over many time scales. The nodal tissue activity, especially that of the sinoatrial node, is governed by a beneficial chaotic behavior. The functioning of nodal cells of the sinoatrial pacemaker is driven by 2 oscillators; the internal cell oscillator characterized by its own pacing rhythm (spontaneous self-excitation) and the external control, mostly nervous.

The electrochemical activity of the heart can be explained, at least partly, by a transport–reaction mechanism. In addition, dynamics and self-organization of cellular metabolism and calcium handling can contribute to the chaotic behavior. Self-organization results from spontaneous positive and negative interactions between system components.

Fractal analysis explores the signal structure; chaos theory examines the organ dynamics responsible for the signal. The chaotic behavior and fractal complexity of the time series reflect different response types of the heart to pharmacological agents. In chronic heart failure, the fractal complexity decreases. The heart chaotic behavior can be lost by bifurcation toward either order or randomness (arrhythmias), especially in the nodal conduction path. Both cases represent a loss in adaptibility to environmental constraints. The ability to adapt is associated with the possibility to generate and transmit new data. Nonfully predictable behavior enables a certain freedom of expression. Furthermore, information generation and storage is defective [26]. A fully random behavior remains meaningless; a regularly periodic behavior repeats the same information.

In summary, fractal scaling of a branched transport circuit is never a prerequisite. It can be used for simple description, as a tool that can simplify calculations, but not for understanding. Furthermore, fractality is usually observed on a limited scaling range, over which the object of interest obeys Eq. 1.2.

1.6.8 Cardiogenesis

> The form, then, of any portion of matter, whether it be living or dead, and the changes of form which are apparent in its movements and in its growth, may in all cases alike be described as due to the action of force.
> D'Arcy Thompson W (1860–1948) [31]

Cardiogenesis (Vol. 5 – Chap. 6. Heart Wall) relies on a set of multipotent cardiac progenitor cells that expand, differentiate, assemble in specific heart compartments, and shape cardiac 3D structures with functional tasks: (1) the pacemaker and conduction circuit (nodal tissue) to trigger contraction; (2) the perfusion and drainage coronary network to feed myocardium and for waste disposal; (3) cardiac chambers to adapt inlet and outlet pressures and propel blood; and (4) heart valves to ensure unidirectional flow (Table 1.12).

1.6.9 Early Stages of Cardiogenesis and Chemical Control

The heart is the first organ to form during embryogenesis. Cardiac precursor cells are detected before gastrulation, the morphogenetic process that leads to the formation of the 3 germ layers (ecto-, meso-, and endoderm). They localize to the lateral posterior epiblast (Table 1.13). Embryonic stem cells undertake genetically and

Table 1.12 Cardiac progenitor cells (Source: [36]; Early cardiogenic precursors generate several cell lineages (Isl1: insulin gene enhancer islet-1, or Isl homeodomain LIM motif-containing transcription factor; MyH11: myosin, heavy chain-11, or smooth muscle myosin heavy chain [smMHC]; NKx2: NK2 transcription factor-related homeobox gene product, NKx2.5 being a cardiac-specific enhancer; TBx18: T-box transcription factor-18; TnT: troponin-T; WT1: zinc finger Wilms tumor transcription factor-1]. Early mesoderm-derived cardiac precursors give rise to progenitors in the first (FHF) and second (SHF) heart fields. The primary heart field (FHF) arises from the anterior lateral mesoderm and forms a cardiac crescent in the early embryo. Later in development, these cells coalesce into the linear heart tube and ultimately give rise to the left ventricle. Marker Isl1 defines SHF progenitors that derive from a cell population in the pharyngeal and splanchnic mesoderm, migrate into the developing heart, and constitute the right ventricle, outflow tract, and parts of the inflow tract. Marker WT1 is associated with proepicardial and epicardial precursor cells, a subset of which differentiates into atrial, ventricular, and coronary smooth myocytes and cardiac fibroblasts. Cardiac progenitor cells from FhF and SHF experience a rapidly changing environment associated with cell migration and changes in the architecture of the primitive heart, hence temporal and spatial variations of signaling molecules (BMP, FGF, TGFβ, and Wnt) released by neighboring endocardial, endothelial, and other mesoderm-derived cells

Precursor	Cell lineages
Isl1+ proepicardial progenitor	Isl1−, WT1+, TBx18+ epicardial progenitor
Isl1−, NKx2.5+ FHF progenitor	
Isl1+ SHF multipotent progenitor	Isl1+, HCN4+ progenitor and nodal cardiomyocyte
	Isl1+, TnT+ progenitor and cardiomyocyte
	Isl1+, MyH11+ progenitor and vascular smooth myocyte
	Vascular endothelial cell

epigenetically regulated evolution toward reversible specification and irreversible determination with differentiation into cells of the endo-, myo-, and epicardium.

Commitment, determination, and differentiation of human embryonic stem cells into a cardiomyogenic lineage require cues from endodermal cells. In other words, the paracrine cardiogenic control exerted by endodermal cells enables segregation of cardiac lineages and their maturation. In addition to genetic control, epigenetic modifications (e.g., DNA methylation, nucleosome positioning, histone methylation and acetylation state by methylases, demethylases, acetyltransferases, and deacetylases) associated with chromatin remodeling participate in cardiogenesis [32].

bHLHc5+ mesodermal cells originated from embryonic stem cells may be the common progenitor for the 2 major cardiac lineages that constitute the so-called first

Table 1.13 Cell fate during cardiogenesis (Source: [32]). A subset of epiblast cells undergoes epithelial–mesenchymal transition and transiently forms the mesendoderm (bHLHa26(27): class-A basic helix-loop-helix protein-26(27) [or HAND2(1): heart and neural crest derivatives expressed protein-2(1)]; bHLHc5(6): class-C bHLH protein-5(6) [or MesP1(2): mesoderm posterior-1(2) homolog]; Fox: forkhead box; Gsc: goosecoid homeobox gene product; MEF: myocyte enhancer factor; Nanog: ever young (Gaelic); NKx2: NK2 transcription factor-related homeobox protein; Oct: octamer-binding protein; Sox: sex-determining region Y (SRY)-box (SOX) gene product; SRF: serum response factor; T: brachyury homolog; TBx: T-box transcription factor). Determination of cardiac cell lineages relies on signaling molecules such as Nodal and bone morphogenetic proteins (BMPs) of the transforming growth factor (TGFβ) family, fibroblast growth factors (FGFs), sonic Hedgehog (SHh), and wingless-type morphogen (Wnt), which cooperate to reprogram epiblast, mesendoderm, and then mesoderm in heart-forming region

Phase	Cell markers (programming messengers)
Epiblast	Oct4+, Sox2+, Nanog+, T+ cell (Nodal, Wnt)
Mesendoderm	Oct4+, Sox17+, FoxH1+, Gsc+, T+ cell (Nodal, BMP2, Wnt)
Mesoderm	bHLHc5/6+, TBx6+, SRF+, T+ cell (BMP2)
Cardiac mesoderm	bHLHa26/27+, GATA4/5/6+, MEF2c+, NKx2.5+, TBx5+ cell (FGFs, BMP2)
Cardiomyocyte	α-Actin+, actinin+, myosin+ cell (FGFs, BMPs, SHh, Wnt)

(Isl1−, TBx5+ cardiac lineage) or second (Isl1+, TBx1+, AlDH1a2+ (aldehyde dehydrogenase-1A2), HES1+, FoxH1+ cardiac lineage) heart fields [32]. Segregation between these 2 pools results from the action of FGF8 factor.

Among developmental messengers, BMP2 is secreted by [32]: (1) visceral endodermal, extraembryonic mesodermal, and promyocardial cells to signal to epiblast-derived cells as well as (2) pharyngeal endoderm in contact with the pharyngeal mesoderm, which creates the second heart field. The crucial balance between Nodal and BMP2 signaling is regulated by morphogens such as Wnt–β Ctn axis in the mesendoderm.

The proximal–distal gradient of Nodal enables the segregation of endoderm and mesoderm as well as that of mesendoderm and cardiac mesoderm. It regulates Oct4 expression. The cardiogenic action of Oct4 involves Sox17, a mesendodermal and endodermal marker.

Determination of the cardiac cell phenotype begins when cells move from the posterior to the anterior region under the influence of instructive factors secreted by both the visceral embryonic endoderm and pharyngeal endoderm [32]. Among mesodermal cells, prospective endo-, myo-, and epicardial cells express DNA sequence GATA-binding transcription factors GATA4, GATA5, and GATA6, heart and neural crest derivatives expressed protein HAND1 and HAND2 (or bHLHa27 and bHLHa26 transcription factors), Wilms tumor protein WT1, and NK2 transcription factor-related homeobox gene product Nkx2-5.

Morphogen Wnt3a may guide the migration of cardiogenic mesodermal cells. It may operate via mesoderm posterior transcription factor MesP1 (or bHLHc5). The latter is required for epithelial-to-mesenchymal transition of a subset of epiblast cells as well as migration of cardiovascular progenitors [32]. Morphogen Wnt2 expressed in the posterior cardiac mesoderm limits the number of VEGFR2+ and hematopoietic cells, but promotes endothelial and cardiac differentiation via a noncanonical pathway. Both Wnt5a and Wnt11 promote both heart fields in a signaling-dependent manner at a given time window during cardiogenesis, favoring the second heart field.

The lateral mesoderm includes progenitors of several cell lineages (hematopoietic, endothelial, and smooth, craniofacial, and cardiac myocytes). An early event segregates the future hemo- and cardiogenic cell populations. An early epiblast progenitor cell may give rise to both VEGFR2+, Brachyury+ and VEGFR2−, Brachyury+ cell populations, under the action of BMP4 secreted by the extraembryonic ectoderm and BMP2 by the visceral endoderm, respectively. The first mesodermal cell lineage emerging from the most posterior mesodermal region is the VEGFR2+ cell population in response to BMP4 [32]. VEGFR2^{high} cells form blood islands; VEGFR2^{low} cells reside in a large zone of multipotent mesoderm.

Hemangioblasts are mesodermal progenitor cells committed to blood, endothelial, and smooth muscle cells. A VEGFR2+ lineage re-emerges from a VEGFR2− cell population and also possibly from the hemangioblast lineage generating the endocardial cell population. A subpopulation (5–10%) of Brachyury+, VEGFR2− cells can re-express VEGFR2 and differentiate into cardiomyocytes upon VEGF stimulation, in addition to endothelial cells, hence generating both myo- and endocardium [32]. The endocardium can originate from both an Isl1− and an Isl1+ lineage (Isl1: insulin gene enhancer islet-1, i.e., Isl homeodomain LIM motif-containing transcription factor). In the cardiac crescent, ETS variant ETV2 of the ETS (E26) family of transcription factors is an NKx2-5 target that serves in endothelial and endocardial specification.

The first identifiable cardiomyocytes are found in the splanchnic mesoderm in the cardiac crescent. As the embryo grows, the crescent fuses to build the primitive heart tube [32]. The primitive heart tube generates the left ventricle, atrioventricular canal, sinus venosus, and major parts of atria. Looping and elongation of the heart tube depends on a second source of cardiac progenitor cells (second heart field) in the pharyngeal mesoderm medially and dorsally to the cardiac crescent. These FGF8+, FGF10+ TBx1+, Isl1+ progenitors form the right ventricule and outflow tract myocardium as well as a minor sleeve of smooth myocytes at the base of large arteries [32].

The epicardium is formed by the outgrowth of proepicardial cells in the proepicardial organ that may arise from the transverse septum and migrate toward the sinus venosus into the pericardial cavity when the heart tube elongates. Epicardium-derived cells can migrate into the myocardium and differentiate into smooth myocytes and fibroblasts. They can also give rise to myocardial cells. A WT1+ mesodermal cell population can create NKx2-5+, Isl1+ cardiac progenitors. Signaling by FGF via MAP2K1 and MAP2K2 is mandatory for the early separation of the epicardial lineage from the precardiac mesoderm [32]. However, FGF is not necessary to induce or maintain expression of TBx18 and WT1 epicardial markers.

1.6.10 Cardiac Progenitor Cells and Repair

New cardiomyocytes originate from division of differentiated previously quiescent cardiomyocytes and differentiation of stem cells. In mammals, enlargement of the heart during embryo- and fetogenesis relies primarily on cell division. Shortly after birth, hypertrophy is mainly related to cardiomyocyte growth. In young adults, the low cardiomyocyte generation rate results from the division of a small proportion of pre-existing cardiomyocytes (<1%) [33]. The cardiomyocyte turnover declines with normal aging.

The heart has a limited intrinsic regenerative capacity. However, cell population regeneration can rise after injury. The minimal renewal of cardiomyocytes in adults heightens near regions of myocardial injury, but the proportion of involved cardiac cells reaches only 3%. Pre-existing cardiomyocytes constitute the main source of cardiomyocyte renewal.

Therefore, boosting the ability of cardiac cell replacement should support proper damaged heart healing. Regenerative medicine is aimed at augmenting the amount of functional cardiomyocytes via the transplantation of progenitor cells that must integrate the myocardium and participate in the controlled electrochemical signal propagation and coordinated mechanical work (Vol. 5 – Chap. 11. Tissue Development, Repair, and Remodeling). Administration of stimulants can also be proposed.

Numerous soluble factors (e.g., neuregulin, periostin, and fibroblast growth factor) stimulate cell division [34]. In addition, certain microRNAs can stimulate division of adult cardiomyocytes, but not cardiofibroblasts. Among microRNAs, 204 augment the rate of division of postnatal rat cardiomyocytes and more than 300 suppress division [35]. Among the 204 cell division-stimulating microRNAs, 40 also function in mice.

Among microRNAs, miR119, miR133, and members of the MIR15 family impede cardiomyocyte proliferation. Exogenous administration of selected microRNAs, such as miR590 and miR199a, stimulates cardiomyocyte proliferation [35]. After myocardial infarction in mice, these microRNAs succor cardiac regeneration.

Cardiac stem cells reside in the myocardium. Embryonic epicardial cells proliferate, migrate, and differentiate into cardiac and vascular smooth myocytes, fibroblasts, and endothelial cells. Epicardial progenitors express TBx18 and WT1

Table 1.14 Cardiac fields and corresponding cell lineages (Source: [32]; MyL2: cardiac, regulatory, slow myosin light chain-2 [MLC2v]; MyL7: regulatory myosin light chain-7 [or MLC2a]; OSR1: odd-skipped-related protein-1)

Tissue	Precursor cells (markers)
Precardiac mesoderm	Common myocardial progenitor (Isl1+, VEGFR2^{low})
	Endocardial progenitor (VEGFR2^{high})
	Epicardial progenitor (WT1+)
First heart field	Atriomyocyte (MyL7+)
	Atrioventricular cell (TBx2+, TBx3+)
	Left ventriculomyocyte (MyL2+)
Second heart field	Right ventriculomyocyte (MyL2+)
	Subaortic myocyte (Sema3c−)
	Subpulmonary myocyte (Sema3c+)
	Atrioventricular cell (TBx2+, TBx3+)
	Venous pole cell (Isl1+, OSR1+), which generates nodal cell (HCN4+), atriomyocyte (MyL7+), and sinus horn cell (TBx18+)

transcription factors, markers of cardiovascular progenitor cells. Terminally differentiated adult epicardium is quiescent.

However, actin-binding thymosin-β4 is able to induce the migration of epicardium-derived progenitor cells from the adult heart [37]. Thymosin-β4 that causes adult epicardial progenitor mobilization is a potent stimulator of coronary vasculo- and angiogenesis,

WT1+ proepicardial cells arising from Isl1+, NKx2.5+ progenitors are cardiomyocyte progenitors that can be used in cardiac regeneration and repair (Table 1.14) [38].[25]

As are mesenchymal, neural, and cancerous stem cells, cardiac stem cells can be isolated according to the high expression of aldehyde dehydrogenase (AlDH) [41]. Atrial appendage, AlDH+, CD34+, PTPRc− stem cells differ from cardiac,

[25]The Wilms tumor Wt1 gene encodes a transcription factor that is inactivated in a subset of Wilms' tumors. Alternative splicing, RNA editing, and the use of alternative translation initiation sites generate multiple isoforms with overlapping and distinct functions during embryo- and fetogenesis as well as in the maintenance of organ function. Protein WT1 is produced in embryos in many tissues, such as the urogenital tract, spleen, some brain regions, spinal cord, mesothelia, diaphragm, limb, proliferating coelomic epithelium, epicardium, and subepicardial mesenchyme [39]. Hypoxia stimulates WT1 expression in the heart. It is produced in the coronary vasculature after local ischemia and may support growth of coronary vessels after myocardial infarction. [40].

multipotent, SCFR+ stem cells. Differentiated cells produce the most important cardiac-specific gene products and have an electrophysiological behavior similar to that of adult cardiomyocytes.

Reprogramming of adult human fibroblasts to a cardiomyocyte requires to determining the optimal combination of factors that are necessary and sufficient to force expression of cardiac transcription factors and/or muscle-specific microRNAs. Four human cardiac transcription factors, GATA4, HAND2, myocardin, and TBx5, as well as 2 microRNAs, miR1 and miR133, activate cardiac marker expression in neonatal and adult human fibroblasts [42].

1.7 Vasculature

The vasculature is characterized by 3 properties: *diversity*, *complexity*, or rather *intricacy* in the restricted sense of its architecture,[26] and *variability*. Diversity means a large between-subject variability in vessel origin, shape, path, and branching. Because the flow dynamics strongly depend on vessel configuration, subject-specific models are required for improved diagnosis and treatment. The variability results from environmental actions and responses.

1.7.1 Systemic and Pulmonary Circulation

The cardiac pump distributes blood to the body's organs and perfuse all constitutive tissues of the organs. Blood is propelled under high pressure, which is greater in the systemic circulation than in the pulmonary circulation.

---

[26]Complexity is a term reserved in this book series for characterizing a behavior of a system composed of many constituent parts in intricate arrangement. The behavior that emerges from a collection of interacting components (i.e., that cannot be reduced to the sum of the contributions of the system's component) is endowed by a set of features, such as adaptivity to environmental (chemical, physical, and mechanical) stimuli and nonlinear interaction with positive and negative feedback, whatever the length scale of the object of interest (i.e., a signaling cascade [Vol. 3 – Chap. 1. Signal Transduction], a cell with its self-organization and -reproduction [Vol. 1 – Chap. 4. Cell Structure and Function], or a physiological apparatus). Complex adaptive systems are complex as their behavior results from dynamical interactions rather than simple summation of relations between entities and adaptive as the individual and collective behavior self-organize when experiencing any event. A complex adaptive system has some or all of the following attributes: the relations between the constituent parts as well as between the system and its environment are nonlinear; it is subjected to feedback; it can adapt to its environment; and it may be sensitive to initial conditions. On the other hand, a complicated system is characterized by linear interactions between its components. When the word "complexity" is related to a structure and architecture as well as even a mechanism (without reference to the properties of a complex system), the noun "intricacy" is used, not as a synonym of subtlety, but with the meaning of "very complicated" (intricate).

The right and left hearts, with their serial chambers, play the role of a lock between low-pressure venous and high-pressure arterial circulation. The atrioventricular coupling sets the ventricle for filling and pressure adaptation, and the ventriculoarterial coupling for ejection.

1.7.2 Blood Circuit

Deoxygenated blood from upper body regions (head, neck, superior limbs, and upper part of the thorax) and from the lower body regions (inferior limbs, and lower torso [lower chest, abdomen, and pelvis]) is brought to the right atrium by the superior (SVC) and inferior (IVC) vena cava.

When the pulmonary valve is open, the right ventricle ejects blood into the pulmonary artery. Pulmonary veins carry oxygen-rich blood from the lungs to the left atrium. The aorta receives blood ejected from the left ventricle.

The aorta generates a set of arteries that themselves divide to give rise to arterial beds with arteries of decreasing size, arterioles, and capillaries that deliver nutrients to tissues. Blood is collected through merging capillaries, which remove catabolites, venules, and small veins. It returns to the heart through the veins under low pressures (Fig. 1.13). Large blood vessels of the upper part of the mediastinum, efferent from and afferent to the heart, are displayed in Figs. 1.14 and 1.15.

1.7.3 Blood Compartments

Each blood circuit in the systemic and pulmonary circulation is composed of 3 main compartments: arterial, capillary, and venous. The arterial tree can be split into a distribution compartment, which includes arteries irrigating main body regions, and a perfusion compartment, which is composed of smaller arteries irrigating tissues.

The main architectural features of the arterial tree are: (1) taper and continuously varying curvature; (2) branchings with a between-branch length of a few vessel radii, and (3) bifurcations after giving birth to 5 to 10 branches. Branching and bifurcations are characterized by their anatomical (branching angles, wall curvature, apex morphology, and area ratio) and functional features such as the flow division ratio that, at least partly, depends on downstream impedances.

The venous collector can also be subdivided into serial compartments. The vascular networks constitute a closed, tortuous, multigeneration circuit of deformable vessels, more or less short, that divide (arteries) and then merge (mainly veins).

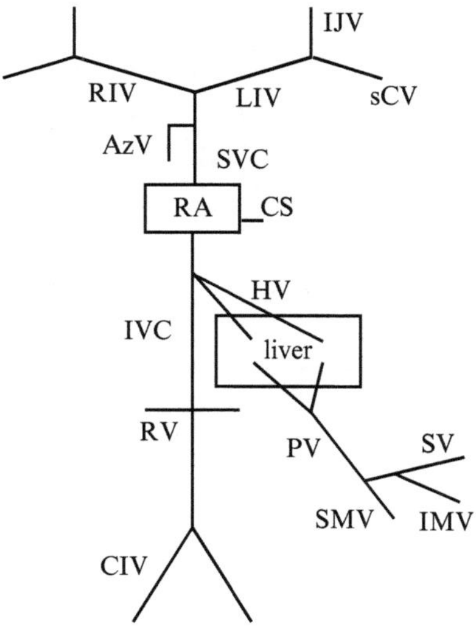

Fig. 1.13 Schematic drawing of the large vein system. The left (LIV) and right (RIV) innominate veins, formed by the junction of the internal jugular (IJV) and subclavian (sCV) veins, unite to form the superior vena cava (SVC). The azygos vein (AzV) enters and ascends in the thorax to finally arch and end in the superior vena cava. Most of the cardiac veins end in the coronary sinus, which goes to the right atrium (RA). The common iliac veins (CIV), formed by the union of the external iliac and hypogastric veins, unite to form the inferior vena cava (IVC). The inferior vena cava receives the renal (RV), suprarenal, inferior phrenic, and hepatic veins. The portal vein (PV) is formed by the junction of the superior mesenteric (SMV) and splenic (SV) veins, which receives the inferior mesenteric vein (IMV)

Table 1.15 Vessel geometry. Approximate magnitude (mm) of caliber d and wall thickness h for different vessel compartments

Vessel	Aorta	Artery	Arteriole	Capillary	Venule	Vein	Vena cava
d	10–25	3–5	0.30–0.01	0.008	0.01–0.30	5	30
h	2	1	0.02	0.001	0.002	0.5	1.5

1.7.4 Local Geometry and Blood Flow

Blood flows depend on vascular architecture and local geometry, i.e., curvature with possible taper and prints of neighboring organs between branching points and branching features (branching angles, hip and apex curvatures, area ratios, and vessel bores and lengths). The order of magnitude in vessel sizes is given in Table 1.15.

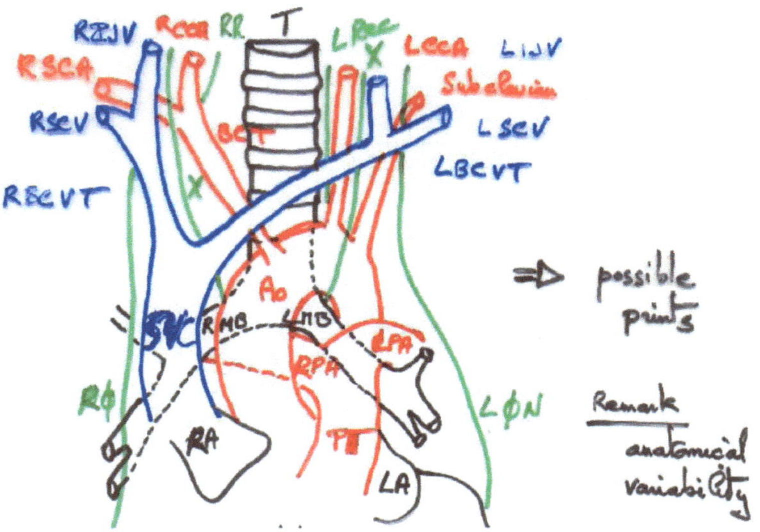

Fig. 1.14 Scheme of large blood vessels of the upper thorax with the airways (coronal view, usual configuration). RA, LA: right and left atria; Ao: aorta (ascending aorta, arch of aorta, and descending aorta); BCT: arterial brachiocephalic trunk; RSCA: right subclavian artery; RCCA, LCCA: right and left common carotid arteries; PT: pulmonary trunk; RPA, LPA: right and left pulmonary arteries; SVC: superior vena cava; RBCVT, LBCVT: right and left brachiocephalic trunks; RSCV, LSCV: right and left subclavian veins; RIJV, LIJV: right and left internal jugular veins; T: trachea; RMB, LMB: right and left main bronchi; RR, L Rec: right and left recurrent nerves; RΦ, LΦN: right and left phrenic nerves; X: vagus nerves

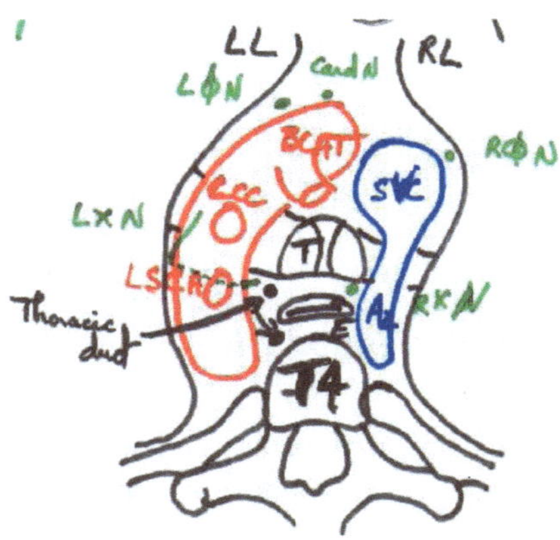

Fig. 1.15 Scheme of large blood vessels of the upper mediastinum (transverse plane). Thoracic (dorsal) vertebra 4 (T4) level. LL, RL, left and right lung parenchymae; Card N: cardiac nerve; RΦ, LΦN: right and left phrenic nerves; LXN, RXN: left and right vagus nerves; E: esophagus; T: trachea; BCAT: brachiocephalic arterial trunk; LCC: left common carotid artery; LSCA: left subclavian artery; SVC: superior vena cava; Az: azygos vein

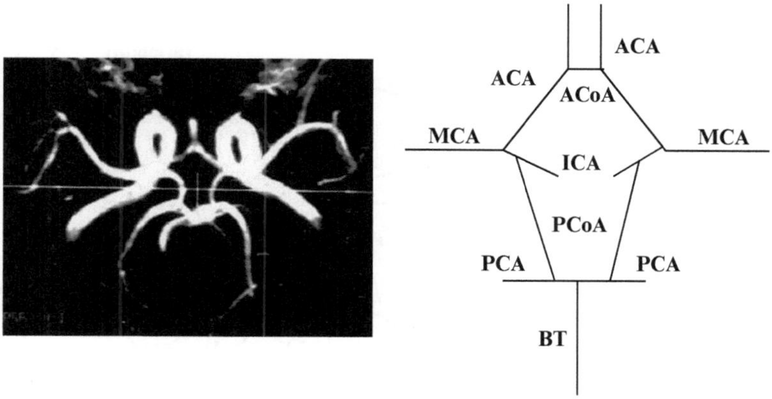

Fig. 1.16 (**Left**) Magnetic resonance image of the Willis circle in a healthy female volunteer. The arteries are characterized by huge curvature and branching. This anastomotic network has various anatomic variations among individuals (frequent absence in communicating arteries, branching variability). (**Right**) Schematic drawing of the cerebral artery network. The internal carotid artery (ICA) gives birth to the middle cerebral (MCA or sylvian artery) and anterior cerebral (ACA) arteries. The anterior communicating artery (ACoA) connects the 2 ACAs. The posterior communicating artery (PCoA) anastomoses ICA with the posterior cerebral arteries (PCA), the terminal branch of the basilar trunk (TB). The basillaris is formed by the union of the right and left vertebralis, largest branches of the subclavian arteries (the arteries merge rather than branch off)

1.7.5 Particular Structures

The vasculature contains some peculiar structures. *Arteriovenous anastomoses* are found in the skin and gut; mural smooth muscles usually close such bypasses.

Arterial anastomoses are frequently observed between the head arteries, in particular between the branches of internal (ICA) and external (ECA) carotid arteries, as well as between the branches of ICA and those of the basilar trunk (an example of artery merger)[27] to form the Willis circle under the cerebral hemispheres (Fig. 1.16).[28]

In the brain, arterial and venous distributions are separate, whereas in other body regions, arteries and veins run together with nerves in a connective sheath as neurovascular bundles.

Portal drainage delivers venous blood directly from a capillary basin to another capillary bed through veins. Portal drainage includes the hepatic and hypophyseal portal system, i.e., blood supply to the liver and pituitary gland from other organs, such as the digestive tract for the hepatic portal system (Fig. 1.17).

[27]The vertebral arteries merge to give birth to the larger basilar trunk.

[28]The large arteries supplying the brain are the carotid and vertebral arteries, leading to the network of pial arteries on the surface of the brain. On the cerebral cortex, the arteries branch into smaller penetrating arteries, which enter the brain.

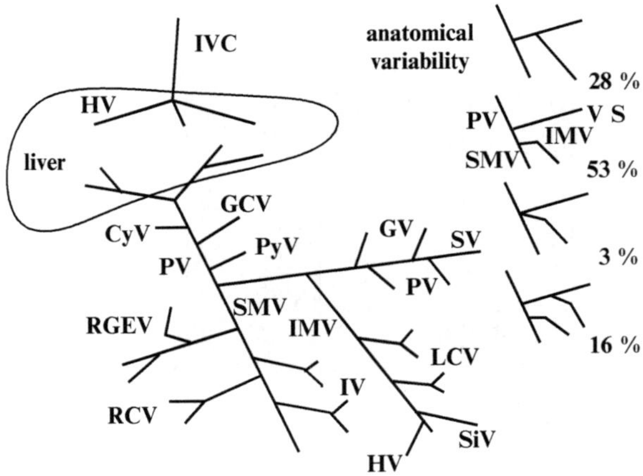

Fig. 1.17 Schematic drawing of the portal system and its anatomical variability

Sinusoids act on sensors such as the *carotid body*, also named the carotid glomus and glomus caroticum. The carotid body detects changes in the composition of arterial blood flowing through it, mainly the partial pressure of oxygen, but also that of carbon dioxide as well as changes in pH and temperature.

Venous sinuses are endothelium-lined spaces in connective tissue where blood is collected (e.g., coronary and dural sinuses and erectile tissue).

Vasa vasorum constitute a microvascular circuit in the adventitia of large blood vessels. These vessels supply oxygen and nutrients to the outer wall region, which is not close to the vessel lumen crossed by blood. However, in the arterial wall, expansion of the vasa vasorum is correlated to progression of atherosclerosis [43].

1.7.6 Microcirculation and Macrocirculation

Different bore (d) values of blood vessels are used in the literature to define the limit beween macro- and microcirculation: 500, 300, or 250 μm. *Microcirculation* starts with arterioles ($10 < d < 250$ μm) and ends with venules. Local microvascular networks have been observed using suitable microscopy[29] in thin tissues, such as the mesentery, the cremaster, etc.

The capillary network is interposed between small arterioles and small venules. Capillaries have the thinnest wall suitable for molecule transfer. In particular, capillaries in the lung parenchyma are closed to the alveolar wall for efficient gas exchanges.

The alveolocapillary membrane, or blood–air barrier, is a double wall (thickness 200 nm–2 μm) formed, at its thinnest part, by: (1) the endothelial glycocalyx; (2) the

[29] A fibered confocal fluorescence microscopy system dedicated to imaging at the microscopic level, with non or minimally invasive access, has been developed by Mauna Kea Technologies (www.maunakeatech.com), in particular for microvascular research.

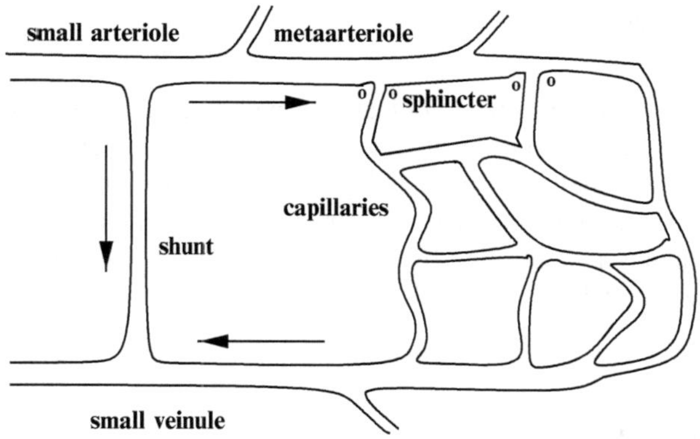

Fig. 1.18 Capillary network interposed between a small irrigation arteriole and a small drainage venule, with an arterioloveinular anastomosis (shunt) and precapillary sphincters

extranuclear part of the continuous capillary endothelium; (3) a fused basement membrane (from those of alveolar epithelium and capillary endothelium); (4) the extranuclear part of the alveolar wall (type-1 and -2 pneumocytes); and (5) a liquid film with surfactant in the alveolar space that lines pneumocytes. Surfactant reduces the surface tension and adapt this tension to the lung volume (Vol. 5 – Chap. 13. Surfactant). The alveolocapillary membrane is permeable to gases (oxygen, carbon dioxide, carbon monoxide, etc.). The alveolar wall is a major barrier to diffusion of small solutes and water. On the other hand, paracrine agents released by the endothelium participate in the control of pulmonary vessel tone and permeability as well as of the conductance to O_2 flux from the alveolus to its uptake by hemoglobin.

A given edge of the pulmonary capillary in the interalveolar septum is related to a given alveolus, the opposite edge to the adjoining alveolus. In connecting parts of the interalveolar septa, the capillary can even be implicated in gas exchanges with more than 2 neighboring alveoli. Flow in the alveolar septum can then be modeled by sheet flow.

A sketch of the microvasculature shows the afferent arteriole and efferent venule with possible arteriolovenular anastomosis (Fig. 1.18).[30] The arteriole gives birth either to metarterioles or directly to capillaries [44]. The entrance segment of the capillary has a precapillary sphincter, which regulates local flow distribution. In some territories, 40–50% of the total blood volume is contained in the microcirculation, the main part being either in the capillaries (heart and lungs) or venules (mesentery) [45].

Taking into account the estimated number of vessels per vascular compartment, the cumulated cross-sectional area is the highest and blood velocity the lowest in the capillaries, the blood–tissue exchange zone. The ratio between the wall thickness and lumen caliber, as well as the resistance are the greatest in the arterioles.

[30] Arteriovenous shunts connect small arterioles and venules.

1.7.7 Adaptation

The vasculature adapts to physiological and pathological situations. The uterine arteries increase in bore during pregnancy and return to their original size after birth (reversible remodeling). Cardiac output and stroke volume also increase during pregnancy. Hypertrophic and hypotrophic remodeling of the uterine arteries during and after pregnancy is controlled by different substances, such as the angiotensin-converting enzyme (Chap. 3) [46].

1.7.8 Fractal Vasculature

Every cell resides close to a blood vessel ($\sim$100μm) to be adequately supplied in oxygen and nutrients. The capillary length was assessed to equal about 600 μm [47] and the capillary density to equal about 250/mm^2 [48]. The number of vessels observed in the field of view of a microscope increases by a given factor when magnification increases. Fractal analysis enables assessment of the architecture of some compartments of the vasculature.

However, the self-organized vascular circuit is not strictly a self-similar branching network. At each bifurcation, the radii of the trunk (R_t) and larger (R_{b1}) and smaller (R_{b2}) branch fulfill the *bifurcation law*, or branching relation:

$$R_t^p = R_{b1}^p + R_{b2}^p, \quad \text{or} \quad d_{\mathrm{h}_t}^p = d_{\mathrm{h}_{b1}}^p + d_{\mathrm{h}_{b2}}^p, \tag{1.5}$$

where d_{h_t}, $d_{\mathrm{h}_{b1}}$, and $d_{\mathrm{h}_{b2}}$ are hydraulic diameters of the trunk (parent [generation i]) and branch (child or daughter [generation $i+1$]) vessels, respectively, and p the bifurcation exponent that specifies the design and head loss, the energy dissipation being minimized. At the network level, the design indices ($\{D_j\}_{j=1}^2$) give the power-law relation between the number of generation i conduits and the hydraulic diameter (d_h) and length (L) [49]:

$$N_i = (d_{\mathrm{h}_i}/d_{\mathrm{h}_0})^{-D_1}, \quad L_i = L_0(d_{\mathrm{h}_i}/d_{\mathrm{h}_0})^{D_2}, \tag{1.6}$$

where d_{h_0} and L_0 are the hydraulic diameter and length of the tree inlet tube. Branching (R_b) and length (R_L) ratios of a branching network are given by

$$\mathsf{R}_b = \frac{N_i}{N_{i+1}}, \quad \mathsf{R}_L = \frac{L_{i+1}}{L_i}. \tag{1.7}$$

The caliber ratio (R_d) and caliber asymmetry ratio (γ_d) are defined as

$$\mathsf{R}_d = d_{\mathrm{h}_t}/d_{\mathrm{h}_{b1}}, \quad \gamma_d = d_{\mathrm{h}_{b2}}/d_{\mathrm{h}_{b1}}. \tag{1.8}$$

The relative change of vessel bore and length from a given generation to the next is nearly constant for numerous vascular circuits (Horton law):

$$\frac{d_{i+1}}{d_i} = k, \quad \frac{L_{i+1}}{L_i} = k. \tag{1.9}$$

Thus vessel length and bore are fractal quantities if vessel generation represents a geometrical or functional scale. The vessel scale can be defined as the volume of tissue supplied by a given artery or drained by a given vein:

$$L_i, d_i = \log V. \tag{1.10}$$

Regional blood flows in organs and tissues are heterogeneous. A power-law function of scale has been shown to govern organ perfusion[31] heterogeneities resulting from scale-independent branching asymmetries,[32] hence difference in tissue volume fed by each branch, and vessel resistance assumed to be a fractal [50]. The degree of perfusion heterogeneity is commonly quantified by measuring *relative dispersion* (RD; or coefficient of variation), which is defined as the standard deviation of perfusion measurements in local volumes of given size divided by the average of these measurements (RD = standard deviation/mean of regional flows). The RD drop with augmented mass (m) of volume elements can be described by a power-law function [50]:

$$RD(m) = m^p \, RD(1), \tag{1.11}$$

where the exponent p is the scale-independent slope of the RD–m graph on a log–log plot. The exponent p was related to the fractal dimension (D_f):

$$p = 1 - D_f. \tag{1.12}$$

It is also related to the correlation coefficient (C) between neighboring perfusion measurements P_1 and P_2:

$$p = \log(C + 1), \quad C = \frac{< (P_1 - < P_1 >)(P_2 - < P_2 >) >}{< (P_1 - < P_1 >)^2 >< (P_2 - < P_2 >)^2 >}. \tag{1.13}$$

The magnitude of spatial heterogeneity characterized by the relative dispersion depends on the length scale,, i.e., on sample size. The relative dispersion of the spatial distribution of flows for a given spatial resolution is given by [53]:

---

[31] Perfusion is the amount of blood delivered to a unit mass of tissue per unit time.

[32] Branching asymmetry (γ_V) is defined as $\gamma_V = V_1/(V_1 + V_2)$, where V_1 and V_2 are tissue volumes fed by daughter branches of a given vessel, with $V_1 < V_2$ (i.e., $0 < \gamma_V \leq 0.5$). For symmetric branching, $\gamma_V = 0.5$.

$$RD(m) = RD(m_{ref}) \frac{m}{m_{ref}}^{1-D_f}, \quad RD(V) = RD(V_{ref}) \frac{V}{V_{ref}}^{1-D_f}, \tag{1.14}$$

where m is the mass of tissue sample (in gram), m_{ref} is chosen to be 1 g, V is the volume of tissue samples, and V_{ref} is an arbitrarily chosen reference volume. The relative dispersion exhibits self-similarity upon scaling with respect to the sample size. Equation 1.14 represents a fractal relation, the parameter D_f, a global measure of heterogeneity, being a fractal dimension ($D_f = 1.0$ indicates uniform blood distribution among organ territories and $D_f = 1.5$ spatial randomness). In the heart, $D_f \sim 1.2$ [53]. In a model that assumed equal terminal flows in normal hearts, the fractal dimension of perfusion heterogeneities ranged from 1.20 to 1.40 [54].

The vessel resistance power law is based on the hypothesis that both vessel bore and length individually follow the power law as a function of volume scale with respective scaling s_d and s_L:

$$d(V) \propto V^{s_d}, \quad L(V) \propto V^{s_L}, \quad \text{therefore,} \quad L \propto d^{s_L/s_d}, \tag{1.15}$$

where the exponent s_L/s_d is experimentally estimated to equal 3/4, the resistance to flow of a given vessel being proportional to its length and inversely proportional to its caliber.

The vasomotor tone was assumed to have a chaotic dynamics due to the interaction of 2 calcium oscillators [26]. The fast oscillator is related to voltage-gated calcium channels and the slow oscillator to calcium-induced calcium release. This chaotic behavior is counteracted by nitric oxide and various vasodilators; conversely, it can result from a sustained reduced perfusion pressure. In the compartment of resistive vessels (small arteries and arterioles), the chaotic behavior was supposed to be regulated by the sympathetic nervous system.

The retinal microvasculature was categorized into 3 ranges of fractal dimension ([1.323–1.427], i.e., sparser microvascular branching pattern [1.428–1.457], i.e., moderately complicated branching pattern, and [1.458–1.506], i.e., denser branching patterns). The higher or the lower the fractal dimension (i.e., suboptimal retinal vascular branching pattern), the stronger the risk of coronary heart disease is [51].

Narrowings (stenoses) or expansions (aneurysms) of the vascular lumen generates abnormal flow patterns. Chaotic convection of particles implies dependence on the initial conditions. Any 2 initially close particles (at a small distance [d_0]) exponentially deviate from each other ($\mathrm{d}(t) = \mathrm{d}_0 \exp\{\lambda t\}$), where λ is the Lyapunov exponent, a measure of the average rate of the exponential path divergence, which occurs along specific curves (unstable foliation, or unstable manifold). Chaotic particle transport causes high stretching and filamentary fractal structures [52]. Platelets can be activated by high shear, e.g., by stretching bonds between the von Willebrand factor and platelet. On the other hand, deposition is associated with stagnant flow regions, i.e., with higher residence time; in these regions, the Lyapunov exponent has lower values.

However, a fractal approach, which may explain some patterns, does not contribute to the knowledge of biological mechanisms responsible for organogenesis and functioning of the physiological apparatus. During embryonic and fetal angiogenesis, blood vessels do not follow a fractal branching pattern [55]. Genetic and mechanical factors such as wall stress field are involved in vascular wall development (Vol. 5 – Chap. 10. Vasculature Growth). Among various isoforms of vascular endothelial growth factor (Vols. 2 – Chap. 3. Growth Factors and 3 – Chap. 8. Receptor Protein Kinases), such as $VEGFa_{121}$ and $VEGFa_{165}$), and placental growth factor (PlGF1 and PlGF2), VEGF homo- and heterodimers and $VEGFa_{121}$–PlGF1 heterodimers increase vascular intricacy, whereas PlGF1 and PlGF2 are not effective [56].

In adult life, nonfractal branching appears also to be the rule, for example, in arteries branching off the aorta. Moreover, arterial anastomoses and interconnected venous circuits yield situations that cannot be modeled by a fractal approach. The vasculature with its finite number of bifurcations is not self-similar throughout the entire network. In particular, the diameter relations vary from large vessels, where area-preserving branching is assumed, to smaller vessels, where cubic branching is postulated [55].

1.7.9 Circulatory Networks

Some circulatory networks are used as illustrations because they are frequently targeted by wall diseases (atheroma, aneurysms, and varicose veins; Vol. 7 – Chap. 2. Vascular Diseases): the aorta and its main branches (Table 1.16); the carotid arteries and their main branches; the cerebral venous circuit; the vein network of the lower limbs; and the portal venous circuit.

1.7.9.1 Aorta

On the anatomy point of view, the aorta is the trunk of the systemic circulation that is directly connected to the left ventricle. It distributes blood to the body's organs via multiple branches. On the fluid mechanics point of view, the aorta is the upstream artery of the systemic circulation that comprises 3 main segments in the thoracic and abdominal cavities: a short entrance segment, the ascending aorta, a large bend, the aortic arch, and a long thoracoabdominal segment, the descending aorta.

Blood Convector

A unidirectional blood jet (peak velocity 1.35 ± 0.35 m/s [16]) originates from the left ventricle into the ascending aorta through the tricuspid (3-cusp) aortic valve during the systolic ejection phase of the cardiac cycle. When the aortic valve

Table 1.16 Large thoracic vessels (Source: [3])

Artery	Origin and branches
Aorta	
Ascending segment	Originates from left ventricle
	Generates left and right coronary arteries
	Aortic sinus (sinus of Valsalva)
	3 sinuses: (left and right posterior and anterior)
	The left posterior and anterior aortic sinus give rise to the left and right coronary artery, respectively
	Noncoronary right posterior aortic sinus
Arch	Brachiocephalic trunk that splits into right common carotid and subclavian arteries
	Left common carotid artery
	Left subclavian artery
Descending thoracic	Visceral branches (bronchial, esophageal, mediastinal, pericardial)
.	Parietal branches (intercostal, subcostal, superior phrenic)
Common carotid artery	Bifurcates into internal and external carotid arteries
	Site of carotid body (chemoreceptor)
	Internal branch has the carotid sinus (baroreceptor)
Subclavian artery	Generates numerous branches, especially vertebral artery that merges with its homologous into the basilar trunk (arterial junction)
	Continuous with the axillary artery

is closed, blood flows backwardly toward the aortic valve, thereby irrigating the coronary arteries during the diastole (perfusion function). The axes of the left and right coronary arteries correspond roughly to the open ends of the left and right leaflets of the aortic valve, the posterior cusp of the aortic valve being the noncoronary leaflet. In addition, the aorta transforms the starting–stopping flow at its entry into a pulsatile flow because of its compliance that enables blood storage during the systole and restitution during the following diastole (buffer function). The aortic wall distend at each systole, storing energy that is liberated during the diastole with the elastic recoil.

Due to its strong nonplanar curvature and the branching set leading to large intrathoracic arteries, time-dependent flow structure is complicated. The distribution of wall shear stress at the top of the aortic arch is characterized by great spatial and temporal variations [57].

Aortic Size and Path

The aorta starts at the upper right corner of the left ventricle (bore 25–30 mm). After ascending for a short distance, it arches backward and to the left, over the main left bronchus (with a possible print). The aortic arch has a variable configuration according to the subject, especially children (Fig. 1.19). It then descends within

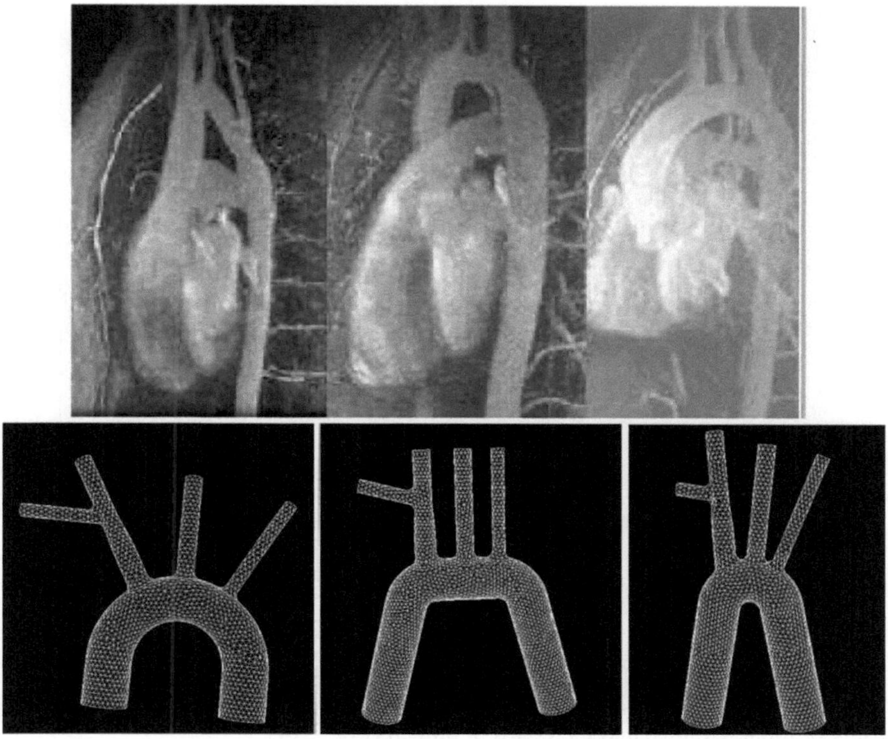

Fig. 1.19 Possible configurations of the aorta arch with its 3 major branches (in most subjects). (**Top right**) Set of radiographies. (**Bottom left**) Mesh of usual shape modeled by a regular circular curvature. (**Bottom mid-panel**) Mesh of aortic arch with a straight segment. (**Bottom right**) Mesh of an angular aortic arch (from M.S. Miette and J. Pichon)

the thorax (bore ~25 mm) on the left side of the vertebral column. It crosses the abdominal cavity (bore 18–20 mm) and ends (bore ~10 mm) by dividing into the right and left common iliac arteries at the lower border of the fourth lumbar vertebra (L4).

The ascending aorta (length ~5 cm) goes obliquely upward, forward, and to the right. Its entrance segment presents 3 mild dilations, the aortic sinuses (Valsalva sinuses), which are bounded by the sinotubular junction. The ascending aorta is contained within the pericardium, together with the pulmonary artery. Its branches are the 2 coronary arteries.

The aorta arch runs at first upward (height 25–40 mm), backward, and to the left in front of the trachea, then directs backward on the left side of the trachea, and finally passes downward at the lower border of the fourth thoracic vertebra (T4).

The descending thoracic aorta localizes to the posterior mediastinum. It ends in front of the lower border of the twelfth thoracic vertebra (T12). It approaches the median line as it descends to be directly in front of the column at its termination. The curved descending aorta has a small tapered shape, as the branches given off are small (Fig. 1.20).

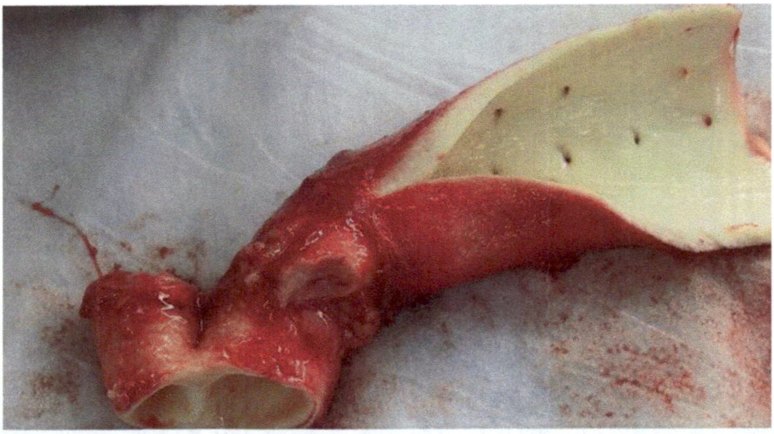

Fig. 1.20 Aorta branchings. The aortic arch downstream from the branching of the brachiocephalic trunk in the calf. A longitudinal incision of the aorta wall, from the dowstream end up to the arterial ligament, exhibits ostia of small branches, such as intercostal arteries

Anatomical Relations

The aorta is in contact with nerves, the left phrenic, the lower of the superior cardiac branches of the left vagus, the superior cardiac branch of the left sympathetic, the trunk of the left vagus, and its recurrent branch (Fig. 1.14).

Below are the bifurcation of the pulmonary artery, the ligamentum arteriosum[33] (Fig. 1.21), which connects the left pulmonary artery to the aortic arch, the left bronchus, and the superficial part of the cardiac plexus.

The descending aorta is in relation to the esophagus and its plexus of nerves, with the hemiazygos veins, azygos vein, thoracic duct, heart, and lungs.

Aortic Branches

The arch gives birth to the brachiocephalic trunk, or innominate (IA), and the left common carotid (LCCA) and left subclavian (LSCA) arteries (Fig. 1.22).

These branches can start from the ascending aorta. In addition, LCCA can arise from the brachiocephalic artery. The left carotid and subclavian arteries can arise from a left brachiocephalic trunk. Conversely, the right carotid and subclavian arteries can branch off directly from the aorta (huge between-subject anatomy variability). In a few cases, the vertebral arteries originate from the arch. The bronchial, thyroid, and internal mammary can also come from the aorta arch.

[33]The ligamentum arteriosum corresponds to the remains of the fetal ductus arteriosus.

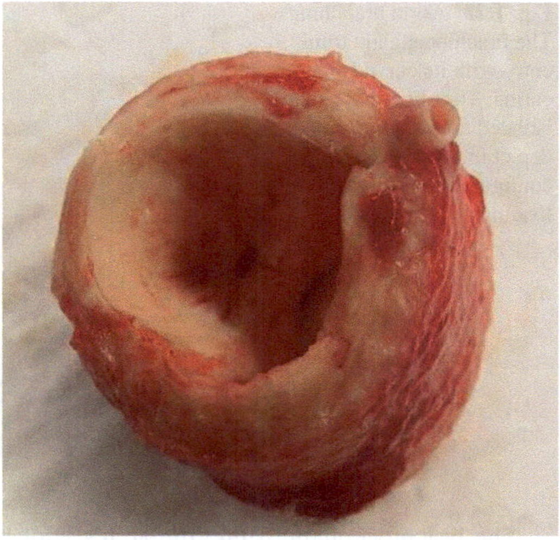

Fig. 1.21 Aorta branching and ligamentum arteriosum: a small side branch (upper right edge) and the closed entry of the large fetal ductus arteriosus (opposite edge). The wall thickness is not uniform, due to wall remodeling in this three-month old calf by flow stresses, according to the nonplanar strong curvature of the aortic arch, using the locus of the ligamentum arteriosum as a landmark

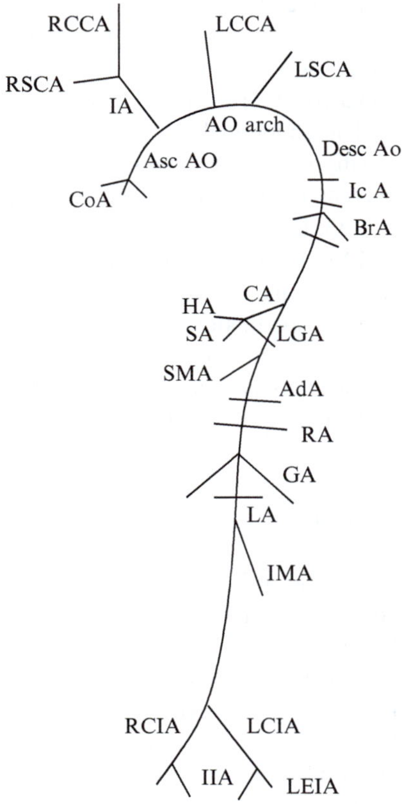

Fig. 1.22 Schematic drawing of the aorta and its main branches. The aorta ascends (Asc Ao) and forms a huge bend, the aorta arch, directed to the left side of the trachea to step over the left main bronchus and then downward, becoming the descending aorta (Desc Ao)

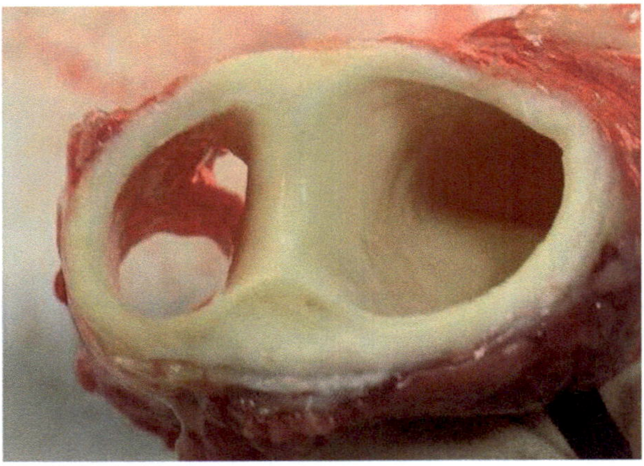

Fig. 1.23 Aorta branchings. The brachiocephalic trunk entry with its curved smooth carina. The slice shape of isolated arteries does not depict the in vivo configuration due to different loadings

The brachiocephalic trunk, the arch's widest branch (length 4–5 cm; Fig. 1.23) ascends obliquely upward, backward, and to the right, and divides into the right common carotid (RCCA) and right subclavian (RSCA) arteries.

The descending aorta gives birth to the costocervical, internal mammary, intercostal (IcA), bronchial (BrA), pericardiacophrenic, mediastinal, esophageal, and epigastric arteries (Fig. 1.22).

The abdominal aorta begins at the aortic hiatus of the diaphragm, diminishes rapidly in size, and gives off many branches (Fig. 1.22). (1) The unpaired branches encompass the celiac artery, which quickly gives birth to 3 branches, the left gastric (LGA), hepatic (HA), and splenic (SA) arteries; the superior (SMA) and inferior (IMA) mesenteric; and middle sacral arteries. (2) The paired branches include on the one hand visceral arteries such as suprarenal, or adrenal (AdA), renal (RA), internal spermatic/ovarian (or genital or gonadal; GA), and, on the other, parietal arteries, such as the inferior phrenic and lumbar (LA) arteries. The aorta finally divides into the 2 common left and right iliac arteries, which divide into the internal (IIA) and external (EIA) branches.

Structure

The aortic wall comprises 3 layers (Vol. 5 – Chap. 7. Vessel Wall). The intima is composed of a thin layer of endothelial cells and a subendothelial connective tissue. It is limited by the internal elastic lamina. The media of the aorta is much thicker than that of other arteries. It is composed of several layers of smooth myocytes immersed in the extracellular matrix separated by lamellae of elastic fibers. It is bounded by the external elastic lamina. The adventitia is made of connective tissue crossed by vasa vasorum and nervi vascularis.

A pressure gradient exists between the aortic lumen and the adventitia. This transmural pressure gradient generates a outwardly radial water flux inside the arterial wall. The hydraulic conductance from the lumen to the adventitia enables transfer of substances that can be retained and processed during their intramural travel.

The intramural mass transfer depends on: (1) permeabilities of the glycocalyx, endothelium, basement membrane, and elastic lamellae; (2) transport properties of the constitutive parietal tissues made of composite poroviscoelastic material; and (3) physicochemical features of moving molecules (mass, charge, hydrophilic properties, affinity for wall components).

Deposition of carried particles can result from: (1) defective washing of blood flow that enables glycocalyx crossing and motion across the intima; (2) increased radial mass transport; and (3) impaired flux out of the adventitia.

Thoracic and Abdominal Aorta

The aorta is a large elastic artery with distinct thoracic and abdominal segments. Heterogeneity between the thoracic and abdominal aorta has developmental origin, as these segments are associated with neural crest cell precursors and mesodermal ancestors, respectively.

In addition, thoracic and abdominal aortic segments differ according to the wall architecture and composition. The thoracic aorta media contains 55 to 60 *medial lamellar units* (concentric bands of elastin filaments and smooth myocytes associated with collagen fibers) that constitute outer vascular and inner avascular (28–30 inner lamellar units close to aorta lumen) zones. The abdominal aortic media is constituted of 28 to 32 units [58]. Media growth after birth is provided by the production of additional lamellar units in the thoracic aorta, whereas lamellar units of the abdominal aorta widen.

Moreover, elastin and collagen content are smaller in the abdominal aorta than in the thoracic aorta. The thoracic aorta then has a greater distensibility than the abdominal aorta.

1.7.9.2 Carotid artery

Blood is supplied to the brain via: (1) 4 main arteries, 2 internal carotid and 2 vertebral arteries and (2) more or less numerous anastomosis according to the subject between the deep and superficial perfusion networks.

The common carotid artery (CCA) bifurcates into the internal (ICA) and external (ECA) carotid arteries. The internal carotid artery supplies the cranial and orbital cavities and the external carotid artery irrigates the exterior of the head, face, and neck.

At the carotid bifurcation, the wall of the carotid sinus has receptor cells innervated by the glossopharyngeal nerve. The carotid glomus lies behind the common carotid bifurcation.

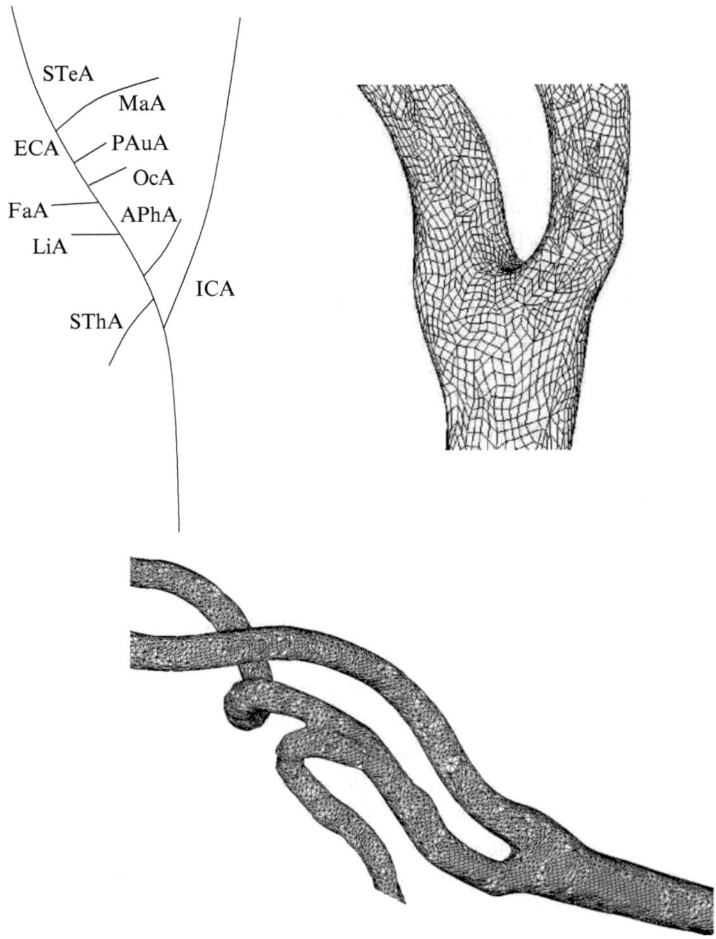

Fig. 1.24 (**Top left**) Carotid arteries and its main branches. (**Top right**) zoom of a quadrangle surface mesh of the carotid bifurcation suitable for fluid-structure interaction computations using MIT shell element. (**Bottom**) 3DR reconstruction and facetization of the regional artery network

Branches of the external carotid artery can be divided into: (1) anterior (superior thyroid [SThA], lingual [LiA], and facial [FaA] arteries); (2) posterior (occipital [OcA] and posterior auricular [PAuA] arteries); (3) ascending (ascending pharyngeal [APhA]); and (4) terminal (superficial temporal [STeA] and maxillary [MaA]) arteries (Fig. 1.24).

The internal carotid artery can be decomposed into 7 successive segments in the streamwise direction: cervical (C1), petrous (C2), lacerum (C3), cavernous (C4), clinoid (C5), ophthalmic (C6), and communicating (C7) [60].

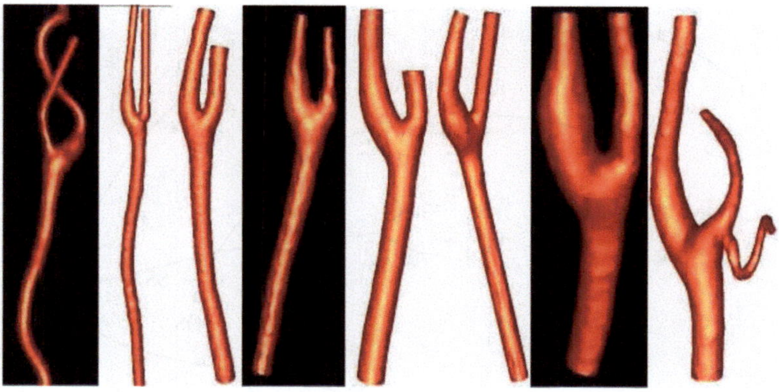

Fig. 1.25 Anatomic variations of the carotid bifurcation [59] (with permission)

1.7.9.3 Cerebral Veins

The cerebral drainage and perfusion circuits use different paths. The cerebral venous network encompasses 2 major circuits [61]: (1) the superficial circuit with large dural veins and sinuses fed by cortical veins and (2) the deep circuit with subependymal, internal cerebral, basal, and great cerebral vein of Galen.

The superficial cerebral venous network drains blood mainly from the cerebral cortex and subcortical white matter; the deep cerebral venous network collects the deep white and gray matter nuclei near the lateral and third ventricles and the basal cistern.

Dural venous sinuses, a.k.a. dural, cerebral, or cranial sinuses, are low-pressure venous ducts. They localize between the periosteal and meningeal layers of the dura mater, or dura, that is the outermost of the 3 layers of the meninges surrounding the brain.

Superficial cerebral veins return venous blood to the superior (SSS) and inferior (ISS) sagittal sinuses. The superior sagittal sinus runs along the midline below the skull in the dura mater at the superior margin of the falx cerebri to reach the confluence of sinuses (Fig. 1.26). Along its path, it receives tributaries, the superficial cortical veins; its caliber then increases. The inferior sagittal sinus courses in the falx cerebri, like its superior counterpart, but along the inferior margin of the falx cerebri, just above the corpus callosum. It empties into the straight sinus, or sinus rectus, at the intersection of the tentorium cerebelli and falx cerebri, at the merging point with the great cerebral vein of Galen. The straight sinus thus drains the deep cerebral venous blood via the vein of Galen as well as the superficial cerebral venous blood via the inferior sagittal sinus.

At the occipital pole, the superior sagittal, straight, and occipital sinuses empty into the confluence of sinuses (CS), a.k.a. confluens sinuum and torcular

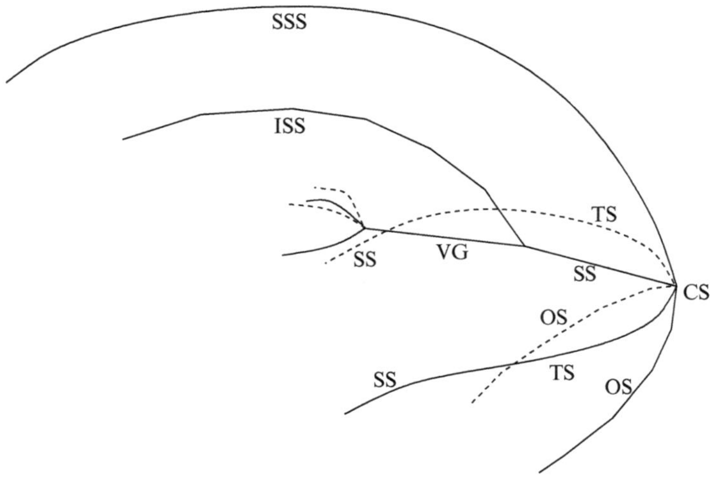

Fig. 1.26 Brain venous network of the dura. The confluence of sinuses (CS) is the connecting point of the superior sagittal sinus (SSS), straight sinus (SS), and occipital sinus (OS). It drains in the left and right transverse sinuses (TS), sigmoid sinus (SS), and then internal jugular veins. The great cerebral vein (vena magna cerebri), or vein of Galen (VG), formed by the merging of left and right internal cerebral and basilar veins, drains in the straight sinus

Herophili,[34] that form 2 transverse sinuses (TS). At each side, the transverse and superior petrosal sinuses, the latter receiving blood from the cerebellum, pons, and rostral medulla, drain into the sigmoid sinus (SS).

The lateral transverse sinus (bore 8–10 mm) drains the superior sagittal sinus. It can collect the superior sagittal sinus in totality in some cases, mostly on the right side, whereas the other can drain into the straight sinus that becomes the left transverse sinus. The transverse sinus can be atretic or even absent on one side.

In addition to the transverse sinus, the sigmoid sinus drains the posterior fossa, as it receives the superior and inferior petrosal sinuses and veins draining the lateral pons and medulla oblongata. The sigmoid sinus has frequent anastomoses with the cutaneous venous network through the mastoid emissary vein.

On both sides and at the base of the brain, the inferior petrosal sinus and the sigmoid sinus join to form the internal jugular vein. The left and right internal jugular veins collect blood from the brain and the superficial regions of the face and neck. At the root of the neck, it unites with the subclavian vein to form the brachiocephalic vein (innominate vein).

Another network of dural sinuses that collects blood includes the cavernous (or lateral sellar compartment), superior and inferior petrosal, and sphenoparietal sinuses. The cavernous sinus, which may be joined by several anastomoses across the midline, collects orbital venous blood from the superior and inferior ophthalmic

[34]Named after the Greek physician Ηρ οφιλὄς (Latinized Herophilus [335–280 BC]).

veins, anterior middle cranial fossa blood via the superior and inferior middle cerebral veins, and blood from the spheno-orbital sinus. From the cavernous sinus, blood flows in the superior petrosal sinus.

Deep sinuses are also interconnected to the basilar venous plexus and the intercavernous sinus. The superior anastomotic vein, or vein of Trolard, is the largest anastomotic cortical vein, which runs to the superior sagittal sinus. The inferior anastomotic vein, or vein of Labbé, together with temporal occipital veins, drains blood from the lateral temporo-occipital cortex and empties into the transverse sinus. The vein of Labbé occasionally drains a part of blood from inferiolateral frontal lobes. Typically, the vein of Labbé runs from the Sylvian fissure toward the transverse sinus, hence creating an anastomosis between the superficial middle cerebral vein, or superficial Sylvian vein, and the transverse sinus before its junction to the sigmoid sinus.

Deep medullary veins originate from the cortex and travel to the subependymal veins near the ventricles. The subependymal veins drain blood from the deeper subcortical structures, basal ganglia, and the dorsal region of the diencephalon to the vein of Galen.

The medullar veins can be subdivided into 2 different sets. The superficial medullar veins collect blood of the white matter. They are connected to the superficial cerebral veins. The deep medullar veins empty into the subependymal veins along walls of the lateral ventricle. The transcerebral veins cross the cerebral hemisphere to reach subependymal veins. Three subependymal veins commonly exist: thalamostriate, septal, and internal cerebral vein. These veins are joined by basal veins of Rosenthal, which drain blood from the basal and medial parts of the frontal lobe, temporal lobe, and basal ganglia, to form the vein of Galen.

Posterior fossa veins are divided into 4 groups: superficial, deep, brainstem, and bridging veins. Brainstem veins are named according to the drainage territory (midbrain, pons, or medulla) and path (transversal or horizontal). In the posterior fossa, venous blood exits through galenic and petrosal circuits and, to a lesser extent, tentorial veins and transverse sinuses. Veins of the upper brainstem, dorsal cerebellum, and vermis merge to the vein of Galen. The anterior pontomesencephalic vein, which lies ventral to the brainstem, drains to the basal vein of Rosenthal or petrosal venous network.

The cerebellum cortex blood is drained by superficial veins: (1) veins of the posterior territory run to the torcular and transverse sinus directly or through medial intratentorial sinuses; (2) veins of the superior territory to the superior petrosal sinus; (3) veins of the anterior territory to the inferior petrosal sinus; (4) veins of the superior aspect to the vein of Galen; (5) veins of inferior vermis to the torcular and/or transverse sinuses. In addition, the precentral vermian vein travels anteriorly to the vein of Galen.

Posterior fossa venous anastomoses include bridging veins between brainstem and cerebellum veins that cross subarachnoid and subdural spaces to reach venous sinuses in the dura. These bridging veins constitute 3 sets: (1) a superior, or galenic, set that targets the vein of Galen, (2) an anterior, or petrosal, set that travels to the petrosal sinuses, and (3) a posterior, or tentorial, set that run to the sinuses converging on the torcula. The petrosal venous circuit drains the inferior anterolateral region of the posterior fossa. The tentorial venous circuit collects blood from the medial superior and inferior cerebellar hemispheres.

1.7.9.4 Limb Veins

The systemic venous network returns deoxygenated blood back to the heart and acts as a blood reservoir (about 75% of the blood volume is contained within the veins). The venous system is a low-pressure system.

The principal veins in legs are the great and lesser saphenous veins and their afferents and in arms the basilic and cephalic veins and their afferents. Two or more deep veins accompany medium-sized arteries. Perforating veins penetrate the deep fascia and connect the superficial veins to the deep veins. Intramuscular venous sinuses are large, very thin walled, valveless veins, mainly within the calf muscles, which connect directly with the deep veins. They act as deformable reservoirs compressed during muscle contraction, which empties the blood out of the venous sinuses into the deep veins.

Thin-walled veins have reduced media. However, the major superficial veins of the limbs have thicker walls than the deep veins. Cutaneous veins virtually bulge during exercise although both venous volume and pressure decrease. But the arterial pressure rises during effort, inducing filtration across the capillary walls. Muscle swelling due to contraction and filtration pushes cutaneous veins toward the skin surface.

Veins have stiffer tiered bulges with valves, most often bicuspid.[35] The resulting sinuses around the valve facilitate valve opening and closing. Venous valves are aimed at preventing retrograde flow when veins are subjected to external compression during muscular contraction. Venous valves are made of connective tissue lined with endothelium.

The limb venous network is composed of a superficial and a deep compartment (Fig. 1.27). The main superficial veins of the lower limbs are the short (SSV or saphena interna) and the long saphenous vein (LSV or saphena externa), which runs from the foot to the knee (saphenopopliteal junction) and from the foot to the groin (saphenofemoral junction), respectively. Venous blood currently also moves from the superficial to the deep venous network via *perforating* or *communicating veins* that cross the deep fascia. The venous system in the leg can thus be divided into superficial veins that lie in the subcutaneous layer, deep veins, perforating veins, and intramuscular venous sinuses.

[35] Valves in the subclavian and internal jugular veins are bicuspid in 90% of humans (100 autopsies of 52 men and 48 women) and unicuspid, unicuspid valves being more common in the internal jugular vein than in the subclavian vein [62]. The average distance from the valve to the junction with the innominate vein is 17 mm for the subclavian vein and 3 mm for the internal jugular vein. Cusp height averages 9 mm.

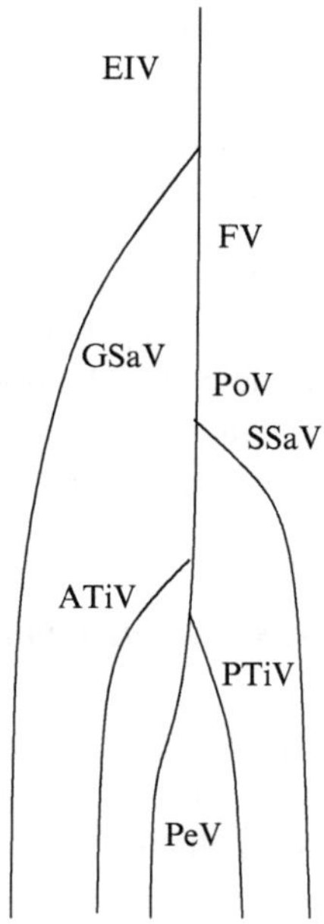

Fig. 1.27 Veins of the lower limbs. The venous network is composed of a superficial (great and small saphenous veins) and a deep circuit (femoral and popliteal veins)

The popliteal vein (PoV) is formed by the union of the anterior tibial vein (ATiV) and the trunk formed by the confluence of the posterior tibial (PTiV) and peroneal (PeV) veins. The popliteal vein becomes the femoral vein. The small saphenous vein (SSaV) crosses the popliteal fossa and drains into the popliteal vein. The great saphenous vein (GSaV), the longest vein, ascends from the foot to the groin to enter into the femoral vein. The femoral vein (FV) becomes the external iliac vein (EIV).

1.7.9.5 Portal System

The portal drainage includes the veins that drain blood from most of the digestive system (Fig. 1.17). The portal vein (PV) is formed by the junction of the superior mesenteric (SMV) and splenic (SV) veins.

The tributaries of the splenic vein are: the gastric (GV), left gastroepiploic (LGEV), pancreatic (PV), and inferior mesenteric (IMV) veins. The inferior mesenteric vein receives the hemorrhoidal (HV), sigmoid (SiV), and left colic (LCV) veins. The superior mesenteric vein receives the intestinal (IV), ileocolic, right colic (RCV), middle colic, right gastroepiploic (RGEV), and pancreaticoduodenal veins. The gastric coronary (GCV), pyloric (PyV), and cystic (CyV) veins end in the portal vein.

1.8 Innervation of the Cardiovascular Apparatus

The sensory-motor nervous system consists of 12 pairs of cranial nerves and 31 pairs of spinal nerves. All spinal nerves are mixed, as they contain both sensory and motor neurons.

The autonomic balance allows adjustment to stresses, in particular on the cardiovascular apparatus. It relies on the mainly excitory influence of the sympathetic and principally inhibitory action of the parasympathetic system. When the sympathetic activation is chronically heightened and parasympathetic activation lowered, the risk of cardiovascular diseases rises.

1.8.1 Parasympathetic and Sympathetic Control

The nervous control of the heart relies on parasympathetic fibers that travel in vagus and sympathetic nerves. Vagus nerves operate as a cardiac inhibitor, whereas sympathetic nerves are cardiac exciters. The sympathetic nervous system supply to the heart leaves the spinal cord at the first 4 thoracic (T1–T4) vertebra.

At the effector organs, sympathetic postganglionic neurons release noradrenaline (or norepinephrine), along with other cotransmitters such as ATP, among others. Noradrenaline binds to adrenergic receptors (Table 1.17). Acetylcholine is the neurotransmitter for both preganglionic sympathetic and parasympathetic neurons as well as for postganglionic parasympathetic neurons. Acetylcholine targets muscarinic receptors on the plasma membrane of effector cells (Table 1.18). At the adrenal medulla, an endocrine gland, presynaptic neurons release acetylcholine that binds to nicotinic receptors, i.e., ligand-gated ion channels. The stimulated adrenal medulla releases adrenaline (or epinephrine) into the blood stream. Adrenaline also targets adrenergic receptors, thereby enhancing sympathetic activity.

Sympathetic and parasympathetic neurons modulate cardiovascular dynamics. Parasympathetic efferent preganglionic neurons in the medulla oblongata project axons via the vagi to intrinsic cardiac parasympathetic postganglionic neurons. Sympathetic preganglionic neurons in the spinal cord send axons to postganglionic neurons in paravertebral ganglia. The interdependent sympathovagal command concept states that the activated sympathetic inhibits the parasympathetic and vice versa. Intrathoracic ganglia process centripetal and centrifugal information using short loops. The main targets of the nervous control are: (1) nodal tissue, (2) cardiomyocytes, and (3) vascular smooth myocytes.

1.8.1.1 Sympathetic Control

The sympathetic system associated with the thoracolumbar outflow consists of cell bodies of general visceral efferent and preganglionic neurons in the lateral horn of the spinal cord from the T1 to L2 level of the vertebral column.

Table 1.17 Receptors of the sympathetic postganglionic neurons (Vol. 3 – Chap. 7. G-Protein-Coupled Receptors)

Type	Effect
Cardiovascular apparatus	
α1a GPCR	Positive inotropy Vasoconstriction
α1b GPCR	Positive inotropy Vasoconstriction
α1d GPCR	Vasoconstriction
α2a GPCR	Sympathetic output reduction
α2b GPCR	Vasoconstriction
α2c GPCR	Venoconstriction
β1 GPCR	Positive bathmotropy, chronotropy, dromotropy, inotropy, lusitropy
β2 GPCR	Positive inotropy, lusitropy
β3 GPCR	Negative inotropy
Respiratory apparatus	
α1 GPCR	Bronchoconstriction (minor contribution)
β2 GPCR	Bronchodilation (major contribution)
Adrenal medulla	
Nicotinic receptor (ion channel)	Secretion of adrenaline and noradrenaline (preganglionic neuron)
Kidney (juxtaglomerular apparatus)	
β1 GPCR	Secretion of renin
Liver	
α1 GPCR	Glycogenolysis, gluconeogenesis
β2 GPCR	Glycogenolysis, gluconeogenesis
Adipose tissue	
β1/3 GPCR	Increased lipolysis

Table 1.18 Receptors of the parasympathetic postganglionic neurons (Vol. 3 – Chap. 7. G-Protein-Coupled Receptors)

Type	Effect
Cardiovascular apparatus	
Muscarinic M_2 GPCR	Decreased cardiac output Attenuated cardiac frequency (negative chronotropy) Reduced contractility (negative inotropy) Decreased conduction (negative dromotropy) Vascular smooth muscle relaxation
Muscarinic M_3 GPCR	Vascular smooth muscle contraction
Muscarinic M_5 GPCR	Vasodilation of cerebral arteries
Respiratory apparatus	
Muscarinic M_3 GPCR	Airway smooth muscle contraction

Preganglionic neurons synapse with postganglionic neurons of paravertebral and prevertebral sympathetic ganglia, in addition to chromaffin cells of the adrenal medulla. Splanchnic nerves such as cervical cardiac and thoracic visceral nerves synapse in the paravertebral chain, whereas thoracic splanchnic nerves synapse in the prevertebral ganglia.

In cats, medullary clusters that project to spinal (T2) sympathetic neurons provide excitatory (cardioacceleration) and inhibitory (cardioinhibition) inputs to spinal cardioacceleratory neurons [63]. The vasomotor center sends multiple excitatory and inhibitory signals from specific reticular nuclei, i.e., cardioacceleration mainly from nucleus lateralis reticularis and parvocellularis and cardioinhibition from nucleus paramedium reticularis, raphe, and medullae oblongatae centralis subnucleus ventralis, to the intermediolateral nucleus.

In cats, clusters of the medullary vasomotor region respond to stimulation of the carotid sinus nerve or aortic depressor nerve. Neurons excited after a short latency (<10 ms) by stimulation of these nerves localize to the nucleus tractus solitarius and medial or paramedian reticular formation [64]. A short-duration (<40 ms) inhibition follows short-latency excitations. Stimulation of the vagus nerve or the dorsolateral column of the spinal cord excite less than half of the neurons of the nucleus tractus solitarius. Long-latency excitatory responses and complicated excitatory–inhibitory patterns are primed in neurons of the medial, paramedian, and lateral–ventral reticular nuclei by stimulation of the carotid sinus nerve and aortic depressor nerve. Most of these neurons are also influenced by stimulation of the vagal nerve and dorsolateral column. Neurons of the lateral ventral reticular region may serve as preganglionic sympathetic neurons in the intermediolateral columns of the spinal cord.

In cats, vasopressor outflow from the medulla and descending tracts of the midcervical spinal cord to the external carotid postganglionic sympathetic nerve is organized into 2 vasopressor pathways [65]: (1) baroreceptor reflex-sensitive, long-onset (>50 ms), slow-conduction sympathetic response (pressor pathway 1) and (2) baroreceptor reflex-insensitive, short-onset, rapid-conduction sympathetic response (pressor pathway 2).

Stimulation of the medullary depressor region inhibits long-latency sympathetic responses from midbrain, medullary, and descending spinal components of the pressor pathway 1, but not short-latency sympathetic responses from descending spinal components of the pressor pathway 2, as does baroreceptor reflex activation [66]. However, unlike baroreceptor reflex activation, stimulation of the medullary depressor region prevents short-latency sympathetic responses from the midbrain and medullary components of the pressor pathway 2. Therefore, 2 distinct sympatho-inhibitory axes can be activated from the depressor region of the medial medulla. The first mimics the baroreceptor reflex acting at a spinal level to inhibit transmission in the pressor pathway 1. The second, which is baroreceptor reflex-independent, acts at a supraspinal level to inhibit the pressor pathway 2.

The medullary depressor region can also cause reflex vasodilatation in skeletal muscles by inhibition of sympathetic vasoconstrictor nervous discharge, which is not associated with excitation of a sympathetic histaminergic vasodilator pathway [67].

1.8.1.2 Parasympathetic Control – Vagus Nerve

The parasympathetic system associated with the craniosacral outflow comprises cell bodies in the brainstem (cranial nerves III, VII, IX, and X) and in the sacral spinal cord. Preganglionic neurons synapse with postganglionic neurons of the parasympathetic ganglia of the head and in or near the wall of innervated organs. In particular, vagus and thoracic spinal accessory nerves innervate the heart and lungs, among other organs.

The *vagus nerve*, or *pneumogastric* nerve, controls many thoracic, abdominal, and pelvic organs (hence its name).[36] It contains axons that run from or to 3 nuclei of the medulla: (1) the *dorsal nucleus of the vagus nerve* that sends parasympathetic output to the viscera; (2) the *nucleus ambiguus* that transmits parasympathetic output to the heart; and (3) the *nucleus of the solitary tract* (nucleus tractus solitarius) that receives afferents from visceral organs.[37] It generates parasympathetic cardiac nerves that form cardiac and pulmonary plexus.

Both ganglia of the vagus nerve are crossed by parasympathetic and possibly some sympathetic fibers. The jugular ganglion is the superior ganglion of the vagus nerve (Fig. 1.28). It contains afferent somatosensory cell bodies that provide sensory information from the auricular and meningeal nerves. Their axons synapse in the sensory trigeminal nucleus. The nodose ganglion, or inferior ganglion of the vagus nerve, is located at the level of the transverse process of the first cervical vertebra (atlas). It chiefly contains visceral afferents, hence receiving sensory cues from the heart, lungs, pharynx, and larynx, in addition to the digestive tract. These visceral afferents synapse in the nucleus of the solitary tract. Preganglionic motoneurons from the dorsal vagal nucleus and some visceral efferents from the nucleus ambiguus that descend to the inferior vagal ganglion form a band skirting the ganglion core.

---

[36]Latin *vagus*: rambling, roving, roaming, strolling, unfixed, unsettled, vagrant, wandering.

[37]The solitary tract (tractus solitarius) extends longitudinally in the posterolateral region of the medulla. Other names encompass round fasciculus (fasciculus rotundus), solitary fasciculus (fasciculus solitarius), solitary bundle (funiculus solitarius), and Gierke and Krause respiratory bundle. It descends to the upper cervical segments of the spinal cord. It is made of primary sensory and descending fibers of the vagus, glossopharyngeal, and facial nerves. It is surrounded by the nucleus of the solitary tract. Inputs to the nucleus of the solitary tract include: (1) taste information from cranial nerves VII and IX; (2) chemoreceptors in the carotid and aortic bodies via cranial nerves IX and X, respectively; (3) arterial baroreceptors of the aorta and carotid arteries; and (4) chemically and mechanically sensitive neurons from the heart, lungs, airways, gastrointestinal system, and liver via cranial nerves IX and X. Outputs are sent to numerous regions of the brain (paraventricular nucleus of the hypothalamus, central nucleus of the amygdala, and other nuclei of the brainstem [parabrachial area and visceral motor and respiratory circuits]).

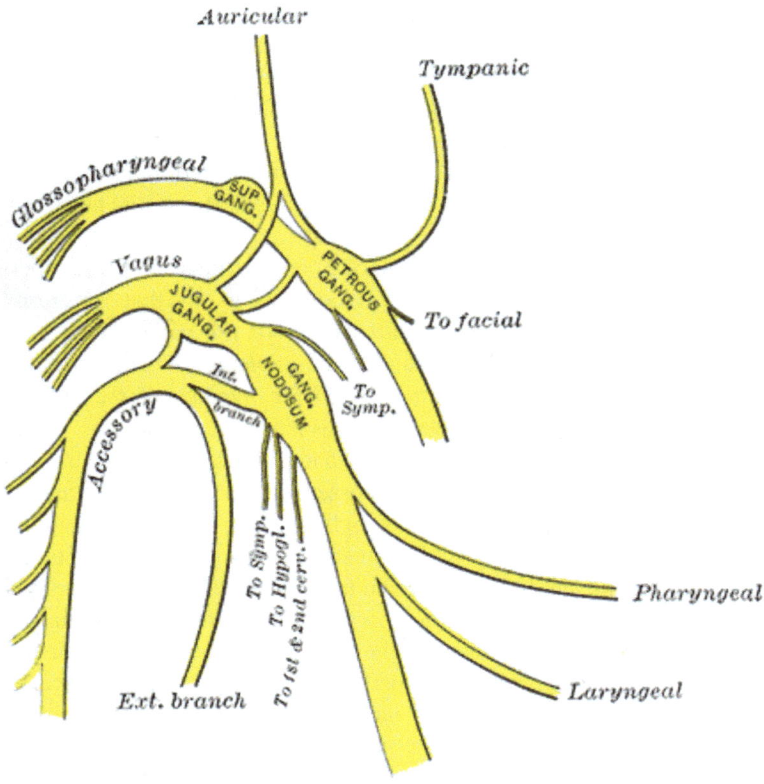

Fig. 1.28 Upper segments of the glossopharyngeal, vagus, and accessory nerves (Source: [3])

1.8.1.3 Glossopharyngeal Nerve

The glossopharyngeal nerve (cranial nerve IX) is a mixed (sensory and motor) nerve[38] that contributes to the pharyngeal plexus (Table 1.19). It results from the merging of 3 to 4 roots emerging from the groove between the olive and inferior cerebellar peduncle. It exits through a foramen of the skull. It is covered with a sheath that emanates from the meninges.

The glossopharyngeal nerve receives: (1) general sensory fibers (ventral trigeminothalamic tract) from the pharynx and posterior region of the tongue (hence its name) as well as the tonsils and middle and outer ear; (2) taste buds from the posterior tongue; and (3) visceral sensory fibers from carotid bodies

[38]Five cranial nerves have mixed sensory, motor, and parasympathetic function: the occulomotor (III), trigeminal (V), facial (VII), glossopharyngeal (IX), and vagus (X) nerves. Three cranial nerves have primarily a sensory function: the olfactory (I), optic (II), and acoustovestibular (VIII) nerves. Four cranial nerves have primarily a motor role: the trochlear (IV), abducens (VI), accessory (XI), and hypoglossal (XII) nerves.

Table 1.19 Role of the glossopharyngeal and vagus nerves. Afferent neurons have their peripheral receptors in walls of organs, such as airways, blood vessels, heart, and lungs; their soma resides in the dorsal root ganglia (or spinal ganglia) or cranial nerve ganglia; their axon ends in the dorsal grey column of the spinal cord or brainstem. Efferent excitatory or inhibitory neurons supply smooth muscles in the walls of hollow organs, airways, blood vessels

Nerve	Number	Special sense	Sensory afferent	Motor efferent	Parasympathetic
Glossopharyngeal	IX	+	+	+	+
Vagus	X	+	+	+	+

and carotid sinuses. These fibers arise from cells of the superior and petrous ganglia and terminate at the medulla oblongata. The glossopharyngeal nerve supplies: (1) parasympathetic fibers to the parotid salivary gland and (2) motor fibers to stylopharyngeus muscle, which dilates the pharynx during swallowing. Motoneurons spring from cells of the nucleus ambiguus of the medulla oblongata and run in the anterior grey column of the spinal cord.

The glossopharyngeal nerve has parasympathetic fibers that innervate the parotid salivary gland. Preganglionic fibers depart the glossopharyngeal nerve as the tympanic nerve and continue to the middle ear where they build a tympanic plexus.

The sympathetic efferent fibers of the glossopharyngeal nerve that originate from the dorsal nucleus are probably both preganglionic motor fibers and preganglionic secretory fibers. The secretory fibers synapse in the otic ganglion; from it secondary neurons travel to the parotid gland.

The glossopharyngeal nerve communicates with the sympathetic system and vagus and facial nerves. Its branches include the auricular, tympanic, carotid, pharyngeal, stylopharyngeus, tonsillar, and lingual nerves.

The sensitive ganglia of the glossopharyngeal nerve include the superior ganglion, or Ehrenritter's ganglion, and inferior ganglion, or Andersch' ganglion, a.k.a. petrous or petrosal ganglion. The sensitive ganglia of the vagus nerve comprise the jugular and nodose ganglion.

1.8.2 Role of Reactive Oxygen Species

Oxidative stress supports sympathetic drive. In other words, reactive oxygen species can provoke an excessive sympathetic nerve activity [68]. In particular, ROS are generated in nodose neurons in culture by stimulated NMDA-type glutamate receptors. Reactive oxygen species can enhance sympathetic activity at the ganglionic level and mediate action of angiotensin in subfornical organs in the hypothalamus. In addition, sustained and augmented ROS levels in baroreceptor neurons may impair the baroreceptor reflex [68]. On the other hand, ROS produced in carotid bodies contribute to an excessive chemoreceptor activation.

1.8.3 Inputs from Vascular Sensors

Most arteries and veins in the body are innervated by sympathetic adrenergic nerves that release noradrenaline. Some blood vessels are innervated by parasympathetic cholinergic or sympathetic cholinergic nerves that secrete acetylcholine. Once it is liberated, the neurotransmitter binds to the adrenergic or cholinergic receptors to activate signaling in effector cells. In this context, receptors are molecules on the plasma membrane of mural cells in relation to nerve endings.

On the other hand, walls of the cardiovascular apparatus (heart, arteries, and veins) contain sensors that are sensitive to blood chemical content (chemoreceptors) or to mechanical stresses (baro- and voloreceptors). In this framework, receptors are cells that communicate with nerve terminals.

1.8.3.1 Chemoreceptors

Peripheral (carotid and aortic bodies) and central (medullary neurons) chemoreceptors primarily function to regulate breathing [69]. Carotid bodies[39] in the carotid artery bifurcations between the external and internal carotid arteries, immediately prior to the origin of the internal carotid artery, are important for eliciting hypoxic ventilatory stimulation in humans [70]. In the absence of carotid bodies, compensatory upregulation of aortic bodies as well as other chemoreceptors contributes to the hypoxic ventilatory response.

Carotid bodies have predominantly ventilatory effects, whereas aortic bodies scattered on the aortic arch and its branches have preponderantly circulatory effects.

Chemoreceptors sense p_{O_2}, p_{CO_2}, and pH in the arterial blood and cerebrospinal fluid. Medullary chemoreceptors primarily respond to changes in p_{CO_2} and concentration of hydrogen ions (Table 1.20). Peripheral chemoreceptors of vascular bodies in the aortic arch and medial to the carotid sinuses are also stimulated by increased p_{CO_2} and reduced pH, but the reflex effect is small compared to the direct effect of hypercapnia and hydrogen ions in the medulla. Carotid bodies monitor chiefly p_{O_2}, whereas aortic chemoreceptors record arterial O_2 saturation [70].

Paraganglia that serve as additional chemoreceptors, i.e., tissues similar to carotid and aortic bodies, also exist in the thorax and abdomen [70]. Chemoreceptors with sympathetic afferent fibers also lodge in the heart. They are activated by ischemia and transmit pain caused by inadequate blood supply to the myocardium.

Peripheral chemoreceptors are critical for ventilatory acclimatization at high altitude [70]. They also contribute in part to exercise-induced hyperventilation,

[39] A.k.a. glomus caroticum and carotid glomus, corpuscule, ganglion, and gland. The carotid body is a small cluster of chemoreceptors and supporting cells located near the bifurcation of the carotid artery. On the other end, the carotid sinus is a localized dilation of the entrance segment of the internal carotid artery.

Table 1.20 Stimuli of baro- and chemoreceptors

Sensor	Stimuli
Baroreceptor	Wall stretch caused by blood (endoluminal) pressure (carotid sinus > aortic arch baroreceptor)
Arterial chemoreceptor	Mainly P_{O_2}
Medullary chemoreceptor	Mainly P_{CO_2}, $[H^+]$

Table 1.21 Arterial baroreceptor innervation (Source: [69])

Sensor	Innervation
Aortic baroreceptor	Vagus nerve (X) via sinus nerve of Hering
Carotid baroreceptor	Glossopharyngeal nerve (IX) via aortic nerve
Aortic chemoreceptor	Vagus nerve (X) via carotid sinus nerve
Carotid chemoreceptor	Glossopharyngeal nerve (IX)

especially with submaximal and heavy exercise. During pregnancy, hypoxic ventilatory sensitivity increases, possibly due to the action of estrogen and progesterone on chemoreceptors.

Chemoreceptory glomus cells[40] give rise to carotid nerve sensory fibers. Afferent nerve fibers from the carotid bodies join with the sinus nerve and then the glossopharyngeal nerve (Table 1.21) [69]. The sinus nerve is formed by nerve fibers that project into the carotid body from the sympathetic superior cervical ganglion and, to a lesser extent, from the intercarotid branch of the glossopharyngeal nerve. In response to hypoxia, chemoreceptor cells of the carotid body release transmitters, which act on the petrosal ganglion neuron terminals to increase afferent discharge.

1.8.3.2 Carotid Glomus

The carotid body is a chemoreceptor organ located in the vicinity of the carotid bifurcation. It is composed of cell clusters formed by chemoreceptor and support cells. Sensory nerve endings of the carotid sinus nerve penetrate these clusters and synapse with chemoreceptor cells.

The carotid body regulates the body's ventilation during hypoxia, hypercapnia, and acidosis. However, in normal humans at rest, hyperoxia that inhibits peripheral chemoreceptor activity decreases sympathetic activity, therefore suggesting that a sympathetic tone exists even in the absence of hypoxia [71].

Once they are detected, chemical stimuli (hypoxia, hypercapnia, and acidosis on glomus cells of the carotid body) provoke an increase in release rate of

[40]Carotid bodies are composed of glomus cells, or carotid body type-1 cells, which are neurons containing various types of neurotransmitters, and sustanticular cells, or S100+ supporting type-2 cells, which resemble glial cells.

neurotransmitters by chemoreceptor cells that heighten action potential frequency in the carotid sinus nerve.

Coupling between glomus cells results from gap junctions. Coupling between glomus cells (GC) and nerve endings (NE) is mostly resistive in the GC-to-NE direction, but mostly capacitive in the opposite NE–GC direction; coupling between nerve endings is mostly capacitive in either direction [72]. Slow chemicoelectrical events originating in glomus cells can be transferred to the nerve endings, but only electrical transients can pass from nerve to glomus cells. Chemoreceptor stimulants (acute and chronic hypoxia, hypercapnia, acidity, cholinergic agents, and dopamine) uncouple most glomus cells, reducing GC–GC signaling via gap junctions, but enhance coupling from glomus cells to nerve endings.

The carotid body possesses 2 concentric bands that contain receptors intermingled with elastin and collagen fibers [73]: (1) a first band close to the adventitia with disperse receptors that have myelinated sensory axons and (2) a second one with circumscribed receptors with myelinated sensory axons that has the highest density of nerves of the entire arterial tree. Types of nerve fibers depend on mammalian species.[41]

Upon glomus cell activation by hypoxia, hypercapnia, or acidosis, afferent carotid nerves prime impulses, thereby triggering respiratory and cardiovascular reflexes. The *chemoreceptor reflex* increases the rate and amplitude of breathing. In addition, as they interact with vasomotor centers in the medulla and via pulmonary stretch receptors, chemoreceptors enhance sympathetic signals to the heart and vasculature by activating the vasomotor centers.

In addition, vasoactive molecules produced in the carotid body can modulate chemoreception, as they control blood flow. In the cat, endothelin-like peptides are produced in the petrosal ganglion and carotid body under normoxia. Endothelin-1 increases basal chemosensory discharges from the carotid body due to local vasoconstriction [75].

Multiple neurohumoral factors, such as catecholamines, angiotensin-2, endothelin, natriuretic peptides, and nitric oxide, influence chemoreceptor activity. Glomus cells synthesize and store multiple neurotransmitters (Table 1.22).

Glomus cells express oxygen-sensitive, Ca^{++}-insensitive, outward voltage-gated K^+ channels ($K_{V_{O_2}}$) constituted by $K_V4.1$ and $K_V4.3$ subunits. At least in rabbits, these cells also produce a $K_V3.4$ subunit that is insensitive to oxygen. Once the $K_{V_{O_2}}$ channel is repressed by hypoxia, an attenuated K^+ current contributes to cell membrane depolarization at rest. This initial depolarization subsequently activates voltage-gated Ca_V channels. Calcium influx allows release of neurotransmitters, such as acetylcholine, ATP, dopamine, serotonin, and substance-P that are produced by chemoreceptor cells to increase discharge in the carotid sinus nerve.

[41] In guinea pigs, 5 different populations of nerve fibers are observed in the carotid body according to their origin and chemical coding [74]. In rats, the innervation pattern of the carotid body differs. Neuropeptide-Y+ fibers act primarily via vascular mechanisms rather than directly as chemosensors. Sensory fibers contain substance-P, calcitonin gene-related peptide (rat and guinea pig), and somatostatin (guinea pig).

Table 1.22 Excitatory and inhibitory neurotransmitters in glomus cells that are present at least in some mammalian species (Source: [76]). Hypoxia releases substance-P using Ca_V1 and $Ca_V2.2$ channels

Biogenic Amines	Acetylcholine, dopamine, noradrenaline, serotonin
Neuropeptides	Atrial natriuretic peptide, Endothelin, enkephalin, substance-P
Gastransmitter processors	Heme oxygenase-2, NADPH–cytochrome-C oxidoreductase

Nitric oxide is synthesized by constitutive nitric oxide synthases NOS1 and NOS3 in nerve fibers and vascular endothelium in the carotid body. Its synthesis requires oxygen. Nitric oxide impedes chemosensor discharge in normoxia by activating calcium-dependent, voltage-gated K^+ channels (K_{Ca}) via a cGMP pathway in glomus cells in rabbits [77]. Potassium flux through K_{Ca} prevents glomus cell membrane depolarization at rest. Nitric oxide thus indirectly precludes Ca_V1 channels in rabbit glomus cells, in addition to direct inhibition via a cGMP-independent mechanism. Angiotensin-2 activates $Ca_V1.2$ channels in vascular smooth myocytes and possibly in glomus cells.

Inwardly rectifying K^+ current carried by the $K_V11.1$ channel in rabbit glomus cells can also influence the resting membrane potential [77].

Nitric oxide in the paraventricular nucleus (PVN) of the hypothalamus suppresses the sympathetic output. Inhibition of NOS in PVN increases sympathetic and phrenic nerve responses to the stimulated carotid body [71].

Endothelin-1 synthesized in the carotid body excites this chemosensory organ. The endothelin receptor antagonist lowers the elevated carotid body discharge induced by chronic intermittent hypoxia in hypoxic as well as normoxic conditions [77].

1.8.3.3 Baroreceptors

Arterial blood pressure is regulated within a narrow range (mean arterial pressure range 11.3–13.3 kPa in adults) to ensure adequate blood flow [69]. A negative feedback from baroreceptors that sense the arterial pressure responds to stretching of the arterial wall. Arterial baroreceptors localize to the carotid sinus at the bifurcation of the common carotid artery into external and internal branches (Table 1.23) and in the aortic arch.

The carotid sinus is the most important cluster of arterial baroreceptors [69]. Baroreceptors in the aortic arch have a higher threshold pressure and are less sensitive than the carotid sinus baroreceptors.

In the sheep, the carotid body (width 1.4–2.6 mm) localizes to the medial surface of the terminal segment of the common carotid artery [78]. It has an oval, circular, or irregular shape. In adult sheep and goats, the internal carotid artery is absent.

Table 1.23 Carotid body and sinus: location and role. The carotid body and sinus are connected to the central nervous system by visceral sensory afferents and visceral motor parasympathetic efferents. The carotid body chemoreceptors strongly influence the magnitude of the sympathetic inhibition produced by activated baroreceptors

Sensor	Location	Role
Carotid body	Carotid artery bifurcation (Common carotid artery)	Chemoreceptor
Carotid sinus	Internal carotid artery	Baroreceptor

The initial part of the occipital artery and the posteriorly adjacent dorsolateral and medial part of the common carotid artery are homologs of the carotid sinus of other animals [78]. The carotid body, represented either by scattered cells or by aggregates around the vessel, is close to the muscular branch of the occipital artery. The sinus lodges at the base of the occipital artery in both species [79]. The carotid body is innervated by the carotid sinus branch of the glossopharyngeal nerve and by branches of the external carotid nerves [78]. The carotid sinus is innervated also by these nerves as well as by a branch of the vagus nerve in 50% of cases and by a branch of the hypoglossal nerve in 23% of cases [78].

When the arterial pressure suddenly rises, i.e., when the arterial wall expands, baroreceptors fire. Conversely, when the arterial pressure suddenly falls, the stretch of the arterial wall decreases and baroreceptor firing falls.

The carotid sinus baroreceptors are innervated by the sinus nerve of Hering, a branch of the glossopharyngeal nerve (Table 1.21) [69]. The glossopharyngeal nerve synapses in the nucleus tractus solitarius in the medulla of the brainstem. This neural nucleus modulates the activity of sympathetic and parasympathetic neurons in the medulla, which in turn regulate the autonomic control of the heart and blood vessels.

The aortic arch baroreceptors are innervated by the aortic nerve, a branch of the vagus nerve that also travels to the nucleus tractus solitarius [69].

1.8.3.4 Reciprocal Interaction Between Baro- and Chemoreceptors

Baroreceptor activation inhibits the chemoreceptor reflex and vice versa. For example, elevated carotid sinus pressure not only reduces vascular resistance, but also precludes vasoconstriction primed by chemoreceptor activation by carotid artery hypoxemia [68]. Conversely, when the baroreceptor activity is attenuated by lowered carotid sinus pressure, vascular resistance increases and vasoconstriction caused by carotid artery hypoxemia is educed. Increased sympathetic nerve activity in a hypertensive subject by hypoxemia is suppressed by activation of carotid baroreceptors stimulated by distension of the carotid sinus region.

Reciprocal dysautonomia, i.e., decreased baroreceptor reflex and enhanced chemoreceptor reflex promote sympathovagal imbalance. Chronic, excessive sympathetic nerve activity in the cardiovascular apparatus that arises from reduced baroreceptor activity, increased chemoreceptor activity, or facilitatory interaction between baro- and chemoreceptors (enhanced chemoreceptor response by reduced baroreceptor activity) is pathological [68].

Electrical stimulation of carotid baroreceptors enhances prognosis in refractory hypertension, as it represses chronic sympathetic nerve activity observed via reduced concentrations of circulating neurohumoral indicators, such as plasma angiotensin-2 and noradrenaline [68]. Vagal nerve stimulation improves long-term survival in rats with chronic heart failure. In addition, inflammatory cytokines (TNF and IL1) of tissue injury activate sensory vagal signals relayed to the nucleus tractus solitarius that activate vagus efferent activity targeting macrophages to inhibit cytokine synthesis through the cholinergic anti-inflammatory pathway.

1.8.3.5 Signal Transduction in Chemoreceptors

Hypoxia may depolarize glomus cells by inhibiting an oxygen-sensitive K^+ channel, such as $K_V11.1$ or the large-conductance K_{Ca} (BK) channel, which acts as a primary oxygen sensor. Plasma membrane depolarization leads to an increase in cytosolic Ca^{2+} level via entry through the Ca_V1 or Ca_V2 channel and subsequent transmitter release [70].

In addition, a heme protein or a redox-sensitive enzyme, such as mitochondrial cytochromes or nitric oxide synthase NOS1 on the one hand or nonmitochondrial heme-containing enzymes NADPH oxidases, NOS1, NOS3, heme oxygenases, or NADPH–cytochrome-C reductase on the other, may act as an O_2 sensor [70]. A change in its redox state may trigger transmitter release.

The carotid body produces many types of neurotransmitters, such as acetylcholine, adenosine triphosphate, and substance-P, which are coreleased during hypoxia. During long periods of hypoxia, inhibitory transmitters may prevent overexcitation of the sensory activity launched by excitatory transmitters.

In mice, hypoxia detection depends on production of hydrogen sulfide and carbon monoxide produced by cystathionine γ-lyase and heme oxygenase-2 [80].

Changes in H^+ concentration caused by acidosis inside the cell stimulates the same pathways involved in p_{CO_2} sensing. Arterial acidosis, either metabolic or from altered p_{CO_2}, impedes acid–base transporters such as Na^+–H^+ exchanger that raise intracellular pH and activates other types of transporters such as Cl^-–HCO_3^- exchanger that decrease intracellular pH. In addition, extracellular pH-sensitive conductances including Cl^- and pH-sensitive potassium TASK-like currents may mediate the depolarization caused by extracellular acidosis [68].

The carotid body produces amiloride-sensitive cation channels ACCN1 and predominant ACCN2 and ACCN3 [68]. The relative expression of various subunits of ACCN channels in glomus cells may determine their activation kinetics and pH sensitivity. Absence of ACCN1 enhances pH sensitivity of ACCN3 isoform. In glomus cells, the ACCN3 channel contributes to pH sensitivity, causing sympathoexcitation.

1.8.3.6 Mechanotransduction in Baroreceptors

Molecules involved in sensory signaling that enable mechanotransduction in baroreceptor neurons and chemotransduction in glomus cells of the carotid bodies include hydrogen ion-sensing ion channels.

Approximately 15 to 20% of arterial baroreceptors are *A-type neurons* with myelinated axons, whereas the remaining 80 to 85% are *C-type neurons* with unmyelinated axons. Among A- and C-type nodose ganglion neurons, in general, C-type, unmyelinated, sensory neurons do not have HCN1; most HCN1+ neurons are myelinated, A-type neurons.

Upon mechanical stimulation, intracellular calcium concentration increases in aortic baroreceptor nodose neurons in rats [68]. Calcium transients that are proportional to the intensity of mechanical stimulation depend on the extracellular Ca^{2+} level. Calcium influx occurs through mechanically gated nonvoltage-gated Ca^{2+} channels.

Neurons of the nodose ganglia connect many types of visceral sensory neurons to the central nervous system. Sensory terminals of nodose neurons can serve as baroreceptors of the aortic arch in the adventitial layer of the arterial wall. All nodose sensory neurons possess hyperpolarization-activated cyclic nucleotide-gated channels HCN2 and HCN4, but only 20% express the HCN1 subtype [81].

Subtype HCN1 characterized by a rapid activation is detected not only in soma, but in mechanosensitive terminals. Both HCN2 and HCN4 responsible for inward current with slow activation localize to mechanosensing terminals of both myelinated and unmyelinated receptor neurons.

Members of the degenerin–epithelial sodium channel (DEG–ENAC) subunit family[42] have a preferential conductance for sodium. They are involved in neurosensory mechanotransduction (Sect. 3.6.1.2).

The epithelial sodium channel (ENaC)[43] is an ion channel permeable for Li^+, H^+, and especially Na^+ ions. It is a component of the mechanosensor in arterial baroreceptors. Pharmacological blockers of ENaC (amiloride and benzamil) suppress mechanically induced Ca2+ transients and depolarizations in isolated aortic baroreceptor neurons as well as the carotid sinus nerve activity and baroreflex in anesthetized rabbits.

Amiloride-sensitive cation channels (ACCN1–ACCN4), or acid-sensing (proton-gated) ion channels (ASIC1–ASIC4), constitute a subfamily of the DEG–ENAC neuronal family of ion channels. They are amiloride-sensitive nonselective cation channels with predominant Na^+ conductance. They have a large extracellular

---

[42]Pickpocket (Ppk) in Drosophila melanogaster; Degenerin-like protein Uncoordinated Unc105 and Degenerin-like protein DeL1 in Caenorhabditis elegans.

[43]A.k.a. sodium channel nonneuronal SCNN1 and amiloride-sensitive sodium channel (ASSC). It is composed of different subunits (SCNN1α–SCNN1δ). It is located in the apical membrane of polarized epithelial cells, particularly in the kidney (primarily in the collecting tubule), lung, and colon. It is involved in the transepithelial Na^+ transport together with the Na^+–K^+ ATPase. In ciliated cells, ENaC localizes along the entire length of the cilium, unlike CFTR channel. It regulates osmolarity of the periciliary fluid.

domain that may tether to extracellular matrix proteins, 2 transmembrane domains, and short intracellular C- and N-termini linked to actin and other proteins that form a mechanically gated complex. Homo- and heteromultimers of various subunits constitute the channels. The ACCN channels are mainly trimeric. Different combinations of subunits constitute ACCN channels. The subunit composition determines the magnitude of H^+ sensitivity. According to the type of sensory terminal in which they reside, ACCN channels mediate different functions (nociception, mechanosensing, thermosensing, salt tasting, pH sensing, and synaptic modulation).

Isoform ACCN1[44] is the least sensitive to H^+ ions. It is involved in mechanosensing in some of the rapidly adapting cutaneous fibers. It acts as a mechanosensory ion channel in baroreceptors. Subtypes ACCN2[45] and ACCN3[46] are very sensitive to H^+ ions. They are implicated in the transduction of nociception in spinal sensory afferents and in pH sensitivity of glomus cells of the carotid body. Another isochannel is the pituitary isoform amiloride-sensitive cation ACCN4 channel.[47]

In mice, 5 ACCN subtypes (ACCN1a–ACCN1b, ACCN2a–ACCN2b, and ACCN3) lodge in nodose ganglia, with prominence of ASIC2b. Isoforms ACCN1a, ACCN1b, ACCN2a, and ACCN2b are splice variants. Subtype ACCN1b is not functional on its own, but modulates activity of heteromultimeric channels as a constituent. Isoforms ACCN1 to ACCN3 exist in various nodose neurons, with varying densities, especially in baroreceptor axons and their terminals in the aortic arch.

Channel ACCN1 is an important determinant of baroreceptor sensitivity and a pressure sensor. Baroreceptor sensitivity decays in ACCN1 null mice [68]. The loss of baroreflex control causes a sympathovagal imbalance with an enhanced sympathetic drive of heart rate and vasomotor tone as well as a reduced parasympathetic control of heart rate. Channel ACCN1 is a mechanosensor in isolated nodose neurons, as mechanically induced depolarization in aortic baroreceptor neurons is correlated to ACCN1 expression. Moreover, aortic depressor nerve activity during the pressor response declines rapidly in knockout mice.[48]

In general, the contribution of the baroreceptor reflex to the pressor response to bilateral carotid occlusion in mice is stronger than that of the chemoreceptor reflex, but the latter becomes very significant with senescence [68]. Both ACCN2 and ACCN3 may contribute to activation of chemoreceptors and their sensitivity may be enhanced when ACCN1 is lacking.

---

[44] A.k.a. ASIC2, brain sodium channel BNaC1, and mammalian degenerin homolog (mDeg).

[45] A.k.a. ASIC1 and brain sodium channel-2 (BNaC2).

[46] A.k.a. ASIC3, neuronal amiloride-sensitive cation channel-3, SLNAC1, testis sodium channel TNaC1.

[47] A.k.a. ASIC4 and BNaC4.

[48] Genetically engineered mice for the loss of a given gene activity, here the Accn1 gene.

Chapter 2
Anatomy of the Ventilatory Apparatus

Anatomy of the ventilatory apparatus refers to its architecture, bulk structure, vascularization, and innervation. Dimensions of tree-like structures, such as the pulmonary arterial, venous, and bronchial circuits, determine the efficiency of fluid distribution and drainage via an optimal structure–function relation.

In adult humans, the chest volume-filling respiratory tree is made of bronchi that connect atmospheric air via cranial and cervical upper airways to alveoli that are the basic respiratory units. A part of walls of these small distal gas-exchange units, or alveolar septa, constitutes the interface between blood and air.[1]

The tracheobronchial tree is constituted of a set of branchings, most often by dichotomy. After each branching, branch dimensions (length and diameter) decay. The human tracheobronchial tree is constituted of conducting ($\mathcal{O}[10^5]$ and respiratory ($\mathcal{O}[10^7]$ airways. In human lungs, the minimum number of bronchial generations is 18, the maximum 33, and the average 23. Walls of the last 8 to 10 bronchial generations include alveoli (Sect. 2.3).

In the human lung, large and small airways differ by their structure, as large as well as mid-sized and small *bronchi* have great and small amounts in cartilaginous rings, respectively. Bronchioles do not possess cartilage. *Terminal bronchioles* form the last purely conductive airways. Last bronchiolar divisions indeed give rise to *respiratory bronchioles* that have their epithelium lining interrupted by alveoli.

The pulmonary alveolus is the principal constituent of the pulmonary parenchyma. This small cavity is situated either along walls of respiratory bronchioles (last bronchiolar generations), which are more and more alveolized distally, traveling away from the last generation of terminal bronchioles, i.e., away from the entry into the pulmonary acinus (for inspiratory air), a subcompartment of the pulmonary lobule, and in alveolar sacs that are closed ends of the airway tree.

Respiratory bronchioles are sites where gas exchange begins. Three generations of respiratory bronchioles exist generally. A single terminal bronchiole with its

[1]Latin *alveolus*: small vase, little tub.

M. Thiriet, *Anatomy and Physiology of the Circulatory and Ventilatory Systems*, BBMCVS 6, DOI 10.1007/978-1-4614-9469-0_2,

Table 2.1 Lung compartments and corresponding irrigating/draining (during inhalation/exhalation) bronchus types with the type of inspiratory air transport (convection or diffusion). The first generation of terminal bronchioles marks the entrance of the tracheobronchial tree into the pulmonary lobule. The first generation of respiratory bronchioles defines the airway entry into the pulmonary acinus

Compartment	Bronchus type
Conductive zone (anatomical dead space) Convective air transport (Navier–Stokes equation) Perfusion by bronchial arteries	
Lobe	Large cartilaginous-walled bronchi
Segment	Mid-sized and small cartilaginous-walled bronchi
Lobule	Terminal bronchioles (wall without cartilage)
Respiratory zone (gas exchange) Multinary diffusive gas transport (Stefan–Maxwell equation)	
Acinus	Respiratory bronchioles (alveolized wall), alveolar ducts (wall completely coated by alveoli), alveolar sacs (closed end), alveoli (basic airspaces)

respiratory bronchioles and 2 to 9 generations of *alveolar ducts* and *alveolar sacs* constitute the basic respiratory compartment of the lung, i.e., the pulmonary *acinus* (Table 2.1). The respiratory tract belongs to the human body's organs with the largest surface area.

The respiratory tract has 2 highly irrigated end regions, the nose and distal alveoli. The nose serve as air conditioner. The lung's major function is the exchange of oxygen and carbon dioxide between air and blood at pulmonary alveolus walls (3.3–$4.8 \times 10^6\,\mu m^3$ [1]; mean volume $4.2 \times 10^6\,\mu m^3$; width $\sim 200\,\mu m$).

Blood and air are delivered to parts of the outer surface and the entire inner surface of alveolar walls—the blood–air barrier—via respiratory and vascular trees, respectively. Whereas about 17 million bronchi carry inhaled air to approximately 480 million alveoli (range 274–790 millions) [1], nearly 73 million precapillary arterioles (bore 10–15 μm) generate directly or indirectly via precapillaries about 280 billion capillaries [2]. The surface area of alveolar septa (surface area $\sim 130\,m^2$, i.e., a tennis court [3]) fits into the lung parenchyma within the chest (air volume 4–6 l according to age, gender, and size).

Three-dimensional reconstruction of airway geometry from image data sets is required for computational biofluid dynamics [4], which are especially carried out to predict sites of inhaled particle deposition, thereby determining inhaled drug delivery protocols. In parallel, large arteries and veins of the pulmonary circulation can also be reconstructed for biomechanical simulations [5]. However, many works focused on upper airway geometry reconstruction fail to provide an anatomically realistic larynx characterized by a ventricle and vestibular and mobile vocal folds that strongly influence air flow. Unlike the nasal fossae, oral cavity, and throat, gaining the geometry of the larynx is not trivial because of the movement of the vocal folds during breathing. Nonetheless, a reconstructed 59-year-old human female airway model from the tracheal inlet down to bronchial generation 5 based on cubic voxel processing of MRI datasets from cadavers was merged to an upper airway model from oral cavity to the larynx derived from casting [6].

2.1 Upper Airways

Two major compartments of the respiratory ducts can be defined: the *upper airways*, in the head and upper neck (Fig. 2.1), and the *tracheobronchial tree*.

Upper airways constitute the proximal compartment with respect to the *airway opening* (facial ends of nose in the case of normal breathing, or mouth during lung function testing). Upper airways include successively the *nose* or the *mouth* according to the breathing mode, the *pharynx*, and the *larynx*. Some upper airway size are given in Table 2.2.

The tracheobronchial tree, below the larynx, corresponds to the distal compartment. The tracheobronchial tree can be subdivided into 2 distinct anatomical regions according to the body's segments, hence subjected to 2 different environments, the extrathoracic trachea (cranial tracheal segment) and the intrathoracic tracheobronchial tree.

2.1.1 Nose

The *nose* is composed of 2 asymmetrical nasal fossae, starting with the naris, or nostril, separated by the nasal septum. The shape of the nasal cavity (cavum nasi) varies between individuals.

The main nasal functions comprise: (1) olfaction,[2] (2) air-conditioning, warming and moistening the inhaled air (30% moisture), (3) cleaning via adsortion and secretion, and detoxification by biochemical action.

---

[2]The olfactory epithelium on the roof of the nasal cavity is made up of 3 layers of cells: supporting cells, bipolar (endowed with 1 dendrite and 1 axon) olfactory receptor neurons, and brush and basal cells. Supporting cells, or sustentacular cells, physically assist olfactory receptor neurons and help produce olfactory mucus, which contains water, mucopolysaccharides, antibodies, enzymes, and electrolytes, in addition to odorant-binding proteins. Brush, microvilli-bearing, columnar cells have a basal surface in contact with afferent nerve endings of the trigeminal nerve (cranial nerve V). Basal stem cells are sources of new olfactory sensory neurons. Olfactory sensory neurons are regularly replaced (life duration 4–8 wk). Olfactory sensory neurons transduce odorant signals from dendrite end knob with immotile cilia that extend through the mucus layer. The plasma membrane contains odorant-binding proteins acting as olfactory receptors. Inhaled odorants are made soluble by the serous secretion from Bowman's glands of the lamina propria of the mucosa. Axons from olfactory sensory neurons pass through tiny holes of the cribriform plate to enter the

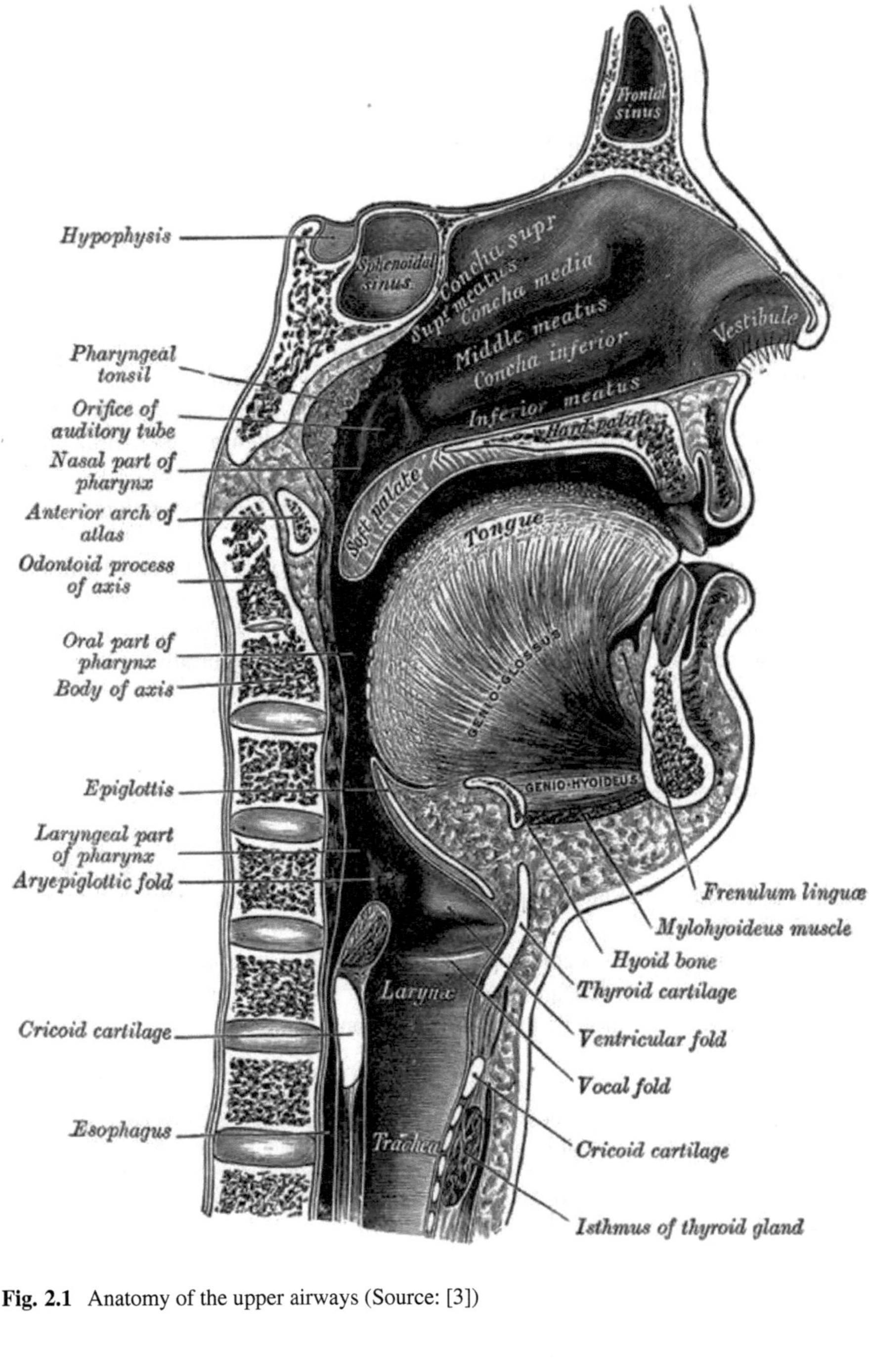

Fig. 2.1 Anatomy of the upper airways (Source: [3])

Table 2.2 Upper airway size (Source: [6]). The larynx length, i.e., the distance between the glottis (more precisely, the splitting apex between the esophagus and larynx) and just above the first tracheal cartilagenous ring equals about 43.8 mm; the sagittal glottis width (left–right axis normal to sagittal plane) equals 17.1 mm (laryngeal [vestibule or tubercle] entry; the sagittal width of the pharynx equals 30.9, 25.5, and 20.2 at its upper, mid, and lower station (at the widest section above the glottis), respectively; the sagittal width of the larynx at the ventricular folds (false vocal cords), at laryngeal ventricules (between false vocal and vocal cords), and at vocal cords equals 9.7, 22.7, and 7.1 mm, respectively; the horizontal distance from the oral cavity aperture to the posterior wall of the oropharynx and to the anterior wall of the larynx in the coronal plane of the abrupt change in curvature of the oropharynx equals 85 and 78 mm, respectively; the anteroposterior distance of the buccal cavity equals 54 mm; the vertical distance from the upper wall of the oropharynx to the upper trachea equals 83.4 mm; the diameter of a circle matching the arc of the oropharynx equals 65.4 mm

Buccal cavity	
Maximum sectional surface area	496–1811 mm^2
Minimum sectional surface area	65–492 mm^2
Nasopharyngeal cavity	
Maximum sectional surface area	174–648 mm^2
Minimum sectional surface area	92–493 mm^2
Laryngopharyngeal cavity	
Maximum sectional surface area	195–667 mm^2
Minumum sectional surface area	62–241 mm^2
Laryngeal cavity	
Maximum sectional surface area	149–328 mm^2
Minimum sectional surface area	30–216 mm^2
Epiglottis tip–soft palate tip distance	13–45 mm

Within the pharynx, the air is characterized by 90% moisture. Air flowing through the nasal cavities is warmed by an extensive capillary network and humidified by nasal goblet cells. Furthermore, the nose keeps heat and moisture from exhaled air. Larger particles (microscopic $\mathcal{O}[10\,\mu m]$) are removed from inhaled air by nose hairs, smaller particles ($\mathcal{O}[1\,\mu m]$) by impaction.

brain via 2 glomeruli (medial and lateral) in the left and right olfactory bulbs over the cribriform plate of the ethmoid bone. Axons of olfactory sensory neurons form a bundle after crossing the cribriform plate to give rise to olfactory nerves (cranial nerve I). The olfactory nerves actually consist of a collection of many sensory nerve fibers that extend from the olfactory epithelium to the olfactory bulb rather than nervous trunks. In addition, in the olfactory bulb, mitral neurons synapse in glomeruli with olfactory receptor neurons and relay signals to some brain regions (piriform and entorhinal cortex and amygdala).

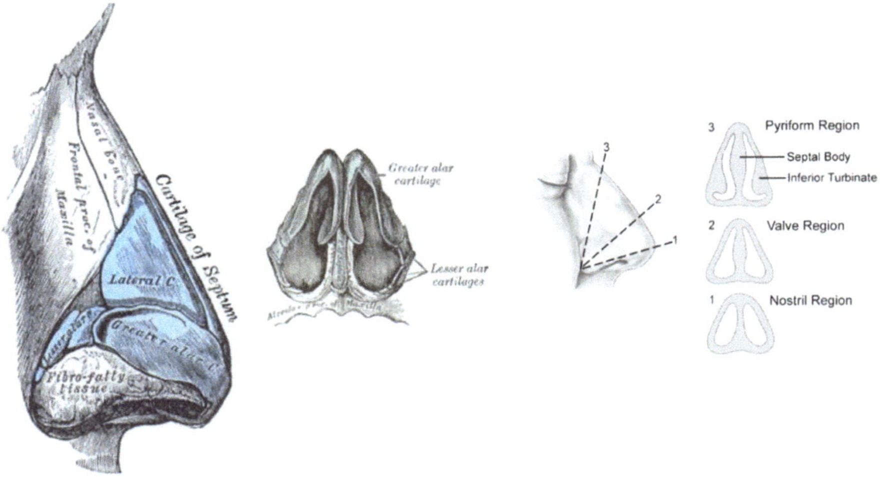

Fig. 2.2 Structure and configuration of nasal fossae (Sources: [3, 8]). (**Left**) Lateral view of nasal bones and cartilages. (**Middle**) Apertures of nasal cavities observed from below with cartilages. (**Right**) Shapes of different sections of the external region of the nasal passage

The 2 nasal cavities have a total surface area of about 160 cm^2 [7]. The nasal passages are flanked by a set of paranasal sinuses (Sect. 2.1.1.3). The nose can be bypassed either transiently (mouth breathing during vigorous exercise and orotracheal intubation) or chronically (tracheostomy).

The nose includes 2 main regions, the *external* and *internal noses*. The external nose is the projecting anterior part of the nose. Inferiorly the external nose open to the atmosphere through 2 *nostrils*, or nares. Nostrils are separated by the nasal septum, which is more or less deviated. The external nose comprizes an anterior cartilaginous part[3] and a posterosuperior bony portion.[4]

2.1.1.1 External Nose

The nasal valve[5] limits a variable area, the nasal vestibule, which depends on the size and shape of the lower lateral cartilage (Fig. 2.2).

The nasal vestibule has compliant walls. It is lined by a skin-like epithelium (stratified, squamous, keratinized epithelium). Behind the nasal pseudovalve, the nasal cavity proper is lined by the respiratory epithelium. The nasal vestibule contains vibrissae, which can stop entrance of large foreign particles during

[3]The nose cartilages include internally the septal cartilage, and, externally, the greater alar or lower lateral cartilage, beneath the upper lateral cartilage, the lesser alar, and the sesamoid.

[4]The paired nasal bones superiorly are attached to the frontal bone and inferolaterally to the ascending nasal processes of the superior maxilla.

[5]The word "valve" is, in the nose, a misnomer, as it does not prevent flow in the 2 opposite directions (exhalation and inhalation).

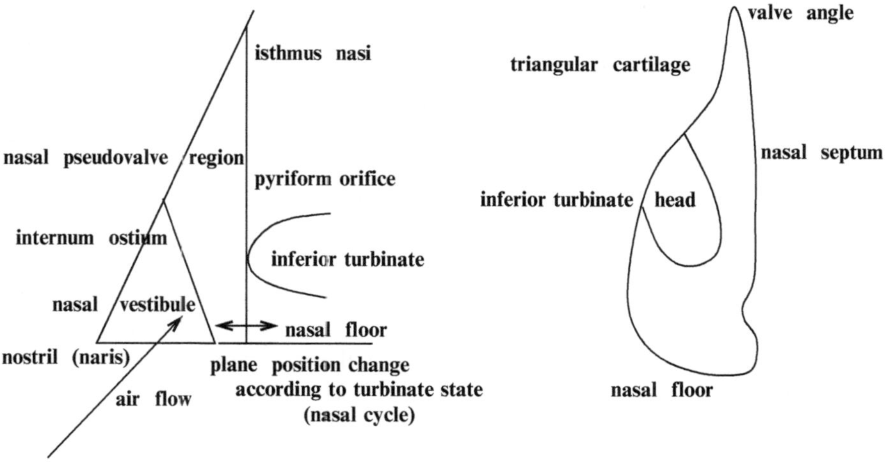

Fig. 2.3 External region of nasal fossa

inspiration. The nasal vestibule also possesses thermoreceptors. The configuration can change during deep breathing. In fact, inward motion of the lower lateral cartilage during forced inspiration is limited.

The nasal narrowing comprises external and internal pseudovalves. The *external nasal pseudovalve* is formed by the fixed columella (end of the nasal septum) medially, the nasal floor, and the nasal rim, or caudal border of the lateral part of the lower lateral cartilage.[6] Dilator naris muscles dilate this part during inspiration, hence counteracting possible wall collapse at high inspiratory flow rates.

The *internal nasal pseudovalve*, or simply the nasal valve, usually the narrowest nasal segment, defines the transition region between the skin and respiratory epithelium. The angle formed between external and internal nasal pseudovalves normally equals 10 to 15 degrees. The nasal pseudovalve lies in the coronal plane at the head of the inferior turbinate.

The anterior nasal aperture, or *pyriform aperture*, is a pyriform, vertical, nasal section with a narrow end upward. Its configuration is determined by the lateral and alar cartilages of the nose. It is bounded above by the inferior borders of the nasal bones, laterally by the thin sharp margins that separate the anterior from nasal surfaces of the maxilla, and below by the same borders that bend medially to join each other at the anterior nasal spine.[7]

---

[6]The lower lateral cartilage, or greater alar cartilage, is a thin flexible plate situated below the lateral nasal cartilage. It bends on itself to form the medial (crus mediale) and lateral (crus laterale) walls of the naris.

[7]Medially, the anterior surface of the maxilla is limited by a deep concavity, the nasal notch, on the margin of which attaches the dilator naris posterior and ends below in a pointed process, which links to that of the opposite bone to form the anterior nasal spine.

The nasal pseudovalve zone is formed by the nasal septum, caudal border of the upper lateral cartilage, head of the inferior turbinate, and isthmus nasi at the pyriform aperture (Fig. 2.3).

The internum ostium is the anterior segment and the isthmus nasi (internal nasal pseudovalve) is the posterior segment of the nasal pseudovalve region. The internal pseudovalve is named a valve because of bulging of the inferior nasal turbinate head mucosa. The external pseudovalve is also named a valve because inferior lateral cartilage movement regulates airflow in the vestibule.

The nasal pseudovalve region can be subdivided into proximal and distal components. The proximal component, or structural compartment, is made of the rigid septal cartilage and inferior portion of the mobile perior lateral cartilage. The distal component, or functional compartment, is the volume-varying tissue of the nasal septum and inferior turbinate head. The inferior turbinate projects forward toward the pyriform aperture and nasal vestibule.

In the normal nose, external and internal nasal pseudovalves are connected by fibrous connections between the caudal border of the superior lateral cartilage and the cephalic border of the inferior lateral cartilage. The lateral wall of the nasal vestibule is concave and corresponds to the internal face of the inferior lateral cartilage lateral crus. The external valve is located in the vestibular border.

The nasal pseudovalve is the narrowest part of the nasal conduit. Therefore, the nasal vestibule is the region of the greatest nasal airflow resistance. In fact, about 2/3 of the total nasal airflow resistance is measured in the bony cavum in the vicinity of the pyriform aperture and about 1/3 in the cartilaginous vestibule [9]. Caval resistance changes proportionately with the degree of mucosal congestion; vestibular resistance changes similarly. This phenomenon results from forward expansion of the anterior end of the inferior turbinates with congestion.

The surface areas of the nostril and of the ostium internum equal about 0.9 and 0.32 cm^2, respectively [10]. The cross-sectional areas of nostrils, ostium internum, and pyriform orifice equal about 1.0, 1.4, and 0.7 cm^2 [11].

2.1.1.2 Internal Nose

The internal nasal valve is formed by the angle between the upper lateral cartilage and the septal cartilage, by the nasal floor, and by the anterior head of the inferior turbinate.

The internal nose is composed of a ceiling, of a floor, of internal and external faces. The nasal floor is formed by the hard palate[8] and, posteriorly, by the soft palate,[9] where the posterior choanae opens into the nasopharynx. The nose ceiling internally is formed by the cribriform plate of the ethmoid[10] and posteriorly by the

[8] The hard palate (palatum durum) is continuous with the soft palate.

[9] The soft palate (palatum molle) is a movable fold between the mouth and the pharynx. It contains myofibers, an aponeurosis, vessels, nerves, adenoid tissue, and mucous glands. Its lower border is free. The lower portion is termed the palatine velum.

[10] The cribriform plate contains numerous tiny orifices for sensory fibers to the olfactory bulbs.

sphenoid. The bony nasal septum is composed superiorly of the perpendicular plate of the ethmoid, posteroinferiorly of the vomer.[11]

The nasal septum is the midline partition that divides the 2 nasal passages in a usual asymmetrical manner. Significant deviation of the nasal septum causes a pronounced narrowing of 1 nasal fossa.

The lateral nasal wall has 3 curved excrescences, the superior, middle and inferior *turbinates*. Each lateral nasal wall indeed contains 3 thin bones: the superior, middle, and inferior *conchae* (Fig. 2.1). The conchae form the bony frameworks of turbinates. Turbinates increase the internal surface area of the nose for heat and water transfer. Posterior to turbinates, where the septul terminates, the left and right nasal cavities merge into the nasopharynx.

Each turbinate extends the corresponding local length of the internal nasal cavity. The inferior turbinate, which is the largest turbinate, runs almost parallel to the nasal floor of the nose. The middle turbinate is located above the inferior turbinate. The superior turbinate is the smallest turbinate, above the middle turbinate.

The *meatus* is the space below the turbinate. There are hence 3 meati: the superior, middle, and inferior meatus. The nasolacrimal duct drains tears into the inferior meatus. The middle meatus contains the semilunar hiatus, with openings of the maxillary, frontal, and anterior ethmoidal sinuses (Sect. 2.1.1.3).

The nasal cavity can be subdivided into 2 compartments: the respiratory and olfactory compartments. The *olfactory cleft* is a narrow cavity on each side of the nasal septum at the top of the nasal cavity coated by the olfactory epithelium.

The respiratory compartment of nasal fossae have a highly vascularized lamina propria. When venous plexus of the turbinate mucosa dilate, congested turbinates restrict air flow and cause air to be directed to the other nasal passage.

The *nasal cycle* (mean duration $\sim$180 mn) is the alternating congestion and decongestion of nasal turbinates due to activation of the autonomic nervous system by the hypothalamus. Turbinates in 1 nasal fossa congest, whereas turbinates of opposite fossa decongest. Changes in local airway caliber result from successive cycles of vasodilation and vasoconstriction in the nasal mucosa, hence tissue swelling and shrinking. Therefore, the 2 nasal cavities work synergistically, but alternatively, to provide constant humidification, warming, and filtering.

The nasal walls are carpeted by a ciliated cylindrical pseudostratified epithelium. A squamous pavement variety in the vestibule makes the transition to the columnar respiratory epithelium. The purification function of the nose is due to ciliary motions and mucus. The mucus is secreted by goblet cells and mucus glands of the wall. The mucus intercepts and excludes at least large (size > 5–$10\,\mu$m) solid particles seeded in inhaled air. The epithelial ciliated cell has 25 to 100 cilia. The cilia undulate from the front to the nasopharynx with a beating frequency of 3 to 20 Hz,[12] driving trapped particles toward the larynx and esophagus.

[11]The vomer participates in the choanal opening of the nasopharynx.

[12]The ciliary beat frequency of a given cell is constant. But the ciliary beat frequency of different cells strongly varies [12].

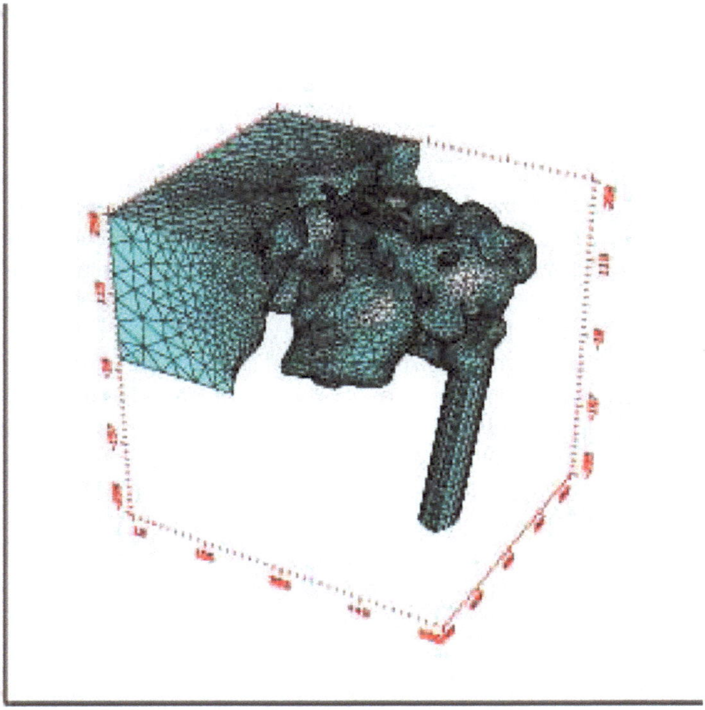

Fig. 2.4 Mesh of a nose and its paranasal sinuses reconstructed from computed tomography with a facial box and a pharyngeal straight distal segment for air flow computations

The nose has its own musculature, which includes 4 principal groups: elevators, depressors, compressor, and dilators. Blood is supplied by branches of both the external and internal carotid arteries. Lymphatics arise from the mucosa and drain posteriorly to the retropharyngeal ganglia and anteriorly to the upper deep cervical ganglia. The nose has sensory trigeminal and motor facial supplies.

2.1.1.3 Paranasal Sinuses

Paranasal sinuses are air-filled cavities within the cranial bones around the nasal fossae (Fig. 2.4). Each sinus is named accordingly to its bone site: (1) a maxillary sinus per fossa, laterally; (2) several small ethmoidal sinuses on the side; (3) a frontal sinus at the anterior angle of the nasal ceiling; and (4) a sphenoidal sinus at the posterior angle of the nasal ceiling, behind the ethmoidal sinuses. Each sinus is connected to the nose by an ostium.

The anterior ethmoidal cells open into the middle meatus. The frontal and maxillary sinuses also drain into the middle meatus. The posterior ethmoidal cells communicate with the superior meatus. The sphenoidal sinus drains in the sphenoethmoid recess, the space between the superior turbinate, the septum, and the sphenoid front wall.

The paranasal sinuses are covered with a ciliated epithelium, which secretes mucus. The paranasal sinuses thereby actively participate in the mucociliary clearance.

2.1.2 *Mouth*

In strong effort, human subjects breath simultaneously through both nose and mouth. During mouth breathing, air is neither filtered nor warmed as efficiently as in nose ventilation. Mouth breathing is often associated with obstruction of upper respiratory tract (e.g., adenoiditis).

The mouth consists of 2 parts: the *outer vestibule*, bounded externally by the lips and cheeks, internally by the gums and teeth, and the *inner buccal cavity* (cavum oris proprium). The vestibule receives the secretion from the parotid salivary glands. The mouth cavity communicates with the pharynx by a constricted aperture, the *isthmus faucium*. It contains the tongue. It receives the secretion from the submaxillary and sublingual salivary glands. A mouthpiece is placed in the vestibule for leakproof flow during spirometry.

2.1.3 *Pharynx*

The posterior wall of the pharynx rests against the cervical vertebrae. The lateral wall has openings of auditory tubes, or *Eustachian tubes*, communicating with the middle ears. The pharynx branches off into 2 conduits, the esophagus and larynx, thereby participating in both the respiratory and digestive functions.

The pharyngeal lumen is lined by a stratified epithelium, either cylindrical in the nasopharynx, or pavimentous.

The pharynx is composed of 3 serial regions (Fig. 2.1). The *nasopharynx* extends from the posterior choanae of the nose to the soft palate. The *oropharynx* is located between the soft palate, superiorly, and the vallecula, inferiorly. Its posterior and lateral walls are formed by the superior and middle pharyngeal constrictors. The *hypopharynx* lies inferiorly from the tip of the epiglottis. Its posterior and lateral walls are formed by middle and inferior pharyngeal constrictors. It leads posteriorly to the cervical esophagus and anteriorly to the larynx.

Table 2.3 Averaged dimensions (mm) of adult larynx (Source: [13]). After puberty, the female larynx does not develop significantly, whereas the male one changes strongly with cartilage enlargments

	Woman	Man
Length	36	44
Width	41	43
Antero-posterior bore	26	36
Circumference	112	136

2.1.4 Larynx

The *larynx* forms the lower part of the anterior wall of the pharynx. It is located between the large vessels of the neck. Its vertical extent corresponds to the fourth, fifth, and sixth cervical vertebrae, but it can have a higher situation in women and childs. Averaged dimensions of the adult larynx are given in Table 2.3.

The larynx is the vocal organ and the conduit that separates food and air. A swallowing reflex prevents foods and liquids from entering the glottal end of the larynx. The epiglottis is a thin fibrocartilaginous lamella at the entrance to the larynx, going upward behind the tongue root. Its free extremity is broad. The lower part of its anterior surface is connected to the upper border of the body of the hyoid bone by the hyoepiglottic ligament. The lateral edges are partly attached to the pharynx wall.

Whereas the pharynx is a soft tissue conduit, the larynx contains bony and 9 cartilaginous structures: the hyoid bone, 3 unpaired cartilages (epiglottic, thyroid and cricoid cartilages), and 3 paired cartilages (cuneiform, arytenoid, and corniculate cartilages; Fig. 2.5). The cartilages are either hyaline (thyroid, cricoid, and most of arytenoid) or elastic (epiglottic, corniculate, cuneiform, and vocal process of arytenoid) in adults. The laryngeal cartilages and hyoid bone are connected by intrinsic and extrinsic ligaments and membranes (Fig. 2.6).

The larynx has an intrinsic musculature (Fig. 2.6). When the false vocal cords are closed by sphincter muscles, the folds form an exit valve acting during coughing, air being explosively released.

Like the pharynx, the larynx possesses anatomical divisions. The *supraglottis* contains the epiglottis, the aryepiglottic folds, the false vocal cord folds and the ventricle.[13] The *glottis* is characterized by the true vocal folds. The *subglottis* is the distal part that leads to the trachea.

The vestibule is the region of the laryngeal lumen above the vocal folds. Its section is wide and has a triangular shape with curved edges, the curved basis corresponding to the anterior wall. The vocal folds have an elliptical lumen.

[13]The laryngeal ventricles are fusiform fossa, situated between the ventricular (false vocal) and vocal folds.

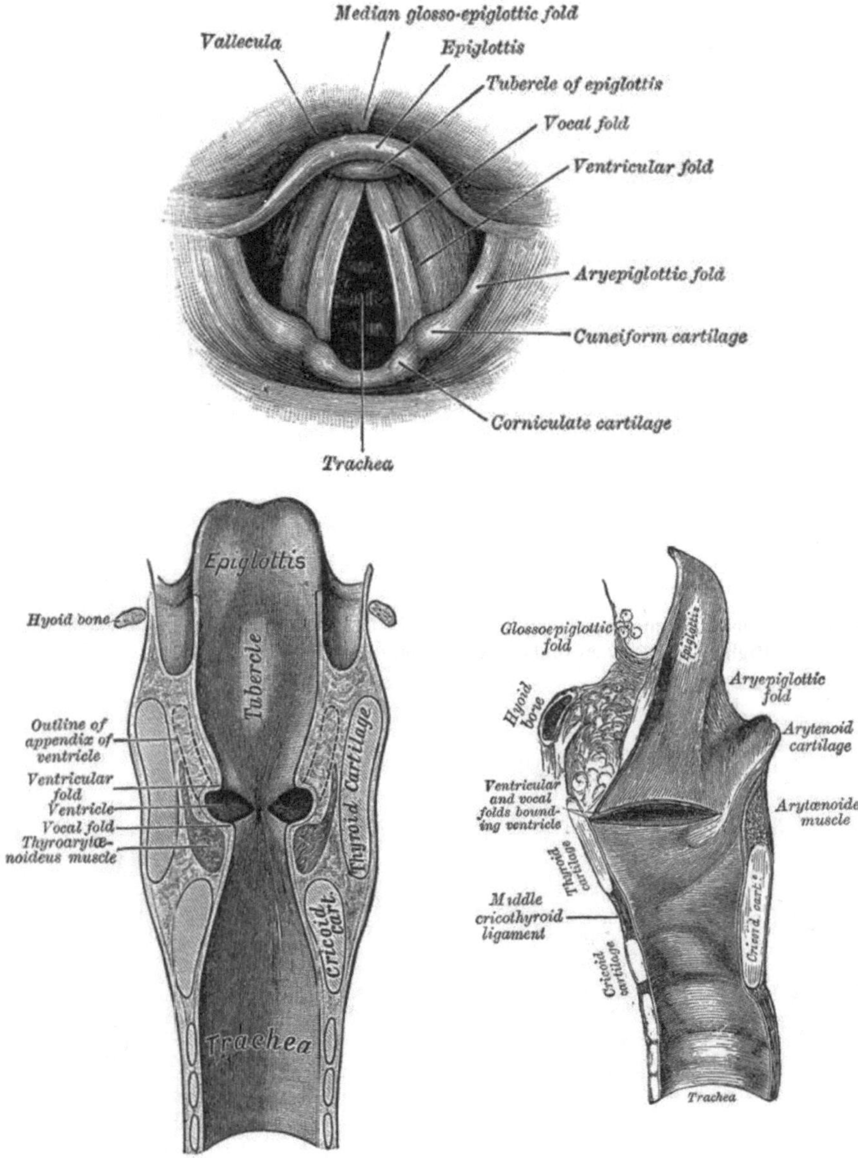

Fig. 2.5 Sections of the larynx (Source: [3]). (**Top**) upper view of the larynx. (**Bottom left**) coronal section of the larynx. (**Bottom right**) sagittal section of the larynx

The *rima glottidis*, the lumen between the vocal folds laterally and the vocal processes of the arytenoid cartilages behind, is the narrowest part of the laryngeal lumen. Its length is about 23 mm and 17–18 mm in men and women, respectively. The size and shape of the rima glottidis vary with the movements of the vocal

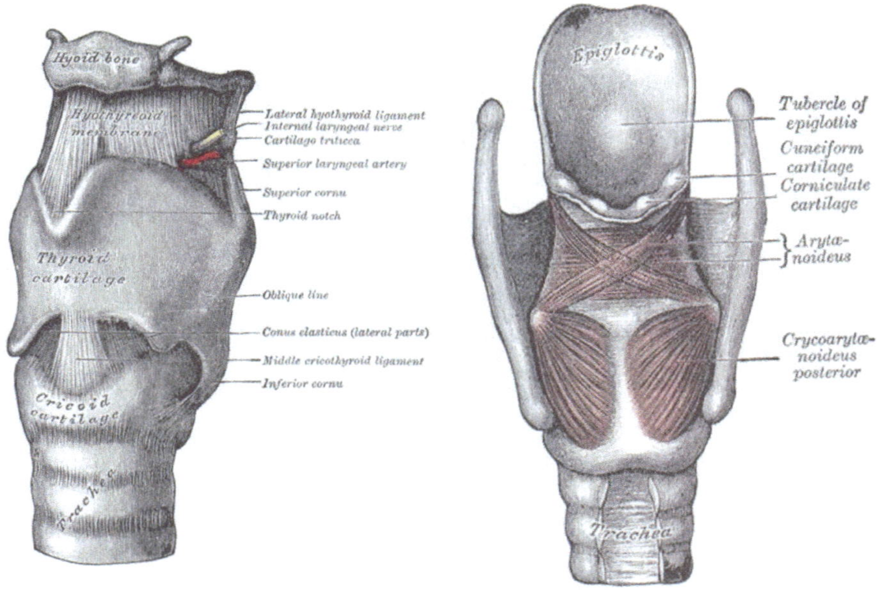

Fig. 2.6 Laryngeal ligaments and muscles (Source: [3]). (**Left**) anterolateral view of the laryngeal ligaments. (**Right**) posterior view of the laryngeal muscles

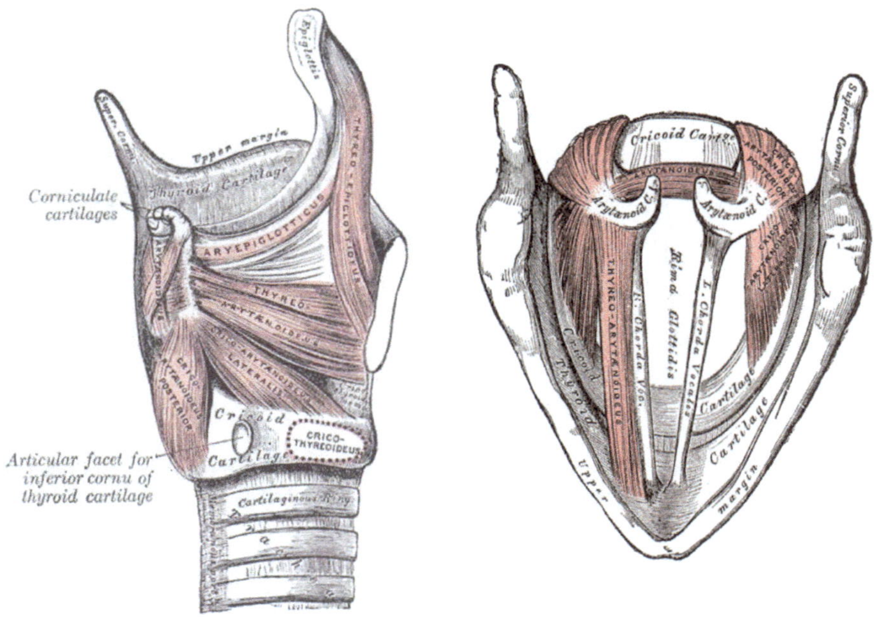

Fig. 2.7 Laryngeal ligaments and muscles (Source: [3]). (**Left**) side view of the laryngeal muscles, with removed right part of the thyroid cartilage. (**Right**) top view of the laryngeal muscles

folds and of arytenoid cartilages during respiration and phonation. At rest, the rima glottidis has a triangular shape with an anterior apex. During extreme adduction of the vocal folds (apposition of the vocal folds), the rima glottidis is reduced to a slit. In extreme abduction of the vocal folds, as in forced inspiration, it nearly becomes lozenge-shaped.

The larynx is the phonation organ. The vocal ligaments within the vocal folds are tensed under the control of cartilage displacements produced by involved muscles (cricothyroid, cricoarytenoid, vocalis muscles). The motion of the surface of the vocal folds produces vocal sounds, which are articulated by the tongue, the lips and the teeth.

The larynx is implicated in throat cleaning and coughing. Foreign matter and mucus raised to the larynx by the tracheobronchial ciliary motion are dislodged by the larynx.

The laryngeal mucosa is continuous with the pharynx one. The larynx wetted surface is covered by a *ciliated epithelium* with some *goblet cells*, except the area that is frequently in contact with food. The latter is lined by a *stratified squamous epithelium*.

The larynx is irrigated by the laryngeal branches of the superior and inferior thyroid arteries. The veins accompany the arteries. They drain either into the superior thyroid vein (then into the internal jugular vein) or into the inferior thyroid vein (then into the innominate vein). The lymphatic vessels consist of 2 sets, superior and inferior.

The larynx receives both motor and sensory innervation. The larynx is innervated by branches of the vagus nerve: (1) the superior laryngeal nerve, with its external laryngeal branch, for sensation of the glottis and supraglottis and motor fibers to the cricothyroid muscle; and, mostly, (2) the recurrent nerve via the inferior laryngeal nerve for sensation of the subglottis and motor fibers to the intrinsic laryngeal muscles.

2.2 Thoracic Respiratory System

The intrathoracic part of the respiratory tract starts (inhalation) or ends (exhalation) at the *superior thoracic aperture*, which is surrounded by the clavicles. The intrathoracic airways and the lung parenchymas are located in the thoracic cage.

2.2.1 Chest Wall

The *thoracic cage*, or chest wall, is formed by: (1) bones and (2) muscles moored to these bones.

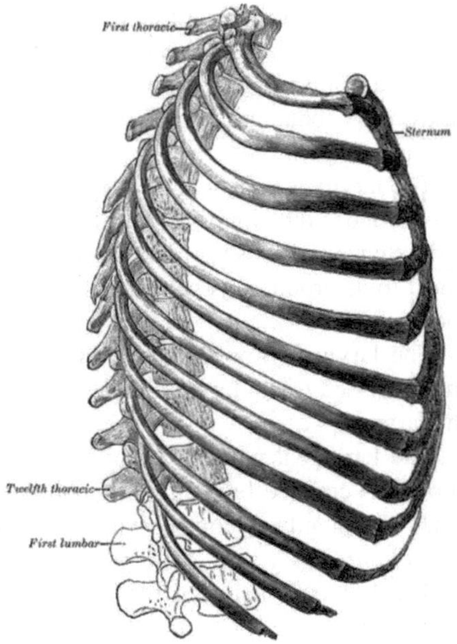

Fig. 2.8 Bones of the thoracic cage (Source: [3])

2.2.1.1 Thoracic Bones

The bones include the *ribs*, *spine column*, with the thoracic vertebrae T1 to T12, and *sternum* (Fig. 2.8).

The 3 types of ribs include: (1) true ribs R1 to R7, with their own costal cartilage; (2) false ribs R8 to R10 that share a common costal cartilage; and (3) floating ribs R11 and R12, without costal cartilage.

The sternum is composed of 3 elements: (1) the manubrium, with the first and second costal cartilages; (2) sternal body (third to seventh costal cartilages); and (3) the xiphoid process.

The skeleton wall is articulated with the *sternocostal and chondrocostal joints* anteriorly, between the sternum and each costal cartilage and between each costal cartilage and respective rib, and the *costovertebral joints* posteriorly, between each ribs and the transverse process of the corresponding vertebra.

2.2.1.2 Thoracic Muscles

The musculature of the thoracic cage comprises the intercostal muscles and the diaphragm. The thoracic cavity is enlarged when air is sucked into the lungs due to the contraction of the inspiratory muscles: the diaphragm dome goes down and the rib cage rises.

The rib cage size increases in the sagittal and transverse directions because ribs R1 to R7 rotate and ribs R9 to R12 bow out laterally and in the axial direction due to the diaphragm contraction causing a caudal displacement. The lungs are stretched.

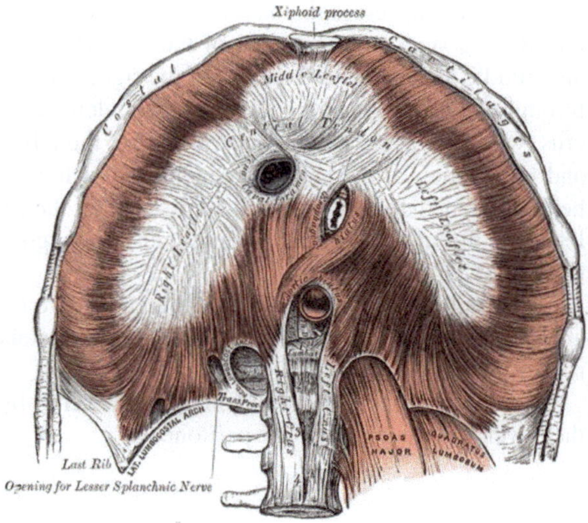

Fig. 2.9 Diaphragm (Source: [3])

The quiet inspiration requires the diaphragm and the external intercostals, with an airway downstream pressure slightly lower than the atmospheric (upstream) one (−0.2 kPa). The quiet expiration is, in general, assumed to be passive, owing to shrinkage of the lung and of the chest wall associated with elastic recoil.

During active inspiration, additional muscles are mobilized, essentially the scalenes and sternomastoids, which elevate the 2 first ribs and sternum. The active expiration requires the abdominal muscles and internal intercostals, which pull ribs downward and inward.

Diaphragm

The diaphragm is the primary muscle for inspiration (Fig. 2.9). This thin dome-shaped muscle separates the thoracic cavity from the abdominal one.

The diaphragm pulls the chest downward during inhalation, displacing the abdominal content downward and frontward. It pushes the lower ribs outside and the lower part of the sternum frontward. Consequently, the thorax height and the tranverse section of the thorax base increase.

The diaphragm pushes upward during exhalation. The diaphragm displacements have a magnitude from about 1 cm at rest to approximately 10 cm during ample breathing.

The diaphragm inserts, peripherally, on the lower ribs, on the posterior face of the xiphoid process, and on bodies of lumbar vertebra via 2 pillars or crura, the left and right crus, on L1 and L2 on the left edge, on L1 to L3 on the right side, and, centrally, on its fibrous central tendon. The central tendon, a thin, strong aponeurosis situated near the center of the vault, is composed of 3 leaflets (right, middle, and left).

Several organs cross the diaphragm via apertures. The inferior vena cava passes through the vena caval foramen of the central tendon, at the junction of the right and middle leaflets, with branches of the right phrenic nerve. The esophagus passes through the esophageal opening, or hiatus, defined by diverging fibers of the right crus, with the vagus (pneumogastric) nerve, the left gastric and esophageal vessels and lymphatics, at the level of the tenth thoracic vertebra (T10). The aorta travels behind the median arcuate ligament, formed by the curved merging of the right and left crus, through the aortic opening with the azygos vein and the thoracic duct, at the T12 level.

The diaphragm is supplied by the phrenic nerve from cervical segments C3, C4, and C5 for the motor inputs and phrenic, intercostals (6–12), and upper 2 lumbar roots for sensory ones.

The upper surface of the diaphragm is covered by the pleura and pericardium, the lower surface partly by the peritoneum.

External Intercostals

The external intercostals, the external layer of the intercostal muscles, go obliquely downward and inward from the inferior border of the ribs to the superior border of the subjacent ribs.

The external intercostals elevate the rib cage. They then increase the lateral and anteroposterior dimensions of the thorax.

The external intercostals receive branches of the intercostal nerves.

Internal Intercostals

The internal intercostals, the internal layer of the intercostal muscles, also link adjacent ribs. They take the opposite path with respect to the external intercostals, going diagonally downward and outward.

The internal intercostals pull the ribs downward and inward, thus decreasing the thoracic volume during ample expiration.

Deep Intercostal Muscles

The *transversus thoracis*, *innermost intercostal*, and *subcostal* muscles constitute the deepest layer of intercostal muscles from the anterior to the posterior region, respectively.

Other Contributing Thoracic Muscles

The *thoracic musculature* involved in respiration includes additional muscles. The *pectoralis minor* are small chest muscles, which pull up on the rib cage, elevating it.

The *serratus posterior superior*, the *levatores costarum*, and the *serratus posterior inferior* attach the ribs to the vertebral column. They assist in respiration, the former two raising and the latter lowering the ribs, when they contract.

Participating Cervical Muscles

Certain cervical muscles are used as rib cage lifters when needed, such as the *sternocleidomastoids* and the *scalenes*. The scalenes and sternocleidomastoids raise and push out the upper ribs and the sternum.

Participating Abdominal Muscles

During strong expiration, abdominal wall muscles are involved, such as the *rectus abdominus*, the *internal and external obliques*, and the *transversus abdominus*. They push up the diaphragm, increasing the pleural pressure, and consequently, the alveolar pressure.

2.2.2 *Mediastinal Airways*

The intrathoracic part of the respiratory conduits can be decomposed into a mediastinal compartment, which comprises the distal (caudal) segment of the trachea and proximal segments of its bifurcation branches, the main bronchi, and the intraparenchymatous respiratory tract down to the alveoli.

2.2.2.1 Mediastinum

The mediastinum is the chest space between the sternum (front), the spine column (rear), and the lungs (sides). It contains the heart and its afferent and efferent vessels, and their branches and tributaries,[14] thymus, trachea and main bronchi, esophagus, lymph vessels and nodes, in particular the *thoracic duct*, the largest lymphatic vessel of the body, as well as nerves and neural plexus.

The mediastinum is decomposed according to the body axis into superior and inferior regions. The superior and inferior mediastina are the region of the thorax above and below the sternal angle, i.e., above and below the junction between the manubrium and the sternum body (above T4–T5), respectively.

[14]The azygos vein, which drains into the inferior vena cava, receives the intercostal veins. The hemiazygos vein drains into the azygos vein.

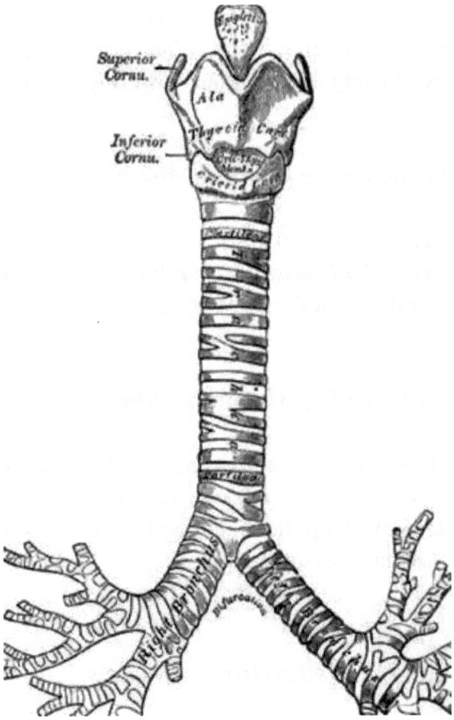

Fig. 2.10 Anterior view on an isolated trachea (Source: [3])

The mediastinum is also subdivided according to the anteroposterior axis (across the sagittal plane) into 3 zones: anterior, middle and posterior. The middle and anterior parts of the inferior mediastinum contain the heart enclosed in its pericardial sac. The esophagus and trachea travel in the posterior mediastinum.

2.2.2.2 Trachea and Main Bronchi

The trachea is a conduit (length 12–14 cm, width 13–22 mm, sagittal size 3–6 mm) supported by U-shaped cartilaginous rings (Fig. 2.10). Sixteen to twenty irregular, eventually branching, dorsally incomplete cartilaginous rings reinforce the tracheal wall that nevertheless remains flexible. Connective tissue and bands of smooth muscle bridge posterior ends of supportive cartilage. Partial cartilaginous rings induce a set of irregularly spaced flanges over the inner and outer tracheal surfaces, as wall narrows in the intercartilaginous segments [14].

The tracheal cranial end, at the level of the sixth cervical vertebra (C6), is attached to the cricoid cartilage. The trachea descends the lower neck, crosses the superior aperture of the thorax and the upper posterior mediastinum, and divides. Both tracheal ends are hence mobile.

The tracheal shape and size vary with head and neck motions. Furthermore, the tracheal configuration and dimensions also change during the respiration, especially during ample thoracic displacements.

The cross-sectional shape of excised tracheas varies throughout the tracheal length [15]. As is the transverse section of any deformable pipe, the tracheal cross section depends on the intrathoracic pressure. During forced expiration [16] and cough, the tracheal lumen, although supported by cartilaginous rings, collapses with a deep invagination of its soft dorsal wall, whereas it widens during deep inspiration. In addition, the tracheal shape can be deformed by possible prints of neighboring organs, such as the aortic arch. Last but not least, the tracheal anatomy changes more or less strongly from one subject to another [15].

The trachea travels with several blood vessels, nerves, and esophagus. It runs near the thyroid gland and the remains of the thymus. Its anterior surface is in relation to the inferior thyroid veins, the anastomosing branches between the anterior jugular veins in the neck, as well as the left brachiocephalic vein and aortic arch with the brachiocephalic and left common carotid arteries, in addition to the deep cardiac plexus in the thorax. Its lateral walls are in relation to the inferior thyroid arteries and recurrent nerves in the neck. Its right side runs with the right vagus, the innominate artery and its left side with the left recurrent nerve, aortic arch, and left common carotid and subclavian arteries.

The tracheal wall is composed of 2 main regions, a posterior fibromuscular membrane and an anterolateral wall with incomplete cartilaginous rings. The cartilaginous rings are incorporated in a connective tissue with elastin and collagen fibers. The pars membranacea is composed of transversally oriented fiber bundles. The myofibers are inserted on the posterior tips of the cartilage. The submucosa contains nerves, lymphatics, blood vessels and glands. The glands, especially located in the intervals between cartilages and externally to the tracheal smooth muscle, secrete mucus.

The tracheal epithelium mainly with ciliated cells and goblet cells ($\sim$1 goblet cell for 5 ciliated cells) lines the luminal surface (Vol. 1 – Chap. 3. Cells of the Ventilatory Tract). *Ciliated cells* (height 40–60 μm) have a row of several hundred cilia at the apical wetted surface. *Goblet cells* are located singly or in groups between ciliated cells. They secrete mucus; their cytoplasma contains mucous vacuoles. *Brush cells* are another cell type with apical microvilli (length $\sim$1 μm, bore $\sim$100 nm) [17]. They are also characterized by apical corpuscules and sometimes few cilia. Brush cells could either be sensors or precursors of ciliated cells.

The tracheal epithelium lies on a basement membrane, and beneath it, on a *lamina propria*. The lamina propria contains nervous and vascular networks. It is pierced by glandular ducts. The lamina propria is composed of collagen fibers and elastin bundles. The submucosa is the middle stratum of the trachea. It is made of connective tissue in which exist tracheal glands, blood vessels, nerves, and lymphatics. The fibrosa is the external layer of the trachea, which contains hyaline cartilage. Connective tissue lies outside the cartilage and extends between the tracheal rings. It contains nerves, blood vessels, and the paratracheal lymph nodes. The cartilaginous partial rings can fuse with adjoining cartilages.

The trachea is supplied by the inferior thyroid arteries. The veins drain in the thyroid venous plexus. The nerves are branches of the vagi and recurrent nerves. The trachea also receives sympathetic inputs. Nerves distribute to the trachealis muscles, gland, and epithelium.

The trachea bifurcates into left (bore 10–14 mm, length ~50 mm) and right (bore 12–16 mm, length ~25 mm) *main bronchi*, or stem bronchi, at the level of the fourth thoracic vertebra or at the upper border of the fifth thoracic vertebra (T4–T5). The division region is characterized by a transition zone to adjust the size of the 2 asymmetrical branches, and a more or less angulated carina. The caudal end is attached to the pericardium and diaphragm by the membrana bronchopericardiaca.

The right main bronchus is shorter than the left one. The left main bronchus branches off at a sharper angle than the right one. Foreign bodies penetrate more easily in the right stem bronchus because of its smaller branching angle. Both stem bronchi enter the respective lungs through the *hilum* with lymph and blood vessels and nerves.

The azygos vein arches over the right main bronchus. The right pulmonary artery lies below and then in front of it. The left main bronchus passes beneath the aortic arch. It is in relation to the thoracic duct and descending aorta. The left pulmonary artery is above, then in front of it.

2.2.3 Lungs

The lung is a cone-like organ on each side of the thorax (Fig. 2.11). A registration-based technique enables the estimation of the local lung expansion from multiple respiratory-gated CT images of the thorax, hence assessing regional pulmonary ventilation and specific volume change [18]. The degree of regional pulmonary expansion is measured using the Jacobian (a function of local partial derivatives) of the registrated displacement field.

2.2.3.1 Lung Configuration

Each lung presents a rounded apex into the root of the neck, an inferior upward convex base, 3 borders (thin inferior and anterior and broad posterior edges), and 2 faces (external costal and internal mediastinal with pericardial impression).

The base of each lung contacts the upper surface of the diaphragm, extending to the level of rib 7 anteriorly and rib 11 posteriorly. The right lung is larger than the left, owing to the presence of the heart.

The lung apex is crossed by the subclavian artery and brachiocephalic vein. The lung base has the same curvature as the diaphragm dome. On its mediastinal face, the lung possesses a hilum enclosed by a reflection of the pleura (Fig. 2.12).

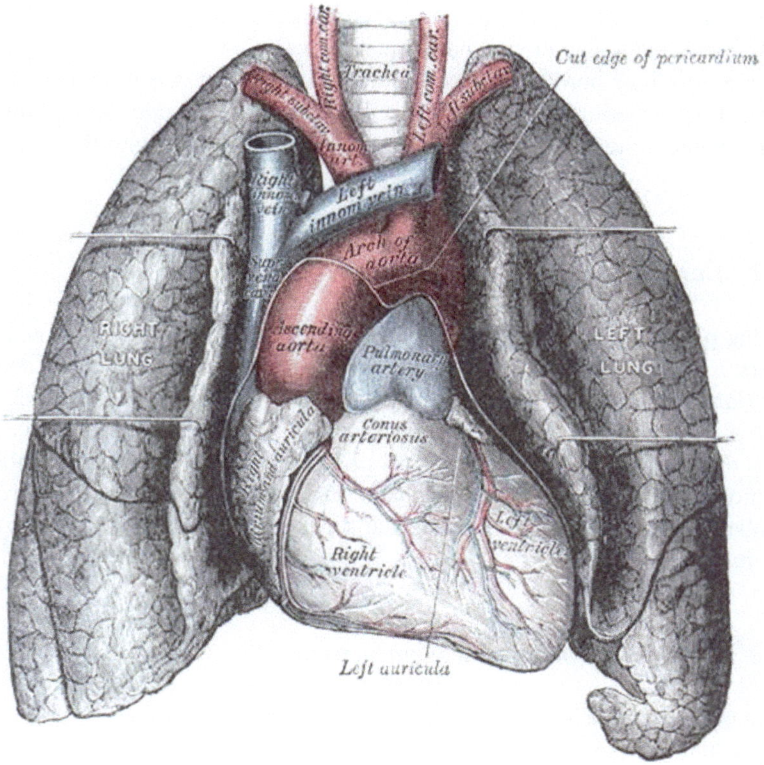

Fig. 2.11 Anterior view on the lungs, trachea, heart, and large thoracic blood vessels (Source: [3])

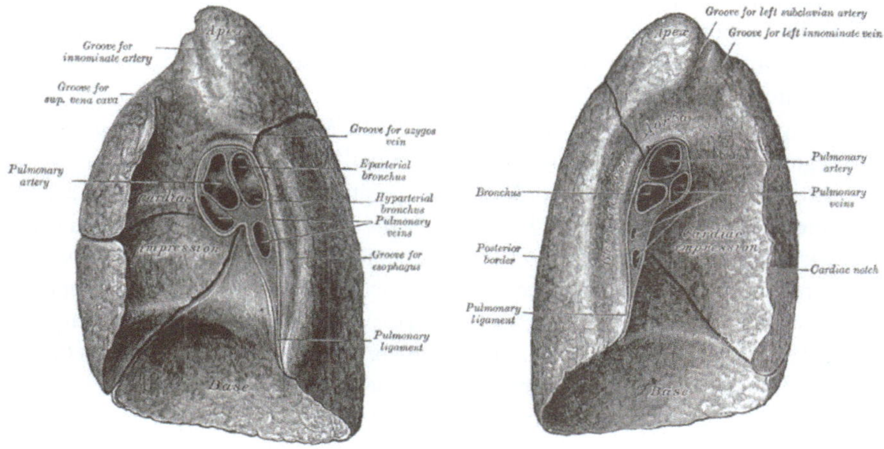

Fig. 2.12 Mediastinal faces of the right (**left**) and left (**right**) lungs with hilum (Source: [3])

The hilum is the passage for airways, entering and exiting blood vessels,[15] lymph vessels connected to ganglions, and pulmonary nerves.

The inferior lung border corresponds to the sixth rib, anteriorly, and the thoracic vertebra T10, posteriorly. The inner edge is delimited by the mid-axillary line. The superior border extends slightly through the thoracic aperture into the base of the neck.

2.2.3.2 Lung Parenchyma

The lung parenchyma can be considered as a deformable three-dimensional tissue network made of interconnected sheaths and septa between pulmonary compartments (*lobes*, *segments*, *lobules*, and *acini*) as well as between alveoli (Sect. 2.2.3.7).

The set of alveolar septa, each septum containing many alveolocapillary membranes, delimits the alveolar spaces, sites of gas exchange. The sheaths embed the respiratory conduits, blood and lymphatic vessels, and nerves. The fissures and interlobar, -segmental, -lobular, and -acinar septa delineate the anatomical structure of the lung parenchyma.

2.2.3.3 Pulmonary Anatomical Segmentation

Both lungs are divided into lobes separated by fissures, 2 horizontal and oblique over the right lung and 1 oblique over the left lung (Table 2.4). The interlobar surfaces are covered with visceral pleura, so that the lobes slide against each other like lungs within the thorax.

The accurate identification of fissures helps in the early detection of pathologies and in the regional functional analysis of lungs. Automatic segmentation of the fissures can be based on information provided by segmentation of airway and vascular trees [19].

The right lung has 3 lobes, the left 2 (Fig. 2.12). Pulmonary lobes are subdivided into segments (Table 2.5), themselves into lobules.

Each segment is named according to its position. The right superior lobe has 3 segments (apical, anterior, and posterior); the right middle lobe 2 (medial and lateral); and the right inferior lobe 5 (superior, anterior basal, lateral basal, posterior basal, medial basal or cardiac). The left upper lobe has 2 compartments: (1) superior, with 3 segments (apical, anterior, and posterior); and (2) inferior or lingular, with 2 segments (superior and inferior). The left lower lobe comprises 5 segments (apical [superior], medial basal, anterior basal, lateral basal, and posterior basal).

The surface of the right lung has a horizontal fissure, which separates the upper lobe from the middle lobe, and an oblique fissure delineating the lower lobe. The

---

[15] The pulmonary artery penetrates between the bronchus, accompanied with the bronchial arteries and veins and pulmonary veins.

Table 2.4 Lung segmentation

Right lung	Upper lobe	Apical segment
		Anterior segment
		Posterior segment
	Intermediate lobe	Medial segment
		Lateral segment
	Lower lobe	Nelson's upper segment
		Lateral basal segment
		Anterior basal segment
Left lung	Upper lobe	Posterior segment
		Apical segment
		Anterior segment
		Upper segment
		Lower segment
	Lower lobe	Nelson's upper segment
		Anterior basal segment
		Lateral basal segment
		Medial basal segment
		Posterior basal segment

Table 2.5 Lobes and segments of the human lungs

right lung		left lung	
Lobe	Segment	Lobe	Segment
Upper	Apical	Upper	Superior, apical
	Posterior		Superior, anterior
	Anterior		Superior, posterior
			Lingular, superior
			Lingular, inferior
Mid	Lateral		
	Medial		
Lower	Superior basal	Lower	Superior basal
	Medial basal		Medial basal
	Anterior basal		Anterior basal
	Lateral basal		Lateral basal
	Posterior basal		Posterior basal

right lung surface has a depression created by the superior vena cava. The left lung surface has a groove made by the aortic arch and the cardiac impression.

2.2.3.4 Pleura

The pulmonary parenchyma is attached to and enclosed by the pleura. The inner surface of the chest wall is covered by the *parietal pleura*. This parietal leaflet is separated from the *visceral pleura* by a thin layer of lubrification fluid. Both pleural leaflets freely slide with respect to each other. The pleural blades fuse at the lung hilum. The pleura below the hilum and behind the pericardial impression forms the pulmonary ligament, which extends between the lower part of the lung mediastinal surface and the pericardium.

The visceral pleura invaginates into the fissures between the lung lobes. The parietal blade is commonly decomposed into 3 parts. The *mediastinal pleura* from the lung dome (or apex) to the basis between the sternum (front) and the spine column (rear). The *diaphragmatic pleura* adheres to a fraction of the diaphragm dome, outside the heart region. The *costal pleura* covers the internal face of ribs and intercostal spaces. The *costodiaphragmatic recess* is the inferior cavity between the lung base and the diaphragm. The *costomediastinal recess* is the junction between the mediastinal and costal pleura.

Pleural Fluid

The 2 layers of pleura are separated by a thin fluid space (thickness of 20–80 μm). The pleural fluid is formed on the parietal pleura at a rate of 7–11 cm^3 per hour (estimated turnover of about 0.15 ml/kg per hour). It is reabsorbed by parietal pleural lymphatics. The pleural liquid volume is 20–25 cm^3 (about 0.3 ml/kg) [20]. The pleural liquid is hypo-oncotic, containing about 1 g/dl of proteins. The pleural space and the lung interstitium behave as functionally independent compartments, because the visceral pleura is only slightly permeable to water and solute.

The flow rate in pleural lymphatics increases when pleural liquid filtration rises, thus controlling pleural liquid volume. A 10-fold increase in filtration rate causes only a 15% increase in pleural liquid volume. Maximum pleural lymph flow reaches about 30 ml per hour. When filtration exceeds maximum pleural lymphatic flow, pleural effusion occurs.

Pleural Structure and Function

The pleura is relatively thick in humans. Mesotheliocytes of the parietal and visceral pleurae are similar in structure and functions. But they act differently in the material resorption from the pleural space. The parietal pleura functions as a barrier for the particle conveyed in the pleural lymphatics on the intercostal muscles. The visceral pleura can resorb inhaled particles.

The mesothelium of the visceral pleura lies on a connective tissue layer bound to a deep fibrous sheet with 1 or 2 strata of collagen fibers and a network of elastin fibers, interwoven with collagen bundles. These fibers adapt to the heterogeneous applied tension during lung expansion. The fibers are connected to the vascular layer[16] and septa of subpleural pulmonary lobules.

[16]The vascular layer lies on the lobular wall. It contains blood and lymph vessels.

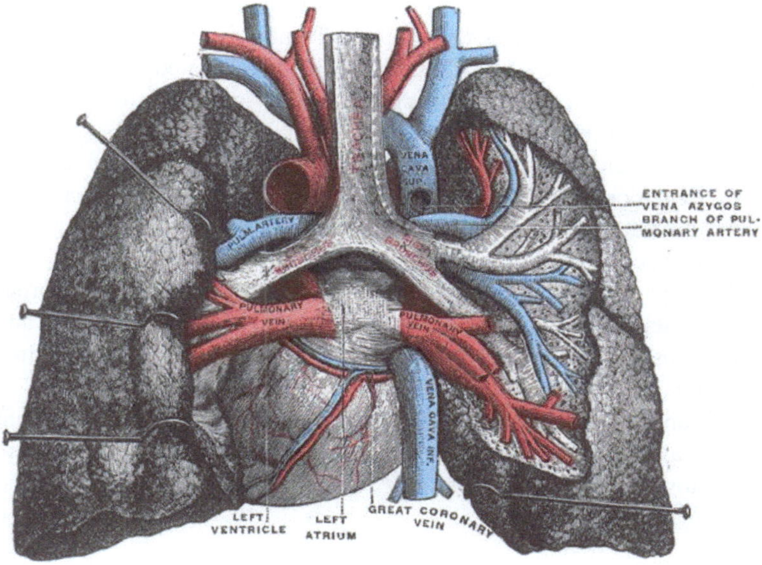

Fig. 2.13 Posterior view on the lungs and pulmonary vessels (Source: [3])

Fibroblasts and macrophages as well as many capillaries and lymphatics reside between the connective tissue fibers beneath the mesothelium.

Pleural Perfusion and Innervation

The pleura is perfused by branches of the intercostal, internal mammary, musculophrenic, thymic, pericardiac, and bronchial vessels. The veins follow the arterial paths. The visceral pleura is irrigated by a coarse mesh of relatively large vessels.[17]

The parietal pleura contains sensory nerve fibers from the phrenic and intercostal nerves. The visceral pleura contains nerve fibers of vagal and sympathetic origin.

2.2.3.5 Pulmonary and Bronchial Vessels

The blood vessels belong to both the *pulmonary circulation* (Fig. 2.13) devoted to gas exchange and the *systemic circulation* dedicated to cell metabolism.

Like the tracheobronchial tree, pulmonary arteries for blood oxygenation and bronchial arteries for bronchus perfusion down to the pulmonary lobule divide to form trees.

[17] The smallest vessels have a relatively great bore ($\sim 30\,\mu m$).

Pulmonary arteries follow bronchioles and alveolar ducts, providing twigs to adjacent alveoli. These arteries then supply alveolar sacs and alveoli. Pulmonary capillaries form the framework of alveolar walls. Oxygenated blood returns to the left atrium via the pulmonary veins using interlobular spaces whereas systemic veins follow the course of their arterial partners.

The systemic circulation arises from the aorta and intercostal arteries. The systemic circulation supplies blood to the pulmonary structures, bronchi down to respiratory bronchioles, and pulmonary vessels. Like alveolar capillaries, bronchial capillaries and peribronchial veins can drain in the pulmonary veins.

The vessel paths and sizes vary during the respiratory cycle, the lung parenchyma and blood vessels being tight by sleeves of connective tissue. Both arteries and veins adjust their position and size according to the lung volume via intertwining fibers between airway and alveolus lattices and vascular sleeves.

In addition, a periarterial plexus of lymph vessels dampens the changes in arterial bore associated with pressure wave propagation. The latter enhances lymph propulsion toward the pulmonary hilum.

2.2.3.6 Vascular and Nervous Plexus

Capillary plexus are observed in bronchi, as well as a capillary network beneath the pleura.

The lung is supplied from the anterior and posterior pulmonary plexus, mainly formed by branches of the sympathetic system and vagus nerves. The nerves are made of fibers associated with the bronchial muscle and mucosa as well as alveolar septa. Small ganglia are associated with these nerves.

2.2.3.7 Lung Interstitium – Fibrous Lung Framework

The interstitium is the supporting lung framework. It contains cells, mainly fibroblasts and pericytes on the capillary outer surface ($\sim$40% of the lung cell population), and fibers. The ground matrix is made of proteins and mucopolysaccharides (or glycosaminoglycans; Vol. 1 – Chap. 8. Cell Environment).

The lung connective tissue is a fibrous continuum that includes: (1) *sleeves* that surround packages of blood and lymph vessels, nerves, and airways; (2) interlobar, interlobular, and between-acinus *septa*; and (3) *visceral pleura.*

The connective tissue represents 20–25% of the lung dry weight. It contains elastin, collagen, proteoglycans, and glycoproteins.

Elastin fibers, the most extensible fibers, are responsible for expiratory recoil and tension transmission throughout the lung parenchyma.

Collagen fibers has a similar pattern than that of elastin fibers. They support the pulmonary architecture and limit the extensibility of airways and alveoli as well as vascular walls. Collagen is a major component of the pulmonary connective tissue, via collagenous and reticular fibers.[18] Collagen 1[19] is found in the bronchovascular sheaths and other parts of the lung parenchyma, collagen 2[20] in the tracheobronchial cartilage, collagen 3 in blood vessels, and collagen 4 in basement membranes.

2.2.3.8 Pulmonary Acinus

The lungs are partitioned into basic ventilatory compartments: the pulmonary acini. Each acinus contains a sequence of branched conduits connected to terminals of the conducting airway tree. The acinus is the smallest respiratory compartment unit of the lung in which all constituent alveoli (basic functional units) are ventilated by the same transitional bronchiole that arise from a given set of terminal bronchioles.

Intraacinar airways include respiratory bronchioles and alveolar ducts coated by gas-exchanging alveoli. Each alveolar duct opens into an alveolar sac,[21] a region into which a group of alveoli that share a common wall at their mouths open. In other words, an alveolar sac is the common space, or atrium, into which the alveolar duct open distally and via which *terminal alveoli* communicate, i.e., receive fresh air and expel air supplied with carbon dioxide. Therefore, an alveolar sac is limited by ends of alveolar walls and mouths of alveolar cavities. The alveolar sac, the terminal generation of the bronchial tree, with its ventilated alveoli represents an airway cul-de-sac. Each alveolus is lined by a single layer of epithelial cells.

Reconstruction from high-resolution (isotropic 2-μm resolution) computed tomography images and quantitative study by image analysis permit to determine the architecture and structure of pulmonary acini, in particular computations of acinar volume and surface area, in addition to estimation of the number of alveoli and capillaries per acinus using a stereologic method [21]. The volume, surface area, and number of alveoli in the acinus are higher in old mice than in young mice. Nonetheless, the surface-to-volume ratio and alveolar density are not significantly different. The acinar volume is 50% greater in old mice than in young mice, whereas total lung volume is 66% larger.

2.3 Tracheobronchial Tree

Within each lung, the main bronchus generates a bronchial tree composed of successive generations of respiratory conduits (Fig. 2.14). The greater the duct generation level, the shorter and narrower the respiratory pipes, and the higher

---

[18] Reticulin is made of collagen and glycoproteins.

[19] Collagen-1 is composed of 2 α1 and 1 α2 chain. It is mostly synthetized by fibroblasts.

[20] Collagen-2, -3, and -4 are made of 3 α1 chains, the α1 chain having a composition that varies according to the collagen type.

[21] Sacculus alveolaris (pl.: sacculi alveolares).

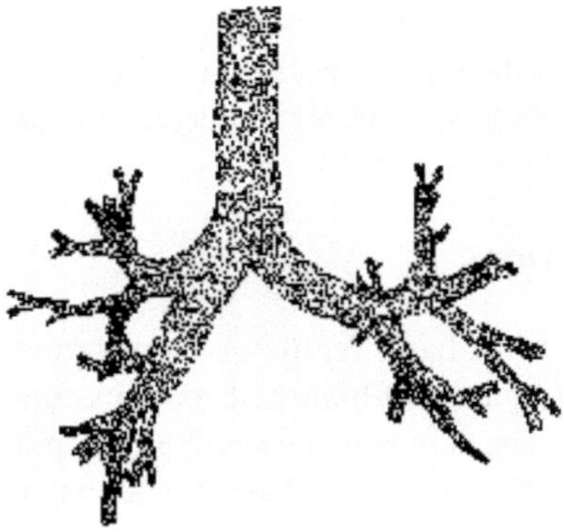

Fig. 2.14 Mesh of the proximal part of the tracheobronchial tree. Functionally, the respiratory tract is composed of 2 main regions: the conducting airway (nasal cavity and associated sinuses, pharynx, larynx, trachea, bronchi, and bronchioles) and the respiratory zone (respiratory bronchioles and alveolar ducts and sacs) within pulmonary acini. The tracheobronchial tree consists of large extra- and intraparenchymal (~5 generations) and small (~15 generations) bronchi. Three-dimensional reconstruction from processing of images acquired by current techniques of medical imaging allows the extraction of a tree down to the generation 6 to 7

the conduit number. From the trachea, the tracheobronchial tree divides into mostly paired branches of unequal length and diameter (*irregular dichotomy*). The maximum number of branches varies according to path, the number of bronchial divisions ranging from 17 to 26.

In adult lungs, the number of intrasegmental cartilaginous-walled generations varies from 8 in the apical segment of the lower lobe to 11 in the lower lingular segment [22]. The number of branchings down to the last cartilaginous bronchial generation changes considerably. Bronchi have cartilaginous walls, but not bronchioles.

Intraparenchymal airways include the distal segment (with respect to the facial airway opening) of the main bronchi, lobar and segmental bronchi (Fig. 2.15), small bronchi, terminal bronchioles that are the last purely conductive airways, respiratory bronchioles, and alveolar ducts that lead to alveolar sacs and alveoli.

The terminal bronchioles correspond to the last bronchial generations that have a complete respiratory epithelium. In adults, the number of terminal bronchioles is estimated at 25,000.

The sizes of intrathoracic airways depend on lung volumes. In mouse airways, the bore and length of small bronchi (caliber at FRC $< 200\,\mu m$) rise to 68.8 and 29.5% in average from functional residual capacity (FRC) to total lung capacity (TLC), respectively, using fast synchrotron radiation CT; the averaged values of the bore and length of large airways (caliber at FRC $> 400\,\mu m$) increase to 45.2 and 22.9% at TLC, respectively [23].

The tracheobronchial tree can be subdivided into 2 main functional compartments: (1) a *conducting zone*, which belongs to the anatomical dead space (Chap. 4); and (2) a *respiratory zone* where gas exchanges occur (Tables 2.6 and 2.7).

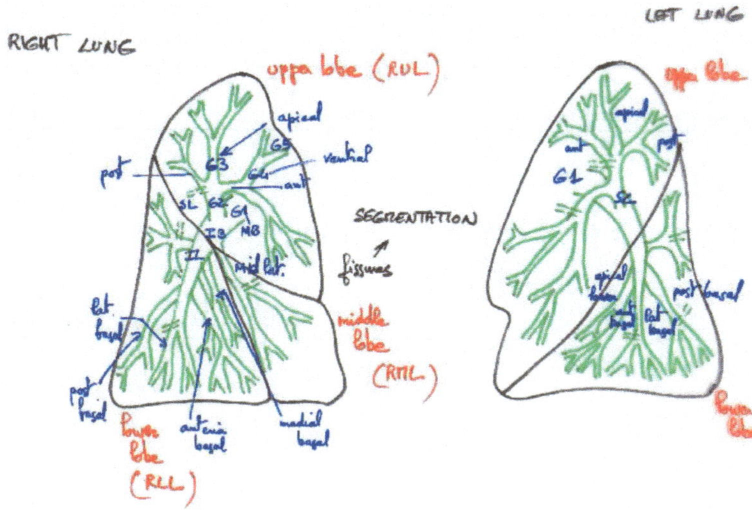

Fig. 2.15 Large bronchi in their respective lung compartment. The trachea divides into 2 main (or stem) bronchi that give rise to intraparenchymal left and right bronchial trees. Inspired air can be still filtered, warmed, and humidified as it passes through this compartment of conducting airways before reaching the respiratory zone, where gas exchange takes place

Table 2.6 The 3 zones of the tracheobronchial tree with the corresponding generations G*i* of a 23-generation model. Small intraparenchymal airways start with approximately generation G10. At the level of G10, the wall cartilage disappears (before G10 in the short bronchial path down to the alveolus level) and the number of smooth myocytes increases

Conduction zone	G0–G16	Gas convection
Trachea	G0	Anatomical dead space
Main bronchi	G1	
Lobar bronchi	G2	
Segmental bronchi	G3	
Subsegmental bronchi	G4	
Bronchi	G5–G8	
Small bronchi	G9–G11	
Bronchioles	G12–G14	
Terminal bronchioles	G15–G16	
Transition zone	G17–G19	Alveolized airway wall
Respiratory bronchioles		Moderate gas exchange
		Gas diffusion
Respiratory zone	G20–G23	Wall wholly dedicated to
Alveolar ducts	G20–G22	gas exchange
Alveolar sacs	G23	

Table 2.7 Geometry of the 6 first generations of the Weibel (regular dichotomy) model of the tracheobronchial tree at 75% of the total lung capacity (Source: [24])

Generation	L (mm)	d_h (mm)	R_h (mm)	L/d_h	A (mm^2)	$\sum_i A$ (mm^2)
0	120	18	9	6.7	254	254
1	47	12	6	3.9	113	226
2	19	8	4	2.4	50	201
3	7.5	5.6	2.8	1.4	25	197
4	12	4.5	2.2	2.7	16	254
5	10	3.5	1.7	2.9	10	308
6	9	2.8	1.4	3.2	6	394

Table 2.8 Geometrical data of the branchings of the tracheobronchial tree. Airway dimensions depend on age, lung size, and lung inflation degree

L/d_t	~3.5
L/d_b	~3.5
d_b/d_t (G0–G16)	0.7–0.8
$2\,A_b/A_t$	1.20–1.25
Branching angle	64–100
R_c	5–10 d_t

The volume of the conducting zone (~150 ml at rest, increasing during deep breathing) is much smaller than the volume of the respiratory zone (~2.5 l). The distance between the terminal bronchiole and alveolus is short (~5 mm) for efficient gas exchanges.

2.3.1 Bronchi

Each bronchial tree is supplied by a suitable number of lobar bronchi. The right pulmonary lobes receives air from the superior, middle, and inferior lobar bronchi, the 2 latter being usually branches of a common stem. The left lung is ventilated from the left main bronchi and its 2 branches, the upper and lower lobar bronchi. Tertiary bronchi distribute air in the different segments of each lobe (segmental bronchi).

Daughter bronchi usually arise from asymmetrical bifurcations. The area ratio, i.e., the ratio between the cross-sectional area (A_t) of the parent (stem) bronchus (the local trunk, subscript t) and the sum of the cross-sectional areas ($\sum A_b$) of the daughter bronchi (local branches, subscript b), is commonly equal to about 1.2 (Table 2.8). Although the airway caliber decreases with branching, the cumulated cross-sectional area quickly increases.

Table 2.9 Bore and length of the generations of the tracheobronchial tree for a respiratory path with a large number of respiratory pipe generations

Generation	d_h (mm)	L (mm)	L/d_h
0	15–22	100–120	6–7.5
1	10–15	30–50	3–4
2	7–11	15–25	2
3–8	4.5–8	10–15	1.4–2.2
8–17	2–3.5	6.5–9.5	2.7–3.2
17–23	1–2	3–6	3
23–35	0.8–0.2	0.5–1	0.6–1.5

Table 2.10 Sizes (mm) of intrathoracic airways in 2 human subjects A and B (Source: [25]; LMB, RMB: left, right main bronchus). In subject B, range of variation of mean values measured in different bronchi of a given generation

	A		B	
Generation	Length	Caliber	Length	Caliber (mean)
0	98.7	13.2–15.2	267.4	14.3
1 – LMB	56.1	9.3–11.6	50.6	8.1
1 – RMB	26.6	10.2–14.2	22.2	13.1
2	18.2	6.4–9.7	10.0–31.8	8.9–10.1
3	12.8	4.5–6.4	9.3–11.0	5.2–7.6
4	14.1	3.2–4.4	6.6–29.8	2.9–7.7
5	17.7	3.6–4.9	7.9–16.8	3.7–4.8
6	22.6	3.7–4.9	5.0–15.4	4.1–4.8
7	13.8	2.8–4.0		

Branches are currently unequal in size. The curved bronchi become shorter with a smaller bore with increased generation rank. The number of generations leading to the pulmonary alveoli varies between the various territories.

Trifurcations and lateral branchings (a small bronchus branches off on the edge of a large one) can be observed. All bronchi are enveloped by a connective tissue sheath, the *peribronchium*. However, bronchioles (bore $\leq 1\,\mu$m) are incorporated into the lung parenchyma (absence of sheath).

The Weibel model [24], in which the arborizing pattern assumes a regular dichotomy, contains 23 airway divisions down to alveolar sacs (Table 2.6). The trachea corresponds to generation zero (G0). Down to generation G11, the lumen bore remains greater than 1 mm and the bronchial wall is composed of muscular and connective tissues with partial cartilage rings for the first generations or complete ones for the following generations. Bronchioles are devoid of cartilage, but have the highest proportion of smooth myocytes relative to their size. Sizes of various bronchial generations are given in Tables 2.9 to 2.15.

Table 2.11 Transverse (Tr) and anteroposterior (AP) calibers (mm) of the trachea and left (LMB) and right (RMB) main bronchi in human cadavers (Source: [6]). Because the airway cross-sectional shape varies along the airway length, 3 sections have been selected to measure the airway caliber (stations in the entry, middle, and exit segments) when the airway length exceeds 40 mm

	Trachea		LMB		RMB	
	Tr	AP	Tr	AP	Tr	AP
Station 1	16.1	14.0	12.9	8.7	15.6	10.3
Station 2	20.0	14.2	10.9	6.53	18.3	8.6
Station 3	21.1	10.3	13.7	6.03		

Table 2.12 Calibers (mm) of left superior (LSLB) and inferior (LILB) lobar bronchi in human cadavers (Source: [6]). Because the airway cross-sectional shape varies along the airway length, 2 sections have been selected to measure the airway caliber (stations in the entry and exit segments) when the airway length exceeds 12 mm

	LSLB		LILB	
	Long	Short	Long	Short
Station 1	9.0	5.4	6.2	4.4
Station 2	10.1	5.4	7.1	4.5

Table 2.13 Caliber, length, cross-sectional area, and perimeter of different bronchi of generations G2 to G5 in human cadavers (Source: [6]). The airway cross-sectional shape varies along the airway length, but the airway length being lower than 12 mm, a single section has been selected (midlength station) to measure the airway caliber. Small bore means edge-to-edge distance perpendicular at midpoint to the largest possible caliber of the section

Generation	Maximal bore (mm)	Small bore (mm)	Length (mm)	Cross-sectional area (mm^2)	Perimeter (mm)
G2	10.4	7.6	18.4	59.6	28.3
(RSLB)	9.0	6.2	9.0	43.5	24.1
(LSLB)	9.6	5.4	11.6	40.9	24.3
(LILB)	6.7	4.5	8.8	21.9	17.5
G3	4.5	3.2	8.5	10.4	12.1
	9.3	4.7	3.8	37.5	29.9
	4.4	3.6		11.7	12.4
	5.4	4.3	10.9	18.5	15.6
	7.5	4.1	4.6	27.9	20.3
	5.9	3.4	7.2	15.6	15.0
(RILB)	5.8	6.0	6.7	29.4	20.1
(RMLB)	5.4	3.8		14.8	14.2

Table 2.14 Caliber, length, cross-sectional area, and perimeter of different bronchi of generations G2 to G5 in human cadavers (Source: [6])

Generation	Large bore (mm)	Small bore (mm)	Length (mm)	Cross-sectional area (mm^2)	Perimeter (mm)
G4	3.4	2.4		6.3	9.6
	3.2	2.4		5.9	8.8
	4.9	3.6		13.0	13.2
	3.9	3.1		9.4	11.2
	4.9	3.4		12.1	12.9
	5.9	3.9	7.1	16.5	15.2
	3.1	2.7		6.1	9.0
	3.9	2.9	8.9	8.9	10.9
	3.0	2.3		5.2	8.4
	3.2	2.6		5.0	8.4
	3.5	2.8		8.1	10.3
	5.7	3.5	8.3	15.2	15.2
G5	4.2	2.4		8.0	10.7
	4.2	3.0		9.9	11.6
	3.2	2.8		7.0	9.6
	2.0	1.4		2.2	5.4
	3.2	2.5		6.3	9.3
	4.4	3.3		11.1	12.2

Table 2.15 Average area, centerline length, and average caliber of the first generations of the tracheobronchial tree (Source: [26]; (IntBr: intermediate bronchus; LLLB: left lower lobe bronchus; LMB: left main [stem] bronchus; LULB: left upper lobe bronchus; RLLB: right lower lobe bronchus; RMB: right main [stem] bronchus; RULB: right upper lobe bronchus [trifurcates])

Generation	Mean area (mm^2)	Centerline length (mm)	Mean caliber (mm)
Trachea	160.9	267.4	14.3
		Left lung	
LMB	59.7	50.6	8.7
LULB (G2)	72.8	10.0	9.6
LLLB (G2)	63.7	12.9	9.0
Right lung			
RMB	134.0	22.2	13.1
RULB (G2)	61.7	16.3	8.9
IntBr (G2)	79.8	31.8	10.1
RLLB (G3)	45.2	11.0	7.6

Table 2.16 Approximated dimensions of lobular and acinar airways, with the airway generations from the Weibel model (the pulmonary acinus can contain up to 12 generations). The airways of the pulmonary acinus have parietal alveoli. In the acinus, communications exist between alveoli and either other alveoli or bronchioles, Kohn pores and Lambert channels

	L (mm)	d_h (mm)
Pulmonary lobule (G5–G16)		
Small bronchi	3–6	1–4
Bronchioles	2–3	0.6–1
Terminal bronchioles	1.5–2	0.5–0.6
Pulmonary acinus (G17–G23)		
Respiratory bronchioles	0.9–1.5	0.4–0.5
Alveolar ducts	0.7–0.9	0.2–0.4
Alveolar sacs		0.15–0.25

2.3.2 Bronchioles

Bronchioles arising from small (terminal) bronchi produce 3 to 4 generations down to the terminal bronchioles (Table 2.16).

In the Weibel model, the terminal bronchioles give birth to 3 successive generations of respiratory bronchioles. However, 2 to 5 generations of respiratory bronchioles can exist.

The last generation of respiratory bronchioles lead to alveolar ducts that convey air to and from a set of alveolar sacs. Alveolar sacs are cavities bounded by a small variable number of alveoli like, in ancient Roman houses,[22] in which atria were surrounded by rooms.

The terminal and respiratory bronchioles belong to the pulmonary *pulmonary lobule*. More precisely, the first generation of terminal bronchioles marks the airway entrance in the pulmonary lobule. A terminal bronchiole commonly divides into 2 branches, but one daughter duct can correspond to terminal bronchiole, whereas the other is a respiratory bronchiole.

The first respiratory bronchiole corresponds to the entrance into the *acinus*, the basic structural compartment of gas exchange. The respiratory bronchioles are the first bronchioles along which alveoli appear for a minimal gas exchange. The nearer the alveolar duct is, the higher the number of alveoli lining the bronchiolar wall. The wall-implanted alveoli are irregularly spaced.

The last in a series of respiratory bronchioles engenders the first generation of alveolar ducts. Alveolar ducts are completely covered by alveoli. The walls of alveolar ducts are, indeed, exclusively constituted of entry tips of alveoli. Five or

[22]in Latin, *domi*.

more ramifications with wall alveoli close to each other have been described. The alveolar sac is the final (distal) closed end airway.

Nervous expansions can be observed in the epithelium, especially in branching segments. Nerves can act on the quality of the mucus coat. Perimuscular plexus can also be found in the airway walls.

2.3.3 Fractal Modeling

Conduit arborescence enables blood and air flow distribution at their common interfaces, the air–blood or alveolocapillary barrier. This interface is characterized by a very large surface area enclosed within a finite thoracic volume (high surface/volume ratio).

The architecture of these trees was described by a *fractal geometry*. The fractal geometry developed by B. Mandelbrot was applied, in particular, to the respiratory tract [27].

Many fractal geometries in biological tissues and physiological apparatus was thought to be aimed at fulfilling the structure–function relation. Given amounts of O_2 and CO_2 are transferred to and from blood by a relatively small inhaled and exhaled volume, respectively. During quiet breathing, the so-called tidal volume (V_T = 500 ml), mix in a total lung capacity (volume [TLC]) of 4 to 6 l (V_T8–13% TLC). Respiratory gases cross the air–blood barrier, the surface area of which has an order of magnitude of 10^2 m^2.

In the lung, the respiratory tract is characterized by successive splittings that give rise to more and more numerous, short, and narrow conduits. Bronchus architecture appears almost similar at different scales. Its configuration at different magnifications indeed resembles its macroscopic structure. In addition to the property of *scale invariance*, or *self-similarity* and the absence of a characteristic scale, the tracheobronchial tree, like a fractal tree, is heterogeneous. Moreover, alveoli of terminal ducts (endpoints) are interconnected. Increase in size with magnification is a property of fractals. The alveolar surface area is assessed to equal 80 m^2 using light microscopy and 130 m^2 at higher magnification using electron microscopy.

On the other hand, a nonfractal theory predicts that anatomical circuits are stochastic networks of branched vessels unaffected by scaling, but only determined by the tube size at each generation.

A simple exponential model was first used to describe a scaling relation between the change in airway generation-averaged length ($\check{L}$) and hydraulic diameter ($\check{d}_h$) and branch order from casts of human lungs, up to generation 10 [28]:

$$\check{d}_{h_i} = \check{d}_{h_0} \exp\{-\kappa i\} = \check{d}_{h_0} S^i, \tag{2.1}$$

where S is a constant scaling (reduction) factor (or size ratio):

$$S = \exp\{-\kappa\} < 1, \quad \kappa = \ln S. \tag{2.2}$$

Measurements in adult lungs of several mammals show that the bronchial length and hydraulic diameter are linearly related over a wide range of generations (i) [29]:

$$\check{L}_i = \lambda_{\text{Ld}}\, \check{d}_{\text{h}_i} \tag{2.3}$$

In humans the average length-to-diameter ratio is $\lambda_{\text{Ld}} \sim 3.25$.

Generation-averaged diameters between parent and child branches are related by the following formula:

$$\check{d}_{\text{h}_{i+1}} = \check{d}_{\text{h}_i}\, S^{-i}\, S^{i+1} = \check{d}_{\text{h}_i}\, S. \tag{2.4}$$

The trunk-to-branch diameter ratio remains nearly similar throughout the tracheobronchial tree. Fractal properties of intrapulmonary airways were then deduced. An idealized model of the tracheobronchial tree was found to be characterized by successive homothetic bronchi with a size ratio equal to $S = 2{-}1/3 \sim 0.79$, at least for the first 10 bronchial generations, i.e., large and small bronchi [30]. The average diameters and lengths of bronchi were better fitted by a power law in bronchial generation number modulated by a harmonic variation than the exponential model [31]:

$$\check{d}_{\text{h}_i} = \check{d}_{\text{h}_0}\, i^{-p}). \tag{2.5}$$

Nonfractal configurations are generally characterized by a simple relation between the surface area and volume.[23] Unlike nonfractal configurations, fractal lungs tackle gas exchanges with a size of order of magnitude of 10 cm.

The fractal theory of a space-filling fractal branching transport structure is based on the principle of self-similarity, which states that the number of generation i conduits is proportional to the dimensionless caliber

$$\text{N}_i \propto \tilde{d}_{\text{h}_i}^{-D} \tag{2.6}$$

where $D \sim p$.

The idealized regular dichotomic model of the intrapulmonary respiratory tract was assimilated to a fractal canopy. A fractal canopy possesses the following 3 properties: (1) the branching angle is invariant throughout the fractal; (2) the length ratio of 2 consecutive generations is constant; and (3) endpoints are interconnected. Intrathoracic airways were thus modeled by a branched fractal structure and sponge-like fractal [32], the fractal geometry being high (can exceed 2).

[23]For a sphere of radius R, the relation between surface area and volume is $3/R$. When R is small, the surface area–volume ratio is large. For a ratio similar to that of the lung, $R \sim 0.1$ mm.

However, the finite nature of the respiratory tract limits the range over which self-similarity scaling characteristics are applicable [29]. A harmonically-modulated inverse power-law scaling was then used to describe tracheobronchial tree dimensions. Compared to several other mammalian species, the human lung differs by the phase of the harmonic modulation for both length and diameter measurements.

The fetal self-similar development of the airway tree was also analyzed quantitatively using fractal theory. Lung development happens in 3 main stages: pseudoglandular, canalicular, and saccular. In the early canalicular stage, during which the mode of growth changes from self-similar branching to surface increasing, the fractal dimension exceeds 2 (2.12 in early canalicular stage and 2.27 in middle canalicular stage) [33]. In the late pseudoglandular stage, the fractal dimension equals 1.73, i.e., similar to that obtained in human adult airways.

According to the Poiseuille law,[24] the flow resistance (R) is proportional to the conduit length and inversely proportional to the fourth power of the diameter. The resistance is thus proportional to S^{-3} and the volume to S^3. Assuming a constant scaling factor, the volume and pressure drop can be written as [30]:

$$V_N = V_0\big(1 + \sum_{r=1}^{N} (2S^3)^r\big),$$

$$R_N = R_0\big(1 + \sum_{r=1}^{N} \frac{1}{(2S^3)^r}\big). \tag{2.7}$$

Resistance minimization in a 10-generation tree enclosed in a finite volume leads to the optimal value $S = 0.79$. Lungs are close to optimality to ensure a safety range and to enable airway caliber regulation [30].

In acinar airways (respiratory bronchioles and alveolar ducts and sacs), oxygen is carried by diffusion. An analytical treatment (rope-walk approximation assuming concentric diffusion space and membrane) and two-dimensional numerical computations of oxygen transport were carried out in fractal, space-filling models of the acinus (Sierpinski model with 3 and 4 branchings) to find whether O_2 reaches the alveolar surface (unscreened transport) or not (diffusion-limited, partly or completely screened transfer according to the travelled distance within the acinus) [34]. Oxygen concentration ($c_{O_2}(\mathbf{x})$) was supposed to obey the Laplace equation in the diffusion space:

$$\nabla^2 c_{O_2} = 0, \tag{2.8}$$

with $c_{O_2} = c_{0_{O_2}}$ at the entrance to the diffusion space (c_0: concentration in a feeding reservoir). At the alveolar membrane surface,

---

[24] The Poiseuille flow, a steady fully developed flow in a straight pipe with a uniform circular cross section, cannot be observed in anatomical conduits. Yet, it is used to assess flow resistance.

Table 2.17 Transport parameters of oxygen flux at 310 K (Source: [35]). Membrane permeability is here given by $\frac{\mathcal{D}_{\rm mb}}{h}\frac{\mathrm{s}_{\rm mb}}{\mathrm{s}_{\rm air}}$ [L·T^{-1}]

Property	Symbol	Value and unit
Diffusivity in air	$\mathcal{D}_{\rm air}$	0.243 cm^2/s
Diffusivity in membrane	$\mathcal{D}_{\rm mb}$	3.08×10^{-5} cm^2/s
Solubility in air	$\mathrm{s}_{\rm air}$	3.89×10^{-10} mol/cm^3/Pa
Solubility in membrane	$\mathrm{s}_{\rm mb}$	1.05×10^{-11} mol/cm^3/Pa
Membrane thickness	h	1110 nm
Membrane permeability	$\mathcal{P}$	7.39×10^{-3} cm^2/s

$$\mathcal{D}\nabla c_{\rm O_2} \cdot \mathbf{n} = \mathcal{P}(c_{\rm O_2} - c_{\rm bO_2}), \tag{2.9}$$

with $\mathcal{D}$ as the diffusion coefficient, $\mathbf{n}(\mathbf{x})$ the surface normal vector pointing into the diffusion space, $\mathcal{P}$ the permeability (number of molecules crossing the membrane per unit time, unit surface area, and unit concentration difference between the 2 sides of the membrane), and $c_{\rm b}$ the concentration in blood. The lung was found to operate in the partial screening regime (almost the unscreened transport) during rest breathing and in the nonscreening regime during exercise. In a subsequent work, oxygen was assumed to travel through 14 generations of bronchi and 9 generations of acinar airways to reach the air–blood barrier of alveoli connected to alveolar sacs [35]. The transition from convection to diffusion then occurs at generation G18, G19, G20, and G21 at rest and moderate, heavy, and maximum exercise, respectively. Oxygen transfer from air to blood was modeled by a 3-step process: diffusion through the acinar airways, diffusion across the membrane (air–blood barrier and plasma), and diffusion and binding to red blood capsules.

$$J = \mathcal{T}_{\rm mb}(p_{\rm air_{O_2}} - p_{\rm b_{O_2}}) = \mathcal{T}_{\rm L}(p_{\rm air_{O_2}} - p_{\rm RBC_{O_2}}), \tag{2.10}$$

where $p_{\rm O_2}$ is the oxygen partial pressure in air (subscript air), blood (subscript b), and red blood capsules (subscript RBC), $\mathcal{T}_{\rm L} = \mathcal{P}A_{\rm eff}$ and $\mathcal{T}_{\rm mb} = \mathcal{D}_{\rm mb}\mathrm{s}_{\rm mb}A_{eff}/h$ are the lung and membrane diffusing capacity (A_{eff}: effective (accessible) surface area of the gas exchanger, a renormalization treatment decomposing the air–blood barrier surface with fractal dimension D_f into regions completely accessible, partly accessible, and completely inaccessible to oxygen (Table 2.17).

However, the architectural intricacy results from asymmetrical branching. In addition to intersubject variability, an intrasubject variability in dimensions exists for any bronchial generation i. Branching is irregular, as: (1) branch diameters ($d_{\rm h_{b1}} = \gamma_d \check{d}_{\rm h_i}$ and $d_{\rm h_{b2}} = [1 - \gamma_d)\check{d}_{\rm h_i}$) and lengths vary and (2) terminal branches are reached after a varying number of generations.

Several types of geometrical asymmetry exist, which can be combined: (1) bifurcation with branches of different cross-sectional areas, and thus given area ratios ($\lambda_{\rm A_i} = A_{\rm b_i}/A_{\rm t}$); (2) bifurcation with branches of similar bores but different branching angles, the branching angle being the angle between the branch and trunk axis (most often, conduit axes are nonplanar); (3) branching that gives rise to a side branch with a smaller caliber ($\lambda_A \ll 1$) and given branching angle, whereas the trunk direction changes only moderately. Three main modes of branching are assumed to include lateral branching and planar and orthogonal bifurcations.

These geometrical variations create asymmetrical flows between branches and, hence, differences in ventilation between pulmonary territories. Size variability between sister bronchi must be optimized according to the length and mechanical properties of bronchial paths from the trachea to alveolar sacs as well as associated vasculature to ensure proper ventilation-to-perfusion ratios.

2.3.4 Bronchial Walls

The bronchus wall consists in 5 layers: a mucosa, muscularis, submucosa, fibrocartilaginous layer, and peribronchial stratum. The structure of the bronchial walls varies with the size. Walls of large bronchi are similar to the tracheal wall. Walls of medium-sized bronchi have irregular, often helical, cartilaginous rings, a muscular layer, and a venous plexus between the muscularis and fibrocartilaginous layer.

The bronchial mucosa is composed of an epithelium, a basal lamina and a tunica propria. The epithelium contains ciliated and goblet cells, as well as Clara cells. The number of goblet cells in the large bronchi is greater than in the trachea. The number progressively decays with a reducing bore. The epithelium protects the lungs. Particles of size greater than 5 μm are stopped in the bronchi. The epithelium has a regeneration capability. The tunica propria has elastin and collagen fibers and capillaries. Glands progressively disappear from the bronchi to the bronchioles.

The submucosa contains bronchial glands. The submucosa progressively becomes distally thinner. The fibrocartilaginous layer possesses discontinuous plates of hyaline cartilage and a fibrous connective tissue. The bronchial cartilage supports are embedded into elastin and collagen fibers. A *tunica fibrocartilaginea* thus results. Bronchi can thereby lengthen and widen or shorten and become narrower during the respiratory cycle, especially when ventilation motions are ample. In large bronchi, elastin and collagen fibers are predominantly arranged in longitudinal bundles. In small bronchi, these bundles are obliquely oriented. At bi- and trifurcations, cartilages form saddles. The peribronchial layer, rich in elastin fibers, separates the wall of the bronchus from the pulmonary parenchyma.

In large bronchi, the muscularis bridges the cartilage tips. In medium-sized bronchi, the smooth muscle detaches from the cartilage. A connective tissue separates the cartilage and the musculature. In small bronchi, the space between the cartilage and the muscularis contains a venous mesh, numerous lymphatics,

Table 2.18 Bronchial and bronchiolar wall structure. In the first generations, the submucosa under the mucosa (h $\sim$70 μm) is thick in the pars membranacea and thin in the cartilaginous pars anulatus

G0–G9	Mucosa, submucosa, tunica fibromusculocartilaginous
G10–G14	Mucosa, submucosa, tunica fibromuscular
G15–G19	Mucosa, submucosa, tunica fibromuscular
G20–G23	Mucosa, submucosa

and thick-walled arterioles. The bronchial muscularis forms bundles rather than a complete layer. The helical disposition of the bronchial musculature allows changes in length and in bore.

2.3.5 *Bronchiolar Walls*

Bronchiolar walls, without cartilage, are characterized by sparse or absent mucus secretory cells. The walls have an outer layer with fibers that are longitudinally oriented. The outer fibers interlace with fibers of the alveolar walls. The wall inner edge is in contact with another layer of fibers with mixed orientation (longitudinal, oblique, and circular).

The fibers of the respiratory bronchioles are connected to the fibrous lattices of the alveolar ducts and alveolar septa, especially entrance rings of alveoli, as well as vascular sheaths. Fiber lattices form dense rings in alveolar ducts.

Three layers can be distinguished in the bronchiolar wall: a mucosa, a muscularis, and a thin outer layer, but this microstructure disappears in small bronchioles. The lamina propria is thin (and thinner in terminal bronchioles than in bronchioles) but rich in elastin fibers. The muscularis is the thickest layer. The connective tissue of the outer layer is continuous with the lung parenchyma

The bronchioles are lined by an epithelium. The bronchiolar epithelium stops at the alveolus entrance. Several cell types can be observed in the bronchiolar epithelium by ultrastructural studies [36]: (1) ciliated cells have apical microvilli between their cylindrical cilia; (2) granulated Clara secretory cells [37]; (3) cuboidal cells in the terminal segment of respiratory bronchioles related to granular pneumocytes; (4) brush cells [17]; and (5) chromaffine-type cells with dense granule along the basal lamina. Ciliated cells progressively disappear as respiratory bronchioles are approached.

In the respiratory bronchiole, 2 epithelia share the same basal lamina: (1) a pseudostratified (bronchial) epithelium with columnar ciliated secretory cells that contain mucins, and columnar nonciliated Clara cells; and (2) a cuboid (alveolar) epithelium with interspersed flat type-1 cells [38]. The cuboid cells are type-2 cells with dense bodies. The columnar and cuboid cell populations distinctly occupy the bronchiolar and alveolar segments.

Table 2.19 Layers of the alveolocapillary barrier. Barrier thickness over the alveolar surface area mostly of 0.5–1.2 μm. The thickness of alveolar and capillary cells is mostly lower than 500 nm. Despite the different components of the alveolocapillary barrier, the gas equilibrium between alveolar spaces and blood conveyed by pulmonary capillaries is quickly reached

Layer	Thickness
Alveolar gas	
Liquid film	∼70 nm
Alveolar epithelium	0.2–3 μm (350 nm in average)
Alveolar basal membrane	50–200 nm
Interstitial fluid	
Capillary basal membrane	50–400 nm
Capillary endothelium	0.2–2 μm
Plasma	∼1 μm
Erythrocyte	

2.3.6 *Pulmonary Alveolus*

The alveolus (size 150–300 μm) is the gas exchange unit of the lung. The shape and dimensions of the alveolus depends on the morphometric method. Technique-associated artifacts can thus be generated.

The air–blood barrier between the alveolar space and pulmonary capillary is extremely thin (thickness 0.2 μm). Oxygen (O_2) and carbon dioxide (CO_2) diffuse through the alveolar epithelium, a thin interstitial space, and the capillary endothelium (Table 2.19). Gas exchange takes approximately 0.25 s, i.e., one-third of the transit time of a red blood capsule. The exchange surface is very high due to the huge number of alveoli. The total surface area of the lung is assessed to equal about 80 m^2 (range of 50–100 m^2).

The interalveolar septa contains a dense capillary network and interalveolar pores. One edge of the capillary faces an alveolus and the opposite side the adjoining alveolus. The supporting lattice is composed of elastin and collagen fibers. Some smooth myocytes exist in the alveolar wall. Slender bundles of smooth myocytes occupy, with elastin and collagen fibers, the knobs of walls between adjacent alveoli, which defines the alveolar entrance rings. Elastin fiber networks are arranged in a helical form around the alveolar ducts and sacs.

The alveolar surface is covered by an epithelium resting on a basal lamina very close to the capillary one. The cell parts with the nuclei bulge into the alveolar space. The space between the alveolar epithelium and the capillary endothelium has a variable thickness. Connective tissue with fibers and cells can, indeed, lodge in the interalveolar septum. In particular, alveolar macrophages (size ∼30 μm) can be observed. They are responsible for uptake of particles (size 0.5–3 μm) which remain in inhaled air.

Alveolar wall epithelium is coated by a continuous layer of type-1 and -2 pneumocytes (Vol. 1 – Chap. 3. Cells of the Ventilatory Tract). Surfactant secreted by type-2 pneumocytes forms a monomolecular film at the air–wall interface. A thin liquid layer—hypophase—separate the surfactant film from the epithelial surface. Within the hypophase reside alveolar macrophages, which are the first line of local defense against inhaled pathogens and harmful nanoparticles.

The alveolar wall is also constituted of a capillary mesh on its outer surface. Pulmonary gas exchanges are achieved by tissular diffusion through the alveolocapillary membrane, the air–blood barrier.

2.3.7 Between-Airway Communications

Collateral ventilation is due to: (1) tiny connections between well-ventilated alveoli and poorly aerated ones via interalveolar pores and (2) tiny conduits between bronchioles and alveoli of neighboring alveolar sacs, the bronchioloalveolar communications.

Interalveolar pores, or pores of Kohn, exist in the interspace of the capillary mesh of the alveolar septum. The caliber varies from 5 to 15 mm, as well as the number (1–6 per septum). Their margins are delineated by fibers. The pores can lead to abnormal fenestrations in diseases. They can be used as routes for alveolar macrophages.

Bronchioloalveolar communications, or canals of Lambert, are built by epithelium-lined tubules (bore few 10 μm). They are observed between alveoli and terminal and respiratory bronchioles with relaxed or contracted wall musculature. These communicating conduits allow aeration of a great amount of alveoli. They can also convey alveolar macrophages.

2.4 Development of Intrathoracic Airways and Airspaces

Branching morphogenesis is used by a group of cells to create cellular and tissular extensions, thereby expanding the surface area in a given volume. Branching can occur in a single cell, such as a neuron when it forms dendrites to communicate with other neurons and axons to relay nerve impulses to remote targets. Branching can happen in a cell pool, such as in the vasculature to deliver oxygen and nutrients and remove metabolic wastes and the respiratory tract to supply oxygen and eliminate carbon dioxide.

Branch pattern is defined by length, caliber, shape, and spacing of conduit generations. Despite many branching modes, tip cells organize the leading front of branch tips. They respond to inductive and guiding signals that initiate cell proliferation and migration, determine cell navigation, and maintain branch outgrowth in a given direction. They drive trailing stalk cells during colonization of their environment.

At each branching, the patterning information must specify [39]: (1) the budding site and the growing direction, and (2) the size and shape of the branch. Furthermore, during each branch development, patterning must control the sprouting space and time of the next branch generation. New branches typically arise as outpouchings of the epithelium, either by migration of a local region of the epithelium or by local, oriented cell divisions.

Iterative models assume that all branching events are under the same genetic control. However, different generations of branches result from distinct cellular mechanisms; they express specific markers and require different sets of genes for their formation. Nevertheless, an iterative process can be repeatedly modified during development to generate different branching patterns.

During embryo- and fetogenesis, sequence of events during branchings and 3D tree patterning is stereotyped. Branching morphogenesis in the mouse lung can be decomposed into 3 simple geometric forms: domain branching and planar and orthogonal bifurcation. At a given position along the craniocaudal axis, branches form in 4 main domains: dorsal and ventral on the one hand (dorsal–ventral axis) and medial and lateral (medial–lateral or left–right axis) [40]. [25] Therefore, a proximal–distal controller comprises a *proximal–distal periodicity generator* that regulates the sequence of branching within each domain, as it coordinates the temporal appearance of subsequent branches (hence between-branch spacing) during the tree development, and a circumferential assigner, or *circumferential domain specifier*, that specifies the positions of domain and the rotation direction. In domain branching, branches can indeed arise at different positions around the circumference of the trunk. As the domains differ in number of branches and the position at which branching is initiated, the proximal–distal controller not only sets the initiating position (*proximal–distal initiator*), but also registers each domain (*proximal–distal register*).

The tracheobronchial tree is then generated by 3 geometrically simple local modes of branching used repeatedly, but in different orders, at least in the mouse lung [40]. During lung morphogenesis, the series of local patterning operations, i.e., the global recursive routine, thus comprise: (1) a proximal–distal periodicity generator that receives inputs from the proximal–distal initiator and register; (2) a circumferential domain specifier that defines the domain number and position; (3) a *branch bifurcator* for planar bifurcation, i.e., a branch tip division subroutine;

[25]For example, in the left (L) primary bronchus lineage, the first generation (LG1L1) buds off the lateral aspect of the founder generation (LG0) using a first branching mode, circumferentially specified budding [40]. The first dorsal branch (LG2D1) buds just distal to LG1. As this domain develops, LG3 begins to sprout from the medial surface, then LG4 from the ventral surface. The latter formed by domain branching switches immediately and permanently to orthogonal bifurcation. Tertiary and later generation branches form by tip expansion and bifurcation (planar bifurcation mode). In a third branching mode (orthogonal bifurcation), branches bifurcate at their tips, as in planar bifurcation, but between each round of branching, a nearly 90-degree rotation in the bifurcation plane occurs, so that the 4 grand daughter branches are arranged in a rosette in a projection plane. These 3 branching modes are used at different times and positions. Domain branching is used first and generates the central scaffold of each lobe; planar bifurcation forms the edges of lobes; orthogonal bifurcation creates lobe surfaces and fills the interior [40]. Branching proceeds at different and variable rates in different lineages.

and (4) a *bifurcation plane rotator* for orthogonal bifurcation, i.e., a rotational orientation subroutine that determines the reorientation of the branch around the axis of the trunk. All subroutines send messages to a *branch generator*.

Subroutine coupling generates 3 main sequences [40]: (1) domain branching and a set of orthogonal bifurcations; (2) domain branching and either sequence 1 (sequence 2a) or planar bifurcation (sequence 2b) and then either sequence 1 (sequence 2b1) or repeat of planar bifurcation (sequence 2b2) (3) domain branching and either sequence 2, sequence 1, or sequence 2b.

The hierarchical and modular program is genetically tractable. Developmental patterning information, i.e., each mode of branching, is controlled by a genetically encoded subroutine regulated by a master routine. Deployment and coupling of branching modes is actually achieved downstream from the dynein, axonemal, heavy chain gene Dnahc11 and the left–right asymmetry pathway [40]. On the other hand, Sprouty-2 restricts the number of branches in the ventral domains. In mice, Sprouty-2 regulates the site of initiation and number of branches in specific domains; Shifty controls the proximal–distal register of entire domains.

2.4.1 Developmental Phases

Lung development proceeds via 2 sequential processes [41]. The sucessive centrifugal, dichotomous branchings of buds from the distal end of the trachea generate a series of tubules of decreasing dimensions, thereby creating an epithelium-coated tree, during the pseudoglandular and canalicular stages (∼5–24 gestational wk in humans). The following development of alveolar sacs and then saccular alveoli occurs along the most distal bronchial generations, beginning at the periphery of the fetal lung, progressing centripetally and centrifugally, during the saccular and alveolar stages (from ∼24 gestational wk to postnatal life in humans).

More precisely, lung development can be decomposed into 4 overlapping phases [42]: (1) *pseudoglandular phase* (gestation wk 5–17); (2) *canalicular phase* (gestation wk 16–26); (3) *saccular phase* (from gestation wk 24–36 until term; and (4) *alveolar phase* (from gestation wk ∼36 to first 3 yr of postnatal life).

The pseudoglandular phase is characterized by preacinar branching of airways and blood vessels. The number of bronchial tubes rises. The epithelial lining is stratified. A ciliated epithelium differentiates between week 11 and 16 of gestation in humans [42]. Cilia develop directly by division of pre-existing centrioles or indirectly from fibrogranular aggregates that give rise to deuterosomes and an aggregate of precentrioles. Presecretory cells appear almost simultaneously as preciliated cells. Intracellular mucus is detected in the human fetal lung at week 13 of gestation. Three types of secretory cells according to granule type can be identified by electron microscopy: cells with electron-dense serous granules, electron-lucent mucous granules, or a mixture of these 2 types. The mesenchyme transforms into cartilage and smooth muscle.

During the canalicular phase, vascularization of peripheral mesenchyme rapidly increases, capillaries move into close contact with the surface epithelium, and connective tissue components are reduced to a minimum. By ingrowth of capillaries between epithelial cells, respiratory regions of the lung appear by conversion of the blind end of the bronchial tree. The last few generations of the bronchial tree form the primitive respiratory bronchioles.

During the terminal sac (saccular) phase, additional respiratory airways develop and acini differentiate. During the alveolar phase, the alveolus number grows.

Accessory communications between distal bronchioles and adjacent alveoli has not been detected in fetal lungs, although in adult lungs, ultimate 3 to 4 generations can have accessory communications [22]. Accessory communications arise mainly in postnatal development during transformation of alveolar ducts into epithelialized bronchioles.

2.4.2 *Development at the Macroscopic Scale*

Development begins as a simple epithelial tube from which new branches successively bud to create a tree-like structure. Additional supporting cell layers then develop around newly formed epithelium-coated tubes. At least for early branch generations, the pattern is highly stereotyped, but not invariant. Continuation of the branching program is not contingent on completion of each prior step, as a branch can originate from a grand parent branch.

In early lung development, endodermal pulmonary buds, or respiratory epithelial tubes, intrude into the mesenchyme covered with pleural cells that define lung borders. These buds then form arborescence by repeated use of morphogenetic processes, such as elongation, lateral budding, and terminal bifurcation [43].

2.4.2.1 Development in Utero

Lung airways develop 22 to 26 days after fertilization as a ventral diverticulum budding from the foregut and lined by epithelium of endodermal origin [42]. In the 5-wk fetus, lobar bronchi exist as small outgrowths from stem bronchi [22]. Segmental bronchi are recognizable from week 6. During the following weeks, the bronchial tree is blind at its distal end and lined by epithelium that is surrounded by mesenchyme. At week 16, the bronchial tree is completely lined with epithelium. At week 24, epithelium is interrupted by ingrowth of capillaries. The alveolar region with respiratory bronchioles arises from transformation of terminal bronchioles. In humans, alveolar development hence starts in utero. However, in rodents, the peripheral part of the acinus is formed by primary saccules; alveoli develop postnatally from saccule walls [44].

In humans, lobar and segmental bronchi appear after week 5 of intrauterine life. Intrasegmental airways quickly divide between weeks 10 and 14. During this period, about 70% of airway generations are formed and at week 16, the airway branching pattern is complete. The longer bronchial paths actually continue to grow until week 16. Bronchial paths of the lingula, middle lobe, and anterior and posterior basal segments include more generations (3–10 additional generations) than those of the apical segments of the upper lobes that contain 14 to 18 generations [22]. In 18-week fetuses, 17 to 19 bronchial divisions can be found between the trachea and terminal bronchioles. After week 24, the total number of fully epithelialized generations is slightly reduced (by 2 to 4 generations) due to the transformation of the last few generations into respiratory bronchioles. Conversely, some respiratory bronchioles transform into fully epithelialized bronchioles between week 24 of intrauterine development and adult life.

2.4.2.2 Postnatal Development

Pre-acinar airways in infants are miniature versions of those in adults [45]. Bronchi grow both in length and bore with lung volume after birth. The tracheobronchial tree of newborns remains a small-scale model of that of adults. Between week 22 of gestation and month 8 of postnatal age, normal infants exhibit a linear increase with age in airway caliber as well as in amount of mural smooth muscles and submucosal glands (except submucosal glands in hilar airways) [46]. Goblet cell number increases with age in the bronchi, but not in distal conduits.

Premature infants have for their postconceptional age normal-sized airways with an increase in amount of bronchial smooth muscle and goblet cells. Lung airway caliber doubles or triples between birth and adulthood [42]. Normal alveolar multiplication is supposed to be complete by between years 2 and 4.

Measurements in bronchial tree casts from children aged from 5 weeks to 17 years show that from 5 to 12 weeks, the caliber ratio is about 1.35, but by month 13, it raises to 1.45 [47]. Peripheral conducting airways are relatively large in diameter at month 1 to 3 and reach adult proportion after year 1. Yet, length ratio does not show trend with age.

2.4.3 Development at the Microscopic Scale

Airway lining and branching requires epithelial growth and morphogenesis. Cell differentiation depends on the amount of available mesenchyme partner [48]. Cell division occurs particularly at tube ends, the so-called *bronchial buds*. Branching depends on adhesion molecules (both intercellular and cell–matrix adhesions), particularly proteoglycans.

In both proximal and distal lung airways, epithelial thickness decreases from early fetal to postnatal life, whereas airway bore increases. All respiratory epithelium cells develop by differentiation and maturation of primitive endodermal cells. Cell maturation begins in large airways and spreads distally.

Lung epithelial branchings result from a repetitive series of cleft (interbud region) and bud formation. Epithelial cells of budding regions of the bronchial tree have at least a 2.5-fold greater proliferation rate than those situated in the adjacent cleft-forming regions [49]. Epithelial proliferation depends on between-epithelial cell contact. Epithelial proliferation in growing lung buds with respect to newly formed clefts may be attributable to the action of specific growth factors, such as TGFβ1, FGF2, EGF, BMP4, and BMP7 proteins.

Airway branching is accomplished by reciprocal morphogenetic interactions between the growing respiratory epithelium and the surrounding mesenchyme, the mesodermal capsule. The airway epithelium is separated from the lung mesenchyme by a basement membrane. Insoluble macromolecules of the extracellular matrix can act as physical barriers and sequester soluble growth factors [41]. The thin basement membrane stretches and is degraded in front of growing buds. Bud outgrowth and epithelial proliferation are correlated with the localization of specific basement membrane components, such as collagen-4, nidogen, laminin-1, and fibronectin [49].

2.4.3.1 Inductive Signals for Branch Initiation

Epithelia and endothelia use growth factors and their cognate receptor protein Tyr kinases as initiating cues to create new branches [50]. Members of the fibroblast growth factor superfamily play a dominant role in branch initiation.

2.4.3.2 Control of Branching Pattern

Formation and subsequent outgrowth of a new branch and its maintenance are under local and global control. Mutual inhibition in branching epithelium upon approaching another branch and self-inhibition may happen in addition to extrinsic factors liberated from the adjoining mesenchyme [50]. In any case, branching pattern is stereotyped, as it is genetically programmed.

2.4.3.3 Coordination of Branching of Nerves, Blood Vessels, and Epithelium

Branching morphogenesis of blood vessels and nerves needs to be coordinated with that of the epithelium for adequate distribution of feeding and nervous control. This coordination relies on communication between tissues, genetic program, and common branch-regulating pathways [50]. Guidance cues of 4 families—Slits, netrins, ephrins, and semaphorins—can attract or repel axons, blood vessels, and epithelium.

Airway epithelia attract endothelial cells and neuronal axons by secreting vascular endothelial and nerve growth factor, respectively. Conversely, endothelial cells contribute to specification and differentiation of the respiratory epithelium. In particular, pulmonary endothelium secretes hepatic growth factor for distal lung epithelial morphogenesis. In addition, endothelial cells produce artemin and neurotrophin-3 to attract axons to travel alongside pioneer blood vessels and nerves produce VEGF to attract blood vessels to their side.

2.4.3.4 Surrounding

Mesenchyme and stroma are major sources of branching regulators that regulate concerted development of tip and stalk cells and eventual rearrangement between these 2 cell types, as well as orchestrate coordinated cell movements [50].

Intercellular and cell–matrix adhesion and cellular protrusions and trailing edge retraction depend on chemical and mechanical interactions between migrating cells and their medium.

2.4.4 Influence Factors

2.4.4.1 Genetic Control, Growth Factors, Hormones, and Morphogens

Embryo- and fetogenesis obey instructions prescribed by the genome. A controlled spatial and temporal expression of genes actually determines the development of complicated, three-dimensional, ramifying circuits of epithelium-lined tubes with successive rounds of branchings that convey biofluids (amniotic liquid before air at birth).[26]

The genetic branching program relies on reiterative use of signaling axes such as the fibrobrast growth factor pathway [39]. The FGF pathway is modified at each stage of branching by feedback to launch distinct branching outcomes. Whereas FGF1 elicits immediate endodermal budding, FGF7 and FGF10 initially cause expansion of the endoderm [51].

During lung morphogenesis, the distal mesenchyme regulates the growth and branching of adjacent endoderm. Molecule FGF10 produced from the earliest stages of lung development diffuses in the mesenchyme and promotes cell proliferation and chemotaxis in lung buds and directs their outgrowth. When a bud is elongating, a FGF10 production peak is formed in the mesenchyme at a single spot near the tip. On the other hand, lateral budding and terminal bifurcation result from 2 separate peaks [43]. The lateral budding mode relies on FGF10 expression localized to spots

[26] Amniotic liquid is, at first, mainly water with electrolytes and, after several weeks, with carbohydrates, lipids, proteins, and other types of organic constituents. It originates from the maternal plasma that crosses the fetal membranes due to osmotic and hydrostatic forces. As placental and fetal vessels develop, amniotic liquid traverses the fetal tissue.

between both sides of the bud stalk where prospective budding occurs and the lung border. In the terminal bifurcation mode, FGF10 expression occurs at 2 split spots located laterally between the bud tip and the lung border.

In addition to FGF and FGFR, other proteins synergistically regulate branching morphogenesis such as Sprouty, Notch, and members of the class-E basic helix–loop–helix family such as Enhancer of Split homolog. Neuronal PAS domain-containing protein-3 (NPAS3), or class-E basic helix–loop–helix PAS protein bHLHe12,[27] is the single regulator of lung branching morphogenesis that is a homolog of Trachealess, a transcription factor involved in tracheogenesis in Drosophila [52]. NPAS3-null newborns die in respiratory distress having diminished alveolarization, decreased SHh, FGF9, FGF10, and BMP4 production, and increased Sprouty homolog Spry2, an inhibitor of receptor protein Tyr kinases upregulated in response to FGF2 in primary dermal endotheliocytes in particular. Hence, FGF signaling is reduced. Exogenous FGF10 rescues branching morphogenesis in NPAS3− mice. On the other hand, $NPAS3^{+/-}$ mice survive postnatally, but develop emphysema and airway hyperreactivity. Factor NPAS3 stimulates SHh and represses Spry2 agent.

Diffusible signaling molecules, such as transforming growth factor-β and sonic Hedgehog, are major negative and positive regulators of FGF10 synthesis [51, 53]. Factors TGFβ and SHh are manufactured in the mesenchyme adjacent to bud stalks and in distal tips, respectively. Factor TGFβ1 reversibly prevents synthesis of nMyC (or bHLHe37) in epithelial cells involved in branching morphogenesis [53]. In the endoderm, SHh precludes FGF10 transcription [51]. On the other hand, SHh via Gli3 controls FoxF1, an activator of FGF10 [54, 55]. Activation of FoxF1 by endodermal signal SHh is counteracted by bone morphogenetic protein BMP4. Both FGF10 and FGF7 impede FoxF1 expression via epithelial BMP4 and inhibition of SHh expression, respectively [54]. Factor Gli3 exists as a full-length ($Gli3_{FL}$) with activator function (Gli3A) that is cleaved into a proteolytic fragment, i.e., a smaller form ($_t$Gli3) with repressor activity (Gli3R) in the absence of SHh ligand. Signaling from SHh prevents Gli3 proteolysis, hence hampering the production of its truncated repressor Gli3R in the developing lung [55]. In fact, both Gli2 and Gli3 have overlapping functions in SHh signaling during foregut (lung, trachea, and oesophagus) development [56]. Long-range activation and short-range inhibition of FGF10 expression by SHh may achieve the spatial FGF10 localization with a peak at an intermediate distance [51].

The spatial organization of cell fates and pattern formation during organogenesis involves morphogen gradients described by the tranport equation with production and degradation terms and defined boundary conditions [57]:

[27] A.k.a. MOP6 and PAS (Per [Period circadian protein]–ARNT [aryl hydrocarbon receptor nuclear translocator]–SiM [Single-minded protein]) domain-containing protein PASD6. The PAS domain regulates heterodimerization, which is required for activity. Protein NPAS3 is the most similar to Trachealess, followed by NPAS1, SiM1, SiM2, hypoxia-inducible factor HIF1α, endothelial PAS domain protein EPAS1, and Per1 proteins.

$$\partial_t c_M = \mathcal{D} \nabla^2 c_M + \mathsf{p}\, c_M - \mathsf{d}\, c_M, \tag{2.11}$$

where c_M is the morphogen concentration, $\mathcal{D}_M$ the morphogen diffusion coefficient, p the morphogen production rate ([T^{-1}]), and d the morphogen degradation rate ([T^{-1}]). The diffusion coefficient and production and degradation rate can dynamically vary because of feedback from triggered signaling axes ($\mathcal{D}_M(\mathbf{x},\{c_{R_i}\})$, $\mathsf{p}(\{c_{R_i}\})$, and $\mathsf{d}(\{c_{R_i}\})$; $\{c_{R_i}\}$: set of concentrations of regulators involved in feedback loops).

Once they are established, morphogen gradients trigger activity of intracellular signaling cascades with their transcriptional networks. Morphogen gradients are transduced nonlinearly into signaling gradients. Signaling effectors activated by morphogens control target genes that cross-regulate each other to form a *gene regulatory network* (GRN), a set of genes that interact with each other indirectly using their RNA and proteic products as well as other substances. The primed gene regulatory network governs transcription rates of network genes. Therefore, the expression of a given target gene is not determined solely by signaling effectors, but also by inputs from members of the gene regulatory network. The cell state depends on the combination of active genes. Furthermore, tissues are composed of many cell types with given combinatorial gene expression that involves numerous genes forming subnetworks linked in a global network [57]. Specific subnetworks may also transduce temporal and spatial derivatives of graded signals.

Gene regulatory networks can be modeled by a set of coupled ordinary differential equations that represent concentrations of contributing substances at time t:[28]

$$\frac{d}{dt} c_i = f_i\big(\{c_\ell\}_{\ell=1}^{N}\big), \tag{2.12}$$

the function (f_i) expressing the dependence of the concentration of substance i (c_i) on concentrations of N molecular species is derived from principles of chemical kinetics. This function is usually represented by a low-order polynomial or Hill function that serves as an ansatz.[29] This model is studied in the framework of nonlinear dynamics.

Morphogen gradients initiate certain gene expression patterns to coordinate events from the nano- to the mesoscale, i.e., molecular, cellular, and tissular processes. Morphogen kinetics often differ according to short and long space (from subcellular to tissular) and time (from second to hour) scales [57]. Morphogen gradients result from: (1) transport associated with cell growth and proliferation; (2) directional transport through a tissue; and, mostly, (3) secretion from a localized

[28]Gene regulatory networks can also be modeled by Boolean and stochastic gene networks.

[29]German *Ansatz*: starting point (in the meaning of *Ausgangspunkt*), attempt, trial, test (in the meaning of *Versuch*), approach, initialization [*die ersten Ansätze*: initial stages]. The initial formulation (starting equation) is an informed guess verified later by results. Once the ansatz is established, the equations in their general formulation are indeed solved.

source, nondirectional spreading, and clearance or local degradation. Changing gradients further complicate patterning. Because of morphogen interactions and feedback, effective degradation rates or diffusion coefficients can depend on morphogen concentration. However, such processes increase the gradient robustness to fluctuations in morphogen production. In addition, morphogen gradient amplitude and duration can also vary during the time course of pattern formation, in particular upon increased production from a growing source or changes in kinetic rates [57].

Airway growth and branching are influenced by hormones (e.g., growth hormone and somatomedin), in addition to growth factors. Among growth factors, some, such as insulin-like or epidermal growth factor, stimulate multiplication of epithelial cells, whereas others inhibit it, such as transforming growth factor-β. Vitamin status participates in lung growth.

Basement membrane components, such as sulfated glycoconjugates and fibronectin, have a stronger concentration in clefts or areas of morphogenetic stabilization than in budding regions [49]. On the other hand, collagen-4, laminin, and nidogen have a uniform distribution at the epithelial–mesenchymal interface. However, transcripts of laminin-α, -β, and -γ have a variable coexpression in budding and cleft-forming regions. Specific laminin isoforms can then achieve variations in epithelial–mesenchymal basement membrane assembly that are permissive to cleft or bud formation.

Transforming growth factor-β1 in developing lung epithelia, a regulator of proliferation and extracellular matrix deposition, and $\alpha_3\beta_1$-integrin, which binds to fibronectin, laminin, and collagen, contribute to the control of lung branching.

The molecular mechanisms that regulate temporal and spatial interactions between the developing respiratory epithelium and mesenchyme depend, at least partly, on HOX genes of branching regions (HOXA1, HOXA3, HOXA5, HOXB3–HOXB4, and HOXB6–HOXB8) [58]. The timing of HOXA1, HOXA4, and HOXA7 peak expression levels is sequential during branchings of the embryonic lung, i.e., during the early pseudoglandular stage, the canalicular stage, and saccular and alveolar stages, respectively. The HOXB3 expression is the greatest during the pseudoglandular stage, whereas that of HOXB4, HOXB6, and HOXB7 reaches a maximun during the terminal stages. In addition, these HOX genes are differentially expressed both between and within lung cell lineages throughout the pseudoglandular, canalicular, and terminal phases. Transcripts of HOXA1 and HOXB3 genes localize thoughout the mesenchyme, whereas those of HOXA4, HOXB5, and HOXB7 genes have a greater spatial restriction, as they reside mainly in the distal bud than in the interbud mesenchyme.

2.4.4.2 Mechanotransduction

Mechanotransduction relies on 3 main coupled aspects of cell and tissue mechanics: cell and tissue rheology; intracellular (endogenous) generation of forces; and response to environmental (exogenous) mechanical stimuli. The mechanical properties of tissues and exerted forces can serve as epigenetic regulators of morphogenesis.

During embryo- and fetogenesis, tissues bend, fold, and twist to elaborate structures of organs. Therefore, developing tissues are subjected to mechanical stress fields. Stretch participates in the control of cell proliferation during development of the airway epithelium.

Mechanical stress fields participate in tissue development. Two main types of mechanical stresses (forces per unit surface area) include pressure, a stress perpendicular to a surface, and shear, stress tangent to a surface caused by friction.

Mechanical stress fields result from contraction of the cytoskeleton during cell motion and tissue deformation that depend on rheology of cells and matrix. Tension can indeed be initiated within the cell, transmitted from the cytoskeleton upon activation of the Rho–RoCK–Dia axis, subsequent myosin phosphorylation, and actin polymerization to prime actomyosin contraction [59]. On the other hand, tension can originate from the extracellular matrix and be transmitted via integrins to the cytoskeleton.

Mechanical properties of the cell's environment and cytoskeletal tension influence, in particular, differentiation of multipotent mesenchymal stem cells. A soft and stiff environment promotes neuronal and osteogenic differentiation, respectively [59].

Mechanical stresses also participate in the regulation of epithelial–mesenchymal transition which can be triggered by numerous soluble cytokines and peptidases during gastrulation and creation of the neural crest. In particular, mechanical stresses are involved in epithelial–mesenchymal transition during heart valve formation [59].

Elevated mechanical stress enhances the nuclear localization of myocardin-related transcription factor MRTFa [59]. This cofactor of serum response factor regulates the transcription of multiple genes involved in cell proliferation and actin cytoskeleton dynamics.

Tension contributes to the regulation of cell turnover (proliferation and death) and differentiation. Tension indeed cooperates with cyclins to enable G1-phase progression upon increased RhoA activity, ubiquitination, and degradation of cyclin-dependent kinase inhibitor CKI1b [59]. Spatial variations in mechanical stresses between epithelial cells and basement membranes may establish local growth signaling differentials. Mechanical stresses from the basement membrane also promote cell survival [59].

Fetal breathing movements induced by the central nervous command of the chest wall via, in particular, phrenic nerves, favors airway development, as stretch is a major stimulator. During fetal breathing, the diaphragm indeed contracts and increases the volume of the thoracic cavity. Resulting stretch promotes airway development. After birth, mechanically ventilated infants have a greater density in mural smooth muscle and goblet cells and an elevated submucosal gland volume [46].

In addition, *airway peristalsis* results from spontaneous contractions of airway smooth myocytes [60]. Waves of contraction launched by calciumwaves travel proximal to distal in freshly isolated fetal lungs from early to late gestation in pigs and rabbits.[30] The behavior of airway smooth myocyte in the fetal lung is phasic, whereas it is tonic in postnatal lungs.

Tension via the Rho–RoCK axis ensures normal differentials in basement membrane thickness and epithelial growth patterns such as bud formation, as well as the presence of normal capillaries [63]. Elevated actomyosin-mediated cell contractility promotes airway branching and increases capillary elongation. *Amount of amniotic liquid* also affects airway development.[31]

2.4.5 Computational Model of Early Intrathoracic Airway Morphogenesis

Morphogenesis deals with the formation of the spatial pattern of structures of biological tissues. Developmental constraints are rules followed by biological patterns or structures. Genes determine the anatomical structure of an organism, as they influence anatomical shapes by determining morphogen production. A set of morphogens produced by genes diffuses and reacts together through a tissue, thereby contributing to developing patterns [64].

Chemical-mediated morphogenesis is modeled by *reaction–transport (diffusion) equations* [64]. Each involved morphogen moves from regions of greater to regions of less concentration (c) at a rate proportional to the gradient of the concentration and also proportional to its diffusivity ($\mathcal{D}$). The reaction rates are assumed to obey the law of mass action applied to the entire set of reactions, including intermediary stages.

Reaction–diffusion equations for a couple of morphogens, an activating morphogen (subscript aM) that stimulates itself and a repressing morphogen (subscript rM) that inhibits itself and exerts a negative feedback on the activating morphogen, are the following:

$$\begin{aligned} \partial_t c_{aM} &= \mathcal{D}_{aM} \nabla^2 c_{aM} + f(c_{aM}, c_{rM}), \\ \partial_t c_{rM} &= \mathcal{D}_{rM} \nabla^2 c_{rM} + f(c_{aM}, c_{rM}), \end{aligned} \tag{2.13}$$

[30]In the pseudoglandular and canalicular stages in fetal pigs, the frequency ranges from 0.4 to 0.6 Hz and lumen caliber reduces from 39 to 46% [60]. In the saccular stage in fetal rabbits, the frequency ranges from 0.2 to 0.2 Hz and the brochus narrowing equals approximately 30%. As terminal tubules of the developing lung relaxes, the liquid flow reverses. Decreasing and increasing lumenal fluid volume causes lung hypo- and hyperplasia, respectively.

[31]After amniocentesis in pregnant macaca fascicularis between gestational days 47 and 64 or 85 and 95, postnatal alveoli are mature, but their number and size are reduced and heightened, respectively [61]. The quantity of airways is also attenuated. These changes occurred regardless of amniocentesis time and removed fluid amount. Simple puncture without fluid removal causes similar changes. Pulmonary hypoplasia that results from chronic amniotic liquid leakage leads to death of respiratory failure within 12 hours of delivery, or in the case of partial pulmonary hypoplasia, to chronic respiratory symptoms in viable infants [62].

where the functions f and g represent the chemical reaction kinetics. A stable homogeneous steady state is defined by $f = 0$ and $g = 0$. With proper reaction kinetics, diffusion within and between cells creates spatial patterns.

The system supposed to be initially in a stable homogeneous condition is disturbed slightly from this state by some slight destabilizing influences, such as Brownian movement, action of neighboring structures, slight irregularities of configuration, and temperature change. However, the system is assumed never to deviate far from the original homogeneous condition (linearity assumption), as patterns produced by nonuniform distribution of morphogens in early stages are expected to to be similar to those in later stages. Morphogens that diffuse and react together undergo an instability that causes spatially varying profiles, or patterns, of concentrations.

In the case of an isolated ring of cells, the morphogen prepattern set up by the reaction–diffusion equations launches a cell-fate program that leads to a branched structure [64]. In addition to diffusion-driven instabilities, instability can arise from interactions between stabilizing system components. At the beginning of instability, a pattern of morphogen concentrations appears which is described by waves.

Data to be specified include [64]: (1) number of involved cells and morphogens; (2) morphogen diffusivities; (3) concerned reactions; (4) kinetic reaction rates; (5) information on random disturbances; and (6) spatial and temporal morphogen distribution. The morphogen permeability of the plasma membrane, the single obstacle to diffusion, equals 5×10^{-6} cm/s.

The changes of state are associated with [64]: (1) variations of position and velocity; (2) stress field depending on material rheology and motion, in addition to osmotic pressures; (3) chemical reactions; and (4) diffusion of chemical substances. Because of mathematical complexity, Turing (1952) focuses on chemical aspects, neglecting the mechanical aspect. The solution of linear differential equations with constant coefficients takes the form of a sum in which each term represents a sinusoidal oscillation:

$$\sum z_1^* \exp\{z_2^* t\}, \tag{2.14}$$

where the quantities $z_1^* = \hat{z}_1 \exp\{\imath\varphi\}$ and $z_2^* = \Im m\, z_2 + \imath \Re e\, z_2$ can be complex numbers.

Auto- and crosscatalysis can account for primary morphogen concentration profiles, or patterns, that determine pattern formation of biological tissues according to short-range activation and long-range inhibition [65]. Activator and repressor concentrations are distinguished from densities of sources (S_{aM}, S_{rM}) and sinks (s_{aM}, s_{rM}) of these regulators. Source densities associated with certain cell types or subcellular structures are expected to change slowly with cell differentiation, whereas regulator concentrations can vary rapidly to achieve primary patterns. Derived inhibition extends into a wider region. The equation for the time rate of change of activator concentration (c_{aM}) that depends on production (p) and degradation (d), both depending on c_{aM} and its space-averaged value $\overline{c_{aM}}$, takes the following general form,

$$\partial_t c_{aM} = \kappa_1 S_{aM} \frac{c_{aM}^{p_1}}{\overline{c_{aM}}^{p_2}} \left(1 - \frac{\kappa_2}{S_{aM}} \frac{\overline{c_{aM}}^{p_3}}{c_{aM}^{p_4}}\right), \tag{2.15}$$

as the production is also proportional to the activator source density, $\{p_j\}_{j=1}^4$ being some powers ($p_1 > p_4$ and $p_3 > p_4 > 0$). The activator–repressor model in which activation and repression of sources are functions of some powers of activator and repressor concentrations becomes:

$$\begin{aligned} \partial_t c_{aM} &= \mathcal{D}_{aM} \nabla^2 c_{aM} + \mathsf{p}_{aM} S_{aM} + \kappa S_{aM} \frac{c_{aM}^{p_5}}{c_{rM}^{p_6}} - s_{aM}(\mathbf{x}) c_{aM}, \\ \partial_t c_{rM} &= \mathcal{D}_{rM} \nabla^2 c_{rM} + \mathsf{p}_{rM} S_{rM} \frac{c_{aM}^{p_7}}{c_{rM}^{p_8}} - s_{rM}(\mathbf{x}) c_{rM}, \end{aligned} \tag{2.16}$$

activator and repressor being cleared by first-order kinetics.

On the other hand, interactions between cells and the extracellular matrix induce a mechanical stress field. In particular, cultured vascular endothelial cells rapidly aggregate into clusters via combined active and passive transports and deform the matrix into a network of cord-like structures. In the absence of cell migration, isotropic, strain-stimulated cell traction is sufficient to predict pattern formation [66].

A two-dimensional mathematical model of FGF10 expression has been developed to predict shapes of lung buds and the lung border, i.e., the mesothelium, a layer of pleural cells, as well as chemical regulation of FGF10 synthesis [43]. Different FGF10 localization modes arise when only geometrical conditions change. A single FGF10 peak occurs when the distance between the tip of lung bud and the lung border is long, leading to elongation of the bud. When the bud tip–lung border distance is short, double FGF10 peaks appear, the location of which depends on the curvature of lung border. The double peaks tend to localize at the more proximal region with increased curvature of the lung border.

Factor TGFβ produced in the mesenchyme around the bud stalk hinders FGF10 expression. Morphogen SHh produced in the mesenchyme near the bud tip inhibits and activates FGF10 production.

The computational domain comprises several subdomains: bud stalk (Ω_s) and tip (Ω_t), with a given uniform curvature, and mesenchyme (Ω_m), as well as boundaries i.e., a straight tube wall (Γ_t) and lung border (Γ_ℓ) represented by the circumference of a semicircle. Its geometry is defined by the length between the bud tip and in the direction of elongation (dimensionless L range [0.0–1.2]), stalk length (dimensionless L_s range [0.0–1.2]) and width (dimensionless $w_s = 1$), tip width (dimensionless $w_t = 1.2$), and curvature of the lung border (dimensionless κ range [0.0–0.6]).

Outside epithelial cells (in Ω_m), the chemical dynamics rely on a feedforward relation. Chemical dynamics are assumed to be much faster than the tissue deformation. The equation of the time rate of change in concentration of molecular species (i) depends on diffusion coefficients ($\mathcal{D}_i$), degradation rates (d_i), transportation rate from cells (t_i), and activation rate (a_i). The concentration of diffusive molecules on Ω_s and Ω_t, respectively, are assumed to be constant (c_{Hh_0} and c_{TGF_0}). The FGF10 concentration equals zero on the tip and stalk, as FGF10 is consumed on the epithelial layer. The Neumann boundary condition (zero flux) is adopted except for FGF10 on the epithelium. The extracellular concentrations of SHh, TGFβ, and FGF10 are functions of space (**x**) and time (t), c_{TGF_i} being the FGF10 concentration in mesenchymal cells:

$$\begin{aligned} \partial_t c_{Hh} &= \mathcal{D}_{Hh} \nabla^2 c_{Hh} - \mathsf{d}_{Hh} c_{Hh}, \\ \partial_t c_{TGF} &= \mathcal{D}_{TGF} \nabla^2 c_{TGF} - \mathsf{d}_{TGF} c_{TGF}, \\ \partial_t c_{FGF} &= \mathcal{D}_{FGF} \nabla^2 c_{FGF} - \mathsf{d}_{FGF} c_{FGF} + \mathsf{t}_{FGF} c_{TGF_i}. \end{aligned} \quad (2.17)$$

Inside epithelial cells,

$$\partial_t c_{TGF_i} = \mathsf{a}_{FGF} \frac{k_1^2}{c_{Hh}^2 + k_1^2} \frac{k_2^2}{c_{Hh}^2 + k_2^2} \frac{k_3^2}{c_{TGF}^2 + k_3^2} - \mathsf{d}_{TGF_i} c_{TGF_i} - \mathsf{t}_{FGF} c_{TGF_i}, \quad (2.18)$$

where k_r ($r = 1, 2, 3$) are the Michaelis constants of the chemical reactions. Although FGF10 diffusivity is influenced by heparan sulfate in the mesenchyme and other matrix components, the spatial distribution of FGF10 is supposed to be given by FGF10 production.

The FGF10 synthesis in the mesenchyme is controlled by SHH both positively and negatively simultaneously. The FGF10 level thus reaches a maximum at an intermediate level of SHh concentration. As the FGF10 expression is repressed by TGFβ around the bud stalk, FGF10 localizes to a single site distal to the bud [43]. When the bud grows and approaches the lung border, the region with intermediate SHH levels splits and primes 2 FGF10 peaks on each side of the axial direction of bud elongation. The splitting of localized FGF10 spots is also observed when the curvature of the lung border is large.

A computed FGF10 gradient vector at each position on the tip surface was fitted to a fourth-order function $\kappa_1 \eta^4 + \kappa_2 \eta^2 + \kappa_3$, where η is the location along the surface of the bud tip ($\eta = 0$ on the direction line of the stalk axis) [43]. A single peak and twin peaks occur when $\kappa_2 < 0$ and $\kappa_2 > 0$, respectively. The distance between peaks $(-2\kappa_2/\kappa_1)^{1/2}$ yields the branching angle.

In addition to geometrical conditions, the localization mode of lateral budding depends on [43]: (1) the balance of activatory and inhibitory SHh signals on FGF10 production, which is controlled by the product of 2 dissociation constants ($k_1 k_2$); (2) the intensity of inhibitory TGFβ; and (3) diffusivities of extracellular SHh and TGFβ.

In summary, lateral budding is observed when: (1) the length between the bud tip and the lung border becomes small; (2) the SHh concentration at the tip becomes high; (3) the SHh diffusion distance becomes long; (4) TGFβ concentration around the stalk becomes low; and (5) the TGFβ diffusion path becomes short.

2.5 Pulmonary Vascularization

Two vasculatures irrigate the lung, the high-pressure bronchial and low-pressure pulmonary arterial circulation. The former supplies airways and lung parenchyma in nutrients; the latter is devoted to gas exchange at the alveolar level.

Pulmonary arteries and veins develop in close connection with the bronchial tree. The intrapulmonary artery tree extends down to generations upstream from terminal bronchioles to give rise to a capillary network that drains into the pulmonary veins.

At the level of terminal bronchioles, small arteries generate arterioles (caliber 100–150 μm). They branch in the pulmonary acinus into thin-walled arterioles that form the pulmonary capillary bed. The latter drains into pulmonary venules and veins.

Conventional arteries that run along bronchi branch with them. They supply the lung parenchyma and airways down to acini. Dorsal bronchial arteries arise from the thoracic aorta. They form a ring in the lung hilum that generates bronchial arteries that accompany bronchi and branch with them as well as subpleural and interstitial branches that run along the pleura and interlobular septa. The bronchial capillary network of large bronchi differs from that of small bronchi and subpleural acini.

Bronchial veins are connected to pulmonary veins. About a third of venous blood is drained by bronchial veins, azygos and hemiazygos (vena azygos minor inferior) veins, and superior vena cava. About two-third of venous blood flows through pulmonary veins into the left atrium.

Additional vessels, the so-called supernumerary arteries, can directly supply adjacent alveoli. In adults, numerous supernumerary arteries represent up to 40% of the cumulated cross-sectional area of all the side branches. The supernumerary vein number is greater than that of supernumerary arteries.

The arterial wall structure has an elastic type in arteries down to a bore of 3 mm (generation 7), a transitional type when the caliber ranges from 2 to 3 mm (generations 7–9), and a muscular type when the bore ranges from 30 μm to 2 mm (i.e., from generation 9 to acinus) [67].

Branches of the bronchial artery travel to lung parenchyma where they form a capillary network that merges with that of the pulmonary artery.

2.5.1 Communications Between Pulmonary and Bronchial Circulation

Two types of anastomoses occur between bronchial and pulmonary arterial vessels [68]: (1) between large vessels that are associated with bronchi of caliber ranging from 1.6 to 3.5 mm and (2) between arterioles within or on the surface of pulmonary lobules.

Table 2.20 Number of arteries in the posterior basal segment of the left lower lobe in fetuses at different ages (week; Source: [69]). Conventional arteries run alongside the airway and branch with it. Supernumerary arteries branch off to go directly into lung parenchyma without accompanying airways. From week 18 onward, the adult number of conventional arteries exists. At week 19, 21 conventional arteries associate with 25 bronchial generations; bronchial generations 5, 9, 16, and 23 are supplied by arterial generations 4, 10, 15, and 21, respectively

Age	12	14	16	17	18	19	20	23	28	36	38
Bronchial generations	15	18	18	18	24	25	20	25	28	26	23
Conventional arteries	11	15	18	16	23	21	19	23	26	24	21
Supernumerary arteries	43	40	46	40	45	58	43	57	75	76	64

In healthy adults, these anastomoses do not participate markedly in the lung function. In pathological conditions, arterial bronchopulmonary shunts can be observed.

On the other hand, the pulmonary capillary bed is connected to the adjacent bronchial capillary network as well as the subpleural bronchial mesh. These connections increase in size and number in lung diseases.

Bronchial and pulmonary veins communicate. This venous drainage is also connected to mediastinal veins.

In normal conditions, the peripheral pulmonary as well as the central part (in large bronchi) of the bronchial vascular bed possess small arteriovenous anastomoses (<2%) of blood circulation. In pathological conditions, these shunts can enlarge.

2.5.2 *Development of Intrapulmonary Blood Vessels*

During development, both conventional and supernumerary arteries as well as veins appear at the same time.

2.5.2.1 Development of Intrapulmonary Arteries

Preacinar bronchial branching is complete at the end of gestational week 16 and main intrasegmental arterial branches can be identified (Table 2.20) [69]. Further intrapulmonary airway changes correspond to an increase in size and appearance of cartilage and glands [22].

The total number of preacinar conventional vessels is present by month 5 of intrauterine life. By day 50, pulmonary arteries are connected to the pulmonary arch. Some respiratory bronchioles and alveolar ducts exist at birth, but alveoli develop mainly after birth. In the later stage of fetal life, intraacinar arteries develop with corresponding airways [69]. Blood vessels grow in bore and length; large vessel size increases faster during fetal life and infancy than childhood.

In human fetuses, artery length and caliber increase linearly with age and increase in lung volume [69]. Artery bore is related to its path length. Although the bronchial tree is still developing, the ratio of supernumerary to conventional arteries is similar to that in adults.

The wall structure of proximal blood vessels is achieved by gestational week 19, as it is the same as in adults despite the smaller vessel size. Near lung hilum, arteries have an elastic wall. The arterial wall transforms in the streamwise direction successively into a transitional, muscular, partially muscular, and finally nonmuscular structure upstream from the capillary bed.

During the fetal life, pulmonary arteries possess a greater amount of smooth muscle than in the adult for a given arterial size. In fetuses, arterioles have thicker walls than in adults. Smooth muscle content progressively increases from week 20 to birth [69]. Proximal arteries grow faster than those that are distal.

2.5.2.2 Development of Intrapulmonary Veins

Drainage pattern of the preacinar region corresponds to that of pulmonary arteries [70]. Growth of acinar veins occurs during childhood. Despite the presence of some vascular smooth myocytes from week 28 of gestation, the muscular layer cannot be detected in the vein wall before birth. The medial thickness remains similar from birth to year 10. In the vein, the muscular layer is thinner than that of arteries. Like arteries, the venous wall structure changes from nonmuscular, partially muscular, muscular, and elastic from venules to large veins.

2.5.3 Pulmonary Circulation

Pulmonary blood volume is shared among the arterial, capillary, and venous compartments (volume $\sim$150, 100, and 200 ml, respectively. Blood flow rates are nearly equal in serial pulmonary and systemic circulation due to permanent adaptation of blood ejections between both hearts.

The circulation time between the pulmonary valve and the left atrium is less than 4 s. The mean capillary circulation time is about 700 ms, whereas the time required for a balance between alveolar gas and blood is lower than 300 ms. At rest in the lying position, $\sim 0.7 < p_{\text{pa}} < \sim 3.3$ kPa and $\sim 0.4 < p_{\text{pv}} < \sim 1.3$ kPa. The pulmonary vessel resistance (PVR) is smaller than 8 kPa/l/s.

The pulmonary blood flow is subjected to cyclic variations of the intrathoracic pressure and indirect upstream (venous return and preload) and downstream effects of the systemic circulation. During inspiration, the intrathoracic pressure decreases and then the transmural pressure exerted at the vascular wall increases and conversely during expiration.

Table 2.21 Mean and range of the diameter and length of branches of a resin cast of a normal pulmonary arterial tree (Source: [2])

	Caliber (μm)		Length (μm)	
Order	Mean	Range	Mean	Range
5	201	140–300	842	210–1600
4	143	80–220	746	90–1900
3	95	60–210	504	20–2010
2	59	40–100	285	10–1200
1	36	30–60		

Pulmonary vasoconstriction is induced by alveolar hypoxia, pH increase, α-stimulating substances, and angiotensin-2, whereas α-blocking molecules, acetylcholine, and bradykinin generate vasodilation.

The pressure range in the pulmonary microcirculation is lower than in the systemic one. Because the pressure drop in the alveolar capillaries is small, RBC transit time rises for efficient gas exchanges. Due to slow travel through the alveolar capillaries, leukocyte concentration in the lung parenchyma is greater than in organs perfused by the systemic circulation.

Heterogeneous distribution of the perfusion in the different lung regions is due to the hydrostatic pressure difference between the lung apex and basis.[32] Ventilation distribution is aimed at partially matching the perfusion variations.

The lung is generally subdivided into 3 zones [144]. Zone 1, where $p_v < p_a < p_A$, is characterized by collapsed pulmonary capillaries. Zone 2, where $p_v < p_A < p_a$ is a transitional region where capillaries may undergo distal partial closure. In zone 3, where $p_A < p_v < p_a$, capillaries are dilated.

2.5.4 *Pulmonary Arterial Circulation*

Large arteries of the low-resistance, low-pressure, pulmonary circulation pertain to the elastic type. These highly distensible vessels, down to generation 7, have a poorly developed mural musculature. In smaller intrapulmonary branches, the amount of smooth myocytes augments. Mid-sized and small arteries as well as large arterioles (caliber 0.1–2.0 mm) are mainly muscular. Capillaries lose their muscular support.

The size and number of branches were measured from a resin cast of a normal pulmonary arterial tree (Table 2.21) [2]. Capillary networks originate from small arteries via various patterns. Most often, short precapillaries (average length ~20 μm) arise at right angles from branches of about 30 μm in caliber.

[32]Hydrostatic pressure decrease with height at a rate of ~100 Pa/cm for arterial (p_a) and venous (p_v) pressure and ~0.1 Pa/cm for alveolar pressure (p_A).

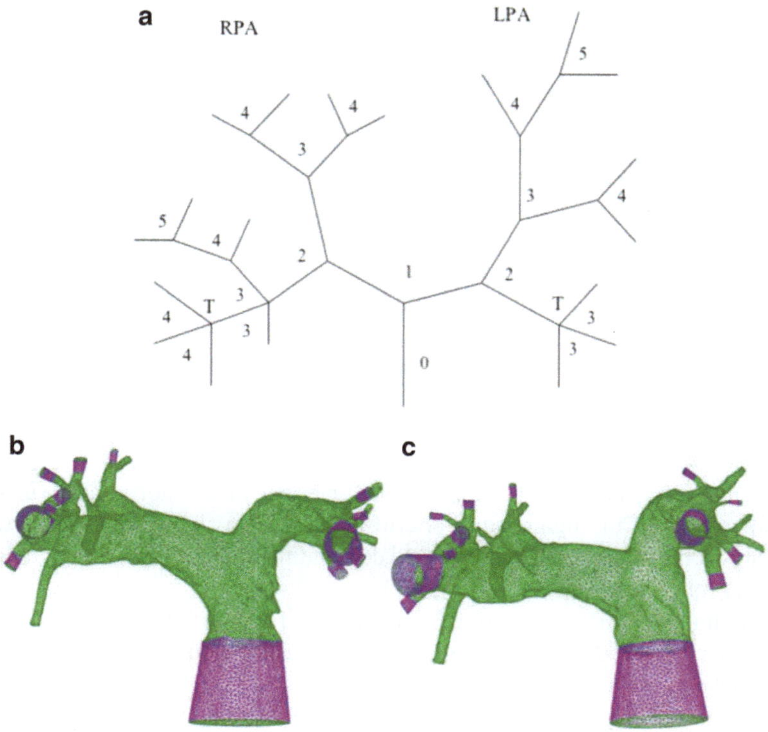

Fig. 2.16 Three-dimensional reconstruction of pulmonary arteries (RPA: right pulmonary artery; LPA: left pulmonary artery; T: trifurcation). (**Top**) Schematic of the investigated pulmonary arterial network with branch generation numbers. (**Bottom**) Computational meshes (courtesy J.J. Christophe)

Arcades join the ends of adjacent precapillaries; they generate capillary networks. Arteries of 10 to 30 μm in diameter can terminate abruptly into capillary loops or can divide to form gradually narrower vessels after 1 to 4 dichotomous divisions and then the capillary network. Capillaries can also arise directly from large arteries (diameter $\geq 100\,\mu$m); they supply alveoli on first- and second-order respiratory bronchioles.

The pulmonary resistance is about one-sixth of the systemic circulation value. The pulmonary pressure is much lower than in that of the systemic circulation (min ~1 kPa; max ~3 kPa; mean ~1.7 kPa). For a peak velocity in the pulmonary trunk of 60 cm/s, a hydraulic radius of 1 cm, and a kinematic viscosity (or momentum diffusivity) of $3 \times 10^{-6}\,m^2/s$, the peak Reynolds number is equal to 2000.

The short and wide pulmonary trunk (pulmonary artery or main pulmonary artery; length ~5 cm; caliber 2–3 cm; Fig. 2.16) arises from the right ventricle base. It has an ascending and posterior orientation, in front of the ascending aorta. It lies within the pericardium. It branches into left and right pulmonary arteries that deliver deoxygenated blood to the corresponding lung.

The bifurcation of the pulmonary trunk has a variable shape (branching angle 100–180 degrees). The area ratio is greater than 1. Both caliber and medial thickness of the pulmonary trunk increase with age [72]. The density of elastin in the pulmonary trunk is not influenced by age and averages approximately 26%, with considerable individual variation.

The caliber of the right pulmonary artery ranges from 17 to 30 mm, i.e., slightly smaller than that of the pulmonary trunk. It passes under the aortic arch and behind the superior vena cava. It runs nearly horizontally to the hilum of the right lung. It then divides into inferior and superior branches. The superior branch supplies the right upper lobe. The inferior branch delivers blood to the right middle and lower lobes.

The short left pulmonary artery runs in a left, cranial, and posterior direction. It then turns caudally to penetrate the hilum of the left lung. It lies in front of the descending aorta, beneath the downstream segment of the aortic arch, to which it is connected by the ligamentum arteriosum. Like the right pulmonary artery, it bifurcates at the hilum into ascending and descending branches that supply the upper and lower lobes, respectively.

The pulmonary microcirculation is characterized by arterioles that rapidly reduce their caliber from 100 to 50 μm. Capillaries (caliber 8–9 μm; length 6–18 μm) have a bore that depends on gravity.

Nutrients are mainly supplied by the bronchial circulation, a component of the systemic circulation. Bronchial capillaries communicate with pulmonary capillaries. The drainage is ensured by both the systemic venous circulation via the azygos vein and pulmonary veins. Therefore, bronchial arteries are unique in that they possess a dual drainage: bronchial and pulmonary veins. Pre- and postcapillary anastomoses exist between the bronchial and pulmonary circulation. In addition, bronchial venous plexus communicate with small rami of pulmonary veins as well as surrounding alveolar capillaries via small venules along bronchi and bronchioles.

From lung hila, bronchial arteries follow 2 distinct directions, the bronchial tree down to terminale bronchioles and the visceral pleura. The pulmonary circulation takes over the nutrition. However, some small bronchial arterioles can irrigate alveolar ducts and, occasionally alveolar septa. In addition, bronchial arteries give rise to vasa vasorum of pulmonary arteries, in addition to vasa nervorum and vasa bronchorum.

2.6 Innervation of Ventilatory Organs

Remote control involves fast signal transmission via nerves that directly signal to target cells as well as the release of hormones from more or less distant endocrine cells that reach their targets by blood circulation (Vol. 2 – Chap. 1. Remote Control). Both signaling types repond to stimuli to adapt ventilation and lung perfusion to the

Table 2.22 Respiratory centers of the central nervous system

Brain region	Respiratory centers	Function
Pons	Pneumotaxic center	Inhibition of inspiratory center (limitation of inspiration duration)
	Apneustic center	Inspiration activation
Medulla oblongata	Inspiratory center	Inspiration Activation of diaphragm and external intercostal muscles
	Expiratory center	Deep and/or forced expiration Activation of abdominal muscles and internal intercostal muscles

body's need. Pulmonary circulation must match ventilation for adequate blood gas content.

Cells can interact with each other via various substances, such as aminergic neurotransmitters,[33] peptides,[34] hormones, growth factors, cytokines, and prostaglandins, among other compounds. Target cells express receptors for these substances.

Cells that participate in the nervous regulation of the ventilatory system include sensors as well as afferent, efferent, and possible relay neurons and glial cells of the central nervous system.

In the central nervous system, 4 main centers regulate pulmonary respiration: inspiratory, expiratory, pneumotaxic, and apneustic centers; Table 2.22).

Inspiration results from a sudden ramped increase in motor discharge to inspiratory muscles. The inspiratory center (dorsal group of respiratory neurons) formed by neurons of the *nucleus tractus solitarius* in the upper region of the medulla oblongata controls the diaphragm and intercostal muscles responsible for inspiration. These inspiratory muscles are connected to the nervous centers via spinal nerves. Contraction of the diaphragm is triggered by 2 phrenic nerves that emanate from cervical spinal nerves 3 to 5. Eleven pairs of intercostal nerves, the anterior branches (rami anteriores) of the thoracic T1 to T11 spinal nerves regulate the activity of intercostal muscles. Before the end of inhalation, motor discharge declines.

[33] All monoamines derive from aromatic amino acids, such as phenylalanine, tyrosine, tryptophan, and thyroid hormones (thyronamines) upon processing by aromatic amino acid decarboxylases. They include histamine, catecholamines (dopamine, noradrenaline, or norepinephrine, and adrenaline, or epinephrine), tryptamines (serotonin and melatonin), and trace amines (β-phenylethylamine, tyramine, tryptamine, octopamine, iodothyronamine, and thyronamines). Various constituents of the aminergic system (adrenaline, dopamine, noradrenaline, and monoamine oxidase) lodge in different regions of the brain (olfactory lobe, hippocampus, cerebellum, pons, medulla, etc.). The synthesis of aminergic neurotransmitters is regulated via the rate-limiting enzyme tyrosine hydroxylase. A vesicular transporter protein in dopaminergic, noradrenergic, and serotonergic neurons in the brain enables accumulation of these 3 amines in synaptic vesicles.

[34] Peptides such as cholecystokinin may act as neurotransmitters.

Table 2.23 Hering–Breuer reflex

Inspiratory center	$\oplus\longrightarrow$	Phrenic and intercostal nerves,
	$\oplus\longrightarrow$	inspiratory muscles,
	$\oplus\longrightarrow$	lung inflation
Pulmonary stretch receptors	$\oplus\longrightarrow$	vagus nerve,
	$\ominus\longrightarrow$	inspiratory center

Exhalation is usually passive, except at high ventilation rates. At rest, expiration results from the relaxation of the diaphragm and intercostal muscles and lung elastic recoil. A feedback control is ensured by central and peripheral chemo- and mechanoreceptors.

When breathing increases, such as during exercise and coughing that helps to remove large foreign bodies from the respiratory tract as well as smaller particles entrapped in the mucus coating the respiratory epithelium, the expiratory center primes the contraction of abdominal muscles and 11 internal intercostals (or intercostales interni).[35] The expiratory center (ventral group of respiratory neurons) formed by neurons of the nuclei ambiguus and retroambiguus is located anteriorly and laterally with respect to the inspiratory center.

The nervous centers change the amplitude and rhythm of respiration upon reception of feedback signals sent by sensory neurons. The Hering–Breuer reflex that uses the vagus nerve prevents lung overinflation (Table 2.23). When lung airways expand during inhalation, the vagus nerve conveys information to the medulla oblongata that, in response, temporarily inhibits the inspiratory center, thereby priming expiration. Once bronchi and bronchioles return to resting size, the vagus nerve ceases to send inhibitory signals, and the inspiratory center reinitiates inspiration.

The pneumotaxic center (or pontine respiratory group) located in the pons provides cyclic inhibitory impulses to the inspiratory center, thereby decreasing the tidal volume. This center, which controls the breathing amplitude and respiratory frequency, is only active during elevated breathing.

The apneustic center, also situated in the pons, activates inspiration at rest. The apneustic center activity is canceled out by the pneumotaxic center when the breathing rate increases. Conversely, the apneustic center delays arrest signals from the pneumotaxic center.

Influence factors of the body's respiration include: (1) voluntary control handled by the cerebral cortex; (2) emotions driven by the hypothalamus and limbic system; (3) body temperature; and (4) mechanically and chemically gated receptors.

[35]External intercostals (or intercostales externi) work in unison during deep inspiration. These thick muscles are directed obliquely downward and laterally. Internal intercostals are also directed obliquely, but in a direction opposite to that of external intercostals. Internal intercostal muscles counteract the action of external intercostal muscles.

Proprioceptive receptors, or proprioceptors, localize to skeletal muscles, tendons, and joints that sense body movements. During exercise, these receptors transmit signals to the respiratory centers to increase breathing frequency and amplitude.

Pain receptors, a major type of nociceptors, operates similarly to provoke hyperventilation. External nociceptors include cutaneous and mucosal nociceptors; internal nociceptors reside in organs (muscle, joint, digestive and urinary tracts, etc.). Cell bodies of nociceptor neurons lodge in the dorsal root ganglia or trigeminal ganglia.

Certain sensory neurons can function as polymodal nociceptors that can respond to both heat and mechanical stimuli. Cutaneous thermoreceptors can also signal to the cerebral cortex, which, in turn, activates respiratory centers, causing hyperventilation.

Irritant receptors are sensory terminals of myelinated afferent fibers that reside in bronchial and bronchiolar walls. Irritant receptors respond to smoke and dust; they are stimulated by histamine, serotonine, and prostaglandins, as well as inflation, deflation, and acute congestion of blood capillaries [73]. Their stimulation causes prolonged inspiratory duration and excitatory responses, such as a cough and gasp.

In addition, juxtacapillary (J) receptors in alveolar walls are innervated by fibers of the vagus nerve. Juxtacapillary receptors, or unmyelinated C-fiber endings,[36] reside in alveolar walls, close to capillaries of pulmonary circulation (hence their name), as well as in bronchial walls, near capillaries of the bronchial (systemic) circulation. They are stimulated by hyperinflation, lung injury, and edema (especially alveolar J receptors), and by exposure to chemicals, such as histamine and serotonine (especially bronchial J receptors) [73]. Their acute and strong stimulation can cause apnea and a fall in systemic vascular resistance. Afferent, unmyelinated C fibers of the somatic sensory system innervate both the pulmonary capillary bed and airways, thereby yielding vagal inputs with different properties from these 2 neighboring structures.

The human body uses 2 major ventilatory control systems to maintain the content of oxygen, carbon dioxide, and protons: the chemical and behavioral control of breathing. Whereas chemical control of breathing relies on arterial and brain partial pressure in O_2 and CO_2 and pH sensed by peripheral and central chemoreceptors, the behavioral control of breathing is based on cues received from different sites (e.g., cortical and subcortical regions and hypothalamus, as well as proprio- and nociceptors). The behavioral control of breathing regulates breathing either by direct control of respiratory motoneurons (corticospinal control of the body's respiration) or by modulation of respiratory centers in thebrainstem via the reticular activating system

[36]C-Fibers are associated with nerves of the somatic sensory system. These unmyelinated afferent fibers (diameter 0.2–1.5 μm) convey inputs from the periphery to the central nervous system with a slow conduction velocity (< 2 m/s). C fibers synapse to neurons in the dorsal horn of the spinal cord. Polymodal C fibers react to various stimuli.

Table 2.24 Variation with age of the breathing frequency (Source: Wikipedia). In a healthy young adult, tidal volume, the volume of air inhaled or exhaled during quiet breathing, equals approximately 500 ml. Eupnea means a normal resting breathing frequency. Tachypnea refers to an elevated breathing frequency and bradypnea to a slow breathing rate. Hyperpnea is an increased amplitude of breathing to match metabolic needs; hypopnea is a decreased breathing amplitude (shallow breathing). Hyperventilation refers to an inappropriate ventilation rate

Age	Breathing frequency (Hz)
Birth–week 6	0.5–1.0
3 years	0.3–0.5
10 years	0.25–0.33
Adults	0.2–0.3

(or extrathalamic control modulatory system). In addition, the behavioral control can modulate the chemical control to optimize breathing for the body's needs.

The average breathing frequency[37] in healthy adults at rest is usually 0.2–0.3 Hz. In fact, resting breathing frequency varies with age (Table 2.24).

2.6.1 *Neurons*

Neurons are electrically excitable cells that generate, receive, process, and transmit information. Neurons are composed of a soma (or cell body), dendrites (thin cellular extensions with many branches), and an axon (thin, long cellular projection).

A majority of neurons receive inputs on the soma and dendrites and transmit output via the axon and discharge of neurotransmitters stored in presynaptic vesicles into synaptic clefts. Synaptic transmission allows the propagation of electrochemical signals, within a millisecond after release, as neurotransmitters activate receptors on target cells and trigger a depolarization wave.

Afferent neurons convey information from tissues to the central nervous system. Efferent neurons transmit signals from the central nervous system to the effectors. Interneurons connect neurons. Excitatory and inhibitory neurons excite and inhibit their targets, respectively. Modulatory neurons are responsible for neuromodulation.

Neuromodulators modulate the release of neurotransmitters (acetylcholine, catecholamines, indolamines, excitatory amino acids such as glutamate, and inhibitory amino acids such as γ-aminobutyric acid) or neuropeptides. Chemokines (interleukins, tumor-necrosis factors, and interferons) that are constitutively expressed in glial cells and neurons can also serve as neuromodulators.

[37] A.k.a. pulmonary ventilation rate and pulmonary respiration rate.

2.6.2 Autonomic Nervous System

The autonomic nervous system is constituted of 2 complementary sympathetic and parasympathetic networks. The autonomic nervous control of breathing and blood circulation enables optimal coupled air and blood flows according to metabolic needs upon signaling from peripheral sensory cells.

2.6.3 Parasympathetic and Sympathetic Control

A stressful event provokes a temporary stimulation of the sympathetic nervous system and a simultaneous reduction in parasympathetic activity. Once this event is ended, parasympathetic activity increases to assist the return of physiological functions to resting levels.

The autonomic nervous system controls delivery of oxygen and nutrients to the body's cells. The sympathetic innervation of the heart and blood vessels is excitatory (positive chronotropy and inotropy as well as vasoconstriction). The sympathetic innervation of airways causes bronchodilation to favor oxygen transfer (Table 2.25). The complementary parasympathetic (vagal) innervation balances sympathetic ouput, thereby preventing excessive activation of the sympathetic nervous system.

An optimal autonomic balance supports the cardiopulmonary function. Blow flow waveform with its amplitude, frequency, and phase is related to breathing rhythm with its frequency and amplitude. Generally, the heart beats faster (e.g., frequency $\sim$1.4 Hz) and slower (e.g., frequency $\sim$1.2 Hz) during inhalation and exhalation, respectively.

2.6.3.1 Nervous Receptors of the Tracheobronchial Tree

More than 5500 afferent nervous fibers from the vagus nerve enter the left pulmonary lobes [74]. Among these fibers, about 9% are myelinated and 91% are unmyelinated.

Sympathetic afferents are characterized by a discharge rate linearly related to transpulmonary pressure and a conduction velocity ranging from 7 to 12 m/s [74]. They may be connected to pulmonary structures and pulmonary veins and arteries. Excitation by some chemical and mechanical stimuli to airways can be mediated by sympathetic afferents.

Stimulation of rapidly adapting irritant and stretch receptors and/or C-fiber receptors in the lungs and airways elicits an *airway reflex defense* (cough, bronchoconstriction, and increased mucus secretion). Tobacco smoke is a potent stimulus for these 2 types of receptors.

Mechanosensitivity of rapidly adapting and slowly adapting low-threshold mechanoreceptors may rely on H^+-gated (acid-sensing) ion channels (ASIC), a

Table 2.25 Effects of the parasympathetic and sympathetic components of the autonomic nervous system on the cardiovascular and ventilatory apparatus. The sympathetic system associated with the thoracolumbar outflow consists of cell bodies of general visceral efferent and preganglionic neurons in the lateral horn of the spinal cord from the T1 to L2 level of the vertebral column. Preganglionic neurons synapse with postganglionic neurons of paravertebral and prevertebral sympathetic ganglia, in addition to chromaffin cells of the adrenal medulla. Splanchnic nerves such as cervical cardiac and thoracic visceral nerves synapse in the paravertebral chain, whereas thoracic splanchnic nerves synapse in the prevertebral ganglia. The parasympathetic system associated with the craniosacral outflow comprises cell bodies in the brainstem (cranial nerves III, VII, IX, and X) and in the sacral spinal cord. Preganglionic neurons synapse with postganglionic neurons of the parasympathetic ganglia of the head and in or near the wall of innervated organs. In particular, vagus and thoracic spinal accessory nerves innervate the heart and lungs, among other organs. The vagus nerve that controls many thoracic, abdominal, and pelvic organs (hence its name) originates in the dorsal nucleus of the vagus nerve and nucleus ambiguus. It generates parasympathetic cardiac nerves that form cardiac and pulmonary plexus (α: α-adrenergic receptor; β: β-adrenergic receptor; M: acetylcholine muscarinic receptor)

	Effects
Parasympathetic	Bronchoconstriction (M_3)
	Negative chronotropy, dromotropy, inotropy (M_2)
	Vasodilation (M_3)
Sympathetic	Bronchodilation (β2 major contribution)
	Positive bathmotropy, chronotropy, dromotropy, inotropy
	Vasoconstriction (digestive tract and skin [α])
	Vasodilation (small coronary arteries [β2])
	Vasodilation (skeletal muscle arteries [β2])

subset of the Degenerin–epithelial Na^+ channel set, which are Na^+-selective or Na^+-preferring cation channels [75].

Slowly Adapting Stretch Receptors

Slowly adapting stretch receptors (SASR) have a long-lasting discharge in response to a maintained lung inflation. Their activity rapidly declines immediately after inflation and slows progressively into a sustained firing, hence their name. They have myelinated axons.

Bronchial and Ttracheal SASRs

Slowly adapting stretch receptors are observed along the tracheobronchial tree down to terminal bronchioles. The SASR concentration is higher in the large proximal airways and progressively declines toward the periphery [74].

Most slowly adapting stretch receptors localize to intrathoracic airways. These receptors regularly increase their rate of discharge in the course of inspiration, reacting to the circumferential tension (*inflation reflex*) [74].

On the other hand, SASRs in the extrathoracic trachea augment their activity during expiration. A proportion of SASRs (27–63%) maintain a discharge at the functional residual capacity. Most SARs are active at zero transmural pressure and respond to both distending and collapsing transmural pressures.

Therefore, activation of extra- and intrathoracic airway SASRs during a breathing cycle is out of phase. During inspiration, activity of the intrathoracic SASRs increases, that of the extrathoracic tracheal SASRs decreases, and conversely during expiration.

Tracheal SASRs reside in the membraneous posterior wall, within the trachealis muscle. When it expands during lung inflation, mural deformable cartilage stretches the highly flexible posterior wall. This mechanical coupling between these 2 wall faces may explain the asymmetrical response curve of tracheal SASRs to positive and negative transmural pressures.

SASR Diversity, Function, and Mechanism of Action

A particular category of slowly adapting receptors that also have myelinated axons and localize to both intrapulmonary and extrapulmonary intrathoracic airways has an expiratory discharge during spontaneous ventilation, at least in some mammal species [74].

Most of the extrapulmonary SASRs are active at FRC; most of the intrapulmonary SASRs have a higher transmural pressure threshold within the tidal volume range [74]. Extrapulmonary SASRs also respond to negative transmural pressure above a given threshold. Type-1 SASR in large airways has a response that tends to saturate above about 1 kPa; type-2 SASR in distal airways has a more linear response.

Like baroreceptors, most SASRs respond not only to the magnitude of the transmural pressure (static response), but also to its rate of change (dynamic response). The receptor fires at a higher rate when the transmural pressure varies than at a steady transmural pressure. The higher the rate of inflation is, the higher the increase in receptor discharge. In addition, the faster the inflation, the lower the recruitment threshold. Moreover, at a given inflation rate, the higher the transmural pressure is, the greater the rise of the firing rate.

Influence Factors

Inhaled air cooling depresses both the static and dynamic responses of SASRs. This effect can explain, at least partly, the bronchoconstrictive action of inhaling cold air.

Inhaled carbon dioxide can either stimulate or inhibit airway SASRs. Inhibition of CO_2 is particularly strong in hypocapnic conditions and weakens at or above normocapnic levels. This gas targets bronchial, but not tracheal SASRs [74]. Hypercapnia and hypoxia can cause bronchoconstriction via chemoreceptor stimulation and then SASRs increase their activity.

Rapidly Adapting Stretch Receptors

Rapidly adapting stretch receptors (RASR) have an irregular and scant activity in eupnea, more often during inspiration; hyperpnea increases their discharge. Their number is smaller than that of SARS. Their conduction velocities are slightly lower than that of SARs [74].

Localization

Rapidly adapting stretch receptors reside in bronchi and bronchioles, but are concentrated in the first generations. In the extrapulmonary airways, the RASRs density increases from the upper trachea to the main stem bronchus. Unlike SARS, they lodge all along the circumference.

Function

They respond to large deformations resulting from positive and negative transmural pressures. Like SASRs, RASRs respond to transpulmonary pressure as well as to its rate of change; However, they do not operate like SARSs. They react to a step change in pressure with a burst of action potentials at irregular intervals that subsides very rapidly [74].

At a constant rate of inflation, RASRs increase their firing as lung volume rises. Their activity depends on flow rate, but this effect tends to saturate, whereas SASRs have an increasing sensitivity to flow rates. An increase in inflation rate lowers the volume threshold at which RASRs are recruited.

Response to Inhaled Agents

Many inhaled irritants, such as the gas sulfur dioxide, aerosols, inert dusts, and fumes, activate RASRs. Bronchoactive substances, such as acetylcholine, bradykinin, histamine, serotonin, substance-P, and prostaglandin $PGf_{2\alpha}$, stimulate RASRs by causing bronchoconstriction [74]. Like for SARSs, CO_2 inhibits RASRs, especially at low levels of hypocapnia.

C-Fiber Receptors

In humans, the alveolar wall contains unmyelinated fibers and afferent terminals; the tracheal epithelium also has unmyelinated fibers with corresponding terminals.

Diversity

Pulmonary C-fiber receptors, or J receptors (Sect. 2.6), lodge in the lung parenchyma, outside the bronchial and bronchiolar walls, unlike bronchial C-fiber receptors. They may reside in the pulmonary or bronchial circulation as well as in alveolar walls. In any case, both types have the same action. They provoke rapid, shallow breathing, bronchoconstriction, cough, mucous secretion, bradycardia, and a decrease in total vascular peripheral resistance [74].

An increase in interstitial pressure such as that during edema stimulates pulmonary C-fiber endings, but has only a weak effect on bronchial C-fiber endings. Pulmonary C-fiber receptors may provide an excitatory input for the control of breathing rate that opposes the inhibitory input from SARs. Bronchial C fibers double their activity when the local temperature rises from 29 to 33°C [74].

Mechanosensory Pulmonary and Chemosensory Bronchial C-Fiber Receptors

Pulmonary C-fiber receptors are mainly mechanosensors; bronchial C-fiber receptors are predominantly chemosensors. Bronchial C fibers respond to humoral mediators of inflammation, such as bradykinin, histamine, prostaglandins, and serotonin. Pulmonary C fibers are stimulated by series-E prostaglandins as well as CO_2 possibly via a vagus-mediated hyperpnea [74].

2.6.3.2 Respiratory Rhythm

Six different sets of neurons are defined according to their firing patterns, membrane potential changes, and synaptic inputs [76]. Four sets of neurons fire during inspiration: (1) preinspiratory neurons (**preI**) that are excited at the transition between expiration and inspiration; (2) early inspiratory neurons (**earlyI**) that send impulses from the beginning to the middle of inspiration; (3) ramp-inspiratory neurons (**rampI**) that discharge throughout inspiration; and (4) late-inspiratory neurons (**lateI**) that are activated at the end of inspiration. During the transition between inspiration and expiration, postinspiratory neurons (**postI**) are activated. Expiratory neurons (**E2**) fire during expiration.

The respiratory outflow in spinal motoneurons supplying respiratory muscles originates from the bilateral dorsal (DRG) and ventral (VRG) group of respiratory neurons within the lower brainstem. Only VRG appears to be essential for rhythm generation.

Respiratory rhythmogenesis relies on a network of mostly antagonistically connected respiratory neurons of the 6 above-mentioned sets [76]. Other groups of neurons shape the activity pattern. Respiratory rhythm results from synaptic interaction and cooperative adjustment of excitability levels via voltage-gated ion channels. Excitatory synaptic interaction operates via glutamatergic synaptic activation through AMPA- and NMDA-type glutamate receptors. Periodic inhibition of synaptic inputs is mediated by GABA and glycine receptors.

2.6.3.3 Reciprocal Interaction Between Baro- and Chemoreceptors

Afferents for both the cardiovascular and ventilatory apparatus originate from baro- and chemoreceptors. Afferents associated with baroreceptors and other mechanoreceptors such as volume receptors project to the nucleus of the solitary tract [221]. Afferents transmitting signals triggered by chemical or physical messengers project to the spinal cord.

Control mechanisms that govern ventilation and blood circulation interact. Hypotension and hypoxemia often happen together, so that baro- and chemoreceptors are activated simultaneously. The corresponding reflexes are then activated synergistically. The vasoconstrictor response to stimulated chemoreceptors is enhanced by hypotension and impeded by transient hypertension, due to, at least partly, a central interaction of chemo- and baroreceptor reflexes [77]. Increased blood pressure in a carotid sinus primes the baroreflex and prevents the vasoconstrictor and ventilatory responses to chemoreflex activation by controlateral chemoreceptors.

In paralyzed anesthetized cats, the fastigial pressor response, the pressor response to electrical stimulation of the fastigial nucleus (or nucleus fastigii) of the cerebellum, negatively interacts with the depressor response to electrical stimulation of the carotid sinus nerve [78]. Blood pressure responses resulting from stimulated fastigial nucleus and carotid sinus nerve are mutually inhibitory. Interneurons that mediate pressor and depressor responses in the paramedian reticular nucleus of the medulla are excited by stimulation of either the fastigial nucleus and carotid sinus nerve. Neurons in the medulla include: (1) *carotid sinus baroreceptor neurons* that fire rhythmically in synchrony with or are modulated by the cardiac cycle; and (2) *carotid body chemoreceptor neurons* that are excited by stimulation of chemoreceptors.

Baro- and chemoreceptor neurons that run via the carotid sinus nerves project either in the intermediate zone of the nucleus of the solitary tract or more medial regions of the medulla, i.e., in the parahypoglossal area or in the dorsal region of the paramedian reticular nucleus [79]. The paramedian reticular nucleus participates in the depressor response to electrical stimulation of myelinated fibers of the carotid sinus nerve and the depressor response to carotid sinus stretch, but precludes the pressor response to chemoreceptor stimulation. It does not intervene in the cardiovagal reflex (heart rate response). On the other hand, the intermediate zone of the nucleus of the solitary tract receives signals from unmyelinated baroreceptor and all chemoreceptor neurons that operate in cardiovascular reflexes (all blood pressure and heart rate responses) to electrical stimulation of the carotid sinus nerves.

Breathing at a slower pace (e.g., frequency $\sim$1.0 Hz) increases the coherence (smoothness) of breath-to-breath heart rate fluctuations and the synchrony (resonance) between respiration and blood circulation parameters. In summary, the breathing tempo is imposed by sympathetic and parasympathetic activity. Conversely, breathing contributes to the short-term modulation of sympathetic and parasympathetic nerve activity.

In addition to short-term between-breath ventilatory–cardiovascular interactions within the phases of ventilation, sustained changes in ventilatory amplitude such as during hyperventilation caused by exercise as well as that resulting from hypoxia or hypercapnia can influence the cardiovascular control in general and the baroreflex in particular. Change in respiratory amplitude influences oscillatory (phasic) features of sympathetic outflow [80]. Sustained isocapnic hyperventilation increases blood pressure and causes tachycardia, but does not change muscle sympathetic nerve activity. The absence of induced cardiac slowing and of sympathetic inhibition in response to higher blood pressure during hyperventilation reflects attenuation of arterial baroreflex sensitivity to oscillations in systolic blood pressure and baroreflex control of both heart rate and sympathetic signaling. Faster breathing rates potentiate the chemoreflex response to hypoxia and hypercapnia [80]. During large inspirations, muscle sympathetic nervous activity changes are followed by an increase in sympathetic signaling during expiration, but integrated sympathetic activity is not influenced by changes in tidal volume. Spontaneous respiratory rate is thus linked to overall sympathetic activity.

During exercise, the baroreflex is overridden, as tachycardia is associated with an increase in blood pressure. A 3-fold rise in tidal volume dampens baroreflex. On the other hand, slow respiratory rate reduces dyspnea and improves exercise performance and baroreflex function in patients with heart failure.

During isocapnic hypoxia, ventilation and blood pressure increase markedly more than during hypocapnic hypoxia, but sympathetic nerve activity rises more during hypocapnic hypoxia than during isocapnic hypoxia [81]. During hyperventilation, the sympathoexcitation of isocapnic hypoxia is attenuated compared to hypocapnic hypoxia, possibly because of the inhibitory influence of activated pulmonary afferents resulting from ventilation increase and of activated baroreceptors due to a greater rise in blood pressure. Consequently, in humans, increased ventilation in response to chemoreceptor stimulation predominates, but it restrains the sympathetic response to chemoreflex stimulation.

Hypercapnia and hypoxia elicit a greater ventilatory response in subjects with a fast breathing rate than in subjects with a slow breathing rate [82]. Spontaneous respiratory rate is linked to chemoreflex sensitivity and muscle sympathetic nerve activity. The faster the respiratory rate is, the greater the sympathetic activity and the higher the chemoreflex response to hypoxia and hypercapnia [82]. The relation between sympathetic output and spontaneous breathing rate is independent of age, body mass (especially amount of body fat), physical activity, and blood pressure [82].

The autonomic nervous system comprises afferent and efferent neurons between the central nervous system and body's organs. The sympathetic and parasympathetic systems control smooth and cardiac myocyte activity as well as glandular secretion. The nervous control of the body's respiration operates via cholinergic parasympathetic neurons and sympathetic efferents.

Cholinergic preganglionic neurons innervate ganglia, glands, or nervous ganglia such as the cardiac ganglionic network. The peripheral ganglia contain motoneurons that control smooth muscles and other visceral targets.

Sympathetic preganglionic cholinergic neurons originate from the spinal cord and travel to pre- or paravertebral ganglia,[38] where they synapse with sympathetic, postganglionic, noradrenergic neurons. Sympathetic postganglionic neurons extend to reach their destinations.

Sympathetic preganglionic neurons receive synaptic contributions from the spinal cord, medulla oblongata, and hypothalamus. Barosensitive sympathetic efferents is mainly fed by the rostral ventrolateral medulla.[39]

Three main classes of sympathetic efferents, baro-, thermo-, and glucosensitive, innervate the heart, blood vessels, kidneys, and adrenal medulla.

Thermosensitive efferents consist of cutaneous vasoconstrictors. They are activated by hypothermia, hyperventilation, and emotions.

Glucosensitive efferents control adrenaline release from adrenal medulla. They are activated by hypoglycemia and exercise.

Barosensitive efferents, the largest class, have a rest activity. They discharge in bursts synchronized with cardiac and respiratory frequency. The barosensitive efferents are responsible for the short-term control of the blood pressure. Barosensitive efferents are subjected to various stimuli, lung stretch-sensitive afferents activated by the ventilation, activated muscle receptors during exercise, activated visceral or cutaneous nociceptors, as well as activated peripheral and central chemoreceptors. The chemoreflex corresponds to the activation of peripheral chemoreceptors by hypoxia and hypercapnia that increase the activity of barosensitive sympathetic efferents.

Like the sympathetic efferent nervous system, the parasympathetic efferent nervous system consists of both pre- and postganglionic neurons. Parasympathetic cholinergic neurons travel to parasympathetic ganglia on or near organs.

The vagus nerves (pneumogastric nerves or cranial nerves X) innervate the thoracic (esophagus, heart, pharynx, larynx, trachea, bronchi, and lungs) and abdominal viscera. Several parasympathetic nerves come off the right and left vagus

---

[38] Axons of preganglionic nerves penetrate in the paravertebral sympathetic chain ganglia, where: (1) they terminate by meeting postganglionic neurons or (2) continue to travel toward: (2.1) a superior or inferior paravertebral ganglion or (2.2) prevertebral ganglion, in both cases to synapse with postganglionic neurons.

[39] The rostral ventrolateral medulla is a set of heterogeneous glutamatergic C1 neurons. A majority of C1 neurons synthesize adrenaline. C1 neurons are connected to the medulla oblongata and pons. Barosensitive C1 neurons also receive inputs from the brain and spinal cord. GABA inputs from the caudal ventrolateral medulla are required in the baroreflex. Brainstem regions, such as the caudal pressor area, midline depressor area, subnuclei of the nucleus of the solitary tract, and gigantocellular depressor area, relay electrical or chemical stimulation aimed at triggering blood pressure adaptation. The rostral as well as the caudal ventrolateral medulla also contains interneurons that contain γ-aminobutyric acid. Many interneurons have a baseline activity.

nerves as they enter the thorax, such as the recurrent laryngeal and cardiac nerves. The latter form cardiac and pulmonary plexus.

The central nervous system is also directly regulated by circulating factors. Hormones and cytokines act via circumventricular organs (subfornical organ, organum vasculosum lamina terminalis, and area postrema). Endothelial cells release regulators (e.g., nitric oxide and prostaglandins) that cross the blood–brain barrier (Vol. 5 – Chap. 7. Vessel Wall) and stimulate the hypothalamus, rostral ventrolateral medulla, and nucleus of the solitary tract.

Changes in ion concentrations are detected in the hypothalamus. Brainstem chemoreceptors sense the level of blood gases. The renin–angiotensin axis also directly acts on the central system. Circulating angiotensin-2 activates angiotensin-2 receptors AT_1 on endothelial cells, releasing nitric oxide.[40]

2.6.4 Evolution of Innervation of the Ventilatory Apparatus

Concentration of certain hormone receptors and β-adrenergic receptors decays during aging, at least in human circulating mononuclear (one-lobed nucleus) leukocytes (i.e., monocytes and lymphocytes) [83]. In fetal rabbits, β-adrenoreceptor density is higher in airways than alveoli [84]. Glucocorticoids increase their number in alveoli, but not in airways and myocardium. In adults, β-adrenoreceptor amount is also higher in airways than alveoli and greater than in fetuses. In rats, β-adrenoceptor density markedly increases in whole lung during postnatal maturation. β-Adrenoceptors are primarily of the β2 subtype [85]. In newborns, β-adrenoceptor density is comparable in bronchial smooth myocytes and alveolar cells, but lower in pulmonary arterial smooth myocytes. In adult (3-month old) rats, β-adrenoreceptor density is similar in pulmonary arterial and bronchial smooth myocytes, but 2-fold greater in alveolar cells.[41] In rats, the postnatal number of cholinergic muscarinic receptors in lungs decays with respect to fetal density [86].[42]

In the human fetal lung, β-adrenoceptors and receptors for vasoactive intestinal polypeptide are present in respiratory epithelium, terminal tubules, and pulmonary arteries at gestational week 14 and their number rises afterward [42]. α1-Adrenergic and muscarinic receptors are detected after gestational week 23. Muscarinic receptors and β-adrenoceptors decay and augment in the first year of life, respectively.

---

[40] Nitric oxide diffuses across capillary walls and potentiates the release of γ-aminobutyric acid.

[41] β-Adrenoceptor density in pulmonary arterial smooth myocytes rises from 1 to 20 days and remains nearly constant thereafter. β-Adrenoceptor density in bronchial smooth myocytes does not change from 1 to 13 days and then increases from 13 to 6 months [85].

[42] The dissociation constant equals 0.25 nmol in 2-month-old rats and does not significantly change during development.

Neuron-specific enolase[43] resides in neuroblasts, neurons, and endocrine cells from gestational week 8, before the appearance of activity of acetylcholinesterase (10–12 weeks) as well as that of dopamine β-hydroxylase, vasoactive intestinal polypeptide, and calcitonin (20 weeks) [87]. Unlike central neurons, peripheral neurons of the lung contain neuron-specific enolase before full maturation and establishment of synaptic contact with target organs.

From the ectoderm, the neural plate and crest develop autonomic nerves that migrate to supply pulmonary structures. Neural crest cells move to the wall of the future trachea before it separates from the esophagus (i.e., at gestational wk 4–5) and lung buds. Main anatomical features of the sympathetic and parasympathetic systems are established at gestational week 6 [42]. Ganglia appear in the trachea by week 7 of fetal life and extend to second generation bronchi. A posterior and inner plexus is observed between the cartilage and epithelium with nerves extending to submucosal glands and tracheal muscle by week 16. At week 16, ganglia develop at bronchial bifurcations and small bronchi. Innervation of adventitia of arteries and veins begins at week 10. Pulmonary nerves mainly synthesize vasoconstrictors.

At birth, sympathetic and parasympathetic nerves extend down to alveolar ducts; both distribution and number of airway nerves are similar to that in adults. Between birth and year 3, the distribution and relative number of nerves does not markedly change. However, the total number of nerves that mainly produce bronchodilators decreases after year 3 [42]. Bronchoconstrictor reactivity also decreases with age during infancy and childhood. Conversely, the relative number of vasoactive intestinal polypeptide-containing (VIP+) nerves that innervate bronchial and bronchiolar smooth muscles is greater in adults (17–24 year old) [88].

2.6.5 Nerve Subpopulations

Nerve subpopulations in airways can be identified by their neurotransmitters and neuromodulators, such as neuropeptide tyrosine (NPY),[44] vasoactive intestine

---

[43]Enolase, a.k.a. phosphoglycerate hydrolase and phosphopyruvate hydratase, is a metalloenzyme that catalyzes the conversion of 2-phosphoglycerate to phosphoenolpyruvate, the final stage of glycolysis. Enolase isozymes, which are homodimers, include: enolase-1 (Eno1), or enolase-α, which resides in various cell types, such as those of the brain, kidney, liver, spleen, and adipose tissue; lung-specific enolase-α (Eno1b); enolase-2 (Eno2), or neuron-specific (NSE) enolase-γ, also detected in neuroendocrine cells; and enolase-3 (Eno3), or muscle-specific enolase-β; in addition to enolase-4 (Eno4). Neuron-specific Eno2 and nonneuronal Eno1 are markers for neurons and glial cells, respectively.

[44]Neuropeptide-Y, a 36-amino acid neurotransmitter (the symbol Y representing tyrosine), shares sequence homology with 2 other peptides: pancreatic polypeptide (PP), secreted by PP cells of the pancreas, and peptide-YY, localized to endocrine cells of the ileum and colon. It is found in noradrenergic ganglion cells, especially nerve fibers in the adventitia of blood vessels and airway smooth muscles, in addition to the brain and heart. During human development, NPY+ nerve fibers are first observed at gestational wk 20 in the trachea; fibers gradually extend down to the

peptide,[45] somatostatin,[46] substance-P,[47] and calcitonin gene-related peptide,[48] as well as tyrosine hydroxylase [88].[49] Many potent bioactive peptides can be detected in the mammalian respiratory tract. Vasoactive intestinal polypeptide and substance-P are found in the lung innervation; bombesin and calcitonin localize to mucosal endocrine cells; cholecystokinin and somatostatin are identified in the respiratory tract [89]. Neuropeptide-Y+ nerves are the most common neuron subpopulation that supplies bronchial smooth muscle, submucosal glands, cartilage, and submucosa.

Nerve types that supply the human pulmonary vasculature can also be identified by neuronal marker (e.g., calcitonin gene-related peptide, neuropeptide-Y, somatostatin, substance-P, synaptophysin [an integral membrane glycoprotein of synaptic vesicles of neuroendocrine cells and almost all neurons of the brain and spinal cord], tyrosine hydroxylase, vasoactive intestinal polypeptide, and ubiquitin C-terminal esterase-L1 [UCHl1]).[50] In normal (from month 1 to year 24) and hypertensive (day 11–year 8) lungs, neuropeptide tyrosine is also the predominant neuropeptide associated with pulmonary vascular nerves [91]. In normal lungs, the density of nerves increases during childhood only in arteries of the respiratory unit.

The density in vasoactive intestine peptide receptors is high in airway epithelium, submucosal glands, bronchial smooth muscle, and alveolar wall, as well as in blood vessels, whereas it is much lower in bronchiolar smooth muscle [92]. In mucoviscidosis, VIP receptor density markedly diminishes, except in bronchial smooth muscle; VIPR decrease is not significant in bronchiectasis; it is not observed in asthma.

2.6.6 *Mechanosensors – Baroreceptors and Voloreceptors*

Fast-responding mechanosensors transduce the stretch undergone by the wall and papillary muscles under increasing luminal pressure and tension exerted by cordae tendinea. They ensure beat-to-beat coordination of the heart rate and contractility.

---

intrapulmonary airways after birth [89]. Neuropeptide-Y causes endothelin-1 secretion in left and right human endocardial (endothelial) cells via activation of the Y_5 receptor and mainly Y_2 and partly Y_5 receptors, respectively [90].

[45]Vasoactive intestinal peptide is produced in many tissues. It has a positive inotropic and chronotropic effect, causes vaso- and brochodilation, and increases glycogenolysis.

[46]Somatostatin is an inhibitory hormone and neuromodulator.

[47]This neuropeptide is a neurotransmitter and neuromodulator. It belongs to the tachykinin family. This potent vasodilator acts via nitric oxide. It is also a bronchoconstrictor.

[48]This peptide is a potent vasodilator and pain transmitter.

[49]Tyrosine hydroxylase, or tyrosine 3-monooxygenase, catalyzes the conversion of ^{L}tyrosine to ^{L}(3,4)-dihydroxyphenylalanine (^{L}DOPA), using oxygen, Fe^{2+}, and tetrahydrobiopterin. Product ^{L}DOPA is a precursor of dopamine, which, in turn, is a precursor of noradrenaline and adrenaline.

[50]A.k.a. ubiquitin thiolesterase and protein gene product-9.5 (PGP9.5). This member of the ubiquitin hydrolase family, an unspecific neural and nerve sheath marker, lodges in neurons and neuroendocrine cells as well as in smooth muscle and germinal centers.

The magnitude of the baroreceptor responses depends on the targeted organ. Signal acquisition depends on receptor sensitivity. Mechanical signals are then transduced into electrochemical events with given neural firing rates.

Arterial baroreceptors are stretch receptors stimulated by intramural tension distortion of the arterial wall when pressure changes. Baroreceptors of the carotid sinus respond to pressures ranging from 8 to 24 kPa. Baroreceptors of the aortic arch, less sensitive than those of the carotid sinus, have a higher threshold pressure.

Baroreceptors are sensitive to changes in both the average blood pressure and rate of change in pressure. The combination of reduced mean pressure and attenuated pulse pressure reinforces the baroreceptor reflex [93]. The background activity can either decay or rise for blood pressure stabilization. The firing rate of the baroceptor nerves increases with blood pressure, and hence wall deformation, from a threshold up to a maximum (saturation).

Activated baroreceptors inhibit the sympathetic nervous system and activate the parasympathetic nervous system, thereby reducing total peripheral resistance and cardiac output. Increased baroreceptor firing actually not only impedes the sympathetic activity to the heart, blood vessels, and kidneys, but also increases vagal tone to the heart. Conversely, a fall in arterial pressure reduces afferent signals, which relieves inhibition of sympathetic tone, increases the peripheral resistances, and restores the cardiac output and subsequently the arterial pressure.

Arterial, high-pressure-sensing baroreceptors localize to the transverse aortic arch and carotid sinuses of the left and right internal carotid arteries (high-pressure circuit). Aortic and carotid baroreceptors monitor the blood pressure delivered to the systemic circuit, except a part of the neck and face, and brain, respectively.

The low-pressure-sensing, less sensitive baroreceptors, or *voloreceptors*, reside in large systemic veins, pulmonary vessels, and walls of the right atrium and ventricles. The stretch receptors in the low-pressure blood circuit are indeed more involved in the regulation of blood volume. These receptors can, in the longer term, change blood circulation pressure.

Aortic baroreceptor neurons exhibit mechanosensitive ion channels that are gadolinium-sensitive and have nonspecific cationic conductances [94]. The fibers of the aortic nerve enter the adventitia, between the left common carotid and left subclavian arteries, and separate into bundles generally containing 1 myelinated fiber and several unmyelinated fibers [95]. When they are close to the aortic media, the myelinated fiber loses its myelin sheath. Both unmyelinated and premyelinated axons branch off. Sensory nerve endings of aortic baroreceptor neurons are located in the adventitia of the aortic arch, between the left common carotid and left subclavian arteries. The basement membrane exists around the sensory terminals. The central axon terminals localizes to the nucleus of the solitary tract.

Once the baroreceptors are stretched, the carotid sinus nerve (or nerve of Hering), a branch of the glossopharyngeal nerve, stimulates inhibitory areas of the vasomotor center (nuclei of the solitary tract and paramedian in the brainstem) [7].[51]

[51] The reticular formation includes the raphe and medial and lateral reticular formation. The cardiac and vasomotor centers of the medulla oblongata reside in the reticular formation.

Baroreceptors of the aortic arch are innervated by the aortic nerve, which then merges with the vagus nerve (cranial nerve pair X), traveling to the brainstem. Efferent limbs are carried through sympathetic and vagus nerves to the heart and blood vessels, controlling the heart rate and vasomotor tone.

In a stress context, the cardiovascular nervous center responds by increasing sympathetic and decreasing parasympathetic outflux. Barosensitive sympathetic efferents control the activity of the heart and kidneys, as well as the release of noradrenaline from adrenal chromaffin cells, and constrict arterioles, except those of the skin.

Multiple transmitters regulate barosensitive neurons. Glutamate, γ-aminobutyric acid,[52] acetylcholine, vasopressin, serotonin, corticotropin-releasing factor, substance-P, oxytocin, and orexin have been found in nerve ends with synapses on pressure-regulating neurons, such as C1 cells[53] and the hypothalamus [221].

2.6.7 *Baroreflex*

The baroreflex, or baroreceptor reflex, maintains the blood pressure. This feedback causes a heart rate decrease, hence a blood pressure reduction, when hypertension happens, or a heart rate increase, thus an elevation of the blood pressure, when hypotension occurs.

2.6.8 *Chemosensors*

Chemosensors, or chemoreceptors, transduce chemical signals into action potentials. Impulses are transmitted via the vagus nerve into the vasomotor centers as well as respiratory centers. These chemosensors detect different chemical stimuli and express several types of ion channels that influence their excitability to trigger action potentials in afferent nervous fibers.

Breathing rate is controlled via (Table 2.26): (1) *central chemoreceptors* located in the ventrolateral surface of medulla oblongata that mainly detect changes in pH of cerebrospinal fluid, but also hypercapnia; and (2) *peripheral aortic and carotid chemoreceptors* (aortic and carotid bodies) that sense changes in blood concentration of oxygen and carbon dioxide, but not of hydrogen ion (pH), and all of these 3 concentrations, respectively. Therefore, hypoxemia is a specific stimulus to arterial chemoreceptors.

[52] γ-Aminobutyric acid signaling from the caudal ventrolateral medulla is important for the baroreflex.

[53] C1 cells belong to a cluster of adrenaline-synthesizing neurons in the rostral ventrolateral medulla. Many C1 cells regulate kidneys.

Table 2.26 Major barosensory neurons monitor arterial wall stretch fluctuations that result from blood pressure variations. Major chemosensory neurons supervise changes in contents of oxygen and carbon dioxide and pH of arterial blood. These 2 sensor types localize near or at the entrance of brain irrigation arteries (carotid sinuses and aortic arch and aortic and carotid bodies, respectively). These neurons generate carotid sinus nerves and aortic depressor nerves. Action potentials generated at their terminals in arterial walls travel to the nodose and petrosal ganglia to the nucleus tractus solitarius. Barosensor, or baroreceptor, and chemosensor, or chemoreceptor, neurons are involved in the baroreflex and chemoreflex, respectively. The baroreceptor reflex is activated during a pressure rise; the chemoreceptor reflex is mainly stimulated by hypoxia and acidosis. The baroreceptor and chemoreceptor reflex is inhibitory and excitatory on the sympathetic nerve activity, respectively. Both hypoxemia and hypercapnia increase ventilation as well as the blood pressure and cardiac frequency. The baroreflex prevents the chemoreflex. Conversely, the chemoreflex may operate on the baroreflex. Increased ventilation yields an additive sympathoexcitation

	Stimuli	Effect
Arterial barosensor	Wall stretch	Sympathoinhibition
(high-pressure baroceptor)	(pressure)	Bradycardia
Volume receptor	Wall stretch	Inotropy ↓
(low-pressure baroceptor)		Vasodilation
Central chemosensor	Acidosis	Sympathoexcitation
	Hypercapnia	Tachycardia
Aortic chemosensor	Hypoxemia	Inotropy ↑
	Hypercapnia	Vasoconstriction
Carotid chemosensor	Hypoxemia	
	Hypercapnia	
	Acidosis	

2.6.9 Arterial Chemoreflex

Arterial chemoreflex generally originates from the carotid body. Glomus cells in the carotid body depolarize in response to hypoxemia and release multiple neurotransmitters that cause impulses in afferent fibers of the *carotid sinus nerve*. Chemosensor afferents project to regions in the pons and medulla, where the primary sites are for respiratory and cardiovascular signal receptors as well as sympathetic integration.

Nevertheless, the hypothalamus modulates the ventilatory and cardiovascular responses to peripheral chemoreflex activation, especially the paraventricular nucleus of the hypothalamus (PVN), where γ-aminobutyric acid receptor-A impedes the chemoreflex that originates from the carotid body. The paraventricular nucleus of the hypothalamus sends efferent neurons to the rostral ventrolateral medulla (RVLM) that relays signals for sympathetic activity.

The inspiratory center in the medulla sends nervous impulses to respiratory muscles via the phrenic nerve to increase breathing rate and lung volume during inhalation. Chemoreceptor-mediated elevation in sympathetic nerve activity is progressively dampened when tidal volume increases because of inhibitory feedback from pulmonary volume receptors.

Central and peripheral chemoreflexes not only control alveolar ventilation to improve O_2 uptake and CO_2 removal, but also blood flow, as gas exchanges require matched ventilation and lung perfusion. A nervous impulse is sent to the cardiovascular centers in the medulla, which then feed back to the sympathetic ganglia, increasing the action potential frequency of the sinoatrial node and sympathetic output to the vasculature. Activated arterial chemoreceptors by hypoxemia increases sympathetic activity to systemic vasculature to compensate for vasodilation caused by hypoxemia on blood vessels and to redistribute blood flow to essential organs. Sympathetic-mediated vasoconstriction exerts particularly on muscular, splanchnic, and renal vascular beds. Moreover, sympathetic activity in adrenal glands augments noradrenaline release, whereas sympathetic activity to brown adipose tissue decays to reduce body temperature and oxygen consumption during hypoxemia [77]. Cardiac sympathetic activity elevates heart frequency and contractility. However, parasympathetic and sympathetic nerves that innervate heart and cerebral vessels are concomitantly activated to limit sympathetic vasoconstriction of coronary and cerebral vessels during hypoxemia and to modulate cardiac chronotropic and inotropic responses. Besides, the sympathetic response depends on ventilation frequency. The higher the breathing rate is, the stronger the chemoreflex responsiveness.

Parasympathetic and sympathetic nerve activity rapidly increase upon exposure to hypoxemia. When hypoxemia is sustained, parasympathetic and sympathetic nerve activity rises and is maintained, whereas the ventilatory rate increases transiently and then progressively declines [77]. When hypoxemia disappears, parasympathetic and sympathetic nerve activity gradually returns to normal level, being maintained during a given period after blood gases and ventilation have been reverted.

Chapter 3
Cardiovascular Physiology

The cardiovascular apparatus supplies blood to the body's organs and responds to sudden changes in demand for nutrients according to the organism's activity. Blood velocity and pressure can be associated with kinetic and potential energy, respectively. The travel time for oxygen delivery between the left atrium and peripheral tissues has a magnitude of $\mathcal{O}[1\ \mathrm{s}]$.

Integrative physiology, which is based on basic biomedical sciences aimed at improving available knowledge on physiological apparatus in normal and pathological conditions, is oriented toward bioengineering applications for enhanced diagnosis, therapy, and prognosis. The latter objectives are supported by the development of computer tools.

When targeting the heart, computational multiscale and multiphysics models are aimed at simulating cardiac electrophysiology, calcium dynamics, contraction–relaxation cycles, electromechanical coupling, cardiac flow in highly deformable chambers, valve dynamics, myocardial perfusion coupled with myocardial mechanics, metabolism, and regulation.

These models integrate adequate image acquisition and processing that determine the computational domain, proper functioning and control (chemically controlled constitutive law of cardiomyofiber mechanics [1], identifiable models of cardiac electrophysiology usually relying on ordinary differential equations, cardiac dynamics [2, 3], excitation–contraction coupling, energetics coupled to perfusion,[1] etc.), and suitable coupled biosolid–biofluid simulation software.[2] High-resolution

[1]Coronary perfusion can be modeled using a poroelastic model for the myocardial microcirculation and a Darcy solver using the arbitrary Lagrangian–Eulerian (ALE) formulation.

[2]A semi-implicit coupling scheme that exhibits a good stability has been proposed [4]. The pressure stress is implicitly coupled with the structure to ensure stability; remaining terms are explicitly treated. The implicit–explicit splitting can be conveniently performed using a Chorin–Temam projection scheme. A stabilized explicit coupling scheme based on Nitsche's method can be used, stability being independent of the fluid–structure density ratio [5].

M. Thiriet, *Anatomy and Physiology of the Circulatory and Ventilatory Systems*,
BBMCVS 6, DOI 10.1007/978-1-4614-9469-0_3,

Table 3.1 Main ion carriers in cardiomyoctes. In the heart, K_{ACh} is activated in atriomyocytes by the parasympathetic (vagal) nerves; flux through the $K_{IR}3.1$ and $K_{IR}3.4$ heteromer rises in hypercholesterolemia [8]

Ca^{2+} carriers	K^+ carriers	Na^+ carriers
$Ca_V1.2$	K_V1	$Na_V1.5$
Na^+–Ca^{2+} exchanger	K_V4	Na^+–Ca^{2+} exchanger
RyR2	K_V7	Na^+–K^+ ATPase
IP_3R	K_V11	Na^+–H^+ exchanger
SERCA2	$K_{IR}2$	
PMCA	$K_{IR}3$	
mtCU	(K_{ACh})	
	$K_{IR}6$	
	(K_{ATP})	

imaging datasets can be used for modeling (e.g., [6]). Input data of various types of models (wall and fluid rheology, electrophysiology, etc.) are extracted from the available literature.

Medical image analysis tackles construction of computational meshes with details on special structures (heart valves that enable unidirectional flow from cardiac cavities,[3] nodal tissue that generates and propagates heart contraction command waves (action potentials), cardiomyofiber orientation that ensures proper heart deformation and synchronized contraction, etc.). Validation focuses on the entire organ rather than nano- and microscopic objects.

The primary function of heart is blood pumping using its striated muscle, the myocardium. Cardiac functioning depends on several factors, such as: (1) ion carriers that determine ion fluxes and intracellular concentrations (Table 3.1; Vol. 5 – Chap. 5. Cardiomyocytes); (2) sarcomere activity, particularly the crossbridge cycling rate; (3) extracellular matrix; (4) wall perfusion responsible for nutrient inputs; and (5) cardiac loads (i.e., flow conditions in up- and downstream vessels).

Ventricle filling and emptying indeed depend on the state of the vascular bed upstream and downstream from the heart. Decrease in intrathoracic pressure during inspiration increases venous return and conversely during expiration. Ventricle filling and emptying are strongly coupled. Ventricle reexpansion helps to refill blood from the atrium. Recoil force due to the downward motion of the beating heart associated with the stretch imposed on upper vessels favors atrium filling.

The autonomous nervous system controls the cardiovascular apparatus and adapts its output to the body's activity and environmental stimuli. The autonomous

[3]Heart valve motions rely on a multibody contact problem with attachment constraints due to chordæ tendinæ [7]. The fluid and immersed valve meshes do not match; the kinematic continuity is imposed using Lagrange multipliers. The method relies on a fictitious domain, which allows very large displacements, combined with the ALE formalism to manage both elastic valve and wall motions. A partitioned fluid–structure algorithm associates independent fluid and structure solvers.

nervous system with its sympathetic and parasympathetic components is made of afferent and efferent neurons that enable communication between the body's organs and the central nervous system.

Estimation of cardiac state variables and parameters relies on inverse problems and data assimilation. The electrochemical activity of the heart is transmitted to electrocardiogram sensors after transformation resulting from traveling through diverse anatomical tissues and possible interferences of other biological signals. Data of interest are original signals and their localization. Once the direct problem that predicts measurements is solved, the inverse problem consists to determining from known outputs their sources that cannot be noninvasively, directly, and precisely measured.

When both direct and inverse modeling are investigated, well-posedness problems must be ensured. The mechanical heart model being too sensitive and large to be well inverted by classical Kalman filters or variational assimilation techniques, robust, low-cost, effective state filters must be developed and combined to parameter estimation procedures. The finite-dimensional state estimator uses the unbiased estimate of the initial condition and corrects the dynamics of the discrete system by a collocated feedback proportional to the measured error. A filter corresponding to a force proportional and opposed to the measured velocity has been applied in an infinite-dimensional mechanical system to estimate loading parameters (fully linear for the whole state-parameter observer system) and stiffness parameters (bilinear observer problem) with volume-distributed measurements of the velocity [9]. In the case of surface measurements, the classical measurement white noise used in Kalman filtering is incompatible with the physical energy space and the state-parameter estimation needs to be reformulated [10].

3.1 Heart

Two heart pumps that propel blood into the pulmonary and systemic circulation are combined into a single muscular organ to synchronously beat. Each cardiac pump is composed of 2 chambers: the atrium and ventricle. Due to pressure differences between the vasculature entry and exit, which drive the blood flow through it, atria are auxiliary pumps that allow rapid ventricle filling, especially at rest when the cardiac frequency (f_c; or heart rate) is low. The atrial chambers generate a blood pressure increment upstream from the ventricle inlet.

The *cardiac output* (CO) is the blood volume pumped by each ventricle that crosses any point in the circulatory system per unit of time, in general taken as the time interval of 1 mn. In a healthy person at rest, the cardiac output ranges from 5 to 6 l/mn.

Cardiac output is determined by multiplying the *stroke volume* (SV), the blood volume pumped by the ventricle during 1 beat by the cardiac frequency. The stroke volume is related to the ventricular contraction force and blood volume.

Table 3.2 Physiological quantities at rest in healthy subjects: cardiac frequency (f_C) decreases and then increases with aging; stroke volume (SV), or systolic ejection volume (SEV), decreases with aging (blood flow rate $q \sim 6.5$ l/mn [~ 110 ml/s] at 30 yr old and $q \sim 4$ l/mn [~ 70 ml/s] at 70 yr old; EDV: end-diastolic volume; ESV: end-systolic volume)

EDV	70–150 ml
ESV	20–50 ml
SEV	50–100 ml
f_C	60–80 beats/mn, 1–1.3 Hz
q	4–7 l/mn (70–120 ml/s)
ejection fraction	60–80%

Table 3.3 Approximative blood compartment volume (%) relative to the systolic ejection volume (SEV 80 ml; total volume 4.4 l)

Pulmonary circulation	16.3
Arteries	5
Capillaries	0.8
Veins	10.5
Systemic circulation	38.7
Aorta	1.3
Arteries	5.6
Capillaries	3.7
Veins	28.1

The stroke volume is the difference between the end-diastolic volume (EDV) and end-systolic volume (ESV). The former is the maximal volume achieved at the end of ventricular filling; the latter the residual volume of the ventricle at the end of a systolic ejection (Table 3.2). The ratios of blood volume to stroke volume in serial compartments of both systemic and pulmonary circulations are shown in Table 3.3.

Various factors determine cardiac output. *Preload* corresponds to a stretching force exerted on the myocardium at the end of diastole, imposed by the ventricular blood volume and *afterload* to the resistance force to ejection. Moreover, heart contractility is affected by different regulatory molecules. As f_C increases, cardiac output rises until a critical f_C is reached; then it decreases. These factors can be combined.

The *cardiac index* (CI) is the ratio between blood flow rate (q) and body surface area ($2.8 < \text{CI} < 4.2$ l/mn/m^2). *Cardiac reserve* refers to the heart's ability to adjust quickly to immediate demands. It is defined by the maximum percentage of cardiac output, which in a healthy young adult is 300 to 400%.

The heart has a chaotic behavior. Its nonperiodic behavior characterizes a pump able to quickly react to any changes of the body's environment. The normal heartbeat indeed exhibits complex nonlinear dynamics. On the other hand, stable periodic cardiac dynamics yield a bad prognosis. A decay in random variability over time, which is associated with a weaker form of chaos, is indicative of congestive

Table 3.4 Duration (ms) of the four main phases of the cardiac (left ventricle) cycle ($f_c = 1.25$ Hz, i.e., 75 beats/mn)

Phase	Cycle timing	Duration	Starting event
IC	0–50	50	Mitral valve closure ECG R wave peak
SE	50–300	250	Aortic valve opening
IR	300–400	100	Aortic valve closure
VF	400–800	400	Mitral valve opening

heart failure [11]. This feature, positive with respect to heart function, is a handicap in signal and image processing, as ensemble averaging is used to improve the signal-to-noise ratio.

Most variability is due to diastole duration changes. The relative difference in mean systolic and diastolic durations reaches values of about 5% and 35% (with a ratio between the standard deviation and the mean of 0.1–0.2), respectively.

3.1.1 *Left Ventricle Cycle*

The cardiac cycle is a misnomer, as a part is taken for the whole (pars pro toto). The cardiac cycle, in fact, represents the cardiac ventricle cycle, more precisely that of the left ventricle, as the latter is the stronger chamber, because the systemic arterial pressure is higher than the pulmonary arterial pressure.

The heart beat carries out a 2-stage pumping action over a period of about 1 s or less: a longer first *diastole* and a *systole*. More precisely, the rhythm of the left ventricle activity consists of 4 main phases: (1) *isovolumetric relaxation* (IR), with closed atrioventricular and ventriculoarterial valves; (2) *ventricular filling* (VF), with open atrioventricular valves and closed ventriculoarterial valves; (3) *isovolumetric contraction* (IC), with closed atrioventricular and ventriculoarterial valves; and (4) *systolic ejection* (SE), with closed atrioventricular valves and open ventriculoarterial valves. Durations of these four phases of the cardiac cycle are given in Table 3.4. Both ventricular filling and systolic ejection are subdivided into accelerating and decelerating phases (rapid [RVF] and slow [SVF] ventricular filling and rapid [RSE] and slow [SSE] ejection). With atrial contraction (AC), mechanical events are divided into 7 phases.

Phase 1 (IC) is the onset of the ventricular systole and coincides with the R-wave peak. When the left ventricle begins to contract, the mitral valve is tightly closed to prevent a backflow of blood. The phonocardiogram shows the first heart sound (S1). Cardiomyocyte tension is developed proportionally to end-diastolic volume. The intraventricular pressure (p_V) quickly increases (Fig. 3.2), whereas both the mitral and aortic valves are closed. The closed ventricle volume is kept constant (incompressible blood) but ventricle shape varies. Isovolumetric contraction creates a rotational motion of the left ventricle wall. As the left ventricle contracts, blood

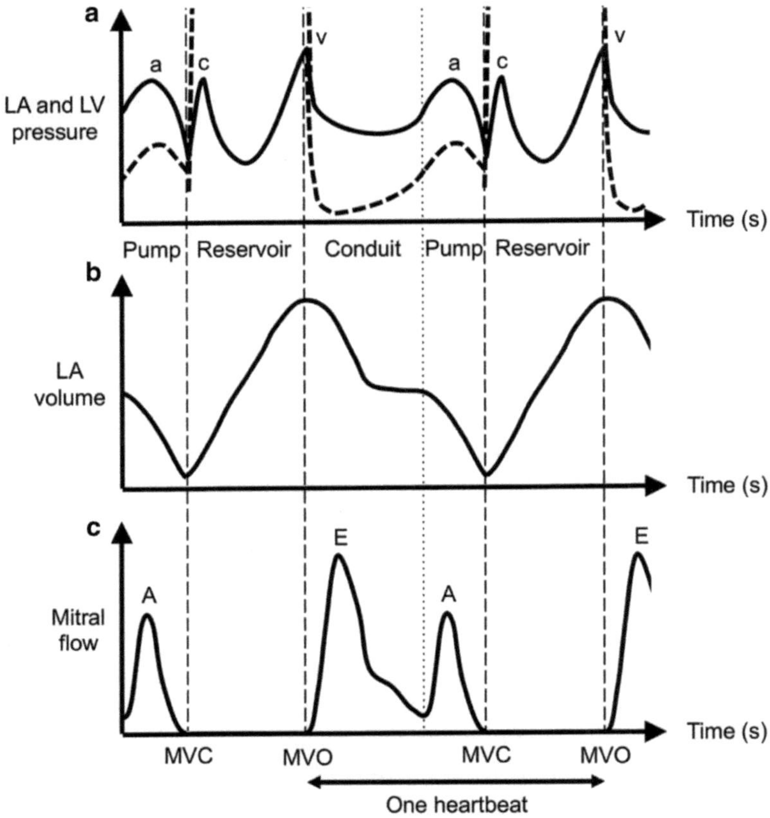

Fig. 3.1 Cyclically evolving left atrial (LA) and ventricular (LV) hemodynamic quantities (Source: [12]). (*Top*) Left ventricular (*dashed line*) and atrial (*full line*) pressures (a, c, and v waves). The a wave of LA pressure corresponds to atrial contraction, c wave to the small increase in LA pressure during early LV isovolumetric contraction, and v wave to the rise in LA pressure associated with LA filling. (*Middle*) left atrial volume. (*Bottom*) Transmitral flow with atrial contraction (A) and early LV filling (E) peaks (MVO[C]: mitral valve opening [closure]). The pulmonary venous blood flow waveform (not displayed) consists of an atrial reversal wave during LA contraction concomitant with A peak, a biphasic S wave during LV systole, and a D wave during opening of the mitral valve concomitant with transmitral blood flow E wave

impacts the closed mitral valve, which bulges into the atria, in its filling process, with a further intra-atrial pressure p_A increment: the c wave (Fig. 3.1). Blood does not regurgitate back into the atria because the parachute-like shape of the closed mitral valve is maintained by papillary muscles and chordae tendineae.[4]

[4]Valve cusps do not evert into the atrium during the ventricular systole by contraction of the papillary muscles, which are related to ventricular myocardium.

Phase 2 (RSE) occurs when p_V overcomes the end-diastolic aortic pressure p_a. The aortic leaflets are pushed open. Traces of p_a and p_V follow each other closely. They reach a maximum. Blood flow in the ascending aorta increases up to a peak, with a delay with respect to the p_V maximum. Aortic blood inflow is greater than the outflow due to the windkessel effect (blood storage in elastic arteries). From the p_V rounded summit, the intraventricular volume reduces quickly. After the c wave, x descent is observed on the p_A plot, which corresponds to the atrium diastole; the x event is imposed by the systolic downward displacement of the right ventricle base.

During phase 3 (SSE), p_a and p_V fall. The aortic blood flow reduces. Meanwhile, the venous return gradually fills the right atrium and the v wave is engendered on the atrial pulse. The beginning of the T wave appears on the ECG trace.

Phase 4 (IR) is characterized by a strong decline in p_V. The second heart sound (S2) and dicrotic notch of p_a trace are due to aortic valve closure. Flow rate in the aorta root exhibits a complete backflow, but blood regurgitation into the ventricles is prevented by VAV closure (both aortic and mitral valves are closed). The constant ventricular volume is the residual volume (ESV). Decline in p_V continues until $p_V < p_A$.

Phase 5 (RVF) starts with mitral valve opening, p_V becoming lower than p_A. The aortic valve remains closed. Atrial pulse tracing shows y descent, which is generated by AVV opening and ventricular filling. The third sound (S3) is recorded. Atrial blood quickly fills the ventricle. Pressures p_A and p_V are similar, with p_A slightly greater than p_V. This stage accounts for 80% of total ventricular filling.

Phase 6 (SVF) is also called *diastasis*. Pressures p_A and p_V rise gradually. The ventricular volume curve increases slowly and slightly. Additional blood is forced from the pulmonary veins into the left ventricle through the opened mitral valve. As blood is collected in the left ventricle, the sinoatrial node sends out the action potential, which leads to atrial contraction. Diastasis sends a small blood volume into the left ventricle (∼15% of ventricular filling) with a slight increase in p_V. Diastasis (duration ∼180 ms) is shortened by f_c increase.

Phase 7 (AC) is the final stage of the cardiac cycle. A large amount of blood (∼70% of the filling capacity) has already filled the ventricles prior to atrial contraction. The atrial systole contributes slightly to ventricular filling at resting f_c, but maintains cardiac output during exercise. The fourth heart sound (S4) recorded by the phonocardiogram and a wave of the atrial pulse result from atrial contraction. This a wave may be elevated by preload augmentation; the a-wave deceleration rate serves as an index of left atrial relaxation. The action potential travels in the ventricular myocardium. The following left ventricle systole begins when the ventricles are full of blood. The cycle begins again.

The heart wall is composed of a matrix with cells, fibers, and blood vessels. Microcirculation represents an important compartment for blood volume storage. Microcirculation filling, enhanced during diastole due to myocardium relaxation, may help the ventricle expansion by a straightening effect.

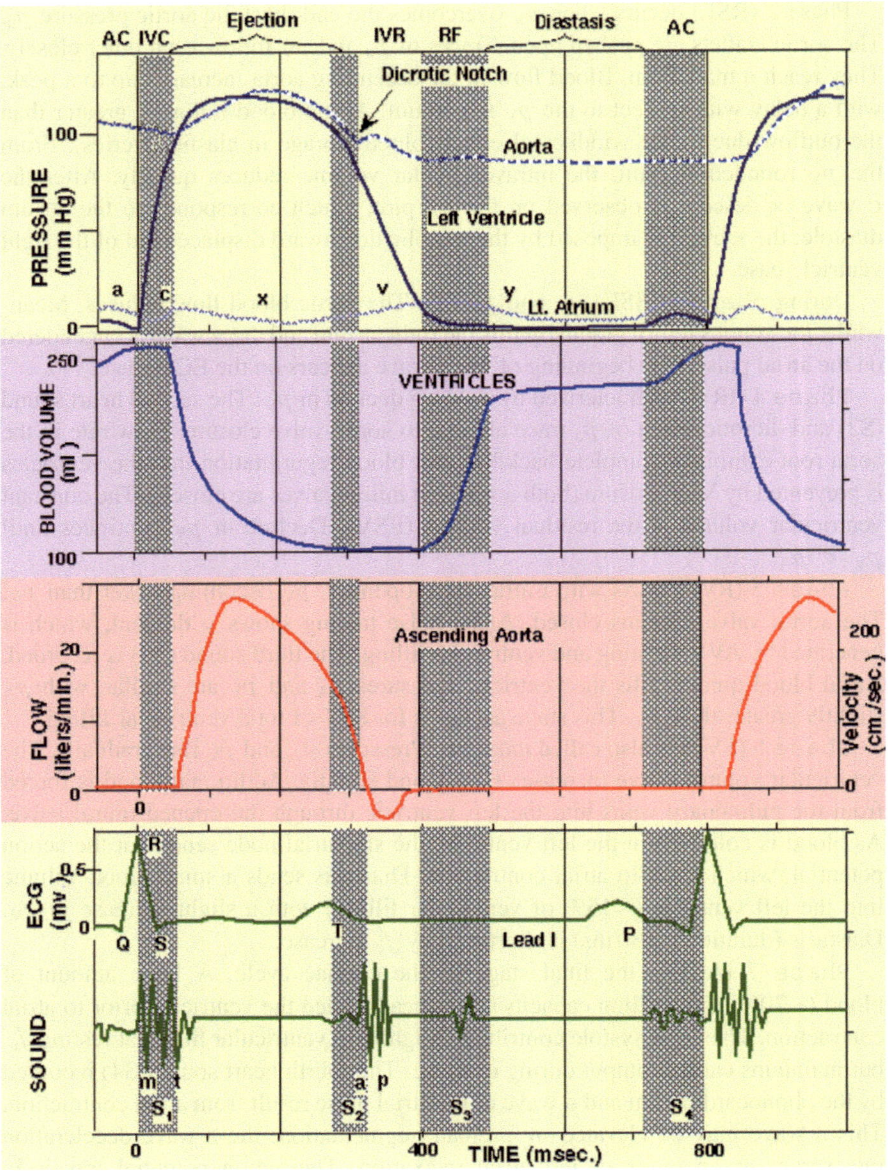

Fig. 3.2 Cardiac cycle. Evolution of the pressure in the left cardiac cavities and aorta (*top*), of left ventricle volume (*second row*), aortic flow (*third row*), and ECG trace and phonocardiogram (*bottom*; from [13])

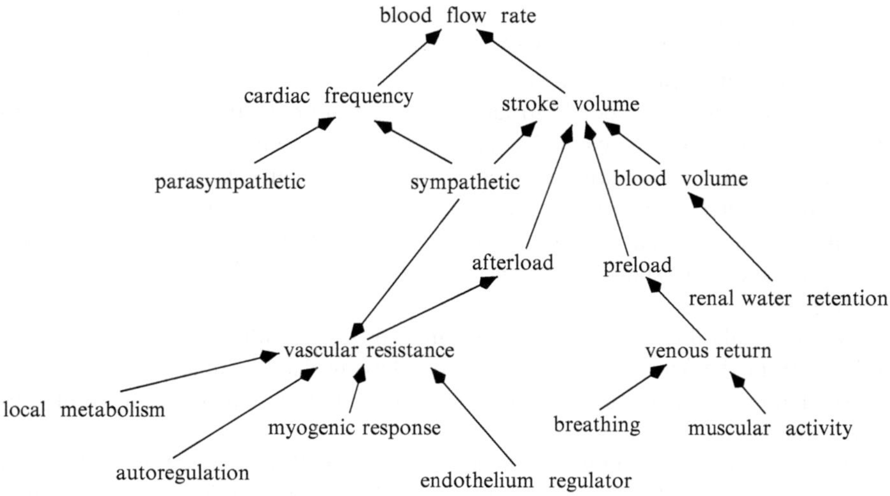

Fig. 3.3 Regulators of blood flow rate

3.1.2 Stroke Volume

Stroke volume can be modified by changes in ventricular contractility (rate of tension development), that is, force generation associated with sarcomere length prior to contraction (*Frank–Starling effect*)[5] and velocity of myofiber shortening (Fig. 3.3). Increased inotropy augments the time gradient of the ventricular pressure and thus the ejection velocity. Inotropy increases causes ESV reduction and SV increase, as displayed by pressure–volume loops. Increased stroke volume causes EDV reduction. Increases in inotropic state maintain stroke volume at high cardiac frequencies, which decreases stroke volume because of reduced time for diastolic

[5]In 1895, Frank found that under isovolumetric conditions, the larger the EDV, the greater the developed tension and pressure. Starling's later experiments demonstrated that the heart intrinsically responds to venous return (to EDV) increases by increasing the stroke volume (heart autoregulation). The relationship between EDV and stroke volume is associated with the relationship between sarcomere length and calcium ion influx and sensitivity. Myofilament length-dependent activation is explained by the separation distance between actin and myosin along the sarcomeric filament axis. The intrinsic ability of the heart to develop greater tension at longer myocardial fiber lengths over a finite range of fiber lengths is due to sliding filament arrangement in cardiomyocytes, with increase of cross-bridge number between actin and myosin filaments. The Frank–Starling mechanism refers to the heart's intrinsic capability of increasing inotropy and stroke volume in response to venous-return increase. The Frank–Starling effect describes static filling mechanisms in an isolated motionless heart. It works for high filling pressure and low flow rate (cardiac failure); however, cardiac functioning is an unsteady phenomenon.

filling and reduced EDV. When afterload increases, ESV initially rises and SV decreases. Afterward, increased ESV raises EDV, if venous return remains constant, and stroke volume can be restored.

The left ventricle responds to an increase in arterial pressure by rising contractility, hence SV, whereas EDV may return to its original value (*Anrep effect*). An increase in cardiac frequency also stimulates inotropy.[6] When the cardiac frequency is high, ion carriers (Vol. 3 – Chap. 3. Main Sets of Ion Channels and Pumps) are not efficient enough to remove all the cytosolic calcium which creates a positive inotropic state (*Bowditch effect*, *Treppe effect*, or frequency-dependent inotropy).[7]

Most signals that stimulate inotropy involve Ca^{2+}, either by increasing Ca^{2+} release from the sarcoplasmic reticulum following Ca^{2+} influx through specific plasmalemmal channels gated by the action potential or sensitizing troponin-C to Ca^{2+} ion.

In particular, calcitonin gene-related peptide (CGRP) increases Ca^{2+} influx and resting intracellular Ca^{2+} concentration as well as end-systolic pressure, ejection fraction, and velocities of contraction and relaxation, whereas it decreases stroke volume and work, and flow rate [14].

The heart is also able to manage excessive blood volume due to increased venous return through sympathetic stimulation with its positive chronotropic (C+) and inotropic (I+) effects.

Stroke work (SW) is depicted by the area inside the pressure–volume curve. It is approximated by the product of mean arterial pressure and stroke volume. Such estimations explain SW differences between left and right ventricles.

Cardiac efficiency is defined as the ratio of stroke work to myocardial oxygen consumption (SW/q_{O_2}). Various factors influence q_{O_2} especially wall tension. Increased inotropy and cardiac frequency augment q_{O_2}.

Systole and diastole are dynamically related. The systolic contraction provides heart recoil and energy that is stored for active diastolic dilation and aspiration [15]. Moreover, heart motion during systole[8] pulls large blood vessels and surrounding mediastinal tissues that react by elastic recoil. Heart diastolic rebound can participate in ventricular filling. The aspiration function of the heart during the diastole can be exhibited by a negative pressure zone on the pressure–volume loop (Fig. 3.4).

---

[6]This frequency-dependent enhanced contractility helps to offset the decreased ventricular filling time at higher cardiac frequencies by shortening the systole time duration, thereby increasing the time available for diastole.

[7]Positive chronotropy (C+) induces positive inotropy (I+).

[8]During systole, the heart moves downward.

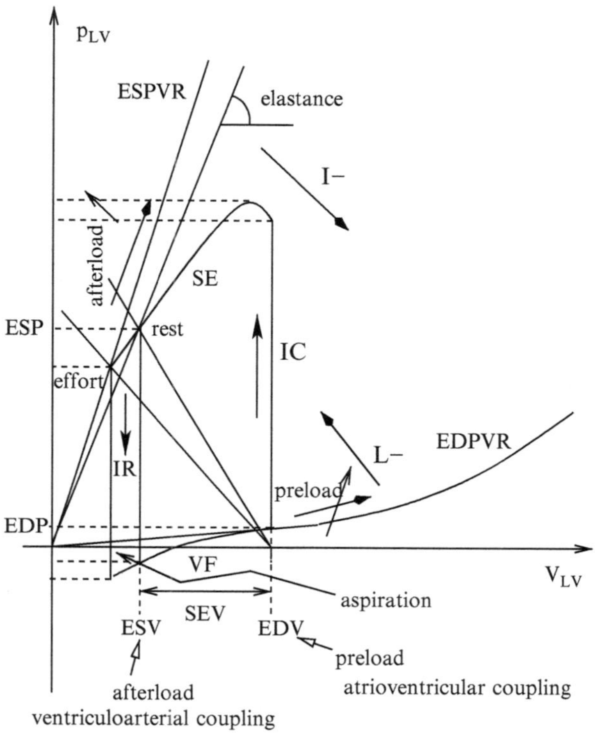

Fig. 3.4 Pressure–volume curve of the left ventricle. EDPVR slope increase is associated with decreased lusitropy (higher pressure with same stroke volume). When preload rises, EDV increases. EDPVR slope is the reciprocal of ventricular compliance. ESPVR gives the maximal pressure generated at a given inotropic state. EDV and EDP are indices of heart preload. ESPVR slope lowering and elevation represent decreased inotropy and increased afterload, respectively (adapted from [15])

3.1.3 Pressure–Volume Curve

At the nanoscale, heart contractility is associated with varying concentrations of Ca^{2+} (systolic calcium entry in the cytosolic and diastolic calcium reuptake) and ATP in the myocyte cytosol in response to the depolarization wave, associated with great and quick sodium ion fluxes (Na^+ do not interfere with intracellular involved processes) through Na^+ channels. Fuel ATP, produced by mitochondrial oxidative phosphorylation (Vol. 1 – Chap. 4. Cell Structure and Function and Sect. 3.1.7), is used to generate the mechanical energy with high efficiency, whatever the loading conditions.

At the macroscale, the physiological concept of heart contractility, the capacity of the myocardium to develp a contraction force whatever the preload or afterload, is depicted by the pressure–volume loops for given cardiac frequencies (Fig. 3.4). Measurements of the cardiac hemodynamics still require invasive techniques, at

Table 3.5 Pressures (kPa) in the cardiac cavities and large vessels (variability of literature data)

Site	Systolic	Diastolic	Mean
RA	0.2–1.3	0–0.5	0–1
RV	0.3–4.0	0–0.9	1.1
PA	1.3–4.0	0.7–2.1	1.1–2.7
LA	0.4–2.2	0.1–1.3	0.3–1.6
LV	10.4–18.2	0.3–1.6	5–5.3
Ao	10.4–18.2	7.8–11.5	11.3–13.3

Table 3.6 Influence factors on pre- and afterload

Heart load	Factors	Indices
Preload	Venous return Blood volume Venous tone LV ejection RA/LA aspiration Muscle contraction (inferior limbs) Respiration Tissue activity	EDV, EDP
Afterload	Downstream resistance Circulating hormones Neural activity Humoral factors (pH, p_{O_2}, p_{CO_2})	$p_{LV_{max}}$, $p_{Ao_{max}}$

least to get good estimates of intracavital pressures (Sect. 5.2.1). The systolic, diastolic, and mean values of the pressure in the 4 heart chambers are given in Table 3.5.

3.1.3.1 Ventricular Pressure–Volume Curve

Ventricular performance is displayed in clinical practice by the pressure–volume diagram (Fig. 3.4). Left ventricular pressure–volume loops illustrate 3 features: myocardial work, myocardial characteristics (inotropy and lusitropy), and blood circulation influences (pre- and afterload, Table 3.6). The volume range corresponds to the stroke volume. The end-diastolic pressure–volume relationship (EDPVR) depicts both venous return and lusitropy. The end-systolic pressure–volume relationship (ESPVR) represents both afterload and ventricle inotropy.

Different types of tests have been carried out: (1) when EDV changes and the afterload is kept constant (constant ESP), ESV remains constant whatever EDV; (2) when afterload varies (varying ESP) at constant EDV (constant EDP), the points {ESP, ESV} run along the ESPVR curve.

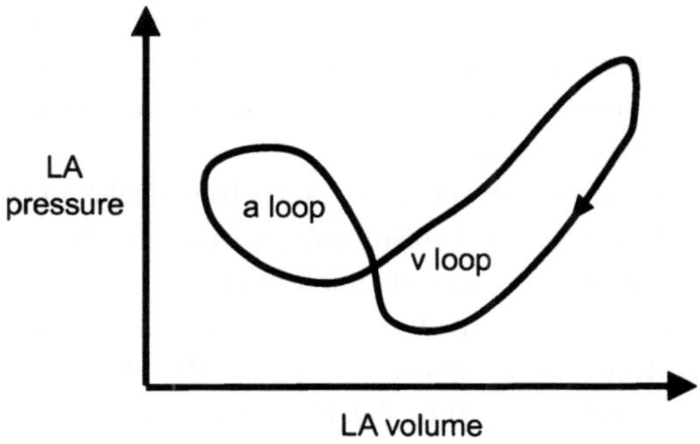

Fig. 3.5 Sketch of a left atrial (LA) 8-shaped pressure–volume diagram (Source: [12]). The left loop (active component [a loop]) that represents active LA contraction proceeds temporally in the trigonometric direction; the right loop (passive component [v loop]) that corresponds to LA reservoir function evolves in a clockwise direction. Left atrial end-diastolic pressure may be defined as the pressure occurring immediately before atrial contraction that corresponds to the LA end-systolic pressure

Left ventricle power can be estimated by the product of the ventricular pressure ($p_V = 13.3$ kPa) by the flow rate:

$$\frac{dV}{dt} \sim VES/T = 8 \times 10^{-5}/8 \times 10^{-1} = 10^{-4}\,\text{m}^3/\text{s}, \tag{3.1}$$

which gives a power of 1.33 W. The right ventricular power is about one-sixth of the left ventricular power.

3.1.3.2 Atrial Pressure–Volume Curve

The left atrium is a (1) reservoir that stores pulmonary venous return during left ventricular systole and isovolumic relaxation (after the closure and before the opening of the mitral valve); (2) conduit for blood flow from pulmonary veins to the left ventricle in early diastole driven by a pressure gradient when the atrioventricular valve is open (passive ventricular filling); and (3) contractile chamber that can actively fills the left ventricle in late diastole when time is available (active ventricular filling), thereby establishing the left ventricular end-diastolic volume.

The atrial pressure–volume plot is composed of 2 lobes (Fig. 3.5) that represent the reservoir and conduit functions at high volumes and active contraction at low volumes [12].

Once end-diastolic (EDP_A and EDV_A) and end-systolic (ESP_A and ESV_A) values identified on the left atrial pressure–volume diagram, left atrial stroke volume ($SV_A = EDV_A - ESV_A$) and emptying fraction (SV_A/EDV_A) can be computed.

After the mitral valve closure, the left atrial pressure and volume progressively increase during the left atrial filling as the left atrium expands (reservoir phase corresponding to the lower part of the A loop and upper portion of the V loop of the left atrial pressure–volume diagram).

The area of the A loop represents active left atrial stroke work [16]. The area of the V loop that represents the total elastic energy stored by the left atrium during the reservoir phase is an index of reservoir function [16].

The left atrial myocardial relaxation and resultant left atrial pressure reduction immediately after atrial systole during early isovolumetric contraction facilitates flow from the pulmonary veins into the left atrium. On the other hand, an end-diastolic retrograde pulmonary venous blood flow from the left atrium is caused by atrial systole. However, it is minimized by the peristaltic-like pattern of atrial contraction and the valve-like anatomy of apertures of the pulmonary veins into the left atrium at normal left atrial pressures [16].

Shortening of the left atrial myofiber depends mainly on preload and inotropic state controlled by the autonomic nervous system. The left atrial myocardium is less sensitive to elevated afterload than the left ventricular myocardium. Left atrial afterload is determined primarily by the left ventricular compliance and pressure.

Left ventricular contraction is a determinant of the early left atrial reservoir function. The cardiac base descends toward the apex during the left ventricular systole, thereby drawing additional blood from the pulmonary venous circulation into the left atrium.[9] Transmission of the right ventricular systolic pressure pulse through the pulmonary circulation, which is responsible for the second peak of the biphasic pulmonary venous S wave, contributes to left atrial pressure and volume elevation later during the reservoir phase.

Left atrial compliance determines reservoir and conduit function, as it supports venous return from the pulmonary circulation. Atrial diseases characterized by an altered compliance are associated with impaired filling. The left atrial appendage is more flexible than the principal atrial compartment [16]. During aging, the left atrial volume increases and its passive emptying decays.

The atrial myocardial β-myosin isoform[10] production is upregulated when left atrial mechanical work rises, thereby enhancing LA emptying fraction, but increasing the left atrial workload due to postload mismatch with the risk of subsequent contractile dysfunction [16].

[9]The mitral annulus descends about 1.3 cm during the left ventricular ejection phase in normal subjects [16].

[10]Cardiac β-myosin is a mechanoenzyme that converts the energy from ATP hydrolysis for muscle contraction. Two cardiac myosin heavy chains (α MHC–β MHC) have different levels of ATPase activity. The β MHC and α MHC subtypes are predominantly expressed in late fetal life and adults, respectively. The former is encoded by the MYH7 gene.

3.1.4 *Interferences Between Cardiac Cells in the Cardiac Function*

The cardiac physiology involves the interaction of various cell populations, cardiomyocytes (atrial and ventricular myocytes with distinct mural types), nodal cells (stretch-sensitive pacemaker [sinoatrial node] and conductors), fibroblasts, endotheliocytes, smooth myocytes, and neurons, as well as those between different compartments (i.e., the extracellular matrix and, at a greater length scale, cardiac parietal tunicae, nodal tissue, blood, lymph, and nerves).

Exosomes are lipidic vesicles (diameter 50–100 nm) implicated in intercellular communication. The protein content of cardiac exosomes differs from that released by other cell types; they contain cytosolic, sarcomeric, and mitochondrial proteins [17]. In particular, cardiomyocytes secrete heat shock protein HSP60 in exosomes. Extracellular HSP60 ouside exosomes causes cardiomyocyte apoptosis via activated Toll-like TLR4 receptor. However, exosomes retain their cargo in different conditions. Reactive oxygen species stimulate exosome production.

The heart is mainly constituted of a syncytium of cardiomyocytes (∼50% of the cardiac cell population) surrounded by fibers and other cell types, among which cardiofibroblasts predominate (40–60% of the cardiac cell population in the heart) [18]. Cardiofibroblast, a spindle-shaped cell with multiple processes, lacks a basement membrane [18].

Interference between intermingled cardiac myocytes and fibroblasts relies on paracrine factors liberated by both cell types as well as direct communication via nanotubes, pannexons, and gap and adherens junctions between both cell types. This crosstalk intervenes in mechanical and electrical function as well as structural remodeling in both normal and pathological hearts [18].

Intercellular communication between atrial and ventricular myocytes and fibroblasts uses hormones (e.g., angiotensin and endothelin-1), growth factors (e.g., CTGF, FGF2, HBEGF, PlGF, TGFβ, VEGF), cytokines (e.g., IL6, IL33, and TNFα), chemokines, and other soluble factors [18]. In addition to structural remodeling, paracrine factors can influence the expression and function of ion channels and gap junctions, hence action potential duration.

Pannexins form gap junction-like structures as well as channels in single membranes, rarely intercellular channels in apposed membranes. They facilitate paracrine signaling by releasing ATP and other small messengers to the extracellular space. Liberation of ATP and UTP from cardiomyocytes via pannexins mediated by subunits of the G12/13 subclass can activate $P2Y_6$ receptor on cardiofibroblasts, subsequently inducing transcription of profibrotic genes [18].

The plasma membrane near gap junctions contains voltage-dependent ion channels and other ion carrier types. In the absence of coupling to cardiomyocytes, cardiofibroblasts act as passive electrical insulators. Coupled with cardiomyocytes, they operate as current sources and/or sinks, thereby modulating excitation threshold,[11] depolarization, action potential duration (shortening in human ventriculomyocyte [18]), and conduction velocity in cardiomyocytes.[12]

Coupling via adherens junctions (i.e., the cadherin–catenin complex linked to actin and intermediate filaments) between cardiomyocytes supports myofibrillogenesis and maintenance of a functional cardiac phenotype. In fibroblasts, the adherens junction promotes the transformation into myofibroblasts. Between cardiac myocytes and fibroblasts, the adherens junction may assist in the activation of mechanosensitive channels and stretch-induced release of paracrine factors from cardiofibroblasts [18].

Long, thin membrane nanotubes, which contain actin and microtubules, enable intercellular communication over long distances between cardiac myocytes and fibroblasts and vice versa. In fact, they permit signal propagation as well as intercellular organelle transfer.

3.1.5 Cardiomyocyte Organelles and Coupled Metabolism, Energetics, Signaling, and Contraction

Transverse (T) tubules (TT) are orderly, deep invaginations of the sarcolemma (cell surface or plasma membrane) perpendicular to the cell surface and longitudinal axis of the cardiomyocyte. They are regularly arrayed along Z discs (longitudinal interval $\sim$1.8 μm; circumferential interval 0.5–1.5 μm), having openings within Z grooves (within 500 nm from the Z disc; Vol. 5 – Chap. 5. Cardiomyocytes). Transverse tubules conduct the action potential into the center of the cardiomyocyte for a rapid spread throughout the cell.

The cardiomyocyte contains about 6000 mitochondria that occupy 30 to 40% of the cell volume. Excitation–contraction coupling consumes large amounts of energy that need to be replenished by mitochondrial oxidative phosphorylation. Mitochondria not only supply locally ATP on demand (Sect. 3.1.7), but are also sources of reactive oxygen species that participate in local Ca^{2+} handling. Calcium ion activates some enzymes of the tricarboxylic acid cycle in the mitochondrial matrix (Tables 3.7 to 3.15).

---

[11]The resting transmembrane potential is higher (less negative) in cardiofibroblasts than that of cardiomyocytes. During action potential upstroke, the transmembrane potential of cardiomyocytes becomes higher than that of cardiofibroblasts, which then can act as current sinks, which slows down cardiomyocyte activation and can attenuate maximum upstroke velocity and peak amplitude of the action potential [18].

[12]The conduction velocity generally increases initially and then decreases when cardiofibroblast density and/or coupling increase.

Table 3.7 Mitochondrial structure and function (AlDH2: mitochondrial aldehyde dehydrogenase-2; CPS: mitochondrial carbamoyl phosphate synthase; MAO: monoamine oxidase; NAGS: ^{N}acetylglutamate synthase; OTC: ornithine carbamoyltransferase; SOD: superoxide dismutase)

Structural compartment	Function
	Mitochondrial envelope
Outer mitochondrial membrane (OMM)	Mass transfer, fatty acid degradation, tryptophan metabolism (kynureninase), monoamine metabolism (MAO)
Intermembrane space	Antioxidative processing $^{(Cu,Zn)}$SOD
	OMM–IMM connection
Inner mitochondrial membrane (IMM)	Oxidative phosphorylation (ATP synthesis), electron and H^+ transport, mass transfer, pyrimidine metabolism (dihydroorotate dehydrogenase)
Mitochondrial matrix	Tricarboxylic acid cycle, fatty acid oxidation, glucose oxidation
	Degradation of some amino acids
	Anaplerosis
	Antioxidative processing (^{Mn}SOD)
	Urea cycle (CSP1, OTC, NAGS)
	Alcohol metabolism (AlDH2)

Table 3.8 Mitochondrial structure and composition. (**Part 1**) Outer mitochondrial membrane (BAK: BCL2-antagonist killer; BCL2: B-cell lymphoma (leukemia) protein; CPT: carnitine palmitoyltransferase; DRP: dynamin-related protein; TOM: translocase of the outer membrane of mitochondria; VDAC: voltage-dependent anion channel). Examples of transported compounds include adenine nucleotide (SLC25a4–SLC25a6 and SLC25a31), citrate (SLC25a1), dicarboxylate (SLC25a10), glutamate (SLC25a22), ornithine (SLC25a2 and SLC25a15), oxoglutarate (SLC25a11), and phosphate (SLC25a3 and SLC25a23–SLC25a25)

Carriers	Enzymes	Miscellaneous
Porin (VDAC)	Mitofusins	DRP1, BAK, BCL2
SLC25A transporters TOM40	(MFn1–MFn2; GTPases), DuSP18, PTPn11, PHLPP1, CPT1	
	Long fatty acylCoA synthase	

Table 3.9 Mitochondrial structure and composition. (**Part 2**) Intermembrane space (BCKDH: branched-chain ketoacid dehydrogenase; PDHP: pyruvate dehydrogenase phosphatase; PInK: phosphatase and tensin homolog deleted on chromosome 10 (PTen)-induced kinase; SOD: superoxide dismutase)

Kinases	Phosphatases	Miscellaneous
PKA, PKCδ, PInK, Src, adenylate kinase creatine kinase	PDHP1, PPM1k, BCKDH	$^{(Cu,Zn)}$SOD

The sarcoplasmic reticulum (SR) occupies 1 to 3% of the cell volume, but creeps throughout the cytosol in the form of longitudinal tubules (free sarcoplasmic reticulum [fSR]) and terminal, pancake-shaped cisterns at Z lines (junctional sarcoplasmic reticulum [jSR]). It serves as Ca^{2+} store (~0.5 mmol; extracellular Ca^{2+} concentration ~1 mmol).

Table 3.10 Mitochondrial structure and composition. (**Part 3**) Inner mitochondrial membrane (ANT: adenine nucleotide translocator; AT: angiotensin-2 receptor; ClC: voltage-dependent chloride channel; CPT: carnitine palmitoyltransferase; G3PDH: glycerol 3-phosphate dehydrogenase; IMAC: inner-membrane anion channel; mtClIC: mitochondrial chloride intracellular channel; mtCU: mitochondrial Ca^{2+} uniporter; mtK_{Ca}: mitochondrial Ca^{2+}-activated K^+ channel; mtK_{ATP}: mitochondrial ATP-sensitive K^+ channel; mtKHE: mitochondrial K^+–H^+ exchanger; mtNCE: mitochondrial Na^+–Ca^{2+} exchanger; mtNHE: mitochondrial Na^+–H^+ exchanger; PiC: mitochondrial inorganic phosphate carrier; PMPc: peptidase, mitochondrial processing; PTP: permeability transition pore; RyR: ryanodine receptor; TIM: translocase of the inner membrane of mitochondria; UCP: uncoupling protein). Mono- (MCC) and tricarboxylate (TCC) carriers (as well as ANTs and UCPs) used for pyruvate ingress and citrate egress, respectively, pertain to the solute carrier superclass, in particular to the SLC25A category

Carriers	Enzymes	Miscellaneous
Electron carriers, H^+ pump, PTP, TIM, IMAC, SLC25A transporters, ANT1–ANT3	F_0F_1 ATP synthase, succinate dehydrogenase, cytochrome-C, ^{ETC}complex	Cardiolipin
mtUCP1–mtUCP4 mtCU, RyR1, $K_V1.3$	NOS1–NOS3	AT_2
mtK_{ATP}	CPT2	
mtK_{Ca}	G3PDH	
mtKHE, mtNCE, mtNHE	PMPcβ	
ClC, mtClIC, PiC, malate–aspartate carrier, glutamate–aspartate carrier, glycerol phosphate carrier		

Table 3.11 Mitochondrial structure and composition. (**Part 4**) Mitochondrial matrix (ETC: electron transport chain; ICDH: Ca^{2+}-activated isocitrate dehydrogenase; KGDH: Ca^{2+}-activated α-ketoglutarate dehydrogenase; PDHc: pyruvate dehydrogenase complex; PDHK: pyruvate dehydrogenase kinase; PDHP: pyruvate dehydrogenase phosphatase; ROS: reactive oxygen species; SOD: superoxide dismutase; TCA: tricarboxylic acid cycle)

Enzymes	Effect, input (i), output (o), regulators
TCA processors	Feeder of ETC for ATP production
	AcetylCoA (i) (glycolysis and fatty acid catabolism product)
	GTP; CO_2; NADH, $FADH_2$ (o)
	Ca^{2+} (ICDH and KGDH activitor)
PDHc	Pyruvate (i); acetylCoA (o)
	Inhibitory PDHK1–PDHK4 (inhibited by pyruvate, CoA, NAD^+, ADP, Ca^{2+})
	(activited by acetylCoA, ATP, NADH)
	(stimulated by high fatty acid levels)
	Activatory PDHP (activited by Ca^{2+} and Mg^{2+})
$^{(Cu,Zn)}$SOD	Limitation of ROS production

Mitochondria are close (40–180 nm) to coupling zones between the junctional sarcoplasmic reticulum and transverse tubule. They are juxtaposed to the junctional sarcoplasmic reticulum, being only separated by a tiny cleft. They can even be tethered to the junctional sarcoplasmic reticulum. Therefore, the trio formed by the

Table 3.12 Major inputs and outputs of the tricarboxylic acid cycle. Each turn of the cycle achieves the complete oxidation of 1 molecule of acetylCoA to form 2 molecules of carbon dioxide, 3 reduced nicotinamide adenine dinucleotide (NADH), 1 reduced flavin adenine dinucleotide ($FADH_2$), and 1 GTP. Both NADH and $FADH_2$ are mainly used in mitochondrial oxidative phosphorylation that yields ATP. Reoxidation of all reduced coenzymes by the respiratory chain yields 9 ATP. Molecule ATP is not produced in the tricarboxylic acid cycle

Input	Output
AcetylCoA	2 CO_2, CoA^{SH}
1 FAD	1 $FADH_2$
1 GDP, 1 P_i	1 GTP
3 NAD^+	3 NADH, 3 H^+

Table 3.13 Regulation of the tricarboxylic acid cycle. In addition, the pyruvate dehydrogenase complex is allosterically inhibited at high [ATP]/[ADP], [NADH]/[NAD^+], and [acetylCoA]/[CoA] ratios

Enzyme	Activators	Inhibitors
Citrate synthase	ADP	Citrate, succinylCoA, NADH, ATP
Isocitrate dehydrogenase	ADP, NAD^+, Ca^{2+}	ATP, NADH
α-Ketoglutarate dehydrogenase	Ca^{2+}	SuccinylCoA, NADH, ATP, GTP
Succinate Dehydrogenase	Succinate, ADP, P_i	Oxaloacetate

junctional sarcoplasmic reticulum, transverse tubule, and mitochondrion, represents a major structural unit of the cardiomyocyte. In particular, it serves as a structural, functional, and regulatory unit of Ca^{2+} and ROS signaling in the cardiomyocyte.

The coregulation of metabolism, energetics, and contraction is linked to transmission of Ca^{2+} signals to mitochondria, that is, change in mitochondrial Ca^{2+} concentration rather than changes in cytosolic Ca^{2+} concentration subsequently to mitochondrial Ca^{2+} uptake [19].

Mitochondria contribute to Ca^{2+} nanodomains at associated contacts between junctional sarcoplasmic reticulum and transverse tubule. Mitochondrial Ca^{2+} uniporter accounts for the rapid beat-to-beat changes. Mitochondrial Na^+–Ca^{2+} exchanger and Ca^{2+}–H^+ antiporter may participate in the slow, pacing-dependent accumulation of Ca^{2+} in the mitochondrial matrix [20]. Mitochondrial Ca^{2+} ion contributes in the regulation of mitochondrial ROS production.

Table 3.14 Mitochondrial enzymes of the tricarboxylic acid cycle used by aerobic organisms to generate energy via oxidization of acetate derived from carbohydrates, fatty acids, and proteins (**Part 1**). Two allosteric enzymes help to controlling the distribution of metabolites and the overall cycle flux. Isocitrate dehydrogenase is inhibited by ATP, NADH, and NADPH, and activated by ADP and Ca^{2+} ions. Oxoglutarate dehydrogenase is also activated by calcium. Involved enzymes are soluble proteins in the mitochondrial matrix, except integral membrane succinate dehydrogenase. The latter uses FAD as a prosthetic group, but the other 3 oxidation steps use NAD as coenzyme. Four dehydrogenase reactions exist. Reoxidation of $FADH_2$ for succinate dehydrogenase by the respiratory chain forms 1.5 ATP/mol. Malate dehydrogenase generate 2.5 ATP/mol when NADH is recycled to NAD by the respiratory chain. Oxoglutatarate and especially isocitrate are even better reducing agents

Enzymes	Substrates	Cofactors	Products
Aconitase	Citrate, H_2O		Aconitate (intermediate), isocitrate, H_2O
Citrate synthase	Oxaloacetate, acetylCoA, H_2O		Citrate CoA^{SH}
Fumarase	Fumarate, H_2O		Malate
Isocitrate dehydrogenase-3	Isocitrate	NAD^+ Mg^{2+}	Oxalosuccinate, NADH, H^+ α-Ketoglutarate, CO_2
Ketoglutarate dehydrogenase	α-Ketoglutarate, CoA^{SH}	NAD^+	SuccinylCoA NADH, H^+, CO_2
Malate dehydrogenase	Malate	NAD^+	Oxaloacetate, NADH, H^+

3.1.5.1 Excitation–Contraction Coupling, Diad, Couplon, and Calcium

Extensive intracellular contacts exist between the sarcoplasmic reticulum and transverse tubules. The TT–jSR coupling is named a *diad*[13] (diadic cleft width 12–15 nm; bore 100–200 nm [21,22]).[14] The diad enables the strong and rapid influx of Ca^{2+} ions in the cytosol of cardiomyocytes.

Excitation–contraction coupling that generates a proper contractile force and propels an adequate blood volume to irrigate the body's tissues indeed relies on localized signaling that results from twinning of clusters of sarcolemmal $Ca_V1.2$ channels of transverse tubules with juxtaposed clusters of sarcoplasmic reticulum ryanodine receptors (RyR), which are in fact ryanodine-sensitive Ca^{2+} channels. Ryanodine receptor undergoes post-translational modifications, such as phosphorylation, nitrosylation, and oxidation. The latter opens the channel pore. Homotetrameric RyR also functions as a stretch sensor and communicator with mitochondria.

[13] δυας: couple, pair, binary number.

[14] In the myocardium, a diad means that a T tubule is associated with a single terminal cisterna. On the other hand, in the skeletal muscle, a triad is formed by a T tubule flanked on either side by the junctional sarcoplasmic reticulum, at the level of the Z line.

Table 3.15 Mitochondrial enzymes of the tricarboxylic acid cycle used by aerobic organisms to generate energy via oxidization of acetate derived from carbohydrates, fatty acids, and proteins (**Part 2**; ETC: electron transport chain; P_i: inorganic phosphate [free phosphate ion]). Water-soluble biotin is also called vitamin-B7. The pyruvate dehydrogenase complex comprises 3 enzymes, pyruvate dehydrogenase (PDH [E1]), dihydrolipoamide acetyltransferase (DLAT [E2]), and dihydrolipoamide dehydrogenase (DLD [E3]), 5 cofactors (CoA, NAD^+, FAD^+, lipoic acid, and thiamine pyrophosphate [TPP]). PropionylCoA is the end product of metabolism of odd-chain and most methyl-branched fatty acids. MethylmalonylCoA derives primarily from propionylCoA from the catabolism of isoleucine, valine, threonine, methionine, thymine, cholesterol, and odd-chain fatty acids

Enzymes	Substrates	Cofactors	Products
MethylmalonylCoA mutase	MethylmalonylCoA	Vitamin-B12	SuccinylCoA
MethylmalonylCoA epimerase	^{R}MethylmalonylCoA	Vitamin-B12	^{S}MethylmalonylCoA
Oxoglutarate dehydrogenase	Oxoglutarate	NAD^+ CoA^{SH}	SuccinylCoA, NADH, CO_2
PropionylCoA carboxylase	PropionylCoA HCO_3^-	Biotin, ATP	MethylmalonylCoA ADP, P_i
Pyruvate dehydrogenase	Pyruvate CoA^{SH}	FAD NAD^+	AcetylCoA, FADH, NADH, H^+
Succinate dehydrogenase (ETC)	Succinate Ubiquinone	FAD	Fumarate, $FADH_2$, ubiquinol
Succinate thiokinase	SuccinylCoA, P_i	GDP Mg^{2+}	Succinate, GTP, CoA^{SH}
SuccinylCoA synthase	SuccinylCoA P_i	GDP	Succinate, GTP, CoA^{SH}

Diads are perpendicular to the TT and jSR membrane plane. They have variable shapes and sizes. Mutually connected $Ca_V1.2$ and RyR2 channels form *couplons*. Array of ryanodine-sensitive Ca^{2+} channels primed by the Ca^{2+}-induced Ca^{2+} release (CICR) mechanism function as intracellular *Ca^{2+} release units* (CRU). Cytosolic Ca^{2+} periodically varies from about 100 nmol at diastole to about 1 μmol [23].

Diads contain RyR2 clusters that range from one to hundreds of tetramers, which can adopt numerous orientations [24]. In some diads, no couplons are formed; in others, only a fraction of the clusters constitutes couplons.

The estimated number of free Ca^{2+} ions in a single diad equals approximately 10 to 100 ions (10^2–10^3 μmol). The Ca^{2+}-induced Ca^{2+} release in a single diad may be mediated by 20 to 50 Ca^{2+} ions [25]. Using a model of stochastic molecular signaling between $Ca_V1.2$ and RyR2 channels with the description of the diad geometry and structure (configuration of main diadic proteins, location of Ca^{2+} binding sites, and membrane surface charges) and movement of individual Ca^{2+} ions within the diad, $Ca_V1.2$–RyR2 signaling was shown to be influenced by both the stochastic dynamics of Ca^{2+} ions in the diad as well as the shape and relative positioning of diad proteins [25]. The relative RyR2 placement and shape enable funneling of Ca2+ ions to RyR2 binding sites, thus increasing the excitation–contraction coupling gain.

Diads differ by their constitutive transporters and exchangers. Other important signaling components are also condensed on T tubules; they include Na^+–Ca^{2+} exchanger operating in reverse or forward mode, voltage-gated sodium channels $Na_V1.1$, $Na_V1.3$, and $Na_V1.6$,[15] inwardly rectifying K^+ channels ($K_{IR}2.1$, $K_{IR}2.2$, and $K_{IR}2.3$ homo- or heteromeric tetramers),[16] Na^+–K^+ ATPase, and β-adrenergic receptor. Moreover, T tubules accumulate numerous effectors of β AR signaling, such as ACases, PDE4, and PKA enzymes as well as AKAP scaffolds that can build signalosomes with $Ca_V1.2$ channels [29].

The junctional membrane compartment is responsible for quasi-autonomous elementary Ca^{2+} release (e.g., Ca^{2+} sparklets and Ca^{2+} sparks in the cytosol as well as Ca^{2+} blinks in the lumen of the SR cistern). Once these signals are synchronized at the cell level by a sarcolemmal excitation (depolarization upon the periodic arrival of an electrochemical impulse, the action potential), the spatial summation of synchronous local Ca^{2+} sparks and rapid diffusion of Ca^{2+} ions manifest as a uniformly distributed Ca^{2+} signal that support excitation–contraction coupling. The organized architecture of the T-tubular network with its array of ion carriers indeed enables the instantaneous and synchronous activation of nearly all Ca^{2+} release units throughout the lenghty ventriculomyocyte [22]. The synchronized Ca^{2+} release allows the coordinated contraction of all contractile (sarcomeric) units within the ventriculomyocyte. The synchronized myofilament contraction of myofibers permits the myocardium contraction.

Transverse tubules extend from the cell surface deep into the cell with branching and axial invaginations of the tubular membrane that all participate in signaling, thereby setting up a three-dimensional membrane network, the *transverse and axial tubular network* (TATN) [21]. The primary transversal tubules aligned with sarcomeric Z discs throughout the entire cardiomyocyte are interconnected with longitudinal branches of T tubules that run between 2 adjoining Z discs. The

---

[15]The principal cardiac pore-forming α subunit isoform $Na_V1.5$ preferentially localizes to intercalated discs, whereas the brain-type α subunit isoforms $Na_V1.1$, $Na_V1.3$, and $Na_V1.6$ reside in transverse tubules [26]. They contribute to the coupling of sarcolemmal depolarization to contraction. On the other hand, $Na_V1.5$ in intercalated discs is primarily responsible for action potential conduction between cardiomyocytes.

[16]Cardiac inward rectifier K^+ currents (i_{K1}) through channels of the $K_{IR}2$ category participate in the maintenance of resting membrane potential as well as late phase repolarization. In rabbits, $K_{IR}2.1$ and $K_{IR}2.2$, which lodge in T tubules, but not $K_{IR}2.3$, are synthesized in ventriculomyocytes [27]. Current i_{K1} is predominantly due to $K_{IR}2.1$–$K_{IR}2.2$ heterotetramers. In mice, i_{K1} also crosses $K_{IR}2.1$–$K_{IR}2.2$ heterotetramers. In guinea pigs, $K_{IR}2.1$, $K_{IR}2.2$, and $K_{IR}2.3$, but not $K_{IR}2.4$ are produced in ventriculomyocytes [28]. Three different inward rectifier conductances are linked to $K_{IR}2.1$, $K_{IR}2.2$, and $K_{IR}2.3$ homotetrameric channels, intermediate-conductance $K_{IR}2.1$ and large-conductance $_{IR}2.2$ being the primary determinants of i_{K1} current with little contribution from low-conductance $K_{IR}2.3$ channel [28]. In humans, $K_{IR}2.1$ resides in ventriculomyocytes, $K_{IR}2.1$ and $K_{IR}2.2$ in atrial cells, and $K_{IR}2.3$ in ventricular cells (cells including not only cardiomyocytes, but also endotheliocytes, vascular smooth myocytes, cardiofibroblasts, and neurons, among others). In the guinea pig heart, $K_{IR}2.1$, $K_{IR}2.2$, and $K_{IR}2.3$ are expressed in both cardiomyocytes and capillary endotheliocytes. Subunit $K_{IR}2.4$ is restricted to cardiac parasympathetic and postganglionic sympathetic neurons as well as sensory nerve fibers [28].

length scale of this membrane network composed of transversal and longitudinal components ranges over several orders of magnitude, from local T-tubule caliber of order 100 nm[17] to cell-wide expansion of 100 μm,[18] which provides thousands of junctional contacts with the sarcoplasmic reticulum in a typical ventriculomyocyte.

Ventriculomyocytes develop a transverse and axial tubular system mainly after birth. During the development of the fetal and neonatal heart, the balance between synthesis and degradation favors synthesis. In the adult healthy heart, the equilibrium maintains the T-tubular network. With increased mechanical strain, oxidative stress, and cardiac injury, degradation overcomes synthesis, thereby causing a progressive loss of T-tubule density and organization.

Atriomyocytes, at least in some species, can have a more or less developed transverse and axial tubular system. Atriomyocytes have a heterogeneous distribution of the tubular system, with the longitudinal components being more prominent [24].

Among nodal cells, Purkinje fibers possess T tubules that form diads [22].

Junctophilin JP2, the single isoform synthesized in the cardiomyocyte among the family members (JP1–JP4), bridges the plasma membrane to the sarcoplasmic reticulum membrane [22].[19] This docking protein localizes near ryanodine receptors in the diadic cleft. This structural factor contributes to the membrane coupling between T tubules and junctional sarcoplasmic reticulum (i.e., diads). Effect of stretch that influences the T-tubule geometry during contraction–relaxation cycle is counteracted by stabilizers of the TT–jSR coupling such as anchoring protein JP2 that maintains a stable positioning of T tubules with respect to the sarcoplasmic reticulum.

Other proteins are implicated in T-tubule formation or remodeling, such as caveolin-3, amphyphisin-2, and telethonin. Caveolin-3 clusters are situated adjacent to RyR2 clusters. Amphyphisin-2 directly binds to membrane phospholipids and supports tubulogenesis. In fact, both amphyphisin-2 and $Ca_V1.2$ channel influence T-tubule formation. Amphyphisin-2 also anchors microtubules to the T-tubule network, hence targeting $Ca_V1.2$ to T tubules [29]. Telethonin is a load-sensitive regulator of T-tubule structure and function [22]. This proteic anchor and mechanical transducer between the myofilament and Z-disc molecular complex also assists transverse tubulogenesis [29].

---

[17]The lumen of T tubules varies within a given mammalian species. Using confocal microscopy, it was assessed to be about 400 nm in humans [22]. However, stimulated emission depletion (STED) imaging that has a better spatial resolution detects a lower caliber.

[18]Although the volume density of the T-tubule network is only 1 to 3%, it represents about one-third of the entire plasma membrane area [22]. Transverse tubule density varies among ventriculomyocytes from different animal species.

[19]$^{P110}\alpha$PI3K and $^{P110}\beta$PI3K are required for the maintenance of the organized network of T tubules, as they regulate junctophilin-2 localization [30]. The 4 junctophilins tether the plasma membrane to the endoplasmic reticulum in excitable cells. The major cardiac junctophilin isoform JP2 has a C-terminal transmembrane domain that anchors the protein in the SR membrane and 8 N-terminal membrane occupation and recognition nexus (MORN) motifs that interact with the plasma membrane to stabilize the junction between the plasma and SR membrane.

Calsequestrin, triadin, and junctin also maintain the structural and functional integrity of the diad. Triadin preserves the structure of the $Ca_V1.2$–RyR2 couplon [31].

Calcium fluxes through $Ca_V1.2$ channels dedicated to contraction are discriminated from those used for other types of Ca^{2+} signaling originating from differently localized $Ca_V1.2$ channels as well as from IP_3 receptors, signal selectively resulting from distinct effectors in different nanodomains such as diads and caveolae [23].

3.1.5.2 Mitochondrial Calcium Nanodomain

During a heartbeat, about 2% of the cellular ATP is consumed. The entire ATP pool of cardiomyocytes is turned over within less than a minute [32]. Various factors, among which Ca^{2+} and ADP, control the mitochondrial redox state and electron flux along the respiratory chain, orchestrating oxidative phosphorylation in response to constantly changing workloads and securing a permanent ATP availability. Mitochondrial Ca^{2+} signaling relies on numerous regulators.

Communication between different organelles and cellular compartments in the cardiomyocytes is organized by nanodomains that enable signaling efficacy. Among organelles, mitochondria represent about a third of the cell volume. They are aligned regularly along the ATP-consuming sarcomeres. They are connected to the sarcoplasmic reticulum (Ca^{2+} store). They are close to diads. Apposition of mitochondria to the endoplasmic reticulum supports communication between these 2 organelles, in particular transfer of lipids and exchange of calcium that regulates endoplasmic reticulum chaperones, mitochondrial ATP production, and apoptosis.

A Ca^{2+} nanodomain is defined by a cellular site close to a Ca^{2+} release zone in which the Ca^{2+} concentration can exceed that in the bulk cytosol by several orders of magnitude for a brief period. Calcium signal peaks are limited spatially and temporarily, as Ca^{2+} ions diffuse away from these spots.

Calcium nanodomains exist between the sarcoplasmic reticulum and T tubule (diad) as well as mitochondria; where Ca^{2+} transfer between organelles is facilitated to match the mitochondrial energetics to the demand generated by cellular processes.

The mitochondrial Ca^{2+} nanodomain refers to the tethering of mitochondria to the sarcoplasmic reticulum that enables high local Ca^{2+} concentrations to overcome the low Ca^{2+} affinity of Ca^{2+} uniporter, which provides the main mitochondrial Ca^{2+} uptake [32]. In cardiomyocytes, the distance between the calcium release unit (RyR2) and the mitochondrial surface ranges between 37 and 270 nm. At this distance, during Ca^{2+} release, Ca^{2+} concentration is about 30 μmol/l.

Mitofusins MFn1 and MFn2 of the outer mitochondrial membrane connect adjoining mitochondria by forming homo- and heterotypic complexes. Mitofusin-2 also localizes to the cardiac sarcoplasmic reticulum. Hence, mitofusin also tethers the mitochondrion to the sarcoplasmic reticulum [32].

Other tethering proteins between the endoplasmic reticulum and mitochondrion include the mitochondrial distribution and morphology MDM complex and phosphofurin acidic cluster sorting protein PACS2.

Mitochondrial Calcium Uptake and Release

Calcium flux across the mitochondrial envelope regulates cellular energetics and signaling and can activate cell death pathways. Various proteins are involved in the transmission of Ca^{2+} between the sarcoplasmic reticulum and mitochondria, among which proteins of the inner (IMM) and outer (OMM) mitochondrial membrane.

Mitochondria take up Ca^{2+} primarily via calcium uniporter (mtCU) in the inner mitochondrial membrane (conductance 5×10^6 Ca^{2+}/s; density 10–40/μm^2) [32]. It oligomerizes and interacts with mitochondrial regulators of calcium uniporter MiCU1 to MiCU3, which have distinct expression patterns [33]. Subtype MiCU1 sets a Ca^{2+} threshold for mitochondrial Ca^{2+} uptake without changing the kinetic properties of mtCU-mediated Ca^{2+} uptake, thereby avoiding Ca^{2+} overload. Another regulator that binds to mtCU is the integral membrane mitochondrial calcium uniporter regulator MCUR1 that supports oxidative phosphorylation [34].

Several other carriers and factors contribute to mitochondrial Ca^{2+} uptake, such as coenzyme-Q10, RyR1, uncoupling proteins UCP2 and UCP3, and leucine zipper EF-hand motif-containing transmembrane LeTM1 protein.

Voltage-dependent anion channel (VDAC) of the outer mitochondrial membrane, which is linked to the endoplasmic reticulum inositol trisphosphate (IP_3)-sensitive Ca^{2+} release channel by 70-kDa heat shock protein HSPa9,[20] participates in Ca^{2+} transfer from the sarcoplasmic reticulum to mitochondria. Subtype VDAC2 interacts with RyR2 isoform, thereby building sarcoplasmic reticulum–mitochondrion junctions.

In cardiomyocytes, the energetically expensive mitochondrial Ca^{2+} efflux primarily is done through dimeric Na^+–Ca^{2+} exchanger with a K_M for an intracellular Na^+ concentration of about 8 mmol (in the physiological level range in cardiomyocytes) [32]. It carries 3 Na^+ or Li^+ in exchange for 1 Ca^{2+} ion. Its action is counterbalanced by mitochondrial Na^+–H^+ exchanger.

Mitochondrial regions located close to diads (distance <500 nm) experience Ca^{2+} flux of 37-nmol cytosolic Ca^{2+} transient, whereas in other mitochondria areas Ca^{2+} concentration increases to 26 nmol per cardiac cycle [32]. Furthermore, the mitochondrial Ca^{2+} concentration reaches a peak earlier in a zone close to the Z line than elsewhere in the mitochondrion, although mtCU is evenly distributed along the mitochondrial envelope.

Two different modes of mitochondrial Ca^{2+} uptake exist: a ^{mtCU}mode 1 that enables a large and rapid transfer of free Ca^{2+} ions, hence controlling the tri-

[20] A.k.a. mortalin Mot2, peptide-binding protein PBP74, mitochondrial stress protein-70, 75-kDa mitochondrial heat shock protein mtHSP75, and 75-kDa glucose-regulated protein GRP75.

carboxylic acid cycle dehydrogenase activation for energy supply-and-demand matching, and a slow ^{mtCU}mode 2 with lower Ca^{2+} affinity that takes up Ca^{2+} amounts leading to relatively small changes of mitochondrial Ca^{2+} concentration [32]. The latter mode is related to a phosphate-dependent buffer.

Mitochondrial Calcium Effect

In the mitochondrial matrix, the tricarboxylic acid cycle produces NADH and $FADH_2$ that deliver electrons into the electron transport chain via ^{ETC}complex-I and -II. Sequential redox reactions along the electron transport chain promote the translocation of H^+ across the inner mitochondrial membrane, thereby creating a proton and electrical gradient that constitute the proton motive force used to regenerate ATP from ADP by the F_1F_0 ATP synthase.

The tricarboxylic acid cycle regenerates NADH required for oxidative phosphorylation at the electron transport chain. On the other hand, NADPH is regenerated by 3 enzymes, substrates of which derive from products of the tricarboxylic acid cycle: isocitrate dehydrogenase (ICDH), mitochondrial $NADP^+$-dependent malic enzyme (ME), and nicotinamide nucleotide transhydrogenase (NNT) [32].

Calcium ion stimulates the electron transport chain, F_1F_0 ATP synthase, and the aspartate–glutamate carrier, thereby further contributing to the acceleration of oxidative phosphorylation caused by ADP molecule.

In the mitochondrial matrix, Ca^{2+} ion stimulates 3 rate-limiting enzymes of the tricarboxylic acid cycle (isocitrate dehydrogenase, α-ketoglutarate dehydrogenase, and the pyruvate dehydrogenase complex) to accelerate the regeneration of oxidized NAD^+ and FAD to reduced NADH and $FADH_2$ form. Calcium uptake in mitochondria also governs mitochondrial ROS synthesis (Sect. 3.1.5.3). Therefore, mitochondrial Ca^{2+} uptake not only allows to match energy supply to demand, but also to keep NADPH in a reduced state to prevent overproduction of hydrogen peroxide (H_2O_2) [32].

3.1.5.3 Signaling by Reactive Oxygen and Nitrogen Species

In the absence of excessive production, reactive oxygen and nitrogen species (Vol. 4 – Chap. 10. Other Major Signaling Mediators)[21] operate as signaling messengers, in addition to other second messengers (e.g., Ca^{2+}, cAMP, IP_3, and arachidonic acid). The intracellular redox balance is safeguarded by ROS scavengers and antioxidants. Molecular players in ROS and RNS generation localize to T tubules, mitochondrial outer and inner membranes, jSR membrane, and cytosol.

[21]Reactive oxygen species comprise free radicals with an unpaired electron (e.g., membrane-impermeable $O_2^{\bullet-}$, and extremely short lifetime $OH^{\bullet}$) and nonradical derivatives (e.g., more stable, membrane-permeable H_2O_2). The main reactive nitrogen species is the free radical $NO^{\bullet}$.

In cardiomyocytes, ROS can be generated by mitochondrial respiratory oxidases of the electron transfer chain (especially ^{ETC}complex-I and -III that generate $O_2^{\bullet -}$ in the mitochondrial matrix and on both sides of the inner mitochondrial membrane, respectively), sarcolemmal (and T-tubular) NADPH oxidases,[22] and cytosolic enzymes of the arachidonic acid pathway (lipoxygenases [LOx] and cyclooxygenases [COx]), cytochrome-P450S, and xanthine oxidase (XOx). Furthermore, translocation of monoamine oxidase (MAO) and NOx4 to mitochondria may contribute to ROS production.

Cardiomyocytes synthesize the 3 nitric oxide synthase isoforms (NOS1–NOS3).[23] The free radical $NO^{\bullet}$ reacts with superoxide ($O_2^{\bullet -}$) to form peroxynitrite ($ONOO^-$), a ROS and RNS species.

Reactive oxygen and nitrogen species may be transferred through mitochondrial channels, such as inner membrane anion channel, mitochondrial permeability transition pore, and voltage-dependent anion channel.

Low ROS concentrations increase Ca^{2+} spark frequency, whereas higher levels suppress Ca^{2+} spark. A moderate increase in ROS production activates RyR channels. At diads, ROS and RNS signaling have different temporal dynamics. A stepwise stretch (static 8% stretch) of cardiomyocyte produces an immediate, short-lived (approximately 10-s duration), ROS transient produced by NOx2 that may be involved in mechanotransduction (^{NOx}ROS transient) [20]. This process sensitizes Ca^{2+}-induced Ca^{2+} release, the stretch increasing the number of Ca^{2+} sparks. On the other hand, a stretch-activated, $NO^{\bullet}$ production can be detected 10 mn after a single-step static 10% stretch.

Cardiac tissue traction and compression generates 2 acoustic wave types, a compressional or tensile (P wave) and a shear (S wave) wave, that propagate at about 1550 m/s and 1 to 5 m/s, respectively [20]. The P wave generated by myocardial deformation during the previous cycle traverses the entire myocardium faster than the propagation of action potential, hence triggering ^{NOx}ROS production that yields an electromechanical coupling gain.

In addition, mitochondria create intermittent, quantal, 10-s $O_2^{\bullet -}$ burst, the so-called *superoxide flash* [20]. Superoxide flashes constitute elemental, local ROS signaling. They are triggered by the opening of a mitochondrial permeability transition pore upon mitochondrial membrane depolarization. They are characterized by a low frequency, brief duration, and spatial confinement, as they do not normally propagate. They can undergo minor-to-moderate changes in amplitude and duration. The resulting ROS nanodomain influences specific effectors at neighboring diads.

Reactive oxygen and nitrogen species set the redox potential of the intracellular medium. They cause specific, usually reversible, oxidative–reductive modification

---

[22]Only Nox1, Nox2, and Nox4 isoforms are synthesized in the heart. Among them, Nox2 localized to T tubules is the predominant isoform in the adult cardiomyocyte.

[23]Subtype NOS3 preferentially lodges in caveolae of the sarcolemma at T tubules and crests. Isoform NOS2 resides in caveolae and SR membrane.

of signaling effectors. Their targets include ion channels and transporters, kinases and phosphatases, and transcription factors, among others.

The diadic domain contains protein kinase-A (PKA) and Ca^{2+}–calmodulin-dependent protein kinase CamK2 as well as protein phosphatases. These kinases and phosphatases can complex with RyR channels. Phosphorylation mostly triggered by PKA and CamK2 can interfere with the ROS- and RNS-mediated redox modifications. Both ROS and Ca^{2+} activate CamK2 enzyme.

Reactive oxygen species also regulate diadic Ca^{2+} signaling on various time scales corresponding to the cardiac period or larger, reshaping Ca^{2+} signal amplitudes and kinetics [20]. Calcium channels and transporters (sarcolemmal $Ca_V1.2$, Na^+–Ca^{2+} exchanger, and plasma membrane Ca^{2+} ATPase as well as on the sarcoplasmic reticulum ryanodine receptors and sarco(endo)plasmic reticulum Ca^{2+} ATPase) are ROS and RNS targets.

The frequency and amplitude of spontaneous Ca^{2+} spark depends on the basal ROS production. ROS transients initially increase and then decrease Ca^{2+} sparks [20]. ROS react with RyRs and affect calmodulin–RyR connection. Calmodulin reduces Ca^{2+} release by RyR channels.

Conversely, Ca^{2+} messenger[24] modulates ROS action. Mitochondrial Ca^{2+} uniporter is involved in the regulation of superoxide flash generation.

3.1.5.4 GPCRs, Phosphorylation, and Sympathetic Control

Adrenoceptor-β1 and -β2 represent the predominant mediators of the sympathetic control of the myocardial function. Selective β1-adrenoceptor stimulation causes cardiomyocyte apoptosis, whereas β2-adrenoceptor activation has an antiapoptotic effect [29].

Cyclic adenosine monophosphate is the predominant positively inotropic and lusitropic effector of Gs-coupled β-adrenoceptors that controls the catecholamine-dependent changes in rate, force, and speed of cardiac contraction. Spatially confined cAMP signals originate from β2-adrenoceptors localized in T tubules and caveolae with molecules which buffer and shape the cAMP signal for fine control [29]. β1AR-mediated cAMP signal propagates throughout large regions of the cell, whereas β2AR-primed signal remains confined in T tubules [29].

In healthy cardiomyocytes, β1-adrenoceptors localized in T tubules and cell crests, but not β2-adrenoceptor, stimulate compartmentalized PKA that then phosphorylates $Ca_V1.2$, RyR2, phospholamban (relieving SERCA inhibition),[25] and

---

[24]Calcium ion signaling ability derives almost entirely from its binding to and unbinding from target proteins and its fluxes through permeable carriers that depolarize the plasma membrane. The cyclic rise and fall of intracellular Ca^{2+} concentration engages and disengages the molecular machinery of contractile myofilaments.

[25]During restoration, cytosolic Ca^{2+} ions reenter the sarcoplasmic reticulum through sarco(endo)plasmic reticulum Ca^{2+} ATPase (SERCA) and is extruded to the extracellular space by sarcolemmal Na^+–Ca^{2+} exchanger (NCX) and plasma membrane Ca^{2+} ATPase (PMCA).

cardiac contractile proteins. Cardiomyocytes synthesize the 2 regulatory subunit isoforms $PKA_{r1\alpha}$ and $PKA_{r1\beta}$ that reside in different subcellular compartments. Activated β2-adrenoceptor produces a cAMP pool that preferentially activates $PKA_{r1\beta}$ [29].

On the other hand, caveolae, tiny invaginations of the plasma membrane (caliber ~50 nm) distributed along the surface of cardiomyocytes, contain signaling components, in addition to β2-adrenoceptor. Among them, Gi subunit reduces and NOS3 antagonizes cAMP signaling.

3.1.6 Influence of Post-Translational Modifications of Regulators

The nitroso–redox balance in the sarcoplasmic reticulum contributes to the cardiovascular function. Ion channels involved in excitation–contraction coupling can also undergo alternative redox-based modifications, such as ^{S}glutathionylation and higher ^{S}oxidations.

3.1.6.1 Glutathiolation

Tachycardia increases calcium release rate from the sarcoplasmic reticulum, hence supporting myocardial preconditioning. Ischemic preconditioning is an adaptation of the heart to short episodes of ischemia. Preconditioning can also be achieved by short periods of tachycardia. Tachycardia-induced preconditioning increases both the density and activity of ryanodine receptor RyR2 and SERCA2a ATPase [35]. In addition, periods of increased cardiac activity may increase ROS generation.

Tachycardia augments NADPH oxidase activity and RyR2 ^{S}glutathiolation in ventriculomyocytes [36]. It also launches association of Rac1 and NADPH oxidase.

3.1.6.2 Nitrosylation–Denitrosylation Cycle

Nitric oxide participates not only in the control of the vasomotor tone by cGMP-dependent signaling, but also in ^{S}nitrosylation (covalent attachment of NO to the thiol group of cysteine residues). Cysteine ^{S}nitrosylation is a post-translational modification by which nitric oxide regulates protein function and signaling.

^{S}Nitrosylation influences systemic peripheral vascular resistance, in addition to cardiac contractility. Constitutive NOS1 and NOS3 subserve endothelium-dependent vasodilation and contribute to the action of β-adrenergic receptors and intracellular calcium handling.

^{S}nitrosoglutathione reductase ($G^{SNO}R$) denitrosylates proteins (removes the NO molecule from the thiol group of cysteine residues in proteins^{SNO}), thereby regulating both vascular tone and cardiac contractility, both under basal conditions and after β-adrenergic receptor stimulation [37]. It raises vascular tone and supports β-adrenoceptor-mediated inotropy associated with denitrosylation of the RyR2 ryanodine receptor.

The control of vascular tone and cardiac contractility results, at least partly, from the balance between nitric oxide synthase and ^{S}nitrosoglutathione reductase (i.e., from the nitrosylation–denitrosylation cycle).

β-Adrenoceptor stimulation relies, at least partly, on the concerted action of NOS1 and $G^{SNO}R$ that enhances Ca^{2+} cycling in the sarcoplasmic reticulum on the one hand and NOS3 and $G^{SNO}R$ that influence peripheral vascular tone.

^{S}Nitrosylation contributes to the regulation of metabolism and mitochondrial energetics as substrate enzymes participate in glycolysis, gluconeogenesis, tricarboxylic acid cycle, and oxidative phosphorylation.

3.1.7 Energetics

A permanent high demand for energy is required by the myocardium to sustain its metabolism and function, that is, ionic fluxes responsible for its electrical activity and continuous series of contraction–relaxation cycles coupled to ionic motions. The myocardium must permanently generates ATP at a high rate. Fuel is mainly supplied by β-oxidation of long-chain fatty acids (50–70% of produced ATP). Fatty acid β-oxidation depends on the myocardial contractility rate, presence of competing substrates (i.e., glucose, lactate, ketones, and amino acids), hormonal milieu, nervous command, and oxygen delivery.

The heart has an average oxygen requirement of 0.10 to 0.13/s (6 to 8 ml/mn) per 100 g at rest. Approximately 80% of oxygen consumption is related to its mechanical work (20% for basal metabolism). Myocardial blood flow must fit this energy demand. The myocardium also uses different substrates for its energy production, mostly fatty acid metabolism, which gives nearly 70% of energy requirements, and glucids.

Myocardium contraction requires a very high ATP/ADP ratio. At low ADP levels, ADP does not diffuse quickly enough back to mitochondria, as the diffusion rate is proportional to the concentration gradient.

Mitochondrial and myofibrillar creatine kinase (CK) and large amounts of creatine and creatine phosphate are in equilibrium with the adenine nucleotide pools. High concentrations of these highly diffusible energy carriers can increase the maximum energy transport rate.

The high cytosolic ATP/ADP ratio implies a low cytosolic AMP concentration, as a result of the myokinase equilibrium.[26] Molecule AMP serve as a signal for ATP supply. It activates glycogen phosphorylase and phosphofructokinase, among others. This nucleotide also generates adenosine that stimulates local blood flow.

Oxidative phosphorylation is a major source of adenosine triphosphate (ATP; Vol. 1 – Chap. 4. Cell Structure and Function). Mitochondria use energy from substrate oxidation to synthesize ATP from ADP and inorganic phosphate. Oxidative energy metabolism relies firstly on a substrate selection. According to circumstances and eventual therapy, carbohydrate oxidation can be favored over long-chain fatty acid β-oxidation for energy production. Energetic substrates used for energy metabolism, such as carbohydrates and long-chain fatty acids, are then diverted away from other metabolic pathways, such as the conversion of long-chain fatty acids to ceramides and diacylglycerides and the processing of glucose into the hexosamine synthesis.

Another pathway of energetic metabolism is anaplerosis that refers to metabolic pathways that replenish the tricarboxylic acid cycle-feeding molecules and intermediates, such as pyruvate, acetylCoA, and citrate.

3.1.7.1 Myocardial Fatty Acid Oxidation

The oxidative phosphorylation P/O ratio of oxidative phosphorylation estimates the number of ATP molecules produced per atom of oxygen reduced by the mitochondrial electron transport chain. It varies according to the energy substrate used for the generation of mitochondrial reducing equivalents (both NADH and $FADH_2$ for fatty acid oxidation and NADH for glycolysis and glucose oxidation, i.e., pyruvate oxidation).[27]

The complete oxidation of 1 palmitate molecule generates 105 ATP molecules and consumes 46 oxygen atoms. On the other hand, the complete oxidation of 1 glucose molecule produces 31 ATP molecules and consumes 12 oxygen atoms [38].

In the normal adult heart, more than 95% of ATP production derives from mitochondrial oxidative phosphorylation. The remainder originates from glycolysis and GTP synthesis in the tricarboxylic acid cycle [38].

The heart has a relatively low ATP content (5 μmol/g wet weight) and high rate of ATP hydrolysis (∼30 μmol/g wet weight/mn at rest) [38]. In normal conditions, the myocardial ATP pool experiences a complete turnover approximately every 10 s.

---

[26]Myokinase converts ATP and AMP into 2 ADP molecules.

[27]Oxidation of NADH at ^{ETC}complex-I is indirectly coupled to ATP production. Oxidation of $FADH_2$ bypasses ^{ETC}complex-I, thereby pumping fewer H^+ across the inner mitochondrial membrane. Therefore, in addition to a greater oxygen requirement than that using glucose as input substrate, fatty acids are less efficient for the generation of ATP than glucose [38].

Mitochondrial ATP synthesis via oxidative phosphorylation depends on the maintenance of an electrochemical proton gradient across the inner mitochondrial membrane generated by the extrusion of protons from the matrix to the intermembrane space by ^{ETC}complex-I, -III, and -IV. Protons reentry into the mitochondrial matrix via F_0–F_1 ATPase drives ATP generation.

Fatty Acid Metabolism

Fatty acids used by the myocardium primarily originate from plasma fatty acids carried by albumin and triacylglycerol of chylomicrons and very-low-density lipoproteins.[28] The concentration of circulating free fatty acids ranges from 0.2 to 0.6 mmol [38].[29]

Fatty acids are taken up by cardiomyocytes in particular via ScaRb3 scavenger receptor (or fatty acid translocase; Table 3.16). In the cytosol, fatty acids bound to fatty acid-binding proteins are esterified to fatty acylCoA by fatty acylCoA synthase (FACS) [38].

Fatty acylCoA can be esterified to triacylglycerol or the acyl group can be transferred to carnitine via carnitine palmitoyltransferase CPT1. Acylcarnitine enters mitochondria, where it is converted back to fatty acylCoA by carnitine palmitoyltransferase CPT2.

The majority of fatty acylCoA that undergoes β-oxidation are mono- (e.g., oleate, the most abundant fatty acid in blood) and polyunsaturated fatty acids. Fatty acid β-oxidation creates acetylCoA, NADH, and $FADH_2$. One cycle of β-oxidation generates acetylCoA that enters the tricarboxylic acid cycle and a 2-carbon-shorter fatty acyl chain.

In some circumstances, mitochondrial thioesterase (mtTE) can cleave long-chain acylCoA to fatty acid anions (FA^-), which can leave the mitochondrial matrix via uncoupling protein [38].

Myocardial Fatty Acid Uptake

Circulating free fatty acids enter cardiomyocytes either by passive diffusion (flip-flop phenomenon due to the lipophilic nature of fatty acids) or mainly via ScaRb3

[28]Most exogenous triacylglycerols derive from chylomicrons; a minor part originates from VLDL particles. A significant proportion of fatty acids from VLDLs is taken up by cardiac VLDL–apoE receptors. VLDL-derived fatty acids are equally distributed between β-oxidation and deposition into intramyocardial lipids.

[29]Concentration of FFAs can vary from very low values in the fetal circulation to more than 2 mmol during myocardial ischemia and uncontrolled diabetes [38]. Activated sympathetic nervous system can rapidly increase circulating FFA concentration, primarily via stimulation by β-adrenoceptors of hormone-sensitive lipase in adipose tissue. Lipoprotein lipase is responsible for most FFAs from chylomicrons used for fatty acid β-oxidation.

Table 3.16 Fatty acid oxidation in cardiomyocytes (Source: [38]; ACC: acetylCoA carboxylase; AcCoA: acetylCoA; AMPK: AMP-activated protein kinase; CPT: carnitine palmitoyltransferase; FA: fatty acid; FABP^{PM}: plasmalemmal fatty acid-binding protein; FACoA: fatty acylCoA; FACS: fatty acylCoA synthase; FATP: fatty acid transport protein; MaCoA: malonylCoA; MCD: malonylCoA decarboxylase; ScaR: scavenger receptor; TCA: tricarboxylic acid; TAG [TG]: triacylglycerol [triglyceride]; $\ominus\longrightarrow$: inhibition). Under certain conditions, mitochondrial thioesterase (mtTE) cleaves long-chain acylCoA to fatty acid anions (FA^-) that can leave the mitochondrial matrix via uncoupling proteins

Compartment	Reaction and Transport
Blood	Fatty acid–albumin
	Triacylglycerol–chylomicron/VLDL
Plasma membrane	ScaRb3, FABP^{PM},
	FATP1–FATP6 (SLC27a1–SLC27a6)
Cytosol	$\text{FA} \xrightarrow{\text{FACS}} \text{FACoA}$
	$\text{FACoA} \rightarrow \text{TAG}$
	$\text{AcCoA} \xrightarrow{\text{ACC}} \text{MaCoA}$
	$\text{MaCoA} \xrightarrow{\text{MCD}} \text{AcCoA}$
	$\text{AMPK}\ominus\longrightarrow \text{ACC}$
	$\text{MaCoA}\ominus\longrightarrow \text{CPT1}$
Mitochondrial	$\text{FACoA} \xrightarrow{\text{CPT1}} \text{acylcarnitine}$
envelope	$\text{Acylcarnitine} \xrightarrow{\text{CPT2}} \text{FACoA}$
	UCP
Mitochondrial	$\text{FACoA} \xrightarrow{\text{mtTE}} \text{FA}^-$
matrix	$\text{FACoA} \rightarrow \text{AcCoA} \rightarrow \text{TCA}$

receptor (50–60% of fatty acid uptake), the plasmalemmal isoform of fatty acid-binding protein (FABP^{PM}), and fatty acid transport proteins (FATP1–FATP6) [38].

Unlike FATPs or FABP^{PM}, ScaRb3 can move between endosomes and the sarcolemma. Both cardiomyocyte contraction and insulin stimulate ScaRb3 translocation to the sarcolemma [38]. Polyubiquination of ScaRb3 causes its proteosomal degradation. Insulin impedes ScaRb3 polyubiquination. On the other hand, fatty acids favor polyubiquination (feedback inhibition).

Myocardial Triglycerid Metabolism

The myocardium has labile stores of triacylglycerols, a source of free fatty acids. Myocardial cytosolic long-chain acylCoA can be converted to triacylglycerol by glycerolphosphate acyltransferase. In healthy people, the intramyocardial triacylglycerol content is low ($\sim$3 mg/g tissue) relative to the FFA uptake rate ($\sim$3 mg/g/h) [38].

Stimulated β-adrenoceptors on cardiomyocytes activates glycerolphosphate acyltransferase and incorporates palmitate into triacylglycerol stores. Simultaneously they elicit intramyocardial triacylglycerol breakdown.

Catabolism of intracellular triacylglycerol is catalyzed by hormone-sensitive lipase, which is activated by cAMP messenger.

Lipoprotein Lipase

A majority of circulating free fatty acids correspond to triacylglycerol contained in lipoproteins. Therefore, triacylglycerol hydrolysis by lipoprotein lipase (LPLase) enables fatty acid uptake and β-oxidation.

Lipoprotein lipase on the capillary endothelial surface is initially synthesized as an inactive monomeric proenzyme in the endoplasmic reticulum of the cardiomyocyte [38]. This proenzyme is processed between the endoplasmic reticulum and Golgi body. Active homodimeric lipoprotein lipase is then secreted. It binds to heparan sulfate proteoglycans at the cardiomyocyte surface. It is subsequently transferred to heparan sulfate proteoglycans of the wetted endothelial surface. Its degradation follows its detachment from the cell surface and release into the blood stream or endocytosis of the HSPG–LPLase complex.

AMP-activated protein kinase may stimulate the transfer of LPLase to the luminal surface of the endothelium. On the other hand, in adipose tissue, LPLase secretion decreases, as angiopoetin-like protein AngL4 favors the conversion of active dimeric LPLase to inactive monomer [38].

The main tissular lipase — adipose triacylglycerol lipase (ATGL) — also contributes to mitochondrial fatty acid uptake.

Mitochondrial Fatty Acid Uptake and Transfer

Mitochondrial carnitine palmitoyltransferase CPT1 converts long-chain acylCoA to long-chain acylcarnitine. Acylcarnitine then moves into the mitochondrial matrix, crossing the inner mitochondrial membrane via the carnitine–acylcarnitine translocase that exchanges carnitine esters, among which acylcarnitine for carnitine, thereby yielding free carnitine for subsequent CPT1 reaction [38]. Once it is in the matrix, acylcarnitine is converted back to long-chain acylCoA by CPT2 on the matrix side of the inner mitochondrial membrane, which is less sensitive to inhibition by malonylCoA. Long-chain acylCoA then enters the fatty acid β-oxidation.

MalonylCoA (turnover half-time $\sim$1.25 mn) inhibits CPT1 enzyme [38]. Myocardial malonylCoA concentration depends on the balance between synthesis from acetylCoA via acetylCoA carboxylase (ACC) and its degradation by tetrameric malonylCoA decarboxylase (MCD).

Two cardiac ACC isoforms exist (ACCα and predominant ACCβ). Activity of ACC enzyme is inversely related to fatty acid β-oxidation [38]. Kinase AMPK phosphorylates (inactivates) both ACCα and ACCβ.

MalonylCoA decarboxylase resides in mitochondria, peroxisomes, and the cytosol. In humans, the Mcd gene has 2 5′ start sites that code for a 54- and 50-kDa protein. Both MCD isoforms are synthesized in the heart [38].

Fatty Acid Regulators

Once in the cytosol, fatty acids are converted into long-chain acylCoA esters by fatty acylCoA synthase (FACS). Long-chain acylCoA can then be used for synthesis of intracellular lipid intermediates (tri- and diacylglycerol and ceramides). The fatty acid moiety can also be transferred to carnitine and taken up by mitochondria. In the normal heart, about 75% of taken up fatty acids are immediately oxidized [38].

Fatty acid are ligands for nuclear receptors NR1c1 to NR1c3 (or transcription factors PPARα to PPARγ) that heterodimerize with NR2b factors (or retinoid X receptors) and translocate to the nucleus, where they bind to their cognate response elements. Numerous transcriptional coactivators, such as PGC1α and PGC1β, participate in NR1c-mediated transcription.

Target genes of NR1c1 encode regulators of [38]: (1) fatty acid uptake (ScaRb3 and FATP1); (2) cytosolic fatty acid binding and esterification (fatty acid-binding protein, fatty acylCoA synthase, glycerol 3-phosphate acyltransferase, diacylglycerol acyltransferase); (3) mitochondrial fatty acid uptake (carnitine palmitoyltransferase CPT1); (4) fatty acid storage (e.g., diacylglycerol acyltransferase (DGAT]); (5) fatty acid oxidation (very-long- [VLCAD], long- [LCAD], and medium [C4 to C12]-chain acylCoA dehydrogenase [MCAD]); (6) malonylCoA metabolism (malonylCoA decarboxylase); (7) mitochondrial uncoupling (mitochondrial thioesterase mtTE1 and uncoupling proteins [UCP2–UCP3]); as well as (8) glucose oxidation (e.g., pyruvate dehydrogenase kinase PDHK4).

Ubiquitous factor NR1c2 (PPARβ) is highly synthesized in the heart. It is involved in the transcriptional control of genes that encode many NR1c1-targeted enzymes [38]. On the other hand, NR1c3 is poorly expressed in the heart.

Upregulation of PGC1 expression by exercise and fasting increase mitochondrial genesis, fatty acid β-oxidation, and oxidative phosphorylation [38].

Fatty acid β-oxidation relies on [38]: (1) oxygen delivery to cardiomyocytes; (2) fatty acid supply to cardiomyocytes as well as their uptake, esterification, and mitochondrial transfer; (3) mitochondrial function (i.e., activity of tricarboxylic acid cycle and electron transport chain activity); (4) transcriptional control of enzymes involved in fatty acid metabolism and mitochondrial genesis; (5) energy demand; and (6) existence of other energy substrates (amino acids, glucose, ketones, and lactate), fatty acids being either major contributors to cardiac oxidative energetic metabolism or minor providers.

Enzymes of Fatty Acid Oxidation

In the mitochondrial matrix, long-chain acylCoA undergoes β-oxidation. Each cycle of fatty acid β-oxidation shortens the fatty acyl moiety by 2 carbons and produces acetylCoA, flavin adenine dinucleotide ($FADH_2$), and nicotinamide adenine dinucleotide (NADH).

Fatty acid β-oxidation involves sequentially 4 enzymes [38]: acylCoA dehydrogenases that constitute a category of enzymes catalyzing the initial step of fatty acid β-oxidation in mitochondria; enoylCoA hydratase (enoylCoA hydratase/3-hydroxyacylCoA dehydrogenase [EHHADH]); (1,3)-hydroxyacylCoA dehydrogenase (HADH); and 3-ketoacylCoA thiolase.

These 4 enzymes lodge in cardiomyocytes with varying fatty acid chain-length specificities. All these enzymes are sensitive to feedback inhibition by their reaction products, including acetylCoA, $FADH_2$, and NADH.

The transcription of fatty acid β-oxidation enzymes is controlled by agonist (eicosanoids and leukotrienes)-activated nuclear NR1c1 receptor (PPARα transcription factor) as well as PGC1α and PGC1β.

At equivalent concentrations, unsaturated fatty acids, such as oleate and arachidonic acid, are oxidized at a similar rate than that of saturated fatty acid palmitate [38]. β-Oxidation of mono- and polyunsaturated fatty acids is facilitated by auxiliary enzymes, such as (2,4)-dienoylCoA reductase and enoylCoA isomerase.

Uncoupling of Mitochondrial H^+ Gradient from ATP Synthesis

Increased delivery of acetylCoA to the tricarboxylic acid cycle and subsequent supply of reducing equivalents ($FADH_2$ and NADH) to the electron transport chain from augmented fatty acid β-oxidation can reduce cardiac efficiency via the activation of uncoupling proteins that dissipate the mitochondrial H^+ gradient and uncouple it from ATP synthesis.[30]

Moreover, uncoupling proteins also contribute to the export of fatty acid anions generated in the mitochondrial matrix during hydrolysis of matrix fatty acylCoA by mitochondrial thioesterases [38].[31]

[30]Uncoupling proteins (UCP1–UCP5) are mitochondrial transport proteins that serve for the reentry of protons from the intermembrane space to the mitochondrial matrix uncoupled to ATP synthesis. Subtype UCP1 is highly expressed in brown adipose tissue, but not in the heart. Ubiquitous UCP2 minimizes generation of mitochondrion-derived reactive oxygen species. Isoform UCP3, a fatty acid anion transporter, is highly produced in the heart, skeletal muscle, and brown adipose tissue.

[31]Fatty acid anions are also generated in the cytosol during hydrolysis of cytosolic fatty acylCoA by cytosolic thioesterases.

Futile Cycling of Fatty Acids

Export of fatty acid anions from the mitochondrial matrix by UCP3 generates a futile cycle [38]. Exported fatty acid anion is converted to an acylCoA ester prior reentry to the mitochondrial matrix for further metabolism via fatty acid β-oxidation. This process requires fatty acylCoA synthase that consumes 2 ATP molecules and releases AMP and pyrophosphate.

When fatty acid utilization increases, cytosolic thioesterases, production of which rises, are committed in the futile cycling of fatty acids.

Adenine nucleotide translocator is inhibited by long-chain acylCoA at either side of the mitochondrial membrane.

Interaction Between Fatty Acid and Glucose Metabolism

In a well-perfused myocardium, 50 to 70% of acetylCoA comes from fatty acid oxidation and 30 to 50% from pyruvate oxidation that derives approximately in equal amounts from glycolysis and lactate oxidation [38].

Pyruvate formed from glycolysis takes 3 main routes (Table 3.17): (1) conversion to lactate; (2) decarboxylation to acetylCoA, an irreversible step in carbohydrate oxidation catalyzed by pyruvate dehydrogenase in the mitochondrial matrix;[32] and (3) carboxylation to oxaloacetate or malate.

Elevated generation of acetylCoA from fatty acid oxidation decreases glucose (pyruvate) oxidation[33] via the activation of pyruvate dehydrogenase kinase and subsequent inhibition of pyruvate dehydrogenase. In addition, augmented supply of fatty acid oxidation-derived acetylCoA to the tricarboxylic acid cycle also lowers glycolysis due to the inhibition of phosphofructokinase PFK1 by cytosolic citrate, a TCA intermediate exported via the tricarboxylate carrier. High rates of fatty acid oxidation inhibit both phosphofructokinase isoforms (PFK1–PFK2), hence glycolysis, via an elevated cytosolic citrate concentration [38].

Conversely, a heightened generation of acetylCoA derived from glucose (pyruvate) oxidation inhibits fatty acid β-oxidation, as 3-ketoacylCoA thiolase is inhibited by acetylCoA [38]. AcetylCoA formed from glycolysis due to carnitine acetyltransferase and subsequent production of acetylcarnitine is also a substrate for carnitine–acetylcarnitine transferase. The latter exports acetylcarnitine to the cytosol, where it can be reconverted to acetylCoA by cytosolic carnitine acetyltransferase. Cytosolic

[32]Pyruvate dehydrogenase is phosphorylated (inactivated) by pyruvate dehydrogenase kinase (PDHK) and dephosphorylated (activated) by pyruvate dehydrogenase phosphatase (PDHP). Among the 4 isoforms (PDHK1–PDHK4), PDHK4 is the predominant cardiac isoform. Pyruvate dehydrogenase kinase is inhibited by pyruvate and reduced acetylCoA/CoA and NADH/NAD^+ ratios [38]. A high level of circulating free fatty acids and intracellular accumulation of long-chain fatty acids support NR1c1-mediated PDHK4 synthesis. Pyruvate dehydrogenase phosphatase is activited by Ca^{2+} and Mg^{2+} ions.

[33]Pyruvate is the usual mitochondrial fuel produced by glycolysis.

Table 3.17 Glucose–fatty acid competition as substrate of cardiac energetics; reciprocal inhibition (Source: [38]) Elevated generation of acetylCoA (AcCoA) from fatty acid oxidation reduces glucose (pyruvate) oxidation via activation of pyruvate dehydrogenase kinase (PDHK) that phosphorylates (inhibits) pyruvate dehydrogenase (PDH). In addition, citrate, a tricarboxylic acid cycle (TCA) intermediate exported to the cytosol via the tricarboxylic carrier (TCC), inhibits phosphofructokinase PFK1, hence glycolysis. Conversely, increased production of acetylCoA from glycolysis inhibits fatty acid oxidation, as ketoacylCoA thiolase is inhibited by acetylCoA. AcetylCoA derived from glycolysis generates acetylcarnitine (AcCarnitine) by activited carnitine acetyltransferase (CAT). Carnitine–acetylcarnitine transferase (CACT) exports acetylcarnitine to the cytosol, where it can be reconverted to acetylCoA by cytosolic CAT. Cytosolic acetylCoA is a substrate for acetylCoA carboxylase (ACC), which catalyzes the generation of malonylCoA, an inhibitor of carnitine palmitoyltransferase CPT1, thereby decreasing fatty acid oxidation when glucose supply rises (ACL: ATP–citrate lyase; LDH: lactate dehydrogenase; MCC: monocarboxylate carrier; $\oplus \longrightarrow$: stimulation; $\ominus \longrightarrow$: inhibition)

Compartment	Reaction and Transport
Cytosol	G6P $\xrightarrow{\text{PFK1}}$ pyruvate
	Lactate $\xrightarrow{\text{LDH}}$ pyruvate
	Citrate $\xrightarrow{\text{ACL}}$ AcCoA
Mitochondrial envelope	MCC (pyruvate entry into matrix)
	TCC (citrate exit into cytosol)
Mitochondrial matrix	Pyruvate $\xrightarrow{\text{PDH}}$ AcCoA
	PDHK$\ominus \longrightarrow$ PDH
	PDHP$\oplus \longrightarrow$ PDH
	AcCoA $\rightarrow$ TCA
	AcCoA $\rightarrow$ FACoA
	AcCoA $\xrightarrow{\text{CAT}}$ AcCarnitine

acetylCoA is processed by acetylCoA carboxylase, which generates malonylCoA, an inhibitor of carnitine palmitoyltransferase CPT1, thereby diminishing fatty acid oxidation when glycolysis rises.

3.1.7.2 AMP-Activated Protein Kinase

AMP-activated protein kinase (AMPK; Vols. 1 – Chap. 4. Cell Structure and Function and 4 – Chap. 5. Cytosolic Protein Ser/Thr Kinases) is a heterotrimer (α catalytic and β and AMP sensitivity-conferring γ regulatory subunits). Numerous isoforms of each subunit exist with a variable tissue distribution pattern. In the heart, $\alpha 1$ and predominant $\alpha 2$, $\beta 1$ and $\beta 2$, and $\gamma 1$ and $\gamma 2$ subunits are synthesized [38].

This sensor of cellular energy status participates in the control of the energetic balance. This kinase integrates nutritional and hormonal signals. It mediates effects of adipokines (e.g., adiponectin and leptin) in the regulation of food intake and glucose and lipid homeostasis [39]. It can enhance insulin sensitivity. It is regulated by kinases such as liver kinase-B (LKB1 or STK11).

Enzyme AMPK responds to depleted cellular ATP level, increased AMP concentration, and elevated creatine/phosphocreatine ratio. Its activation usually requires changes in AMP/ATP or Cr/PCr ratio. Nonetheless, insulin inhibits myocardial AMPK when these ratios do not change.

Protein AMPK regulates not only fatty acid β-oxidation, but also glucose uptake and glycolysis. Kinase AMPK increases and decreases fatty acid β-oxidation during augmented and lowered energy demand, respectively. It operates via acetylCoA carboxylase activity and malonylCoA concentration [38].

During ischemia, the upstream kinase AMP2K of the signaling cascade that activates AMPK is neither LKB1, nor CaM2Kβ, but may be myosin light chain kinase (MLCK) [38].

3.1.7.3 Mitochondria

Mitochondria constitute approximately 30% of the myocyte volume. Mitochondria operate in the energetic metabolism, as they generate ATP, the ultimate energy transfer molecule, thermogenesis, signaling, as they produce reactive oxygen species in particular, maintenance of calcium homeostasis, and apoptosis, the latter being preceded by mitochondrial fission apoptosis.

Mitochondria are dynamic organelles that reorganize, reshape by fusion and fission, and relocalize within cells. Cyclic changes in fusion and fission involve regulators such as transmembrane GTPases [40].

Mitochondrial fusion enables the exchange of soluble matrix proteins and complementation between mitochondrial DNA molecules, modulates mitochondrial dimensions and shape, and determines the size and dynamics of the mitochondrial Ca^{2+} pool. The mitochondrial transmembrane GTPases, mitofusin MFn1 mediates the mitochondrial fusion, whereas MFn2 supports clustering of mitochondria [41].

Nuclear-encoded, mitochondrial, 120-kDa dynamin-like molecule optic atrophy protein OPA1[34] is involved in normal mitochondrial morphology and division. Cytoplasmic dynamin-related protein DRP1 contributes to mitochondrial fission, as it accumulates and assembles on the outer mitochondrial membrane at sites of membrane fission.

In addition to the nucleus, the mitochondrion is a source of DNA in cells. The mitochondrial encodes 13 constituents of the electron transport chain, the remaining

[34]Mutations in the OPA1 gene cause autosomal dominant type-1 optic atrophy. Optic atrophy protein OPA1, a.k.a. mitochondrial genome maintenance protein MGM1 and mitochondrial nucleoid protein, contains a mitochondrial targeting signal, hence localizing to mitochondria.

Table 3.18 Stages and components of the citric acid cycle, also called the tricarboxylic acid (TCA) cycle and Krebs cycle. Both NADH and FADH2 supply electrons from metabolic substrates. Mitochondrial oxidative phosphorylation relies on the transfer of electrons from electron donors to electron acceptors to produce ATP via multimeric complexes in the mitochondrial inner membrane called the electron transport chain (^{ETC}complex-I–^{ETC}complex-IV). Both NADH and succinate produced in the TCA cycle donate electrons to ^{ETC}complex I (NADH dehydrogenase) and -II (succinate dehydrogenase). These electrons are transferred to ubiquinone and delivered to ^{ETC}complex-III and, then, ^{ETC}complex-IV. As electrons are transferred, hydrogen ions are pumped to the intermembrane space to create a gradient and the mitochondrial membrane potential. ATP synthase (^{ETC}complex-V) uses achieved proton gradient to convert ADP to ATP. The TCA cycle operates in 2 spans, acetylCoA to α-ketoglutarate and α-ketoglutarate to oxaloacetate, the second span with succinate, fumarate, and malate intermediates yielding anaerobic ATP

Stage	Substrate	Enzyme
Input	Pyruvate (cytosol)	Enzymes of glycolysis
	AcetylCoA (oxidative decarboxylation in mitochondrial matrix)	Pyruvate dehydrogenase (PDH)
	Tricarboxylic acid cycle	
1	Oxaloacetate	Citrate synthase (CS)
2	Citrate	Aconitase (Aco)
3	Aconitate	Aconitase
4	Isocitrate	Isocitrate dehydrogenase (ICDH)
5	Oxalosuccinate	Isocitrate dehydrogenase
6	Ketoglutarate	Ketoglutarate dehydrogenase (KGDH)
7	SuccinylCoA	SuccinylCoA synthase (SCoAS)
8	Succinate	Succinate dehydrogenase (SDH)
9	Fumarate	Fumarase (Fum)
10	Malate	Malate dehydrogenase (MDH)

79 polypeptides being encoded in the nucleus. These polypeptides combine to create the ^{ETC}complexes required for ATP generation. Other nuclear-encoded proteins comprise members of the tricarboxylic cycle and transporters for protein shuttling between the mitochondrion and cytosol, among others. Protein translocation through mitochondrial membranes requires conditioning mediated by chaperone and motor proteins.

Source of Energy – Tricarboxylic Acid Cycle and Electron Transport Chain

Cardiomyocytes have a large population of mitochondria that produce ATP from products of the tricarboxylic acid cycle, or *citric acid cycle* as well as fatty acid and amino acid oxidation (Vol. 1 – Chap. 4. Cell Structure and Function; Table 3.18).

At the inner mitochondrial membrane, electrons from NADH and succinate pass through the *electron transport chain* to oxygen, which is reduced to water. The electron transport chain comprises a set of electron donors and acceptors. Each electron donor passes electrons to a more electronegative acceptor. Molecular oxygen is the most electronegative, hence the terminal electron acceptor in the chain.

Passage of electrons between donor and acceptor releases energy, which is used to generate a proton gradient across the mitochondrial membrane by active proton pumps into the intermembrane space, producing a thermodynamic state that has the potential to do work. A given Gibbs free energy is available to do work.

Although the electron transport chain efficiently utiizes O_2 in the controlled oxidation of redox carriers, superoxide anion ($O_2^{\bullet -}$) is an unavoidable by-product of oxidative phosphorylation.

Transcription of Mitochondrial DNA

Many proteins of the respiratory chain and ATP synthase are translated by mitochondrial ribosomes. Oxidative phosphorylation capacity then depends on the regulation of mitochondrial DNA (mtDNA) expression and mitochondrial ribosome genesis. Mitochondrial DNA is a double-stranded circular molecule that encodes 22 tRNAs, 2 rRNAs, and 13 subunits of the respiratory chain.

Transcription is initiated from 2 sites, the light- (LSP) and heavy-strand (HSP1) promoters [42]. A second transcription initiation site (HSP2) downstream from HSP1 also exists. Transcription from LSP also produces the RNA primers required for initiation of mtDNA replication at the origin of the heavy strand.

The machinery responsible for transcription initiation in mitochondria consists of mitochondrial RNA polymerase, mitochondrial transcription factor mtTFb2 encoded by the TFB2M gene, and mitochondrial transcription factor mtTFa, a high-mobility group-box (HMG) protein encoded by the TFAM gene. The latter binds a specific sequence immediately upstream from mitochondrial gene promoters (i.e., upstream from HSP1 and LSP transcription start sites). Factor mTFa recruits the transcription machinery [42]. In addition, mtTFa is a DNA-packaging factor that enables mtDNA coating and packaging into nucleoids, as it can bind, wrap, and bend DNA [42]. In humans, mitochondrial, DNA-directed, RNA polymerase mtRPol encoded by the POLRMT gene recognizes promoter elements. Transcription termination is also regulated by binding of mitochondrial factors such as transcription termination factor mtTerF1 of the mitochondrial transcription termination factor (MTERF) family (mtTerF1–mtTerF4).

Methylation of rRNA supports ribosome genesis. In mitochondria, the dimethyladenosine transferase, mitochondrial transcription factor mtTFb1 encoded by the TFB1M gene, dimethylates 2 adenines of 12S rRNA to permit genesis of the small ribosomal subunit [43]. Member mtTerF4 recruits NSUn domain-containing RNA methyltransferase NSUn4 to the large ribosomal subunit [44].[35]

[35] Methyltransferases of the NOL1–NOP2–SUn (Sad1P [Schizosaccharomyces pombe] and Unc84 [Caenorhabditis elegans)] domain-containing protein family catalyzes the methylation of cytosine to 5-methylcytosine. The proteic domain refers to ribosomal RNA methyltransferase nucleolar protein homolog NoL1, also called NoP2 and NSun1, which is found in archaeal, bacterial, and eukaryotic proteins.

Mitochondrion Displacements

Mitochondrial transport within cells relies mainly on microtubules, whereas actin may mediate short-range mitochondrial displacements and anchoring. Filamentous actin enhances mitochondrial size and number, whereas globular actin causes mitochondrial fragmentation and reduces mitochondrial density [45]. The cellular ^{F}actin/^{G}actin ratio regulates action of the serum response factor. The latter and actin modulate mitochondrial dynamics via actin-severing cofilin and slingshot phosphatase. Actin polymerization and depolymerization increases and reduces SRF activity, respectively. Activity of SRF is not only adjusted by actin dynamics. Conversely, SRF adapts mRNA levels of actin isoforms (ActA–ActC) and actin binding proteins, such as myosins, tropomyosins, gelsolin, and actinin, via cofactors of the myocardin-related transcription factor (MRTF) family. In addition, phosphorylation of the actin-severing protein cofilin rises. Cofilin activity is inhibited by LIMK and assisted by slingshot phosphatases.

In addition, redistribution of mitochondria in subcellular compartments can influence cell signaling. In particular, hypoxia alters the microtubule-dependent transport of mitochondria. Consequently, these organelles accumulate in the perinuclear region. The resulting elevated intranuclear ROS concentration stimulates the transcription of the hypoxia-sensitive VEGF gene by oxidizing DNA sequences in the hypoxia response element of the VEGF promoter [46].

Metabolism Controller

Defective mitochondrial oxidative metabolism of fatty acids is associated with the development of the metabolic syndrome (i.e., diet-induced obesity and insulin resistance in adipose tissue and skeletal muscle). Obese individuals have mitochondria with impaired energetic capacity.

Mitochondrial genesis, thermogenesis, and glucose and fatty acid metabolism are regulated by peroxisome proliferator-activated receptor-γ coactivator PGC1β, a transcriptional coactivator that interacts with numerous transcription factors. It is encoded by the PPARGC1B gene and preferentially expressed in tissues with high mitochondrial content, such as myocardium, slow skeletal muscle, and brown adipose tissue.

The Pgc1b gene encodes 2 microRNAs, miR378 and miR378*, from a common hairpin RNA precursor. These transcript inhibitors counterbalance the metabolic actions of PGC1β factor [47]. These microRNAs target transcripts of carnitine ^{O}acetyltransferase (CRAT), a mitochondrial enzyme involved in fatty acid metabolism, and Med13, a component of the Mediator complex that controls nuclear hormone receptor activity.[36] Therefore, both miR378 and miR378* participate in the regulation of mitochondrial metabolism and energy homeostasis.

[36]Subunit Med13 is also targeted by miR208a and other muscle-specific miRNAs encoded by introns of myosin heavy-chain genes.

Signaling Node

Mitochondria serve as nodes in numerous signaling pathways. They operate especially in metabolism and apoptosis. Mitochondria can modulate calcium-induced signal transmission. They may support stabilization of oxygen-sensitive transcription factors.

In addition to being major sites of energetics control, mitochondria constitute sources of reactive oxygen (ROS) and nitrogen (RNS) species that react with mitochondrial molecules (Sect. 3.1.5.3).

A small percentage of electrons exits from the electron transport chain. Electron leakage predominantly occurs at ^{ETC}complex-I and -III, leading to a partial reduction of oxygen. These electrons then form the free radical superoxide, a highly reactive molecule that contributes to redox state-associated signaling (Vol. 4 – Chap. 10. Other Major Signaling Mediators). Manganese superoxide dismutases, ^{Mn}SOD in the mitochondrial matrix and $^{(Cu,Zn)}$SOD) in the intermembrane space, convert superoxide into hydrogen peroxide. The latter can then be fully reduced to water by antioxidant enzymes, such as glutathione peroxidase and catalase (in mitochondria of heart and liver cells).

Therefore, the balance between ROS generation and the antioxidant activity controls the oxidative status. Manganese addition, hyperglycemia, and leptin support superoxide production. Increased levels of oxygen and decreased electron flow through the electron transport chain augment ROS generation. Calcium regulates ROS production via the TCA cycle and cytochrome-C loss [48].

At normal concentrations, reactive oxygen species are implicated in mitogenic signaling. Superoxide increases MAPK activity and stimulates vascular smooth myocyte proliferation. In addition, hydroxyl radical activates guanylate cyclase, thereby producing cGMP that mediates vascular smooth myocyte relaxation.

Higher ROS levels cause strong oxidative modifications, mitochondrial dysfunction, and oxidative stress. In addition, superoxide can also combine with nitric oxide to produce reactive peroxynitrite.

Mitochondrial Ion Carriers

Mitochondrial ion-selective or nonselective channels control the ionic balance and volume of the mitochondrial matrix (Tables 3.19 to 3.21). Ion transfer is restricted at the mitochondrial envelope by maintenance of charge neutrality and limited changes in the matrix osmolarity [49].

Symporters cotransport different ion types, or an ion and a metabolite, in the same direction across the membrane (e.g., mitochondrial P_i–H^+ cotransporter). Antiporters exchange ions on different sides of the membrane; they are either electroneutral (e.g., mitochondrial Na^+–H^+ antiporter) or electrogenic (e.g., Na^+–Ca^{2+} exchanger), when the ion flux is driven by both the electrochemical gradients

Table 3.19 Mitochondrial inner membrane carriers: role and features (Source: [49]; ANT: adenine nucleotide translocase; mtAC: mitochondrial apoptosis-induced channel; mtPBR: mitochondrial peripheral benzodiazepine receptor [primary component: 18-kDa isoquinoline carboxamide-binding protein (IBP)]). Translocase of the outer membrane of mitochondria (TOM) and its partner on the inner membrane, TIM, can form large conductance channels enabling the transfer of mitochondrial preproteins encoded in the nucleus. The first step of steroid synthesis is the conversion of cholesterol transported through PBR to pregnenolone catalyzed by the cytochrome-P450 side-chain cleavage enzyme (CyP450scc) located on the mitochondrial inner membrane. Voltage-dependent anion channel (VDAC) that abounds in the mitochondrial outer membrane enables entry and exit of metabolites and ions; it is a component of the mitochondrial contact site between the outer and inner membranes

Type	Role and regulation
mtAC	Cytochrome-C release (apoptosis)
	Voltage independent, cation selective
mtPBR	Cholesterol transport, steroid genesis
TOM40	Protein transport
VDAC (porin)	Metabolite transport (ATP, ADP, P_i),
	Free diffusion of cations (Ca^{2+}, K^+, Na^+)
	PTP complex component ANT and ATP synthase (F_1F_O ATPase) cytochrome-C release (apoptosis [?]) binding site for hexokinase-2
	Phosphorylation by PKA, PKCϵ, Tyr kinases ↓ ADP flux by NADH and Mg2+-NAD(P)$H^{Mg^{2+}}$

of the transported ions and the membrane potential. In a uniporter, ions flow electrophoretically down their electrochemical gradient (e.g., mitochondrial Ca^{2+} uniporter).

The entry of a cation depolarizes the mitochondrial membrane and stimulates proton pumping to balance the entering positive charge. This increases the pH gradient component of the *proton-motive force*. When the cation accumulation in the matrix is accompanied by either the movement of a permeant weak acid such as acetate or the electroneutral uptake of inorganic phosphate (P_i) driven by the proton gradient, the pH change is prevented with a net increase in matrix osmolarity and mitochondrial volume. Mitochondrial swelling can be counteracted by cation–proton antiporters. Potassium influx through mitochondrial ATP-sensitive K^+ channel and efflux through K^+–H^+ exchanger can regulate the mitochondrial volume [49]. Anion efflux through an inner membrane anion channel (e.g., inner membrane anion channel) coupled with cation efflux can cause mitochondrial shrinkage.

A rapid cation uptake is done by the K^+ uniporter. Calcium uptake decrease surrounding pH, stimulates cell respiration, causes a cycle of swelling and contraction of the mitochondrion, oxidizes NADH, and alters the redox state of electron transport chain carriers. If the mitochondrial permeability transition pore (PTP) is not activated, mitochondria can take up large amounts of Ca^{2+} ions, acting as a Ca^{2+} sink, especially during the diastole.

Table 3.20 Mitochondrial inner membrane carriers: role (Source: [49]; ClC: voltage-dependent chloride channel; mtClIC: mitochondrial chloride intracellular channel; IMAC: inner-membrane anion channel; mtCU: mitochondrial Ca^{2+} uniporter; PTP: permeability transition pore). Mitochondrial ATP synthase transfers protons down the electrochemical gradient. Mitochondrial uncoupling protein has 3 isoforms (mtUCP1–mtUCP3). Partial uncoupling may protect by modifying the production rate of free radicals; that of UCP2 and/or UCP3 may optimize the efficiency of energetic metabolism. Ryanodine receptor RyR1 resides in the cardiac mitochondrial inner membrane, where it can participate in Ca^{2+} influx or efflux according to the orientation of the Ca^{2+}-triggering site. Mitochondrial K_{ATP} channel protects the heart during ischemia, unlike sarcoplasmic K_{ATP} channel that plays a minor role, despite its greater conductance. Activated mitochondrial K_{Ca} channel may cause a partial depolarization that may decrease the driving force for Ca^{2+} entry under inotropic stimulation or ischemia

Type	Role and features
IMAC (ClC?)	Volume regulation
	Inorganic anion permeability: NO_3^- > Cl^- > Pi
	Organic anion permeability: oxaloacetate^{2-} > citrate^{3-} > malate^{2-} > ATP^{4-}
	Temperature-sensitive inhibition by H^+ and Mg^{2+}
	Activation by ROS-induced ROS release
K_V1.3	Cell death
mtClIC (?)	Apoptosis
mtCU	Ca^{2+} uptake (flux ~20.10^3 Ca^{2+}/s) (dependent on ATP, P_i, and Mg^{2+})
mtK_{Ca}	Volume regulation, protection against ischemia PKA substrate
mtK_{ATP}	Volume regulation, protection against ischemia GSK3β substrate
mtUCP	Thermogenesis H^+ and Cl^- conductance (H^+ leak route)
RyR1	Ca^{2+} flux
PTP	Nonselective transport of ions and metabolites
	Opening by Ca^{2+} overload, depolarization, P_i, ROS
	Inhibition by adenine nucleotides, Mg^{2+}, matrix H^+ complex formed by ANT, VDAC, cyclophilin-D, F_1F_0 ATPase, hexokinase

Calcium

Mitochondrial calcium uptake through Ca^{2+} uniporter relies on both the concentration gradient and mitochondrial membrane potential. Calcium extrusion through the Na^+–Ca^{2+} exchanger maintains the mitochondrial Ca^{2+} level. The mitochondrial permeability transition pore also enables calcium efflux [48].

Cardiac Mitochondrial NOS

Nitric oxide is a free radical and gaseous uncharged 30-Da regulator, which is highly diffusible in aqueous and lipid phases, but travels over short distances due to the high reactivity. Nitric oxide is involved not only in the regulation of coronary vasomotor tone, but also in cardiac metabolism, ion channel activity, energy production, autocrine regulation of myocardial contraction and relaxation, O_2 consumption, apoptosis, and hypertrophy.

Table 3.21 Mitochondrial inner membrane carriers: substance fluxes (Source: [50]; ANT: mitochondrial adenine nucleotide translocase; mtKHE: mitochondrial K^+–H^+ exchanger; IMAC: inner-membrane anion channel [permeable to various anion types and small anionic metabolites, such as Cl^-, malate, oxaloacetate, malonate, citrate, ATP]; mtCU: mitochondrial Ca^{2+} uniporter; mtK_{Ca}: mitochondrial Ca^{2+}-activated K^+ channel; mtK_{ATP}: mitochondrial ATP-sensitive K^+ channel; mtNCE: mitochondrial Na^+–Ca^{2+} exchanger; mtNHE: mitochondrial Na^+–H^+ exchanger; PiC: mitochondrial inorganic phosphate carrier; PTP: mitochondrial permeability transition pore)

Type	Substance fluxes
ANT	Outward ATP and inward ADP fluxes
IMAC	Outward anion [$O_2^{\bullet -}$] flux
mtCU	Inward Ca^{2+} flux
mtK_{Ca}	Inward K^+ flux
mtK_{ATP}	Inward K^+ flux
mtKHE	Outward K^+ and inward H^+ fluxes
mtNCE	Outward Ca^{2+} and inward Na^+ fluxes
mtNHE	Outward Na^+ and inward H^+ fluxes
PiC	Inward P_i transfer
PTP	Opening by ROS, Ca^{2+}

Its signaling is compartmentalized. It activates soluble guanylate cyclase that produces cGMP, which stimulates protein kinase-G and cGMP-regulated phosphodiesterases. However, it operates via cGMP-independent pathways.

In cardiac endothelial cells and myocytes, nitric oxide synthase isoforms (NOS1–NOS3) are confined to distinct sites. Inducible NOS2 transiently lodges in the cytosol during abnormal conditions.

Isoform NOS1 resides in the sarcoplasmic reticulum membrane close to ryanodine-sensitive Ca^{2+} carriers. Synthesized NO promotes RyR opening (positive inotropy).

Subtype NOS3 localizes to the plasma membrane, more precisely to caveolae (with caveolin-3, β-adrenoceptors, and $Ca_V1.2$), including T tubules, close to diads.

Mitochondrial NOS (^{C}mtNOS), mostly NOS1 (but also NOS3 and ephemerally NOS2), is situated in the inner mitochondrial membrane, like angiotensin-2 AT_2 receptor, close to the electron transport chain.

Activity of ^{C}mtNOS is regulated by Ca^{2+}, H^+, O_2, and ^{L}arginine, as well as the mitochondrial transmembrane potential and mitochondrial metabolic state [51]. Extramitochondrial free Mg^{2+}, a mitochondrial Ca^{2+} uniporter blocker, precludes NO production. On the other hand, elevated levels of extramitochondrial Ca^{2+} stimulate ^{C}mtNOS activity

In mitochondria, at normal nanomolar concentrations, NO targets heme irons, $O2_2^{\bullet -}$, and thiols, that is, at the macromolecular scale, mostly the electron transport chain (ETC) and permeability transition pore (PTP). It has a high affinity for cytochrome-C oxidase (^{ETC}complex-IV), reversibly preventing its activity [51].

In addition, it reacts with ubiquinol–cytochrome-BC$_1$ complex (^{ETC}complex-III), thereby inhibiting electron transfer and enhancing production of superoxide anion. Nitric oxide also causes ^{S}nitrosylation of NADH dehydrogenase (^{ETC}complex-I), hence its inhibition. Therefore, NO diminishes production of mitochondrial energy, but supports that of reactive oxygen species.

In the mitochondrial matrix, manganese superoxide dismutase forms uncharged hydrogen peroxide (H_2O_2). Nitric oxide also reacts with superoxide formed by the mitochondrial electron transport chain at a rate constant of about 2×10^{10}/mol/s to create peroxynitrite ($ONOO^-$) [51].

3.1.7.4 Mitochondrial Electron Transport Chain or Respiratory Chain

The electron transport chain, or respiratory chain, couples electron transfer between an electron donor such as NADH and an electron acceptor such as O_2 with the transfer of H^+ ions (or protons) across a membrane. The resulting electrochemical proton gradient is used to generate chemical energy in the form of adenosine triphosphate (ATP). In mitochondria, the conversion of oxygen to water, NADH to NAD^+, and succinate to fumarate is required to generate the proton gradient.

The electron transport chain is a set of mitochondrial enzymes and redox carriers that ferry reducing equivalents (i.e., a set of chemical species that transfer the equivalent of one electron in redox reactions, from substrates to oxygen). In the heart under normal conditions, a major source of reducing equivalents is fatty acid oxidation.

The electron transport chain consists of a spatially separated series of redox reactions. The underlying force driving these reactions is the *Gibbs free energy* of reactants and products.

3.1.7.5 Oxidation–Reduction Potential

Oxidation–reduction potential, or redox potential, yields a measure of oxidizing or reducing power. It actually measures the tendency of a chemical species to acquire electrons, thereby being reduced. Reduction potential is measured in volts (V). Oxidizing agents such as molecular oxygen have positve redox potentials; reducing agents have negative redox potentials. Electrons move spontaneously toward compounds with the more positive redox potentials. The effective redox potential depends on the proportion of oxidized and reduced forms (*Nernst equation*):

$$E = E_0 + \frac{R_g T}{z\mathcal{F}} \ln\left(\frac{[\text{oxidized form}]}{[\text{reduced form}]}\right), \tag{3.2}$$

where E_0 is the standard reduction potential measured with 50% oxidized and 50% reduced form (equimolar level of each contributor) for the compound of interest (a partial pressure of 1 atm for each gas component of the reaction), at 25°C, R_g the universal gas constant (~8.31 J/K/mol), T the absolute temperature, z the number of moles of transferred electrons in the reaction, and $\mathcal{F}$ the Faraday constant (~96,485 C/mol). The NADPH–NADP couple is a much better reducing agent than the NADH–NAD pair, although their redox potentials are identical.

Similarly, the *Henderson–Hasselbach equation* relates the pH of a buffer solution to the ionization constant for the buffer and the fraction of protonated and nonprotonated forms:

$$pH = pKa + \log\left(\frac{[\text{deprotonated form}]}{[\text{protonated form}]}\right), \tag{3.3}$$

where pK_a that reflects the intrinsic affinity of the buffer for protons is the base-10 logarithm of the association constant ($K_a = [BH]/([B^-] \times [H^+])$; B: base, buffer).

The Nernst equation in biology also relates the membrane potential ($\Delta\Psi$) to concentrations (internal $[c_i]$ and external $[c_e]$) of a diffusible ion in equilibrium with the potential on each side of the membrane:

$$\Delta\Psi = \frac{R_g T}{z\mathcal{F}} \ln\left(\frac{c_i}{c_e}\right) = 2.303\,\frac{R_g T}{z\mathcal{F}} \log\left(\frac{c_i}{c_e}\right). \tag{3.4}$$

The *Gibbs free energy* (ΔG) measures the work for a chemical reaction. It reflects the displacement from equilibrium ($\Delta G = 0$).

When an oxidizing reagent interacts with a reducing agent, the difference between their respective redox potentials is related to the Gibbs free energy using the following reaction,

$$\Delta G = -z\mathcal{F}\Delta E, \tag{3.5}$$

where ΔG and ΔE are measured in joules and volts, respectively.

3.1.7.6 Mitochondrial Redox Carriers

Electron carriers associated with the respiratory chain are arranged in the approximate order of their redox potentials: the best reducing agents couple at the substrate end and the best oxidizing molecule at the oxygen end. Along the chain, the difference in redox potential between adjacent carriers provides the driving force to pump protons out of the matrix space and into the cytosol.

Energy obtained via electron transfer is used to pump protons from the mitochondrial matrix into the intermembrane space, creating an electrochemical proton gradient across the mitochondrial inner membrane. The latter allows ATP synthase to use H^+ flux through it back into the matrix to generate ATP from ADP and inorganic phosphate.

Table 3.22 Molecular complexes and components of the mitochondrial electron transport chain. Mitochondrial oxidative phosphorylation relies on the transfer of electrons through multimeric complexes to produce ATP (^{ETC}complex-I–^{ETC}complex-V; NAD [NAD^+]: nicotinamide adenine dinucleotide coenzyme [oxidizing agent, i.e., electron acceptor that then becomes reduced on receiving electron]; NADH: reducing agent, or electron donor; NADP [$NADP^+$]: nicotinamide adenine dinucleotide phosphate; NADPH: reduced form of $NADP^+$). Despite their equivalent redox potentials, cells maintain the mitochondrial NADPH–NADP couple about 500 times more reduced than the mitochondrial NADH–NAD pair. In other words, the NADPH–NADP pair is nearly entirely (∼100%) in the reduced form, whereas the NADH–NAD couple exists partly (<1/3) in the reduced form. In general, substrates of enzymes that react with NADP are often better reducing agents than those of enzymes that react with NAD cofactor. Most dehydrogenases are specific for either NAD or NADP. A pyridine nucleotide transhydrogenase catalyzes the chemical reaction: $NADPH + NAD^+ \rightleftharpoons NADP^+ + NADH$. Energy-linked transhydrogenase catalyzes the reversible transfer of reducing equivalents between the mitochondrial NADPH–NADP and NADH–NAD pool with simultaneous pumping of 2 protons and 2 charges across the inner mitochondrial membrane

Respiratory chain components	Location	Role
NADPH–NADP pair	Matrix	Mobile carrier
NADH–NAD pair	Matrix	Mobile carrier
Transhydrogenase	Membrane	Proton pump
NADH dehydrogenase (I)	Membrane	Proton pump
Succinate dehydrogenase (II)	Membrane	
Ubiquinol–ubiquinone pair	Membrane	Mobile carrier
ubiquinol–cytochrome-c reductase (III)	Membrane	Proton pump
Cytochrome-C (IV)	Intermembrane space	Mobile carrier
Cytochrome-C oxidase (IV)	Membrane	Proton pump
F_0F_1 ATPase	Membrane	ATP synthase

Four molecular complexes contain the principal electron transport chain carriers, mitochondrial ^{ETC}complex-I to -IV, which have one segment into the mitochondrial matrix and another into the intermembrane space (Table 3.22).

^{ETC}Complex I, or NADH coenzyme-Q reductase, accepts 2 electrons from electron carrier nicotinamide adenine dinucleotide (NADH) and transfers it to the lipid-soluble carrier coenzyme-Q, or ubiquinone. The reduced product, ubiquinol, freely diffuses within the membrane. ^{ETC}Complex I translocates 4 H^+ from the mitochondrial matrix to the intermembrane space across the inner membrane. ^{ETC}Complex-I contains 46 subunits.

Ubiquinone also receives electrons from ^{ETC}complex II, or succinate dehydrogenase. ^{ETC}Complex-II and enzymes of the electron transfer flavoprotein system share ubiquinone as the final electron acceptor in the inner mitochondrial membrane. ^{ETC}Complex-II is anchored to the membrane and oxidizes succinate to fumarate. The electron transfer flavoprotein system is composed of the heterodimeric soluble electron transfer flavoprotein that shuttles electrons from dehydrogenases to the membrane-associated electron transfer flavoprotein: ubiquinone oxidoreductase.

Ubiquinone passes electrons to ^{ETC}complex III, or cytochrome-BC_1, a dimer formed by 11 different subunits per monomer. Cytochrome-BC_1 sends 2 electrons to the hydrosoluble electron carrier cytochrome-C in the intermembrane space.

Cytochrome-C transmits electrons to ^{ETC}complex IV, or cytochrome-C oxidase. The latter uses the electrons and hydrogen ions to reduce molecular oxygen to water. Four electrons are removed from 4 molecules of cytochrome-C and transferred to molecular oxygen, hence producing 2 molecules of water. Simultaneously, 4 H^+ are translocated across the membrane, contributing to the proton gradient.

Three H^+ pumps include mitochondrial ^{ETC}complex-I, -III, and -IV. Mitochondrial ^{ETC}complex-III uses the Q cycle, a sequential oxidation and reduction of the lipophilic electron carrier, ubiquinol–ubiquinone pair. Oxidation of ubiquinol (reduced coenzyme-Q) to ubiquinone (oxidized coenzyme-Q) leads to the transfer of 4 protons into the intermembrane space across the inner mitochondrial membrane. Therefore, except succinate dehydrogenase (^{ETC}complex-II), all other complexes of the electron transport chain in the inner mitochondrial membrane pump protons from the matrix space as they transfer reducing equivalents, either hydrogen atoms or electrons, from one carrier to the next.

3.1.7.7 Oxidative Phosphorylation and ATP Synthesis

Oxidative phosphorylation refers to ADP phosphorylation to ATP using the energy of hydrogen oxidation carried out in many steps. The electron transport chain and oxidative phosphorylation are indeed coupled by the proton gradient across the inner mitochondrial membrane. Free energy available from substrate oxidation is employed for ATP synthesis.

The F_0F_1 ATP synthase consists of at least 12 different subunits, several of which are present in multiple copies. The F_1 head group of ATP synthase contains 3 nucleotide-binding sites. The F_0 base is embedded in the inner mitochondrial membrane. The F_0 component acts as an ion channel that enables H^+ flux back into the mitochondrial matrix. This reflux releases free energy produced during the generation of the oxidized forms of the electron carriers NAD^+ and ubiquinone. The free energy is used for ATP synthesis catalyzed by the F_1 component of ATP synthase.

Synthesis of ATP that requires energy (7.3 kcal/mole) occurs not only in mitochondria by cellular respiration, but also in the cytosol by glycolysis:

$$\mathrm{ADP} + \mathrm{P_i} \quad \rightarrow \quad \mathrm{ATP} + \mathrm{H_2O}, \tag{3.6}$$

where P_i is inorganic phosphate (free phosphate ions $H_2PO_4^-$ and HPO_4^{2-}). The total number of ATP obtained after complete oxidation of one glucose by glycolysis, citric acid cycle, and oxidative phosphorylation is estimated to be about 30 [54].

3.1.7.8 Post-Translational Modifications of Mitochondrial Proteins

Post-translational modifications, such as acetylation, glutathionylation, nitration, ^{S}nitrosylation, and phosphorylation, are implicated in the regulation of mitochondrial metabolism. Unlike signaling mediators, such as kinases, acetylases, deacetylases, phosphatases, and ion channels, for which the modification of a small fraction can influence signaling, a significant fraction of target proteins must be modified to have an impact on the metabolic rate [55].

Mitochondrial matrix proteins involved in energy metabolism include membrane transporters, intermediary metabolism enzymes, and complexes of oxidative phosphorylation.

Mitochondrial protein kinases of the intermembrane space include PInK, PKA, PKCδ, and Src [55]. Protein phosphatases also lodge in mitochondria, such as Ca^{2+}-sensitive pyruvate dehydrogenase phosphatase PDHP1, the major isoform in the heart, branched-chain ketoacid dehydrogenase (BCKDH) phosphatase, and PPM1k. Other phosphatases localize to the outer membrane or intermembrane space, such as DuSP18, PTPn11, and PHLPP1 of the PPM1 family [55].

Phosphorylation of Inner Membrane Carriers

Protein phosphorylation is a classical regulatory mechanism. In the cardiac mitochondrial matrix and respiratory complexes, it controls mitochondrial oxidative phosphorylation, hence energy metabolism. Phosphorylation of enzymes, such as mitochondrial pyruvate dehydrogenase and glycogen phosphorylase,[37] modulates enzyme translocation and activity to match energetic needs imposed by cardiac workload.

Numerous protein Ser/Thr kinases, such as PKA, PKB, PKC, and JNK, localize in mitochondria. They are implicated in the regulation of metabolism or apoptosis, as they target matrix and inner and outer membrane components, such as ATP synthase, porin, BCL2 antagonist of cell death (BAD), and BCL2-associated X protein (BAX).

The respiratory chain creates both a pH differential and a voltage gradient across the inner mitochondrial membrane. Proteic carriers control fluxes of materials across the membrane (Table 3.23). The majority of inner membrane carriers are antiporters that exchange one molecule for another. Symporters carry 2 molecules in the same direction. Electrogenic uniporters ferry charged ions such as Ca^{2+} ion.

[37] Pyruvate dehydrogenase enables entry from glycolysis to oxidative phosphorylation. Glycogen processing by glycogen phosphorylase is the early metabolic response to cardiac work change.

Table 3.23 Examples of transmembrane tranfer of molecules (Source: [56]). In neutral solution, ATP is ionized mostly as ATP^{4-} with a small proportion of ATP^{3-} ion

Substance	Mechanism
Phosphate	Electroneutral exchange of $H_2PO_4^-$ for OH^- (phosphate–hydroxyl carrier)
Dicarboxylate	Electroneutral exchange of malate^{2-} for succinate^{2-} or $H_2PO_4^-$ (malate–phosphate carrier) (SLC25a10: mitochondrial dicarboxylate transporter)
Tricarboxylate	Electroneutral exchange of citrate^{3-} and H^+ for malate^{2-} (SLC25a1: mitochondrial citrate transporter)
Oxoglutarate	Electroneutral exchange of malate^{2-} for oxoglutarate^{2-} (SLC25a11: mitochondrial oxoglutarate–malate carrier)
Adenine nucleotides	Electrogenic ATP^{4-}–ADP^{3-} exchange (adenine nucleotide [ATP–ADP] translocase [antiporter]) Electrogenic exchange of ADP^{2-} for ATP^{3-} Exchange of ATP^{4-} for ATP^{3-}
Glutamate	Electroneutral exchange of glutamate$^-$ for OH^- (glutamate–hydroxyl antiporter)
Aspartate	Electrogenic exchange of glutamate$^-$ and H^+ for aspartate$^-$ (SLC25a12: mitochondrial aspartate–glutamate carrier-1) (SLC25a13: mitochondrial aspartate–glutamate carrier-2) (SLC25a18: mitochondrial glutamate–H^+ symporter) (SLC25a22: mitochondrial aspartate–H^+ symporter-1)
Calcium	Electrogenic uniport for Ca^{2+} Electroneutral exchange of Ca^{2+} for $2H^+$
Calcium–sodium	Electroneutral exchange of Ca^{2+} for $2Na^+$
Sodium–proton	Electroneutral exchange of Na^+ for H^+ (Na^+–H^+ antiporter)

The mitochondrial ATP^{Mg}–P_i carrier modulates the matrix adenine nucleotide pool (ATP, ADP, and AMP).[38] It is the single carrier regulated by Ca^{2+} ion. This carrier is present in renal and hepatic mitochondria, but not in the heart [53].

Communication between the mitochondrion and cytosol across the outer and high-resistance inner mitochondrial membrane relies on numerous transporters. Cytosolic ADP is exchanged with matrix ATP generated by oxidative

[38] Adenine nucleotide transport occurs as an electroneutral divalent exchange of $^{Mg}ATP^{2-}$ for HPO_4^{2-} anion. In biology, phosphorus is found as a free phosphate ion in solution and is called inorganic phosphate (generally denoted P_i) to distinguish it from phosphates bound in various phosphate compounds, especially adenosine phosphates (AMP, ADP, and ATP). In a dilute aqueous solution, phosphate exists in 4 forms. The phosphate ion (PO_4^{3-}) is the conjugate base of the hydrogen phosphate divalent ion (HPO_4^{2-}) that is the conjugate base of the dihydrogen phosphate monovalent ion ($H_2PO_4^-$), which in turn is the conjugate base of phosphoric acid (H_3PO_4). In humans, H^+- and Na^+-dependent renal type-2a sodium–phosphate cotransporter also has a preferential affinity for the inorganic phosphate species HPO_4^{2-} [52].

Table 3.24 Members of the mitochondrial adenine nucleotide translocator family, or solute carrier set SLC25A

Subtypes	Other aliases
ANT1	SLC25a4
ANT2	SLC25a5
ANT3	SLC25a6
ANT4	SLC25a31

phosphorylation using adenine nucleotide (ADP–ATP)[39] translocase (ANT1–ANT3; Table 3.24). Phosphorylation by intermembrane space Src kinases is required for maximal translocase function [55].

Voltage-dependent anion channel, or mitochondrial porin, permits metabolite exchange across the outer mitochondrial membrane. It can be controlled by cytosolic kinases (GSK3β, PKA, and NEK1) and phosphatases [55].

Translocases of the outer membrane (TOM) and of the inner membrane (TIM) are also phosphorylated, hence leading to activation and inhibition. The former regulates exchange of proteins between the cytosol and mitochondrial intermembrane space. Many other carriers exist (Table 3.25).

Phosphorylation of Enzymes and Complexes of Electron Transport Chain

Acetyl groups are produced in the mitochondria by pyruvate dehydrogenase and are transported to the cytoplasm. Citrate synthase converts acetylCoA and oxaloacetate to citrate. Citrate exits the mitochondria using the citrate–malate antiporter to be used for fatty acid synthesis. In the cytoplasm, citrate is cleaved to regenerate acetylCoA and oxaloacetate. Cytosolic oxaloacetate is reduced to malate by NADH from glycolysis; malate is supplied to citrate–malate antiporter. In the cytosol, malate is converted to oxaloacetate for gluconeogenesis. Succinate can target mitochondrial ^{ETC}complex II (succinate dehydrogenase); additional electrons are delivered into the quinone pool originating from succinate and transferred via FAD to ubiquinone.

In addition, NADH generated in the citric acid cycle may donate its electrons in oxidative phosphorylation, hence for ATP synthesis. Moreover, reduced form of flavin adenine dinucleotide ($FADH_2$) is covalently attached to succinate dehydrogenase, an enzyme functioning both in the citric acid cycle and mitochondrial electron transport chain.

The phosphorylation state of pyruvate dehydrogenase (PDH) depends on the balance between PDH kinase (PDHK) and phosphatase (PDHP). Another enzyme,

[39]Electrogenic ATP^{4-}–ADP^{3-} exchange. The charge imbalance associated with the adenine nucleotide carrier leads to a large difference in the ATP/ADP ratio between the mitochondrial matrix space and cytosol. The transport of ATP, ADP, and phosphate across the inner mitochondrial membrane costs additional energy (about 1/3 of the minimum required for ATP synthesis within the mitochondrial matrix) supplied by the respiratory chain.

Table 3.25 Other members of the SLC25A set, each encoded by a gene of the human genome

Member	Name
SLC25a2	Mitochondrial ornithine transporter
SLC25a3	Mitochondrial phosphate carrier
SLC25a7	Mitochondrial uncoupling protein UCP1
SLC25a8	Mitochondrial uncoupling protein UCP2
SLC25a9	Mitochondrial uncoupling protein UCP3
SLC25a14	Mitochondrial uncoupling protein UCP5
SLC25a15	Mitochondrial ornithine transporter
SLC25a16	Mitochondrial carrier
SLC25a17	Mitochondrial carrier; peroxisomal membrane protein
SLC25a19	Mitochondrial thiamine pyrophosphate carrier
SLC25a20	Mitochondrial carnitine-acylcarnitine translocase
SLC25a21	Mitochondrial 2-oxodicarboxylate carrier
SLC25a23	Mitochondrial calcium-binding phosphate carrier (ATP^{Mg}–P_i; SCaMC3)
SLC25a24	Mitochondrial calcium-binding phosphate carrier (ATP^{Mg}–P_i; SCaMC1)
SLC25a25	Mitochondrial calcium-binding phosphate carrier (ATP^{Mg}–P_i; SCaMC2)
SLC25a26	Mitochondrial ^{S}adenosylmethionine transporter
SLC25a27	Mitochondrial uncoupling protein UCP4
SLC25a28	Mitochondrial iron transporter-2, mitoferrin-2 (MFrn2)
SLC25a29	Mitochondrial carnitine-acylcarnitine carrier
SLC25a30	Kidney mitochondrial carrier protein-1
SLC25a32	Mitochondrial folate transporter
SLC25a33	Mitochondrial carrier
SLC25a34	Mitochondrial carrier
SLC25a35	Mitochondrial carrier
SLC25a36	Mitochondrial carrier
SLC25a37	Mitochondrial iron transporter-1, mitoferrin-1 (MFrn1)
SLC25a38	Mitochondrial carrier
SLC25a39	Mitochondrial carrier
SLC25a40	Mitochondrial carrier (various metabolic intermediates)

branched chain α-ketoacid dehydrogenase (BCKDH), is involved in the metabolism of 3 essential amino acids: isoleucine, leucine, and valine. It is targeted by BCKDH kinase and phosphatase. In addition, most of the enzymes of the citric acid cycle (succinylCoA ligase, citrate synthase, isocitrate dehydrogenase, mitochondrial malate dehydrogenase, fumarate hydratase, and aconitate hydratase) can be phosphorylated. However, phosphorylations of these enzymes may fail to modify their activity [55].

Whereas pyruvate dehydrogenase regulates the entry of reducing equivalents from glycolysis, carnitine palmitoyltransferase is involved in an another source of reducing equivalents, fatty acid oxidation. In the outer mitochondrial membrane, carnitine palmitoyltransferase CPT1, the muscle isoform of which is predominant in the heart, can be phosphorylated. Its sensitivity to malonylCoA changes, but not its catalytic activity [55]. AcylCoA dehydrogenases catalyze the initial step in each cycle of fatty acid oxidation in mitochondria.

Phosphorylation regulates the cytochrome chain. Both PKA and PKCϵ may target ^{ETC}complex-IV in particular [55].

Dephosphorylation

Protein phosphatases PP1 and PP2 localize to mitochondria. Ungrouped mitochondrial protein Tyr phosphatase PTPMt1 lodges specifically in the mitochondrial matrix; it acts in both ATP production and insulin secretion.

Phosphatase PPM1k is a soluble resident of the mitochondrial matrix that belongs to the PP1M family of Mg^{2+}- or Mn^{2+}-dependent protein Ser/Thr phosphatases. These members (at least 17) are monomeric with 2 structural domains: a common N-terminal catalytic domain and a variable C-terminal region among isoforms. Subtype PPM1k participates in the control of cell survival and mitochondrial permeability transition pore opening, but does not contribute to mitochondrial oxidative phosphorylation and respiratory chain activity [57].

Mitochondrial Protein Tyrosine Nitration

Mitochondria support formation of reactive oxygen and nitrogen species. Mitochondrial-derived, short-lived reactive species either contribute to signaling or cause mitochondrial dysfunction via oxidation according to their concentrations. Among oxidative modifications in mitochondria, peroxynitrite and other nitrating species enable the nitration of tyrosine into 3-nitrotyrosine, a post-translational oxidative modification that can alter protein structure and function [58].

Mechanisms of Protein Nitration in Mitochondria

Tyrosine nitration is achieved by a 2-step process: (1) oxidation of tyrosine to yield tyrosyl radical by 1-electron oxidants and (2) addition of nitric oxide ($NO^{\bullet}$) or dioxide $NO_2^{\bullet}$ to tyrosine. Oxidants include carbonate radical ($CO_3^{\bullet-}$), hydroxyl radical ($^{\bullet}OH$), nitrogen dioxide ($^{\bullet}NO_2$), and compound-1-like peroxidases.

Heme-containing peroxidases can oxidize diverse substrates by reacting with hydrogen peroxide (H_2O_2) as electron acceptor. Most peroxidase substrates are small aromatic molecules. However, cytochrome-C peroxidase in the mitochondrial electron transport chain that has cytochrome-C as its redox partner is an exception. The catalytic cycle of heme peroxidases is composed of several stages. In a first step, the resting ferric state (Fe^{3+}) is converted to an oxyferryl state ($Fe^{4+} = O$) with a simultaneous formation of a radical ($[Fe^{4+} = O]–R_1$), or compound 1. One oxidizing equivalent resides on iron, giving the oxyferryl intermediate, whereas in many peroxidases the porphyrin (R) is oxidized to the porphyrin π-cation radical (R_1). Compound-1 then oxidizes an organic substrate to give compound-2 ($[Fe^{4+} = O]–R_2$) and a substrate radical:

$$Fe^{3+} + H_2O_2 \rightleftharpoons [Fe^{4+} = O]R_1 + H_2O;$$
$$Fe^{4+} \rightleftharpoons OR_1 + \text{substrate} \rightarrow [Fe^{4+} = O]R_2 + \text{oxidized substrate};$$
$$Fe^{4+} \rightleftharpoons OR_2 + \text{substrate} \rightarrow Fe^{3+} + H_2O + \text{oxidized substrate};$$

The oxyferryl intermediate in cytochrome-C peroxidase is referred to as compound ES. In cytochrome-C peroxidase, the radical center is Trp191 connected to the heme via the hydrogen-bonded His–Asp–Trp triad, which is relatively easy to oxidize [59]. However, Trp191 residue is dispensable for the electron-transfer pathway, as alternative pathways exist.

Lipid peroxidation refers to the oxidative degradation of lipids. Free radicals that accept electrons from lipids of cell membranes can cause cell damage. Free radical chain reaction most often targets polyunsaturated fatty acids, because they contain multiple double bonds. In membranes, lipid peroxyl radicals ($LOO^{\bullet}$) can result from the peroxidation of polyunsaturated fatty acids by lipoxygenase. Antioxidants protect polyunsaturated fatty acids from peroxidation, as they scavenge peroxyl radicals. Vitamin-E phenoxyl radicals arise from scavenging peroxyl radicals.

Peroxynitrite anion ($ONOO^-$) and its conjugate acid (ONOOH), collectively termed peroxynitrite, are the main nitrating agents in mitochondria. Peroxynitrite anion is formed by the diffusion-controlled reaction of $NO^{\bullet}$ with $O_2^{\bullet -}$. Mitochondrial respiration is a major source of $O_2^{\bullet -}$ formation, mainly by ^{ETC}complex-I and -III, as 0.2 to 2% of oxygen consumed is converted to $O_2^{\bullet -}$ in normal conditions. Produced $O_2^{\bullet -}$ is released mostly into the matrix, but also into the intermembrane space.

Nitric oxide synthesized by a mitochondrial NOS competes with oxygen interacting with the Cu^{+}–Fe^{2+} center of cytochrome-C oxidase, thereby modulating oxygen consumption and ATP and $O_2^{\bullet -}$ creation. Increased $O_2^{\bullet -}$ concentrations shift the reactivity of $NO^{\bullet}$ toward $O_2^{\bullet -}$, the reaction kinetics becoming relevant.

Proton-catalyzed homolysis of peroxynitrite leads to $OH^{\bullet}$ and $NO_2^{\bullet}$. When the concentration of CO_2 is high (~1.4 mmol), free radicals from peroxynitrite[40] arise after the formation of the nitrosoperoxocarboxylate adduct[41] ($ONOOCO_2^-$), which undergoes homolysis[42] to form a pair of radicals, carbonate radical ($CO_3^{\bullet -}$) and nitrogen dioxide ($NO_2^{\bullet}$). Most often, these 2 radicals recombine to form carbon

[40] I.e., $ONOO^-$ reacts nucleophilically with carbon dioxide. In vivo, the concentration of CO_2 is about 1 mmol; its reaction with $ONOO^-$ occurs quickly to form nitrosoperoxycarbonate ($ONOOCO_2^-$).

[41] An adduct (Latin *adductus*: strict) is a product of a direct addition of at least 2 distinct molecules forming a single reaction product that contains all atoms of all components.

[42] Dissociation of a molecule generating 2 free radicals.

dioxide and nitrate. Nonetheless, they sometimes escape the solvent cage[43] and become free radicals, hence causing peroxynitrite-related cellular damage.

Nitrotyrosine can also result from peroxidase-catalyzed oxidation (e.g., myeloperoxidase or eosynophil peroxidases) of nitrite (NO_2^-) and tyrosine by hydrogen peroxide (H_2O_2).[44] Peroxidase compound 1 can also oxidize free or protein-bound Tyr to tyrosine.

Mitochondrial cytochrome-C promotes H_2O_2-mediated oxidation of different substrates. Peroxidase activity associated with cytochrome-C is enhanced by nitration of its own Tyr and by its interaction with cardiolipin, an abundant phospholipid of the inner mitochondrial membrane. Catalase peroxidase can also cause tyrosine nitration.

Effects of Protein Nitration

Protein tyrosine nitration does not necessarily have an effect on the function of target proteins. Nitration of mitochondrial aconitase does not affect its catalytic activity. Nonetheless, in many cases the nitration of some Tyr residues alters the structure and function (mostly inhibition) of mitochondrial targets of nitrating species. In particular, ^{Mn}SOD superoxide dismutase is inactivated by Tyr34 nitration [58]. On the other hand, nitrated mitochondrial cytochrome-C loses both its ability to transport electrons by the respiratory chain and to activate the apoptosome, but gains an augmented peroxidase activity upon Tyr74 or Tyr97 nitration.

Protein nitration can influence the protein turnover. Although polymerization of nitrated proteins limits proteasomal degradation, protein nitration can accelerate proteolysis. Protein nitration may also alter signaling. Protein nitration is reestablished during the reoxygenation.

Modulation of Protein Tyr Nitration

Mitochondrial antioxidant enzymes such as ^{Mn}SOD protect and support cell survival. Mitochondria also possess peroxiredoxin-3, which is restricted to mitochondria, and -5, which also lodges in peroxisomes; they catalyze peroxynitrite reduction to NO_2 molecule [58].

---

[43]In chemistry, the cage effect describes influence of its surroundings on properties of a molecule. In a solvent, a molecule is often supposed existing in a cage of solvent molecules, the so-called solvent cage. Reactions occur when a molecule occasionally exits and meets another molecule.

[44]Hydrogen peroxide carries out a 2-electron oxidation of heme-peroxidase produced compound-1 (obtained by monovalent reaction from NO_2^- into $NO_2^\bullet$) and compound 2 (result of another $NO_2^\bullet$ molecule forming a second molecule of $NO_2^\bullet$). Nitrogen dioxide can promote nitration of free and protein tyrosine, performing Tyr oxidation to tyrosine, followed by the addition of a second $NO_2^\bullet$ molecule [58]. However, nitrogen dioxide alone is an inefficient nitrating catalyst, as 2 $NO_2^\bullet$ molecules are needed to nitrate one tyrosine and because $NO_2^\bullet$-mediated oxidation of tyrosine is slow w.r.t. oxidation of thiols.

Glutathione (concentration in the millimolar range in mitochondrial matrix) is a peroxynitrite scavenger, but, as the reaction rate constant is quite low, alone it is not a major detoxifier of nitrating species. Nevertheless, glutathione is a major recycler of thiols in proteins, thereby protecting Tyr from nitration.

Glutathionylation

Many proteins contain free thiols that can undergo a reversible ^{S}thiolation, that is, the formation of mixed disulfides with low-molecular-weight thiols mainly via the interaction of protein thiols with glutathione. Protein glutathionylation enables defense against oxidative damage, in addition to redox state-associated signaling.

As mitochondria are major contributors of redox state and oxidative damage, glutathionylation of mitochondrial proteins has an important impact [60].

^{S}Nitrosylation

Nitric oxide is an important messenger of cardioprotection. It not only activates guanylate cyclase, but also signals by post-translational modification of protein thiol groups (^{S}nitrosylation or ^{S}nitrosation, i.e., covalent attachment of an NO group to protein thiols). ^{S}*Nitrosylation* refers to the addition of NO without a change in the formal charge of the substrate; ^{S}*nitrosation* is defined as the formation of a covalent bond between a nitrosonium equivalent (NO^{+}) and a nucleophilic amine.

Reversible ^{S}nitrosylation allows regulation of cell apoptosis, modulation of G-protein coupled receptor signaling, mediates excitation–contraction coupling, and ensures cardioprotection [61]. ^{S}Nitrosylation occurs upon stimulation that initiates NOS activation. ^{S}Nitrosylated proteins can specifically bind to other proteins and by transnitrosylation can ^{S}nitrosylate other proteins.

Role of ^{S}Nitrosylation

Reversible ^{S}nitrosylation influences the structure and activity of proteins (Tables 3.26 to 3.28). This post-translational modification is very labile, thereby causing transient affects. It protects from irreversible oxidation by shielding thiol groups.

Cardioprotection is ensured by NOS that converts substrates oxygen and arginine into NO and citrulline with cofactor tetrahydropterin (BH4). The latter has decayed concentration with aging. Uncoupled NOS, which occurs more frequently with aging or during some diseases, generates superoxide. ^{S}Glutathionylation uncouples NOS3 enzyme [61].

Numerous mitochondrial proteins undergo ^{S}nitrosylation, especially during preconditioning (∼52% ^{S}nitrosylated proteins) [61]. ^{S}Nitrosylation targets selective

Table 3.26 Effect of ^{S}nitrosylation on protein structure and function of some proteins of interest in the cardiovascular apparatus. (**Part 1**) Ion carriers (Source: [61])

Protein	Effect of ^{S}nitrosylation
Connexin-43	Increase in open probability
$K_V7.1$	Activation
$Ca_V1.2$	Inhibition or activation according to NO level and redox state
NCX1	Stimulation
RyR1/2	Increase in open probability
SERCA	Increased activity
TRP	Activation

Table 3.27 Effect of ^{S}nitrosylation on protein structure and function of some proteins of interest in the cardiovascular apparatus. (**Part 2**) Signaling effectors that intervene in cell fate for which mitochondria play a role (Source: [61]; GRK: G-protein-coupled receptor kinase). Caspase-3 (Casp3: cysteine-aspartic acid peptidase-3), an effector caspase for apoptosis is activated by extrinsic (death ligand) and intrinsic (mitochondrial) pathways. Glyceraldehyde 3-phosphate dehydrogenase (GAPDH) catalyzes step 6 of glycolysis and can initiate apoptosis [62] and control transcription of the histone gene [63]. Transglutaminases form cross-linked, generally insoluble proteic polymers that can create barriers and stable structures. Extracellular tissue transglutaminase TG2, rather than intracellular TG2, is enzymatically active. Nitric oxide promotes TG2 externalization [64]. Multifunctional, Ca^{2+}-dependent TG2 mediates cell death. However, nuclear localization of TG2 generally protects against cell death [65]

Protein	Effect of ^{S}nitrosylation
β-Arrestin	Promotion of binding to clathrin and receptor internalization
Caspase-3	Inhibition
COx2	Increased activity
Creatine kinase	Inhibition
GAPDH	Inhibition
	Increased binding to Seven in absentia (Ub ligase)
	Increased nuclear localization and degradation of nuclear proteins
GRK2	Decreased phosphorylation of β-adrenoceptors
HIF1	Stabilization
HSP90	Inhibition of ATPase activity
Myosin	Inhibition
NFκ B	Inhibition
NOS3	Inhibition
PTen	Ubiquitination and degradation
PTPn1	Inhibition
Transglutaminase	Inhibition

Table 3.28 Effect of ^{S}nitrosylation on protein structure and function of some proteins of interest in the cardiovascular apparatus. (**Part 3**) Mitochondrial constituents (Source: [61]; ALDH2: mitochondrial aldehyde dehydrogenase-2). Parkin encoded by the PARK2 gene, mutations of which cause a form of autosomal recessive juvenile Parkinson disease, is a component of an ubiquitin ligase. Together with PTen-induced putative kinase PInK1 is involved in the protection of mitochondrial structure and function. Kinase PInK1 regulates the localization of Parkin to the mitochondria [66]. Thioredoxin that resides in cytoplasm and mitochondria contributes to the defense mechanism with thioredoxin reductase against oxidative stress in mitochondria, where reactive oxygen species are primarily produced. Mutations in the PARK7 gene cause recessively inherited Parkinson disease. Multifunctional Park7 has antioxidant and transcription modulatory activity [67]. Upon oxidant challenge, cytoprotector translocates to mitochondria. Dynamin-related protein DRP1, a mitochondrial fission protein that localizes to mitochondria, contributes to mitochondrial division [68]

Protein	Effect of ^{S}nitrosylation
Aconitase	Inhibition
ALDH2	Inhibition
^{ETC}Complex-I	Inhibition
^{ETC}Complex-IV	Inhibition
DRP1	Activation of GTPase activity, increased mitochondrial fission, decreased mitochondrial fusion
F_1ATPase	Inhibition
Park7	Impairment of protein dimerization
Parkin	Initial increase then repression of ubiquitin ligase
PKB	Inhibition
Thioredoxin	Increased reductase activity Increase transnitrosylase activity

cysteine residues of specific mitochondrial proteins. Scaffold proteins facilitate interaction of NOS and proteins. Transnitrosylation also assists ^{S}nitrosylation of selected proteins by binding of proteins^{SNO}.

3.1.7.9 Mitochondrial Dysfunction

Mitochondrial damage and/or dysfunction occurs during aging and in calcium dysregulation and impaired oxidative phosphorylation with increased ROS generation, as well as atherosclerosis. Altered mitochondrial dynamics is a cause of mitochondrial dysfunction.

Mitochondrial dysfunction can also result from DNA damage. Unlike nuclear DNA, mitochondria lack protection ensured by histones and is closer to generators of free radicals. Basic DNA repair, such as mitochondrial base excision repair, removes smaller adducts incorporated by alkylation, deamination, or oxidation. Smoking can induce DNA damage and impede DNA repair [48].

Mitochondrial deletion occurs at sites of misrepaired mitochondrial DNA damage. Exposure to ROS increases mtDNA oxidative lesions and reduces mitochondrial protein concentrations and ATP production. Mitochondrial DNA damage precedes atherogenesis and is exacerbated by impaired antioxidant activity [48].

Defective oxidative phosphorylation, altered electron transport chain, especially ^{ETC}complex-I and its associated cardiolipin in the inner mitochondrial membrane, increase ROS production, hence creating DNA damage and oxidation of mitochondrial lipids and proteins, in particular adenine nucleotide translocase.

Calcium dysregulation affects several Ca^{2+}-sensitive mitochondrial enzymes, such as pyruvate, oxoglutarate, and sodium isocitrate dehydrogenases, thereby reducing substrate provision for oxidative phosphorylation, thereby promoting apoptosis. On the other hand, reduced mitochondrial ATP generation can disrupt cellular Ca^{2+} ATPase activity.

3.1.7.10 Extramitochondrial Carbonic Anhydrase

Carbon dioxide is produced abundantly in cardiac mitochondria by the tricarboxylic acid cycle as well as pyruvate dehydrogenase (~3 mmol CO_2/l/mn at rest for a myocardial O_2 consumption of 0.1 ml/mn/g) [69]. The production of CO_2 fluctuates according to the mitochondrial activity regulated by Ca^{2+} transients and the redox state.

Carbon dioxide exit from the mitochondrial matrix to the cytosol and then the extracellular space, crossing inner and outer mitochondrial membranes and then sarcolemma. Molecular hindrance restricts moderately CO_2 diffusion. Mitochondrial membranes are significantly more permeable to CO_2 gas than H^+ and HCO_3^- ions, particularly when they are enriched with gas channels (Vol. 3 – Chap. 4. Membrane Compound Carriers).

Carbonic anhydrase (CA) lodges in various sites in the cell. The 13 members of the CA family are categorized into *cytosolic* (CA1–CA3, CA7, and CA13), *mitochondrial* (CA5a and CA5b), *secreted* (CA6), and *membrane-associated* (CA4, CA9, CA12, and CA14–CA15). Ventriculomyocytes express several types of carbonic anhydrases that reside in the cytosol (CA2), mitochondria (CA5), sarcolemma (CA4 and CA14), and sarcoplasmic reticulum membrane (CA4, CA9, and CA14).

Carbonic anhydrase is involved in mitochondrial CO_2 clearance, as it catalyzes CO_2 hydration to H^+ and HCO_3^- ions as well as the reverse reaction. The activity of extracellular-facing CA isoforms accelerates CO_2 exchange between the intra- and extracellular compartments [69].

Mitochondria are surrounded by a carbonic anhydrase-rich domain that contains cytosolic CA2 and sarcoplasmic reticulum-associated CA14 isoform [69]. Extramitochondrial CA activity improves the efficiency of mitochondrial respiration and alkalinizes the mitochondrial matrix. The inner membrane pH gradient contributes to Ca^{2+} extrusion by Ca^{2+}–H^+ exchanger, phosphate uptake by P_i–H^+ cotransporter, and pyruvate ingress by H^+–monocarboxylate carrier, thereby avoiding a reduced ATP production.

3.1.7.11 Cardiac Anaplerosis

The tricarboxylic acid cycle (TCA; Vol. 1 – Chap. 4. Cell Structure and Function) generates NADH and $FADH_2$ for the electron transport chain (ETC). Adequate NADH and $FADH_2$ production requires a constant supply of acetylCoA and a constant pool of TCA substrates that possess the acetyl groups targeted for oxidation.

Most intermediates are also involved in other metabolic pathways (molecular synthesis as well as metabolic signal transmission) in mitochondria and cytosol after influx from mitochondria through appropriate transporters (*cataplerosis*). Conversely, entry of intermediates into the TCA by *anaplerosis* balance cataplerosis, thereby maintaining TCA substrate concentrations for optimal TCA activity. Anaplerotic substrates, particularly pyruvate, ensure cardioprotection. Amino acids, such as aspartate, glutamate, and glutamine, increase anaplerosis.

Carbohydrates and long-chain fatty acids are used for energy production. Insulin stimulates anaplerosis. Glucose is an important source of pyruvate. Pyruvate (plasma concentration 0.05–0.20 mmol) and lactate (plasma concentration 0.5–2.0 mmol) produced by glycolysis are substrates for energy metabolism and anaplerosis. Pyruvate is partitioned between decarboxylation and carboxylation.

In the myocardium of healthy humans, mitochondrial citrate efflux is correlated with nutrient uptake, positively with long-chain fatty acids, but negatively with glucose. Cytosolic citrate restricts utilization of both glucose, as it inhibits phosphofructokinase, and long-chain fatty acids, as it inactivates carnitine palmitoyltransferase-1 upon its cleavage by ATP–citrate lyase to acetylCoA, a precursor of malonylCoA [70].

Mechanisms and objectives of entry (anaplerosis) and exit (cataplerosis) of TCA substrates include [70]: (1) pyruvate carboxylation, a major anaplerotic pathway; (2) citrate removal; (3) succinate removal, which rises during oxygen deprivation; (4) malate decarboxylation to pyruvate; (5) transamination between oxaloacetate and α-ketoglutarate and their corresponding amino acids, aspartate and glutamate, respectively, α-ketoglutarate removal being able to balance anaplerosis rate; (6) formation of fumarate; (7) anaplerotic production (low in normal conditions) of succinylCoA from propionylCoA precursors, such as branched-chain amino acids, propionate (using propionylCoA carboxylase and methylmalonylCoA racemase and mutase), propionyl ^{L}carnitine, odd-carbon ketone bodies, and odd-carbon medium-chain fatty acids (e.g., heptanoate); (8) ATP formation by succinylCoA using substrate phosphorylation and succinylCoA synthase; (9) ketone body metabolism via 3-oxoacidCoA transferase (SCOT); (10) heme synthesis from succinylCoA due to the activation of heme oxygenase in response to oxidative stress; and (11) mitochondrial alterations, particularly opening of the permeability transition pore.

A restricted flux at any reaction of the tricarboxylic acid cycle, e.g., a loss of activity of aconitase or α-ketoglutarate dehydrogenase, may increase the entry of substrates downstream from the defective reaction, e.g., α-ketoglutarate or succinylCoA. The participation of amino acids to oxidation, such as glutamate or glutamine, may also control anaplerotic and cataplerotic pathways.

In the myocardium, TCA substrate concentrations are very low (total substrate level <2 μmol/g tissue; oxaloacetate level: 5–10 nmol/g). In the TCA, the rate of acetylCoA oxidation ranges from 0.1 to 4 μmol/mn/g [70]. The turnover time of TCA intermediates is very rapid, albeit variable, ranging from less than 1 s to 1 mn. The turnover rate for oxaloacetate reaches values up to 800 mn.

3.1.8 Ketone Bodies

Two predominant circulating ketone bodies include acetoacetate and β-hydroxybutyrate. The latter is oxidized to the former using mitochondrial β-hydroxybutyrate dehydrogenase BDH1.

Hepatic ketogenesis produces only ^{D}β-hydroxybutyrate, which is the single stereoisomer that is a BDH substrate. Cardiomyocytes, neurons, and skeletal myocytes are the 3 greatest consumers of ketone bodies.

Each enzyme involved in ketone body oxidation catalyzes a reversible reaction [71]. Therefore, ketone bodies can be synthesized and oxidized in the same cell.

3.1.8.1 Ketogenesis

Following lipolysis from triacylglycerols (or triglycerides), transport to and across the hepatocyte plasma membrane, and then cytosolic transfer to mitochondria via carnitine palmitoyltransferase, fatty acid β-oxidation creates acetoacetylCoA and acetylCoA that are the primary ketogenic substrates. The major hormonal regulators are glucagon and insulin.

Ketogenesis predominantly occurs in the liver mitochondria. Ketone bodies are released by the liver via solute carriers of the SLC16A category (SLC16a1 and SLC16a6–SLC16a7). They then circulate to be taken up by extrahepatic cells, where they can undergo a terminal oxidation [71]. Carriers of the SLC16A category do not have stereoselectivity for β-hydroxybutyrate, transferring both ^{D}β OHB and ^{L}β OHB, at least in rats.

Ketone bodies are synthesized in the liver from acetylCoA derived primarily from fatty acid oxidation (Table 3.29). They are then carried to other organes for terminal oxidation, hence contributing to energy metabolism, especially during the neonatal period and after exercise. The concentration of circulating ketone bodies

Table 3.29 Ketone body metabolism (Source: [71]; AACS: cytoplasmic AcAcCoA synthase; AcAcCoA: acetoacetylCoA; AcCoA: acetylCoA; ACC: acetylCoA carboxylase; CoA^{SH}: free coenzyme-A; Cs: cholesterol; FAO: fatty acid β-oxidation; FAS: fatty acid synthase; HMGCL: HMGCoA lyase; HMGCoA: 3-hydroxy methylglutarylCoA; HMGCR: HMGCoA reductase; HMGCS1: cytoplasmic HMGCoA synthase; HMGCS2: mitochondrial HMGCoA synthase; MaCoA: malonylCoA; NAD^+ (NADH): nicotinamide adenine dinucleotide oxidized (reduced); PDH: pyruvate dehydrogenase; TCA: tricarboxylic acid; SCOT: succinylCoA:3-oxoacidCoA transferase). Six thiolases exist (acetylCoA acyltransferases ACAA1–ACAA2, acetylCoA acyltransferases ACAT1–ACAT2, and hydroxyacylCoA dehydrogenases HADHa–HADHb; c[m]Tase: cytosolic [mitochondrial] thiolase). Cytosolic lipogenesis and cholesterol synthesis constitute nonoxidative metabolic axes of ketone bodies. The predominant circulating ketone bodies, acetoacetate (AcAc) and β-hydroxybutyrate (β OHB), are substrates for lipogenesis, the latter being oxidized to AcAc using mitochondrial β OHB dehydrogenase (BDH1). Conversely, acetoacetate is transformed to β OHB and NAD^+ using NADH. Acetoacetate can also be nonenzymatically decarboxylated to acetone and CO_2 in the mitochondrion

Ketogenesis in hepatocyte mitochondria
$\text{AcylCoA} \xrightarrow{\text{FAO}} \text{AcCoA} \xrightarrow{\text{mTase}} \text{AcAcCoA}$
$\text{AcAcCoA} \xrightarrow{\text{HMGCS2}} \text{HMGCoA} \xrightarrow{\text{HMGCL}} \text{AcAc} \xrightarrow{\text{BDH1}} \beta\ \text{OHB}$
Terminal ketone body oxidation in extrahepatic mitochondria
$\beta\ \text{OHB} \xrightarrow{\text{BDH1}} \text{AcAc} \xrightarrow{\text{SCOT}} \text{AcAcCoA} \xrightarrow{\text{mTase}} \text{AcCoA} \rightarrow \text{TCA}$
Lipogenesis and cholesterol synthesis in the cytosol
$\text{AcAc} \xrightarrow{\text{AACS}} \text{AcAcCoA} \xrightarrow{\text{HMGCS1}} \text{HMGCoA} \xrightarrow{\text{HMGCR}} \text{mevalonate} \rightarrow \text{Cs}$
$\text{AcAcCoA} \xrightarrow{\text{cTase}} \text{AcCoA} \xrightarrow{\text{ACC}} \text{MaCoA} \xrightarrow{\text{FAS}} \text{lipogenesis}$

increases from about 50 μmol in the normal fed state to 7 mmol and to 1 mmol after 16 to 20 h of fasting in healthy adult humans as well as 20 mmol in diabetic ketoacidosis [71].

3.1.8.2 Ketone Body Oxidation and Nonoxidative Catabolism

Ketolysis happens either in the mitochondrion or cytosol leading to terminal oxidation (ketone body oxidation) or lipogenesis and sterol synthesis (nonoxidative catabolism). Ketone body metabolism changes during aging. It is disturbed in type-1 and -2 diabetes and heart failure.

Ketone body catabolism generates acetylCoA that can be terminally oxidized within the tricarboxylic acid cycle as an alternative, glucose-sparing fuel source, especially in the heart and skeletal muscles. Moreover, neurons that do not effectively generate high-energy phosphates from fatty acids oxidize ketone bodies during starvation as well as in the neonatal period.

Ketone bodies are extracted from the blood circulation and taken up by cells via SLC16a1 and SLC16a7 transporters [71]. Within mitochondria, β-hydroxybutyrate dehydrogenase BDH1 catalyzes the oxidation of β OHB to acetoacetate. The latter is transformed into acetoacetylCoA by succinylCoA:3-oxoacidCoA transferase (SCOT), which is synthesized in all cell types endowed of mitochondria, except hepatocytes. It exchanges coenzyme-A between succinate and acetoacetate. The free energy released by hydrolysis of acetoacetylCoA is greater than that of succinylCoA. Therefore, the reaction favors the formation of acetoacetate.

AcetylCoA is also rapidly used by citrate synthase to feed the tricarboxylic acid cycle and competes with acetoacetylCoA formation. A reversible reaction by acetylCoA thiolase between acetoacetylCoA and acetylCoA yields 2 molecules of acetylCoA that can enter the tricarboxylic acid cycle. On the other hand, succinylCoA:3-oxoacidCoA transferase enables ketone body oxidation. Ketone bodies are oxidized in proportion to their delivery until saturation of uptake or oxidation [71]. Synthesis of the CoA transferase SCOT and activity diminish during states of sustained ketosis. In the absence of SCOT, ketone bodies cannot be oxidized.[45]

On the other hand, ketone bodies are also used for sterol synthesis and lipogenesis. This *nonoxidative metabolism* happens particularly in the developing brain, liver, and lactating mammary gland [72].[46] It is initiated by cytoplasmic acetoacetyl-CoA synthase that converts acetoacetate to acetoacetylCoA. The latter serves as a direct substrate for cytoplasmic HMGCoA synthase HMGCS1 that catalyzes the committing step of sterol synthesis. Lipogenesis starts with acetoacetylCoA processing by a thiolytic cleavage to yield acetylCoA. The latter is transformed to the lipogenic substrate malonylCoA upon carboxylation by acetylCoA carboxylase.

3.1.8.3 Role of Ketone Bodies

Ketone bodies support energetic homeostasis, particularly in the brain, heart, and skeletal muscle during ketogenic diet (low-carbohydrate, high-lipid diet, i.e., when carbohydrates are lacking, but fatty acid are available).

Glucose metabolism accounts for less than 1% of circulating ketone bodies in the state of low-carbohydrate intake, because pyruvate predominantly enters the hepatic tricarboxylic acid cycle via carboxylation to oxaloacetate or malate rather

[45] SCOT-knockout mice develop hyperketonemic hypoglycemia and die within 48 h of extrauterine life [71].

[46] D3-Hydroxy[3-14C]butyrate is incorporated into lipid in lactating mammary glands of rats, a major site of ketone body utilization. This incorporation decreases in short-term insulin deficiency (2 h) and starvation (24 h), but increases again on refeeding (2 h) [72]. The activity of cytosolic acetoacetylCoA synthase follows changes in nutritional state, but is not affected by short-term insulin deficiency.

Table 3.30 Role of ketone body metabolism (Source: [71]; GPCR: G-protein-coupled receptor; HDAC: histone deacetylase; ROS: reactive oxygen species). Ketone body oxidation represents the contribution of acetylCoA derived from ketone bodies to the tricarboxylic acid cycle. This contribution in extrahepatic cells via citrate synthase is called terminal oxidation of ketone bodies; it converts 1 mol of acetylCoA to 2 mol of CO_2 and 2 mol of NADH. Like niacin, $^{D}\beta$-hydroxybutyrate is a ligand for GPR109a

Process	Effect and target
	Metabolism of ketone bodies
Ketogenesis (hepatic)	Use of acetylCoA derived from fatty acid β-oxidation
Oxidation	Improvement of ATP production rate per oxygen
	Reduction of ROS generation
Catabolism	Lipogenesis and sterol synthesis
	Signaling by ketone bodies
GPCR	GPR109a (Gi-coupled niacin receptor NiacR1)
	GPR41 (Gi-coupled short-chain fatty acid receptor)
Membrane excitability	Inhibition of K_V4 by $^{L}\beta$ OHB in ventriculomyocytes (transient outward K^+ current) (action potential prolongation)
Oxidative stress	HDAC-dependent regulation of antagonist mediators

than decarboxylation to acetylCoA [71]. Amino acid catabolism yields a small percentage of circulating ketone bodies, leucine catabolism generating up to 4% of circulating ketone bodies.

Ketone bodies serve in the synthesis of lipids, especially in the developing brain, lactating mammary gland, and liver. The metabolism of ketone bodies interferes with [71]: (1) β-oxidation of fatty acids;[47] (2) the tricarboxylic acid cycle and mitochondrial electron transport chain; (3) lipogenesis[48] and sterol synthesis; (4) glycolysis and gluconeogenesis; and (5) cell signaling (Table 3.30).

After a meal, excess fuel substrates are processed and stored as triglycerides in adipocytes. During fasting, triglycerides are hydrolyzed to free fatty acids and glycerol (lipolysis); the former are released into the blood stream and subsequently used by cells as energy substrates.

In humans, the 3 main hormones that regulate lipolysis are insulin that inhibits lipolysis, as well as catecholamines (adrenaline and noradrenaline) and glucagon

[47]β-Oxidation is the sequential derivation of 2 carbon units in the form of acetylCoA from fatty acyl chains. It mainly occurs within the mitochondrial matrix and sometimes in peroxisomes.

[48]I.e., synthesis of new lipids from acetylCoA using acetylCoA carboxylase and fatty acid synthase.

that stimulate lipolysis.[49] Like niacin,[50] ${}^{D}\beta$-hydroxybutyrate is a ligand for GPR109a receptor that can hinder lipolysis in adipocytes, hence limiting delivery of nonesterified fatty acids to the liver [71].[51] In addition, GPR109a promotes reverse cholesterol transport in macrophages.

In mice, β-hydroxybutyrate decreases sympathetic outflow, as it antagonizes the Gi/o-coupled receptor GPR41 for short-chain fatty acids (scFA; e.g., acetate, propionate, and butyrate) that abounds in sympathetic ganglia [71].

${}^{D}\beta$-Hydroxybutyrate also inhibits class-1 histone deacetylases, thereby increasing transcription of genes that encode mediators of resistance to oxidative stress.

---

[49]In obesity, lipolysis is deregulated. The basal lipolysis rate rises. The concentration of circulating free fatty acids augments, hence provoking insulin resistance. Stimulation of lipolysis by catecholamines as well as its inhibition by insulin are precluded.

[50]Nicotinic acid, also termed niacin, vitamin-B3, and vitamin-PP, is used in the treatment of dyslipidemia. It increases the concentration of high-density lipoproteins and decreases that of very-low-density and low-density lipoproteins [73]. Nicotinic acid also lowers plasma concentrations of free fatty acids and triglycerides. It decreases lipolysis in adipocytes, as it impedes cAMP accumulation via the Gi–ACase axis and hormone-sensitive triglyceride lipase [74].

[51]Two subtypes of GPR109 exist (GPR109a–GPR109b) that are also called nicotinic acid and niacin receptor-1 and -2 (NiacR1–NiacR2) as well as protein upregulated in macrophages by interferon-γ (PUMaγ) in mice and HM74a and HM74b (HM74) in humans. NiacR1 and NiacR2 are high- and low-affinity receptor for nicotinic acid, respectively [73]. Receptor NiacR1 is highly expressed in adipose tissue and spleen. In fact, NiacR1 is also detected in the lung and trachea, whereas NiacR2 resides in the lung, adipose tissue, and spleen, as well as leukocytes. The ligand-inducible nuclear receptor NR1c3 (or PPARγ) is necessary and sufficient for adipogenesis, as it controls the differentiation, maintenance, and function of adipocytes. It also regulates numerous genes of the adipocytic phenotype, such as those involved in lipid uptake (e.g., lipoprotein lipase, scavenger receptor ScaRb3, and oxidized LDL receptor) and synthesis, lipid storage and lipid droplet stabilization (e.g., perilipin), glycerol and fatty acid recycling (i.e., reesterification of fatty acids and glycerol to triglycerides), and fatty acid oxidation [75]. The transcription factor NR1c3 has 2 isoforms, NR1c3a (PPARγ1) and NR1c3b (PPARγ2). The latter that has a longer N-terminus is restricted to adipocytes. The former is more widely distributed (e.g., adipocytes, enterocytes, monocytes, and macrophages). Polyunsaturated fatty acids and eicosanoids activate NR1c3 factor. Factor NR1c3 heterodimerizes with retinoic acid X receptors (NR2b) and binds to PPAR-responsive elements (PPREs). The human *antilipolytic G-protein-coupled receptors* comprise GPR81, GPR109a, and human-specific GPR109b. The genes that possess a NR1c–NR2b-binding site in their promoter, such as those encoding GPR81, GPR109a, and GPR109b, contribute to the reduction of circulating free fatty acids [75]. The Gpr81, Gpr109A, and Gpr109B genes colocalize in chromosome 12q24.31. In addition, certain aromatic ^{D}amino acids, including ^{D}phenylalanine, ^{D}tryptophan, and the metabolite of the latter, ^{D}kynurenine, connect to GPR109b, which abounds in human neutrophils, but not GPR109a [76]. ^{D}Isomers may operate as hormonal, antibacterial, or modulatory peptides of immunity, or neuropeptides. They serves as chemoattractants for neutrophils via activated GPR109b. They diminish the activity of adenylate cyclase and elicit a transient influx of calcium ions. The potent chemotactic factor for eosinophils and neutrophils, 5-oxo-eicosatetraenoic acid, links to GPR48 receptor, which has high sequence similarity to GPR109b. Other chemotactic GPCRs comprise GPCRs for leukocyte chemoattractants CXCL8, C5a, ^{N}formyl methionyl-leucyl-phenylalanine (fMLP), platelet-activating factor, and leukotriene-B_4), as well as human, neutrophil-specific, Gi/o- and Gq-coupled GPR43 receptor for short chain fatty acids, such as sodium acetate and sodium propionate, which are by-products of anaerobic bacteria.

Table 3.31 AcetylCoA (AcCoA) in mitochondria and the cytosol and ketogenic flux between these 2 compartments; monocarboxylate and tricarboxylate carriers (Source: [71]; ATP, adenosine triphosphate; OAA: oxaloacetate; PEP: phosphenol pyruvate; PEPCK: phosphenolpyruvate carboxykinase; PK, pyruvate kinase). AcetylCoA (AcCoA) derives from pyruvate (Pyr) using pyruvate dehydrogenase or fatty acid oxidation (FAO). It is converted either by citrate synthase to citrate that feeds the tricarboxylic acid cycle (TCA) or acetoacetylCoA (AcAcCoA). Pyruvate (Pyr) either enters the tricarboxylic acid cycle by anaplerosis that preserves terminal oxidation of acetylCoA. The primary anaplerotic reaction is catalyzed by pyruvate carboxylase, which generates oxaloacetate from pyruvate. Oxidation of odd-chain fatty acids and amino acid catabolism also yield anaplerotic substrates. The tricarboxylic acid cycle coupled to the mitochondrial electron transport chain (ETC) produces ATP. Citrate can exit the mitochondrion using the citrate-dependent tricarboxylate carrier and, in the cytosol, be processed by ATP–citrate lyase to create acetylCoA. AcetoacetylCoA is processed by succinylCoA:3-oxoacidCoA transferase (SCOT) using succinate into acetoacetate (AcAc) and succinylCoA. Acetoacetate crosses the mitochondrial enveloppe using the monocarboxylate carrier. In the cytosol, acetoacetate can be processed by thiolase or acetylCoA carboxylase into acetoacetylCoA or malonylCoA, thereby leading to lipogenesis or cholesterol synthesis

Glucose (Glc) metabolism	
Glycolysis: Glc $\rightarrow$ PEP $\xrightarrow{\text{PK}}$ pyruvate	
Pyr $\rightarrow$ AcCoA or direct entry into TCA	
Gluconeogenesis: OAA $\xrightarrow{\text{PEPCK}}$ PEP $\rightarrow$ Glc	
AcetylCoA	ATP production (TCA–ETC)
	Citrate production followed by entry into TCA or exit from the mitochondrion using the tricarboxylate carrier
Fatty acid (FA) metabolism	
Fatty acid $\rightarrow$ acetylCoA and acetoacetylCoA	
AcetylCoA	ATP production (TCA–ETC)
	AcetoacetylCoA–acetoacetate axis (using thiolase and SCOT successively)
	Acetoacetate exit from the mitochondrion using the monocarboxylate carrier (ketogenic flux)

The mitochondrial pool of acetylCoA controls and is regulated by ketogenesis (Table 3.31). AcetylCoA activates [71]: (1) pyruvate carboxylase that increases anaplerotic entry of metabolites into the tricarboxylic acid cycle and (2) pyruvate dehydrogenase kinase, which phosphorylates and inhibits pyruvate dehydrogenase, thereby further augmenting flux of pyruvate into the tricarboxylic acid cycle. Cytoplasmic acetylCoA inhibits fatty acid oxidation. AcetylCoA carboxylase catalyzes the conversion of acetylCoA to malonylCoA, the lipogenic substrate and inhibitor of mitochondrial carnitine palmitoyltransferase-1, thereby decreasing delivery of acyl chains to the mitochondrial matrix for terminal oxidation [71].

3.1.8.4 Energetics of Myocardial Ketone Body Oxidation

In the normal adult heart, more than 95% of ATP needed for its metabolic and signaling activities as well as electrical and mechanical functions is supplied by mitochondrial oxidative phosphorylation. Fatty acid oxidation yields up to 70% of ATP, whereas the rest derives from the metabolism of glucose, lactate, amino acids, and ketone bodies.

The myocardium is the highest ketone body consumer per unit mass [71]. The myocardium oxidizes ketone bodies at the expense of fatty acid oxidation. In the myocardium, anaplerosis (replenishment of TCA intermediates) enables maintenance of energetics. SuccinylCoA:oxoacidCoA transferase that transforms acetoacetate to acetoacetylCoA sequesters free CoA; hence, CoA precursors or anaplerotic substrates are needed to maintain anaplerosis.

Ketone body oxidation is more energetically efficient than terminal fatty acid oxidation. In addition, unrestrained mitochondrial fatty acid oxidation may augment the generation of reactive oxygen species to a level that exceeds scavenging. Ketone bodies diminish ROS formation with respect to fatty acid oxidation, as they support free radical scavenging and maintain ubiquinone in the oxidized state [71]. In cardiomyocytes, when the ability to switch between fatty acid and glucose sources is altered, contribution of ketone bodies to the tricarboxylic acid cycle may be adaptive.

During ketogenic diet, the myocardium engages a transcriptional program that modulates myocardial utilization of ketone bodies, as it suppresses OXCT1 expression, hence SCOT synthesis [71].

3.1.9 Matching of ATP Supply to Demand

Cells need mechanisms that adapt the rate of formation of ATP via mitochondrial oxidative phosphorylation to energetic requirements. Synthesis of ATP can be modified by change in activity of mitochondrial dehydrogenases (pyruvate, isocitrate, and oxoglutarate dehydrogenase) that generate NADH and $FADH_2$ for the mitochondrial electron transport chain upon elevated mitochondrial free Ca^{2+} concentration as well as augmented [ADP]/[ATP], [NAD^+]/[NADH], and/or [CoA]/[acetylCoA] ratio.

Among its manifold functions (e.g., muscle contraction trigger and second messenger of many hormonal signals), Ca^{2+} stimulates glycogen catabolism and regulates the tricarboxylic acid cycle, as it activates Ca^{2+}-dependent dehydrogenases. Therefore, the same signal stimulates muscle contraction and ATP production to sustain contraction using many feedback loops that couple NADH reoxidation and ATP production to oxygen consumption and energy expenditure.

The firing rate of spontaneous action potential of sinoatrial cells involves activity of Ca^{2+}–calmodulin-activated adenylate cyclases, cAMP-activated protein kinase PKA, and Ca^{2+}–calmodulin-dependent protein kinase CamK2 that phosphorylates sarcoplasmic reticulum and plasma membrane ion carriers. In response to β-adrenergic receptor activation, elevated firing rate of spontaneous action potential increases the demand for ATP delivery. Augmented ATP production accompanied by an increased O_2 consumption must then match required ATP demand. This matching is achieved via a greater mitochondrial Ca^{2+} contribution and enhanced Ca^{2+}–cAMP–PKA–CamK2 signaling to mitochondria [77].

In addition to activation of soluble guanylate cyclase, nitric oxide modulates transiently and reversibly the oxygen consumption, as this messenger generated in small quantities is a competitive inhibitor of cytochrome oxidase [78].

Mitochondrial free calcium activates certain TCA dehydrogenases, thereby increasing the rate of oxygen consumption on the one hand, and mitochondrial nitric oxide synthases, thus decreasing the rate of oxygen consumption on the other. Activated Ca^{2+}-activated dehydrogenases, firstly α-ketoglutarate dehydrogenase and, with higher Ca^{2+} concentrations, pyruvate and isocitrate dehydrogenases that have a lower affinity for Ca^{2+}, and then stimulated constitutive forms of mtNOS (mtNOS1 and mtNOS3) modulates the tricarboxylic acid cycle activity [78]. When oxygen supply is limited, nitric oxide slows down the mitochondrial metabolism.

Calcium activates α-ketoglutarate dehydrogenase, thereby processing α-ketoglutarate and subsequently relieving inhibition of aspartate aminotransferase that stimulates the malate–aspartate carrier, which is driven by the mitochondrial proton motive force. Resulting aspartate egress from mitochondria is associated with glutamate ingress in mitochondria through Ca^{2+}-stimulated aspartate–glutamate carrier. The cytosolic ratio between lactate and pyruvate concentration declines, promoting pyruvate and glucose oxidation [78].

3.1.10 Myocardial Oxygen Consumption

The total oxygen consumption (2–10 ml/mn/100 g) depends on the subject and age. The heart has the highest arteriovenous O_2 difference. A large ATP amount in the cardiomyocyte originates from aerobic metabolism.

Myocardial contraction accounts for at least about 75% of the cardiac oxygen consumption (MV_{O_2}). The myocardial oxygen consumption can be assumed to be an affine function of the heart's mechanical work (stroke work W) done during contraction and basal oxygen consumption ($MV_{O_2}b$):

$$MV_{O_2} = \kappa W + MV_{O_2}b. \tag{3.7}$$

Blood is supplied to the myocardium by the coronary arteries. The entry orifices of the coronary arteries localize to the upper part of the Valsava sinuses of the aortic root at the upper edges of open valve cusps. The coronary blood flow is mainly produced by 2 successive fluid–structure interaction phenomena: (1) systolic blood ejection produced by cardiac pump contraction and (2) diastolic recoil of elastic arteries, mainly the aorta. The latter produces a backflow toward the closed aortic valve with additional input in coronary arteries.

Coronary arteries run on the heart surface. The main arteries branch off in several large vessels that remain on the epicardium, the path of which experiences a deformation during the cardiac cycle. Large coronary branches give birth to a set of small arteries that penetrate into the cardiac wall. Both the downstream compartment of the coronary macrocirculation and the microvasculature bear strong variations of the transmural pressure caused by the working myocardium. Intramural blood vessels are more or less compressed during the myocardial contraction. Therefore, intramural perfusion is more or less hampered by the systolic compression. On the other hand, the coronary drainage is transiently enhanced by the compression of the upstream compartment of the venous network, like the blood return in valved veins of inferior limbs improved by the contraction of surrounding muscles, hence external temporary compression. Conversely, intramural artery expansion during diastole produces a beneficial aspiration effect associated with stretching of the superficial coronary bed.

Oxygen extraction in the capillary bed is more effective during diastole because capillaries, which cross the relaxed myocardium, are dilated. The coronary blood flow is equal to about 5% of the cardiac output. Whenever O_2 demand increases, various substances promote coronary vasodilation: adenosine, lactate, nitric oxide, K^+ ions, and prostaglandins.

The coronary perfusion pressure depends on the aorta root pressure, which varies strongly during the cardiac cycle. Activated sympathetic nerves innervating coronary arteries cause transient vasoconstriction mediated by α1-adrenergic receptors. A brief vasoconstriction is followed by a vasodilation due to augmented vasodilator production and β1-adrenergic receptor activation. In addition, parasympathetic stimulation of the heart provokes a slight coronary vasodilation.

Energy consumption by the sarcomere is determined by the kinetics of cross-bridging. The number of cross-bridges and the rate of cross-bridge recruitment is determined by the kinetics of calcium–troponin bonds and sarcomere length (i.e., the number of available myosin-binding sites on actin). The *Fenn effect* is related to the adaptation of energy liberation by the myocardium contraction when the load undergone by the myocardium varies during contraction.

Oxidative phosphorylation, which depends on oxygen transfer, is regulated to adapt to fluctuations of either oxygen supply or needs. A normal heart can rise its oxygen use about 20 times. Oxygen consumption increases particularly with the cardiac frequency, but electron transfer in the respiratory chain can be reduced due to insufficient oxygen supply.

Oxygen availability in mitochondria depends on its transport from adjoining coronary capillaries across the extracellular medium and cardiomyocyte. However, both the extra- (<2 μm) and intracellular (<10 μm) spaces represent a very short transport distance. Furthermore, a mitochondrium population is located near the sarcolemma and myoglobin carry oxygen in the cytosol down to mitochondria in the cell core. Yet, the intracellular transport is not fast enough to match the rate of consumption. Consequently, gradients in oxygen concentration can be observed in the cytosol of cardiomyocytes, although the myoglobin near the sarcolemma is almost fully saturated with oxygen [79]. Nevertheless, intrinsic respiratory regulation in mitochondria of the cell center can compensate for relatively slow oxygen transport within cardiomyocytes, electron transfer being sustained despite reduced oxygen supply. The electrochemical gradient across the mitochondrial membrane is thereby maintained for suitable mitochondrial functioning.

Assessment of oxygen usage can be performed with positron emission tomography using labeled acetate (or acetic acid; ^{11}Cacetate) as a positron-emitting radionuclide (tracer) [80].[52] Acetate is quickly incorporated and processed into acetylCoA in cells. AcetylCoA can enter into the tricarboxylic acid cycle, thereby being involved in energy production. This catabolism predominates in the myocardium. It can also participate in lipid synthesis and cellular membrane building. This anabolism dominates in tumors.[53] In the tricarboxylic acid cycle, acetate is oxidized and the radioactivity is transferred to carbon dioxide. Myocardial clearance of the radiochemical acetate relies on mitochondrial oxidative flux (i.e., on oxygen consumption).

3.1.11 Cardiac Electrophysiology

Cardiac rhythm is achieved by electrochemical activity generated and propagated in the nodal tissue by a precise temporal control of opening and closing of ion carriers of nodal and common cardiomyocytes (Vols. 3 – Chap. 3. Main Classes of Ion Channels and Pumps and 5 – Chap. 5. Cardiomyocytes). These regular impulses, or action potentials, spread throughout the myocardium to trigger its contraction and propel blood in the systemic and pulmonary circulation.

Mechanisms of sinoatrial node pacemaking (i.e., pacemaker spontaneous depolarization) rely on: (1) activity of the hyperpolarization-activated cyclic nucleotide-gated cation (HCN) channel (depolarizing pacemaker "funny" i_f current) that prevents excessive hyperpolarization and slow pacemaking; (2) activation of $Ca_V1.2$ channel ($i_{Ca,L}$ current) that can generate a pacemaker activity; (3) deactivation of the delayed rectifier K^+ channels ($i_{K,r}$ and $i_{K,s}$ currents) that modulate action potential amplitude and frequency; (4) sustained (steady) inward current ($i_{Na,sus}$) essentially through Na^+ channels that counteracts hyperpolarization; and (5) spontaneous Ca^{2+} release from sarcoplasmic reticulum and subsequent activation of Na^+–Ca^{2+} exchanger (i_{NaCaX} current).

The generation and propagation of the electrochemical command is disturbed in the case of channelopathies. The latter often result from gene mutations. In animals, molecular technology enables genome transformation by removing, inserting, substituting, inhibiting, or overexpressing genes to test the function of target proteins. However, the genotype does not directly determine the phenotype

---

[52] Acetate is rapidly taken up by cells and is transformed to acetylCoA in both the cytosol and mitochondria by acetylCoA synthase. AcetylCoA is a common metabolic intermediate for synthesis of cholesterol and fatty acids, which are incorporated into membranes. In mitochondria, acetylCoA is also oxidized in the tricarboxylic acid cycle to carbon dioxide and water. The radiochemical acetate is used as a positron emission tomography (PET) tracer for studying myocardial oxidative metabolism and regional myocardial blood flow.

[53] In tumoral cells, overexpressed fatty acid synthase converts most of the acetate into fatty acids that are incorporated into intracellular phosphatidylcholine-based membrane nanodomains.

because of intermediary events. In addition, the existence of a dysfunctional protein can be associated with changes in production and/or regulation of multiple other proteins used in compensatory mechanisms.

3.1.11.1 Adaptors and Scaffolds

Ion channels (Vols. 3 – Chap. 3. Main Sets of Ion Channels and Pumps and 5 – Chap. 5. Cardiomyocytes) are components of molecular complexes embedded in membrane lipids. The cardiomyocyte cytoskeleton modulates ion channel function, in particular Na^+ channel kinetics. Binding proteic partners serve in feedback ion sensing, enzymatic regulators, and adaptors and scaffolds. Among the latter, A-kinase-anchoring proteins and ankyrins recruit components of molecular complexes associated with ion carriers.

Ankyrins

The ankyrin family members are encoded by 3 genes (ANK1–ANK3) that produce ankyrin-1,[54] ankyrin-2,[55] ankyrin-3,[56] respectively. They contain 3 functional domains: N-terminal membrane-binding, spectrin-binding, and C-terminal death domain. They thus bind to cell membranes and the spectrin-based cytoskeleton. They coordinate localization and assembly of interacting proteins, hence ensuring proper insertion of ion carriers into appropriate cellular membranes and their clustering.

In ankyrin-$2^{-/-}$ ventriculomyocytes, i_{Na} current is smaller than in wild-type myocytes, the former exhibiting a hyperpolarizing shift in activation and inactivation kinetics [81]. Mutations of the ANK2 gene disrupt the cellular organization of ankyrin-2-binding ion carriers, such as sodium–potassium pump, sodium–calcium exchanger, and inositol trisphosphate receptors. The densities of these proteins in transverse tubules as well as their overall levels is reduced [82]. Mutations of the ANK2 gene also alter Ca^{2+} signaling in cardiomyocytes, hence provoking extrasystoles, in addition to type-4 long-QT arrhythmia (LQT4; Vol. 7 – Chap. 1. Heart Pathologies). Moreover, protein phosphatase-2 regulatory subunit $PP2_{r5\alpha}$ is a binding partner for ankyrin-2 [83].

AKAPs

A-kinase anchor proteins constitute a set of proteins with diverse structure that share binding to the regulatory subunit of protein kinase-A, thereby confining

[54] A.k.a. ankyrin-R, R standing for restricted expression and RBC-related isoform.

[55] A.k.a. ankyrin-B referring to broad expression and brain-related isoform.

[56] A.k.a. ankyrin-G (general expression).

the holoenzyme to adequate sites within the cell. In addition, AKAP6 binds to RyR2, AKAP7α to $Ca_V1.2$, AKAP7δ to the SERCA–phospholamdan complex, and AKAP12 to β-adrenoceptor.

Scaffold AKAP9 connects to α ($K_V7.1$ encoded by the KCNQ1 gene) and β (minimal K^+ channel MinK encoded by the KCNE1 gene) channel subunits that build the slowly inactivating K^+ channel, which causes the slow delayed rectifier K^+ current ($i_{K,s}$) and is strongly regulated by β-adrenergic receptors. The latter mediate sympathetic regulation of cardiac action potential duration. The anchoring protein AKAP9 recruits PKA and PP1, thereby controlling the phosphorylation state of $K_V7.1$ and modulating its activity [84].

3.1.11.2 Enzymatic Regulation

The cardiac ryanodine receptor required for excitation–contraction coupling tethers to multiple regulatory subunits, such as calmodulin and calstabin-1 and -2 (FKBP12 and FKBP12.6), kinases, such as PKA and CamK2, and phosphatases, such as PP1 and PP2, the latter via $PP2_{r3\alpha}$ [85]. These partners modulate RyR2 function in response to extracellular signals. Phosphorylation of RyR channels by PKA reduces the affinity for calstabin, thus activating channels and increasing calcium release from intratracellular stores.

Calcium–Calmodulin-Dependent Protein Kinase CamK2

Calcium–calmodulin-dependent protein kinase CamK2 directly and indirectly modulates the function of depolarizing, voltage-gated $Na_V1.5$ and $Ca_V1.2$ channels, repolarizing $K_V4.3$ channels that cause transient outward K^+ currents ($i_{K,to}$), $K_{IR}2.1$ channels responsible for inwardly rectifying K^+ current (i_{K1}), ryanodine receptor RyR2, sarco(endo)plasmic reticulum Ca^{2+} ATPase, and phospholamban, as well as gene transcription (Table 3.32). Increased CamK2 synthesis and activity such as in heart failure destabilizes excitation–contraction coupling. Overall changes to Ca^{2+} handling during CamK2 overexpression result from interactions between CamK2 substrates.

Protein Kinase-A

Activated β-adrenergic receptor increases levels of cAMP that stimulates PKA kinase. The latter then phosphorylates numerous substrates, such as $Na_V1.5$, $K_V7.1$, RyR, SERCA, Na^+–K^+ ATPase, troponin-I, and glycolytic enzymes (Table 3.33). In addition, PKA phosphorylates HDAC5 that then inhibits cardiac fetal gene expression and cardiomyocyte hypertrophy [87].

Table 3.32 Calcium–calmodulin-dependent protein kinase in cardiomyocyte (Source: [86]). This kinase contributes to the regulation of the cardiac function, as it targets ion carriers of the sarcolemma and the sarcoplasmic reticulum. In the nucleus, it phosphorylates HDAC5 histone deacetylase (MEF: myocyte enhancer factor)

Substrate	Role
Ion carriers	
$Na_V1.5$	Depolarization
$Ca_V1.2$	Electromechanical coupling
$K_V4.2/3$	Transient outward K^+ current ($i_{K,to}$)
$K_{IR}2.1$	Inwardly rectifying K^+ current (i_{K1})
RyR2	Electromechanical coupling
SERCA	Cardiomyocyte relaxation
Phospholamban	SERCA inhibitor (unphosphorylated state)
	PKA substrate (sympathetic stimulation)
Gene transcription regulators	
HDAC5	MEF2 repressor

Table 3.33 Protein kinase-A in cardiomyocyte (CMC; PFK2/F(2,6)BPase: bifunctional regulator with both kinase [PFK2: phosphofructokinase-2] and phosphatase [F(2,6)BPase: fructose (2,6)-bisphosphatase] activity)

Substrate	Role
$Na_V1.5$	Depolarization
Na^+–K^+ ATPase	Repolarization
$Ca_V1.2$	Electromechanical coupling
$K_V7.1$	Slow delayed rectifier K^+ current ($i_{K,s}$)
RyR	Eexcitation–contraction coupling
SERCA	CMC relaxation
Phospholamban	Weakening of inhibition of SERCA2a
Troponin-I	CMC contraction
PFK2/F(2,6)BPase	Glycolysis regulation

In cardiomyocytes, PKA phosphorylates ryanodine-sensitive calcium channel (Ser2809), thereby increasing the open RyR probability and Ca^{2+} release from the sarcoplasmic reticulum.

Activated PKA heigthens Na^+ current through the $Na_V1.5$ channel involved in the initiation and conduction of action potentials. It is thus responsible for the action potential upstroke that initiates excitation–contraction coupling in the heart. In fact, $Na_V1.5$, the pore-forming α-subunit of the cardiac voltage-dependent Na^+ channel, has numerous binding partners.[57] In humans, reduced cardiac sarcolemmal Na^+

[57] Interacting proteins include PKA, CamK2, methylase, cytosolic protein Tyr phosphatase PTPn3, glycerol-3-phosphate dehydrogenase 1-like protein (GPD1L), α-actinin, ankyrin-3, calcium ion, calmodulin, caveolin-3, N-cadherin, connexin-43, disc large homolog DLg1, fibroblast growth

current causes bradyarrhythmias and sick sinus and Brugada syndrome as well as progressive cardiac conduction defect. Protein kinase-A increases $Na_V1.5$ transfer to the plasma membrane from both the perinuclear and cortical regions [88]. It may also modulate cardiac Na^+ flux [89].

It also phosphorylates phospholamban (Ser16), thereby relieving SERCA inhibition and causing faster kinetics of cytosolic Ca^{2+} decay. Phosphorylation of phospholamban by PKA actually causes its dissociation from SERCA, increasing the rate of Ca^{2+} reuptake into the sarcoplasmic reticulum.

Protein kinase-A constitutes molecular complexes, the localization and assembly of which are mediated by AKAP anchors. In ventriculomyocytes, AKAP5 scaffold recruits PKA and PKC kinases, PP3 phosphatase, and adenylate cyclase to a caveolin-3-associated complex that also contains $Ca_V1.2$ channel and β-adrenoceptor [90]. The AKAP5–PKA-containing complex controls Ca^{2+} uptake, as it phosphorylates (activates) $Ca_V1.2$ channels. Afterward, Ca^{2+} binds to PP3 and calmodulin. Activated PP3 then dephosphorylates the channel, in addition to PP2 phosphatase.

A PKA-containing complex also promotes Ca^{2+} release from the sarcoplasmic reticulum, as it activates ryanodine Ca^{2+} channels coupled with the Ca^{2+}-binding junctin–triadin–calsequestrin complex, as well as Ca^{2+} storage in the sarcoplasmic reticulum, as it stimulates SERCA pumps [91].

When located at the cytoskeleton, a PKA-containing complex regulates mitochondrial activity necessary for contraction.

In addition to its role as an intermediate in the citric acid cycle, succinate acts as a signaling molecule in cardiomyocytes. It modulates global Ca^{2+} transient and cell survival via a PKA pathway [92]. It first targets succinate receptor-1 (SucnR1), a G-protein-coupled receptor (GPR91) expressed by cardiomyocytes.[58]

3.1.11.3 Calcium Influx

Calcium activates and regulates multiple processes in every cell type. In the cardiomyocyte, cyclic fluctuations in cytosolic Ca^{2+} concentration trigger and regulate the strength of cardiac contraction (contractile Ca^{2+} influx). Changes in Ca^{2+} level can regulate myocyte growth and viability (cell fate-regulating Ca^{2+} influx). Different sources of Ca^{2+} signals may be involved in maladaptive hypertrophy and impaired contractility in cardiac diseases [93].

---

factor FGF12, Nedd4-2 ubiquitin ligase, plakophilin-2, RanGRF, syntrophin-α1, -β1, -β2, and -γ2, telethonin, and 14-3-3η [89]. N-glycosylation of $Na_V1.5$ associated with glycogenes (i.e., glycosyltransferases, glycosidases, and sugar nucleotide synthesis and transporters) modulates its electrical signaling. It is phosphorylated by PKA, PKC, CamK2, Fyn, and dephosphorylated by PTPn3 [89].

[58]In addition to SucnR1, extracellular succinate is actively transported through sodium–dicarboxylate cotransporters. These cotransporters are not produced in the heart.

The source of cell fate-regulating Ca^{2+} influx that activates the PP3–NFAT cascade is $Ca_V3.2$ channels during pressure overload-induced hypertrophy [94]. These carriers are expressed during embryo- and fetogenesis in ventriculomyocytes, but are undetectable in adult ventriculomyocytes. They are reexpressed in cardiac maladaptive hypertrophy and failure.

Muscle-specific microRNA, miR1, is implicated in arrhythmia and heart failure. Overexpression of miR1 in rat ventriculomyocytes causes a marked increase in the amplitude of inward Ca^{2+} current, flattening of Ca^{2+} transient voltage dependence, enhanced frequency of spontaneous Ca^{2+} sparks, and reduced sarcoplasmic reticulum Ca^{2+} content [95]. Overexpression of miR1 raises phosphorylation of ryanodine receptor RyR2 at Ser2814 (a target of CamK2), but not at Ser2808 (at target of PKA). Hyperphosphorylation of both RyR Ser2814 and Ser2808 by CamK2 and PKA produce alterations in RyR behavior that underlie arrhythmogenic Ca^{2+} release from stores [93]. MicroRNA-1 increases phosphorylation of $Ca_V1.2$ and RyR2 channels, as it suppresses transcripts of the regulatory $PP2_{r5\alpha}$ subunit of PP2 phosphatase, hence reducing PP2 activity to these channels.

3.1.11.4 Sympathetic Control

Signaling primed by β-adrenoceptors regulates not only myocardial function (Sect. 3.6.2), but also metabolism (glycolysis) and gene regulation.

Stimulated β1-adrenoceptor raises contraction frequency (positive chronotropy) and contractility of cardiomyocytes (positive ionotropy). β2-Adrenoceptor produces only modest positive chronotropy. The minor β3-adrenoceptor also expressed in the myocardium modulates myocyte function. Unlike the action potential with its all-or-none response, the force of cardiac contraction is graded according to the amplitude of the global Ca^{2+} transients.

In normal sympathetic activity, catecholamines bind to β-adrenergic G-protein coupled receptors and signal via the Gs–cAMP–PKA pathway. The effect of the β1-adrenoceptor is mediated by PKA, which phosphorylates in particular troponin-I and phospholamban for increased inotropy and lusitropy. Kinase PKA phosphorylates other substrates. It can elicit cell growth and death.

Among PKA substrates, some are shared with CamK2 kinase. During β-adrenoceptor stimulation, CamK2 and PKA phosphorylate numerous mediators of Ca^{2+} handling with similar effects ($Ca_V1.2$, ryanodine receptors, and phospholamban). They can synergistically increase contractility of cardiomyocytes. Myocardial contractility is actually modulated by the phosphorylation–dephosphorylation cycle of $Ca_V1.2$, RyR2, and phospholamban by PKA and CamK2 kinases as well as PP1 and PP2 phosphatases.

Acute β-adrenergic signaling pertains to the fight-or-flight response. On the other hand, persistent β-adrenergic signaling primes maladaptive hypertrophy, fibrosis, and heart failure.

3.1.11.5 Modeling

Computational models based on a proper modeling in the context of a selected theory integrate accumulated knowledge. Experimental and clinical observations are used to yield input data. Numerical simulations of the direct problem are aimed in particular at predicting and demonstrating the role of influence parameters, using adequate boundary conditions.

On the other hand, the goal of solving the inverse problem is the assessment of the role and magnitude of factors that cannot be directly measured.

After a phenomenological analysis, computational models combine relevant spatial and temporal scales. Computational physiology and pathophysiology indeed may integrate functional and structural information from multiple spatial (nm–m) and temporal (ps–wk or more [mo or yr]) scales.

Computational physiology relies on a 4-step work that comprises modeling, data extraction, possibly using data mining, simulation, and comparison to experimental data [96, 97]. Experimental and clinical data enable model validation. Models can then generate novel predictions and hypotheses as well as new experimental design. Model predictions are further compared to new observations.

Sensitivity analysis can unravel causes and consequences of variability in specific properties. Error estimation in both collected simulation and experimental data improves validation.

Computational cardiac electrophysiology investigates the electrical activity of the heart. This multiscale modeling type integrate information across molecular, cellular, tissular, and organ levels with a large range of temporal scales.

Mathematical models of the cardiac electrical activity have been carried out to mimic interplay between structural, electrical, and mechanical mechanisms in normal and abnormal conditions. In particular, computations are performed to explore links between sympathetic (β-adrenergic) control, excitation–contraction coupling, genetic mutations that alter ionic fluxes and action potentials, and arrhythmogenesis.

The model type (i.e., the number of equations and parameters) is determined by the problem of interest, desired level of detail, and quality of available experimental data.

The earliest models of cardiac electrophysiology describes currents across the cell membrane through a few carriers. For example, the cardiac cell model carried out by D. Noble in 1960 is a very simplified representation of actual physiology with only 3 ion fluxes [97]. It comprises a single voltage-gated inward current accounting for both Na^+ and Ca^{2+} fluxes (the latter being discovered only in 1967 by H. Reuter) and 2 K^+ effluxes through 2 types of channels, the conductance of which either falls with membrane depolarization or slowly rises. Nevertheless, it reproduced roughly the action potential shape in Purkinje nodal cells.

Sophisticated cellular models incorporate relevant ion fluxes, ion buffers, and metabolism, including their spatiotemporal effects, as well as regulation by circulating factors such as hormones, cell signaling (especially via CamK2 and PKA) with its control mechanisms (pathway interference, feedback, and spatial compartmentation), genetic mutations of cardiac genes, and anatomical heterogeneities (anisotropy, myofiber orientation,[59] and specific electrophysiological characteristics of distinct cardiomyocyte types according to their location in and across the cardiac wall).

Modeling of Cardiac Signaling

Computational models aimed at studying cell signaling networks are based on the law of mass action and Michaelis–Menten enzyme kinetics that represent each involved chemical reaction. These models require a large number of parameters. Cardiac excitation–contraction coupling is controlled by Ca^{2+}–calmodulin-dependent protein kinase CamK2 and protein kinase-A. Some models integrate many signaling pathways, such as those mediated by Ca^{2+} messenger and PKA, PKC, CamK2 enzymes.

On the other hand, several modeling approaches more closely based on network topology have been elaborated, such as discrete-level (Boolean) and fuzzy logic modeling. These approaches used for qualitative predictions require few parameters and facilitate analysis of systems properties such as feedback loops. A logic-based ordinary differential equation modeling of cell signaling networks based on normalized Hill activation–inhibition functions controlled by logical operators was developed to characterize signaling crosstalk with integrin-mediated mechanotransduction [98]. The cardiac β1-adrenergic signaling network comprises 36 reactions and 25 species. Sensitivity analysis points out the role of PKA negative feedback on upstream signaling and the importance of phosphodiesterases as negative regulators of the network.

In a myocyte action potential model with Na^+ and Ca^{2+} transport coupled to a Markovian Na^+ channel model with CamK2-dependent changes. Kinase CamK2 shifts Na^+ current (i_{Na}) availability to more negative voltage, enhances intermediate inactivation, and slows recovery from inactivation, but also enhances late noninactivating i_{Na} current [99]. This kinase also increases cardiac Ca^{2+} (i_{Ca}) and K^+ ($i_{K,to}$) currents. Effect of CamK2 on i_{Ca} and $i_{K,to}$ prolongs and shortens the action potential duration, respectively. The combined effect of CamK2 on all 3 currents causes action potential shortening [99].

In a model representing CamK2–RyR interaction in the canine ventriculomyocyte, phosphorylation of $Ca_V1.2$ by CamK2 has a greater effect on RyR flux than phosphorylation of RyR channel due to shift in $Ca_V1.2$ gating and the apparent faster recovery from inactivation [100]. Phosphorylated $Ca_V1.2$ channel increases action potential duration. Phosphorylation of $Ca_V1.2$ by CamK2 has a greater effect on RyR-mediated Ca^{2+} release than RyR phosphorylation.

[59] Current models use local tensors representing electrical conductivity and mechanical stiffness parameters in 3 orthogonal directions (myofiber, in-sheet transverse, and sheet-normal axis) in few parietal layers.

Compartmentation and dynamics of the cAMP–PKA axis determine signaling specificity. Cardiac effect depends on spatiotemporal variations in cAMP–PKA signaling. Combining Förster resonance energy transfer (FRET) imaging in neonatal rat ventriculomyocytes to a β AR signaling model, spatial PKA gradients can be explained by restricted cAMP diffusion, phosphodiesterase-mediated cAMP degradation, and PKA-mediated cAMP buffering [101]. Accumulation and compartmentation of cAMP can be responsible for the rate-limiting PKA activation and subsequent phosphorylation activity downstream from the stimulated β-adrenoceptor. On the other hand, prostaglandin-E1 primes higher PKA activity in the cytosol than at the sarcolemma.

A computational model was carried out to demonstrate the importance of cAMP compartmentation during activation of β1-adrenergic and M_2 muscarinic receptor in ventriculomyocytes [102]. The model consists of 2 plasmalemmal, caveolar and extracaveolar, and a single bulk cytosolic subdomain. Both caveolar and extracaveolar subdomains contain β1 AR and M_2R, stimulatory Gs and inhibitory Gi subunits, and phosphodiesterase isoforms (PDE2 and PDE4). The caveolar subdomain possesses adenylate cyclases inhibited by Gi (AC5 and AC6) as well as an additional PDE (PDE3). The extracaveolar subdomain has adenylate cyclases stimulated by Gi (AC4 and AC7). The cytosol subdomain contains PDE2, PDE3, and PDE4 isozymes inside and AC4 and AC7 at its boundary. Messenger cAMP diffuses between the 3 compartments.

In a sinoatrial node model coupled with a β1-adrenergic signaling, increased firing frequency in sinoatrial nodal cells results from a combination of changes in $i_{Ca,L}$ ($Ca_V1.2$), $i_{s,i}$ (sustained inward Na^+ and K^+ current), and i_{ha} (hyperpolarization-activated nonselective Na^+ and K^+ currents [103]. In addition, $i_{K,s}$ current (slow component of the delayed rectifier K^+ current), one of the major targets of the β1-adrenergic cascade, contributes markedly to counterbalancing $i_{Ca,L}$ and i_{NaCaX} (Na^+–Ca^{2+} exchanger) current during β-adrenoceptor stimulation.

The Ca^{2+}–calmodulin-dependent protein kinase CamK2 and the cAMP-dependent protein kinase-A cascades interfere. Some models incorporate PKA- and CamK2-mediated modulation of calcium handling with dual phosphorylation of $Ca_V1.2$, RyR, and phospholamban, in adddition to phosphorylation of substrates by either kinase in the rabbit ventriculomyocyte [104]. The model consists of 3 main modules: (1) activation of CamK2 and phosphorylation of components of excitation–contraction coupling; (2) β-adrenergic activation of PKA and phosphorylation; and (3) the cardiac excitation–contraction coupling model of the rabbit ventricular myocyte. The computational domain is composed of dyadic (with $Ca_V1.2$, RyR, CamK2, PKA, PP1, PP2, Cam, and Ca^{2+}), subsarcolemma (with CFTR, $Ca_V1.2$, Na_V, K_V7, NCX, CamK2, PKA, PP1, Cam, and Ca^{2+}), cytosolic subdomains (with SERCA, Plb, CamK2, PKA, PP1, PP2, Cam, I1, TnI, and Ca^{2+}). During overexpression simulations, active CamK2 also targets fast and late sodium ($i_{Na,f,\ell}$) and transient outward K^+ ($i_{K,to}$) currents. Other PKA substrates include inhibitor-1 (I1), troponin-I (TnI), $K_V7.1$–minK ($i_{K,s}$), and cystic fibrosis transmembrane conductance regulator (CFTR). Phosphatases PP1 and PP2 oppose phosphorylation by either kinase. Synergy between individual CamK2 activities and between CamK2 and PKA may contribute to full adrenergic responses. Kinase CamK2 participates in a CamK2–Ca^{2+}–CamK2 positive feedback loop that is potentiated by an enhanced Ca^{2+} flux primed by a PKA signal.

Cell-signaling networks modulate excitation–contraction coupling that translate Ca^{2+} influx to contraction. Accurate representation of myocardial Ca^{2+} dynamics is crucial. Arrhythmia (delayed and early afterdepolarizations, premature ectopic beats, and initiation of ventricular tachycardias and fibrillation) can arise from altered Ca^{2+} handling. A positive feedback results from activation from local Ca^{2+} release through RyR units of neighboring units. A negative feedback is primed by released Ca^{2+} of $Ca_V1.2$ channels. A positive nonlinear relation exists between the released Ca^{2+} amount and sarcoplasmic reticulum load.

Simulations have investigated the effects of [86]: (1) the surface charge on local calcium concentration within the dyad; (2) spontaneous Ca^{2+} release from the sarcoplasmic reticulum; (3) relation between sarcoplasmic reticulum load, Ca^{2+} spark amplitude, and RyR sensitivity; and (4) relation between dyadic volume, $Ca_V1.2$ number, and excitation–contraction coupling. Concentration and speed of Ca^{2+} diffusion within the sarcoplasmic reticulum influence Ca^{2+} wave propagation.

Modeling of the Cardiac Electrochemical Activity

The heart electrical activity emerges from multiple interactions of numerous molecular components of cardiomyocytes. The mathematical formulation of transfer kinetics relies on existing knowledge, background assumptions, and hypotheses on biophysical features of ion carriers, as well as assumptions on best fit with experimental data.

Modeling of the electrical activity of a cardiomyocyte that incorporates changes in ionic concentrations and mechanisms that regulate the influx and efflux of ions through cellular membranes, essentially the plasma and endoplasmic reticulum membrane defies simple mathematical analysis. Therefore simplified mathematical models only keep essential elements. Usually, these models are based on a set of ordinary differential equations of the transmembrane voltage and some ionic currents.

The cell membrane is modeled as a capacitor connected in parallel with variable resistances and capacitances representing different ionic fluxes through channels, exchangers, and pumps.

The electrophysiological behavior of a single cell was described by Hodgkin and Huxley with the following differential equation [105]:

$$\mathrm{u}_t = -\frac{i_{\mathrm{stim}} + i_{\mathrm{ion}}}{C_{\mathrm{mb}}}, \tag{3.8}$$

where u is the transmembrane voltage, t the time, i_{ion} the sum of transmembrane ionic currents, i_{stim} externally applied stimulus current, and C_{mb} is cell capacitance per unit surface area.

Each current type (i_k) is given by:

$$i_k = \mathtt{G}_k(\mathtt{u}_{mb} - \mathtt{u}_k) \tag{3.9}$$

where $\mathtt{G}_k$ and $\mathtt{u}_k$ are the membrane conductance and constant membrane equilibrium potential for the corresponding ion k.

The membrane conductance is expressed by a constant conductance and activation gate ($m(t)$) and inactivation functions $h(t)$ and $n(t)$ for Na^+ and K^+, each depending on the voltage-dependent opening and closing rates. The function $m(t)$ corresponds to a fast scale, whereas $h(t)$ and $n(t)$ are slow-scale functions, which, after simplifications, lead to a system with slow–fast dynamics. A system of 4 ordinary differential equations, 1 for the current and 3 for the 3 gating variables, is solved for the membrane potential as it varies with time.

A 2D sheet of cardiomyocytes was modeled as a continuous domain (ignoring discrete subdomains of cells) with the following partial differential equation,

$$\mathtt{u}_t = -\frac{i_{stim} + i_{ion}}{C_{mb}} + \frac{1}{\mathtt{r}_x SVR_x C_{mb}} V_{xx} + \frac{1}{\mathtt{r}_y SVR_y C_{mb}} V_{yy}, \tag{3.10}$$

where $\mathtt{r}_{x(y)}$ is the electrical resistivity of the cell in the x and y directions, SVR_x and SVR_y the surface-to-volume ratio in the x and y directions, and i_{ion} is the sum of all transmembrane ionic currents [106]:

$$\begin{aligned} i_{ion} &= i_{Ca,L} + i_{Ca,p} + i_{Ca,b} + i_{Na} + i_{Na,b} + i_{NaCaX} + i_{NaK_{ATP}} \\ &\quad + i_{K,p} + i_{K,b} + i_{K1} + i_{K,to} + i_{K,r} + i_{K,s}, \end{aligned} \tag{3.11}$$

where $i_{Ca,p}$ and $i_{K,p}$ and $i_{Ca,b}$ and $i_{K,b}$ are plateau and background Ca^{2+} and K^+ currents, respectively. In a 1D model, or isotropic 2D model, the cell capacitance per unit surface area $C_{mb} = 2\mu F/cm^2$, resistivity $\mathtt{r} = 180\,\Omega \times cm$, and surface-to-volume ratio $SVR = 0.2\mu m^{-1}$ [107].

A 3-gate formulation was introduced for i_{Na} [108]:

$$i_{Na} = (\mathtt{G}_{Na} m^3 hj + \mathtt{G}_{Na,b})(\mathtt{u} - \mathtt{u}_{Na}), \tag{3.12}$$

where m is the activation gate, h a fast inactivation gate, j a slow inactivation gate, and $\mathtt{u}_{Na}$ the sodium reversal potential (50 mV). The Beeler–Reuter model incorporates 4 currents: 2 voltage- and time-dependent inward currents (the excitatory inward Na^+ current and a slow inward current primarily associated with calcium ions); a time-independent outward K^+ current exhibiting inward rectification; and a voltage- and time-dependent outward current mainly carried by K^+ ions.

The simplest ionic model used to describe cardiac electrophysiology is the following,

$$\begin{aligned} \mathtt{u}_t &= g\, i_{in}(\mathtt{u}) - i_{out}(\mathtt{u}) + i_{stim}, \\ g_t &= G(V, g), \end{aligned} \tag{3.13}$$

where i_{stim} is the stimulation current, $i_{\text{in}}(\text{u})$ and $i_{\text{out}}(\text{u})$ the ionic in- and outflux that require energy ($i_{\text{out}}(\text{u}, e_{\text{ATP}})$; e_{ATP}: energy consumption).

The following model,

$$\text{u}_t = i_{\text{in}}(\text{u}) - i_{\text{out}}(w) + i_{\text{stim}}; \quad w_t = W(\text{u}, w), \tag{3.14}$$

does not correspond to ionic current models; no switch controls stimulus-induced membrane potential bifurcation. Eikonal models cannot differentiate between uncontrolled and controlled bifurcations (i.e., $g \times i_{\text{in}}(\text{u})$ and $i_{\text{in}}(\text{u})$) and repolarization mode (i.e., $i_{\text{out}}(\text{u}, e_{\text{ATP}})$ and $i_{\text{out}}(w)$). This kinematic model does not incorporate the coupling dynamics of chemical reaction and electrical conduction. Both $i_{in}(V)$ and $i_{out}(V)$ can be decomposed in specialized ionic currents, especially when realistic action potential duration and conduction speed are required.

Models that take into account continuously varying parameters within a region, keeping the same number of parameters, may be more realistic than considering uniform parameters for each region.

The most appropriate model must be selected according to the goal of numerical experiments, taking into account the well-posedness of data-assimilation problems and real-time constraints. A 2 state-variable model of cardiac action potentials such as the Mitchell–Schaeffer model [109] has an ionic current interpretation relevant for inverse problems.

Ionic current models use the transmembrane potential as a variable of state, with binary gate variables (g) that quickly commute between 2 values (0 and 1; i.e., open and close gate; switching needing only a small amount of energy, which is neglected) and ion concentrations.

A hierarchy of models of growing complexity exists, starting from eikonal models of depolarization front and evolving with a series of ionic models. They are commonly simplifications of complex ionic models of cardiomyocyte plasma membrane that aim at reproducing quantitatively many of the dynamical characteristics of nodal and common cardiomyocytes. Certain models can be processed analytically.

Among others, this series encompasses (Table 3.34): (1) the phenomenological Aliev–Panfilov reduced bioelectric model [110], which is not usually well suited for cardiac electrophysiology (i.e., for restitution of action potential duration and conduction velocity); (2) the Mitchell–Schaeffer 2-variable ionic model (possibly with incorporation of a pacemaker function to simulate ectopic activation sites); (3) the Fenton–Karma 3-currents (loosely corresponding to sodium, calcium, and potassium) model [111], which enhances the Mitchell–Schaeffer model, but simplifies the Luo–Rudy model [112] of the cardiac membrane; and (4) the Djabella–Sorine 8-variable model [113].

The Fenton–Karma model contains 3 variables: transmembrane normalized potential ($\text{u} = 0$: rest voltage; $\text{u} = 1$: peak voltage) and 2 gating parameters (fast $[f]$ and slow $[s]$ function). It takes into account essential ionic currents associated with calcium concentration. This simplified ionic model approximates well the restitution properties and spiral wave behavior of more complex ionic models of

Table 3.34 Examples of ionic flux models (AP: action potential; APD: action potential duration; cb: calcium buffer; gv: gate variable; if: ionic flux; tmp: transmembrane potential; $\Delta[Ca^{2+}]$: change in intracellular calcium concentration)

Ionic current model l	Variable number	Features
Panfilov	6 (tmp+6 gv)	Incorporates initial fast repolarization during which a rapid–slow effect occurs Omission of $\Delta[Ca^{2+}]$
Fenton–Karma (1998)	6 (tmp+3 if +2 gv)	3 currents (Ca^{2+}, K^+, Na^+) Absence of Ca^{2+} dynamics
Mitchell–Schaeffer (2003)	2 (tmp+gv)	Enables ECG modeling, reasonable APD restitution 2 currents only Absence of Ca^{2+} dynamics
ten Tusscher (2004)	17 (tmp+3 if +12 gv)	Ca^{2+} dynamics (minimal model) Validation
Djabella–Sorine (2007)	8 tmp+3 if +2 cb +2 gv)	Adequate AP shape

cardiac action potential such as the Beeler–Reuter kinetics model [108]. Numerical tests were carried out in a parallelepipedal myocardial domain of varying wall thickness and fiber rotation rate (degree/mm), as action potential propagates in an anisotropic myocardium due to intramural rotation of myofibers.

The transmembrane voltage change that triggers ionic currents is given by the relation [111]:

$$\mathrm{u}_t = -J_{\text{fast}} + J_{\text{slow}} + J_{\text{ung}} + J_{\text{stim}}, \tag{3.15}$$

where J_{fast}, J_{slow}, J_{ung}, and J_{stim} are the fast and slow inward, ungated, and stimulation current, respectively. The first 3 ionic fluxes represent Na^+, Ca^{2+}, and K^+ currents, respectively. The stimulus current J_{stim} is an external current applied during experiments.[60]

Ionic flux $J_{\text{fast}} = -f\,D(\mathrm{u})/\tau_{\text{fast}}$ depends on a characteristic current time (τ_{fast} [0.25 ms]) and a voltage dependence ($D(\mathrm{u})$):

$$D(\mathrm{u}) = \begin{cases} (\mathrm{u} - \mathrm{u}_{\text{crit}})(1 - \mathrm{u}) & \text{if } \mathrm{u} > \mathrm{u}_{\text{crit}} \\ 0 & \text{otherwise.} \end{cases}$$

The gate variable f evolves according to:

$$f_t = [f_\infty(\mathrm{u}) - f(\mathrm{u})]/\tau_f(\mathrm{u}), \tag{3.16}$$

[60]Typically, it consists of a periodic train of brief pulses (duration 1 ms; magnitude $\sim$ twice the amplitude required to excite fully recovered tissue).

where the voltage-dependent functions on the right-hand side of Eq. 3.16 are step functions given by:

$$f_\infty(\mathrm{u}) = 0 \text{ and } \tau_f(\mathrm{u}) = \tau_{f_{\mathrm{close}}} \text{ if } \mathrm{u} > \mathrm{u}_{f_{\mathrm{gate}}},$$
$$f_\infty(\mathrm{u}) = 1 \text{ and } \tau_f(\mathrm{u}) = \tau_{f_{\mathrm{open}}} \text{ if } \mathrm{u} < \mathrm{u}_{f_{\mathrm{gate}}}, \tag{3.17}$$

where $\tau_{f_{\mathrm{close}}}$ and $\tau_{f_{\mathrm{open}}}$ equal typically 10 and 18 ms, respectively.

Similarly, the slow inward current density $J_{\mathrm{slow}} = -sS(V)/\tau_{\mathrm{slow}}$ depends on a characteristic current time (τ_{slow} [127 ms]) and a sigmoid function:

$$S(\mathrm{u}) = (1 + \tanh[\kappa(\mathrm{u} - \mathrm{u}_{\mathrm{sig}})])/2. \tag{3.18}$$

The gate variable s evolves according to:

$$s_t = [s_\infty(\mathrm{u}) - s(\mathrm{u})]/\tau_s(\mathrm{u}), \tag{3.19}$$

where the voltage-dependent functions on the right-hand side of Eq. 3.19 are given by:

$$s_\infty(\mathrm{u}) = 0 \text{ and } \tau_s(\mathrm{u}) = \tau_{s_{\mathrm{close}}} \text{ if } \mathrm{u} > \mathrm{u}_{s_{\mathrm{gate}}},$$
$$s_\infty(\mathrm{u}) = 1 \text{ and } \tau_s(\mathrm{u}) = \tau_{s_{\mathrm{open}}} \text{ if } \mathrm{u} < \mathrm{u}_{s_{\mathrm{gate}}}, \tag{3.20}$$

where typical values of $\tau_{s_{\mathrm{close}}}$ and $\tau_{s_{\mathrm{open}}}$ are 1 s and 80 ms, respectively.

The ungated current takes the form

$$J_{\mathrm{ung}} = P(\mathrm{u})/\tau_{\mathrm{ung}}, \tag{3.21}$$

where τ_{ung} is the characteristic ungated current time (130 ms) and the piecewise linear voltage dependence $P(\mathrm{u})$ is given by [114]:

$$P(\mathrm{u}) = \begin{cases} 1 & \text{if } \mathrm{u} > \mathrm{u}_{\mathrm{out}} \\ \mathrm{u}/\mathrm{u}_{\mathrm{out}} & \text{if } \mathrm{u} < \mathrm{u}_{\mathrm{out}}. \end{cases}$$

The Djabella–Sorine 10-variable model is a reduced order version of the ten Tusscher–Noble–Panfilov 17-variable model [106]. The ten Tusscher–Noble–Panfilov model takes into account the dynamics of the main ionic currents (fast sodium, L-type calcium, transient outward, rapid and slow delayed rectifier, and inward rectifier currents) through specific carriers as well as main calcium buffers in the bulk cytosol and sarcoplasmic reticulum. It improves calcium dynamics w.r.t. the 6-variable Bernus–Panfilov γ-model, that restitutes action potential duration and conduction velocity and reproduces the main properties of human epi-, midmyo-, and endocardial cells by modifying selected ionic currents [107]. The ten Tusscher–Noble–Panfilov model realistically models epi-, midmyo-, and endocardial action potential, differences being explained by variations of transient outward and slow delayed rectifier currents, as well as conductances, current gating, and intracellular calcium transients. Due to its asymptotic behavior without drifts of the state, this model can be used in multibeat simulations from the cell to the heart scale. The model has been also reduced and extended to represent pacemaker or Purkinje cells.

The Djabella–Sorine differential model of a cardiac pacemaker cell has 10 state variables and comprises 2 subdomains: (1) the cell membrane with capacitance, voltage-dependent ion channels, electrogenic pump and exchanger; and (2) a lumped compartmental model that accounts for intracellular changes in concentrations of Na^+ and K^+ and the main processes that control intracellular calcium concentration, that is, the release and uptake by the sarcoplasmic reticulum and buffering in the sarcoplasmic reticulum (calsequestrin) and in the cytosol (calmodulin and troponin). Three variables describe the membrane (membrane potential and 2 gate variables of ionic channels), taking into account the dynamics of 5 major ionic currents through channels (inward sodium, L-type calcium, and outward potassium), Na^+–Ca^{2+} exchangers, and Na^+-K^+ pumps. Seven variables are related to Ca^{2+} binding to calmodulin and troponin and intracellular concentrations of free calcium, sodium, and potassium. The control by the calcium of the pacemaker activity is also considered.

Three-dimensional activity in the electrical component of the cardiac model is governed by the parabolic partial differential monodomain equation. This equation describes the flow of current from cell to cell through low-resistance gap junctions of intercalated discs and relates the transmembrane potential to the transmembrane current.

When it is necessary to explicitly account for the extracellular current, the bidomain model, a coupled system of partial differential equations, is solved rather than the monodomain model equation.

Modeling of Electrocardiograms

Numerical experiments that model the electrocardiogram relies on the simulation of electrochemical impulse propagation. Ionic flux models preferentially are based on phenomenological models such that inverse problems can be more easily solved. Transmembrane kinetics are described by a set of ordinary differential and algebraic equations.

Numerical simulations of electrocardiograms (ECG) using a PDE- and ODE-based mathematical model are aimed at yielding realistic 12-lead ECGs using the bidomain equations coupled to a Mitchell–Schaeffer phenomenological ionic model in the heart and a generalized Laplace equation in the extracardiac regions of the thorax (torso) [115].

Other features of modeling the electrical ventricular activity comprise adequate heart–torso transmission, anisotropy, cell heterogeneity (hence transmural action potential heterogeneity), and His bundle incorporation to display proper amplitudes, shapes, and polarities for the entire set of standard skin electrodes.

The equation set made of the heart and torso models is solved using a finite element method and a second order semi-implicit time marching scheme. The coupling conditions at the heart–torso interface are enforced by a Dirichlet–Neumann domain decomposition algorithm. This numerical implementation offers a good compromise between accuracy, stability, and efficiency.

At the heart level, the collection of cardiomyocytes, connected by specialized end-to-end or side-to-side junctions, immersed in the extracellular fluid and ground matrix, is modeled as a periodic array that leads to a *homogenization procedure*, the smallest unit is a group of cardiomyocytes. The heart is assumed to be a continuum with 2 electrical domains, the intra- and extracellular media connected by plasma membranes. The extracellular milieu represents the intercellular space. The plasma membrane acts as an electrical isolator with a given surface area-to-volume ratio (A_m) and capacitance per unit area (C_m). It hosts transmembrane proteic ion carriers, such as ion channels, exchangers, and pumps.

The electrical propagation is often simulated using a entire-heart reaction–diffusion model, that is, the *bidomain model* or simplified formulations such as the monodomain and eikonal models.

The local tissue conductivity tensor is obtained from the local fiber orientation and sheet arrangement as well as values of the myocyte-specific conductivity associated with the local myocardial orthotropic electrical properties.

At the heart level, parameters include the intra- and extracellular conductivity tensors. Two elements of information are then required: myofiber architecture and conductivity. Fiber architecture yields the local direction of the conductivity tensor. It is extracted from diffusion tensor-MRI images. A 3D reconstruction of the heart and meshing determines the geometry of the computational domain.

The bidomain model consists of 2 partial differential equations that relate intracellular (u_i) and extracellular (u_e) electrical potential to current at each node. The transmembrane potential (u_m) is the difference $\mathrm{u}_i - \mathrm{u}_e$. Each medium is modeled as an anisotropic network of resistors. Two fundamental laws are used: *Ohm's law* that relates electrical potential to ionic fluxes (transmembrane and intra- and extracellular currents) and *Kirchhoff's law* (i.e., the conservation of charge). Ohm's law states that:

$$J_i = -\mathbf{G}_i \nabla \mathrm{u}_i; \quad J_e = -\mathbf{G}_e \nabla \mathrm{u}_e, \tag{3.22}$$

where J_i and J_e are the averaged intra- and extracellular current densities and $\mathbf{G}_i$ and $\mathbf{G}_e$ the intra- and extracellular conductivity tensors.

The averaged intra- and extracellular current densities, conductivity tensors, and electrical potentials are defined in the whole heart volume (Ω_H). The electrical charge conservation becomes:

$$\nabla \cdot (J_i + J_e) = 0, \quad \text{in } \Omega_\mathrm{H}. \tag{3.23}$$

At the cell level, the plasma membrane is considered as a capacitor in parallel with a nonlinear term accounting for the conductivities of transmembrane proteic carriers. The homogenized equation of the electrical activity of the plasma membrane is the following,

$$\mathsf{A}_{\text{mb}}\left(\mathsf{C}_{\text{mb}}\frac{\partial \mathrm{u}_{\text{mb}}}{\partial t} + i_{\text{ion}}\right) + \nabla J_i = \mathsf{A}_{\text{mb}} i_{\text{app}} \text{ in } \Omega_{\text{H}}, \tag{3.24}$$

where $i_{\text{ion}}(\mathrm{u}_m, w)$ represents the transmembrane ionic current per unit membrane area, w being an ionic variable that generally satisfies a system of ODEs:

$$w_t + g(\mathrm{u}_{\text{mb}}, w) = 0 \text{ in } \Omega_{\text{H}}, \tag{3.25}$$

(g and i_{ion} depend on the considered cell ionic model), and i_{app} a given applied current stimulus per unit membrane area.

A perfect electric transmission is generally assumed at the interface (Γ_{HT}) between the heart and torso (domain Ω_{T}):

$$\begin{aligned} \mathrm{u}_e &= \mathrm{u}_{\text{T}} \\ \mathbf{G}_e \nabla \mathrm{u}_e \cdot \mathbf{n} &= \mathbf{G}_{\text{T}} \nabla \mathrm{u}_{\text{T}} \cdot \mathbf{n} \text{ on } \Gamma_{\text{HT}}, \end{aligned} \tag{3.26}$$

where u_{T} is the torso potential and $\mathbf{G}_{\text{T}}$ the conductivity tensor of the torso.

The torso can be assumed to be a passive conductor. Therefore, the torso potential satisfies the generalized Laplace equation:

$$\nabla \cdot (\mathbf{G}_{\text{T}} \nabla \mathrm{u}_{\text{T}}) = 0 \text{ in } \Omega_{\text{T}}. \tag{3.27}$$

As the handling of modeling ionic currents is not trivial, an alternative approach consists of modeling cardiac action potential dynamics using phenomenological models. This type of tractable model is aimed at capturing shape and rate-dependent properties of action potentials at given locations within the heart (pacemaker, transmission nodal cells, and atrio- and ventriculomyocytes in different myocardial layers) without including a detailed description of implicated ionic currents. Moreover, the reduced number of parameters and equations enables identification and inverse problem solving.

The sensitivity analysis demonstrates that the most critical parameters of the bidomain model are A_m, C_m, the angular velocity of the activation wave, and the transverse conductivities $\mathbf{G}_{i_t}$ and $\mathbf{G}_{e_t}$. In addition, the ECG trace is sensitive to the ionic model parameters.

3.1.12 Calcium Handling During Exercise

Sarcalumenin, a Ca^{2+}-binding protein of the longitudinal sarcoplasmic reticulum, regulates Ca^{2+} reuptake into its store, as it interacts with the SERCA2a pump [116]. It may serve in adaptation of the Ca^{2+} storage to exercise. Endurance exercise (1 h/d during 12 wk) significantly upregulates SERCA2a expression.

Furthermore, the production of other Ca^{2+}-handling proteins, such as calsequestrin-2, phospholamban, ryanodine receptor RyR2, and Na^+–Ca^{2+} exchanger NCX1, also rises after training [116].

3.2 Large Blood Vessels

The blood pressure maintains a suitable blood flow, which is distributed among the different compartments of the body (Table 3.35). Blood irrigates the heart pump, brain (control center), endocrine organ (remote regulation), bone marrow (source of blood cells), lungs of the strongly energy-consuming human body, muscles, ligaments and bones (life gestures), digestive tract (nutrient input), kidneys, liver, and so on.

The main organs regulating blood volume are the kidneys. The kidneys receive about 20% of the cardiac output for blood processing by filtration and reabsorption. The liver acts as a filter (detoxifying various substances), a storage (for glucids, vitamins, etc.), and an excretory gland (bile).

Beyond large arteries, blood pressure abruptly falls and the systolo-diastolic pressure difference decays. The pressure drop occurs mostly in arterioles and moderately in capillaries. In the venous bed, most of the pressure decrease is observed in the venules with little further drop in large veins. Therefore, a small pressure difference is sufficient to fill the atrium.

Blood velocity decreases from arteries, with a scale (peak value) of $\mathcal{O}[10\,\text{cm/s}]$, to the capillaries with a magnitude of $\mathcal{O}[0.1\,\text{mm/s}]$. Pressure in the pulmonary circulation is much lower than that in the systemic circulation. Although the right pump is weaker, flow rates are identical in both circulations, because the pulmonary resistances are smaller.

Pressure variations follow external constraints. The main factors that affect pressure values are blood volume, ejection volume, and vascular resistance. Blood volume mainly depends on Na^+ concentration and cardiac function, and vessel tone on K^+ concentration.

3.2.1 *Arterial Circulation*

Left ventricle pressure soars during isovolumetric contraction to reach a value close to the minimum of the aortic pressure. During systolic ejection, the pressure of the left ventricle is slightly higher than that of the aorta. Both pressures reach their maxima and decrease.

Table 3.35 Estimated blood flow distribution (%) among body parts (datum variability)

Compartment	[117]	[118]	[119]
Heart	3	4	4
Brain	14	13	13
Skeletal muscle	15	21	21
Kidneys	22	19	20
Other abdominal organs	27	24	24
Skin	13	10	
Lungs, pelvis, etc.	6	9	

Table 3.36 Blood flow quantities in proximal arteries (Sources: [122, 123])

Artery	Velocity (cm/s)	Pressure (kPa)
Aortic arch	−20–60	10.7–16.0
Descending thoracic aorta	−10–60	9.3–15.3
Abdominal aorta	−10–60	9.3–15.3
Common iliac artery	−7.5–60	8.7–13.3

In certain subjects who have thick-walled hearts, the pressure wave displays 2 peaks. Several indices can then be calculated. The *augmentation index* (AIx), derived from the ascending aortic pressure waveform, is the difference between the first and second systolic pressure peaks (the first peak being lower than the second), expressed as a percentage of the pulse pressure. (Pulse pressure is defined as systolic pressure minus diastolic pressure.) This index has been used as a measure of systemic arterial stiffness and additional load imposed on the left ventricle. An increased augmentation index is related to the risk of coronary heart disease.

The so-called *time to reflection* (TR) is the time from the foot of the pressure wave to the first systolic peak. The *diastolic pressure time interval* (DPTI) is the time from the foot of the pressure wave to the dicrotic notch. The *systolic pressure time interval* (SPTI) is the time from the dicrotic notch to the end of the waveform. The *subendocardial viability ratio* (SEVR) is the ratio of the diastolic pressure time interval to the systolic pressure time interval. Applanation tonometry assesses the augmentation index and pulse wave velocity, but the use of this technique to derive the central waveform from noninvasively acquired peripheral data needs to be validated [120].

The arterial pressure (p_a) evolves between its systolic p_s and diastolic p_d values during the cardiac cycle (Table 3.36). The systolic pressure reflects the cardiac output and distensibility of elastic arteries, whereas p_d is an index of the peripheral vessel state.

The pulse pressure is the difference between p_s and p_d. It mainly depends proportionally on stroke volume and is inversely proportional to arterial compliance.

The mean arterial pressure (mAP) is currently estimated by $p_d + (p_s - p_d)/3$.[61] It decreases from the aorta (~13 kPa) to arterioles (~5 kPa). Arterioles are the main site of blood flow resistance—systemic vascular resistance (SVR)—and control blood input into capillaries.

The mean arterial pressure can be considered in a first approximation as the CO × SVR product. Arterial pressure has a circadian pattern. Pressure values increase progressively with aging up to 120% after 60-years old. Vessel pressure variations along the circulatory network are given in Table 3.37.

[61] The mean arterial pressure is underestimated using 0.333 as a multiplier rather than 0.412 [121]:

$$\text{mAP} = p_d + 0.412(p_s - p_d).$$

Table 3.37 Indicative pressure (kPa) evolution in the vasculature

	Mean	Systolic	Diastolic
Compartment	Systemic circulation		
Aorta	12.7	16	11.2
Artery	12	16.8	10.6
Arteriole	9.1	12.5	7.7
Capillary	3.5		
Vein	1.5		
Right heart			
RA	0.9	1.2	0.4
RV	1.1	3.9	0.8
Pulmonary circulation			
Artery	1.7	2.9	1.2
Arteriole		2	0.7
Capillary		1.2	0.5
Vein		1	0.2
Left heart			
LA	1	1.9	0.3
LV	5	16	0.5

Table 3.38 Flow features in the femoral artery in a healthy volunteer at rest (Source: [124])

Accelerating flow duration	90–100 ms
Latency between ECG R wave and systolic ejection	190 ms
Decelerating flow duration	140–190 ms
q_{smax}	40 ml/s
q_{min}	−12 ml/s
q_{dmax}	6 ml/s
$\bar{q}$	4 ml/s

Pressure variations drive the unsteady blood flows, the flow time variations being complex (multiharmonic signal; Sect. 5.2). In addition, the waveform varies with the vessel station. Ranges of cyclic variations in blood velocity in several arteries are given in Table 3.36. Values of the flow rate extrema in the femoral artery and duration of the acceleration and deceleration phases are provided in Table 3.38.

3.2.2 *Venous Return*

Veins are distensible; the venous compartment is used as blood storage. Veins are collapsible, especially thin-walled superficial veins. Gravity causes blood pooling in the legs in the upright position. Veins provide heart filling flow (venous return) with a lower velocity than in arteries, the right heart pumping blood into the pulmonary circulation.

Table 3.39 Influence agents on venous tone

Constriction	Deep inspiration
	Hypoxemia
	Hypopressure in carotid sinus
	Central hypercapnia
	5HT
	Cold
Dilation	Rest
	Hyperpressure in carotid sinus
	ACh

Table 3.40 Influence factors on the venous pressure

Vis a tergo	Capillary pressure
Vis a fronte	Cardiac function
Vis a latere	External pressure
	Muscle contraction (limbs)
	Breathing (abdomen, thorax)
	Vein tone
Vis a parte interiore	Vein volume
	Energy dissipation

Venous return is the blood volume reaching the right atrium, which equals the cardiac output. The venous return is inversely proportional to the central venous pressure (CVP $\sim$0.25–0.80 kPa for CO of 5 l/mn). The central venous pressure decreases with inspiration, thereby increasing venous return, due to negative intrathoracic pressure at inspiration and increased intra-abdominal pressure.

Flow in large veins experiences cardiac activity. The jugular vein pulse is characterized by an a wave generated by atrial contraction, which is followed by an x valley associated with the peak jugular velocity, due to atrial relaxation and descent of the atrioventricular floor during ventricular ejection, the downslope being interrupted by a c wave induced by the carotid artery pulse (acceleration phase of the arterial flow) [128]. The following v wave occurs during the ventricular relaxation, and the y valley, shallower than that of the x one, during rapid ventricular filling.

Several agents act on venous return (Tables 3.39 and 3.40). The main factors affecting venous flow include: (1) the pressure difference between venules and the atrium, (2) muscular tone of the venous wall, (3) venous compliance and venous blood volume, (4) external pressure, in particular pressures in the abdominal and thoracic cavities, which vary during the breathing cycle, and (5) massage by skeletal muscle contraction.

Muscle contractions help venous return by compressing the surrounding veins equipped with valves. Both intra-abdominal and intrathoracic pressures decrease during inspiration (lung inflation) and increase during expiration [129]. Respiration-induced flow rise and decay in the inferior vena cava are observed with a phase lag in the thoracic segment with respect to the abdominal one. Moreover, the hepatoportal contribution explains the quantitative difference between these 2 IVC segments.

The respiratory cycle affects venous return via the systemic and hepatoportal venous supplies in the opposite way [129]. The systemic venous flow rises, whereas the hepatoportal venous contribution decays during inspiration.

Like arteries, the venous tone is governed by the autonomous nervous system. Venous flow is thus a time-dependent flow, especially in veins close to the heart due to rhythmic activity of the right cardiac pump and breathing, as well as in leg veins due to muscle contraction-assisted flow. In leg veins, the peak velocity $\widehat{V}_q = 20$–30 cm/s and the mean velocity $\overline{V}_q = 2$–4 cm/s.

3.2.3 Metabolic Blood Flow Regulation

Blood flow regulation is controlled by various local and remote mechanisms. Arterioles dilate or constrict in response to changing intravascular pressure (myogenic response), to mechanical stress via endothelial release of nitric oxide (mechanical stress-dependent response), to signal transmission both upstream and downstream along vessel walls (propagated or conducted response),[62] and local metabolism.

Oxygen exchange between blood and tissue cells occurs primarily in the microcirculation. In the microcirculation, blood flow is regulated by the local metabolic demand. Oxygen supply must match oxygen need. Regulatory messengers manufactured and released from a site of metabolic activity influence the tone of blood vessels. Active hyperemia is the increase in blood flow in regions where the metabolic activity rises.

Metabolic activity controls local blood flow via synthesized catabolites, such as CO_2, ADP, extracellular K^+ ions, particularly those released by contracting cardiac and skeletal muscles, hydrogen ion, and organic acids such as lactic acid, a carboxylic acid product of anaerobic metabolism generated from pyruvate by lactate dehydrogenase, and inorganic phosphate liberated by the hydrolysis of adenine nucleotides. These catabolites directly prime vasodilation of local arterioles, thus increasing blood flow. Other substances such as nitric oxide are involved in coupling metabolic demand to local and regional blood flows.

Metabolic blood flow regulation involves ATP release by red blood capsules that triggers arteriolar vasodilation. Red blood capsules respond to a lower oxygen level such as that happening during exercise by releasing ATP at a rate that depends on their oxyhemoglobin saturation level [125]. Secretion of ATP through SLC29a1 nucleoside transporter is electrically balanced by the simultaneous influx of extracellular chloride and/or bicarbonate across the erythrocyte membrane via SLC4a1 anion exchanger. Once released from RBCs, ATP binds to P2Y receptors on the luminal surface of the endothelium. Liberated ATP initiates a conducted response that triggers arteriolar vasodilation [126, 127].

---

[62] Acetylcholine primes vasodilation of both local and regional arterioles and arteries.

3.2.4 *Coronary Circulation*

The left and right coronary arteries and their main epicardial branches function as distribution vessels. Epicardial arteries branch into smaller arteries, which dive into the myocardium and give birth to the microvasculature with its resistance vessels and dense capillary network along cardiomyocytes.

Tissue perfusion for nutrient transport and drainage for catabolite removal are effected by a complex vasculature. The microcirculation with arterioles, capillary bed, and small and large draining venules, is wholly intramural. Large venules merge to form the venous collector. Large veins are epicardial like large arteries.

3.2.4.1 Coronary Capillary Density

The capillary density (1.3–5.0/mm^2) with respect to the cardiomyocyte population minimizes the transport distance. A close contact between capillaries and cardiomyocytes is necessary to supply oxygen for suitable ATP synthesis, a major factor of myocardium contraction, whereas calcium ion is a major determinant of excitation–contraction coupling.

3.2.4.2 Coronary Blood Volume

Blood volume in epicardial coronary arteries has been estimated to equal about 1.6 ml/100 g of left ventricle tissue for a perfusion pressure of 13.3 kPa, the right coronary bed usually contributing about 15% of the volume in the common structure of the coronary tree [130]. The left coronary artery irrigates both ventricles; the coronary flow that perfuses the right ventricle is about 65% of that of the left ventricle. Blood volume in the intramyocardial bed is assessed to be 4–5 ml/100 g of left ventricle tissue.

3.2.4.3 Coronary Perfusion Pressure

The perfusion pressure that drives the blood from the entrance of the coronary arteries to the right atrium varies from a minimum of 10 to 10.5 kPa to a maximum of 14 to 15 kPa. However, coronary flow responds not only to upstream coronary perfusion pressure, but also to the energetic needs of the myocardium. Myocardial perfusion is mainly adjusted to oxygen extraction by changes in coronary resistance. Coronary flow is proportional to the oxygen supply when the arterial oxygen content is normal [130]. Oxygen demand is more relevant to heart perfusion than the blood driving pressure.

3.2.4.4 Coronary Reserve

Coronary reserve at any perfusion pressure is expressed by the difference between autoregulated and maximally vasodilated flow. Coronary reserve is hence the ability to increase flow for maximal extraction of oxygen from the coronary blood. Normal hearts have a 4- to 5-fold reserve.

3.2.4.5 Coronary Flow Resistance

Flow resistance associated with friction is composed of 2 varying components: (1) resistance associated with the autoregulated vasomotor tone for flow adaptation to the myocardium demand and coronary flow reserve, and (2) transient additional systolic resistance induced by the contracting myocardium.

Melatonin causes either vasoconstriction (e.g., rat cerebral arteries) or vasodilation (e.g., rat and rabbit aorta, basilar iliac, and renal arteries) according to the artery type [131]. In other vascular beds (e.g., rat caudal arteries), melatonin does not have any direct action, but potentiates that of other vasoconstrictors. In humans, melatonin increases blood flow in some vascular territories (e.g., forearm) and decreases it in others (e.g., kidney). Both melatonin receptor types (MT_1–MT_2) reside in the vasculature. The vasomotor response to melatonin depends partly on the relative distribution and effect of melatonin receptor subtypes. In coronary arteries, melatonin binds to the MT_2 receptor and impedes nitric oxide-induced vasorelaxation. In vascular smooth myocytes, it increases the phosphorylation (activation) by PKG1[63] of phosphodiesterase PDE5,[64] which degrades cGMP, thereby precluding cytosolic cGMP accumulation in response to nitric oxide [131].[65]

3.2.4.6 Coronary Arterial Flow During the Ventricular Cycle

Blood flow at the entrance of the coronary arteries deeply varies during the left ventricle cycle. The entry segment of the coronary arteries is perfused during ventricular ejection. They receive a portion of the flow that crosses the open aortic orifice during the systole, the valve reaching at most about the level of the artery axis (Fig. 3.6).

The coronary circulation has 2 special time-dependent features. Epicardial vessels undergo strong path deformation over the beating heart during the cardiac

---

[63] Subtype PKG1 is the predominant vascular isoform.

[64] A major subtype in smooth myocytes.

[65] On the other hand, PKG activated by the NO-cGMP axis phosphorylates numerous ion carriers, thereby reducing the cytosolic Ca^{2+} level as well as causing a membrane hyperpolarization, hence a relaxation of arterial smooth myocytes.

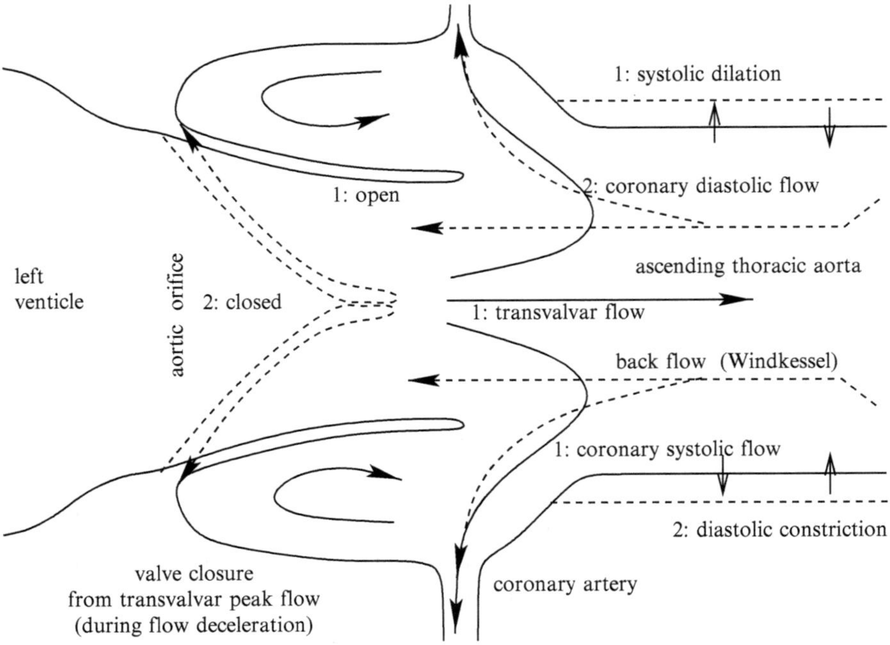

Fig. 3.6 Coronary arteries receive blood during systolic ejection from the left ventricle and during the diastole from restitution of stored blood by the elastic arteries (strong windkessel effect). Aortic valve leaflets begin to close from the peak transvalvar flow, thus reducing the coronary input. The first inflow is followed by a second distolic inflow (Fig. 3.7), characterized by a second peak flow

cycle (Vol. 8 – Chap. 1. Hemodynamics). Time-dependent vessel curvature is associated with a variable motion amplitude due to changes in heart cycle period.

Intramural vessels can be compressed during systole by contracted cardiomyocytes and expand during myocyte relaxation. Capillary bore varies from 4.5 to 9 μm, especially due to myocardium activity. Systolic perfusion is then restricted, whereas drainage is transiently supported by collapse of the upstream venous circuit.

Arterial inflow reaches its greatest values during diastole, at least in the larger left coronary artery (Fig. 3.7), and venous outflow during systole [132]. During systole, blood flow rate is greater in deformable epicardial arteries than in intramural arteries, and conversely during diastole. The amplitude of cyclic variations of pressure is much higher in intramural than in epicardial arteries.

Myocardial contraction reduces the caliber of penetrating arteries (diastolic bore range: 140–650 μm) to about 17% of end-diastolic values, whereas the bore of epicardial coronary vessels increases during systole to about 8% of the end-diastolic value [134]. The compression degree is stronger in the deep wall than in the superficial wall. The longitudinal dimension increases up to about 5% of the end-diastolic value.

Conversely, luminal expansion of intramural arteries associated with a curvature increase of epicardial arteries during ventricular myocardium relaxation enhances

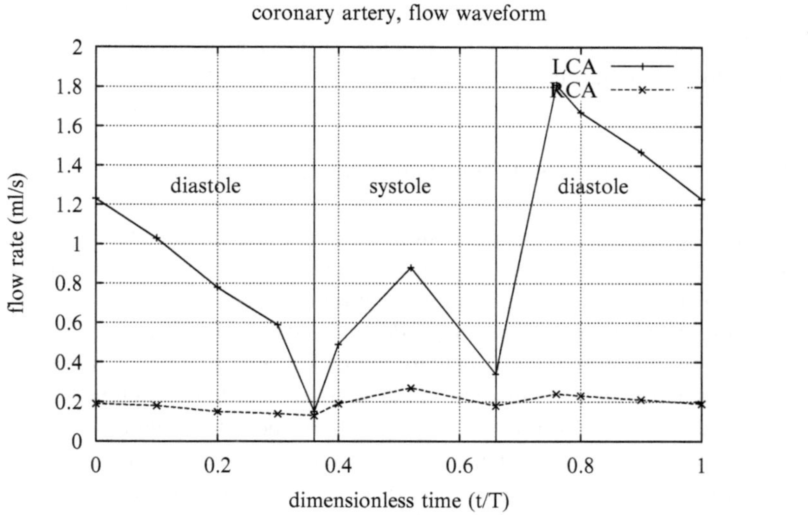

Fig. 3.7 Flow rates in the left (LCA; solid line) and right (RCA; dashed line) coronary artery during the cardiac cycle with a phasic flow pattern (T: period; Source: [133]). During systole, the coronary flow is lowered due partial hindrance of its coronary artery inlet that are partly hidden by the aortic valve cusps and compression of intramural vessels by the contracting myocardium. During diastole, the coronary vessels can be irrigated, although the pump is not propelling blood, because of the windkessel effect associated with the large proximal elastic arteries, mainly the aorta, is responsible of a backflow toward the coronary arteries and closed aortic valve

coronary filling. After the peak aortic valve flow, the pressure difference between the aorta and any capillary bed reaches its highest values and favors tissue perfusion. When the aortic valve closes, the elastic arteries shrink and the left ventricle enters its isovolumetric relaxation phase. The pressure decays abruptly in the left ventricle and more slowly in the arteries due to the windkessel effect, whereas it decreases much more in the coronary circulation owing to myocardium relaxation. Coronary perfusion is particularly enhanced during this time interval of the cardiac cycle.

A retrograde flow can occur in the arteries of the deep and middle compressed myocardial layers toward the arteries of the subepicardial layer [130]. However, the retrograde flow is mostly concealed by distensible epicardial arteries. Distensible epicardial arteries can indeed store about 75% of input stroke volume during the systole. During the following diastole, the blood stored in the epicardial arteries is discharged with the flow incoming from the aorta into the intramural arteries. The flow remains anterograde in arteries of the subepicardial layer as well as in veins of all layers.

Certain small coronary arteries of the midmyocardium may convey retrograde flow during systole. Cardiomyocyte sets form branched bundles, along which are located capillaries. The cardiomyocyte shortening direction corresponds to the main axis of the cardiac myofiber. The distance of the between-cardiomyocyte space decays. The capillary between 2 myofibers can be more or less compressed.

Nevertheless, collagen struts between cardiomyocytes and capillaries may adapt the capillary path and cross-section, avoiding strong narrowing of the capillary lumen and reduction in needed oxygen supply.

3.2.4.7 Coronary Blood Flow Autoregulation

Autoregulation of the coronary flow refers to the maintenance of myocardium perfusion despite changes in perfusion pressure when myocardial metabolism is kept constant (Sect. 3.6.1.8). Autoregulation extends for a given range of perfusion pressure. The autoregulation range depends particularly on mammal species and age. Coronary autoregulation usually works for perfusion pressures between 8 and 20 kPa. Autoregulation fails first in the deep myocardium when perfusion pressure reaches the threshold (either upper or lower) of the autoregulation range [130].

Autoregulation is set up before any usual regulatory effect occurs. Otherwise, the vasomotor tone of the coronary arteries is regulated by sympathetic nerves and several vasoactive substances (acetylcholine, adenosine, prostaglandins, leukotrienes, histamine, serotonin, atrial natriuretic peptide, angiotensin, endothelial-derived vasoactive molecules, etc.).

Blood flow autoregulation depends on myogenic (response to intraluminal pressure) and metabolic responses [135]. The combined action of the metabolic and myogenic responses may overcome the effect of the endothelial-mediated (shear-dependent) vasodilation, thereby achieving autoregulation of flow with increasing arterial pressure. Agent ATP released by red blood capsules that can act as oxygen sensors may trigger an upstream-conducted arteriolar vasodilation [127].

Several factors are involved in the metabolic response. Hypoxia causes vasodilation. Carbon dioxide production is coupled with oxygen consumption and ATP production, but is washed out by venous drainage. When cardiac frequency increases, the rise in interstitial K^+ level precedes the change in coronary vascular resistance [136]. However, K^+ released from myocytes during cell membrane depolarization reenters the cell during recovery and vasodilation persists. Potassium may only participate in the early phase of hyperemia. Adenosine also contributes to myocardial reactive hyperemia. Adenosine concentration in the interstitium is regulated by uptake by cardiocytes and degradation to inosine and hypoxanthine removed in the coronary venous effluent. In addition, prostacyclin is a powerful vasodilator in the microcirculation.

3.2.4.8 Coronary Transit Times

Phasic vasomotor activities may explain heterogeneity in transit times of blood particles between myocardial regions. Coronary flow is mainly characterized by a heterogeneous regional distribution of blood from deep to superficial myocardial layers, that is, from the inner part (subendocardium) to the outer one (subepicardium). The pressure within the heart wall, indeed, decreases from the endocardium to the epicardium, at least during the systole [130]. A triple-layer model then includes the subendo-, midmyo-, and subepicardial strata.

3.2.4.9 Coronary Blood Flow Distribution Control

Blood flow distribution can be centrally controlled at the local level, particularly targeting arterioles, independently of neural and hormonal signals. Increase in cardiac activity augments coronary blood flow. Activation of sympathetic nerves of the coronary vasculature induces transient and slight vasoconstriction mediated by $\alpha 1$-adrenoceptors. Vasoconstriction is followed by vasodilation caused by vasodilators associated with $\beta 1$-adrenoceptor activation of the myocardium to increase the coronary flow for an elevated metabolic activity. Parasympathetic stimulation generates a small coronary vasodilation. Additionally, endothelial production of nitric oxide mediates coronary vasodilation.

3.2.4.10 Coronary Adaptation to Exercise

Static, 20-s duration handgrip at 70% of maximal voluntary contraction causes coronary vasoconstriction (elevation in mean blood pressure without change in coronary blood velocity), but systemic hypoxia induces coronary vasodilation (augmentation in coronary blood velocity without variation in mean blood pressure) to fit metabolic demand [137]. A concordant increase in coronary blood velocity and blood pressure occurs during the cold pressor test. The activity of the sympathetic nervous system activated by physiological stress is modulated by mechanisms independent of autonomic influences.

Skeletal muscle perfusion response to exercise involves reflex and metabolic vasodilation, changes in oxygen extraction, capillary recruitment (increase in the total number of perfused capillaries), and anaerobic glycolysis [136]. On the other hand, the myocardium adjusts to changing loads mainly by vasodilation. Coronary vascular resistance is correlated with the oxygen consumption rate. Basal oxygen extraction is actually high at rest as the heart has a constant activity. The basal oxygen usage is assessed to equal 15 nmol/g of dry weight/mn [136]. In addition, the anaerobic capacity of myocardium delivers only a minor fraction of needed energy. Capillary recruitment and nervous reflexes have a less important role in the myocardium than in skeletal muscle.

3.2.5 Cerebral Circulation

The brain has a high metabolic demand and thus requires adequate nutrient input. The average distance from capillary to neuron ranges from 8 to 20 μm. The microvasculature of the central nervous system maintains the homeostasis of the internal medium, as a constant environment is required for neuronal activity.

3.2.5.1 Blood–Brain Barrier

The blood–brain barrier (Vol. 5 – Chap. 7. Vessel Wall) corresponds to a specialized endothelium of cerebral capillaries that avoids any fluctuation in substance concentration in the extracellular space, especially ions, whereas blood supplies the brain with required nutrients.

3.2.5.2 Functional Hyperemia

In the brain, neuronal activity, energy metabolism, and blood flow are strongly coupled. Cerebral circulation adjusts to brain stimulation (functional hyperemia). The gray matter is among the highest oxygen consuming body's tissues. Local changes in neuron metabolism and blood supply (especially glucose and oxygen input) are ensured by neurovascular coupling. Glucose oxidation provides the major part of energy to support neuronal activity. The magnitude of changes in oxygen consumption depends on stimulation features and brain region (i.e., the amplitude of oxidative metabolism in neuronal circuits). Most of ATP consumption in the brain serves for ion pumps. Hypermetabolic regions correspond to areas in which both excitatory and inhibitory neurons are active. However, synaptic inhibition requires less energy than excitation.

Excitatory activation in postsynaptic cells may be the main determinant of changes in cerebral blood flow associated with neuronal activity, which depends on metabolism adaptation. Glutamatergic postsynaptic neurotransmission influences both hemodynamic and metabolic responses to brain activation. In the cerebellum, an increase in oxygen consumption due to excitatory input is reduced by synaptic inhibition [138]. γ-Aminobutyric acid modulates activity-dependent oxygen consumption triggered by synaptic excitation.

Brain microcirculation responds to neuronal activation by capillary dilatation and flow increase [139]. Heterogeneity in capillary perfusion decreases, as small capillaries dilate more than large capillaries. In addition, capillaries covered by a discontinuous layer of pericytes dilate more than the medium-size vessels wrapped by a discontinuous strip of smooth myocytes. Functional hyperemia can sometimes be associated with focal redistribution of microcirculatory flow.

3.2.5.3 Anticipation of Nervous Activity

Elevation in local blood flow in the cerebral circulation is only partly correlated with the neuronal activity in the corresponding brain region. Electrochemical signals sent by brain cells mediate locally augmented delivery of blood. However, cerebral blood flow can locally rise in anticipation of neural events.

In alert animals, hemodynamic signals can be decomposed into 2 distinct components: (1) a component is predictable from neuronal responses generated by visual input, and (2) another component of nearly equal amplitude independently of visual input or standard neural predictors of hemodynamics [140]. An anticipation mechanism brings additional arterial blood to the cerebral cortex in preparation of expected tasks, whereas neurons can remain nearly unresponsive.

3.2.5.4 Cerebral Autoregulation

Cerebral blood flow is regulated via changes in cerebrovascular conductance in response to changes in cerebral perfusion pressure (cerebral autoregulation; Sect. 3.6.1.6) as well as sympathetic activity. The latter depends on fluctuations in blood gas concentrations, blood flow rate, and regional metabolism, which causes functional hyperemia. These factors are not taken into account when only pressure and flow velocity are measured.

Cerebral autoregulation is displayed by a plateau with a slight slope and variable upper and lower limits of arterial pressure. Cerebral arteries are actually able to keep brain perfusion almost constant over a wide range of arterial pressures by means of myogenic, neurogenic, or metabolic mechanisms.

Chronic hypertension shifts upper and lower limits of autoregulation toward higher values [141]. Cerebral blood flow is maintained despite gradual, long-term, or acute changes in arterial pressure by adaptating cerebrovascular resistance.

Classical studies of cerebral autoregulation are carried out in steady state, as pressure change occurs gradually up to the new stable value. Transcranial Doppler ultrasonography and servocontrolled finger photoplethysmography allow us to investigate dynamical pressure–flow relations during a sit-to-stand procedure, for example. A period of reduced flow in the cerebral artery precedes the return to flow baseline after hypotension.

Although in mild or moderate hypertension dynamical cerebral autoregulation is not substantially altered, it is impaired during sudden hypotension episodes in the elderly with carotid stenosis. Postprandial or orthostatic hypotension thereby can cause syncope in elderly individuals.

Moreover, low-frequency (0.07–0.15 Hz) blood pressure changes preserve dynamical cerebral autoregulation, but not high-frequency changes whatever the age. In addition, cerebral autoregulation features depend on the type of cerebral artery.

Cerebral autoregulation is often assessed in both healthy subjects and patients using 3 widespread methods: (1) the rate of regulation (RoR index); (2) the autoregulatory index (ARI); and (3) transfer function analysis. The rate of regulation index quantifies the rate of change in beat-to-beat cerebrovascular conductance, or resistance, after an acute transient hypotension. The autoregulatory index is derived by fitting the arterial pressure and cerebral blood flow velocity within a 30-s window after a transient hypotension to a mathematical model that incorporates both the latency and gain of cerebral autoregulation. Transfer function analysis on blood pressure relates dynamic changes in arterial pressure and cerebral blood flow velocity in the middle cerebral artery, for example, in the frequency domain. This approach yields 3 parameters that describe the magnitude of flow changes associated with arterial pressure (gain) as well as the timing (phase) and linearity (coherence) of the relationship. Higher values of coherence and gain and lower values of phase are generally considered as reflecting weaker cerebral autoregulation. These indices are generally unrelated to one another or show only weak to moderate correlations [142]. Interpretation of these measures must thus be cautious.

3.2.5.5 Changes During Aging

Composition, geometry, and mechanical properties of human cerebral arteries vary during aging. Intima and media thicken, but outer diameter remains relatively constant with age [143]. Elastin structure changes from a fiber network primarily oriented in the circumferential direction to a less organized mesh, especially at the intima. Cerebral arteries stiffen with age due to dysfunctional elastin with aging, as the volume fraction of elastin remains nearly constant.

3.2.6 Pulmonary Circulation

Pulmonary blood volume is shared among the arterial, capillary, and venous compartments with volume of about 150, 100, and 200 ml, respectively. Blood flow rates are nearly equal in serial pulmonary and systemic circulation due to permanent adaptation of blood ejections between both hearts.

3.2.6.1 Pulmonary Circulation Time

The circulation time between the pulmonary valve and the left atrium is less than 4 s. The mean capillary circulation time is about 700 ms, whereas the time required for a balance between alveolar gas and blood is lower than 300 ms.

3.2.6.2 Pulmonary Circulation Pressure and Resistance

The pressure range in the pulmonary microcirculation is lower than in the systemic circuit. Because the pressure drop in the alveolar capillaries is small, RBC transit time rises for efficient gas exchanges. Due to slow travel through the alveolar capillaries, leukocyte concentration in the lung parenchyma is greater than in organs perfused by the systemic circulation.

At rest in the lying position, pulmonary blood pressure ranges in arteries $\sim 0.7 < p_{pa} < \sim 3.3$ kPa and veins $\sim 0.4 < p_{pv} < \sim 1.3$ kPa. The pulmonary vessel resistance (PVR) is smaller than 8 kPa/l/s.

The pulmonary blood flow is subjected to cyclic variations of intrathoracic pressure and upstream (venous return and preload) and downstream effects of the systemic circulation. During inspiration, the intrathoracic pressure decreases and the vessel transmural pressure increases, and the reverse occurs during expiration.

Pulmonary vasoconstriction is induced by alveolar hypoxia, pH increase, α-adrenoceptor stimulators, and angiotensin-2. On the other hand, α-adrenoceptor blockers, acetylcholine, and bradykinin generate vasodilation.

3.2.6.3 Heterogeneous Distribution of the Pulmonary Perfusion

A heterogeneous distribution of the perfusion in different lung regions according to their vertical situation results from the hydrostatic pressure difference between the lung apex and basis.[66]

Ventilation distribution is aimed at partly matching the perfusion variations. The lung is generally subdivided into 3 zones [144]: *zone 1*, where $p_v < p_a < p_A$ is characterized by collapsed pulmonary capillaries; *zone 2*, where $p_v < p_A < p_a$ is a transitional region in which capillaries may undergo distal partial closure; and *zone 3*, where $p_A < p_v < p_a$, hence the capillaries are dilated.

3.3 Microcirculation

The microcirculation, with its 4 main duct components: arterioles, capillaries, venules, and terminal lymphatic vessels, regulates blood flow distribution within organs, transcapillary exchanges, and removal of cell wastes.

Arterioles are small precapillary resistance vessels. They are richly innervated by sympathetic adrenergic fibers and highly responsive to sympathetic vasoconstriction via both α1 and α2 postjunctional receptors. It is thus a major site for the regulation of systemic vascular resistance. The primary function is flow regulation, thereby determining nutrient delivery and catabolite washout. They regulate capillary entry hydrostatic pressure and fluid exchanges.

In some organs, precapillary sphincters can regulate the number of perfused capillaries. Venules are collecting vessels. Sympathetic innervation of larger venules can alter venular tone which controls capillary exit hydrostatic pressure.

Large lymphatic vessels have muscular walls. Spontaneous and stretch-activated vasomotion in terminal lymphatic vessels helps to convey lymph. Lymph circulation is mainly under local control. Smooth myocytes of the stretched lymph vessels, indeed, rhythmically contract at low frequency due to lymph accumulation. Sympathetic nerves cause contraction. Valves direct lymph into the systemic circulation via the thoracic duct and subclavian veins. Lymph flow is very slow. Lymph has a composition similar to plasma but with a small protein concentration.

---

[66]The hydrostatic pressure decreases with height at a rate of ~100 Pa/cm for an arterial (p_a) and venous (p_v) pressure and ~0.1 Pa/cm for an alveolar pressure (p_A).

Table 3.41 Main features of the capillary circulation

Length	0.2–0.4 mm
Radius	4 μm
Between-capillary distance	10–30 μm
Capillary density	300–5700 (heart)
Total surface	60–1100 mm^2/mm^3 of tissue (heart)
Volume	10^6 μm^3 (∼500 RBCs in a capillary)
Compartmental blood volume	∼300 ml (∼5% of total blood volume)
Transit time	0.6–3 s
Pressure	3.5–4 kPa (arteriolar side)
	1.5–2.9 kPa (venular side)
Capillary set flow	100 ml/s
Mean velocity	≤1 mm/s
Hematocrit	<Ht in large vessels (Fahraeus effect)
Viscosity	$\mu = \mu(R_h, Ht)$ (Fahraeus–Lindqvist effect)
Diphasic flow of deformed cells	
Plasma skimming and cell screening	
Flow regulation by recruitment (precapillary sphincter) and possible shunt	

The protein concentration in the lymph is about half that of plasma. Lymphatics have important roles in tissue clearance of molecules, removal of cells from tissues after inflammation, and interstitial fluid homeostasis.

3.3.1 Capillary Flow

The capillary circulation is characterized by: (1) a low flow velocity (quasi-steady Stokes flow) and (2) a short distance between the capillary lumen and cells.

The capillary wall is adapted to molecular exchanges (Table 3.41; Vol. 5 – Chap. 7. Vessel Wall). It is constituted by: (1) the glycocalyx; (2) the endothelium; and (3) a basement membrane.

Endothelium has a variable structure and function according to the type of organ. It is characterized by either pores or intercellular cleft sealed by tight junctions according to the perfused territories.

Large pores represent the vesicular transport through endotheliocytes, sites of apoptotic endotheliocytes, and loci of lost glycocalyx associated with discontinuous tight junctions, in addition to fenestrations without glycocalyx.

The capillary flow locally depends on upstream resistance in arterioles and downstream resistance in venules. Local control of substance transport to the tissue is done by: (1) recruitment of terminal arterioles (the higher the number of open capillaries is, the greater the solute delivery and waste removal); (2) possible autoregulation (maintenance of a constant flow despite changing vascular pressure) associated with the cellular activity; and (3) vascular permeability.

3.3.2 *Capillary Recruitment*

Blood vessel recruitment can refer to angiogenesis that ensures normal tissue growth. Most often, blood vessel recruitment corresponds to the contribution of previously unperfused capillaries to increase the total surface area of mass exchange between the perfused tissue and blood.

3.3.2.1 Pulmonary Capillary Recruitment

The elevated surface area at the interface between the pulmonary capillary blood and alveolar air by capillary recruitment enables the maintenance of an adequate oxygenation when the metabolic demand rises. Hypoxia augments the cumulative length of perfused capillaries in the pulmonary capillary bed due to the recruitment of previously unperfused capillaries [145].

In the pulmonary circulation, recruitment has an exclusive capillary origin, as pulmonary arterioles and venules do not participate in the vascular recruitment [146].

The pulmonary capillary recruitment is modulated by pressures in alveoli and up- and downstream vessels, walls of which contain smooth muscle and are innervated, as well as intrinsic factors. The distribution of blood flow within the lung is indeed related to the relation between alveolar, arterial, and venous pressures.

The precapillary pressure influences capillary recruitment. When the postcapillary pressure increases temporarily, the volume of alveolar vessels that constitute an important volume storage heightens [147] and pulmonary vascular resistance decays, as both the small and large vein resistances are reduced [148].

Intrinsic factors of capillary recruitment rely on activity of local mural cells. Endotheliocytes produce enzymes, such as angiotensin-converting enzyme, 5′-nucleotidase, carboxypeptidase-N, and aminopeptidase-P, that regulate the action of circulating hormones. In particular, circulating angiotensin-1 and bradykinin are processed by enzymes on the luminal surface of pulmonary endothelial cells [149]. Angiotensin-1 is converted into vasoconstrictory angiotensin-2, whereas vasodilatory bradykinin is inactivated. Endothelial cells also synthesize nitric oxide, among other vasoactive substances, to modulate the vasomotor tone, hence the vessel lumen caliber and blood flow resistance.

3.3.3 *Transcapillary Transfer of Materials*

The capillary endothelium forms a stable anti-inflammatory, antithrombotic, and antiadhesive interface between circulating blood components and cells and the body's tissues.

Three fluid compartments exist in series: blood, interstitial fluid, and lymph. According to pressure inside these compartments, water flows continuously from blood to lymph, as walls (i.e., mainly continuous endothelia) of capillaries and postcapillary venules operate as semipermeable membranes. Lymph drains back into the blood circulation via the major veins at the base of the neck.

Water flow across the capillary wall depends on the balance between the osmotic absorption pressure of plasma proteins, or colloid osmotic pressure, and the capillary pressure generated by the cardiac pump.

3.3.3.1 Microvascular Permeability

The permeability of the microvasculature is a property of its endothelium to govern exchange between blood and perfused tissue. The motion of water and solutes (nutrients and wastes) is driven by a pressure difference balance (difference between the hydrostatic pressure in the capillary lumen and in the interstitium and difference between colloid osmotic pressure in plasma and interstitium [subglycocalyx layer]) for the water convection that conveys molecules as well as concentration gradient for solute diffusion.

The permeability coefficients relate the net fluxes of fluid (water; J_w) and solute (J_s) driven by concentration (c) and pressure (p) differences. Four coefficients are involved: hydraulic conductivity ($\mathcal{G}_\mathrm{h}$), diffusional permeability ($\mathcal{P}$), solvent drag coefficient (κ_d), and osmotic reflection coefficient (κ_o) [163].

Mass Transfer Factors

Mass transfer in a microvascular unit between the flowing blood and any irrigated tissue is regulated by [150]: (1) mechanisms that control the number and type of perfusing microvessels; (2) the balance between adhesion and contraction in endotheliocytes that determine the conductance of intercellular spaces to water and small solutes; (3) pressure and chemical potential gradients that govern mass transfer; and (4) organization of barriers to macromolecules in the endothelial glycocalyx.

Capillaries are the primary site of exchange for fluid, electrolytes, gases, and macromolecules by filtration and absorption that result from endocytosis as well as diffusion due to the concentration difference and convection due to the pressure difference between the vascular lumen and tissue.

Obviously, fenestrated capillaries have a higher permeability than continuous capillaries. In most capillaries, there is a net filtration of fluid by the capillary endothelium (filtration exceeds reabsorption). Excess fluid within the interstitium is removed by the lymphatic system.

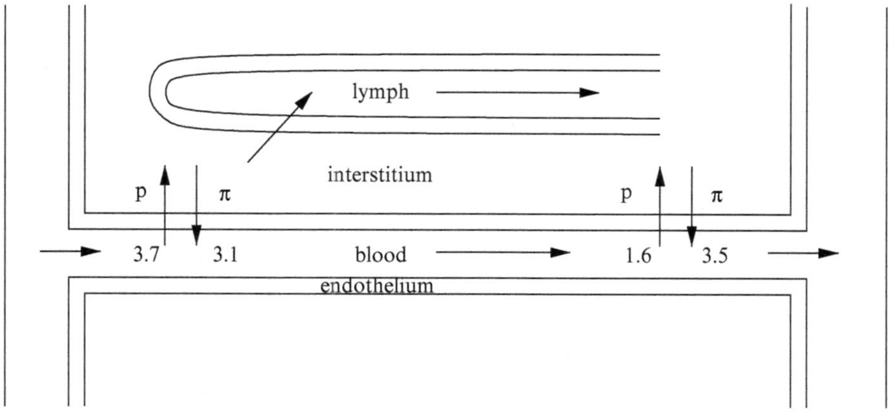

Fig. 3.8 Capillary exchanges associated with hydrostatic and osmotic pressure value (kPa) distribution

Hydrostatic and Osmotic Pressure

The transport of water and solutes (electrolytes and small molecules) between blood and tissue fluid is mainly determined by 2 opposing forces, *hydrostatic* and *osmotic* transendothelial pressures, as well as endothelium permeability (Fig. 3.8).

Water flux results from the imbalance between hydrostatic and osmotic pressures for given features of the vessel wall and exchange surface area. A net motion of water out of (positive water flux) or into (negative water flux) the vessel lumen leads to filtration and absorption, respectively. The net driving pressure for fluid motion is determined by the sum of the hydrostatic and osmotic contributions.

The capillary hydrostatic pressure is normally much greater than tissular hydrostatic pressure. The net resulting hydrostatic pressure gradient is positive from the lumen to the tissue in the upstream capillary segment. It drives fluid out of the capillary into the interstitium.

The plasma osmotic pressure is usually much greater than the interstitial osmotic pressure. The resulting osmotic pressure gradient favors fluid reabsorption from the interstitium into the capillary. The osmotic pressure difference must be multiplied by a reflection coefficient associated with capillary permeability to proteins responsible for the osmotic pressure.

In the traditional theory, plasma is filtered out of the capillary entry segment (arterial side) and reabsorbed back into capillary exit segment (venular side). The ultrafiltrate has the same ion concentration as plasma and does not contain large molecules.

Steady-state equilibrium between the filtration of the arterial side of the microvasculature and reabsorption in venular microvessels (the latter being previously assumed to balance the former) usually does not occur. The balance between the hydrostatic and oncotic pressures is modulated by the control of the interstitial pressure by fibrocytes in particular. Underhydrated glycosaminoglycans can take up fluid and swell, thereby reducing the interstitial pressure [153].

Water Hydraulic Conductivity

The hydraulic conductivity determines the fluid flow per unit pressure gradient across the barrier. In continuous capillaries, water is transported by convection through intercellular clefts of the microvascular endothelium (paracellular route) and, in a smaller, but significant proportion (up to 20%), across the plasma membrane via aquaporins and through the cell (endo- and transcytosis; intracellular route). In fenestrated capillaries (e.g., renal glomerulus), the fenestra is the predominant path that is characterized by a high hydraulic conductivity and larger cross-sectional area than that of the intercellular clefts.

A mixture of low- and high-molecular-weight molecules exerts an osmotic pressure across a semipermeable membrane that is proportional to the size of the channels through which molecules travel. The driving pressure is thus the weighted difference between the hydrostatic pressure difference and colloid osmotic pressure difference between the capillary lumen and the interstitium, or subglycocalyx layer.

Solute Permeability

The solute permeability determines the solute flux by diffusion per unit concentration gradient and per unit surface area. In fact, a given solute is transported by both diffusion and convection by water flow driven by a pressure difference between the microvascular lumen and interstitium.

When a porous medium flow theory is used to model the transcapillary exchange and lymphatic removal or the transport of water and solute into biological tissues, the transfer of molecules from the microvascular endoluminal blood, actually depends on: (1) the intersticial void fraction that varies with the local hydrostatic (interstitial) pressure and the tissue compliance; (2) molecule void fraction; (3) solute diffusivity; (4) tissue hydraulic conductivity; and (5) solvent drag coefficient.

Osmotic Reflection Coefficient

The degree of permeability to a given solute can be quantified by the normalized (ranging 0 to 1 from entirely permeable to wholly impermeable) Staverman osmotic reflection (selectivity) coefficient (κ_o). In other words, the osmotic reflection coefficient of a porous membrane is a measure of the selectivity of the membrane to the solute of interest. "Ideal" solutes have an osmotic reflection coefficient equal to that of the solvent drag.

The *sieving coefficient* (s) is the inverse of the osmotic reflection coefficient. The sieving coefficient determines the ratio of molecules that is sieved across the membrane. The reflection coefficient (κ_r) is defined as the complementary of the sieving coefficient with respect to unity ($s = 1 - \kappa_r$) [151]. Two reflection and sieving coefficients are then considered, the osmotic or solvent drag sieving coefficient ($\kappa_d = 1 - s_d$) and the frictional sieving coefficient ($\kappa_f = 1 - s_f$).

The higher the osmotic reflection coefficient is, the stronger the colloid osmotic pressure in the entrance compartment. An endothelium permeable to water and impermeable to plasma proteins is thus associated with a plasma that exerts a great osmotic pressure that antagonizes the hydrostatic pressure; in the interstitium both the hydrostatic and osmotic pressures are smaller.

Solvent Drag Coefficient

The capillary blood–tissue interstitium exchange (i.e., the movement of water and proteins through the interstitium by convection and diffusion) depends on the solvent drag effect on solutes. Mass transfer theory is associated with several assumptions: (1) dependence of water flow into tissue on the hydrostatic and osmotic pressure gradient between microvascular blood and tissue; (2) extracellular space changes with variations in local interstitial pressure; (3) water convection-driven motion of proteins through the tissue at a velocity proportional to the water velocity. Macromolecular (protein) permeability in large pore theory relies on the solute retardation factor, or solvent drag coefficient, in addition to the osmotic reflection coefficient.

If λ_p denotes the ratio between the molecule radius and pore radius and $\phi = (1 - \lambda_p)^2$ the reduction of pore cross section available to the solute, the osmotic reflection and solvent drag coefficient in the pores are computed using the following equations [151]:

$$\kappa_{o,p} = (1 - \phi)^2, \qquad (3.28)$$

and

$$\kappa_{d,p} = \frac{16}{3}\lambda_p^2 - \frac{20}{3}\lambda_p^3 + \frac{7}{3}\lambda_p^4. \qquad (3.29)$$

3.3.3.2 Endothelial Barrier

The endothelium and its associated structures, e.g., the glycocalyx, basement membrane, junctions between apposed endotheliocytes, and supporting pericytes, form the primary barrier to water and plasma protein motion, thereby maintaining the plasma volume and venous return. Nevertheless, it enables transvascular exchange to fulfill the local metabolic needs.

Adherens and Tight Junctions

The tight junction is a nearly continuous band that consists of transmembrane proteins junctional adhesion molecule JAMa, claudin and occludin. In the cytosol, tight junction proteins are connected to actin filaments via zona occludens ZO1 and ZO2 proteins.

These junctional strands, especially in the inner segment (near the wetted surface) of the intercellular cleft, limit outward water flow and create a difference between colloid osmotic pressures in the subglycocalyx space and interstitial fluid.

In fact, the endothelial tight junction strand is not continuous, but contains infrequent breaks (length 100–400 nm). These breaks occur at 2 to 4 μm intervals, thereby forming large pores and, hence, unrestricting passage of molecules through intercellular spaces. Therefore, water as well as solutes such as plasma proteins is funneled through these breaks.

Adherens junctions strengthen intercellular adhesion by a homophilic Ca^{2+}-dependent transinteraction of cadherin-5 (or vascular endothelial [VE]-cadherin) as well as junctional adhesion molecules (JAMa–JAMc). In the cytosol, adherens junction proteins are anchored to cortical actin filaments that reside around the perimeter of each cell via catenins (Ctnnα–Ctnnδ).

Therefore, cell adhesion molecules, such as cadherin-5 and occludin, stabilize the endothelial barrier. On the other hand, both adherens and tight junctions are tethered to the cortical actin cytoskeleton via several adaptors, such as catenins and zona occludens proteins, respectively (Vol. 1 – Chap. 7. Plasma Membrane).

Furthermore, vascular endotheliocytes are also anchored to the extracellular matrix via focal adhesions supported by integrins and actin-linking proteins, such as paxillin, talin, and vinculin.

Endotheliocyte Adhesion and Contraction

The endothelial barrier relies on a balance between intercellular and cell–matrix adhesions that maintain the integrity of the barrier and stress fiber contraction within endotheliocytes that opens intercellular gaps.

Myosin light chain kinase (MLCK)[67] phosphorylates myosin light chain, principally MLC2, in response to Ca^{2+} binding to MLCK-associated calmodulin, thereby supporting actomyosin contractility, weakening interendothelial cell adhesion, and causing vascular hyperpermeability via gap formation [152].

Once it is activated by Ca^{2+}–calmodulin or phosphorylation by protein Tyr kinase (Tyr464 and Tyr471), MLCK phosphorylates MLC (Ser19 and subsequently Thr18) [152].

On the other hand, myosin light chain phosphatase (MLCP) dephosphorylates MLC, hence relaxing the cytoskeleton. Activated RhoA GTPase stimulates RoCK kinase that subsequently phosphorylates (inhibits) MLCP, thereby increasing endothelial permeability.

[67]Four MLCK types exist: skeletal (skMLCK), cardiac (cMLCK), and smooth muscle (smMLCK) isoforms encoded by the MYLK2, MYLK3, and MYLK1 genes, respectively, as well as nonmuscle MLCK (nmMLCK) with 4 high-molecular-weight isoforms (MLCK1–MLCK4) that are splice variants translated from the MYLK1 gene. Isoforms MLCK1 and MLCK2 are the most highly expressed subtypes in the vascular endothelium.

Endothelial hyperpermeability mediated by MLCK in response to histamine, thrombin, and VEGF relies on PKC activation as well as NO production and subsequent priming of the sGC–cGMP–PKG–cRaf–ERK1/2 pathway [152].

Cortical Actin Cytoskeleton

Cellular junctional complexes are stabilized by inhibition of cofilin and recruitment of ubiquitous cortactin. Unphosphorylated cofilin binds to ^{F}actin and promotes filament severing, thereby destabilizing the cortical actin cytoskeleton. Kinase LIMK phosphorylates (inactivates) cofilin.

On the other hand, cortactin is implicated in cortical actin assembly, reorganization, and strengthening. Cofilin phosphorylation regulates cortical actin fibers.

Glycocalyx

The glycocalyx layer at the interface between blood and the vessel wall constitute the first barrier that is permeable to water and solutes, but retains plasma proteins and inflammatory leukocytes in the vascular space, before any trans- or paracellular transfer.

The glycocalyx excludes circulating blood cells and macromolecules. *Hindrance* of fluid and solute radial transfer, or conversely *permeation* of proteins into the glycocalyx, depends on the molecule size and charge as well as on the glycocalyx fiber size, orientation, and spacing, limiting accessibility to small plasma molecules. The glycocalyx is able to restrict access of bloodborne agents by both size and electrostatic exclusion. It reduces albumin permeability from a magnitude order of 10^{-6} cm/s in the absence of the glycocalyx to that of 10^{-7} cm/s (albumin reflection coefficient close to 0.9) [150].

The *Fahraeus effect* (i.e., a microvascular hematocrit lower than the macrovascular hematocrit) associated with a central flow where red blood capsules accumulate and a peripheral plasma layer (width 0.4–0.6 μm) in contact with the glycocalyx that moves with a slower velocity than the mean RBC velocity, can be explained not only by phase separation of RBCs and plasma at upstream bifurcations, but also exclusion of circulating RBCs and plasma by the glycocalyx.

The glycocalyx can be considered as a negatively charged gel-like mesh of glycosaminoglycans (glypican, hyaluronan, and syndecan-1) and sialoglycoproteins (Vol. 1 – Chaps. 8. Cell Environment and 5 – Chap. 9. Endothelium).[68] These

[68]Glycosaminoglycans are linear heteropolysaccharides, combination of which create different GAG types, such as heparan (50% of the total GAG pool at the endothelial surface), chondroitin, and dermatan sulfate, and hyaluronic acid or hyaluronan. Membrane-bound glypicans with their heparan sulfate chains localize to caveolae. Transmembrane syndecans cluster in the outer edge of caveolae. They connect to the cytoskeleton. Hyaluronan is a very long glycosaminoglycans that is not sulfated. It is not attached to a core protein. Transmembrane epican, or heparan sulfate proteoglycan, can contain chondroitin and heparan sulfate as well as oligosaccharides. It localizes to caveolae.

molecules extend from 40 to 600 nm from the endothelial luminal membrane. The scaffold of the glycocalyx is formed by anionic polysaccharides (glycosaminoglycans, proteoglycans, and glycoproteins).[69]

Heparan sulfate proteoglycans contain abundant binding sites for plasma proteins. Their core proteins, syndecans and glypicans, can function as signal transducers. The glycosaminoglycan hyaluronan, which is not sulfated and lacks binding sites for proteins, contributes significantly to the volume of the glycocalyx due to its hydration capacity.

The endothelial glycocalyx contains a relatively thin layer of membrane-bound macromolecules and a relatively thick layer of adsorbed plasma components. Plasma proteins, endothelial-derived growth factors, enzymes, and water are adsorbed within the polysaccharidic mesh.

The glycocalyx coats the inner wetted surface of capillaries, including the entrance of the intercellular cleft and most fenestrations. It operates as both a determinant of vascular permeability and mechanotransducer.[70] A sustained physiological shear participates in the maintenance of a stable endothelial phenotype.

Intercellular Cleft

Endothelial cells overlap along their perimeter by a distance of 0.1 to more than 1 μm. Therefore, the lateral edge is not perpendicular to the wetted surface, but oblique.

The cleft between apposed endotheliocytes follows a long, narrow, tortuous path. The cleft is sealed by porous tight junctions. The width of the interendothelial cleft (18–22 nm) is determined by transmembrane adhesion proteins such as cadherin-5 that provides a homophilic link between adjoining cells.

The paracellular passage is the primary route for water and solute exchange across the microvascular wall when gaps form between apposed cells. In the cardiac microvasculature, venules possess tight junction strand breaks as well as much smaller gaps [150]. The permeability of these venules reaches the upper values reported in mammalian microvessels.

Transcytosis

The capillary endothelium contains plasmalemmal vesicles (size 60–80 nm). Caveolae invaginate from the luminal plasma membrane and form vesicles carried along

[69] A glycoprotein has short oligosaccharide branched chains. Glycoproteins encompass many receptors on the cell surface, such as integrins, selectins, and members of the immunoglobulin superfamily.

[70] After removal of the glycocalyx by heparinase, cultured endotheliocytes do not align in the streamwise direction and can proliferate after 1 d of experiencing flow.

the cytoskeleton that fuse with the abluminal plasma membrane (Vol. 1 – Chap. 9. Intracellular Transport). Caveolae are coated by the integral membrane protein caveolin-1 on their cytoplasmic face.

Other vesicular transport means can also contribute to transendothelial transfer of materials.

Intertitium

Blood capillaries are surrounded by a loose connective tissue that contains a given interstitial fluid volume.[71] The extracellular matrix consists of 4 main components (Vol. 1 – Chap. 8. Cell Environment): (1) a fibrous collagen network that serves as a scaffold; (2) fibrous elastin meshwork; (3) ground substance formed from proteoglycans composed of glycosaminoglycans linked to proteins as well as the free glycosaminoglycan hyaluronan.

Intertitium acts as a storage space for the fluid filtered across the capillary wall. The interstitial fluid is an ultrafiltrate of plasma with a total protein concentration of 50 to 60% of plasma level and with an electrolyte composition similar to that of plasma.

The composition of the interstitial fluid depends on the transcapillary fluid flow, itself function of the quality of the capillary wall barrier, and molecule size and charge.

Several matrix components (e.g., chondroitin and heparan sulfate and hyaluronan) are cleared out from the interstitium by an elevated fluid flow. The turnover rate that depends on clearance and synthesis rate varies according to the organ type.

Intertitium is a dynamical structure that can participate in transcapillary water filtration via a modulation of the interstitial fluid pressure. The latter is the filling pressure for initial segments of lymphatics.

Vasodilation

Increased nutrient exchange relies on the regulation of both perfusion and permeability. In fact, it depends more on vasodilation that raises the exchange surface area and associated augmented water-driving pressure (endoluminal and hence transparietal pressure), and elevated concentration differences between blood and tissue than a marked increase in permeability [153].

Vasodilation increases the number of microvessels perfused during exposure to an inflammatory agent as well as the permeability on blood–tissue exchange of fluid and solutes. Nitric oxide elevates permeability of the microvascular bed. On the other hand, vasoconstriction attenuates these effects.

[71] The skin and skeletal muscles jointly contain around two-thirds of the extracellular fluid.

The heparan sulfate-rich glycocalyx layer limits perfusion volume in capillaries and hinders transparietal fluid and solute transport. Vasoactive substances can alter glycocalyx exclusion ability [154]. Vasodilators such as adenosine and bradykinin operate at the level of the glycocalyx rather than by increasing the local blood flow. Adenosine provokes shedding of hyaluronan from the glycocalyx [154].

Monomeric GTPases in the Maintenance of the Endothelial Barrier

Cell adhesion stabilizers target small (monomeric) GTPases (Vol. 4 – Chap. 9. Guanosine Triphosphatases and Their Regulators). These effectors contribute to the organization of the cortical actin cytoskeleton and distribution of constituents of endothelial adherens junctions. Small GTPases are activated by guanine nucleotide-exchange factors (GEF) and inactivated by GTPase-activating proteins (GAP).

The maintenance of the microvascular barrier depends on various types of small GTPases, in particular Rac1, CDC42, and RhoA proteins. Both Rac1 and CDC42 are the main monomeric GTPases that protect the endothelial barrier [155]. They are antagonized by RhoA regulator.

Effect on Cytoskeleton Dynamics

Both Rac1 and CDC42 GTPases activate via PAK1 and PAK4 LIMK enzyme [155]. On the other hand, the RhoA effector RoCK kinase inhibits LIMK, hence favoring cofilin-mediated ^{F}actin depolymerization.

In addition, RhoA GTPase destabilizes the endothelial barrier via RoCK kinase that inactivates myosin light chain phosphatase (MLCP) as well as phosphorylates MLC, hence increasing MLC phosphorylation and provoking actin–myosin fiber contraction.

However, both Rac1 and RhoA GTPases can also have opposite effects. For example, the microvascular endothelial barrier is destabilized by Rac1 during angiogenesis. In addition, Rac1 acts via PAK to enhance MLC phosphorylation in macrovascular endothelial cells. On the other hand, RhoA can strengthen endothelial cortical actin via its effector Dia in a profilin-dependent manner.

Microtubule Participation

Microtubule polymerization stimulates RhoA, thereby increasing permeability of endothelial cell monolayers. Once it is uncoupled from microtubules, the microtubule-associated protein RhoGEF2 targets RhoA, a mechanism used by by thrombin [155].

Effect on Cell Junctions

IQ motif-containing GTPase-activating protein IQGAP1 impedes α-catenin–β-catenin binding, hence preventing ^{F}actin linkage to adherens junctions [155]. Both Rac1 and CDC42 may regulate cadherin-mediated cell adhesion by sequestering IQGAP1 inhibitor.

A loss of cell adhesion can result from phosphorylation mainly by SRC family kinases and endocytosis of cadherin–catenin complex components [155]. Several barrier-destabilizing mediators, such as VEGF, thrombin, and TNFSF1, signal via small GTPases such as Rac1 and provoke phosphorylation of cadherin-5 and β- and δ1-catenin. Activator RhoGEF2 connects to the tight junction complex via the adaptor cingulin; it augments endothelial permeability.

cAMP Mediator

Monomeric Rac1 GTPase activated by sphingosine 1-phosphate enhances peripheral localization of cadherin-5 as well as α- to γ-catenins as well as reduces stress fiber formation, but promotes cortical actin assembly [150]. Sphingosine 1-phosphate elevates the cytosolic concentration of cAMP messenger.

Prostaglandin-E_2 and -I_2 as well as atrial natriuretic peptide ANP elevates cytosolic cAMP concentration, hence promoting Rac1 and CDC42 activation via the RapGEF3–Rap1 pathway and TIAM1 and Vav2 GEFs [155].

The barrier protector cAMP operates via Rap1-mediated activation of Rac1 and CDC42 as well as RhoA inhibition. Messenger cAMP acts via RapGEF3 and RapGEF4 that stimulate Rap1 and indirectly activate Rac1 GTPase. The cAMP–RapGEF–Rap1 axis stabilizes the peripheral actin band.

Mediator cAMP activates not only Rac1, but also synergistically CDC42 via Vav2 protein [155]. Stabilization of the endothelial barrier by oxidized phospholipids relies also on CDC42 GTPase. In addition, CDC42 is a major GTPase involved in barrier restoration.

Molecule cAMP represses effect of TGFβ that disorganizes the microtubular mesh and, hence, increases endothelial permeability likely via RapGEF3 activator.

Nitric Oxide

The path through small pores of tight junctions can be modulated, in particular by nitric oxide [150]. Once it is synthesized by nitric oxide synthase-3 and released in response to inflammatory agents, such as bradykinin, platelet-activating factor (PAF) and vascular endothelial growth factor (VEGF), nitric oxide increases microvascular permeability [156].

The signaling pathway involved in hyperpermeability comprises soluble guanylate cyclase (sGC), cGMP, protein kinases PKC and PKG as well as ERK1 and ERK2.

Platelet-activating factor that activates NOS3 increases endothelial permeability, but does not cause vasodilation. On the other hand, some vasodilators such as acetylcholine (ACh) activate NOS3, but does not alter the microvascular permeability [156]. Both PAF and ACh separate NOS3 from caveolin-1, but ACh promotes preferential movement of NOS3 to the Golgi body, whereas PAF favors its motion in the cytosol.

In endotheliocytes, NOS3 localizes mainly to the plasma membrane and Golgi body, but also distributes in the cytosol. Its localization to plasmalemmal caveolae relies on ^{N}myristoylation and palmitoylation. It translocates from the plasma membrane to subcellular compartments upon demyristoylation and depalmitoylation, thereby relieving the inhibitory connection with caveolin-1 [156]. It is internalized with caveolae.

However, both membrane-bound and cytosolic NOS3 are able to release nitric oxide, the former form releasing more basal NO than cytosolic form [156]. Its relocalization to specific subcellular compartments determines its function as well as access to arginine pools and scavengers of reactive nitrogen species, thereby ensuring protection, as NO is also a radical.

Its activity is determined by multiple mechanisms (Vol. 4 – Chap. 10. Other Major Signaling Mediators): (1) activating and inactivating phosphorylation and dephosphorylation of specific NOS residues (Ser116, Ser617, Ser635, Ser1179, Thr497, and Tyr83); (2) ^{S}nitrosylation that prevents NOS3 activity and reduces sensitivity of sGC to NO activator; (3) interaction with different proteins: (4) and specific subcellular localization.

Activation of NOS3 in response to proinflammatory stimuli responsible for microvascular hyperpermeability proceeds from: (1) translocation from the plasma membrane to cytosol via caveolae; (2) dissociation from caveolin-1, association with heat shock protein HSP90,[72] Ca^{2+}–calmodulin binding, phosphorylation (Ser1177), and dephosphorylation (Thr495) [156].

Insulin-like and vascular endothelial growth factor as well as hemodynamic stress can increase permeability of venules via protein kinase-B that phosphorylates (activates) NOS3 (Ser1177). Bradykinin stimulates calmodulin-dependent kinase CamK2 that also phosphorylates NOS3 (Ser1177).

VEGF

Vascular endothelial growth factor (VEGF), previously called vascular permeability factor and vasculotropin, regulates vascular permeability (Vols. 2 – Chap. 3. Growth

[72]Chaperone HSP90 binds to both NOS3 and sGC and facilitates their interaction, stabilizing sGC and enhancing cGMP production.

Factors, 3 – Chap. 8. Receptor Kinases, and 5 – Chap. 10. Vasculature Growth). In humans, the VEGF family includes 5 gene products (VEGFa–VEGFd and placental growth factor [PlGF]).[73]

It signals via VEGFR2, calcium influx through Ca^{2+} store-independent transient receptor potential (TRP) channels TRPC3 and TRPC6 as well as Ca^{2+} store-dependent TRPC1, phospholipase-Cγ, Src kinase, protein kinase-Cα, the cRaf–MAP2K1/2–ERK1/2 module, nitric oxide synthase, guanylate cyclase, the Rho–Rac axis, cadherin, zona occludens, and occludin [157].

It can act as a stimulator or inhibitor of endothelial cell proliferation. It promotes transfer by vesiculovacuolar organelles and reduces the cell adhesion molecule content at endothelial junctions, which causes a loss of tight and adherens junctions, thereby enhancing mass transfer by both trans- and paracellular routes. In addition, it supports the formation of fenestrations [157].

Subtype VEGFa binds to VEGFR1 and VEGFR2. It favors albumin extravasation. Proangiogenic splice variants of VEGFa (e.g., $VEGFa_{165A}$) increase also the hydraulic conductivity [157]. Antiangiogenic VEGFa isoforms such as $VEGFa_{165B}$ can also raise the hydraulic conductivity via VEGFR1 receptor.

Subtype VEGFc binds to VEGFR2 and VEGFR3. It augments the hydraulic conductivity acutely via VEGFR2 receptor.

The other VEGF family member, placental growth factor (PlGF), binds only to VEGFR1 receptor. It does not to increase the hydraulic conductivity acutely [157].

3.3.3.3 Mass Transfer

Materials that cross the vessel wall include respiratory gases, water, ions, amino acids, carbohydrates, proteins, lipid particles, and cells. The capillary permeability is high for water and moderate for ions, lipids, and proteins. Fluid can move from the intravascular compartment to the extravascular space composed of cellular, interstitial, and lymphatic subcompartments.

Transcapillary fluid exchange through fenestrated capillaries and across the glycocalyx and within the intercellular cleft in continuous capillaries can rapidly adjust.

As the tissular fluid volume is 3- to 4-times larger than the plasma volume, the tissular fluid serves as a reservoir that can supply additional fluid to the circulatory apparatus or draw off excess.

[73] Isoform VEGFa, or VEGF, is synthesized in almost all cells subjected to hypoxia or other stress types. These proteins signals upon binding to their cognate receptors, in particular VEGFR1 to VEGFR3, once they homo- and heterodimerize. They operate as a potent (but not very powerful) endothelial growth factor, a powerful vascular permeabilizing agent, and a potent vasodilator.

3.3.3.4 Inflammation and Edema

Inflammatory and allergic reactions are characterized by an altered endothelial barrier. During inflammation, microvascular permeability is selectively increased in small postcapillary venules.

During inflammation, the glycocalyx and intercellular junctions are acutely modified. Inflammatory mediators increase vascular permeability primarily by the formation of intercellular gaps in venules, once adherens and tight junctions open. Inflammatory gaps (size $<1\ \mu m$) form at intercellular sites in venules [150].

Alteration of Glycocalyx

The near-wall flow region in the microvasculature can be decomposed into: (1) the glycocalyx, a carbohydrate-rich, blood-excluding intraluminal layer that contributes to a low capillary hematocrit; and (2) a cell-free plasma layer. The core region carries blood cells.

Exclusion by the endothelial glycocalyx of flowing cells and molecules is reduced by adenosine and other vasoactive substances [154].

Vesicular Transfer

Increased endothelial permeability caused by platelet-activating factor (PAF) is counteracted by NOS inhibition and reduced NO production. This mechanism explains inhibition of caveolin-1 that directly interacts with NOS3 enzyme. Caveolin-1 action on microvascular permeability is antagonized by an elevated endothelial cytosolic Ca^{2+} concentration [158]. Caveolin-1 and endothelial Ca^{2+} level antagonistically regulate Ca^{2+}–calmodulin-dependent NOS3 activity.

Degradation of Adhesion Sites

Microvascular permeability is modified by altering adhesion sites between apposed endotheliocytes as well as between them and the extracellular matrix.

Histamine, thrombin, and VEGF disturb the endothelial barrier [159]. Thrombin disrupts the Cdh5–Ctnn complex in adherens junctions. Cadherin-5 and occludin undergo marked widespread rearrangement with a local loss of adhesion proteins at intercellular junctions.

The bacterial endotoxin lipopolysaccharide disrupts endothelial tight junctions. Cytokine TNFSF1 breaks the endothelial barrier at cell adhesion supported by cadherin-5 [155].

Intercellular Gap Formation

Whereas RoCK kinase, myosin light chain kinase (MLCK), and myosin ATPase are implicated in cell contraction during formation of endothelial gaps in cell cultures, occurrence of gaps in vivo in mesenteric venules in response to bradykinin or platelet-activating factor does not depend on this mechanism [150].[74]

Transmembrane glycoproteic adamlysin ADAM15 is involved in vascular inflammation. Its production is upregulated upon exposeure to proinflammatory cytokines. This inflammatory mediator increases endothelial permeability in response to thrombin via the Src–ERK1/2 pathway that supports endothelial cytoskeleton contraction and, hence, intercellular gap formation [160]. This effect is independent of FAK activation as well as disruption of adherens junctions and focal adhesions (i.e., connections between integrins and matrix contituents) and shedding of ectodomain of cell adhesion molecules (e.g., cadherin-5 and integrin). Enzyme ADAM15 promotes also neutrophil transendothelial migration [160]. Neutrophils migrate across the endothelium via the paracellular route maintained by junctional molecules (e.g., cadherin-5, PECAM1, and JAMa).

Reduction of cAMP Concentration

Inflammatory mediators raises endothelial permeability. They disrupt the endothelial barrier, at least partly, as they reduce near-membrane cAMP concentration. Messenger cAMP actually enhances the barrier function, because it activates Rac and Rap1 GTPases to stabilize the barrier, independently of PKA kinase [153]. On the other hand, RhoA GTPase promotes cytoskeleton contractility, thereby augmenting endothelial permeability.

The cAMP concentration is regulated by phosphodiesterases, especially PDE2a and PDE3a. The other messenger cGMP inhibits PDE3a, but activates PDE2a, hence its bimodal regulation [150]. At low concentrations (<100 nmol), cGMP inhibits PDE3a without activating PDE2a, thereby raising cAMP level and lowering endothelial permeability. At higher concentrations, cGMP activates PDE2a, overcoming effects of PDE3a inhibition, thereby diminishing local cAMP level and increasing endothelial permeability.

---

[74]In cultured endothelial monolayers most often stimulated by thrombin as an acute inflammatory stimulus, paracellular gap formation is blocked upon inhibition of RoCK or MLCK, but not in situ. Thrombin launches an active cell RhoA-dependent contraction that creates large gap formation and, hence, an acute increase in endothelial barrier permeability. Some of the effects of thrombin on vascular permeability result from the release of other inflammatory mediators from mastocytes or neurons. Thrombin primes a contraction-dependent increase in permeability [150]. On the other hand, platelet-activating factor rises endothelial permeability in a cell contraction-independent manner in inflamed rat mesentery venules. In addition, thrombin activates platelets that contribute to elevated endothelial permeability.

Similarly, at a low concentration, atrial natriuretic peptide triggers synthesis of cGMP at a low intracellular level and attenuates endothelial permeability, whereas a higher ANP concentration increases endothelial permeability [150].

Activation of Phospholipases

Upon receptor binding, proinflammatory agents activates phospholipases PLA2, PLC, and PLD, as well as Ca^{2+} channels [156].

Activated PLA2 via the lipoxygenase axis leads to the formation of leukotrienes, particularly LTc_4, that activate NOS enzyme. Via the cyclooxygenase axis, PLA2 activates thromboxane synthase (hence vasoconstriction caused by platelet-activating factor).

Phospholipase-C catalyzes the synthesis of inositol trisphosphate and diacylglycerol, hence releasing Ca^{2+} from intracellular stores and activating protein kinase-C. The latter phosphorylates (inactivates) NOS3 (Thr495). In addition, PKC may phosphorylate (activate) cytoskeletal proteins, such as caldesmon and vimentin, as well as myosin light chain kinase. Moreover, an imbalance in activity of PKCβ2 and PKCδ contributes to the endothelial hyperpermeability [156].

Reactive Oxygen Species

In addition, products of activated leukocytes such as reactive oxygen species increase the permeability. Platelet adhesion to sites where the endothelial barrier is opened initiates the inflammatory cascade, the strong interaction of platelets with leukocytes enhancing activation of leukocytes. In fact, endotheliocytes, leukocytes, platelets, macrophages, and mastocytes interfere via their products (cytokines, chemokines, and ROS) [153].

Hyperpermeability and Edema

Once the control of the microvascular permeability is lost, the endothelium becomes hyperpermeable. During inflammation, capillaries become even leaky; water volume within interstitium thus increases and leads to tissue swelling, or edema.

In addition, the interstitial fluid pressure lowers down to -1.33 kPa (-10 mmHg; the normal capillary filtration pressure ranging from 65 to 133 Pa [0.5 to 1 mmHg]), thereby increasing the capillary filtration [169].

Inflammatory agents regulate the microvascular permeability using endothelial-dependent and -independent mechanisms, as they act also on fibroblasts and lymphatics of the microvascular unit [150].

Excessive accumulation of water in interstitium is due to a lack of reabsorption into capillaries and/or in filtration into lymphatics. In fact, the edematous interstitium can result from: (1) an increased capillary and venous hydrostatic pressure associated with heart failure or venous obstruction; (2) decreased plasma osmotic pressure; (3) increased capillary permeability caused by proinflammatory mediators and damaged leaky capillaries; and (4) lymphatic obstruction.

Microvascular Permeability and Leukocyte Adhesion

Protein leakage from plasma to tissue is temporally and spatially dissociated from leukocyte adhesion and endothelial transmigration during inflammation [161]. Leukocyte adhesion and migration are thus uncoupled from microvascular permeability changes. After exposure to inflammatory stimuli, most of the adherent or migrating leukocytes that take the paracellular route (i.e., not through activated endothelial cells) are observed at intercellular junctions of postcapillary venules distinct from endothelial gaps. Both solute leakage and leukocyte adhesion occur in the early phase, but leakage lasts longer than the transient leukocyte adhesion. Although both platelet-activating factor and leukotriene-B_4 cause leukocyte adhesion, only the former augments the fluid filtration, as LTb_4 exposure does not alter endothelial barrier [161]. In addition, leukocyte transmigration involves only a transient opening of adherens junctions. Altered cadherin-5 during cell migration quickly diffuse back to reseal the gap.

On the other hand, production primed by cytokines and release of reactive oxygen species from stimulated neutrophils increases the microvascular permeability.

In addition, during inflammation, platelets support leukocyte recruitment and elevated microvascular permeability. Platelet interaction with venular endothelial cells is initiated by endothelial gap formation [161]. In inflamed venules, platelet–endothelium interaction mediated by P-selectin recruits additional leukocytes. Consequently, ROS production is amplified and enables a sustained increase in microvascular permeability.

3.3.3.5 Microvascular Permeability

Exchange vessels aim at supplying dissolved gases, ions, and solutes to cells, after transferring these low-molecular-weight molecules across the vessel wall.

Except liver, adrenal, and bone marrow sinusoids, in which the endothelium has large pores, the vascular endothelium constitutes a selective barrier between blood and tissue. A filter at the luminal entrance to endothelial clefts provides low permeability to macromolecules. In most microvessels, the macromolecular transport is done by transcytosis (vesicular mass transfer) and not through porous clefts. Microvascular exchange is mainly passive.

Endothelial cells have tight junctions that limit the vascular permeability. The strength of these intercellular connections is controlled by various messengers to modulate the vascular permeability, in addition to vascular growth and repair. The stability of the tight junction depends in particular on RhoA guanosine triphosphatase, itself activated by Rho guanine nucleotide-exchange factors.

Angiopoietin-1 and vascular endothelial growth factor VEGFa strengthen and disrupt tight junctions, respectively. They determine the localization of the RhoA-specific guanine nucleotide-exchange factor PlekHg5[75] to control RhoA activity at tight junctions, hence endothelial permeability [162]. Angiopoietin-1 supports PlekHg5 recruitment to tight junctions, thereby stabilizing cell adhesion. Protein PlekHg5 is recruited to cell junctions by adaptor and scaffold members of the Crumbs polarity complex[76] at sites of intercellular contacts (adherens and tight junctions as well as synapses) and promotes junction integrity by activating Diaphanous. Diaphanous inhibits cadherin-5 endocytosis and Src activity.

On the other hand, VEGFa activates protein kinase-D1 that phosphorylates (Ser806; inactivates) PlekHg5, thereby reducing PlekHg5 tethering to its junctional anchors and removing it from cell junctions [162].

3.3.3.6 Exchanges of Water and Hydrophilic Solutes

The *hydraulic conductivity* measures the porosity of the capillary wall for the water flux. The *filtration coefficient* is the hydraulic conductance per unit exchange surface area.

Nutrients and metabolic wastes are transported between blood and cells by convection associated with fluid motions and diffusion according to the pressure and concentration gradients, respectively. Molecular transfer is facilitated by the very slow blood flow in the capillary (i.e., by a high residence time).

Diffusion from a compartment of higher concentration to a region of lower concentration across a membrane of infinitesimal thickness is governed by the Fick law. Brownian motion states that the substance flux is proportional to the solute permeability coefficient, which depends on the diffusivity and solubility of the investigated solute, transport surface area, and concentration gradient.

The selectivity coefficient, or reflection coefficient, measures the probability of solute penetration across the vessel wall. A selectivity coefficient equal to 1 means that the vessel wall is impermeable to the molecule. Most plasma proteins does not penetrate (reflected off) the barrier; they are responsible for the osmotic pressure. A selectivity coefficient equal to 0 means that transport is not restricted.

---

[75] A.k.a. synectin-binding RhoA exchange factor (Syx).

[76] In Drosophila melanogaster, the Crumbs polarity complex includes: (1) type-1 transmembrane Crumbs (Crb) and (2) Stardust (Sdt), a scaffold protein of the MAGUK (membrane-associated guanylate kinase) family. In humans, it comprises: (1) Crumbs homologs (Crb1–Crb3) encoded by 3 genes and (2) members of the P55-like (P55 Stardust) MAGUK subfamily, or membrane protein, palmitoylated (MMP1–MMP7), among which MPP5, or protein associated with Lin7 PALS1, is most similar to Stardust; (3) protein (PALS1) associated with tight junctions (PATJ) and its related molecule multiPDZ domain-containing protein MuPP1; and (4) Lin7 homolog-C (Lin7a–Lin7c; Abnormal cell lineage Lin7 in Caenorhabditis elegans) encoded by 3 genes.

Small solutes, such as salts and glucose (selectivity coefficient 0.01), are easily transferred. In the range bounded by these 2 values, the wall layer is semipermeable to solute with some amount that hardly penetrate the barrier (strong reflection).

In microvessels with a continuous endothelium, the main route for water and solutes is the endothelium cleft, except when tight junctions exist. Transcapillary water flows and microvasculature transfer of solutes, from electrolytes to proteins, in both continuous and fenestrated endothelium, can be described in terms of 3 porous in-parallel routes: (1) a water pathway across endothelial cells; (2) a set of small pores (caliber 4–5 nm); and (3) a pool of larger pores (bore 20–30 nm) [163].

The estimated exchange area between apposed cell is on the order of 0.4% of the total capillary surface area. The array of junctional strands between endothelial cells is interrupted at intervals, allowing water and solute fluxes. Loss of the glycocalyx in regions of discontinuous tight junctions between adjacent cells can form localized nonselective regions (large pores) without widening of intercellular junctions.

The *pore theory* simulates the cleft passages between adjacent endothelial cells in continuous endothelium. The pore density is defined by the effective fraction of the cleft width used for molecular exchange. The pore size is determined by the interfiber spacing in a fiber matrix at the cleft entrance.

The associated fiber matrix model of capillary permeability associated with the endothelium glycocalyx (thickness $\sim$100 nm; fiber spacing 7 nm) provides a basis for molecular size selectivity, especially in a fenestrated endothelium. A thin matrix at the cleft entrance can indeed be a major determinant of the permeability properties of the capillary wall. Such a fiber matrix model can also be applied to the cleft itself, because large portions of the cleft contain matrix components. Models have predicted that the fiber layer (typical thickness 100 nm), which extends from the endothelium surface into the cleft entrance region, sieves solutes [164].

The entire membrane permeability coefficient is the sum of the different route coefficients. The membrane selectivity coefficient is the sum of the individual coefficients of the in-parallel paths weighted by the fractional contribution of each path to membrane hydraulic conductivity. Despite its limitations,[77] the pore theory is a useful pedagogical tool. Moreover, any fiber matrix-based modeling must not only take into account fiber size and volume, but also matrix organization [163].

The *effective pore radius*, a selectivity measure, is determined by interstices in the fiber matrix, which depend on matrix composition and arrangement. The *effective pore number* is determined by the size and frequency of endothelium passages in the intercellular spaces as well as through cells.

Water and small solutes can indeed cross endotheliocytes, using specific channels. Aquaporins (Aqp) are membrane water-transport proteins [165]. Subtype Aqp1 is found in endothelia. However, the contribution of the transcellular transfer to the net flux is supposed to be small. In most microvessels, aquaporins contribute to

[77]Macromolecular transport is not always coupled with water flows. Furthermore, capillaries without large pores exist.

less than 10% of the hydraulic conductance, except in the blood–brain barrier and possibly skeletal muscle capillaries [163]. Aquaporins intervene when interstitial osmolarity increases such as in renal descending vasa recta.

An elevated transmural pressure on endothelial cell culture on a porous rigid matrix increases the endothelial hydraulic conductivity [166]. Augmented shear stress in endothelial cleft resulting from heightened transmural flow raises the hydraulic conductivity via a NO–cAMP-dependent mechanism.

3.3.3.7 Macromolecule Permeability

Macromolecules can cross the endothelium between cells (paracellular transport) or preferentially through endothelial cells (transcytosis), using proper receptors, specific or not, and vesicles (Vol. 1 – Chap. 9. Intracellular Transport).

Vacuole-like structures can be observed, isolated inside the endotheliocyte, as luminal membrane invaginations, or connected to the abluminal compartment [167]. Both luminal and abluminal surfaces of the capillary endothelium are dynamical with invaginations and protrusions associated with environmental stimuli. Moreover, vesicle translocation between luminal and abluminal membranes is accelerated when the transendothelial pressure rises. The endothelial permeability can increase by vesiculovacuolar organelles without formation of trans- or intercellular gaps.

The macromolecule flux from blood to interstitium can mainly occur in postcapillary venules because their endothelial cells have simple intercellular junctions.

Microvascular wall models comprise pores for small and intermediate-sized molecules and transendothelial channels for macromolecules. The macromolecular transport, which is independent of convection, needs endothelial vesicles, a mass transfer mechanism that use cytoskeletal tracks and their coupled nanomotors, a more efficient process than diffusion.

3.3.3.8 Transport Laws

The fluid volume flux per unit endothelial area J (or J_w; subscript w: water), also called volumetric flow rate per unit area and volume filtration rate per unit area, across the endothelial barrier that is modeled by a membrane is given by the *Starling relation*:

$$J(\kappa_o = 0) = \mathcal{G}_h(\Delta p - \Delta \Pi), \tag{3.30}$$

where $\mathcal{G}_h$ (dimension: $M^{-1}.L^2.T$) is the hydraulic conductivity, or capillary filtration coefficient, $\Delta p = p_\ell - p_{\text{int}}$ the difference in hydrostatic pressures between capillary lumen (p_ℓ [3.3–3.7 kPa (25–28 mmHg)]) and interstitial fluid (p_{int} [2.0–2.2 kPa]), $\Delta \Pi = \Pi_\ell - \Pi_{\text{int}}$ the difference in osmotic (oncotic) pressures between capillary lumen (Π_ℓ) and interstitial fluid (Π_{int}), and κ_o osmotic selectivity (reflection) coefficient of the vessel wall for the chemical species of interest ($\kappa_o = 0$ if the endothelial membrane is fully permeable to transported molecular species and $\kappa_o = 1$ if the membrane is impermeable).

Therefore, volumetric filtration rate per unit endothelial area increases when: (1) the capillary hydrostatic pressure rises; (2) the interstitial hydrostatic pressure lowers; (3) the capillary colloid osmotic pressure decays; and/or (4) the interstitial colloid osmotic pressure augments. The transcapillary water flow contributes to transendothelial solute flux.

Because both the plasma and interstitial fluid contain many solutes (molecular species $_i$), the water movement across the capillary wall can be written as [170]:

$$J_w = \mathcal{G}_h(\Delta p - \sum \kappa_{o_i} \Delta \Pi_i). \tag{3.31}$$

In most microvascular beds, only concentrations of macromolecular solutes differ between the plasma and interstitial fluid. Hence, J_w can be described approximately by the difference between the effective osmotic pressure exerted by macromolecules in plasma and interstitial fluid.

The Kedem–Katchalsky equations incorporate convection and diffusion:

$$J_w = \mathcal{G}_h(\Delta p - \kappa_o \Delta \Pi), \tag{3.32}$$

$$J_s = \mathcal{P}\Delta c + J_w(1 - \kappa_d)c_*, \tag{3.33}$$

where subscripts w and s stand for water and solute, respectively, $\mathcal{P}$ (dimension: L.T^{-1}) is the membrane diffusional permeability for the solute, Δc the difference in solute concentration across the endothelium (between the plasma and interstitial fluid just beneath the endothelium), c_* the averaged solute concentration within the pore that represents the endothelial cleft, and κ_d the dimensionless solute drag reflection coefficient that depends on membrane permeability for both water and solute.

The quantities Δp and $\Delta \Pi$ are driving and hindering forces for water flow (i.e., volume transport that convects a given solute), respectively. The quantity Δc is the driving force for diffusive solute transport. The quantities $\mathcal{G}_h$ (order 10^{-7} cm/s/cm H_2O both in vivo and in vitro conditions), $\mathcal{P}$ (values of which varies between in vivo [order 10^{-8}–10^{-7} cm/s for albumin] and in vitro [order 10^{-6} cm/s for albumin; order 10^{-7} cm/s for 70-kDa dextran] microvascular endothelium [168]), κ_o, and κ_d are transport properties. The endothelial transport properties are sensitive to both the chemical and mechanical environment.

The relative contribution of convection and diffusion is given by the dimensionless transport parameter ($\mathcal{T}$) [168],

$$\mathcal{T} = \frac{\mathrm{Pe}}{\exp \mathrm{Pe} - 1}, \tag{3.34}$$

where Pe is the Péclet number that is defined by:

$$\mathrm{Pe} = (1 - \kappa_{\mathrm{d}})\frac{J_w}{\mathcal{P}}. \tag{3.35}$$

The Péclet number is the ratio of the velocity at which the macromolecular species is washed by convection (solvent drag) to its diffusive velocity in the endothelial barrier.

Therefore, the solute transport equation can be written as:

$$J_s = \Big(\mathcal{P}\mathcal{T} + (1 - \kappa_{\mathrm{d}})J_{\mathrm{w}}\Big)\Delta c. \tag{3.36}$$

When the solute transport is dominated by convection, $\mathrm{Pe} \gg 1$, $\mathcal{T} \rightarrow 0$, and $\mathrm{Pe} \rightarrow (1 - \kappa_{\mathrm{d}})J_{\mathrm{w}}$. When the solute transport is dominated by diffusion, $\mathrm{Pe} \ll 1$, $\mathcal{T} \rightarrow 1$, and $\mathrm{Pe} \rightarrow \mathcal{P}$.

The fluid volume in the interstitial space is assumed to be regulated within a narrow range by readjustment of the interstitial hydrostatic and colloid osmotic pressures according to capillary and lymphatic conditions. However, the interstitial fluid pressure (p_{int}) is an active force that create a fluid flux across the capillary wall particularly in inflammation.

3.3.3.9 Limitations of the Starling Law

The traditional form of the Starling principle governing microvascular fluid exchange between plasma and the interstitial fluid across the endothelial barrier related to the balance between hydrostatic and oncotic pressures must be modified due to variations of interstitial pressures and the semipermeable glycocalyx [169, 170].

The classical filtration–reabsorption model that relies on a spatial fluid balance does not describe fluid exchange in the real microcirculation, because: (1) both p_{int} and Π_{int} change when J_w varies; and (2) the composition of the subglycocalyx fluid diverges from that of the bulk interstitial fluid [170].

The tissue fluid balance in the lung, muscle, skin, among others, is not achieved by an equilibrium between upstream filtration and downstream reabsorption in the steady state. The drainage of the capillary filtrate by lymphatic vessels is thus the dominant factor of the interstitial volume homeostasis. However, the globally averaged net filtration constraint is much smaller than the hydrostatic–osmotic pressure balance.

The arteriolar vasomotion cycles may support the lymphatic drainage. Arteriolar contraction lowers the capillary pressure and promotes transient fluid absorption, whereas the arteriolar relaxation favors filtration, thereby creating a temporal fluid balance [170]. In addition, the net filtration force across the capillary endothelium depends more on the colloid osmotic pressure below the endothelial glycocalyx than that in the bulk intertitium.

Filtration Rate–Osmotic Pressure Relation

Microvascular absorption is only transient in capillaries with a low blood pressure (i.e., in most tissues); a slight filtration can prevail even in venules [170]. Starling's relation states that the filtration rate drops when the capillary concentration of plasma proteins rises or when the interstitial concentration of plasma proteins diminishes, for a given difference in transendothelial fluid pressure.

In some specialized organs (e.g., kidney and intestine), fluid absorption is associated with local epithelial egress of plasma proteins from the interstitium to the lymphatic system (Sect. 3.3.3.10).

Absorption in capillaries with a low blood pressure is observed only transiently. The concentration of plasma proteins in the interstitial fluid (plasma protein filtrate concentration; c_{int}) is inversely related to the capillary filtration rate, because the interstitial fluid is continually renewed and the interstitial concentration of plasma proteins is determined by the rate of solute influx (J_s) relatively to the rate of water flow influx (J_w; extravascular dilution curve) [169, 170]:

$$c_{\text{int}} = J_s / J_w. \tag{3.37}$$

When the capillary filtration rate rises, interstitial and lymph macromolecules are diluted, hence their concentration falls. The pericapillary colloid osmotic pressure is thus inversely related to the filtration rate.

The low rate of filtration and lymph formation in most tissues can be explained by the plasma protein concentration gradient within the intercellular cleft of continuous capillaries and around fenestrations. the subglycocalyx space.

Subglycocalyx Space

The interstitial fluid at the glycocalyx abluminal side is separated from the pericapillary interstitium by tortuous intercellular clefts with gaps through tight junctions. A gradient of plasma protein concentration can arise along the cleft. Therefore, the difference of colloid osmotic pressure across the capillary barrier in Starling's law is defined by that between the capillary lumen and the subglycocalyx space rather than the difference of colloid osmotic pressure between the the capillary lumen and the bulk interstitial fluid:

$$J_w = \mathcal{G}_h\big[(p_\ell - p_{\text{int}}) - \kappa_o(\Pi_\ell - \Pi_{\text{sg}})\big], \tag{3.38}$$

where Π_{sg} is the colloid osmotic pressure in the subglycocalyx space.

The colloid osmotic pressure in the subglycocalyx space of the ultrafiltrate sieved by the glycocalyx can be small because κ_o is high and the outward ultrafiltrate flow through the narrow breaks in the intercellular junctional strands prevents protein diffusion equilibrium between the subglycocalyx and pericapillary fluid [170]. Differences between plasma protein concentrations in the subglycocalyx space and pericapillary medium depend on the magnitude of the capillary filtration pressure (p_ℓ). The effective osmotic pressure difference opposing fluid filtration may equal Π_ℓ rather than $\Pi_\ell - \Pi_{int}$ term of Starling's law, when Π_{sg} is very low. Therefore, the basal volumetric filtration rate per unit endothelial area is balanced by the lymph drainage rate [170].

Varying Interstitial Pressure

In addition to adjustability of the gradients of plasma protein levels across the glycocalyx and within the intercellular cleft in continuous capillaries as well as around fenestrations, the extracellular matrix and interstitial fluid pressure influence the transcapillary exchange.

Cells exert a tension on the extracellular matrix constituents. The interstitial fluid pressure (p_{int}) depends on the interstitial volume and bulk modulus,

$$B = \frac{\Delta p_{int}}{\Delta V_{int}/V_{int}}, \tag{3.39}$$

where V_{int} is the interstitial volume. A decreased interstitial fluid pressure increases the transcapillary fluid flow that, in turn, augments the solute convection across the capillary wall.

In turn, an elevated capillary filtration heigthens the interstitial fluid pressure and reduces the interstitial colloid osmotic pressure, thereby limiting further filtration augmentation. This automatic readjustment is called *capillary filtration autoregulation* by the interstitial pressure [169].

The interstitial fluid pressure acts as the filling pressure that increases lymph flow in initial lymphatics. In addition, the *interstitial exclusion*, that is, the fact that plasma proteins do not have access to the full interstitial volume due to steric exclusion by structural matrix molecules, influences the time at which a new steady state is reached [169]. A higher exclusion lowers the interstitial content of plasma proteins, thereby reducing the time at which a new steady state occurs. Moreover, lymph flow increases until a steady state is achieved. The enhanced removal of protein associated with the lymph clearance reduces the protein amount.

Cellular tension on the extracellular matrix in cooperation with matrix fiber networks restrains expansion of the glycosaminoglycan gel, thereby keeping the interstitial fluid pressure at a relatively high value.

Upon removal of the cellular tension exerted on the collagen and microfibrillar networks in the connective tissue via the collagen-binding β_1-integrins, glycosaminoglycans of the ground substance that are normally underhydrated can expand and take up water [169].

On the other hand, several growth factors and cytokines such as PDGFbb (but neither PDGFaa, nor FGF) can cause the contraction of fibroblasts and restore the matrix compaction and a high interstitial fluid pressure via $\alpha_V\beta_3$-integrin as well as phosphoinositide 3-kinase and protein kinase-C [169].

Insulin and prostaglandin-$F_{2\alpha}$ can also renormalize the interstitial fluid pressure. On the other hand, proinflammatory substances, such as IL1 and TNFSF1, as well as prostaglandins PGe1, PGe2, and PGi2 (prostacyclin), attenuate the matrix cell contraction and collagen gel compaction and decrease the interstitial fluid pressure.

In the skin, macrophages of the connective tissue exerts a control of the interstitial fluid volume using tonicity-responsive enhancer-binding protein (TonEBP) [169]. The latter binds to the promoter of the VEGFC gene, thereby leading to VEGFc secretion by macrophages and avoiding interstitial hypertonic volume retention during high-salt diet. Factor VEGFc may act as an osmosensitive regulator that contributes to the handling of sodium balance.

3.3.3.10 Epithelial Mass Transfer in the Kidney, Lymph Node, and Gut

In peritubular capillaries of the renal cortex and ascending vasa recta of the renal medulla, in intestinal mucosal capillaries, and in lymph nodes, fluid is absorbed continuously [170]. In the kidney and intestine, the interstitial fluid is renewed by a release of protein-free fluid by the epithelium.

In the rat ileum, the pressure in fenestrated capillaries of the mucosa is much lower than that in continuous capillaries of circular and longitudinal smooth muscles. In cats, the intestinal lymph flow increases and its protein concentration decreases as fluid is absorbed from the small intestine lumen. During the absorption of digested food, the epithelial absorption of glucose and amino acids is confined to the outer third of the villus, fluid absorbed through epithelium may be taken up into the extensive capillary bed of the inner two-thirds of the villus and lacteals.

In the renal cortex, the low interstitial plasma protein concentration and, hence, interstitial colloid osmotic pressure is maintained by the low macromolecular permeability of the fenestrated peritubular capillaries and the fast production of a protein-free interstitial fluid by the tubular epithelium, which flushes the relatively small amount of plasma protein entering the cortical interstitial fluid into lymphatics (efficient lymphatic drainage) [170].

In the renal medulla, fluid absorbed by the collecting duct from the nascent urine is cleared continuously from the medullary interstitial fluid to ascending vasa recta. Fenestrated ascending vasa recta, unlike fenestrated cortical peritubular capillaries, have a relatively high permeability to plasma proteins. Albumin is cleared from the medullary interstitial fluid and directly taken up into medullary capillary blood (countercurrent exchange) by protein convection [170].

In lymph nodes, the interstitial fluid is continuously replaced by the inflowing prenodal lymph of low protein concentration [170]. The rapid renewal of the interstitial fluid counteracts the inverse relation between the capillary filtration rate and colloid osmotic pressure.

3.4 Rhythmicity

The internal clock observed in flowers by the Swedish botanist K. von Linné is estimated by the opening and closing time of a certain species of flowers. Similarly, animals possess an internal circadian clock described by endogenous self-sustained oscillations of physiological and metabolic processes (Vol. 2 – Chap. 5. Circadian Clock).

The cardiovascular system is subject to the circadian rhythm imposed by molecular clock gene activities (Clock, Bmal1, PER1–PER2, CRY1–CRY2, FBXL3, NR1D1, NR1F1, CSNK1D, and CSNK1E),[78] as well as to shorter rhythms imposed by the autonomous nervous system and adrenal glands. The circadian rhythm is characterized by temporal variations in the release of hormones, such as catecholamines and corticosteroids.

Vasopressin is synthesized and released with an approximate 5-fold increase during daylight hours. In addition, exposure to environmental stress leads to adrenal responses that vary during the day.

In humans, concentrations of several peptide hormones exhibit daily variations: the leptin level rises during the evening and that of growth hormone and prolactin during the night. Plasma concentrations of glutathione and amino acids also bear daily changes, some with a peak at around midnight (glutamine, isoleucine, leucine, methionine, phenylalanine, proline, threonine, tryptophan, and valine), others in the evening (glycine) [171]. Blood levels of ornithine, citrulline, and arginine reach a peak after midnight. Plasma concentrations of creatine, dimethylglycine, sarcosine, and creatinine (creatine pathway) successively peak before and slightly after midnight.

The molecular clock relies on positive and negative feedback loops with a period of about 24 h that govern circadian gene expression [172] (Vol. 2 – Chap. 5. Circadian Clock). In particular, genes of the molecular clock regulate enzymes of the synthesis of catecholamines.

The core of the circadian clock is composed of 2 major transcription factors CLOCK and BMAL1. Upon heterodimerization, the CLOCK–BMAL1 complex primes positive (BMAL1) and negative (Per1–Per3, Cry1–Cry2, nuclear receptor NR1d1, and basic helix–loop–helix proteins bHLHb2 [or bHLHe40], and bHLHb3 [or bHLHe41]) loops. The former stimulates clock output genes, among which many encode transcription factors, whereas the latter represses transcription by the CLOCK–BMAL1 heterodimer. The circadian clock yields anticipation, as it enables cells to perceive day instant and respond to periodic environmental stimuli (e.g., feeding–fasting cycles, awakening habit [physical activity or feeding], etc.).

This autonomous molecular mechanism of the circadian clock exists in almost every cell type. Molecular clocks actually lodge not only in the suprachiasmatic and dorsomedial nuclei of the hypothalamus (central pacemaker) but also in

[78] The CSNK1D and CSNK1E genes encode casein kinases CK1δ and CK1ϵ, respectively.

most tissues, especially aorta, liver, kidney, and heart for cardiovascular system functioning.[79] The clock gene PER1 is active in the heart, veins, and arteries. Peak PER1 activity occurs during the late night in cultured heart tissues and all arterial samples, whereas the phases of the rhythms in veins vary according to the anatomical location [174].

Hormones control the expression of vascular clock genes, such as CLOCK and neuronal PAS domain protein NPAS2 (or bHLHe9) via nuclear receptors RARα (NR1b1) and RXRα (NR2b1) [175].

The suprachiasmatic nucleus clock entrains the phase of peripheral clocks via chemical cues, such as rhythmically secreted hormones. However, circadian gene expression in peripheral cells can be uncoupled from cyclic gene expression in the suprachiasmatic nucleus [176]. For example, the glucocorticoid hormone analog dexamethasone transiently changes the phase of circadian gene expression in liver, kidney, and heart, but does not affect cyclic gene expression in neurons of the suprachiasmatic nucleus [177].

The circadian pacemaker synchronized with the sleep–wake cycle influences the body's functioning (e.g., body temperature and heart rate). Time variations in cardiac frequency and arterial blood pressure are the most well-known cardiovascular rhythms. Nocturnal heart rate and blood pressure can decay down to 0.5 Hz and 4 to 7 kPa. Stroke volume and cardiac output undergo circadian rhythms even at rest, but they can be less easily monitored. Blood flow peaks are observed in early afternoon [178]. Vascular reactivity to adrenaline is greater in the early morning (3–4 h). Sympathetic tone and cathecolamine concentration have higher values during early day time hours than during the night. Blood volume increases during the evening and decreases at midnight, in association with circadian renal activity. Plasma levels of hormones (renin, angiotensin, aldosterone, and atrial natriuretic peptide), as well as of plasma proteins (clotting and fibrinolytic factors)[80] and hemoglobin content bear daily rhythms. Consequently, the circadian clock can explain fluctuations in hemodynamic values.

In addition to modulation of myocardial metabolic activity in response to numerous environmental stimuli, heart functioning exhibits diurnal variation including responsiveness to external fluctuating stimuli that influence myocardial contractility. Some environmental factors indeed fluctuate in a predictable manner according to the moment during the day and year (circadian and seasonal oscillations).

The circadian clock of cardiomyocytes regulates myocardial metabolism both directly by controlling triglyceride and glycogen turnover and indirectly by modulating response to workload, hormones, and nutrients [179]. The cardiomyocyte circadian clock thus allows the anticipation of some environmental stimuli, as it modulates myocardial metabolism. Heart working periods must be matched

---

[79]The circadian clock in the suprachiasmatic nucleus (Bmal1-based oscillator) is entrained by light. Another circadian clock in the dorsomedial nucleus of the hypothalamus (also Bmal1-based oscillator) is primed by food [173].

[80]Fibrinolytic activity falls during early morning hours.

by elevated rates of oxidative and nonoxidative metabolism. Generation of ATP in cardiomyocytes relies on both exogenous (glucose, lactate, and fatty acids) and endogenous (glycogen and triglyceride) substrates. Many genes of atrial and ventricular myocytes that target carbohydrate and lipid metabolism show circadian oscillation in activity in wild-type mice compared to cardiomyocyte clock mutant mice [179].[81]

Regulation of endogenous fuel turnover by the circadian clock allows cardiomyocytes to anticipate diurnal variations in recurrent environmental stimuli, but uptake and metabolism of exogenous substrates are influenced by sudden unpredictable factors. Nonetheless, the circadian clock of the cardiomyocyte mediates diurnal variations in the responsiveness of the heart to increased workloads, thereby also anticipating diurnal variations in certain neurohumoral factors such as insulin [179].

Seasonal variations in lipid concentrations have been observed. Total cholesterol level, as well as LDL and HDL cholesterol levels, are higher in the winter in most studies (spring in some investigations) and lower in the summer (autumn in certain works), with substantial monthly fluctuations. The change magnitude and month for high and low levels can depend on the subject, especially on the subject's age, gender, and ethnic group. Other biomarkers of cardiovascular risk have been less explored.

Sleep deprivation, work shift, and jet lag disrupt biological rhythms. Circadian clock disorganization alters cardiovascular and renal functioning [180]. Loss of synchronization between the heart and its environment predisposes the heart to metabolic maladaptation and subsequent myocardial dysfunction.

The circadian clock also influences cardiovascular diseases. The clinical onset of both myocardial infarction and stroke occurs more frequently in the early morning than at other times of day. Endogenous rhythms of numerous physiological parameters (cardiovascular, pulmonary, hematologic, and endocrine) determine timing of medications. β-Blockers and calcium channel blockers lower blood pressure and cardiac frequency more during the day than during the night owing to the time variation in sympathetic tone, metabolism, and pharmocokinetics [178]. Other cardiovascular drugs also exhibit daily variations in activity. Treatment timing coordinated with body clock hence increases its efficiency.

Cardiovascular vulnerability has a daily pattern with a morning peak that has been observed later, that is, myocardial infarction, stroke, and sudden cardiac death have a peak occurrence at a time window of 9 to 11 h [181]. Static and dynamic measures of locomotor activity and cardiac frequency fluctuations exhibit circadian rhythms. The suprachiasmatic nucleus of the hypothalamus influences the physical activity regulation that in turn adjusts cardiac dynamics.

---

[81] Several proteins encoded by these genes tune lipogenesis and lipolysis, as well as lipid droplet stability (e.g., adiponutrin, 1-acylglycerol-3-phosphate O-acyltransferase, and diacylglycerol O-acyltransferase-2 that are involved in lipogenesis) on the one hand, and glycogenolysis and glycolysis (e.g., protein kinase-A, protein phosphatase-1, and phosphofructokinase) [179]. Moreover, the transcriptional activity of peroxisome proliferator-activated receptors PPARα (NR1c1) that activates genes for fatty acid oxidation is controlled by the circadian clock of the cardiomyocyte.

Chronotherapy exploits individualized internal body time to maximize potency and minimize toxicity during drug administration. Internal body time can be assessed by timetables of plasma concentrations of the body's clock-controlled metabolites [171]. Circadian blood metabolomics[82] measures personalized body time that accounts for genetic background, gender, age, and living habits, as well as jet-lag-induced circadian rhythm disturbances.

3.5 Convective Heat Transfer

Convective heat transfer by the blood acts in homeothermy of the human body. Circulatory heat exchangers and conservers allow the heat or cooling of the body, depending on environment and body conditions, heat being convected by blood circulation. During exercise, the body temperature rises.

The main exchange surfaces (with a given thermal conductivity) are the lungs and skin. Blood heat is transferred to alveolar air to be exhaled to environmental air across the exchange barrier, the thin alveolocapillary membrane, or the skin with its numerous cell layers.

Heat exchangers can be bypassed, especially in the limbs with 2 venous networks, the superficial under the skin and the deep accompanying arteries, with anastomoses between them.

Cancer treatments can use thermal ablation,[83] which is less invasive than surgery. Therapy efficiency depends on the local blood flow.

Heat transfer models use heat source(s) and sink(s) and effective conductivity to describe the thermal influence of blood flow. A continuum model cannot account for the local thermal impact of the vasculature, with countercurrent vessel segment pairs. The vasculature must also be modeled, down to a certain caliber, and be combined with a continuum model for heat transfer in the irrigated tissues.

The bioheat transfer equation proposed by Pennes in 1948 is used in physiological heat transfer modeling dealing with the microcirculation [182]. The Pennes model provides the rate of heat change in a given body tissue from the sum of

---

[82]Metabolomics technology relies on method coupling (mass spectrometry, gas and liquid chromatography, and capillary electrophoresis. Hepatic lipase mRNA in the mouse liver has a peak expression slightly before that of lysophosphatidylcoline [171].

[83]Image-guided radiofrequency ablation treats cancers particularly localized to the liver, kidney, and adrenal glands by heating. One or more radiofrequency needles are inserted into the tumor. Cryotherapy uses gas-refrigerated cryoprobes, which are inserted inside the tumor, initiating the formation of ice balls to destroy cancerous cells by freezing and thawing processes. One of the main difficulties is the determination of the optimal position of the probes and treatment duration for complete destruction of cancerous cells without damaging too many surrounding normal cells. Another kind of tumor therapy consists of thermal and mechanical exposure to high-frequency focused ultrasound (HIFU). Tumor antigens and other compounds released from destroyed cells can stimulate antitumoral immunity. The optimal exposure time is an important parameter to avoid damage of normal cells, especially walls of neighboring blood vessels.

Table 3.42 Mean and standard deviation (SD) of thermal conductivity (mW/cm/C) and diffusivity ($\times 10^3$ cm2/s) of aortas and atherosclerotic plaques at 35 °C (Source: [183])

	Thermal conductivity		Thermal diffusivity	
Tissue	Mean	SD	Mean	SD
Normal aorta	4.76	0.41	1.27	0.07
Fatty plaque	4.84	0.44	1.28	0.05
Fibrous plaque	4.85	0.22	1.29	0.03
Calcified plaque	5.02	0.59	1.32	0.07

the net heat conduction into the tissue, metabolic heat generation, and heating (or cooling) effects of the arterial supply.[84] The convective heat transfer term associated with blood flow is given by: $q_b c_b (T - T_a)$ (q_b: organ perfusion rate per unit volume of tissue; c_b: specific heat of blood; T_a: arterial temperature; and T: local tissue temperature). The energy field in the perfused tissue domain is given by the bioheat transfer equation:

$$\rho_{tis} c_{tis} \frac{\partial T}{\partial t} = \mathcal{G}_{T_{tis}} \nabla^2 T + q_b c_b (T - T_a) + q_{met} + q_{so}, \tag{3.40}$$

where c_{tis} is the specific heat of the explored biological tissue, ρ_{tis} its density, $\mathcal{G}_{T_{tis}}$ its thermal conductivity, q_{met} the metabolic heat source (rate of energy deposition per unit volume assumed to be homogeneously distributed throughout the tissue of interest, but usually neglected), and q_{so} a possible heat source (e.g., in the case of thermal ablation).

This equation is coupled with the energy field equation in the flowing blood domain, which includes: (1) a directional convective term due to the net flux of equilibrated blood $\rho_b c_b (\mathbf{v} \cdot \nabla T)$; $\mathbf{v}$: blood velocity, which can be composed of 2 terms, a hemodynamic and acoustic streaming component generated by mechanical effects of high-frequency focused ultrasound when this therapeutic procedure is used to destroy the tumor); (2) the contribution of the nearly equilibrated blood in a tissue temperature gradient $\mathcal{G}_{T_b} \nabla^2 T$ ($\mathcal{G}_{T_b}$: blood perfusion thermal conductivity); and (3) heat deposition q_h due to an imposed source:

$$\rho_b c_b \frac{\partial T}{\partial t} = \mathcal{G}_{T_b} \nabla^2 T - \rho_b c_b (\mathbf{v} \cdot \nabla T) + q_h. \tag{3.41}$$

Values of the thermal conductivity and diffusivity of cardiac and arterial walls are given in Tables 3.42 and 3.43.

[84] Pennes underestimated the magnitudes of the conduction and convection terms in the energy balance, using inappropriate values of tissue thermal conductivity and tissue perfusion rate.

Table 3.43 Mean and standard deviation (SD) of thermal conductivity (mW/cm/K) and diffusivity ($\times 10^3$ cm2/s) of myocardia at 37 C (Source: [183])

Thermal conductivity		Thermal diffusivity	
Mean	SD	Mean	SD
5.31	0.37	1.61	0.20

The heat transfer coefficient for blood in a vessel (d: vessel bore; L vessel length) can be evaluated from the *Sieder–Tate equation*, when $q_m \mathrm{c_p}/(\mathcal{G}_\mathrm{T} L) > 6$:

$$h_\mathrm{T} d/\mathcal{G}_\mathrm{T} = 1.75(q_m \mathrm{c_p}/(\mathcal{G}_\mathrm{T} L))^{0.33} \times (\mu_\mathrm{pl}/\mu)^{0.14}, \tag{3.42}$$

where h_T is the heat transfer coefficient, $\mathrm{c_p}$ isobar heat capacity, $\mathcal{G}_\mathrm{T}$ the thermal conductivity, q_m the mass flow rate, and μ_pl the near-wall plasma viscosity at wall temperature.

Spatial variations in temperature distribution have been computed in a 3D muscle vascular model [184]. The tissue domain is composed of a twin artery and vein, which give birth to dichotomic trees of arterioles and venules, with eight generations of paired, closely spaced vessels, assuming property constancy at each generation. The arteriovenous spacing, vessel bore and density, and flow rate depend on tissue depth. An efficiency function, which depends on the volumic tissue blood perfusion rate and radial coordinate, is proposed to improve the bioheat equation. Using an appropriate procedure to analyze Pennes data, the data support the theory [185]; but the Pennes model, as in new bioheat transfer models, lacks experimental validation and reliable evaluation of tissue properties [186].

3.6 Regulation of the Circulation

The heart adjusts the body requirements of nutrient supply by increasing the ejection volume and its beating frequency. Stroke volume can rise from about 70 ml to 100 ml (i.e., an increase of almost 50%) and cardiac frequency from about 1 to 3 Hz or more (i.e., a 3-fold augmentation). The blood flow rate, or cardiac output, is adjusted using mainly cardiac frequency rather than stroke volume. Stroke volume depends on heart inotropy and on pre- and postload.

The postload, or afterload, is determined by the arterial resistance, which is mainly controlled by the sympathetic innervation. The higher the resistance and arterial pressure, the smaller the ejected volume. The preload influences diastolic filling, and consequently, end-diastolic values of the ventricular volume and pressure.

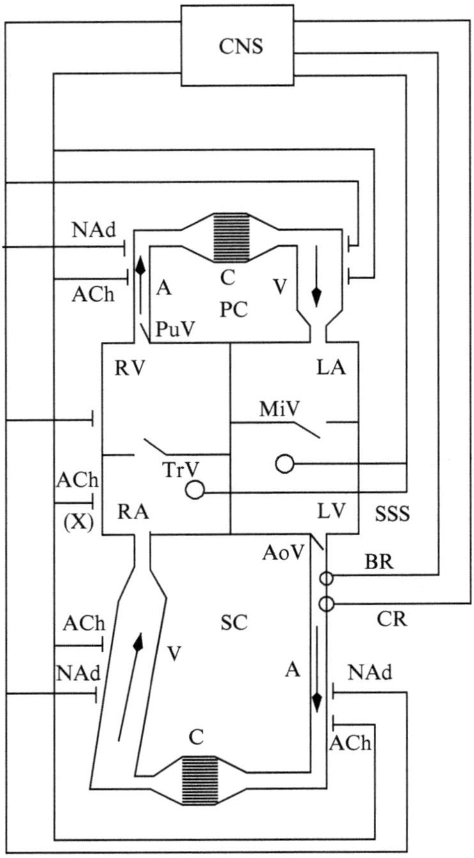

Fig. 3.9 Neural control of the blood circulation (SSS: stretch-sensitive sensor; BR: baroreceptor, which responds to p_a; CR: chemoreceptor, which reacts to pH, p_{O_2}, and p_{CO_2})

Blood circulation is controlled by a set of regulatory mechanisms: (1) remote control that involves the central nervous system via action potentials (Fig. 3.9) and endocrine organs via hormone release that act like very-fast-wired and slower wireless communications, respectively (Table 3.44); and (2) local phenomena (mechanotransduction; Vol. 5 – Chaps. 8. Smooth Muscle Cells and 9. Vascular Endothelium). Neural and endocrine signals allow the cardiovascular system to adapt to environmental changes, regulate, and synchronize the functions of more or less autonomous cell sets.

When hypotension occurs, the baroreceptor reflex induces a vasoconstriction in the systemic vasculature. Blood flow to the brain and myocardium does not significantly decay because of autoregulation, at least if the arterial pressure does not decline below the autoregulatory range. A partial occlusion of large distribution arteries elevates the local flow resistance, but can reduce pressure in downstream irrigating arteries. The downstream resistance arteries dilate in response to attenuated pressure and blood flow to avoid local hypoxemia.

Lymphatic vessels also adapt to pressure variations. Acute and prolonged increases in transmural pressure increase lymphatic vessel contractility [188].

Table 3.44 Additional vasoactive substances from remote, regional, or local origin (VIP: vasoactive intestinal peptide; CGRP: calcitonin gene-related peptide; NMJ: neuromuscular junction; VD: vasodilation; VC: vasoconstriction; Source: [187])

Molecule Vasodilation	Effect
Acetylcholine	NO release
Adrenaline	α (VC) and β (VD) effects
Bradykinin	NO release
CGRP	
Dopamine	Inhibition of NAd release
Enkephalin	
Histamine	Inhibition of NAd release from arteriole NMJ
	NO release
Noradrenaline	Major α1 and moderate β1 effects
	NO release
Serotonin	NO release
	Arteriole vasodilation
	Enhancement of action of NAd and ATn2
	Artery vasoconstriction
VIP	Enhancement of ACh release and effects
Vasoconstriction	
Angiotensin-2	PLC activation
Vasopressin	PLC activation

3.6.1 Autoregulation in the Arterial Bed

The primary function of local blood flow control is the provision of an adequate supply of nutrients to match tissue activity, especially to maintain equilibrium between oxygen delivery and consumption. Blood flow distribution can be controlled by the central nervous system as well as locally, particularly in arterioles, independently of remote signaling (neural impulses and hormonal cues).

In 1902, W. Bayliss discovered that an elevated transmural pressure causes a paradoxical vasoconstriction and arterial resistance augmentation. This pressure-induced myogenic response is also called the *Bayliss effect* [189]. The myogenic response is the mechanism of blood flow autoregulation that enables matching of blood supply to tissue nutrient and energy demand over a given pressure range (8–20 kPa [60–150 mmHg]).

The *myogenic response*, or myogenic autoregulation (Vol. 5 – Chaps. 8. Smooth Muscle Cells and 9. Endothelium), is related to the intrinsic ability of resistance arteries to respond to a transmural pressure change, hence a type of *mechanotransduction*. Endoluminal pressure imposes a circumferential and axial stretch in the vessel wall and a shear stress at the wetted surface of the endothelium that are sensed by parietal cells. The myogenic response can change quantitatively (i.e., change in magnitude) and qualitatively (i.e., change in mediating mechanism) in response to environmental conditions.

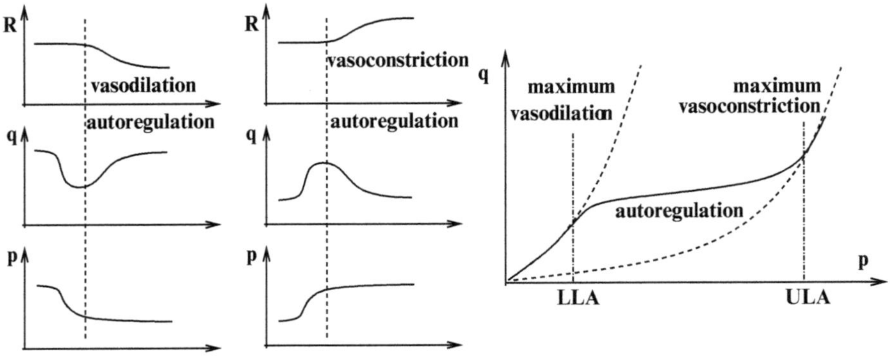

Fig. 3.10 Autoregulation of local blood flow. (**Left**) Perfusion of certain organs is autoregulated when the blood driving pressure either decays or rises, the local blood flow being maintained almost constant by either vasodilation or vasoconstriction of the organ vasculature. (**Right**) Autoregulation occurs for a pressure range, between the lower (LLA) and upper limit of autoregulation (ULA)

Table 3.45 The vascular myogenic response refers to the intrinsic ability of a blood vessel to constrict in response to an increase in intraluminal pressure or dilate with a decrease in intraluminal pressure. Autoregulation aims at maintaining a constant blood flow rate (q) when the local pressure (p) changes . When the pressure decreases, a paradoxical vasodilation (VD) widens the local vessel caliber, hence reducing the local resistance ($\mathtt{R} = \Delta p/q$), but potentiating the pressure drop. Conversely, when the pressure decreases, a paradoxical vasoconstriction (VC) narrows the vessel lumen, thereby elevating the local resistance, but strengthening the pressure rise

$\downarrow$p $\rightarrow$ VD $\rightarrow$ $\downarrow$R $\rightarrow q$ = cst.
$\uparrow$p $\rightarrow$ VC $\rightarrow$ $\uparrow$R $\rightarrow q$ = cst.

Organ perfusion is autoregulated by blood-driving pressure variations. A nearly constant flow is maintained although the blood pressure varies in a given range. The pressure–flow relation exhibits a very small slope (a quasi-plateau) in the autoregulatory range, and steep slopes outside the autoregulatory range (Fig. 3.10).

An artery is maximally vasodilated below the lower limit and maximally vasoconstricted above the upper limit of autoregulation. Below the autoregulatory range, the pressure–flow relation of a given vessel matches the nonlinear pressure–flow curve of a maximally dilated vessel after vasodilator infusion into the explored organ, which disables autoregulation. The artery dilates with increasing pressure, hence reducing flow resistance and augmenting flow rate (Table 3.45). When the pressure rises beyond the autoregulatory range, the pressure–flow relation of the vessel follows the pressure–flow curve of a maximally constricted vessel.

Autoregulation thus fails whether the perfusion pressure is too low, dropping below the lower limit of autoregulation, local blood flow falling with decaying blood pressure, or too high, rising above the upper limit, the local blood flow heightening with elevating blood pressure.

Autoregulation is aimed at ensuring an adequate blood flow and oxygen delivery to essential organs. The cerebral, cardiac, and renal circulation exhibit a strong autoregulation. The skeletal muscular and visceral circulation show a moderate autoregulation. The cutaneous circulation displays a slight autoregulation.

3.6.1.1 Autoregulation and Vascular Resistance

Autoregulation is based on modified vascular resistance, the ratio between perfusion pressure and blood flow:

$$\mathrm{R} = \Delta p/q. \tag{3.43}$$

When blood pressure increases, stretched arteriole walls contract, their lumens narrow, and subsequently the local blood pressure further rises to maintain the blood flow rate at a nearly constant value. Usually, elevation in intracellular pressure in an elastic pipe leads to its expansion. Resistance blood vessels, small arteries and arterioles, thus behave oppositely. Vasoconstriction in autoregulatory segments of the arterial bed is aimed at keeping blood flow rate invariant. Conversely, when blood pressure decays, resulting vasodilation decreases vascular resistance; subsequently the local blood pressure parodoxically declines, and the local blood flow remains almost constant.

The vascular myogenic response, an inherent property of smooth myocytes in walls of small arteries and arterioles, not only stabilizes the local blood flow rate, but also the capillary filtration pressure when blood pressure changes.

Smooth muscle tone is controlled not only by hemodynamic stress applied by flowing blood to the vessel wall, but also by metabolic and neural mechanisms. These control mechanisms are locally integrated [190]. The resulting vascular tone can thus be elevated or attenuated by nerve transmitters, autacoids, and hormones.

3.6.1.2 Myogenic Signaling

The myogenic response relies on regulatory effectors of several signaling cascades that operate either in series or in parallel and cooperate or exert a mutual exclusion (Table 3.46) [191]: (1) sensors, such as stretch-sensitive G-protein-coupled receptors, stretch-activated cation channels, membrane-bound enzymes (e.g., MMP2 and MMP9), cell adhesion molecules such as integrins, and cytoskeleton; as well as (2) cytosolic messengers that are either calcium-dependent (e.g., phospholipase-C and diacylglycerol, phospholipase-A2 and arachidonic acid metabolites, ryanodine- and inositol trisphosphate-stimulated intracellular calcium release channels) or calcium-independent (e.g., protein kinase-C, RhoA GTPase, and RoCK kinase).

All myogenic mechanisms are potential contributors to the overall response with variable impact depending on the circumstances. Two organizational, Ca^{2+}-dependent and -independent mechanisms are integrated and modulate Ca^{2+} concentration levels and sensitivity. Both the sensitivity to transmural pressure

Table 3.46 Sensors and mechanotransduction mediators in the myogenic response (Source: [191]; ASIC: acid-sensing ion channel; PKC: protein kinase-C; PLA2: phospholipase A2; PLC: phospholipase C; S1P: sphingosine 1-phosphate; TREK: TWIK-related potassium channel; TRP, transient receptor potential)

Function	Molecule types
Sensors, transducers	GPCRs
	TRPs, TREK1, ASIC2, Piezo-1/2
	Integrins
Messengers, intracellular mediators	S1P
	PKC, Src, PLA2, PLC
	Vinculin, actin

(i.e., the magnitude of the myogenic response) and the selection and weighting of contributors (i.e., the orchestration of calcium-dependent and -independent signaling cascades that mediate the myogenic response) are under a temporal control.

The participation of various mechanisms can be modified, hence changing [191]: (1) the magnitude of the myogenic response with the recruitment of proximal vascular segments, dormant processes being activated; and/or (2) the pattern of the operating mode without necessary variation of the magnitude of the response.

Among mechanically activated proteins that initiate a signaling cascade, some sense directly the membrane stretch, others are stimulated by a mediator downstream from force sensors. Nevertheless, both types are indiscriminately termed mechanosensitive. In any case, a mechanosensitive protein must react according to the application speed and amplitude of the mechanical stimulus. The activation time of mechanically stimulated proteins is very short: lower than 50 μs for membrane-tethered enzymes, than 5 ms for ion channels, and than 500 ms for G-protein activation [192]. In addition, a given protein may not be mechanosensitive, but may react to mechanical cues in conjunction with other proteins within a macromolecular complex. In other words, mechanical signals can be sensed and transduced by mechanosensory proteic complexes rather than a single molecule.

Mechanosensitive G-Protein-Coupled Receptors

G-protein-coupled receptors not only sense and transmit physical (photons, temperature, and olfactory and gustatory stimuli) and chemical cues (hormones and neurotransmitters), but also convert mechanical strains and stresses into chemical and electrical signals (mechanosensation, or mechanoreception, and mechanotransduction). Vascular endotheliocytes, smooth myocytes, and cardiomyocytes, in addition to nephron epitheliocytes and neurons, are equipped with mechanosensitive GPCRs. For example, mechanically activated angiotensin-2 type-1 receptor (AT_1), a component of the renin–angiotensin regulator, operates in myogenic vasoconstriction as well as cardiac hypertrophy.

Table 3.47 Mechanosensitive G-protein-coupled receptors (Source: [192]; CMC: cardiomyocyte; EC: vascular endotheliocyte; HEK: human embryonic kidney cell vSMC: vascular smooth myocyte). Activated subunits of the $G\alpha_s$ subclass stimulates adenylate cyclases that forms cAMP, whereas subunits of the $G\alpha_{i/o}$ inhibit these enzymes. Activated subunits of the $G\alpha_{q/11}$ subclass stimulates phospholipase-Cγ, hence producing diacylglycerol (DAG) and inositol trisphosphate (IP_3)

Receptor	Coupling	Expression
AT_1	Gq/11, Gi/o	vSMC
		CMC
B_2	Gq/11	vEC
D_5	Gs	vEC
ET_A	Gs, Gq/11	HEK, neutrophils
H_1	Gq/11	HEK
M_5	Gq/11	HEK
PTH_1	Gs, Gq/11	HEK
V_{1A}	Gq/11	HEK

Several GPCRs, such as Gq-coupled acetylcholine M_5, Gq/11- and Gi/o-coupled angiotensin AT_1, Gq/11-coupled bradykinin B_2, Gs-coupled dopamine D_5, Gs- and Gq/11-coupled endothelin ET_A, Gq-coupled histamine H_1, Gs- and Gq/11-coupled parathyroid hormone PTH_1, Gq-coupled vasopressin V_{1A}, and Gi/o-coupled formylpeptide FPR_1 receptors are activated by stretch (Table 3.47) [191, 192]. However, they must be associated to Ca^{2+} mobilization and/or sensitization. In addition, GPCRs such as Gs-coupled β2-adrenergic (noradrenaline) receptor, which is not mechanosensitive, mediates the control of the nervous system.

Mechanical forces stretch cell membranes, thereby modifying the configuration of integral membrane proteins. Plasmalemmal mechanosensitive proteins respond to mechanical stresses and strains, as the configuration change opens an effector site of the protein, rendering it active. An increase in intraluminal pressure heightens axial and circumferential wall tension. Augmented wall tension is associated with the maintenance of myogenic tone in a given blood pressure range. Myogenic vasoconstriction lowers the intraparietal azimuthal tension. Reduced wall tension below a threshold also maintains the myogenic tone. Myogenic vasodilation increases the intramural circumferential tension.

G-protein-coupled receptor and chemosensor OlfR78 (human ortholog OR51e2)[85] is expressed not only in the olfactory epithelium, where it detects odorants, but also in vascular smooth myocytes, especially those of the renal afferent arteriole of the juxtaglomerular apparatus. It participates in the regulation of the

[85] Receptor OR51e2 may be activated by several androgens.

glomerular filtration rate and launches renin secretion. It responds to vasodilatory short-chain fatty acids (SCFA),[86] such as acetate and propionate.[87]

In addition, other SCFA receptors, such as Gi/o-coupled receptors GPR41 and GPR43 expressed in adipocytes and neutrophils as well as the sympathetic ganglia and major arteries, in addition to the kidney intervene in adiposity and inflammation. Both Gs-coupled OlfR78 and Gi/o-coupled GPR41 are produced in smooth myocytes of small resistance vessels, where they mutually antagonize each other. Propionate modulates blood pressure via OlfR78 and GPR41 receptors [193].

Mechanically Activated Ion Channels

Mechanically activated ion channels open directly by membrane stretch.[88] Their opening is triggered directly by mechanical stimuli rather than via a second messenger stimulated by another mechanically sensitive molecule. The stretch is transmitted by constituents of the extracellular matrix, integral membrane proteins, and the cytoskeleton. It can be sensed by a proteic mechanotransducer associated with the ion channel that can be considered as a subunit or directly transmitted by plasmalemmal lipid tension. However, cell membrane stretch activates or inactivates mechanosensitive ion channels.

In fact, ion channels can respond not only to stretch, but also to strain such as membrane deformation (curvature). When a volume change (associated with a dilution or concentration of cytosolic ions and macromolecules) is the stimulus rather than a mechanical force, the term volume-sensitive channel is employed. In addition, activity of voltage- and ligand-gated channels can be modulated by mechanical stimuli.

The *stretch activation model* states that tension in the lipid bilayer triggers conformational changes which open the channel. The *spring-like tether model* assumes that tethers are attached to the channel and are connected directly or not to the cytoskeleton and the extracellular matrix. Whereas the cytoskeleton can act as a tether connecting mechanosensitive proteins, the extracellular matrix can serve as an anchor relaying external relative movements to integral membrane proteins. Tethers act like springs of a shutter.

---

[86] Short-chain fatty acids are terminal products of fermentation by the gut microbiota that enter the blood circulation.

[87] Receptor OR51e2 is unresponsive to other SCFAs. Renin release by juxtaglomerular apparatus cells depends on calcium-inhibitable adenylate cyclase AC5 and/or AC6 isoforms. On the other hand, AC3 produced in macula densa cells participates in renin secretion by juxtaglomerular apparatus cells as a paracrine factor.

[88] Stretch-activated currents have been described in various mammalian cell types, such as vascular smooth myocytes, renal epithelicytes, somatosensory dorsal root ganglion neurons, and inner ear hair cells.

Once they are gated, mechanosensitive ion channels open in a pore-like manner. They carry a mechanically gated transduction current rather than creating a favorable environment for transduction. These mechanotransducers generate ion fluxes and associated electrical signals such as pressure-induced depolarization.

Stretch-activated channels respond typically to a mechanical stress in the temporal magnitude order of millisecond ($\mathcal{O}$[1 ms]; voltage-gated ion channel gating $\mathcal{O}$[1 μs]).

Mechanosensitive channels vary in selectivity for the permeating ions from mild cation selectivity for Ca^{2+}, K^+, and Na^+ to highly selective K^+ channels in eukaryotes.

Mechanogated ion channels play a role in cardiac and arterial myocytes. Opening of the nonselective stretch-activated cation channels in smooth myocytes is involved in the pressure-dependent myogenic constriction of resistance arteries (Sect. 1.8.3.6).

Transient Receptor Potential Superfamily

Channels of the transient receptor potential superfamily are categorized into 7 families: TRPA (ankyrin-like), TRPC (canonical), TRPM (melastatin-related), TRPML (mucolipin-related), TRPN (no mechanoreceptor), TRPP (polycystin-related), and TRPV (vanilloid; Vol. 3 – Chap. 2. Membrane Ion Carriers). They tetramerize and generally form nonselective cation channels. They are involved in vision, hearing, taste, olfaction, thermosensation, and mechanosensation. Some are activated directly by sensory stimuli, others by second messengers.

Certain cation channel of the transient receptor potential superfamily are activated by mechanical stimuli such as osmotic stress, membrane stretch, and shear. Multiple TRP channels are mechanosensitive (TRPA1; TRPC1, TRPC3, TRPC5–TRPC6; TRPM3–TRPM4, TRPM7; TRPN1; TRPP1–TRPP2; TRPV1–TRPV2, TRPV4; and TRPY1) [194, 195]. Several mechanosensitive TRP channels are synthesized in cardiomyocytes, vascular smooth myocytes, endotheliocytes, and vasosensory neurons.

Both TRPC6 and TRPM4 are depolarizing cation channels involved in the constriction of resistance arteries [192]. Activity of TRPM4 channel, which is activated to membrane stretch, is regulated by free intracellular Ca^{2+} ion and protein kinase-C.

Thermosensitive TRPV2 channels may also serve as mechanosensors in vascular smooth myocytes. The TRPV4 channel serves as a mediator of a signaling pathway that encompasses PLA2 and cytochrome-P450 epoxygenase rather than being directly gated by a mechanical stress.

Whereas TRPP1 homomer encoded by the PKD2[89] gene is not directly mechanosensitive, TRPP1 complexed with TRPV4 or PKD1[90] may be a constituent of a mechanosensitive complex in the vasculature [192].

TREK1

In the heart, several types of stretch-activated channels that reside in atrial and ventricular myocytes include mechanogated ATP-sensitive K_{ATP}, acetylcholine-sensitive K_{ACh}, and Ca^{2+}-activated K_{Ca} channels, as well as TREK1, TREK2, and TRAAK channels, which are members of the tandem 2-pore domain (2P) K^+ channel category (encoded by the KCNK2, KCNK10, and KCNK4 gene, respectively) [196].

The TREK1 channel ($K_{2P}2.1$) causes outwardly rectifying K^+ current. Two pore-domain background potassium channels are regulated by mechanical stretch, in addition to oxygen, hydrogen ion, and G proteins (Vol. 3 – Chap. 3. Main Classes of Ion Channels and Pumps).[91] Membrane stretch activates TREK1 channel with half-maximal activation at about 5.3 kPa (40 mmHg) [192]. It is involved in pressure-induced vasodilatation in the skin.

Piezo

In addition to 2-pore potassium channels and some members of the epithelial sodium channel (ENaC) family, and the large hydrophobic ion channels Piezo-1 and -2 are also mechanosensitive [197].

Both Piezo-1 and -2 are pore-forming subunits of mechanosensitive ion channels or indispensable accessory subunits of a larger mechanosensitive proteic complex [192].

ASIC2

Mechanosensitive ASIC2 heterotrimeric channel of the ASIC (acid-sensing ion channel) family of the ENAC (epithelial sodium channel)–DEG (degenerin) superfamily lodges in aortic baroreceptor neurons in the nodose ganglia and their terminals, where it participates in the arterial mechano- and chemosensory reflex (Vol. 3 – Chap. 3. Main Sets of Ion Channels and Pumps).

---

[89] Polycystic kidney disease-2.

[90] Previously called polycystin-1.

[91] Carbon monoxide is a messenger and regulator of TREK1 as well as epithelial Na^+ (ENaC) channels, in addition to calcium-activated K^+ (BK), $K_V2.1$, Ca_V1, and ligand-gated ionotropic P2X (e.g., $P2X_2$ and $P2X_4$).

Integrins and Plasma Membrane-Associated Proteic Complexes

Several types of mechanosensitive proteins synergistically operate with GPCRs in the initiation, maintenance, and termination of the myogenic tone.

In arterioles, plasma membrane-associated proteic complexes (*dense bands*) that consist of α-actinin and vinculin connected to integrins and Src kinase are enriched in the abluminal region. However, these complexes may serve as stabilizers of vascular smooth myocytes rather than mechanosensors.

Nonetheless, the actin cytoskeleton play a role in mechanotransduction, as it links mechanosensitive proteins between them and adapts the cell configuration to the local stress field.

Integrins are plasmalemmal heterodimeric glycoproteins that link components of the extracellular matrix (e.g., fibronectin, laminin, and collagen) with intracellular structural proteins (e.g., dense band constituents). Integrins can act either as direct mechanosensors or as mechanotransduction modulators. In addition, integrins influence activity of ion channels, hence playing an additional role in mechanosensation.

Sphingosine 1-Phosphate

Sphingosine 1-phosphate (S1P) is an important determinant of the myogenic response that serves as both an intracellular messenger and extracellular ligand of cognate receptors ($S1P_1$–$S1P_3$ on vascular cells). In mesenteric arteries in evolving heart failure, the relative S1P contribution can increase, thereby enhancing the myogenic reactivity.

It is synthesized by 2 rate-limiting kinases, sphingosine kinases SK1 and SK2, that phosphorylate the highly abundant cell membrane sphingosine (Vol. 4 – Chap. 2. Signaling Lipids). Localization of SK1 to the plasma membrane, especially to membrane rafts, enables S1P signaling initiation and, hence, calcium mobilization and sensitization. Myristoylation and palmitylation of SK1 create a distinct S1P pool acting on calcium mobilization, but not calcium sensitivity.

Sphingosine 1-phosphate targets multiple signaling mediators (e.g., RhoA and Rac GTPases, phospholipase-C, and adenylate cyclase). It is involved in both Ca^{2+} mobilization and sensitization, hence vascular smooth myocyte contractility. Translocation of S1P is launched by hemodynamic stress and ensured upon ERK-mediated phosphorylation [191].

Different S1P pools exist whether S1P resides in the intra- or extracellular medium and whether it triggers a Ca^{2+}-dependent and -independent vasoconstriction. Conversely, Ca^{2+} that do not intervene in SK1 translocation and activation can operate via calmodulin and calcium- and integrin-binding protein CIB1 (CIBP or calmyrin) as a SK1 myristoylation switch.

Protein Kinase-C

The 12 isozymes of the PKC family may represent a second myogenic regulatory node, as they modulate both intracellular Ca^{2+} concentration via several ion channels and calcium sensitivity [191]. Signaling mediated by PKC is controlled in space and time.

Distinct PKC isoforms may be used as integrators in the myogenic response. Subtype PKCα and PKCδ serve in Ca^{2+} sensitization and influx, respectively [191]. Moreover, a developmental shift in swine mesenteric arteries has been observed. Activity of PKC is required for the myogenic response in arteries isolated from 1-day postnatal animals, but is dispensable in arteries isolated from 10-day postnatal animals characterized by a reduced production of PKCα and PKCϵ [191].

Other Mediators of Mechanosensation and Mechanotransduction

Protein Tyr kinases of the SRC family participate in the mechanotransduction as they influence the actin cytoskeleton function via integrins. However, their action in mechanosensation remains to be demonstrated.

Membrane-associated phospholipases PLA2 and PLC are implicated in mechanosensation in different cell types (e.g., skeletal myocyte and mesangial cell, respectively) [192]. In the blood vessels, PLC isoforms are nodes of mechanotransduction signaling pathways in the myogenic response. The stretch-induced receptor activation launches a Gq/11-mediated signaling that targets PLC; subsequently, diacylglycerol stimulates TRPC channel and, hence, sodium and calcium influx. On the other hand, PLA2 may potentiate TRPC6 activity via ω-hydroxylase and 20-hydroxyeicosatetraenoic acid [192].

3.6.1.3 Arteriolar Network

Elements of an arteriolar tree may react sequentially, but independently, to changes in blood pressure to regulate capillary pressure [190]. Small reductions in perfusion pressure may provoke a dilation of the generation-1 arteriole, thereby preventing or minimizing downstream myogenic changes in the microcirculation. A sufficient perfusion pressure drop causes a maximal myogenic dilation of the parent arteriole. Any further decrease may cause myogenic dilation of generation-2 arterioles. Once these segments are maximally dilated, any further pressure decrease may be experienced by generation-3 arterioles, and so on. Therefore, a progressive reduction in perfusion pressure may lead to sequential dilation of successive arteriolar generations. Axial progression of the myogenic response initiated upstream support spreading of myogenic dilation or constriction from one generation to the next.

However, this process works only for certain vascular beds. The behavior of other microvascular circuits does not support the concept of a *series-coupled myogenic effector* with a vasodilation that progresses gently downstream, from parent to child arterioles. All arteriolar generations can dilate simulaneously. The smaller the arteriolar caliber, the higher the myogenic sensitivity, except in small arterioles with a discontinuous layer of smooth myocytes [190]. This feature hence explains the homogeneous pattern of dilation in response to progressive blood pressure reduction. Moreover, a third type of arteriolar network does not exhibit autoregulation.

3.6.1.4 High Intravascular Pressure

The mechanical stretch of the vascular wall acts as a stimulus on medial smooth myocytes to elicit myogenic contraction. Autoregulatory contraction of arterioles controls local blood flow in perfusion compartments.

In vivo, arteriolar blood pressure at rest, at least in rat cremaster muscle, is assessed to range from 9.3 to 10.6 kPa [198]. In fact, an acute increase in intraluminal pressure from 4.0 to 5.3 kPa can prime vasoconstriction [190]. This type of myogenic response may be antagonized by mechanical stress-dependent, endothelium-derived, NO-induced vasodilation.

The myogenic behavior (myogenic vasoconstriction, maintenance of basal myogenic tone, and myogenic vasodilation) relies on mechanosensors in vascular smooth myocytes and subsequent transduction to adapt states of actin polymerization and contractility. Different mechanisms occurs during initial and sustained phases of the myogenic response of rat small mesenteric arteries. Both phases depend on calcium influx through voltage-dependent calcium channels [199]. However, only the sustained phase is mediated by a cytochrome-P450 ω-hydroxylase metabolite such as 20-hydroxyeicosatetraenoic acid and calcium-activated K^+ channel.

In fact, numerous signals participate in myogenic behavior, in addition to ion channels and cytosolic calcium [200], such as the RhoA–RoCK axis [201], protein kinase-C [202], and arachidonic acid derivatives [203]. Epoxyeicosatrienoic acids are released from the endothelium in response to acetylcholine. They hyperpolarize smooth myocytes, open K_{Ca} channels, and relax arteries [204]. Another major metabolite of arachidonic acid, 20-hydroxyeicosatetraenoic acid is a potent vasoconstrictor. In normotensive salt-sensitive rats, myogenic activation of skeletal muscle resistance arteries does not depend on cytochrome-P450 ω-hydroxylase. On the other hand, in hypertensive salt-sensitive rats, myogenic activation of these vessels partly relies on 20HETE production, this substance inhibiting K_{Ca} channels via PKCα, at least in canine basilar artery [204]. Inhibition of the BK channel sensitizes the canine basilar artery to mechanical stretch.

Mechanotransduction in vascular smooth myocytes can result from myogenic, neurogenic, or hemodynamic stress, or metabolic processes. The pathway starts at the plasma membrane and cell cortex to target the contractile cytoskeleton.

The myogenic contraction-triggering cascade relies on a connection between the extracellular matrix, cell adhesion sites, and cytoskeleton. The latter depends

Table 3.48 Integrins of vascular endothelial and smooth muscle cells (Source: [205]). Cells of vascular walls connect their cytoskeleton to the extracellular matrix via integrins. The latter operate as assembly points for numerous cytoskeletal and signaling components. These cell-surface receptors communicate with the extra- and intracellular media. They transmit forces from both sides and inside–out and outside–in signals used to control vascular tone, permeability, and wall remodeling. These mechanosensors decipher the pressure-dependent myogenic response and serve as detectors of certain molecules. Integrins of both cell types are enmeshed in a meshwork of interacting matrix proteins, such as collagen-1 and -3 to -6, elastin, several types of laminin, fibro- and vitronectin, osteopontin, tenascin, and thrombospondin. Integrins on vascular smooth myocytes and ablumenal surface of endothelial cells participate in controlling vascular tone

Endothelial cell	Smooth myocyte
$\alpha_1\beta_1$	$\alpha_1\beta_1$
$\alpha_2\beta_1$	$\alpha_2\beta_1$
$\alpha_3\beta_1$	$\alpha_3\beta_1$
$\alpha_5\beta_1$	$\alpha_4\beta_1$
$\alpha_6\beta_1$	$\alpha_5\beta_1$
$\alpha_6\beta_4$	$\alpha_6\beta_1$
$\alpha_V\beta_3$	$\alpha_7\beta_1$
$\alpha_V\beta_5$	$\alpha_8\beta_1$
	$\alpha_9\beta_1$
	$\alpha_V\beta_1$
	$\alpha_V\beta_3$
	$\alpha_V\beta_4$
	$\alpha_V\beta_5$

on integrins (Table 3.48) for both outside–in and inside–out signaling. Integrin mechanosensors are associated with the regulation of arteriolar caliber and hence vascular myogenic response, in addition to vascular permeability and remodeling [205].

Cell adhesion and corresponding forces result from interactions between fibronectin and $\alpha_5\beta_1$-integrin [206]. Both $\alpha_5\beta_1$- and $\alpha_V\beta_3$-integrins are needed for myogenic constriction [207]. These integrins act as mechanosensors of vascular smooth myocytes that enable these cells to detect and respond to changes in intraluminal pressure.

Vascular smooth myocyte contraction is accompanied by a remodeling of both contractile and noncontractile components of the actin cytoskeleton and adhesion junctions. Integrin molecule reconfiguration and integrin population rearrangement contribute to reset arteriole reference state.

Table 3.49 Integrins on vascular myocytes and vascular tone. Once they are exposed to matricryptic site-containing peptides, for example, when a mechanical stress field is applied or upon matricryptin production, activated $\alpha_V\beta_3$-integrin inhibits the $Ca_V1.2b$ channel, thereby promoting vasodilation. On the other hand, $\alpha_4\beta_1$- and $\alpha_5\beta_1$-integrin support $Ca_V1.2b$ gating, via Src kinase and possibly protein kinase-A, which phosphorylate (activate) $Ca_V1.2b$ channel, thereby causing vasoconstriction in response to matricryptic site-containing peptides (Source: [205])

Type	Effect (activatory matrix ligand)
$\alpha_4\beta_1$	Vasoconstriction
$\alpha_5\beta_1$	(fibronectin)
	(matricryptins)
$\alpha_V\beta_3$	Vasodilation
	(vitronectin, fibronectin)
	(matricryptins)

Numerous vascular integrins recognize the RGD matricryptic,[92] integrin-binding motif of vascular wall matrix proteins such as collagen, fibronectin, and vitronectin.

On endothelial cells, $\alpha_5\beta_1$- and $\alpha_V\beta_3$-integrin respond to RGD matricryptic site-containing peptides and mechanical forces by launching production of vasoactive autacoids.[93] On vascular myocytes, integrin-$\alpha_4\beta_1$ and -$\alpha_5\beta_1$ cause vasoconstriction in response to LDV and RGD matricryptic site-containing peptides, respectively, whereas integrin-$\alpha_V\beta_3$ provokes vasodilation in response to RGD matricryptic site-containing peptides (Table 3.49) [205].

Some synthetic peptides containing the RGD integrin-binding sequence repress spontaneous myogenic tone in isolated arterioles via $\alpha_V\beta_3$-integrin. On the other hand, activated $\alpha_4\beta_1$- and $\alpha_5\beta_1$-integrin support myogenic tone. Certain RGD-containing peptides can attenuate the myogenic response to step increase in intravascular pressure [205].

In addition, shear-mediated endothelial cell response is triggered upon activation of integrins and inwardly rectifying $K_{IR}2.1$ channel within seconds, thereby causing endothelial cell hyperpolarization. The signaling cascade includes the integrin–SHC connection, activation of FAK, JNK, ERK1, and ERK2 kinases, as well as that of SREBP1 and Jun factors, and modulation of Rho activity [205].

---

[92]Matricryptic sites are active sites, such as Arg–Gly–Asp (RGD) and Leu–Asp–Val (LDV), are hidden in the mature secreted form of matrix molecules. They become exposed upon conformational changes caused by oligomerization, adsorption, and mechanical stress. Matricryptic sites contribute in particular to acute changes in vascular permeability and fibrin–fibronectin polymerization in wound repair. Matricryptins are active fragments of matrix molecules with exposed active matricryptic sites. They can bind to specific integrins and regulate arteriolar tone using calcium and potassium channels.

[93]Hormone-like substances that act locally and briefly.

Moreover, in cooperation with connexins, integrins may contribute to signal propagation that enables remote regulation of resistive vessels. A local stimulation of integrins can actually initiate a remote change in vascular caliber. Last, but not least, upon interaction with their ligands, integrins can operate via the calcium channel and cytosolic influx in both endothelial and smooth muscle cells.

Mechanics of the fibronectin–integrin–actomyosin cytoskeleton complex can be modulated by G-protein-coupled receptors. Stimulated G-protein-coupled receptors change the cortical elasticity as well as cell adhesion strength [208]. Vascular smooth myocyte relaxation reduces coupling to extracellular matrix proteins, whereas contraction elicits a tight coupling. Integrin-mediated adhesion of vascular smooth myocytes to the extracellular matrix is reinforced during contraction and attenuated during relaxation. A vasodilator (e.g., adenosine) softens the plasma membrane–cytoskeleton complex and reduces the membrane adhesion strength. On the other hand, a vasoconstrictor (e.g., angiotensin-2) stiffens the complex and heightens the membrane adhesion strength.

Cortical stiffness can result from rearrangement and cross-linking of the actin mesh, generation of actin bundles (actin stress fibers), actin binding to integrins, interaction of actin with cortical myosin, creation of new actin fibers, and stretch of cortical actin by connected actin of the contractile compartment [209].

3.6.1.5 Low Intravascular Pressure

Vasodilation can result from hyperpolarization generated by activity of endothelial Ca^{2+}-activated K^+ channels $K_{Ca}2.3$ (SK or SK_{Ca}) and $K_{Ca}3.1$ (IK or IK_{Ca}), among other mechanisms such as vasodilation dependent on nitric oxide or prostacyclin (Vol. 5 – Chap. 9. Vascular Endothelium).

This hyperpolarization, the predominant endothelium-dependent mechanism in small resistive arteries and arterioles, spreads both radially and longitudinally. It is transmitted to adjacent smooth myocytes via endothelial projections through the basement membrane and internal elastic lamella toward smooth myocytes. Calcium-activated K^+ channels localize to myoendothelial microdomains between endothelial and adjoining smooth muscle cells.

These 2 cell types generate elementary and spatially restricted Ca^{2+} events spontaneously. In particular, spontaneous, rapid, localized (subcellular) Ca^{2+} sparks through activated ryanodine-sensitive calcium channels on the sarcoplasmic reticulum of cerebral artery myocytes attenuate the vascular tone, as they activate BK channels [210]. The latter hyperpolarize and dilate pressurized arteries.

In fact, 3 types of nanodomains on projections can be defined: (1) myoendothelial gap junction that facilitates the intercellular transfer of ions (e.g., Ca^{2+}) and small molecules (e.g., inositol trisphosphate [IP_3]); and (2) cluster of K_{Ca} channels and (3) nonselective cation transient receptor potential (TRP) channels. The 2 latter nanodomains can merge.

An intercellular functional unit is composed of: (1) connexins that form gap junctions; (2) endothelial IP_3 receptor (IP_3R) and $K_{Ca}3.1$ channel; and (3) inward rectifier $K_{IR}2.1$ channel and voltage-dependent $Ca_V1.2$ channel on smooth myocyte.

In particular, Ca^{2+} ion can also be locally released through IP_3 receptors of endothelial endoplasmic reticulum to create spontaneous, localized, and asynchronous calcium pulsars in myoendothelial junctions. Calcium pulsars encode signals between vascular endothelial and smooth muscle cells. They differ from Ca^{2+} sparks generated by ryanodine receptors. One target of calcium pulsars is the $K_{Ca}3.1$ channel in endothelial projections to relax the adjoining smooth myocyte. Exiting K^+ ion targets the myocyte $K_{IR}2.1$ channel to trigger hyperpolarization of the myocyte plasma membrane.

Endothelial, mechanosensitive TRPV4 channels cluster on the myoendothelial junction. They are selectively activated at low intravascular pressure [211]. Intraluminal pressure below 6.7 kPa causes TRPV4 channel opening and 2- to 3-fold increase in the frequency of endothelial Ca^{2+} sparklets. These localized Ca^{2+} sparklets on the endothelial plasma membrane activate endothelial IK, but not SK channels. In addition, Ca^{2+} sparklets may enhance the frequency of Ca^{2+} puff and pulsar, as Ca^{2+} can stimulate clustered IP_3Rs on the myoendothelial junction.

3.6.1.6 Cerebral Arterioles

The pressure-dependent activation of vascular smooth myocytes of cerebral arterioles is the main regulator of blood flow in the brain (Sect. 3.2.5.4). The greater the wall stretch is, the stronger the vasoconstriction.

Increased vessel wall stretch caused by rising perfusion pressure activates phospholipase-C that produces, via cytochrome-P450-4A, 20-hydroxy eicosatetraenoic acid, which in turn, activates protein kinase-C [212]. Protein kinase-C inhibits potassium channels, especially calcium-activated potassium channels, modifying the membrane potential[94] of the smooth myocytes and inducing vasoconstriction.

The metabolic activity of the nerve cells also intervenes. Glutamic acid is released from metabolically active neurones, stimulating astrocytes for arachidonic acid production. Arachidonic acid is converted by cytochrome-P450-2C11 to epoxyeicosatrienoic acid, which stimulates calcium-activated potassium channels, antagonizing 20-hydroxyeicosatetraenoic acid. Cerebral autoregulation thus results from a balance between vasoconstriction generated by 20-hydroxyeicosatetraenoic acid and vasodilation induced by epoxyeicosatrienoic acid. Astrocytes also release other vasoactive molecules, such as thromboxane-A2, prostacyclin, and prostaglandin-E and -F. The calcium-activated potassium channels can be affected by various mediators (nitric oxide, adenosine, and prostacyclin).

---

[94]The membrane is maintained in a relatively depolarized state, partially because of inhibition of K^+ channels.

3.6.1.7 Renal Arterioles

Renal autoregulation is mainly located in glomerular arterioles. Stretch-activated vasoconstriction and tubuloglomerular feedback maintain a nearly constant blood flow. The tubuloglomerular negative feedback deals with increased delivery of water and salts to the distal tubule, which causes vasoconstriction. This feedback can be affected by angiotensin-2, nitric oxide, thromboxane, 20-hydroxyeicosatetraenoic acid, and adenosine triphosphate.

3.6.1.8 Coronary Arterioles

Coronary blood flow is tightly coupled with oxygen demand, as myocardium has a very high basal oxygen consumption. Increased cardiac activity augments coronary blood flow. Coronary autoregulation works for perfusion pressures between 8 and 26 kPa (Sect. 3.2.4.7).

Coronary circulation produces vasoactive substances. Endothelial production of nitric oxide mediates coronary vasodilation. Coronary endothelial cells can also produce epoxyeicosastrienoic acids, which cause coronary vasodilation. In addition, ATP-gated potassium channels and adenosine are implicated.

Ischemic myocardium releases increased amounts of adenosine, which regulates coronary flow. Oxygen is a major mediator. Myocardial oxygen consumption increases when coronary blood flow augments, antagonizing autoregulation. In abnormal states, coronary autoregulation can change. Hypoperfused human epicardial coronary arteries change their autoregulatory responsiveness. Vasoconstriction distal to the site of coronary angioplasty results from altered autoregulation [213].

3.6.1.9 Role of Vascular Smooth Myocytes

Autoregulation is mediated by vascular smooth myocytes independently of endothelial cells. Vasoconstriction that results from an increase in pressure inside most small resistance arteries involves stretch-induced activation of plamalemmal, nonselective cation channels of vascular smooth myocytes. This activation causes membrane depolarization, calcium influx through $Ca_V1.2$ channels, and smooth muscle contraction [214].

Stretch-induced channel activity originates from interaction between mechanosensitive angiotensin Gq-coupled AT_2 receptor and cation TRPC6 channel [215]. The former can be activated by membrane stretch in the absence of its ligand [216]. Excited Gq protein activates phospholipase-C that hydrolyzes the membrane phospholipid phosphatidylinositol (4,5)-bisphosphate into diacylglycerol and inositol trisphosphate. Diacylglycerol stimulates several members of the TRP family, such as TRPC3, TRPC6, and TRPC7 [217]. In fact, PIP_2, DAG, and IP_3 are all able to prime activity of certain TRP channels [218].

3.6.1.10 Blood Flow Modeling in Arterioles

Some mathematical models of blood flow regulation describe the mechanics of the vascular wall exposed to a pressure field and its reaction by varying the vessel tone. The model of thin-walled, elastic, cylindrical vessels is based on Laplace's law, this hypothesis being valid for a ratio of a wall thickness to hydraulic diameter much smaller than 1. Simple constitutive laws relate wall shear stress to vessel wall tension.

Changes in hematocrit (Ht) influence blood pressure. A small hematocrit elevation is associated with decreased blood pressure and increased blood flow rate. Augmented hematocrit and resulting rise of blood viscosity heighten wall shear stress (WSS), thereby increasing nitric oxide production in the endothelium and subsequently causing vasodilation resulting from stress-induced nitric oxide release from the endothelium.

Arteriolar mechanics, which is responsible for autoregulation of blood flow, was modeled incorporating elastic (passive) and myogenic (active) responses modulated by wall shear stress effect for small changes in luminal pressure (p_i) and hematocrit [219]. This model considers a one-dimensional, steady, laminar, fully developed flow (Poiseuille flow) in a thick-walled, elastic, cylindrical arteriole (straight duct of circular cross-section; unstressed internal radius $R_i = 10\,\mu$m, endothelial thickness 1 μm, and wall thickness 10 μm). Wall shear stress is given by $c_w = (\Delta p_i/L)R_i/2$ and the volumetric flow rate by:

$$q = \frac{\pi R_i^4}{8\mu_{\rm pl}} \frac{\Delta p_i}{L} \Big(1 + \lambda_{\rm co}^4 \Big[\frac{\mu_{\rm pl}}{\mu_{\rm co}} - 1\Big]\Big), \tag{3.44}$$

where $\mu_{\rm co}$ and $\mu_{\rm pl}$ are the blood core and plasma layer viscosity, respectively.

The 2-layer blood domain in the arteriolar lumen consists of a cell-rich core and a plasma layer (Vol. 8 – Chap. 1. Hemodynamics). The thickness of the plasma layer is assumed to be related to Ht by the following linear relation,

$$h_{\rm pl} = \kappa_1 Ht + \kappa_2, \tag{3.45}$$

κ_3 and κ_4 being fitting parameters. Then the normalized thicknesses of blood core ($\lambda_{\rm co}$) and plasma layer ($\lambda_{\rm pl}$) in an arteriole are given by:

$$\lambda_{\rm pl} = h_{\rm pl}/R; \quad \lambda_{\rm co} = 1 - \lambda_{\rm pl}.$$

The blood core viscosity ($\mu_{\rm co}$) is related to the blood core Ht ($\mathrm{Ht_{co}}$) by another linear relation:

$$\mu_{\rm co} = \kappa_3 Ht_{\rm co} + \kappa_4, \tag{3.46}$$

where κ_3 and κ_4 are fitting parameters ($\kappa_3 = 1.68 \times 10^{-4}$ and $\kappa_4 = 43.48 \times 10^{-4}$ Pl) The core hematocrit is related to the measured Ht by:

$$Ht = 2\, Ht_{\rm co}\lambda_{\rm co}^2 \Big(1 - \lambda_{\rm co}^2 + \lambda_{\rm co}^2 \frac{\mu_{\rm pl}}{2\mu_{\rm co}}\Big) \times \Big(1 - \lambda_{\rm co}^2 + \lambda_{\rm co}^2 \frac{\mu_{\rm pl}}{\mu_{\rm co}}\Big)^{-1}. \tag{3.47}$$

The blood vessel is supposed to be a tethered cylinder; axial strain is neglected and the relation between radial strain ($\mathbf{e}_r$) and radial ($\mathbf{c}_r$) and circumferential ($\mathbf{c}_\theta$) stress vector is assumed to be linear:

$$\mathbf{e}_r = \frac{r}{R_i}[\mathbf{c}_\theta(1 - \nu_P^2) - \nu_P \mathbf{c}_r(1 + \nu_\text{P})], \tag{3.48}$$

where $E = 1 \times 10^4$ N/m^2 is the elastic modulus of blood vessel supposed to be constant and $\nu_\text{P} = 0.5$ the Poisson ratio (incompressible material).

The active (myogenic) and passive responses of vessel wall to blood flow are the sums of their respective passive ($\mathbf{c}_r^\text{p}$) and $\mathbf{c}_\theta^\text{p}$) and active ($\mathbf{c}_r^\text{a}$) and $\mathbf{c}_\theta^\text{a}$) components. The passive components are derived from Lame's equations:

$$\mathbf{c}_r^\text{p} = \frac{R_i^2 p_i}{R_e^2 - R_i^2}\left(1 - \frac{R_e^2}{r^2}\right); \quad \mathbf{c}_\theta^\text{p} = \frac{R_i^2 p_i}{R_e^2 - R_i^2}\left(1 + \frac{R_e^2}{r^2}\right). \tag{3.49}$$

The active components are assumed to be linearly proportional to the tension at the surface of the vessel wall ($p_i R_i/2$), that is, the myogenic response is a simple function of wall tension:

$$\begin{aligned} \mathbf{c}_r^\text{a} &= \frac{R_i^2}{R_e^2 - R_i^2}\left(1 - \frac{R_e^2}{r^2}\right)\gamma_\text{WSS}\kappa_5 p_i \frac{R_i}{R_\text{ref}}; \\ \mathbf{c}_\theta^\text{a} &= \frac{R_i^2 p_i}{R_e^2 - R_i^2}\left(1 + \frac{R_e^2}{r^2}\right)\gamma_\text{WSS}\kappa_5 p_i \frac{R_i}{R_\text{ref}}, \end{aligned} \tag{3.50}$$

where $\gamma_\text{WSS} = -\tanh(\kappa_6/c_\text{w})$ is the modulation of the myogenic response by the wall shear stress normalized with a fitting parameter (κ_6), R_ref the vessel radius at baseline pressure, and $\kappa_5 = 8$ another fitting parameter using experimental observations.

The maximum NO synthesis rate ($\mathsf{p}_{\text{NO}_\text{max}} = 150\,\mu$mol/s) is supposed to evolve linearly with the wall shear stress. Nitric oxide production follows Michaelis–Menten kinetics (with Michaelis–Menten constant $\text{K}_\text{M} = 4.7$); hence the corresponding reaction rate $\mathcal{R}$ is given by:

$$\mathcal{R} = \frac{\mathsf{p}_{\text{NO}_\text{max}} p_{\text{O}_2}}{p_{\text{O}_2} + \text{K}_\text{M}}, \tag{3.51}$$

where p_{O_2} is the oxygen partial pressure.

In the cell-rich core, NO concentration (c_NO) is governed by the following steady-state reaction–diffusion equation,

$$\frac{\mathcal{D}_{\text{NO}_\text{f}}}{r}\frac{\partial}{\partial r}\left(r\frac{\partial c_\text{NO}}{\partial r}\right) = \mathsf{d}_\text{f} c_\text{NO}, \tag{3.52}$$

where $\mathcal{D}_{NO_f} = 3300\,\mu m^2/s$ is the diffusion coefficient of NO in the cell-rich core and $\mathsf{d}_f = 382.5/s$ the scavenging rate of NO by red blood capsules, which can be assumed to remain constant, whatever the hematocrit.

In the plasma layer and glycocalyx (subscript f), the transport of NO and O_2 is assumed to satisfy the following steady-state diffusion equations,

$$\frac{\mathcal{D}_{NO_f}}{r}\frac{\partial}{\partial r}\left(r\frac{\partial c_{NO}}{\partial r}\right) = 0, \tag{3.53}$$

and

$$\frac{\mathsf{s}_{O_2}\mathcal{D}_{O_{2f}}}{r}\frac{\partial}{\partial r}\left(r\frac{\partial p_{O_2}}{\partial r}\right) = 0, \tag{3.54}$$

where $\mathsf{s}_{O_2} = 0.01\,\mu mol/Pa$ is the solubility of O_2 in the fluid phase and $\mathcal{D}_{O_{2f}} = 2,800\,\mu m^2/s$ is the diffusivity of O_2 in the vessel lumen.

In the endothelium (susbscript w), the rate of O_2 consumption is assumed to be twice the rate of NO production. The transport equations become:

$$\frac{\mathcal{D}_{NO_w}}{r}\frac{\partial}{\partial r}\left(r\frac{\partial c_{NO}}{\partial r}\right) + \mathcal{R} = 0 \tag{3.55}$$

and

$$\frac{\mathsf{s}_{O_2}\mathcal{D}_{O_{2w}}}{r}\frac{\partial}{\partial r}\left(r\frac{\partial p_{O_2}}{\partial r}\right) - 2\mathcal{R} = 0, \tag{3.56}$$

where $\mathcal{D}_{NO_w} = 1,400\,\mu m^2/s$ and $\mathcal{D}_{O_{2w}} = 1,650\,\mu m^2/s$ are the diffusivities of NO and O_2 in the arteriolar wall.

In the remaining part of the arteriolar wall,

$$\frac{\mathcal{D}_{NO_w}}{r}\frac{\partial}{\partial r}\left(r\frac{\partial c_{NO}}{\partial r}\right) - \mathsf{d}_w c_{NO} = 0, \tag{3.57}$$

and

$$\frac{\mathsf{s}_{O_2}\mathcal{D}_{O_{2w}}}{r}\frac{\partial}{\partial r}\left(r\frac{\partial p_{O_2}}{\partial r}\right) - \mathcal{R} = 0, \tag{3.58}$$

where $\mathsf{d}_w = 1/s$ is the rate of consumption of NO in the arteriolar wall.

The steady blood flow rate through a single arteriole was thus shown to depend on NO production and availability and the balance between passive and active (myogenic) mechanical responses of the arteriole wall. Transient effects in both the viscoelastic response of the arteriole wall and NO production as well as wall anisotropy were neglected, but these hypotheses should not change the conclusion.

3.6.2 Neural Effects on Heart and Blood Vessels

The central nervous system is the center of the regulation loop. It receives signals from peripheral organs via afferent nerves and sends cues to visceral effectors through efferent nerves. The brain gets signals from sensors and adjusts the circulatory parameters to match the body's needs.

3.6.2.1 Nervous Regulation Loop

The nervous control of the circulation operates via: (1) afferents from the cardiovascular apparatus and other endocrine glands; (2) interconnecting neurons; and (3) efferents from the nervous centers. These neurons are either peripheral or central, with excitatory and inhibitory synapses.

Afferent neurons are constituted of 4 main types: (1) *barosensitive* (i.e., afferents from the arterial baroreceptors); (2) *thermosensitive* (i.e., cutaneous vasoconstrictors activated by hypothermia, emotions, and hyperventilation);[95] (3) *glucosensitive* (adrenaline release from the adrenal medulla stimulated by hypoglycemia and physical exercise); and (4) afferents from the heart, blood vessels, kidneys, and adrenal medulla.

3.6.2.2 Parasympathetic and Sympathetic Nervous System

The main targets of the nervous control are, in the heart, nodal cells and cardiomyocytes and, in the vessel wall, smooth myocytes.

The nervous regulation of the blood circulation is composed of cholinergic preganglionic neurons of the central nervous system, which lead to peripheral control nodes, such as para- or prevertebral sympathetic ganglia, endocrine glands, especially adrenal glands involved in the catecholamine secretion, or visceral ganglionic networks.

Postganglionic neurons include 2 types: (1) parasympathetic cholinergic neurons, which mainly innervate the heart, and (2) sympathetic noradrenergic neurons, which innervate the heart and vessel walls.[96]

Both sympathetic and parasympathetic nerves act synergetically, exerting a mutual inhibition. Whenever the sympathetic system is activated, the parasympathetic activity lowers and conversely.

---

[95] Skin circulation is mostly regulated via the rostral ventromedial medulla and medullary raphe [221].

[96] When it is not caused by vascular or renal disorders, hypertension can be due to a strong sympathetic tone.

Table 3.50 Neural activation of circulatory organs

Effect	Sympathetic	Parasympathetic
Heart		
Bathmotropy (CMC excitability)	B−	B+
Chronotropy (emission frequency of action potential)	C+ (major)	C− (major)
Dromotropy (conductibility)	D+ (moderate)	D− (major)
Inotropy (CMC contraction force)	I+ (major)	I− (minor)
Tonotropy (distensibility)	T+	T−
Blood vessels		
Resistance	+ (major)	Nonsignificant
Capacitance	− (minor)	Nonsignificant

Table 3.51 Response speeds of various commands of the cardiovascular system (f_C: cardiac frequency; SV: stroke volume)

Command	Response
Parasympathetic	Very quick (f_C)
Sympathetic	Quick (f_C, SV)
Biochemical	Slower (f_C, SV)

Activation of sympathetic efferent nerves to the heart increases the cardiac frequency (positive chronotropy [C+]), contractility (positive inotropy [I+]), and conduction velocity (positive dromotropy [D+]; Table 3.50).

Parasympathetic nerves have mostly negative effects, but quicker than those of the sympathetic nerves (Table 3.51). Parasympathetic effects on inotropy are weak in ventricles and significant in atria.

α1-Adrenergic receptors elicit a negative inotropic effect in 68% of cardiomyocytes of the mouse right ventricle and in 36% of those in the left ventricle [220]. Negative and positive inotropy is associated with increased and decreased Ca^{2+} transients, respectively. Diminished Ca^{2+} transients is linked to an elevated Ca^{2+} export from the cell and reduced Ca^{2+} content of the sarcoplasmic reticulum.

Most blood vessels in the body do not have parasympathetic innervation. On the other hand, both the heart and blood vessels are effectors of excitatory sympathetic fibers. In blood vessels, sympathetic activation constricts arteries and arterioles. Vasoconstriction causes an increase in resistance and pressure, and a decrease in distal blood flow. Sympathetic-induced constriction of capacitance veins decreases venous blood volume and increases venous pressure.

The overall effect of sympathetic activation is to increase the cardiac output (blood flow rate), systemic vascular resistance, and arterial blood pressure.

3.6.2.3 Cardiovascular Nervous Centers

Nervous signals are integrated in cardiovascular centers localized in the brainstem. Various neuron clusters form: (1) the *cardiac inhibitory*, (2) *cardiac excitatory*, and (3) *vasomotor* centers. Cardiac centers maintain a balance between inhibitory parasympathetic nerves and excitatory sympathetic nerves.

The cardiac inhibitory center sends an inhibitory efferent pathway to the sinoatrial node via vagal parasympathetic fibers. The other 2 centers project sympathetic fibers to the sinoatrial node and the myocardium on the one hand, and smooth muscles of blood vessel walls.

Several regions in the brain, the rostral ventral lateral medulla (RVLM), nucleus of the solitary tract (NTS), and parabrachial nucleus (PBN) in the dorsolateral pons,[97] are integrative centers of cardiovascular cues involved in the regulation of sympathoexcitatory reflexes. The nucleus of the solitary tract receives inputs from the cardiac vagal and sympathetic afferents.

Limbic, cortical, and midbrain structures function for the short-term regulation of blood pressure by the sympathetic tone. The background activity of the sympathetic tone for long-term control of blood pressure is driven by neurons of the rostral ventrolateral medulla, spinal cord, hypothalamus, and nucleus of the solitary tract (Fig. 3.11) [221].

The nucleus of the solitary tract is an integrative center for blood circulation control. It directly receives cues from baroreceptors, voloreceptors, and chemoreceptors, as well as many synaptic inputs.

The spinal cord receives chemical (tissue oxygen content) and physical (tissue stretch) outputs.

The hypothalamus, with its paraventricular and dorsomedial nuclei, is another integrative center for the regulation of blood circulation. The dorsomedial nucleus is implicated in environmental stresses. Neurons of the paraventricular nucleus are affected by blood volume, pressure, and osmolality.

The regulation of *renal sympathetic activity* by arterial baroreceptors uses the rostral ventrolateral medulla. The renal nerve response is also induced by hepatoportal osmoreceptors and voloreceptors, in addition to arterial baroreceptors.

The response to atrial voloreceptors involves the nucleus of the solitary tract and the paraventricular nucleus of the hypothalamus to regulate sodium reabsorption by the kidney, and subsequently blood volume. Peripheral and brain osmoreceptors and hypothalamic sodium receptors, particularly in the median preoptic nucleus, affect renal sympathetic activity. Other central osmoreceptors and sodium receptors are located in circumventricular organs.

Circulating hormones such as aldosterone control blood circulation via circumventricular organs (subfornical organ, organum vasculosum lamina terminalis, and area postrema). Circulating hormones such as angiotensin-2 also directly affect the sympathetic ganglia. Angiotensin-2 acts on the median preoptic nucleus, nucleus of the solitary tract, rostral ventrolateral medulla, and paraventricular nucleus of the hypothalamus.

[97]The parabrachial nucleus is separated by the brachium conjunctivum into 2 main regions: the lateral and medial parabrachial nuclei. The lateral parabrachial nucleus is connected to the rostral ventral lateral medulla and nucleus of the solitary tract. The lateral parabrachial nucleus can be further subdivided to the dorsal (dPBN), ventral (vPBN), central (cPBN), and external lateral (elPBN) parabrachial nuclei.

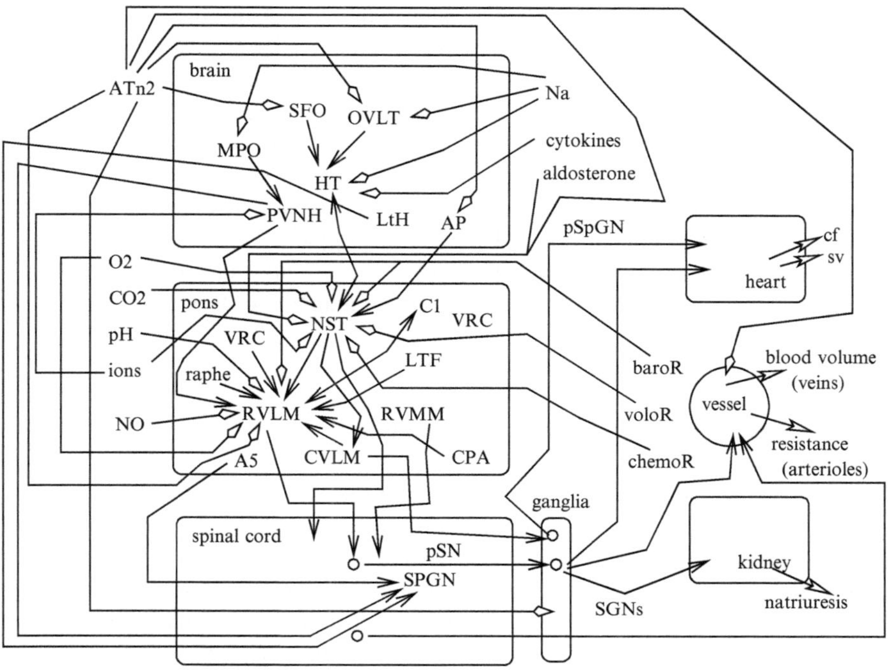

Fig. 3.11 Nervous regulation of the blood circulation, especially sympathetic activity and its influence agents (Source: [221]). An increase in blood pressure activates baroreceptors and inhibits cardiac, renal, and vascular sympathetic efferents. The baroreceptor reflex dampens short-term fluctuations of blood pressure. However, it can be reset via both neural and humoral mechanisms such that the operating range is shifted without reduction in reflex sensitivity. The nucleus of the solitary tract (NST), hypothalamus (HT), with its paraventricular nucleus (PVNH) and median preoptic nucleus (MPO), are integrative centers for the regulation of blood circulation. The interneurons (releasing γ-aminobutyric acid) of the caudal ventrolateral medulla (CVLM) inhibit barosensitive neurons of the rostral ventrolateral medulla (RVLM). RVLM neurons coexist in a pons region with the ventral respiratory column (VRC) neurons for coordination of respiration and circulation, and a cluster of adrenaline-synthesizing neurons (C1). Circumventricular organs are implicated, such as subfornical organ (SFO), organum vasculosum lamina terminalis (OVLT), and area postrema (AP). Efferent neurons include sympathetic preganglionic (SPGN), sympathetic ganglionic (SGN), parasympathetic (pSN), and parasympathetic postganglionic neurons (pSpGN). Other involved nervous structures comprise the caudal pressor area (CPA), lateral hypothalamus (LtH), lateral tegmental field (LTF), and a noradrenergic cluster located at the pontomedullary junction (A5). The cutaneous circulation is mainly regulated via the rostral ventromedial medulla (RVMM) and raphe. Stimuli of neuronal activity include aldosterone, angiotensin-2 (ATn2, which acts on median preoptic nucleus, PVNH, NTS, RVLM, and SPGN), ions (especially sodium and pH), and blood gas (cf: cardiac frequency, sv: stroke volume)

The central nervous system has its own receptors, responding to changes in blood gas (O_2 and CO_2) levels via brainstem chemoreceptors, and in sodium and osmolality via hypothalamic receptors.

Exercise activates the sympathetic system, thus increasing blood pressure and flow to skeletal muscles, whereas it reduces the blood flow to other body organs, such as the kidneys. Blood flow during exercise is regulated by both the central nervous command and local process. Exercise elevates blood flow and stress exerted on arterial walls, thereby causing shear-mediated release of nitric oxide and vascular lumen dilation. The brachial artery response to shear increase during the forearm exercise with handgrip protocol shows that vascular endothelium transduces the mean time-dependent shear [222]. However, retrograde diastolic flow is more or less noticeable depending on the mode of exercise and workload; it influences the hemodynamic behavior, hence shear stress magnitude as well as amplitude variations [223].

3.6.2.4 Neurotransmitters and Cotransmitters

Perivascular adrenergic and cholinergic nerves release many types of neurotransmitters, usually cotransmitters, such as peptides, purines, and nitric oxide, for blood flow regulation on the regional or general scale.

Messenger ATP is released as a cotransmitter with noradrenaline for sympathetic vasoconstriction in small arteries and arterioles. Vascular endothelium is engaged in blood flow control on a local scale. Strong, local blood flow variations as well as hypoxia lead to changes in vascular tone mediated by endothelium. However, both regulator types interact.

Nitric oxide synthase NOS3 in the rostral ventrolateral medulla reduces blood pressure. NOS+ (NOS1+–NOS3+) neurons in the rostral ventrolateral medulla receive inputs from baroreceptors and are activated during cardiac sympathoexcitatory reflexes.[98]

Postganglionic sympathetic vasoconstrictor fibers use noradrenaline as a neurotransmitter. The resting frequency of 1 to 4 Hz can increase to 10 Hz. Inactive units may be recruited.

Vasodilatory fibers are not involved in the bulk regulation of peripheral resistances, but locally increase blood flow. *Sympathetic vasodilatory fibers* arise from the cerebral cortex, and, after a synapse in the hypothalamus, run through the medulla to alter the activity of preganglionic vasodilator fibers. Postganglionic fibers of the sympathetic vasodilator system are cholinergic. *Parasympathetic vasodilatory fibers* participate in autonomic reflexes such as digestive secretion.[99]

---

[98]In anesthetized cats, mean arterial pressure and renal sympathetic nerve activity rely on NO, especially that synthesized by NOS2 in the rostral ventrolateral medulla [224].

[99]Two groups of parasympathetic vasodilatory fibers originate from the chorda tympani nerve, a branch of the facial nerve, and from the trigeminal portion of the distal lingual nerve (probably via the glossopharyngeal nerve) [225].

A population of neurons in the arcuate nucleus express estrogen receptor ERα (NR3a1), neurokinin receptor NK_3, kisspeptin-1,[100] neurokinin-B,[101] and dynorphin[102] (KND+ neurons). They project to gonadotropin-releasing hormone[103] (GnRH)+ terminals in the median eminence and preoptic region that regulate body temperature. KND+ neurons promote *cutaneous vasodilation*, a heat dissipation effector, via projections to rostral hypothalamic structures that control thermoregulatory effectors, such as the medial preoptic area, which contains NK_3+, warm-sensitive, gabaergic neurons, and median preoptic nucleus, which receives information from warm-sensitive, cutaneous thermoreceptors [228]. They participate in 17β-estradiol effect on body temperature during the light phase, but not during the dark phase.

Sympathetic nerves express both neurotrophin protein Tyr receptor kinase NTRK1 and Sema3a receptor neuropilin-1. Semaphorin-3A promotes aggregation of neurons into sympathetic ganglia during early embryogenesis. Cardiomyocyte-released chemoattractant nerve growth factor is required for sympathetic axon

---

[100] Formerly known as metastin, as it was originally identified as a human metastasis suppressor. Its alias is formed by 2 groups of letters "Ki" refering to the location of its discovery, Hershey, Pennsylvania, home of Hershey Chocolate Kiss and "SS" for suppressor sequence. Kisspeptin is encoded by the KISS1 gene. The gene product is a 145-amino acid precursor that is cleaved to 54-amino acid peptide, which can be further truncated to 14-, 13-, or 10-amino acid C-terminal fragments, the kisspeptins. Kisspeptin_{145} represents the precursor, kisspeptin_{54} (or $\text{kisspeptin}_{(68--121)}$) the peptide, and kisspeptin_{14} (or $\text{kisspeptin}_{(108--121)}$), kisspeptin_{13} (or $\text{kisspeptin}_{(109--121)}$), and kisspeptin_{10} (or $\text{kisspeptin}_{(112--121)}$) the C-terminal fragments. Kisspeptin signaling in the brain mediates the negative feedback action of sex steroids on gonadotropin secretion, generating the preovulatory GnRH–LH surge, triggering and guiding the tempo of sexual maturation at puberty, controlling seasonal reproduction, and restraining reproductive activity during lactation [226]. The KISS1 gene expression is regulated by estradiol (E_2) in the hypothalamus. Kisspeptin-1 is also synthesized in the neocortex of fetal adrenal glands. It stimulates secretion of aldosterone [226]. It is also produced in pancreatic β cells, where it can stimulate insulin release (auto- and paracrine action). Kisspeptin_{54}, -_{13}, and -_{10} are potent vasoconstrictors. Kisspeptin-1 is a G-protein-coupled receptor ligand for GPR54 (or Kiss1R). The kisspeptin-Kiss1R complex initiates secretion of gonadotropin-releasing hormone (GnRH) at puberty. Synthesized in the brain, kisspeptin unleashes hormones that stimulate the production of estrogen or testosterone in ovaries and testes, starting the physical transformations of puberty. Menopause is characterized by ovarian follicle depletion, reduction of ovarian steroids, compensatory gonadotrophin hypersecretion, and hypertrophy of neurons expressing neurokinin-B (NKB), kisspeptin-1, and estrogen receptor-α within the hypothalamic infundibular (arcuate) nucleus.

[101] Neurokinin-B is encoded by the TAC3 (tachykinin-3) gene. On the other hand, the other human tachykinin gene, the TAC1 gene, encodes neurokinin-A (or substance-K), neuropeptide-K (or neurokinin-K), neuropeptide-γ, and substance-P.

[102] Dynorphin, an opioid peptide derived from the prodynorphin gene product, inhibits the reproductive axis. The dynorphin gene expression decreases in postmenopausal women. The secretion of luteinizing hormone can then rise [227].

[103] Secretion of GnRH into portal capillaries stimulates luteinizing hormone (LH) secretion from the anterior hypophysis, which stimulates the secretion of 17β-estradiol from the ovary. The negative feedback primed by 17β-estradiol via NR3a1 reduces the plasma LH level and decreases neurokinin-B and kisspeptin synthesis in KND+ neurons.

growth and innervation of the heart. The neural chemorepellent Sema3a is abundantly expressed in the trabecular layer in early-stage embryos, then restricted to Purkinje fibers after birth. Semaphorin-3A builds a transmural sympathetic innervation patterning, characterized by an epicardial-to-endocardial innervation gradient [229]. Alterations in Sema3a expression trigger various kinds of arrhythmias.

Glutamate is a ubiquitous excitatory neurotransmitter in the central nervous system. It is used in processing cardiovascular excitatory reflex inputs from cardiac sympathetic afferents to initiate sympathoexcitatory reflex responses. Glutamate in the dorsal, ventral, and central lateral parabrachial nuclei decreases blood pressure and cardiac frequency. Reflex inputs from the heart are transmitted mainly through sympathetic afferents and processed in the external lateral parabrachial nucleus in the pons [230].

The external lateral parabrachial nucleus processes inputs from cardiac sympathetic afferents to trigger sympathoexcitatory cardiac reflexes using AMPA- and NMDA-type ionotropic glutamate receptors (Vol. 3 – Chap. 2. Ion Carriers) [230]. During myocardial ischemia, the ischemic metabolites bradykinin and thromboxane-A2 stimulate cardiac spinal sympathetic sensory nerve endings and afferents that excite elPBN neurons.

Other neurotransmitters, such as GABA and ACh, in the external lateral parabrachial nucleus can also process cardiac sympathoexcitatory reflexes [230]. The former causes a depressor response with an associated decrease in renal sympathetic nerve activity. Cholinergic neurons from the external lateral parabrachial nucleus that project to the rostral ventral lateral medulla contribute to elevation in blood pressure.

3.6.2.5 Exercise – Muscle Metaboreflex and Mechanoreflex

Two main mechanisms increase sympathetic nerve activity: (1) the exercise pressor reflex and (2) the central command. The exercise pressor reflex arises from chemo- (muscle metaboreflex) and mechanoreceptors (muscle mechanoreflex) of skeletal muscles, which stimulates the nervous centers via afferent fibers. The central command simultaneously activates the locomotor and cardiovascular systems.

Muscle metaboreflex desensitization and mechanoreflex sensitization are observed after myocardial infarction. During exercise in heart failure, exaggerated sympathetic activation occurs with augmented muscle sympathetic signaling and excessive renal vasoconstriction.[104] Moreover, elevation of muscle blood flow decays with respect to normal vasculature. The renal and lumbar sympathetic responses associated with the central command increase in heart failure, causing excessive peripheral vasoconstriction [231].

---

[104]Renal vasoconstriction reduces the blood supply, causing excessive renin secretion and inappropriate salt and water retention.

Table 3.52 Rhythmic processes associated with blood flow

Process	Frequency (Hz)
Heart cycle	1
Respiratory cycle	0.3
Vessel myogenic activity	0.1
Neurogenic control activity	0.04–0.4
Endothelium metabolic activity	$\leq$ 0.01

3.6.2.6 Regulation of Cardiac Frequency

The heart is able to beat independently. The cardiac frequency is higher than rhythms of the main processes involved in the regulation of blood flow (Table 3.52). Adaptation needs a slight delay.

Nervous Control

The nervous system regulates cardiac frequency (f_c), in superimposition of heart automatism, to adapt f_c to the changing needs of the body. However, the frequency increase is bounded by the necessary *diastolic filling* associated with venous return and *diastolic perfusion* of the coronary arteries. Various sensors of the circulatory apparatus send messages to the cardiac centers that, in turn, send messages to the heart.

Respiratory Sinus Arrhythmia

Breathing modulates autonomic nervous outflow from the brainstem, thereby causing fluctuation of the cardiac cycle period measured by the R–R interval on ECG traces that are synchronous with respiration. Respiratory sinus arrhythmia (RSA) is a vagal- or combined vagosympathetic-mediated increase and decrease in cardiac frequency concomittant with inspiration and expiration, respectively. In other words, respiratory sinus arrhythmia is an alternation between cardiac acceleration and deceleration exhibited by inspiratory R–R interval shortening and expiratory R–R interval lengthening. Breathing frequency is related to RSA amplitude.

At least in the upright position, cardiac sympathetic activity may modulate both RSA amplitude and the phase relation between respiration and respiratory sinus arrhythmia [232]. At least in the supine position, cardiac sympathetic activity attenuates RSA amplitude, but does not modulate the frequency-dependent breathing–RSA phase relation. Cardiac sympathetic tone is lower in the supine posture. Changes in breathing–RSA relation at different breathing frequencies primarily reflect variations of cardiac vagal activity due to central and/or baroreflex interactions. Yet, mechanical factors associated with fluctuations in intrathoracic pressure during breathing can also contribute to respiratory sinus arrhythmia.

The time interval between R–R interval maximum and expiration onset remains similar (∼2.5–3.0 s) whatever the breathing frequency, whereas the R–R interval minimum progressively shifts from expiratory onset into midinspiration with slower breathing (1:1 inspiration-to-expiration ratio without pause) [232]. The maximal R–R interval happens simultaneously with inspiratory onset only during 0.20-Hz breathing, whereas during 0.15- and 0.10-Hz breathing, the R–R interval maximum begins before inspiratory onset. During slow (0.10-Hz) breathing, 2 stages of cardiac acceleration can be observed; the rate of cardiac acceleration before inspiration is lower than that following inspiratory onset. However, pre- and postinspiratory cardiac accelerations differ during slow (0.10 Hz), but not fast (0.20 Hz) breathing. Yet, the time interval from inspiration onset to systolic blood pressure (SBP) minimum and from expiration onset to SBP maximum remain unchanged whatever the breathing frequency. The relation between breathing and RSA is thus nonlinear.

In addition, the inspiratory-to-expiratory duration ratio influences respiratory sinus arrhythmia. A short inspiration relative to expiration enhances RSA amplitude and causes stronger cardiac acceleration with respect to regular and relatively long inspirations. Moreover, both end-inspiration and end-expiration pauses affect the cardiac deceleration component of respiratory sinus arrhythmia. Furthermore, in the absence of V_T control, the temporal relation between expiration and R–R minimum changes when breathing slows from 0.20 to 0.10 Hz, that is, significant phase shifts are observed between respiration and respiratory sinus arrhythmia [232].

Other Factors

Hormones

Additional factors such as hormones and body temperature also influence the cardiac frequency. Under stresses, *cathecolamines* are released from the adrenal medulla into the circulation to produce an increase in heart rate. *Thyroid hormones*—thyroxin (T_4) and triiodothyronine (T_3)—accelerate the cardiac frequency (f_c) and modulate myocardial contraction.

Ion Concentrations

Ion concentrations in the extracellular environment may have a significant influence on cardiac functioning. Potassium excess in the extracellular medium reduces cardiac frequency and contractibility, as well as calcium level reduction. Excessive Na^+ concentration depresses the cardiac function, whereas Na^+ deficiency in the extracellular space leads to cardiac fibrillation.

Body Temperature

Elevation of body temperature is associated with an increase in cardiac frequency. Conversely, hypothermia is accompanied by a reduction of cardiac frequency.

Exercise

Exercise modulates the cardiac chronotropic response and heart rate variability. At 3 pedaling frequency (70, 80, and 90 rpm), a pedaling frequency component in heart rate variability can be continuously monitored over the time course of exercise [233]. At peak exercise, amplitude of pedaling frequency component accounts for about 43, 39, and 49% of the total heart rate variability at 70, 80, and 90 rpm, respectively. The pedaling frequency component and respiratory sinus arrhythmia overlap at low workloads.

3.6.2.7 Regulation of the Arterial Pressure

At the nanoscopic scale, messengers transmit signals between cells, thereby controlling blood pressure. These messengers interact with receptors at the cell surface. Upon agonist binding on its extracellular side, the receptor experiences a conformational changes and becomes activated.

At the macroscopic scale, blood flow rate depends on cardiac frequency and stroke volume:

$$q = f_c \times \text{SV}.$$

The cardiac frequency is mainly regulated by the nervous system (both parasympathetic and sympathetic components), as well as by hormones. Stroke volume is controlled by myocardial contractility and its regulating factors (sympathetic control), pre- and afterload, with their sympathetic command, and circulating regulators. Arterial pressure depends on both blood flow rate and vascular resistance:

$$p_a = f_c \times \text{SV} \times \text{SVR}. \quad (3.59)$$

Blood pressure is regulated by short-, mid-, and long-term mechanisms (Fig. 3.12). Both blood pressure and sympathetic activity undergo a circadian rhythm.

A beat-to-beat arterial blood pressure variability can be analyzed using a multivariate parametric model of interactions between systolic and diastolic blood pressure, pulse pressure, cardiac frequency, and breathing cycle [234]. Systolic blood pressure can be defined as the sum of the preceding diastolic blood pressure and pulse pressure. Diastolic blood pressure is supposed to account for arterial baroreflex; pulse pressure reflects changes in stroke volume related to cardiac and breathing frequency and afterload. Fast, beat-to-beat responses are observed in systolic and pulse pressure total power, diastolic pressure low- and high-frequency power, pulse pressure very-low-, low-, and high-frequency power.

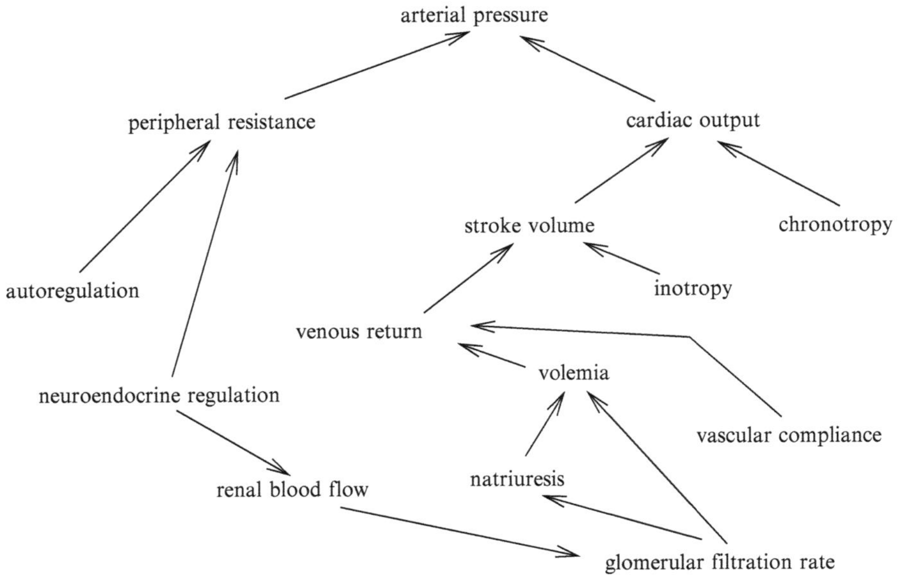

Fig. 3.12 Influence factors and regulation of arterial pressure. The arterial pressure is determined by the cardiac output and total systemic peripheral resistance. Cardiac output depends on cardiac functioning (chronotropy, inotropy, etc.) and venous return, itself particularly affected by volemia and vascular compliance. Blood pressure thereby is controlled by the autonomic nervous system, each factor being regulated by either the sympathetic system or both the sympathetic and parasympathetic systems. The sympathetic baroreflex is a feedback loop, with afferents from baroreceptors. The distention of the arterial wall following an increase in blood pressure activates the baroreceptors, and subsequently inhibits cardiac, renal, and vasomotor sympathetic efferents to restore the blood pressure. With the parasympathetic counterpart, the sympathetic baroreflex is first aimed at dampening short-term fluctuations of blood pressure with appropriate resetting when the blood pressure adapts to the imposed conditions, such as during exercise. The new operating range then does not disturb reflex sensitivity. The sympathetic activity fluctuates because of the time-dependent nature of arterial baroreceptor activity, polysynaptic transmission through the baroreflex loop, and breathing. These factors are also regulated by endocrine (renin–angiotensin–aldosterone system, cathecolamines, vasopressin, and natriuretic peptides) and paracrine factors. The total peripheral resistance is mainly determined by the activity of the sympathetic system and local autoregulations. The kidney is the primary regulator of extracellular fluid volume and electrolyte balance. The kidney participates in blood pressure regulation via the relationship between the renal blood pressure and natriuresis. Any increase in sodium retention produces an initial blood volume expansion, then an increase in blood pressure associated with a rise in cardiac output. Subsequent tissue overperfusion leads to an increase in peripheral resistance, and cardiac output returns to its resting values. Renal blood flow and glomerular filtration are controlled by the renal sympathetic nerves, concentrations of circulating hormones, and renal paracrine and autocrine factors (ATP, nitric oxide, etc.)

Short-Term Nervous Control

The central nervous system integrates afferent signals from the heart, vasculature, and kidney and sends efferent impulses to these organs (Sect. 3.6). Systemic vascular resistance depends mainly on arteriolar vasoconstriction state.

Baroreceptors

In the short-term control, the arterial pressure is monitored by suitable receptors, mainly baroreceptors. Baroreceptors have a rest activity associated with the sympathetic tone of the vascular smooth myocytes. They react with bursts synchronized with the arterial pressure pulse and respiration. Adjustments are made via neural mechanisms that change the cardiac output and peripheral resistances.

Orthostasis and Position Changes

Humans experience a ceaseless orthostatic stress, orthostasis meaning standing in an upright (or orthostatic) position supported only by feet and relying on a dynamic balance. The base of support is narrow, delimited by 2 feet. During standing, a gravitational fluid shift toward the lower part of the body, particularly lower limbs, is compensated to avoid postural hypotension. Orthostatic hypotension refers to an unusually low blood pressure when the patient is standing up. Arterial baroreflex, a feedback loop from baroreceptor tension sensing to arterial pressure control, is the major compensatory mechanism. Defective baroreceptor afferents cause postural hypotension.

In addition, arterial pressure is maintained during frequent body position changes. An upright tilt resets baroreflex control of sympathetic nerve activity to a higher level and, hence, renal and cardiac sympathetic nerve activities [235].

At rest, in the absence of posture changes, spontaneous blood pressure oscillations with breathing can be explained by intrathoracic pressure variations that introduce fluctuations in venous return and stroke volume. A significant augmentation of systolic blood pressure fluctuations happens during slow (0.10 Hz) breathing [232]. In fact, mechanical causes with elevated tidal volumes during slow breathing and a hybrid central nervous control and baroreflex processing can generate blood pressure Mayer waves.

Long-Term Control

The long-term control of blood pressure involves indirect monitoring of blood volume. Hormonal mechanisms restore the blood volume and indirectly the blood

Table 3.53 Handling of the water and salt balance enables adjustment of blood volume and pressure

↑ Blood volume ⟶ ↑ cardiac flow rate ⟶ ↑ vascular resistance ⟶ ↑ arterial pressure ⟶ ↑ natriuresis.

pressure. The long-term regulation involves mainly: (1) the renin–angiotensin pathway,[105] (2) natriuretic peptides, and (3) antidiuretic hormone, or vasopressin (Sect. 3.6.7).

Kidney

Blood volume and, subsequently, blood pressure are controlled by fluid and electrolyte (particularly sodium and potassium) excretion by the kidney (Sect. 3.6.7). Sodium ion is the predominant cation in the extracellular fluid. The kidney is responsible for sodium and fluid balance handling, hence blood volume and pressure adjustment (Table 3.53).

The regulation of sodium reabsorption depends, at least partly, on the predominant $\alpha 1$-Na^+–K^+ ATPase and additional $\alpha 2$- or $\alpha 3$-isozyme expressed in the kidney.[106]

Most high blood pressure disorders result from either disturbed regulation of peripheral vascular resistance or defective salt and water reabsorption by the distal nephron. Defective salt reabsorption in a portion of the distal nephron leads to hypertension, a major risk factor for cardiovascular disease. Reduced renal Na^+ excretion causes extracellular volume expansion and subsequently augmented blood volume. Salt increases arterial constriction, hence peripheral vascular resistance and blood pressure.

A high-salt diet is implicated in hypertension. However, the sensitivity of blood pressure to salt varies among individuals. Salt-dependent, sustained elevation of blood pressure results from an increased sympathetic nerve activity with resetting of

---

[105] In 1898, R. Tigerstedt and P. Bergman discovered that a cortical extract of rabbit kidney causes vasoconstriction when injected intravenously [236]. They isolated the substance that was named renin.

[106] The Na^+–K^+ ATPase is a heteromeric enzyme that comprises a catalytic α and a glycosylated β subunit. The central isoform-specific region is targeted for protein kinase-C activation. This pump catalyzes the export of 3 Na^+ ions and import of 2 K^+ ions at the expense of 1 ATP molecule. $\alpha 4$-Isoform lodges exclusively in the sperm tail. In the heart, the cytosolic Ca^{2+} concentration increases during contraction and decreases during relaxation via sarco(endo)plasmic reticulum Ca^{2+} ATPase and plasmalemmal Ca^{2+} ATPase and Na^+–Ca^{2+} exchanger. Tha latter uses the Na^+ gradient established by Na^+–K^+ ATPase. Inhibition of Na^+–K^+ ATPase raises cytosolic Na^+ concentration. Therefore, Na^+–Ca^{2+} exchanger augments cytosolic Ca^{2+} concentration, thereby causing a positive inotropic effect. $\alpha 2$-Isozyme is a main regulator of calcium in the myocardium, its inhibition increasing calcium influx and contractibility [237]. $\alpha 1$-Isoform depresses cardiac contractility via excess extracellular K^+ level without apparent changes in intracellular calcium handling.

the baroreflex that triggers vasoconstriction and attenuated endothelium-mediated vasodilation, as well as, in the long term, arterial structural remodeling (collagen deposition, wall thickening, and lumen narrowing) [238]. Vascular wall remodeling may result from angiotensin-2-triggered inflammation and oxidative stress.

Like aldosterone, ouabain[107] is both a hypothalamic and adrenal hormone and growth factor that can promote cell growth and proliferation. It can thus participate in the arterial wall remodeling.

Hypothalamus

High dietary salt (NaCl) raises Na^+ concentration in blood and cerebrospinal fluid. Sodium-sensing hypothalamic circumventricular organs such as the subfornical organ then provoke a central sympathetic excitation, thereby elevating sympathetic nerve activity. The hypothalamic signal involves aldosterone, epithelial Na^+ channel, ouabain secreted by hypothalamic cells, ouabain-sensitive Na^+–K^+ pump, and angiotensin-2, hence priming a direct, rapid, short-term, angiotensin-2-mediated sympathetic nerve activity as well as a slower neuromodulatory axis [238]. Unlike the fast synaptic transmission, the much slower modulation of synaptic transmission depends on protein phosphorylation as well as changes in protein synthesis (e.g., increase in angiotensin-converting enzyme, angiotensin AT_1 receptor, and NADPH oxidase; decrease in nitric oxide synthase NOS2) in the hypothalamus.

Aldosterone raises the ouabain content in the hypothalamus via mineralocorticoid receptor. A high-salt diet heightens hypothalamic aldosterone content. Aldosterone regulates the activity of the ubiquitin ligase, neural precursor expressed, developmentally downregulated protein NEDD4-2, which labels the ENaC channel for retrieval from the plasma membrane and degradation [238].

Ouabain targets several types of ouabain-sensitive Na^+–K^+ ATPases in the central nervous system and peripheral organs. Arterial myocytes and endothelial and glial cells as well as some neuron types possess both α1- and α2-Na^+–K^+ ATPases (ratio ∼4:1); most neurons express α1- and α3-Na^+–K^+ ATPases [238].

Ouabain augments hypothalamic production of AT_1 receptor and NADPH oxidase, but reduces that of nitric oxide synthase NOS2 [238].

Cotransmitters ATP and neuropeptide-Y can increase sympathetic activity. In addition, they both modulate the effect of noradrenaline on smooth muscle. Moreover, neuropeptide-Y is a mitogene and growth regulator that can intervene in vascular wall remodeling [238].

---

[107] Somali waabaayo: arrow poison. Ouabain is also called G-strophanthin. It is found in ripe seeds of African plants Strophanthus gratus and the bark of Acokanthera ouabaio. It is a poisonous cardiac glycoside at high (micromolar to millimolar) concentrations. It is structurally related to digoxin, another lipophilic cardiac glycoside. It binds to and inhibits plasmalemmal Na^+–K^+ ATPase (sodium pump). At low (nanomolar and subnanomolar) concentrations, at least in guinea pig ventriculomyocytes, this endogenous substance stimulates the Na^+–K^+ ATPase (α1–α3-isoforms) [239].

Sympathetic Nerves

In addition to its effects in the central nervous system, ouabain synergistically operates on peripheral sympathetic neurons. It increases spontaneous neurotransmitter release from postganglionic sympathetic nerve terminals and both short- and long-term potentiation of synaptic transmission in sympathetic ganglia [238].

Adrenal Gland

Adrenal glands (or suprarenal glands) are endocrine glands above the kidneys. They release hormones in response to stress. Cortical cells produce corticosteroids such as androgens (zona reticularis or inner cortical layer), cortisol (zona fasciculata), and aldosterone (zona glomerulosa or outer region), which controls blood osmolarity. The adrenal medulla secrete catecholamines, such as adrenaline (~80%) and noradrenaline (~20%).

Ouabain is also liberated by the adrenal cortex and supports vasoconstriction via the Na^+–K^+ ATPase–Na^+–Ca^{2+} exchanger-Ca^{2+} pathway. In addition, circulating ouabain, the plasma level of which augments, activates the Na^+–K^+ pump–Src kinase axis, which heightens the NCX–TRP signaling in arterial smooth myocytes, but diminishes endothelium-mediated vasodilation.

Plasma ouabain is mostly originated from the adrenal cortex. Ouabain secretion from adrenocortical cells is stimulated by ACTH, angiotensin-2, and catecholamines. Released ouabain suppresses its own secretion (feedback) [238].

In arterial smooth myocytes, Na^+–Ca^{2+} exchanger NCX1 and canonical transient receptor potential channel TRPC3 and/or TRPC6 are upregulated in rodent hypertension models and in pulmonary artery myocytes in human primary pulmonary hypertension. Consequently, Ca^{2+} signaling is enhanced. This ouabain-induced upregulation may result from a Na^+–K^+ ATPase–Src–MAPK cascade (Na^+ pump signalosome). Other Na^+–K^+ ATPase-mediated cascades involve PI3K, PKB, and PKD kinase.

Arterial constriction is caused by α2-Na^+–K^+ ATPase and NCX1 exchanger. Arterial constriction can be secondarily sustained (but not initiated) by the activation of the RhoA–RoCK pathway and consequent phosphorylation (inactivation) of MLCP phosphatase [238]. Venous smooth muscle contraction is also influenced by Na^+ transport, in particular Na^+–Ca^{2+} exchanger. Pulmonary vessels are also targeted.

Endothelial cells also express α2-Na^+–K^+ ATPase, NCX1, and some TRPC channels. Arterial constriction activates Ca^{2+} signaling in endothelial cells that stimulates NOS3 activity and hence NO production, which limits vasoconstriction. Endothelium-derived hyperpolarizing factor is another antagonizing vasodilatory mechanism. Vasodilation can be primed by acetylcholine and bradykinin. In cultured endothelial cells acutely exposed to nanomolar ouabain, excited bradykinin causes cytosolic Ca^{2+} transients and enhances NO release [238]. However, in endothelium of descending vasa recta microvessels in the renal outer medulla, where

NO can impede Na^+ reabsorption in adjacent nephrons, unlike acute exposure, chronic exposure of ouabain attenuates acetylcholine- and bradykinin-induced cytosolic Ca^{2+} transients as well as NO release, thereby potentiating sympathetic-caused vasoconstriction.

Renin–Angiotensin Pathway

Dysregulation in the renin–angiotensin pathway is often implicated in arterial hypertension. Angiotensin-2 increases blood pressure via the angiotensin AT_1 receptor. Renal AT_1 is a primary determinant of hypertension [240].[108] Angiotensin-2-mediated aldosterone responses are not involved.

Subcutaneous Lymphatic Circuit

The subcutaneous lymphatic circuit collects fluid, proteins, and cells from the interstitial extracellular space and transports them back into the circulation, thereby acting in particular in immunological surveillance. In addition, this fluid buffer blunts blood pressure rise during excessive salt intake. After an increase in dietary salt, the excess sodium is indeed stored under the skin on interstitial proteoglycans. Osmotically inactive Na^+ ions then cause a local hypertonicity sensed by macrophages infiltrating the interstitium of the skin. Consequently, nuclear factor of activated T-cells NFAT5, or tonicity-responsive enhancer-binding protein (tonEBP), activates the VEGFC gene [241]. Macrophages then release vascular endothelial growth factor-C that provokes lymphatic growth, thus raising lymph capillary density and enlarging the local lymph compartment. Lymphatic vessels drain the interstitial compartment, in addition to blood capillaries. Osmotic differences regulate fluid fluxes between the interstitial fluid and lymphatic and blood vessels. The lymphatic compartment can buffer the increased Na^+ level and associated fluid volume. Factor VEGFc also promotes NOS3 expression, thereby vasodilation, during extracellular Na^+ and fluid volume retention.

Modeling of Arterial Pressure Regulation

Short-term regulation (<20 s) of blood circulation, such as those primed by the body's position and ventilatory rhythm changes, by the autonomous nervous system relies on behavior and representation models. In most models, the regulated variable is the arterial pressure.

---

[108]Receptor AT_1 is expressed by epithelialocytes throughout the nephron, in the glomerulus, and renal blood vessels. Once it is activated, it promotes sodium reabsorption by stimulating both sodium–proton antiporter and sodium–potassium ATPase on the apical (luminal) and basolateral plasma membrane, respectively, in the proximal tubule of the nephron. It stimulates epithelial sodium channels in the collecting ducts. Furthermore, activated vascular AT_1 induces vasoconstriction, which subsequently reduces renal blood flow and sodium excretory capacity.

The autonomous nervous system controls activity of the heart, vasculature, and lungs. Organ sensors include baroreceptors sensitive to local, pressure-induced, mural stress field, chemoreceptors that analyze blood gas, thermoreceptors, and mechanoreceptors such as those sensitive to rib-cage stretch. These sensors send information via afferent nerves that are processed and integrated by the central nervous system. This controller then responds by messages transmitted via efferent sympathetic and parasympathetic nerves. Response type depends on the vagosympathetic balance. The body's respiration influences blood flow, especially the venous return, via the intrathoracic pressure.

Spectral analysis exhibits 2 important peaks in short-term regulation: (1) the first one (~0.1 Hz) is associated with sympathetic control of arterial pressure and (2) the second evolves in the range of the ventilatory frequency (0.15–0.40 Hz).

Models are thus defined by: (1) input parameters, such as ventilatory frequency that can be represented by an oscillator with a given frequency and amplitude; (2) blocks or compartments; (3) transfer functions between inputs and outputs (i.e., between ventilation and blood circulation parameters as well as between blood circulation parameters), often assuming linear relationships between physiological signals.

Behavior models do not differentiate various anatomical constituents, but use interaction blocks. For example, the cardiovascular system can be constituted by 3 blocks: a first one related to the effect of cardiac rhythm on arterial pressure, a second to arterial compliance, peripheral resistance, venous return, and nervous control of blood circulation (on both heart and vessels), and a third to nervous response to baroreceptor signals that result from a change in arterial pressure.

Representation models include sensors, nervous command, and receptors of physiological components (heart, blood vessels, lungs, and central nervous system). Each compartment can contain several submodels. Essential components encompass heart, vasculature, baroreceptors, central nervous system, and ventilatory apparatus.

Effects of influence parameters can be yielded by measurments of physiological signals such as the R–R interval. Effects on cardiac function encompass chronotropy (i.e., effect on cardiac frequency), dromotropy (i.e., effect on nodal conductivity), inotropy (i.e., effect on myocardium contractility), bathmotropy (i.e., effect on myocardium exitability), and lusitropy (i.e., effect on myocardium relaxation). These effects can be modeled by differential equations that incorporate sympathetic and parasympathetic cues.

The vascular network is defined by its properties over which the central nervous system operates (resistance, compliance, and inertance).

Baroreceptors record the pressure signal. An augmented activity attenuates the sympathetic command and enhances the parasympathetic signaling.

The central nervous system integrates circulatory and ventilatory signals and reacts via its 2 sympathetic and parasympathetic constituents.

The respiratory apparatus is represented by its sensors that influence the central command.

Table 3.54 Vessel innervation

	Adrenergic receptor	Muscarinic receptor
Artery	$\alpha 1$	M3
Vein	$\alpha 1$, $\alpha 2$, β	M3

3.6.2.8 Other Regulations

The CVS adjusts the blood volume and distribution to provide nutrients to the working muscles involved in postural changes and locomotion. Anticipatory responses are obvious when body motion is associated with any emotional context [242]. However, postural changes are not necessarily preceded by preparatory cardiovascular actions.

Other autonomic regulation mechanisms include thermoregulation and respiration frequency-dependent respiratory sinus arrhythmia (RSA)[109] which, together with baroreflex regulation, affect cardiac frequency. Respiratory sinus arrhythmia depends on respiration frequency and amplitude. It can require several mechanisms. Pressure variations experienced by thoracic blood vessels during respiration, hence changes in afterload, can influence mechanoreceptors of vasculature walls and mediate a baroreflex. Coupling between the respiratory and circulatory autonomic nervous centers can also be involved. The respiratory sinus arrhythmia can also contribute to respiratory arterial pressure fluctuations [243]. Thermoregulation operates at very low frequencies (below 40 mHz), whereas baroreflex regulation and RSA are low- (40–150 mHz) and high-frequency (150–400 mHz) components, respectively.

3.6.3 Adrenergic and Cholinergic Receptors

The action of autonomic nerves are mediated by the release of neurotransmitters that bind to specific receptors in the heart and blood vessels (Table 3.54). These receptors are coupled with signal transduction pathways. In the heart, catecholamines, such as adrenaline (Ad; or epinephrine) or noradrenaline (NAd; or norepinephrine), released by sympathetic nerves preferentially bind to $\beta 1$-adrenergic receptors, causing I+, C+, and D+ effects (Table 3.55). Gs subunit of G-proteins are activated and, in turn, stimulate adenylate cyclase.

Stimulated $\beta 2$-adrenergic receptor that has similar cardiac effects have an increasing role in heart failure as $\beta 1$-adrenergic receptor expression is downregulated. Noradrenaline can also bind to $\alpha 1$-adrenergic receptor on cardiomyocytes, increasing contractility.

[109] Respiratory sinus arrhythmia is the variation in heart rate occurring simultaneously with respiration. On ECG traces, it induces fluctuations of the R–R interval series.

Table 3.55 Types and responses of adrenergic receptors. Adrenaline is released by the adrenal medulla, and noradrenaline is secreted by the nerves and adrenal medulla (Source: [119])

	α1	α2	β1	β2
Ligand	Ad			
Second messenger	IP3		cAMP	
Vessel	Vasoconstriction		Vasodilation (β1 < β2)	
Heart			C+, D+, I+, B+	
Lipolysis		Decreased	Increased (β1 < β2)	
Glycogenolysis	Increased (β1 < β2)			
Insulin release		Inhibition	Stimulation (β1 < β2)	
Synapse		Inhibition of ACh and NAd release		
Micellaneous		TC aggregation	Stimulation of renin release	

In blood vessels, noradrenaline preferentially binds to α1-adrenergic receptor inducing vasoconstriction. Response to activated α-adrenergic receptors is linked to intracellular calcium stores. Similar responses occur when noradrenaline binds to postjunctional α2-adrenergic receptors located on some blood vessels.

Noradrenaline can also bind to postjunctional β2-adrenergic receptors that causes vasodilation. β-Adrenergic receptors are linked to adenylate cyclase. Relaxation can be mediated by cAMP-dependent phosphorylation (inactivation) of myosin light chain kinase.

Noradrenaline can also regulate its own release by acting on prejunctional α2- (inhibition) and β2-adrenergic (stimulation) receptors. Circulating adrenaline binds to β2adrenergic receptors, causing vasodilation in some organs.

The stimulation of β-adrenergic receptors increases cytosolic Ca^{++} concentration and, hence, cardiac contraction, whereas excessive activation of β-adrenergic receptors induces myocardial hypertrophy and dysfunction in the case of infarction.

Ca^{++}–calmodulin-dependent protein kinase CamK2 is an effector og the β-adrenoceptor signaling cascade associated with maladaptive myocardial remodeling [244]. Inhibition of CamK2 may hamper such a pathological remodeling.

The myocardium also contains muscarinic receptors associated with adenylate cyclase and a K^+ channel in the sarcolemma. Acetylcholine (ACh) released by parasympathetic nerves binds to muscarinic receptors. Acetylcholine reduces cytosolic cAMP concentration and increases K^+ current. It thus produces I–, C–, and D– effects.

In blood vessels, muscarinic receptors are coupled with the formation of nitric oxide that causes vasodilation. Activation of prejunctional muscarinic receptors inhibits NAd release. Therefore, vagal stimulation overrides sympathetic stimulation in the heart.

Arteries in skeletal muscle are innervated by sympathetic nerves that release ACh to induce hyperemia, particularly at the onset of exercise.

The vasomotor tone can be locally regulated. It is also controlled by the autonomic nervous system (ANS), which acts by the nervous[110] and humoral path.[111] Adrenoceptors stimulated by NAd localize to arterial, arteriolar, and venous smooth myocytes. They induce slow depolarizations that last for several seconds. Agents ATP and NAd are costored in synaptic vesicles in sympathetic nerves. When they are coreleased, they act postjunctionally for contraction of the vascular smooth muscle.

Nucleotide receptors activated by ATP generate fast depolarizations. Conversely, muscarinic receptors in arteries and veins are inhibitory. Five distinct, but related muscarinic receptors have been identified [245]. The G-protein-coupled M_3 receptor localizes to the surface of the endotheliocyte.

Heart failure is characterized by cardiac overstimulation of sympathetic nerves for compensation of decreased cardiac function associated with an increased blood concentration of catecholamines. During heart failure, α2-adrenergic receptor in chromaffin cells of the adrenal medulla are disturbed by increased activity of G-protein-coupled receptor kinase GRK2, contributing to an elevated blood level of catecholamines [246]. Normally, α2-adrenoceptors generate via the Gi/o protein an autocrine feedback inhibition of catecholamine secretion induced by activation of cholinergic nicotinic receptors. During heart failure, α2-adrenoceptors of the adrenal gland lose their inhibitory function on the sympathetic system. Inhibition of GRK2 in adrenal glands during heart failure restores the inhibition of catecholamine release by activating α2-adrenoceptors. In the heart, GRK2 expression is also upregulated, thereby attenuating the ventricular function. In the cardiomyocyte, GRK2 phosphorylates and desensitizes β-adrenoceptors, thus lowering catecholamine-induced signaling via the Gs–ACase–PKA pathway, causing reduced contractility.

3.6.4 Circulation Sensors

Mechano- (Sect. 2.6.6) and chemosensor cells (Sect. 2.6.8) in walls of the heart and blood vessels, especially in diverse cardiac regions as well as in coronary, large intrathoracic (particularly inner aortic arch and bases of both venae cavae) and cervical blood vessels, continuously record the hemodynamic regime, transduce signals, and feed the information to corresponding afferent neurons.

Chemosensor, or *chemoreceptor* cells, transduce a chemical signal into an action potential. Vascular chemoreceptors localize to the carotid glomus at the common carotid artery bifurcation and aortic body of the aortic arch (i.e., close to baroreceptor cell sites such as the carotid sinus).

[110]Nerve fibers in adventitia act by electrochemical stimulation at neuromuscular junctions and biochemical processes (release of neurotransmitters) preferentially at external layers of the media.

[111]Flowing vasoactive hormones act after transmural migration up to internal layers of the media.

Table 3.56 Baroreceptors and chemoreceptors

Receptor	Signal	Cardiac effect	Vessel effect
Baroreceptor	p	$p \searrow \Rightarrow \Sigma c \oplus$, $p\Sigma c \ominus$	$p \searrow \Rightarrow \Sigma c \oplus$
(threshold ~8 kPa)	Wall deformation	(I+, C+)	(vasoconstriction)
Chemoreceptor	p_{O_2}	$p_{O_2} \searrow \Rightarrow p\Sigma c \oplus$	

The chemoreceptory carotid body contains clusters of excitable glomus cells. The main stimulus for glomus cells is diminished arterial partial pressure of oxygen, whereas chemoreceptors in the central nervous system are highly sensitive to partial pressure of carbon dioxide. Reduced O_2 and increased CO_2 and H^+ concentrations stimulate chemoreceptors (Table 3.56).

Sensing is also done by *baroceptor* cells in the high-pressure circulation (aorta and carotid sinuses) and *voloreceptors* in the low-pressure circulation (pulmonary artery, atria, ventricles, and venae cavae).

Sympathetic nerve activity is inhibited by activated lung stretch receptors and carotid and aortic baroreceptors. Baroreceptors enable quick control by the central nervous system to adjust the arterial blood pressure and maintain it at physiological values (baroreflex negative feedback).

3.6.5 Short-Term Control of the Circulation

The control of the circulation here deals with overall circulation (rather than local controls of blood flow in the skeletal muscles and head). The time scale of the short-term regulation of the circulation is $\mathcal{O}[1\,\text{s}]$ to $\mathcal{O}[1\,\text{mn}]$, whereas for the long term, it is $\mathcal{O}[1\,\text{h}]$ to $\mathcal{O}[1\,\text{d}]$.

The short-term control includes several reflexes, which involve the following inputs and outputs: arterial pressure, cardiac frequency, stroke volume, and peripheral resistance and compliance. So the autonomic nervous system can receive complementary information from the circulation and has several processing routes.

The importance of a given feedback loop with respect to the various other reflexes can become primary in certain circumstances or secondary in others. Nervous control of the circulation must therefore take into account the whole set of involved factors.

The control of the peripheral resistance and compliance is slower than the command of the heartbeat period and stroke volume. Because the cardiac frequency can be noninvasively measured using the R–R interval of ECG records, most investigations consider the relation between arterial pressure and cardiac frequency.

The short-term control of circulation requires: (1) many receptor varieties, (2) nervous signaling, and (3) corresponding feedback loops. Among several types of mechanosensitive receptors of blood circulation, baroreceptors act as starting elements in the reflex regulation of arterial pressure.

Both arterial pressure and R–R interval vary from one heartbeat to another. Arterial pressure and R–R interval are considered baroreflex input and output, respectively. The effect of arterial pressure on R–R intervals and reciprocally of the R–R interval on arterial pressure (closed-loop interaction between the 2 physiological signals) is described by 2 transfer functions [247].

The power spectrum of the R–R interval serves as a measure of its neural modulation. The relationship between arterial pressure and R–R interval via the baroreflex is exhibited by low- (LF) and high- (HF) frequency spectral components. Normalized indices of LF and HF components of spectral analysis of R–R interval variability are used, dividing the quantities by the difference between power variance and power of the very-low-frequency (VLF) component. Low-frequency oscillations in the R–R interval not only depend on the baroreflex control loop, but also on central rhythmic modulation of neural autonomic activity.

Mayer waves, low-frequency arterial blood pressure oscillations, are observed in response to decreased central blood volume, seen in upright posture [248]. The stability of the arterial baroreflex feedback is then reduced.

During exercise, the mesencephalic locomotor region inhibits the baroreceptor reflex by activating interneurons of the nucleus tractus solitarius (NTS), which inhibit NTS cells receiving baroreceptor input [249].

3.6.6 Response to Stress

The reestablishment of homeostasis after disturbances involves the coordinated activation and control of the autonomic nervous and neuroendocrine systems. The autonomic nervous system provides the most immediate response to stress (potential or actual threat) exposure ("fight or flight" response). The main circuits involved in stress response include the *sympatho–adrenomedullary* and *hypothalamo–pituitary–adrenocortical* axes [250].

Different types of stressors trigger distinct responses. Physical stressors (e.g., blood loss and trauma) rapidly implicate the brainstem and hypothalamus, whereas psychological stressors (e.g., deadlines) primarily engage brain regions that subserve emotion (amygdala and prefrontal cortex), learning and memory (hippocampus), and decision making (prefrontal cortex) [251].

Many molecule types influence duration and magnitude of the response to stress. Various substances released in response to stress influence diverse neuronal circuits. Stress mediators act on specific neuronal populations to adapt to the changing environment [251]. Stress mediators, neurotransmitters (e.g., noradrenaline, dopamine, and serotonin), neuropeptides (e.g., corticotropin-releasing hormone [CRH], dynorphin, galanin, ghrelin, neuropeptide-Y, orexin, oxytocin, substance-P, urocortin, and vasopressin), and steroids (e.g., cortisol), have overlapping origins.

Table 3.57 Examples of anatomical compartments of the central nervous system (brain and spinal cord). Both psychogenic and systemic stimuli are processed in limbic and associated forebrain structures (amygdala, hippocampus, and prefrontal cortex). These regions receive information from subcortical and cortical regions involved in sensory processing (e.g., olfactory nuclei and piriform and insular cortex), memory (medial septum and entorhinal and cingulate cortex), and attention and arousal (e.g., locus coeruleus and raphe nuclei). The limbic system devoted in particular to behavior and long-term memory is mainly composed of the amygdala, hippocampus, parahippocampal and cingulate gyri, hypothalamus, thalamus, fornix, and other parts of the limbic cortex

Compartment	Components
Forebrain	
Cerebrum	Neocortex, basal ganglia, amygdala, hippocampus, rhinencephalon (olfactory nuclei, piriform cortex, etc.)
Interbrain	Thalamus, epithalamus, hypothalamus, subthalamus, pretectum, epiphysis (pineal gland), hypophysis (pituitary gland)
Brainstem	
Hindbrain	Cerebellum, pons, medulla
Midbrain	Cerebral peduncles, tectum, pretectum
Spinal cord	
Sensory afferent pathways	Dorsal column-medial lemniscus tract, spinocerebellar tract, anterolateral circuit
Motor efferent pathways	Pyramidal and extrapyramidal tracts

Response to stress also generates interoceptive cues, such as blood volume and osmolarity. Entailed nervous circuits are overlapping networks of the *limbic forebrain*,[112] *hypothalamus*, and *brainstem* (Table 3.57) [250].

Neurons of the medulla and spinal cord inform autonomic integrative sites in the interbrain (e.g., dorsomedial hypothalamus), hindbrain (e.g., raphe pallidus and lateral parabrachial nucleus), and midbrain. These structures modulate the autonomic response according to messages from the hypothalamus and limbic forebrain.

3.6.6.1 Sympatho–Adrenomedullary Axis

The sympathoadrenomedullary component, that is, preganglionic sympathetic neurons of the intermediolateral column (T1–L2) of the thoracolumbar spinal cord that project to pre- or paravertebral ganglia innervating chromaffin cells of the adrenal medulla, rapidly increase cardiac frequency and contractibility and blood pressure following vasoconstriction of some territories. Concentrations of circulating adrenaline and noradrenaline primarily liberated from adrenal medulla and sympathetic nerves, respectively, rise.

[112] Latin limbus: border, edge, fringe, hem, selvage.

3.6.6.2 Parasympathetic Activity

Excitation of the autonomic nervous system quickly wanes following activation of the parasympathetic reflex mediated by the nucleus ambiguus and the dorsal motor nucleus of the vagus nerve, possibly via inputs from the nucleus of the solitary tract. Activated craniosacral preganglionic nuclei activates postganglionic nuclei located in or near innervated organs. Coordinated action of the parasympathetic system thus controls the duration of autonomic responses.

Parasympathetic outflow uses the circuit from the dorsal motor nucleus of the vagus nerve and nucleus ambiguus [250]. It is influenced by the parasympathetic prelimbic cortex and mixed paraventricular nucleus. The parasympathetic anterior *bed nucleus of the stria terminalis* also participates via relays in the paraventricular nucleus and nucleus of the solitary tract. However, it has numerous subregions that contribute differently in stress signal integration [250]. Its anterolateral domain contains CRH+ neurons that project to the paraventricular nucleus. On the other hand, its posteromedial domain that receives predominantly gabaergic inhibitory axons from the paraventricular nucleus prevents the action of the hypothalamo–pituitary–adrenocortical axis.

The nucleus of the solitary tract receives inputs from neurons in the sympathetic infralimbic cortex and central amygdala as well as the paraventricular nucleus. Gabaergic axons from the central amygdala contribute to regulate responses to systemic stress.

3.6.6.3 Hypothalamus

The *paraventricular nucleus* of the hypothalamus is a major integrator of stress signals. It houses distinct populations of neurons that project to the median eminence and to autonomic centers in the brainstem and spinal cord (parabrachial nucleus in the pons, dorsal motor nucleus of the vagus nerve and nucleus of the solitary tract in the medulla, and intermediolateral cell column in the spinal cord) [250].

Some nuclei of the autonomic nervous system possess both sympathetic and parasympathetic synapses. In other words, parasympathetic and sympathetic neurons are intermingled in these nuclei. In particular, the paraventricular nucleus is connected to both sympathetic and parasympathetic nuclei, such as the mixed nucleus of the solitary tract,[113] parasympathetic dorsal motor nucleus of the vagus nerve, and sympathetic intermediolateral column of the spinal cord, locus ceruleus, and ventrolateral medulla [250]. The paraventricular nucleus receives inputs from the dorsal raphe, tuberomammillary nucleus, supramammillary nucleus, and spinal cord, among others.

[113]The hypothalamic paraventricular nucleus receives direct noradrenergic, adrenergic, and peptidergic innervation from the nucleus of the solitary tract [250].

The paraventricular nucleus is heavily innervated by gabaergic inhibitory neurons, among which some reside in the periparaventricular nucleus region. The latter is targeted by several limbic regions [250]. The *medial preoptic hypothalamus*, the action of which is modulated by the medial amygdala and hippocampus, provides gabaergic innervation to the paraventricular nucleus.

The *suprachiasmatic nucleus*, the coordinator of circadian rhythms, has few direct projections to the paraventricular nucleus, but heavily innervates its surrounding region.

Some ascending brainstem and possibly spinal circuits project to the *parvocellular group* of the paraventricular nucleus. The dorsal region of the medial parvocellular paraventricular nucleus is inhibited by numerous hypothalamic regions, such as the medial preoptic area, ventrolateral component of the dorsomedial hypothalamus, and local neurons in the periparaventricular nucleus region. The parvocellular group also receives serotonergic innervation from the median raphe nuclei of the midbrain. Serotonin of serotonergic fibers targets serotonin-2A receptors on neurons of the paraventricular nucleus and its surrounding. It then activates the hypothalamo–pituitary–adrenal axis.

The rostral ventrolateral medulla, locus ceruleus, and paraventricular nucleus innervate the intermediolateral column. The mixed *dorsomedial hypothalamus* is linked to the paraventricular nucleus. The dorsomedial hypothalamus and the arcuate nucleus excite the hypothalamus. The former regulates autonomic response to psychogenic stimuli. Its stimulation (but not hypovolemic stimulion) indeed increases cardiac frequency, blood pressure, and the hypothalamo–pituitary–adrenal axis. In fact, the dorsomedial hypothalamu contains segregated neuronal populations that activate or inhibit the hypothalamo–pituitary–adrenal axis.

3.6.6.4 Circumventricular Organs – Signal Integration

Components of the lamina terminalis of the forebrain (median preoptic nucleus, subfornical organ, and organum vasculosum of the lamina terminalis) respond to perturbations in fluid and electrolyte balance. They are involved in the regulation of blood pressure by angiotensin-2 and AT_1 receptor. The subfornical organ has direct angiotensin-2-containing axons to the medial parvocellular paraventricular nucleus, where it stimulates the hypothalamo–pituitary–adrenal axis. The lamina terminalis also projects to other hypothalamic structures (anteroventral preoptic nucleus, dorsomedial hypothalamus, and preautonomic paraventricular nucleus.

3.6.6.5 Hypothalamo–Pituitary–Adrenal Axis

The brainstem rapidly excites not only the autonomic nervous system, but also the hypothalamo–pituitary–adrenal axis, as its neurons project to hypophysiotrophic neurons and to preganglionic autonomic neurons such as sympathetic neurons of the medulla (e.g., rostral ventrolateral medulla) and intermediolateral cell column of the spinal cord.

The activated hypothalamo–pituitary–adrenocortical axis elevates the circulating glucocorticoid level with a peak about 10 mn after stress onset. Activated hypophysiotrophic neurons of the hypothalamic paraventricular nucleus secrete hormones, such as corticotropin-releasing hormone and vasopressin, into the portal circulation of the median eminence. These hormones act on the anterior hypophysis, thereby triggering release of adrenocorticotropic hormone, which in turn acts on the inner adrenal cortex (zona fasciculata) to initiate secretion of glucocorticoid hormones (cortisol). Circulating glucocorticoids then foster the mobilization of stored energy and potentiate numerous sympathetic effects such as vasoconstriction. Moreover, the adrenal cortex innervated by the sympathetic nervous system regulates corticosteroid release.

Once they are released in the blood circulation, glucocorticoids trigger within minutes a negative feedback on the hypothalamo–pituitary–adrenal axis. Glucocorticoid receptors in the forebrain are sufficient to initiate the termination of the stress response [250]. However, the importance of limbic glucocorticoid signaling regulates response of the hypothalamo–pituitary–adrenal axis to psychogenic, but not systemic stress. On the other hand, glucocorticoids bind to nuclear receptors, relatively low-affinity glucocorticoid receptor (NR3c1), and, in some tissues, high-affinity mineralocorticoid receptor (NR3c2), which are ligand-activated transcription factors, to prime long-latency and -acting gene transcription.

3.6.6.6 Stress Mediators

Neurons in the basolateral amygdala, which are excited by noradrenaline, dopamine, CRH, and corticosteroids, support sympathetic and neuroendocrine activation via their output to the central amygdala, fine-tuning of arousal via the locus ceruleus, and processing of emotional and contextual aspects of stress via the prefrontal cortex and hippocampus [251].

Neurons of the hippocampus express receptors for monoamines, neuropeptides, corticosteroids, and opioids. They enable fine-tuning of neuronal firing in response to signaling [251].

Three main classes of stress mediators that are characterized by their spatial niches and temporal windows contribute to the nervous response to stress [251]. Two temporal waves of rapid and delayed effects of stress mediators occur. The first includes dopamine, noradrenaline, serotonin, and corticotropin-releasing hormone. These stress mediators promote vigilance, appraisal of the situation, and the choice of an optimal strategy to face the challenge. Furthermore, monoamines and neuropeptides can regulate transcription factors, such as CREB and AP1 (second wave). A sustained stress response results from gene expression (classical second wave) primed by corticosteroids acting via glucocorticoid receptors that are transcription factors. Corticosteroids can also initiate rapid change in the brain functioning via nongenomic signaling, especially in the hypothalamus and hippocampus [251].

Table 3.58 Examples of stress-activated neuropeptides in the central nervous system (Source: [251]; ACTH: adrenocorticotropic hormone; BNST: bed nucleus of the stria terminalis; CRH: corticotropin-releasing hormone; CRHR: CRH receptor; EWN: nonganglionic Edinger–Westphal nucleus; Ox: orexin receptor; V_1: vasopressin receptor-1)

Type	Location	Receptor	Function
CRH	Amygdala	CRHR1/2	Emotional memory, anxiety
	BNST	CRHR1/2	Stress-related anxiety
	Hippocampus	CRHR1	Long-term potentiation, stress-related learning and memory
	Hypothalamus	CRHR	Neuromodulation
	Locus ceruleus	CRHR1	Interaction with noradrenergic circuits
Dynorphin	Hippocampus	Opioid receptors	Stress-related dysphoria
Orexin	Hypothalamus	$Ox_{1/2}$	Stress-related energy
Urocortin (Ucn1–Ucn3)	EWN	CRHR1/2	Stress adaptation
Vasopressin	Hypothalamus, BNST, amygdala	$V_{1A/B}$	Emotional memory ACTH release from hypophysis

Monoamines (dopamine, noradrenaline, and serotonin) are released within minutes after the stress onset by specific neuronal populations. Their liberation is triggered by nervous circuits devoted to stress evaluation or indirectly by an activated sympathetic nervous system. They generally act via G-protein-coupled receptors and prime fact action, supporting behavioral strategies.

Neuropeptides are released by stress in specific neuronal populations and activate multiple receptor types (Table 3.58). They mediate the stress response or antagonize it under specific contexts [251].

Threat triggers a large burst of corticosteroids, supporting central and peripheral aspects of the stress response via gluco- and mineralocorticoid receptors. The latter has a 10-fold higher affinity. It abounds in neurons of the hippocampus and lateral septum. It is moderately expressed in the amygdala, paraventricular nucleus, and locus coeruleus. The distribution of the mineralocorticoid receptor overlaps that of corticotropin-releasing hormone CRH_1 receptor. On the other hand, the glucocorticoid receptor, which is ubiquitous in the brain, abounds in the hippocampus, lateral septum, and paraventricular nucleus. In the hippocampus, the activated mineralocorticoid receptor maintains stress signaling, whereas the activated glucocorticoid receptor has suppressive and normalizing effects, as it yields a negative feedback on behavioral aspects of the stress response [251]. However, in the basolateral amygdala, the latter has excitatory effects.

Different types of stress mediators interact. In particular, noradrenaline, corticotropin-releasing hormone, opioids, and corticosterone cooperate in the locus coeruleus and its projection regions, such as the hippocampus and amygdala to optimize the stress response [251].

3.6.6.7 Serotonin

Serotonin, or 5-hydroxytryptamine, intervenes in the control of blood pressure, as it targets its receptor in the heart, vasculature (vascular serotonergic system [uptake, synthesis, and metabolism]), adrenal gland, kidney, blood cells, and central nervous system (serotonergic nervous system). In many cell types, serotonin is taken up by the serotonin transporter (SerT) and other amine transporters. It is involved in orthostatic hypotension[114] and hypertension [252]. In blood, serotonin exists as a free circulating form (65–250 ng/ml) and stored form in platelets (but not synthesized in platelets; 3.4–3.5 nmol/10^9 platelets). The platelet actually serves as a serotonin store.

In fact, cardiovascular responses to serotonin comprise brady- or tachycardia, hypo- or hypertension, and vasodilatation or -constriction. This monoamine actually targets different receptor types in various target cells, such as that of the central nervous system, autonomic ganglia, and postganglionic nerve endings, as well as vascular smooth muscle and endothelial and cardiac cells.

Most serotonin receptors are G-protein-coupled proteins, except $5HT_3$ ion channels. All serotonin receptors ($5HT_{1A}$–$5HT_{1F}$, $5HT_{2A}$–$5HT_{2C}$, $5HT_3$–$5HT_4$, $5HT_{5A}$–$5HT_{5B}$, and $5HT_6$–$5HT_7$; Tables 3.59 and 3.60), except $5HT_6$, are involved in the regulation of blood flow. Central $5HT_{1A}$, $5HT_3$, and $5HT_7$ receptors participate in the control of cardiovascular reflexes, especially the parasympathetic (vagal) drive [253].

Serotonin influences blood pressure, as it acts on central neural networks that control the magnitude of the sympathetic tone to cardiovascular cells, hence blood pressure. Major neural circuits that control blood pressure and cardiac frequency include (Table 3.61): (1) sympathetic ganglion cells innervating the heart, blood vessels, adrenal medulla, and kidney; (2) sympathetic preganglionic neurons in the intermediolateral nucleus of the thoracolumbar spinal cord; (3) spinal interneurons; (4) glutamatergic supraspinal populations of sympathetic premotor neurons, that is, cardiovascular regulatory neurons in the rostral ventrolateral medulla connected to pontine, hypothalamic, and limbic areas, as well as cardiovascular baro- and chemoreceptor reflex circuits.

The baroreceptor reflex depends on the nucleus of the solitary tract that excites neurons in the caudal ventrolateral medulla, which, in turn, inhibits sympathoexcitatory neurons of the rostral ventrolateral medulla. Reduced baroreceptor activity launches the sympathetic activation that increases cardiac frequency and contractility, venous stiffness, and vasoconstriction to compensate for the decrease in blood pressure that occurs upon standing when gravity pulls blood toward the lower extremities. Baroreceptor sensory neurons in the nucleus of the solitary tract provide the parasympathetic premotor drive to the cardiac vagal preganglionic neurons in the medulla that drive cardiac vagal nerve activity to lower the heart activity.

[114]Postural hypotension is defined as at least a 2.66-kPa decrease in systolic blood pressure or 1.33-kPa decrease in diastolic blood pressure upon standing.

Table 3.59 Serotonin receptors in the cardiovascular apparatus as well as central (CNS) and peripheral (PNS) nervous system (**Part 1**; Source: [252]). Sympathetic nerve terminals release sympathetic neurotransmitters (noradrenaline, neuropeptide-Y, and ATP) to contract vascular smooth myocytes. Serotonin binds to the inhibitory $5HT_{1A}$ receptor on sympathetic premotor neurons and stimulatory $5HT_{2A}$ receptor on sympathetic preganglionic neurons. Stimulation of the central $5HT_{1A}$ receptor decreases vasoconstrictory sympathetic nerve activity and increases cardiac vagal nerve activity. Receptors $5HT_{1A}$, $5HT_{1B}$, and $5HT_2$ to $5HT_4$ exist in the nucleus of the solitary tract (NTS), the site of termination of sensory neurons (from baro- and chemoreceptors and other types of cardiopulmonary receptors)

Subtype	Location	Function
$5HT_{1A}$	CNS	Lowers or raises blood pressure
	Kidney	
$5HT_{1B}$	SMC	Vasoconstriction
	CNS	Lowers or raises blood pressure (CNS)
	PNS	Inhibition of NAd release in sympathetic nerve terminals and transmission in sympathetic ganglia
	Kidney	
$5HT_{1D}$	SMC	Vasoconstriction
	PNS	Inhibition of NAd release in sympathetic terminals
	Kidney	
$5HT_{1E}$		
$5HT_{1F}$	SMC	Vasoconstriction
	CNS	Trigeminal nerve
	Kidney	

3.6.7 Delayed Control of the Circulation

Delayed mechanisms involve circulating hormones and regulatory peptides and gases, such as catecholamines, angiotensin, endothelin, prostaglandins, and nitric oxide, among others.

3.6.7.1 Nephron

Urination removes water, certain electrolytes, and certain wastes from the body.[115] Urine is produced in the nephrons (Fig. 3.13) by 3 regulated processes: filtration, reabsorption, and secretion.

Renal blood is first filtered from the glomerulus, a capillary ball formed from an afferent arteriole and leading to a narrower efferent arteriole, by the *Bowman capsule* (glomerular filtration).

[115]Nitrogenous wastes are excreted as ammonia, urea, or uric acid.

Table 3.60 Serotonin receptors in the cardiovascular apparatus as well as central (CNS) and peripheral (PNS) nervous system (**Part 2**; Source: [252]; CTGF: connective tissue growth factor; TGF: transforming growth factor)

Subtype	Location	Function
$5HT_{2A}$	Platelet	Platelet aggregation
	SMC	Vasoconstriction
	Heart	Tachycardia, contraction
	CNS	Sympathetic outflow
	Kidney	ERK1/2 activity (mesangial cells), stimulation of CTGF and TGF-β
		Stimulation of podocyte VEGF expression
	Adrenal	Adrenaline release
$5HT_{2B}$	EC, SMC	Vasodilation and -constriction
	Cardiac valves	Growth (mitogen)
	Kidney	
$5HT_{2C}$	Kidney	
$5HT_3$	Vagus, ganglia	Bradycardia
	NTS	
$5HT_4$	CMC	Positive chronotropy, inotropy, and lusitropy
	Adrenal	Stimulation of cortisol
	NTS	
$5HT_6$		
$5HT_7$	CMC,	Positive inotropy and lusitropy
	SMC	Vasodilation
	Adrenal	Aldosterone production

Table 3.61 Central autonomic circuits targeted by serotonin. Excitatory (+), inhibitory (−), and mixed (±) connections between involved autonomic nuclei (Source: [252]; BAT: brown adipose tissue; BR: baroreceptor; CVLM: caudal ventrolateral medulla; DMH: dorsomedial hypothalamus; IML: intermediolateral nucleus (thoracolumbar spinal cord); LTF: lateral tegmental field; NA: nucleus ambiguus; NTS: nucleus of the solitary tract; RVLM: rostral ventrolateral medulla)

		Inputs	Outputs
Neuron cluster	Presence of 5HTRs	(Connection type)	
DHM	N		Raphe (+)
Raphe	Y	DHM (+)	IML, NTS (+, +)
IML	Y	Raphe (+)	Cutaneous vessels, BAT, heart
		RVLM (+)	Visceral and muscular vessels, heart
NTS	Y	BR (+)	CVLM, LTF, NA (+, ±, +)
CVLM	Y	NTS (+)	RVLM (−)
RVLM	Y	CVLM, LTF (−, +)	IML (−)
LTF	Y	NTS (±)	RVLM (+)
NA	Y	NTS (+)	Heart

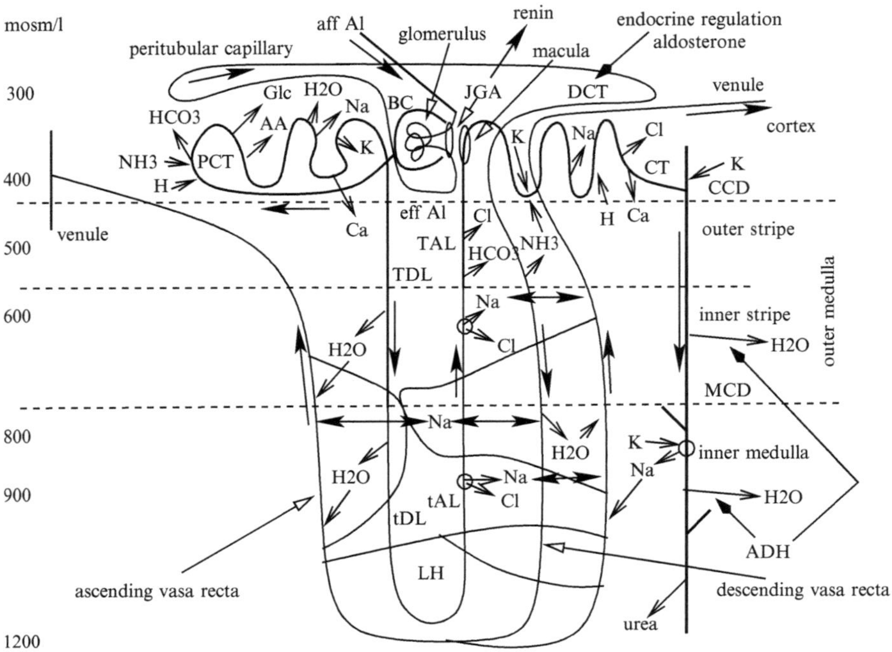

Fig. 3.13 The juxtamedullar nephron with a long Henle loop (Source: [254]). The glomerulus with its afferent arteriole (aff Al) and efferent arteriole (eff Al), and with the Bowman capsule (BC). Water, electrolytes (Na, K, Cl, etc.), glucose (Glc), amino acids (AA), urea, and other filtered chemicals form the glomerular filtrate. The tubule is composed of several segments: (1) proximal convoluted tubule (PCT); (2) loop of Henle (LH), with its descending (thick [TDL] in the outer medulla and thin [tDL] in the inner medulla) and ascending (thin [tAL] in the inner medulla and the inner stripe of the outer medulla, thick [TAL] in the outer stripe of the outer medulla) limbs; (3) distal convoluted tubule (DCT); (4) connecting tubule (CT); and (5) cortical (CCD) and medullary (MCD) segments of the collecting duct. The juxtaglomerular apparatus (JGA) secretes renin. The macula densa regulates arteriolar resistance. Interstitial osmolarity of 300 mosm/l is observed at the level of cortical nephrons with short loops and ordinary peritubular capillaries. At the hairpin curve, the interstitium osmolarity is equal to 1200 mosm/l or more

The glomerular filter consists of 3 layers: (1) the fenestrated endothelium, (2) glomerular basement membrane, and (3) interdigitated podocyte extensions, which completely enwrap the glomerular capillaries.

The filtration barrier restricts the passage of molecules according to their size, shape, and charge. Water, electrolytes, glucose, amino acids, wastes (urea), and other filtered chemical species form the glomerular filtrate. The juxtaglomerular apparatus of the arteriole walls contains granular cells that secrete renin.

Molecules (water, glucose, amino acids, ions, and other nutrients) are reabsorbed from the renal tubules back into the peritubular capillaries, which drain into a venule.

The tubule is composed of several segments, (1) the *proximal convoluted tubule*; (2) *loop of Henle*, with its descending limb that consists of a thick (in the outer medulla) and thin (in the inner medulla) segments and ascending limb constituted of a thin (in the inner medulla and inner stripe of the outer medulla) and thick (in the outer stripe of the outer medulla) segments; (3) *distal convoluted tubule*; (4) *connecting tubule*; and (5) *cortical* and *medullary* segments of the *collecting duct*.

About 60% to 70% of salts and water are reabsorbed in the proximal convoluted tubule. The descending limb of the loop of Henle is permeable to water, but impermeable to salts. The ascending limb of the loop of Henle is impermeable to water; the active pumping of sodium concentrate salts in the hypertonic interstitium, with Na^+–Cl^- cotransporters in particular. The distal convoluted tubule secretes hydrogen and ammonium.

Molecules (hydrogen and potassium ions, urea, and ammonia) are secreted from peritubular capillaries into the distal and collecting tubules via either active transport (Na^+–K^+ pumps) or diffusion. The collecting duct is permeable to water owing to antidiuretic hormone and urea in its downstream segment.

The distal nephron is the site of endocrine regulation. Mineralocorticoid receptors are located in the Henle loop, distal convoluted tubules, connecting tubules, and collecting tubules and ducts [255, 256].

The *macula densa* is the specialized area of the downstream segment of the thick ascending limb and the upstream segment of the distal tubule in the neighborhood of the afferent and efferent arterioles of its own glomerulus. The macula densa regulates arteriolar resistance.

Solute and water transport between ascending and descending urine and blood pipes are not only governed by osmotic forces and interstitial hydrostatic pressure, but also by active processes and membrane permeability. An osmotic gradient is caused by Na^+ and Cl^- transport out from the ascending tubule to the interstitium. An interstitial osmolarity of 300 mosm/l is observed at the level of cortical nephrons with short loops and ordinary peritubular capillaries. Interstitial hyperosmolarity pulls water from descending vasa recta and descending tubules. In the medulla, the interstitial sodium level balances sodium concentrations in both vasa recta. Water from the descending tubule and descending vasa recta enters in the interstitium and then the ascending vasa recta. At the hairpin curve, interstitium osmolarity reaches 1200 mosm/l or more (urea leaving the collecting duct intensifies the osmotic gradient) [254].

In the nephron, sodium uptake is associated either by chloride reabsorption or potassium secretion. Simultaneous sodium and chloride ion reabsorption by sodium–chloride cotransporters is an electroneutral process. Electrogenic sodium reabsorption via the sodium channel generates a negative charge in the urinary lumen and subsequent secretion of potassium via potassium channels, such as the renal outer medullary potassium channel (ATP-dependent $K_{IR}1.1$). Sodium carriers are found in the distal convoluted tubule. When sodium–chloride cotransporter activity is augmented, the sodium channel function decays and vice versa.

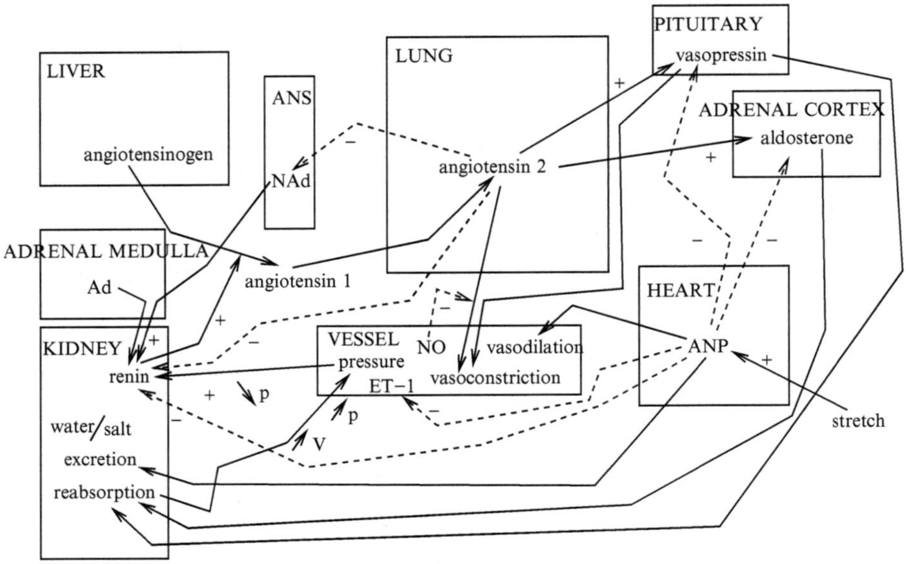

Fig. 3.14 Atrial natriuretic peptide (ANP) and the renin–angiotensin axis. Angiotensin-2 induces vasoconstriction associated with changes in renal hemodynamics, vasculature remodeling, and release of aldosterone by the adrenal cortex and vasopressin by the pituitary gland. Vasopressin increases water retention by the kidneys. Its effect on vasomotor tone at physiological concentrations is negligible. ANP generates vasoconstriction. In the kidneys, it inhibits renin release and decreases sodium reabsorption. In the adrenal cortex, it inhibits aldosterone release

Table 3.62 Renin–angiotensin–aldosterone axis and cardiac natriuretic peptides (Source: [257])

	Angiotensin-2	ANP
Artery	Vasoconstriction	Vasodilation
Myocardium	I+, C+	
Adrenal medulla	Aldosterone release	Aldosterone release inhibition
Adrenal cortex	Cathecolamine release	
Kidney	Glomerulus filtration reduction	Glomerulus filtration increase
	Water and salts reabsorption	Water and salt excretion
	Renin release	Renin release inhibition
Pituitary (hypophysis)	ADH release	ADH release inhibition
	ACTH release	
	Prolactin release	
Sympathetic	NAd release	

3.6.7.2 Renin–Angiotensin–Aldosterone System

Late-adaptive mechanisms are provided by the kidneys, which control the volemia through Na^+ and water reabsorption under action of the renin–angiotensin–aldosterone axis (RAAS; Fig. 3.14 and Table 3.62).

Sympathetic stimulation via β1-adrenoceptors, renal artery hypotension, and decreased Na^+ delivery to the distal tubules stimulate the release of renin by the kidney. Renin cleaves angiotensinogen (ATng) into angiotensin-1 (ATn1). Angiotensin-converting enzyme (ACE) produces angiotensin-2 (ATn2). The latter constricts the arterioles, thereby raising the systemic vascular resistance and arterial pressure. It acts on the adrenal cortex to release *aldosterone*, which increases Na^+ and water retention by the kidneys. Angiotensin-2 stimulates the release of *vasopressin* (or antidiuretic hormone, ADH) from the posterior pituitary, which also increase water retention by kidneys. It favors noradrenaline release from sympathetic nerve endings and impedes noradrenaline reuptake by nerve endings, hence enhancing the sympathetic activity.

Angiotensin-converting enzyme regulates blood pressure. It cleaves small peptides, such as angiotensin-1 and bradykinin. It also sheds various glycosylphosphatidylinositol-anchored proteins from the plasma membrane. This activity is enhanced by the membrane raft disruptor *filipin* [258].

Kinase WNK4[116] regulates the sodium–chloride cotransporter of the distal convoluted tubule, hence switching the balance from electroneutral to electrogenic sodium reabsorption in the distal convoluted tubule (Fig. 3.15) [259]. Kinase WNK4 inhibits the sodium–chloride cotransporter and renal outer medullary potassium ROMK1 channel. Kinase WNK1 may inhibit WNK4 kinase.

Aldosterone is released by the adrenal gland either by volume loss stimulated by angiotensin-2,[117] or hyperkalemia. Aldosterone favors sodium reabsorption via the sodium channel in the mineralocorticoid-sensitive segments of distal nephron. Kinase WNK4 can mediate the renal response to aldosterone.

Acute small changes in plasma sodium concentration within the physiological range between 135 and 145 mmol, in the presence of aldosterone, increase endothelial stiffness by about 10% within mn [262]. Hence, endothelial cells are less responsive to rhythmically applied stresses. This reduced impact could modify the release of vasoactive substances. In addition, salt loading upregulates the endogenous inhibitor dimethyl ᴸarginine of nitric oxide synthase. High extracellular sodium concentration resulting from aldosterone activity not only stiffens endothelium, but also reduces nitric oxide release from the vascular endothelium. The lack of nitric oxide leads to vasoconstriction and hypertension.

In addition to the endocrine function of the circulating renin–angiotensin pathway, local renin–angiotensin axes have an auto- and paracrine role. The brain, like several organs, expresses all components of the renin–angiotensin cascade. Angiotensin peptides may also serve as neurotransmitters.

[116]Protein Ser/Thr kinases of the WNK (with no lysine [Lys or K]) set are characterized by the absence of lysine usually found in the catalytic domain of all other protein serine/threonine kinases. Kinases WNK1 and WNK4 are expressed in the distal convoluted tubule, connecting tubule, and collecting duct of the nephron (Vol. 4 – Chap. 5. Cytoplasmic Protein Ser/Thr Kinases).

[117]Angiotensin-2 also directly stimulates renal sodium reabsorption, independently of aldosterone.

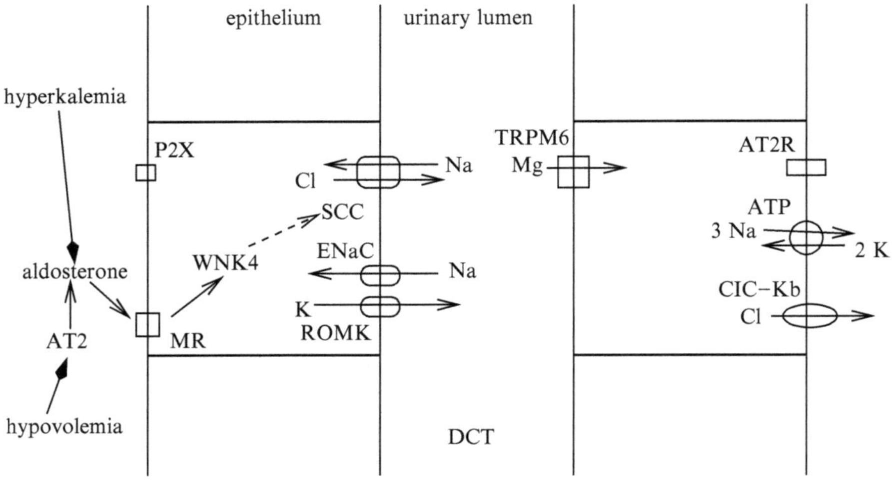

Fig. 3.15 The distal convoluted tubule (DCT) with certain ion carriers and receptors, as well as aquaporin and kinase WNK4, a possible effector of aldosterone (Sources: [259–261]). Ion carriers in epithelial cells of the distal nephron include inward rectifier renal outer medullary potassium channel (ROMK), sodium–chloride cotransporter (SCC), and epithelial sodium channel (ENaC). Others such as Na^+–K^+ ATPase, Na^+–Ca^{++} exchanger, Na^+–H^+ exchanger, cation–chloride (Na^+–K^+–$2Cl^-$) cotransporter, amiloride-sensitive Na^+ channel, Ca^{++}-sensitive K^+ channel, pH-sensitive K^+ channel, epithelial Ca^{++} channel, plasma membrane Ca^{++}-ATPase, calbindin-D28k, Ca^{++}-dependent Cl^- channel, ATP-sensitive Cl^- channel, and H^+ ATPase, are not represented here. Mg^{++} is transported transcellularly by TRPM6 (it is reabsorbed paracellularly in the thick ascending limb of the Henle loop. Plasmalemmal receptors include the mineralocorticoid receptor (MR), activated by aldosterone, angiotensin-2 receptor (AT2R), and nucleotide receptors P2X4, P2X5, and P2X6, among others

The renin–angiotensin axis is involved in the regulation of fluid balance and energy metabolism, in particular, in the control by the central nervous system of energy balance. In addition, angiotensin-2 increases synthesis of uncoupling protein UCP1 (or thermogenin)[118] and β3-adrenergic receptor in brown adipose tissue and that of β3-adrenoceptor in white adipose tissue [263]. Delivery of angiotensin-2 to the brain heightens energy expenditure. Eventual associated anorexia is correlated to increased levels of hypothalamic agouti-related protein, proopiomelanocortin, adrenocorticotropic hormone, and thyrotropin-releasing hormone as well as suppression of hypothalamic neuropeptide-Y and orexin expression [263, 264]. However, overexpression of the renin–angiotensin axis that excites sympathetic nerve activity in brown adipose tissue and energy expenditure does not reduce the daily food intake [265]. Sympathetic activity mediates thermogenic metabolism and contributes to the weight-reducing action of leptin.

[118]Uncoupling protein-1 in mitochondria of brown adipose tissue produces heat by nonshivering thermogenesis, a primary process of heat generation in human infants.

3.6.7.3 Leptin

Leptin, the renin–angiotensin axis, and the sympathetic nervous system are implicated in obesity-associated hypertension. A positive peripheral leptin–angiotensin interaction occurs, as plasma and pulmonary angiotensin-converting enzyme synthesis and activity are supported by leptin.

In addition to the classical circulating renin–angiotensin pathway, tissue-specific renin–angiotensin axes exist, such as those of the brain and adipose tissue, among others.

The adipose renin–angiotensin axis operates via an increased production of adipose tissue angiotensinogen that promotes adipose tissue development and raises circulating angiotensin-2 level. Moreover, angiotensin-2 produced by adipocytes increases leptin release from adipocytes.

The brain renin–angiotensin axis contributes to the regulation of sympathetic nerve activity and arterial pressure. Both leptin and angiotensin-2 receptors AT_{1A} are observed in several regions of the central nervous system involved in cardiovascular and metabolic control, such as the nucleus of the solitary tract, subfornical organ, and the ventromedial and dorsomedial hypothalamic nuclei, which are involved in angiotensin signaling, as well as the arcuate nucleus, a prominent site of leptin action [266]. In the brain, the leptin–renin–angiotensin axis facilitates the sympathetic nerve activity in the kidney and thermogenic brown adipose tissue, but does not operate on leptin-induced decrease in food intake.

Under physiological conditions, leptin provokes endothelium-dependent vasorelaxation by stimulating nitric oxide and endothelium-derived hyperpolarizing factor [267]. Leptin activates nitric oxide synthase NOS3 via the AMPK–PKB axis (NOS3 phosphorylation at Ser1177). Leptin also stimulates endothelial NOS1 production that can compensate, at least partly, a lack of NO production by NOS3 to maintain endothelium-dependent relaxation [268]. Under pathological conditions with chronic hyperleptinemia (obesity and metabolic syndrome), the NO-mediated vasodilation caused by leptin is impaired.

3.6.7.4 Vasopressin (Antidiuretic Hormone)

Neurohypophyseal vasopressin exerts its regulation via 3 receptors: vascular V_1, renal V_2, and pituitary V_3 receptors.[119] These receptors interact with PKC and GRK5 kinases. Vasopressin, via V_1 receptor, stimulates steroid secretion in adrenal glands. Cardiovascular effects of this 9-amino acid peptide hormone are low, even at high concentrations.

Vasopressin regulates blood pressure, especially in pathophysiological conditions, such as severe hypovolemia. Vasoconstriction elicited by the V_1 receptor increases blood pressure (Fig. 3.16). Conversely, V_2 receptor-mediated release of

[119]Receptors V_1 and V_3 are also called V1Ra and V1Rb receptors, respectively.

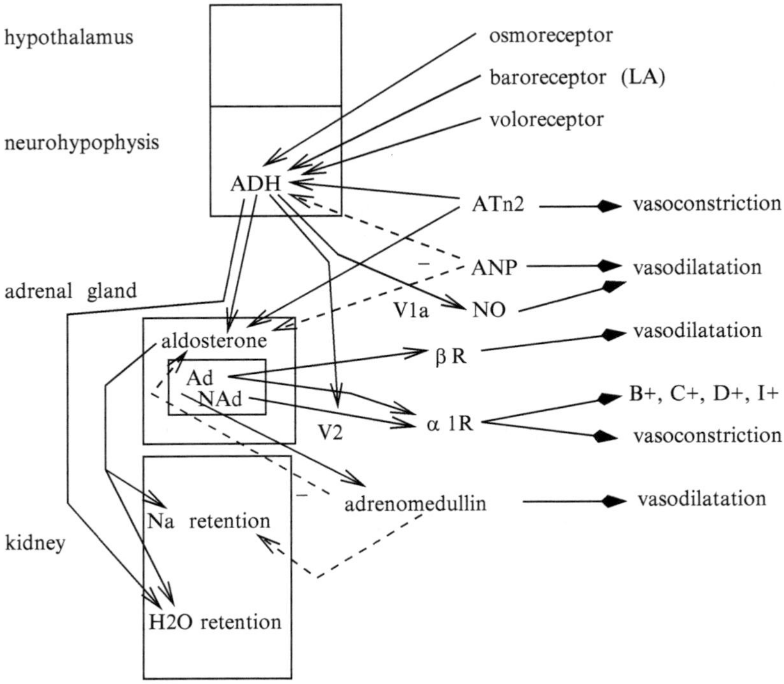

Fig. 3.16 Interactions of hormones (angiotensin-2 [ATn2], atrial natriuretic peptide [ANP], vasopressin [ADH], aldosterone, and cathecolamines) and effects on blood circulation

nitric oxide from the vascular endothelium decreases blood pressure via vasodilation. Therefore, the magnitude of blood pressure changes due to vasopressin results from a summation of vasoconstriction triggered by V_1 receptors and vasorelaxation primed by V2-receptors.

Moreover, vasopressin acts on the brain. Its indirect vasodilator effect is caused by inhibiting sympathetic efferents and enhancing the baroreflex [269].[120] Receptor V_1 maintains blood pressure at physiological levels, not via direct vasoconstriction, but by regulating the neural and hormonal actions of vasopressin [270]. Receptor V_1 is strongly expressed in the nucleus of the solitary tract. It is thus involved in the regulation of the baroreflex control.

Vasopressin reduces water excretion in the kidney by producing aquaporin-2 after binding V_2 receptors in the collecting duct cells and regulates the total

[120]Reflex control of the cardiovascular system mainly involves baroreceptors, afferents to the central nervous system, cardiovascular centers, and sympathetic and parasympathetic efferents to the heart and vasculature. The potentialization of baroreflexes is done via central action, activating V1-receptors in the area postrema, and sensitization of the arterial baroreceptors, as well as cardiac afferents.

solute concentration (osmolality) of the blood.[121] Vasopressin yields excess water reabsorption from prourine to blood in heart failure.

Collecting duct cells correspond to the site of vasopressin-regulated ion transport and water absorption as well as to the preferential intrarenal zone of bacterial initiation of inflammatory response. Supraphysiological vasopressin levels dampen the inflammatory response of collecting duct cells triggered by receptors to lipopolysaccharides on the bacterial surfaces [271]. Possible suppression in aquaporin-2 expression caused by inflammation and lipopolysaccharide-induced inhibition of V2 receptors can attenuate the anti-inflammatory action of vasopressin [272].

Vasopressin acts jointly with noradrenaline to produce vasoconstriction. ADH potentiates NAd action on vascular smooth myocytes. Therefore, significant ADH effects can be observed. In addition, electrolytes in the blood and extracellular fluid modify the surface polarity of smooth myocytes and force. Blood concentrations of O_2 and CO_2 also affect the force developed by smooth myocytes.

3.6.7.5 Natriuretic Peptides

The endocrine heart acts as a modulator of the activity of the sympathetic nervous system and the renin–angiotensin–aldosterone system (Fig. 3.14) [273]. Natriuretic peptides, or natriuretic factors (Vols. 2 – Chap. 1. Remote Control and 3 – Chap. 6. Receptors), control the body fluid homeostasis, hence blood volume and pressure.

Natriuretic Peptide Receptors

Natriuretic peptide acts by binding to 3 receptors: NPR_1,[122] NPR_2,[123] and clearance receptor (C-receptor). Both ANP and BNP bind specifically to NPR_1 receptor. C-type natriuretic peptide (CNP) synthesized by endothelial cells, chondrocytes, and brain cells, acts as a local regulator and is a selective ligand for the NPR_2 receptor. Natriuretic peptides exert their activities by increasing the intracellular concentration of cyclic guanosine monophosphate.

A- and B-type Natriuretic Peptides

A- (ANP [atrial]) and B-type (BNP [brain]) natriuretic peptides are synthesized by cardiomyocytes as preprohormones, which are processed to yield prohormones and ultimately hormones. Natriuretic peptides are released into the circulation at a

---

[121] Lack in vasopressin or in V_2 receptor in the collecting ducts is responsible for central and nephrogenic diabetes insipidus, respectively.

[122] A.k.a. NPRa and GCa.

[123] A.k.a. NPRb and GCb.

basal rate. Augmented secretion follows hemodynamical or neuroendocrine stimuli. In cardiomyocytes, activated PKB increases the expression of atrial natriuretic peptide using the phosphatidylinositol-3 kinase pathway [274]. Atrial natriuretic peptide exerts its activity not only as a circulating hormone, but also as an auto- and paracrine factor.

Natriuretic peptides relax vascular smooth myocytes. They also regulate vascular smooth myocyte proliferation. They prevent apoptosis and hypertrophy of cardiomyocytes. They repress proliferation of cardiac fibroblasts. They impede endothelin synthesis and sympathetic nerve activity, and reduce baroreflex activity.

Natriuretic peptides have direct and indirect renal actions. Protein ANP increases renal blood flow. It inhibits renin release by the kidney, raises the glomerular filtration rate, and decreases the tubular sodium reabsorption (diuresis and natriuresis). In the adrenal cortex, natriuretic peptides preclude aldosterone synthesis and release (Table 3.62).

C-type Natriuretic Peptide

The endothelial production of C-type natriuretic peptide (CNP) is stimulated by TGFβ and TNFα [275]. Endothelial CNP can regulate the local vascular tone (relaxation) and growth via cGMP production by vascular smooth myocytes. The diuretic and natriuretic effects of CNP are much weaker than those of ANP and BNP peptides.

Urodilatin

Urodilatin is a peptide synthesized in the kidney by a different processing of the atrial natriuretic peptide prohormone. Whereas in the heart and other tissues ANP prohormone is cleaved between amino acids 98 and 99 to form kaliuretic peptide, in kidneys, ANP prohormone is cleaved between amino acids 94 and 95 to produce urodilatin. Therefore, 126-amino acid ANP prohormone gives rise to 4 peptide hormones in the kidney: (1) long-acting natriuretic peptide (amino acids 1–30 of proANP); (2) vessel dilator (amino acids 31–67 of proANP); (3) kaliuretic peptide (amino acids 79–94 of proANP); and (4) urodilatin (amino acids 95–126 of proANP).

Urodilatin regulates renal sodium and water excretion, as it increases renal blood flow. It is secreted in response to increased arterial pressure and blood volume from cells of the distal tubule and collecting duct. As it causes diuresis, it can serve in acute and chronic renal failure.

Urodilatin as well as 4 cardiac peptide hormones that regulate blood pressure and maintain plasma volume, that is, atrial natriuretic peptide, vessel dilator, long-acting natriuretic peptide, and kaliuretic peptide, have potent anticancerous effects, especially against renal carcinoma [276].

3.6.7.6 Ghrelin

Circulating ghrelin, a 28-amino acid acylated peptide hormone, reduces blood pressure. Ghrelin-producing cells contain ghrelin ^{O}acyltransferase that acylates ghrelin before its secretion. Both ghrelin and its active unacylated form, desacyl ghrelin, travel in the blood circulation (plasma concentration ~3 nmol in the fasted state and ~500 pmol in the fed state in humans).

Ghrelin targets growth hormone secretagogue receptor (GHSR), but not desacyl ghrelin. Transcript Ghsr generates a full-length G-protein-coupled receptor GHSR1a and a short spliced variant GHSR1b that terminates after the fifth transmembrane domain. Growth hormone secretagogue receptor GHSR1a is detected in renal distal tubules, hypothalamic neurons, and sympathetic preganglionic neurons such as those of the vasoconstrictor axis. For example, in the celiac ganglion, terminals surround all neurons, including somas of postganglionic vasoconstrictor neurons in the periphery of the ganglion.

Ghrelin reduces sympathetic nerve activity, but does not have direct vasodilator effects on blood vessels, as GHSR1a does not reside in arteries [277]. The depressor effect of ghrelin is mediated by a decreased sympathetic vasoconstriction.

3.6.7.7 Vasoactive Intestinal Peptide

Numerous peptide-containing neurons exist in the brain, spinal cord, and peripheral nervous system. These peptides include substance-P, vasoactive intestinal polypeptide, enkephalin, and somatostatin. Substance-P is observed in the central nervous system, in primary sensory neurons, sensory neurons in the vagus nerve, and taste buds, as well as intestinal neurons.

Vasoactive intestinal peptide lodges in cholinergic neurons innervating exocrine glands. It may be responsible mainly for vasodilation, whereas acetylcholine mainly causes secretion [278]. Combined action of both substances in very low doses provokes a marked potentiation of secretion and vasodilation.

Cholecystokinin-like peptide resides in dopaminergic neurons projecting mainly to limbic areas. In central CCK+, dopamine+ neurons, both molecules operate as cotransmitters.

3.6.7.8 Cholecystokinin

The gastrointestinal circulation is controlled not only by the enteric nervous system, but also by the sympathetic nervous system, especially in the postprandial period.

Cholecystokinin (CCK) is released from enteroendocrine cells of the intestinal mucosa in response to feeding. It causes a gastrointestinal vasodilation by supressing the sympathetic vasomotor drive [279]. Activated CCK_1 receptor on abdominal vagal afferent neurons stimulates an intramedullary circuit analogous to that of other sympathetic cardiovascular reflexes. Abdominal vagal afferent neurons then influence the discharge rate of sympathetic vasomotor neurons in the rostral ventrolateral medulla.

Both cholecystokinin and leptin lower blood pressure, as they act on endings of subdiaphragmatic vagal afferent neurons of vasomotor control circuits. Excitation of vagal afferents inhibits splanchnic sympathetic nerve activity.

3.6.7.9 Catestatin and Sympatheticoadrenal Axis

The sympatheticoadrenal axis that comprises sympathetic postganglionic neurons and adrenal chromaffin cells release adrenaline, noradrenaline, and dopamine into blood upon activation of neuronal nicotinic receptors by acetylcholine. Noradrenaline is the neurotransmitter of sympathetic postganglionic neurons, whereas adrenaline is the predominant catecholamine released from chromaffin cells.

Catecholamines increase vascular resistance and systolic blood pressure, as well as cardiac frequency and output. Noradrenaline can augment diastolic blood pressure, whereas adrenaline lowers it. Dopamine regulates epithelial sodium transport, vascular smooth myocyte contractility, and production of reactive oxygen species. Dopamine also interacts with the renin–angiotensin and sympathetic nervous systems.

Catestatin results from cleavage of chromogranin-A. Both catestatin and its precursor chromogranin-A are stored and released with catecholamines from the storage vesicles in chromaffin cells of the adrenal medulla and adrenergic neurons [280].

Catestatin inhibits catecholamine release by acting on the neuronal nicotinic receptor (negative feedback). On the other hand, it stimulates the release of vasodilator histamine from mastocytes via inhibitory Gi protein. Catestatin is more efficient than other histamine-releasing peptides, such as substance-P and neurotensin.

Therefore, catestatin has a hypotensive effect via: (1) autocrine inhibition of catecholamine release from adrenal chromaffin cells and sympathetic adrenergic neurons and (2) paracrine stimulation of histamine release from peritoneal and pleural mastocytes.

An additional mechanism of action of catestatin is related to a catestatin isoform. This isoform has a central effect, as it acts on nicotinic synapses of the cardiovascular center of the nucleus of the solitary tract, where it elevates baroreceptor sensitivity and cardiac parasympathetic activity and reduces cardiac sympathetic activity [280]. Consequently, both arterial blood pressure and cardiac frequency decrease.

Chapter 4
Physiology of Ventilation

To learn, we must necessarily reason about what we have observed, compare the facts and judge them by other facts used as controls.

(C. Bernard [1813–1878] [1])

Breathing is a vital function under both autonomous and voluntary control. Breathing is a continuous process during wakefulness and sleep that can transiently stop during speech, defecation, and emesis.

4.1 Main Features and Input Data of the Body's Respiration

The main function of the respiratory tract is gas transport during breathing, also called the body's ventilation and body's respiration. Breathing enables regulated exchanges of 2 gases—oxygen (O_2 supply) and carbon dioxide (CO_2 removal)—in lungs between air in pulmonary alveoli and blood in pulmonary capillaries of alveolar septa to support activity of body's cells and maintain pH homeostasis.

4.1.1 Oxygen Input and Cell Metabolism

The body's cells need oxygen used to generate the main immediate energetic source, adenosine triphosphate (ATP). The latter donates energy by splitting a third phosphate group (P_i) and creating adenosine diphosphate (ADP). The latter can regenerate ATP by phosphorylation.

The cell metabolism relies on (Table 4.1): (1) aerobic and anaerobic glycolysis that forms ATP, NADH, pyruvate, and lactate; (2) tricarboxylic acid cycle (Vol. 1 – Chap. 4. Cell Structure and Function) that uses pyruvate produced by glycolysis and acetyl coenzyme-A from catabolism of fatty acids and generates NADH,

M. Thiriet, *Anatomy and Physiology of the Circulatory and Ventilatory Systems*, BBMCVS 6, DOI 10.1007/978-1-4614-9469-0_4,

Table 4.1 Cellular energetic metabolism (Source: [2]; ND: not described). Cellular respiration, a combustion reaction chain, is the set of catabolisms that convert chemical energy from nutrients (sugars, amino acids, and fatty acids) into adenosine triphosphate (ATP), a cellular fuel, using an oxidizing agent (electron acceptor), molecular oxygen (O_2)

	Carbohydrates Glucose Aerobic	Lipids Palmitic Anaerobic	Peptides acid	Alanine
Energy produced (kJ/mol)	2880	140	9790	1320
ATP synthesis (mol/mol)	36	2	130	ND

the latter being reoxidized to NAD^+ by the respiratory chain; and (3) oxidative phosphorylation in which hydrogen atom collected from NADH is transferred to molecular oxygen to create ATP, a cellular fuel used in material synthesis and transfer across cell membranes.

The electron transport chain pumps hydrogen ions into the space between the mitochondrial inner and outer membranes, hence creating a concentration gradient used to synthesize adenosine triphosphate. At the end of the electron transport chain, electrons are donated to oxygen (O_2), which is reduced to form water.

Carbon atoms of glucose give rise to carbon dioxide (CO_2). Glycolysis supplies 2 molecules of pyruvate (C_3) per molecule of glucose (C_6). Pyruvate produced by glycolysis enters the mitochondrion by active transport and is converted to acetylCoA (C_2) that is oxidized to CO_2 whereas NAD is reduced to NADH. In the mitochondrion, pyruvate is oxidized in the tricarboxylic acid cycle that generates CO_2, H_2O, and ATP. The net gain from one cycle is 3 NADH, 1 $FADH_2$, and 1 GTP for a single pyruvate molecule. Subsequently, GTP can create ATP.

The electron carriers NAD^+/NADH or FAD/$FADH_2$ that participate in the tricarboxylic acid cycle bring electrons to the electron transport chain. Their reducing potential enables the formation of further ATP via the electron transport chain with oxygen as the terminal electron acceptor.

Glycolysis can occur with or without oxygen. In the presence of oxygen, glycolysis, or usage of energy stored in the chemical bonds of glucose ($C_6H_{12}O_6$), is the preferred first stage of cellular respiration (which comprises 3 main stages: glycolysis, tricarboxylic acid cycle, and mitochondrial electron transport).

In the absence of oxygen, glycolysis produces a much lower ATP amount, more precisely 2 ATP, that is, 5% of the production of aerobic glycolysis. Anaerobic glycolysis requires NAD^+, which is regenerated from NADH%. Electrons from NADH are indeed added to pyruvate to produce lactate. In the absence of oxygen, pyruvic acid remains in the cytoplasm, where it undergoes fermentation, which produces lactic acid.

Myocytes require sudden contraction, that is, to use energy from ATP very rapidly. This fuel is supplied from a high-energy organic phosphate reservoir, creatine phosphate (C^P). Long-term energy is supplied by efficient oxidative phosphorylation that depends on a continuous O_2 consumption and generates 36 mol ATP. Short-term energy is supplied by glycolysis (2 mol ATP/mol Glc) that releases ATP rapidly. Maximal oxygen consumption (aerobic capacity) is achieved during exercise such as running. Excessive work can rely on anaerobic glycolysis, which raises lactate concentration in the blood, hence causing acidosis.

4.1.2 Gas Mixture

The *mole* is a measure of the amount of a substance, defined as the amount of any substance that contains as many elementary entities (e.g., atoms, molecules, ions, electrons) as there are atoms in 12 grams of pure carbon-12, the isotope of carbon with relative atomic mass 12, that is, the *Avogadro constant* ($\mathcal{A}$ ~6.022×10^{23} /mol) elementary entities of the substance [3]. If a substance contains N entities, its amount is $N/\mathcal{A}$.

An *extensive property* depends on the amount of substance in a sample. An *intensive property* does not depend on the amount of substance.

The *molar mass* (M) of a substance, or mass per mole of molecule, is the mass of 1 mol of the substance, that is, the average molecular mass of the molecule in the substance multiplied by the Avogadro constant. The *molar volume* is the volume per mole.

The *number density* (n_i [/m^3]) is the ratio of the number (N_i) of gas molecules of a constituent i in a gas mixture to the volume of the gas mixture:

$$n_i = N_i/V. \tag{4.1}$$

The *mass density* (ρ_i; [kg/m^3]) is the product of the number density and the ratio of the Avogadro constant to the molecular mass (M_i; kg/mole):

$$\rho_i = n_i\, M_i/\mathcal{A}. \tag{4.2}$$

The concentration of a gaseous component in a gas mixture can be defined in various ways. The *mass concentration* (ρ_i; kg/m^3 [M/L^3]) of a constituent i in a gas mixture is the ratio between its mass (m_i) and the volume of the mixture (V):

$$\rho_i = m_i/V. \tag{4.3}$$

The mass concentration is equivalent to the mass density of a component in a mixture.

The *molality* (mol/kg) of a solute is defined as the number of moles (n_{mol_i} of solute i divided by the mass (kg) of the solvent (m_{solvent}; not the mass of the solution).

The *molar concentration* (c_i; mol/m^3), or *molarity*, is defined as the ratio of the number of moles (n_{mol_i} of a constituent i to the volume of the mixture V:

$$c_i = n_{\text{mol}_i} / V. \tag{4.4}$$

The *number fraction* (f_{N_i} [dimensionless]) is defined as the number of molecules (N_i) of a constituent i divided by the total number of all molecules (N_{tot}) of the mixture:

$$f_{N_i} = N_i / N_{\text{tot}}. \tag{4.5}$$

The *volume fraction* (f_{V_i} [dimensionless]) is defined as the ratio between the volume (V_i) of a constituent i and the volume of the mixture (V):

$$f_{\text{V}_i} = V_i / V. \tag{4.6}$$

The *mass fraction* (f_{M_i} [dimensionless]) is the ratio of the mass (m_i) of a constituent i to the gas mixture mass (m_{tot}):

$$f_{\text{M}_i} = m_i / m_{\text{tot}}. \tag{4.7}$$

The *molar fraction* (f_i) is the ratio of number of moles (n_{mol_i} of a constituent i to total number of moles ($n_{\text{mol}_{\text{tot}}}$) of the mixture:

$$f_i = n_{\text{mol}_i} / n_{\text{mol}_{\text{tot}}}. \tag{4.8}$$

It can be calculated from masses (m_i) and molar masses (M_i) of the gas components: $f_i = (m_i / M_i)/(\sum_k m_k / M_k)$.

... like the kinetic theory of gases conceive pressure by considering it as a cluster of collisions.

(G. Bachelard [1884–1962] [1])

In a gas mixture, the *partial pressure* (p_i) of a gas constituent i is its virtual pressure assuming it occupies alone the volume of the mixture at the same temperature. The total pressure (p_{tot}) of an ideal gas mixture is the sum of the partial pressures of each gas component (*Dalton law*):

$$p_{\text{tot}} = \sum_i p_i. \tag{4.9}$$

Given the molar fraction (f_i) of a gas constituent i in a mixture,

$$p_i = f_i \, p_{\text{tot}}. \tag{4.10}$$

In air,

$$p_{\text{atm}} = p_{N_2} + p_{O_2} + p_{CO_2} + p_{H_2O}, \tag{4.11}$$

and

$$p_{\text{gas}_i} = f_i(p_{\text{atm}} - p_{H_2O}). \tag{4.12}$$

4.1.3 Gas Transfer

The flux (J [kg/s/m^2]) of a substance i is defined as the rate of mass transfer across a surface of unit area:

$$J_i = dm_i/dt; \quad \mathbf{J}_i = c_i \langle \mathbf{v}_i \rangle; \quad \mathbf{J}_i = \sum_{k=1}^{N} \kappa_{ki} \langle \mathbf{f}_k \rangle, \tag{4.13}$$

where m is the solute mass on one side of the surface, c its concentration, $\langle \mathbf{v} \rangle$ the average velocity, $\langle \mathbf{f} \rangle$ a set of generalized forces, and κ_{ki} phenomenological coefficients relating applied forces on component k to the resultant flux of component i.

Diffusion[1] is a random transport phenomenon that causes the gradual mixing of content of apposed compartments in the absence of bulk motion (i.e., convection or advection). Diffusive equilibrium is reached when the concentration of the diffusing substance in the 2 compartments becomes equal (complete mixing). Diffusion of a substance through a semipermeable membrane is called osmosis. Effusion is the process in which molecules flow through a hole without collisions between them. According to *Graham's law*, the rate of gas effusion is inversely proportional to the square root of its molecular mass.

4.1.3.1 Diffusion in the Gas Phase

The *first Fick law* yields the steady diffusive flux per unit area and unit time (J [mol/m^2/s]) resulting from a concentration gradient (from a region of high concentration to a compartment of low concentration) using a diffusion coefficient, or diffusivity ($\mathcal{D}$ [m^2/s]):

$$J = -\mathcal{D}\nabla c. \tag{4.14}$$

[1]Latin, *diffundo*: to spread out, to pour.

The temporal variation of the concentration can be computed using the *second Fick law*:

$$\frac{\partial c}{\partial t} = \mathcal{D} \nabla^2 c. \tag{4.15}$$

When the diffusion coefficient varies within the domain, the second Fick law becomes:

$$\frac{\partial c}{\partial t} = \nabla \cdot (\mathcal{D} \nabla c). \tag{4.16}$$

However, the Fick law is only valid for a binary gas mixture. The Stefan–Maxwell diffusion is a model of gas diffusion in multicomponent mixtures [4]:

$$-\nabla f_i = \sum_{i \neq j} \frac{f_j J_i - f_i J_j}{\mathcal{D}_{ij}}. \tag{4.17}$$

The binary mutual diffusivity depends on the molecular mass, the diffusion medium, and the temperature. In kinetic theory applied to Brownian motion, the Einstein–Smoluchowski relation yields that diffusion coefficient as

$$\mathcal{D} = \mathcal{M} k_B T, \tag{4.18}$$

where $\mathcal{M}$ is the mobility (i.e., the ratio of the particle terminal velocity upon an applied force), k_B the Boltzmann constant (1.38×10^{-23} J/K), and T the absolute temperature.

The Stokes–Einstein equation is related to the diffusion of spherical particles through a liquid with a low Reynolds number:

$$\mathcal{D} = \frac{k_B T}{6\pi \mu R_p}, \tag{4.19}$$

where μ is the viscosity of the medium and R_p is the radius of the spherical particle.

4.1.3.2 Diffusion in the Liquid Phase

Molecular diffusion is also related to the thermal motion of liquid particles at a given temperature, because particles in a solution can move. The transfer rate depends on the temperature, viscosity of the fluid, and particle size and molecular mass.

In solutions, diffusing solutes (ions and molecules) down to a concentration gradient (from an area of high concentration to a zone of low concentration), a simple equation is used to approximate the diffusion time (t) taken by a given molecule to diffuse an average distance ($\langle L \rangle$) in a given direction (*Einstein approximation equation*) [3]:

$$\langle L \rangle = 2\Big(\frac{\mathcal{D} t}{\pi}\Big)^{1/2}, \quad t = \frac{\pi}{4} \frac{\langle L \rangle^2}{\mathcal{D}}. \tag{4.20}$$

Table 4.2 Diffusivities of some ions and small and large molecules in aqueous solutions (Sources: [5–7])

Molecule	Molecular mass (g/mol)	Diffusivity (cm^2/s)
H^+	1.0	$\sim 9.3 \times 10^{-5}$
Na^+	23.0	$\sim 1.3 \times 10^{-5}$
K^+	39.1	$\sim 2.0 \times 10^{-5}$
Ca^{2+}	40.1	$\sim 0.8 \times 10^{-5}$
Cl^-	35.5	$\sim 2.0 \times 10^{-5}$
H_2	1	$\sim 4.5 \times 10^{-5}$
NH_3	17	$\sim 1.5 \times 10^{-5}$
O_2	32	$\sim 2.1 \times 10^{-5}$
CO_2	44–48	$\sim 1.9 \times 10^{-5}$
Glycine	75	$\sim 0.9 \times 10^{-5}$
Glucose	180–192	$5.0\text{-}6.6 \times 10^{-6}$
Insulin	5,730	$\sim 2.1 \times 10^{-6}$
Albumin	66,500	$\sim 6.0 \times 10^{-7}$
Hemoglobin	68,000	$\sim 6.9 \times 10^{-7}$
Fibrinogen	330,000	$\sim 2.0 \times 10^{-7}$
Myosin	493,000	$\sim 1.1 \times 10^{-7}$

The diffusion coefficient of a solute in free solution depends on the physical characteristics of both the solute and the medium. It is inversely related to the molecular mass of the solute. In the case of ions, the size of the hydration shell influences the value of the diffusion coefficient. In the case of large molecules, molecular shape influences the diffusivity. The molecular mass and diffusion coefficient of some substances in dilute aqueous solution are given in Table 4.2.

The diffusion of a gas in a liquid depends on its solubility. The Henry's law states that, at a constant temperature, the amount of a given gas that dissolves in a given type and volume of liquid is directly proportional to its partial pressure (p) in the gas in equilibrium with the liquid. In other words, the Henry parameter (H; solubility) of a gas species in a solution is proportional to its partial pressure in the gas in contact with the surface of the solution:

$$p = \mathsf{H} c, \tag{4.21}$$

where c is the concentration of the solute (H: kPa/mol/l). In fact, various forms of Henry law exist according to the variable used for the amount of gas in solution (molar fraction, molarity, etc.). Consequently, the Henry parameter can be expressed with different units. In the alveolocapillary barrier, the solubility of carbon dioxide in contact with plasma is about 22 times that of oxygen.

When a gas is dissolved in a liquid, its relative rate of diffusion is proportional to its solubility in the liquid and inversely proportional to the square root of its molecular mass (Graham's law). Carbon dioxide has a greater molecular mass (44) than that of oxygen (ratio 44/32 $\sim$1.4). The rate of diffusion of CO_2 is then about 19 times that of oxygen.

Table 4.3 Composition of inhaled and exhaled air (%), neglecting the contribution of argon and other inert gases that exist in small amounts. Cells of the human body consume oxygen to convert fuel (carbohydrates, lipids, and proteins) into energy and heat. The rate of cellular metabolism depends on the degree of physical activity

Gas	Inhaled air	Exhaled air
Nitrogen	78	78
Oxygen	21	16 (14–18)
Carbon dioxide	0.04	4 (4–5)
Water vapor	0.96	3 (3–4)
Temperature	Variable	310 K

4.1.4 Inhaled Air

The composition of the atmospheric air has fascinated philosophers of old civilizations. Anaximenes of Miletus (−585−−528), a Greek pre-Socratic philosopher, asserted that air is the fundamental substance from which all other things are made.[2] For him, natural forces constantly act on the air and transform it into other materials, which come together to form the organized world. Based on rarefaction and condensation, air causes successive changes of matter.

Air, the natural breathing gas, is a mixture of molecular nitrogen (dinitrogen [N_2]; ~78.08%), molecular oxygen (dioxygen [O_2]; ~20.95%), carbon dioxide (~0.04%), water vapor (variable amount), inert gases (argon, helium, hydrogen, krypton, neon, xenon, etc.), and traces of harmful gases (e.g., carbon monoxide [0.00001%]; Tables 4.3 and 4.4).

The main air component is molecular nitrogen, an inert gas. Molecular oxygen, which is required for life and combustion, is the second next most abundant air component. These 2 gases constitute 99% of atmospheric air. The remaining gases are found in much smaller proportions.

The level of carbon dioxide in the atmosphere is balanced by several processes, such as photosynthesis, plant death, plant and animal respiration, and volcanic activity, as well as combustion and anthropogenic emissions.

[2]On the other hand, other inhabitants of Miletus, Thales and Anaximander, thought that water and unspecified material, respectively, are the source of all things.

Table 4.4 Average composition of air at the Earth surface. The volume mixing ratio is the mole fraction (defined as the number of moles of a trace gas per mole of air [unit mol/mol]). Air is a gas mixture, the composition of which varies with the altitude at which air is sampled. Argon is the most abundant member of the rare gas family

Component	Fraction
Water (H_2O)	0.1–1.0
Dry air	
Nitrogen (N_2) (inert gas)	~78.1
Oxygen (O_2) (cellular fuel)	~20.9
Total	99
Rare inert gas family	
Argon (Ar)	0.934
Helium (He)	5×10^{-4}
Hydrogen (H_2)	5×10^{-5}
Neon (Ne)	18×10^{-4}
Krypton (Kr)	1×10^{-4}
Xenon (Xe)	9×10^{-6}
Carbon dioxide (CO_2)	0.03–0.04 (biological and industrial origin)

4.1.4.1 Atmospheric Reactions and Air Content

Early changes in the atmosphere were characterized by an increase in the proportion of oxygen because of photosynthesis by plants and decrease in the proportion of carbon dioxide because it was entrapped in sedimentary rocks and fossil fuels, absorbed by plants for photosynthesis, and dissolved in oceans.[3]

Earth experiences several interacting cycles that transfer and exchange matter and energy, such as atmospheric circulation, ocean currents, global heat conveyor (distributor of ocean heat), hydrologic cycle (evaporation of ocean water, transport by the atmosphere, and fall as rain or snow), and the rock cycle (erosion, lava flow, and transformation into igneous, metamorphic, or sedimentary rocks). The atmospheric circulation relies on the motion of air warmed near the equator that rises and flows toward Earth's poles, cools, returns to the surface, and flows back to the equator. The atmospheric circulation combined with the Earth's rotation distributes heat and moisture and creates weather patterns.

Air surrounds Earth and is progressively impoverished with the distance from the surface. The atmosphere has no definite outer edge but fades gradually into space. The lowest layer of the atmosphere is the *troposphere*, which is in constant motion.

[3]About 70% of Earth's surface is covered by water, mostly oceans.

Sunlight heats the Earth's surface. Warm air rises, expands, and cools as air pressure decreases, the change of atmospheric pressure with altitude being given for about the first 10 km of altitude in the troposphere by:

$$p_{\text{atm}}|_{\text{h}} = p_{\text{atm}}|_0 \cdot \left(\frac{288 - 6.5\,h}{288}\right)^{5.2558}. \tag{4.22}$$

The upper atmosphere expands during the day and contracts during the night. On the other hand, cool denser air sinks and warms near the Earth's surface.

Water vapor, carbon dioxide, and other gases in the atmosphere trap heat originated from the sun, warming Earth (greenhouse effect). The hydrosphere (oceans) stores the major part of heat received from the sun.

Above the troposphere (about 50 km from Earth's surface), the stratosphere contains the ozone layer. Ozone prevents most of the sun's harmful ultraviolet beams from reaching Earth's surface.

Oxygen is the most abundant element (∼47%) in minerals and rocks in Earth's lithosphere (crust).[4]

Human activities and effects on the environment change the Earth's atmosphere (e.g., burning fossil fuels and deforestation; Vol. 7 – Chap. 3. Pathologies of the Respiratory Tract; Table 4.5). Burning of fossil fuels adds carbon dioxide to the atmosphere at a faster rate than its removal rate.

Air also contains numerous natural pollutants, such as dust and volcanic ash, and anthropogenic harmful pollutants, which have a given mean residence time (Table 4.6).

The airborne particle mix with its chemical and physical characteristics depends on the range of sources and their proportional contribution to the mix. The particle mix is composed of primary (emitted) and secondary (formed in the atmosphere) particles, such as sulfates and nitrates.

Particles, or particulate matter, are defined by their aerodynamic diameter (μm) and classified into PM_{10}, $PM_{2.5}$, and $PM_{0.1}$, the subscript indicating the higher size threshold. Particles derived from combustion sources (vehicles, power plants, etc.) are generally small, whereas those coming from abrasion (road dust and wind-blown soil) are often large. Smaller particles tend to homogeneously spread over large areas, and penetrate indoors.

Chemical reactions in the atmosphere represent a source of atmospheric trace molecules and pollutants. Trace gas concentrations in polluted air can be higher than those in unpolluted air, as the reaction rate can be elevated.

The tropospheric secondary pollutant ozone is an oxidant formed by solar beams in the presence of primary pollutants, mainly nitrogen oxides (especially NO) and

[4]The Earth's crust (continental and oceanic crust thickness ∼40 and ∼8 km) also contains silicon (∼27%), aluminum (∼8%), iron (∼5%), calcium (∼4%), sodium (∼2%), potassium (∼2%), magnesium (∼2%), and other elements such as carbon.

Table 4.5 Change in air composition at the Earth's surface due to human activities (data from various sources). Sunlight drives photochemistry of important atmospheric species, as photons break bonds and create free radicals. Molecule photodissociation occurs above the dissociation energy. Resulting radicals can then react with many molecules. Earth's surface ozone is highly reactive and toxic. Stratospheric ozone-depleting gases, such as chlorofluorocarbons (CFC), hydrochlorofluorocarbons (HCFC), and hydrofluorocarbons (HFC) were commonly used as refrigerants in air conditioners, dehumidifiers, freezers, and refrigerators, as well as aerosol propellants and solvents (and now a priori replaced). On the other hand, nitrogen dioxide generates ozone in the troposphere, in addition to contributing to air pollution

Component	Fraction
Ammonia (NH_3)	2×10^{-7}
Bromomethane (CH_3Br)	1×10^{-9}
Carbon monoxide (CO)	1×10^{-5}
Carbon tetrachloride (CCl_4)	6×10^{-8}
Chloromethane (CH_3Cl)	6×10^{-8}
Dichlorodifluoromethane (CCl_2F_2)	3×10^{-8} (aerosol spray propellant)
Hydrogen chloride (HCl)	$\sim 1 \times 10^{-10}$
Hydroxyl radical (OH)	$\sim 1 \times 10^{-11}$
Ozone (O_3)	7×10^{-6}
Methane (CH_4)	2×10^{-4}
Nitric acid (HNO_3)	$\sim 1 \times 10^{-9}$
Nitrous acid (HNO_2)	$\sim 1 \times 10^{-10}$
Nitric oxide (NO)	14×10^{-6}
Nitrogen dioxide (NO_2)	2×10^{-6}
Nitrous oxide (N_2O)	3×10^{-5}
Peroxyacetyl nitrate	$\sim 1 \times 10^{-9}$
Sulfur dioxide (SO_2)	1×10^{-4}
Sulfuric acid (H_2SO_4)	1×10^{-6}
Tetrafluoromethane (CF_4)	7×10^{-9}
Trichlorofluoromethane (CCl_3F)	17×10^{-9} (refrigerant)
Trichloroethane ($C_2H_3Cl_3$)	98×10^{-10} (solvent)

volatile organic compounds. Its short-term effects comprise reduced pulmonary function and increased airway responsiveness and airway inflammation.

Ammonia (NH_3), a soluble (in water) and reactive gas, the most abundant alkaline component in the atmosphere, is an important element of the nitrogen cycle. It is oxidized in the environment by microorganisms (nitrification). Environmental ammonia results from agricultural, municipal, industrial, and natural processes. Concentration of ammonia in the air is greater in regions of emissions from

Table 4.6 Gas components in atmospheric air and their mean residence time (Source: [8])

Gas	Fraction	Mean residence time
Argon	0.9	
Hydrogen	0.00006	
Water vapor	0–3	8–10 d
Carbon dioxide	0.035	50–200 yr
Ammonia	10^{-8}–10^{-6}	3 d
Carbon monooxide	4–20.10^{-6}	5 mo
Freons (CFC)	10^{-7}	50–150 yr
Methane	17.10^{-5}	7–10 yr
Nitrogen oxides	10^{-8}–5.10^{-5}	3 d
Nitrous oxide	33.10^{-6}	130 yr
Ozone		
tropospheric	10^{-6}–10^{-5}	wk–mo
stratospheric	10^{-5}–5.10^{-5}	mo
Peroxyacetylnitrate (PAN)	10^{-7}–5.10^{-6}	
Sulfur dioxide	10^{-7}–5.10^{-5}	3 d
Volatile organic compounds (VOC)	10^{-5}–10^{-4}	

livestock farming (poultry, pig, cow, sheep, goat, horse, and deer), more precisely from cattle urine and manure, as well as some fertilizers and crops, in addition to industries (fertilizer manufacture, sugarbeet processing, ceramics and cement works, iron and steel mills, and oil refineries, among other combustion sources). Ammonia concentration depend on near-surface winds, atmospheric stability, and Earth surface roughness, among other factors. Ammonia can react with other chemicals in the atmosphere to produce fine particles that contain ammonium ion (NH_4^+; particulate ammonium). Most of these particles are removed from the atmosphere by rain, whereas gaseous ammonia is absorbed by soil, water, and vegetation. A large fraction of NH_3 (up to 40%) can deposit near its source. Nevertheless, ammonia compounds can be transported in the atmosphere over long distances.

Nitrogen species, especially nitrogen oxides (NO_x), are sources of photochemical reactions. Nitric oxide (NO), or nitrogen monoxide, is a free radical and a by-product of combustion of substances in air as well as a natural product of electrical discharges of lightning during thunderstorms. In air, nitric oxide is rapidly oxidized to nitrogen dioxide (NO_2), a major air pollutant, as it interacts with ozone. (The reaction of NO with O_2 is slow.) Nitrogen dioxide results mainly from vehicles. In the presence of sunlight, NO_2 participates with hydrocarbons and oxygen in the formation of ozone and other secondary photochemical oxidants. It also reacts with aerosols to form secondary, often acidic particles. Both NO and NO_2 participate in stratospheric ozone layer depletion. In air, NO also converts to nitric acid (HNO_3) implicated in acid rains.

In general, concentrations are inversely proportional to rates of photodissociation and reaction in the atmosphere. Among nitrogen species, N_2O and HNO_3 are extremely photostable, whereas NO_2, NO_3, dinitrogen pentoxide (N_2O_5), and peroxynitric acid (HO_2NO_2) are easily degradable.

Sulfur dioxide (SO_2) is released by volcanoes and industrial processes. Its oxidation, usually in the presence of a catalyst such as NO_2, forms sulfuric acid (H_2SO_4). In addition to acid rains, emission of SO_2 is a precursor to particulates in the atmosphere. Sulfur dioxide blocks nervous signals from pulmonary stretch receptors, thereby abolishing the Hering–Breuer inflation reflex.

Carbon monoxide is mainly produced by incomplete combustion of carbonaceous fuels, such as gasoline and natural gas, hence having outdoor and indoor sources. Its toxicity is attributed to its high affinity with hemoglobin and myoglobin. Its affinity to hemoglobin is 200 to 250 times that for oxygen, forming carboxyhemoglobin (COHb).

Hydroxyl radical attacks various pollutants in air, such as carbon monoxide (CO)), methane (CH_4), and acetaldehyde (CH_3CHO). The resulting methyl radical (CH_3) is involved in reactions that oxidize NO in NO_2 and regenerate OH radicals simultaneously.

Photochemical reactions can create a smog characterized by suspended particulate materials. The condensed phase is represented by organic and inorganic (metal oxides and salts of acids) substances. Salts and acidic droplets result from the combination of solid particles with nitric and sulfuric acids. Acidic droplets can rapidly react with ammonia to form ammonium nitrate and sulfate aerosols.

Liquid droplets are sites of reactions such as oxidation of sulfur dioxide (SO_2) in sulfuric acid (H_2SO_4). In a polluted atmosphere, high levels of metal ion catalysts, hydrogen peroxide (H_2O_2), and ozone support this oxidation.

4.1.4.2 Biological Carbon Dioxide

Photosynthesis processes carbon dioxide and produces oxygen, whereas small amounts of carbon dioxide are produced in darkness. The carbon cycle results from photosynthesis and respiration. The reaction of photosynthesis is the following,

$$6\,CO_2 + 6\,H_2O \quad \rightarrow \quad C_6H_{12}O_6 + 6\,O_2. \tag{4.23}$$

The respiration reaction is as follows,

$$C_6H_{12}O_6 + 6\,O_2 \quad \rightarrow \quad 6\,CO_2 + 6\,H_2O. \tag{4.24}$$

Carbon dioxide is emitted from hot springs and geysers; it is also liberated by breathing of humans and animals, human activities (combustion of fossil fuels, coal, and hydrocarbons), and fermentation of liquids.

During cellular respiration, especially during the citric acid cycle (Krebs or tricarboxylic acid cycle) oxidation of organic substances, mainly lipids and carbohydrates, which are mainly composed of carbon with some hydrogen and oxygen, produces CO_2 and water.

Carbon dioxide is related to pH of blood and other body fluids via the *carbonate buffer*, which is composed of bicarbonate ions, dissolved carbon dioxide, and carbonic acid. Carbonic acid and bicarbonate ion neutralize hydroxide and hydrogen ions.

Carbon dioxide changes water pH, as it dissolves slightly to form a weak carbonic acid (H_2CO_3):

$$CO_2 + H_2O \quad \rightarrow \quad H_2CO_3. \tag{4.25}$$

Carbonic acid reacts slightly and reversibly in water to form a hydronium cation (H_3O^+) and bicarbonate ion (HCO_3^-):

$$H_2CO_3 + H_2O \quad \rightarrow \quad HCO_3^- + H_3O^+. \tag{4.26}$$

Carbon dioxide also stabilizes transmission of nervous signals. This transmission arises when the transmembrane voltage is higher than a given threshold that is sensitive to the local CO_2 content to prevent accidental impulses.

Moreover, CO_2 promotes O_2 release from hemoglobin in capillaries that supply blood to cells (Bohr effect), hence tissue oxygenation. Carbon dioxide also provokes local vasodilation. In the respiratory tract, a normal CO_2 level favors bronchodilation. Last, but not least, it influences the nervous control of the cardiovascular and respiratory apparatus. It is sensed by central chemoreceptors in the medulla oblongata and peripheral chemoreceptors of the aortic and carotid bodies.

4.1.5 Breathing Modes

Rest breathing remains nearly invariant throughout healthy adulthood. However, the breathing frequency adapts to change in posture and exercise as well as to sleep.

Eupnea corresponds to a normal resting breathing frequency; bradypnea and tachypnea refer to elevated and slow breathing frequency, respectively. Minute ventilation during normal breathing in healthy subjects equals about 6 l/mn (range 4.5–8.2 l/mn according to age and gender).

Hyperpnea is an increased amplitude of breathing to match metabolic needs; it corresponds to respiratory compensation to correct a metabolic acidosis. Hypopnea is a decreased breathing amplitude (shallow breathing).

Sensu stricto, hyperventilation is an increased inappropriate breathing; it is also called overbreathing in the music field. Breathing disorders result mainly from lung and cardiovascular diseases as well as dysfunctions of the neural control of the body's ventilation.

The wetted surface of the lung parenchyma provides a large exchange area, particularly for gas, such as oxygen and carbon dioxide, owing to a huge quantity of tiny alveoli connected to the tracheobronchial tree downstream (during inspiration) from a craniocervical respiratory conduits for air conditioning and purification.

4.1.6 Functions of the Body's Respiration

The external respiration (in opposition to the cell's internal respiration) includes ventilation and lung perfusion with gas exchange between pulmonary alveoli and blood capillaries. Blood gas content depends on the local ventilation–perfusion ratios in lungs. Perfusion of lung territories must match local ventilation. Lung ventilation and perfusion are indeed coupled for efficient gas transport toward and from the body's tissues, including those of lungs and associated structures.

Breathing quality depends on: (1) respiratory mechanics that involves airway walls, lung parenchyma, and chest wall; (2) air distribution among lung territories; and (3) control of the respiration according to the body's tissue needs.

The lung has additional functions associated with the functioning of its cell population, such as water and material transfer regulation especially for stability of the airway surface fluid, maintenance of surfactant, and immune defense of the monolayer of respiratory epithelium[5] that ensures the airway surface fluid proximally and blood reoxygenation for the body's survival distally.

Mucosal protection in the respiratory tract among others is achieved by the existence of an epithelial barrier with tight junctions and secretion by goblet and other specialized secretory cells of a mucus enriched with antimicrobial glycoproteins, immunoglobins-A, cytokines, and chemokines to defend against the invasion of pathogens. In addition, the viscous mucus layer prevents the adherence of inhaled particles or microorganisms on the respiratory epithelium that are then expelled by ciliary movement.

Lungs achieve oxygen uptake from the inhaled air and carbon dioxide removal from the venous blood. The extent of gas transport depends on tissue metabolism. Gas transport between the atmosphere and body tissues in the 2 coupled, serial physiological apparatus, ventilation and blood circulation, uses both active and passive processes.

Gas transport between the atmosphere and blood is ensured by convection in large respiratory conduits and diffusion in the last generations of bronchioles and alveoli. Air convection is generated by alternate positive (expiration) and negative (inspiration) alveolar pressure with respect to the atmospheric pressure. The transalveolar pressure (transmural alveolointrathoracic pressure $[p_A - p_e]$)

---

[5]Epitheliocytes line and protect all the body's tissues as they regulate or prevent the passage of exogenous matter (Vol. 1 – Chap. 3. Cells of the Ventilatory Tract).

Table 4.7 Oxygen transport from atmosphere to the cell. Subscripts $\bullet_A$, $\bullet_a$, $\bullet_{ACM}$, $\bullet_{aw}$, $\bullet_b$, $\bullet_{cy}$, $\bullet_E$, $\bullet_{ECF}$, $\bullet_I$, $\bullet_{mt}$, and $\bullet_v$ stand for alveolar, arterial, alveolocapillary membrane, airway, blood, cytosol, expiration, extracellular fluid, inspiration, mitochondrial, and venous, G_{cy} is proportional to the number of electron transport chains (ETC; i.e., to the number of mitochondria), $p_{mtO_2} = 0$, and $k = 1/6$

Airway	$\dot{V}_{O_2} = G_{aw}(p_{IO_2} - p_{EO_2})$.
Alveolocapillary membrane	$\dot{V}_{O_2} = G_{ACM}(p_{AO_2} - p_{bO_2})$.
Blood	$\dot{V}_{O_2} = G_b(p_{aO_2} - p_{vO_2})$.
Intertitium	$\dot{V}_{O_2} = G_{ECF}(p_{bO_2} - p_{cyO_2})$.
Mitochondrial sink	$\dot{V}_{O_2} = G_{cy}(p_{cyO_2} - p_{mtO_2})$.
	$\dot{M}_{mtO_2}) = k\dot{M}_{ATP}$,

decays and then rises with collapsing and inflating air spaces, respectively, due to the muscular activity of the chest wall that directs the evolution of the intrathoracic pressure.

When the thoracic cavity expands during inspiration by contraction of the diaphragm, the principal inspiratory muscle, and external intercostal muscles, lungs dilate and air flows in the respiratory tract at a rate that depends on airway resistance. Expiration is passive at rest. The lung and chest elastic recoils to their equilibrium positions indeed participate in air exhalation.

The functioning of the respiratory apparatus depends on multiple parameters, local and global, intrinsic and environmental. Integrative approaches of a complex dynamical system are based on knowledge accumulated in various involved fields, from genomic and proteomic information to histological, anatomical, physiological, and mechanical data.

4.1.7 Body's Oxygen and Carbon Dioxide Content

An adequate amount of oxygen is supplied to cells via respiratory and circulatory networks (Table 4.7). The former is subdivided into 3 compartments: (1) external medium (i.e., atmospheric air); (2) conduction airways; and (3) gas exchange compartment with pulmonary alveoli. Blood carries oxygen from respiratory alveoli to cells. Due to mass conservation,

$$\dot{V}_{O_2} = \dot{V}_E(f_{IO_2} - f_{EO_2}) = \mathcal{T}_{LO_2}(p_{AO_2} - p_{\bar{v}O_2}) = q(c_{aO_2} - c_{\bar{v}O_2}), \tag{4.27}$$

where $\dot{V}_E$ is the ventilation rate, $\mathcal{T}_L$ the transfer capacity of the alveolocapillary membrane, and q the blood flow rate and subscripts $\bullet_A$, $\bullet_a$, $\bullet_E$, $\bullet_I$, and $\bullet_{\bar{v}}$ stand for alveolar, arterial, expiration, inspiration, and mixed venous, respectively.

Table 4.8 Examples of values of the physiological function of body's respiration and blood circulation (Source: [2])

Quantity	Value
f_R	0.25 Hz
V_T	480 ml
V_D	170 ml
$\dot{V}_E$	7200 ml/mn
$\dot{V}_D$	2550 ml/mn
$\dot{V}_A$	4650 ml/mn
$\dot{V}_{O_2}$	270 ml/mn
f_C	1 Hz
VES	85 ml
q	5100 ml/mn
$\dot{V}_A/q$	0.91

In the adult human body, the total oxygen quantity is equal to nearly 1.5 l (~70 mmol). The total carbon dioxide quantity is equal to about 6.2 l (~280 mmol).

At rest, the oxygen consumption ($\dot{V}_{O_2}$) of a 70-kg adult human male ranges from 250 to 300 ml/mn (3–4 ml/kg/mn), that is, approximately 4 to 5 ml/s (~0.183 mmol/s; Table 4.8). The carbon dioxide production ($\dot{V}_{CO_2}$) amounts to about 200 ml/mn, that is, nearly 3.3 ml/s (~0.15 mmol/s). Therefore, the resting oxygen consumption per unit time is 1/6 to 1/5 of the oxygen body store and the carbon dioxide production is about 1/30 of the carbon dioxide body capacity.

A low content of blood oxygen for more than a few minutes causes irreversible brain damage. Due to the limited O_2 reservoir, breathing rises rapidly to satisfy the metabolic demands of sustained body motion. The body's respiration also accommodates slower body changes during development, pregnancy, and aging, as well as diseases.

During exercise, the oxygen consumption is hence greater than the available oxygen quantity. During a modest effort such as walking, human O_2 consumption triples to nearly 800 ml/mn. During exercise, $\dot{V}_{O_2}$ can increase more than 20-fold with respect to its basal value until its maximal value ($\dot{V}_{O_{2max}}$). Production of CO_2 ranges between 70 and 100% of oxygen consumption.

4.1.8 Respiratory Epitheliocytes

Airway epitheliocytes (Table 4.9) participate in the regulation of lung homeostasis. They serve as a barrier that excludes incoming environmental antigens. They modulate the activity of adjacent cell populations, such as airway smooth myocytes [9].

They secrete into the overlying fluid a range of mediators involved in defense against pathogens (Sect. 4.10) [9]. They synthesize nitric oxide, a potent inhibitor of the activation of lung dendritic cell and memory T lymphocytes. On the other hand, they secrete cytokines and chemokines, in particular, granulocyte (CSF3)

Table 4.9 Fraction of various coating cell types in the lung parenchyma

Cell type	Proportion
Pneumocyte-1	0.08
Pneumocyte-2	0.16
Interstitial cells	0.36
Alveolar macrophage	0.10
Endotheliocyte	0.30

and granulocyte–macrophage colony-stimulating (CSF2) factors, type-1 and -2 interferons, as well as adhesion molecules such as ICAM1 to enhance the local survival of recruited inflammatory cells. They also modulate airway B-lymphocyte accumulation and activation.

4.1.9 Respiratory Cilia

Motile cilia (length 6–7 μm; caliber ~0.2–0.3 μm [10]) sense their environment and propel periciliary fluid (with momentum transfer to the mucus layer) and directly, but more transiently, mucus to clear mucus and entrapped debris from airways. Its structure is characterized by a $9+2$ axoneme arrangement[6] (Vol. 5 – Chap. 12. Airway Surface Liquid and Respiratory Mucus). Dynein arms (dynein heavy chains), radial spokes, and one pair of interdoublet links are associated with cilium microtubules. Cilia regulate their bending by means of inner dynein arms and their frequency (12–15 Hz) using outer dynein arms [10].

Motile cilia undergo a 3-phase motion: (1) *effective stroke* in a plane perpendicular to the cell surface, reaching a maximum velocity of 1 mm/s at its tip and describing an arc of approximately 110 degrees; (2) *brief rest*; (3) *recovery stroke* of almost 180 degrees backward close to the cell surface and then extending almost fully, approximately 2 to 3 times slower than the effective stroke.

When ciliary beat frequency increases, all 3 phases of the beat cycle are shortened, especially the resting period. Force adjustment may be mediated by the transient receptor potential vanilloid TRPV4 channel that increases intraciliary calcium, hence axonemal beating (within 80 ms) [10]. Augmented cytosolic Ca^{2+} concentration increases ciliary beat frequency.

Ciliary dynein arms move microtubules with a slow or fast mode. Changes in dynein arm movement state (i.e., frequency adjustments) result from phosphorylation or dephosphorylation of axonemal substrates. Phosphorylation is ensured by PKA, PKC, and/or PKG kinase [10].[7] The predominant mechanism relies on

[6]The ciliary array of microtubules consists of 9 peripheral doublets of microtubules (A- and B-tubules) and 2 central single microtubules. Each A-tubule possesses inner and outer dynein arms. The microtubule is a heterodimer of α- and type-4 β-tubulin.

[7]Kinase activation happens downstream from stimulated G-protein-coupled receptors (e.g., $P2Y_2$ [ATP], adenosine, bradykinin, and muscarinic acetylcholine receptors).

cAMP-dependent protein kinase-A.[8] Both cGMP and cGMP-dependent protein kinase-G contribute to the regulation of ciliary beating and potentiate PKA effect. On the other hand, protein kinase-C slows ciliary beating [10].

Whereas large hyaluronan (~800 kDa) does not affect ciliary beat frequency, small hyaluronans (50–200 kDa) increase it via the hyaluronan-mediated motility receptor (HMMR)[9] possibly associated with macrophage-stimulating-1 receptor (MSt1R, or Ron) at the apex of ciliated airway epitheliocytes [10].

Ciliary beat frequency decreases when extracellular pH decreases below 7.0 in bronchi and below 5.0 in bronchioles [10]. Intracellular alkalization directly causes faster ciliary beating; intracellular acidification lowers ciliary beat frequency in human tracheobronchial epitheliocytes.

4.1.10 Deformation of the Lung Parenchyma

The lung generates a stabilizing prestress [11]. The residual filling pressure after expiration stiffens basement membranes, collagen fibers, elastin bundles that surround each alveolus, and resist surface tension acting on the epithelium. Expiration and inspiration that shorten and lengthen alveolar septa bend and buckle as well as extend collagen fibers. Breathing also shrinks and expands intercellular spaces between epitheliocytes without compromising the tissue structural integrity. This reversible mechanical deformation might activate intracellular signaling.

4.2 Lung Volumes

The volume distribution of lung compartments in rats is given in Table 4.10. Alveolar septa have been estimated to constitute 15% of lung volumes, but this value is overestimated because section thickness exaggerates volume proportion of tissue [12]. Morphometric data from 8 normal human lungs (6 male and 2 female 10–40-yr-old lungs) at about 75% of total lung capacity are given in Table 4.11).

There are several lung volumes that can be measured directly or not (Chap. 5). The *tidal volume* (TV or V_T) is the volume inhaled and exhaled during quiet breathing (average ~500 ml). During quiet breathing, the bronchus bore is commonly assumed to be quasi-constant.

[8]The main intracellular second messenger cAMP can regulate axonemal bending and ciliary beating. Sources for cAMP are G-protein-responsive, Mg^{2+}-sensitive, transmembrane adenylate cyclases at the apical membrane. In addition, G-protein-insensitive, Mn^{2+}-sensitive, soluble adenylate cyclases in the axoneme are activated by HCO_3^- in a pH-independent manner and Ca^{2+} in the presence of HCO_3^- ions.

[9]A.k.a. receptor for hyaluronic acid-mediated motility (RHAMM).

Table 4.10 Volume distribution (approximative values) of lung compartments in rats (Source: [12])

Lung compartment	Volume fraction
Alveolar space	O.53
Alveolar duct lumen	O.24
Alveolar wall	O.15
Conducting airway lumen	O.03
Conducting airway wall	O.01
Blood vessels	O.03
Connective tissue	O.01

Table 4.11 Morphometric data from 8 normal human lungs at about 75% of total lung capacity (Source: [12])

Morphometric data	Value
Total lung volume	4.35 l
Alveolar surface area	140 m^2
Capillary surface area	130 m^2
Capillary lumen volume	210 ml
Air–blood barrier thickness	
Arithmetic mean	2.2 μm
Harmonic mean	0.6 μm

4.2.1 *Pressure–Volume–Temperature Relations of an Ideal Gas*

Some physical laws on gas are widely used in respiratory physiology. The *combined gas law* states that:

$$\frac{pV}{T} = k, \tag{4.28}$$

where p is the gas pressure, V the gas volume, T the gas temperature (K), and k a constant.

In thermodynamics, the *internal energy* of a system is expressed in terms of pairs of conjugate variables such as the thermal parameters, temperature and entropy (Ts conjugate pair), and the mechanical parameters, pressure and volume (pV conjugate pair).

A temperature difference provokes a change in entropy (hence temperature is thermodynamically conjugate to entropy), and their product corresponds to the energy associated with heat. A pressure change causes a volume change, and their product corresponds to the energy associated with mechanical work.

The *entropy* can be defined as a measure of energy dispersal at a specific temperature or as the ineffective thermal energy per unit temperature (unavailable energy for carrying out a mechanical work, i.e., energy loss). The entropy of an isolated system never decreases, because this system spontaneously evolves toward a thermodynamic equilibrium (i.e., a state of maximum entropy). The entropy of any system only decreases by increasing the entropy of another connected system.

The *heat capacity* (C) of a material is the ratio between the amount of thermal energy (Q) transferred to/from the material and the resulting increase/decrease in temperature (ΔT) of the material:

$$\mathrm{C} = \frac{Q}{\Delta T}. \tag{4.29}$$

The heat capacity is defined at constant volume or constant pressure. The *specific heat capacity*, or thermal capacity, at constant volume or constant pressure is the amount of heat required to change the material temperature per unit mass to raise the material by 1 degree for 1 kg of matter:

$$\mathrm{c} = \frac{\mathrm{C}}{m} = \frac{\mathrm{C}}{\rho V}, \tag{4.30}$$

where ρ is the material density.

The *internal energy* (u) is the sum of all forms of energy intrinsic to a thermodynamic system, such as potential (e.g., chemical and deformation energy) and kinetic energy, in addition to external energy provided to the system. The change in internal energy is related to heat flow (Q_{T}) and change in mechanical work (change in volume due to an external pressure) in the absence of other perturbations.

The internal energy of an ideal gas is proportional to the amount of gas substance, or number of moles n (i.e., the mass [m] divided by the molar mass [M]) and temperature T:

$$u = \mathrm{c_V} n T, \tag{4.31}$$

where $\mathrm{c_V}$: isochoric (isovolumetric) heat capacity.

The gas temperature and pressure are the partial derivatives of the internal energy w.r.t. entropy and volume, respectively:

$$T = \frac{\partial u}{\partial s}, \quad p = -\frac{\partial u}{\partial V}, \tag{4.32}$$

The equation of state of an ideal gas, or the *ideal gas law*, states that:

$$pV = R_{\mathrm{g}} n T = N k_{\mathrm{B}} T, \tag{4.33}$$

where R_{g} is the universal (ideal or molar) gas constant (8.31446 J/[mol.K]; the symbol R standing for Regnault constant), N the number of gas particles, and k_{B} the Boltzmann constant. The specific gas constant ($R_{\mathrm{g_{spec}}}$) of a gas (gas mixture) is obtained by dividing R_{g} by the molar mass (M) of the gas (gas mixture) as well as by dividing the Boltzmann constant by the molecular mass of the gas (m):

$$R_{\mathrm{g_{spec}}} = \frac{R_{\mathrm{g}}}{M} = \frac{k_{\mathrm{B}}}{m} = \mathrm{c_p} - \mathrm{c_V}. \tag{4.34}$$

In a closed thermodynamic system, the fundamental thermodynamic relation when the volume is the single external parameter and the process reversible is expressed as

$$du = Tds - pdV. \tag{4.35}$$

The *Boyle–Mariotte law* relates the gas pressure to the gas volume when the temperature and amount of gas remain constant in a closed system (i.e., isothermal change) when internal energy does not vary, but external mechanical work (kinetic energy) is applied to the gas:

$$pV = k. \tag{4.36}$$

The change in entropy (s) during isothermal expansion or compression of an ideal gas from an initial volume V_0 and pressure p_0 to a final volume V and pressure p is given by

$$\Delta s = nR_g \ln \frac{V}{V_0} = -nR_g \ln \frac{p}{p_0}, \tag{4.37}$$

where n is the number of moles of gas and R_g is the gas constant.

The *Charles law* when pressure and amount of gas remain constant is related to isobaric change, the variation in volume and temperature being associated with a change in external mechanical work (which is related to changes in volume due to an externally exerted pressure) and internal energy (i.e., in molecular velocity):

$$\frac{V}{T} = k. \tag{4.38}$$

The change in entropy during heating or cooling of a gas at constant pressure from an initial temperature T_0 to a final temperature T is:

$$\Delta s = nc_p \ln \frac{T}{T_0}, \tag{4.39}$$

where c_p is the isobar heat capacity.

The *Gay–Lussac law* when the mass and volume of the gas are constant describes isovolumetric change (i.e., in the absence of mechanical work), the variation in temperature being associated with a change in pressure and internal energy (i.e., in molecular velocity):

$$\frac{p}{T} = k. \tag{4.40}$$

The change in entropy during heating or cooling of a gas at constant volume from an initial temperature T_0 to a final temperature T is:

$$\Delta s = n c_v \ln \frac{T}{T_0}. \tag{4.41}$$

An *adiabatic process* occurs in a very short time scale without exchange of heat (heating or cooling) with its environment, the gas domain being limited by a boundary impermeable to heat approximated by a thermally insulated barrier. The amount of energy is kept constant. A reversible adiabatic process is an *isentropic process* (isoentropic process); an irreversible adiabatic process that extracts no work is an *isenthalpic process*.

In an isentropic process (in the absence of dissipative effect), the change in enthalpy ($h = u + pV$) is given by

$$dh = du + pdV + Vdp, \tag{4.42}$$

as

$$du = nc_v dT = -pdV, \quad dh = nc_p dT = Vdp. \tag{4.43}$$

The heat capacity ratio (γ), which is constant for an ideal gas, can be written as

$$\gamma = \frac{c_p}{c_v} = -\frac{dp/p}{dV/V}. \tag{4.44}$$

Therefore,

$$p\,V^{\gamma} = \text{constant}. \tag{4.45}$$

4.2.2 Measurement of Pulmonary Volumes and Respiratory Flow Rates

Spirometry, or spirography, is the basic respiratory functional examination to assess mobilizable lung volumes inhaled and exhaled at rest and during deep breathing maneuvers that are performed quietly or rapidly such as forced expiration (carried out with maximal amplitude and maximal speed).

Using a pneumotachograph (flowmeter), the flow–volume curve during forced inspiration and expiration can be drawn. Inhaled and exhaled air volumes can then be obtained by integration.

The residual volume that cannot be extracted from lungs is currently measured using the low-cost *helium dilution method*. Helium is an inert gas that does not cross the alveolocapillary membrane. The subject is connected to a spirograph (volume V_S) containing a given helium molar concentration (c_{He_S}) at the end of a rest exhalation (at FRC). Helium is then distributed throughout the lung (volume FRC = ERV + RV) with a steady-state concentration (c_{He_L}). The amount of helium reamining constant,

$$V_S \times c_{He_S} = (V_S + \text{FRC}) \times c_{He_L}, \tag{4.46}$$

$$\text{RV} = \text{FRC} - \text{ERV} = V_S\Big(\frac{c_{He_S}}{c_{He_L}} - 1\Big) - \text{ERV}. \tag{4.47}$$

Plethysmography is easy, but expensive. The subject seated in a plethysmograph breaths through a mouthpiece endowed with a pneumotachograph and obturator. Air flow is interrupted whereas the subject continues to breathe, hence compressing and dilating air in the respiratory tract as well as (but in the reverse order) between his body and the impermeable box, the pressure being measured near the mouth and/or in the plethysmograph. After mouthpiece closure, during following inhalation, the lung volume increases and pressure in airways decreases. Assuming that the Boyle–Mariotte law is valid ($pV = \text{cst.}$):

$$p_1\text{FRC} = p_2(\text{FRC} + V_T); \quad \text{FRC} = \frac{p_2}{p_1 - p_2}\Delta V = p_2\frac{\Delta V}{\Delta p}. \tag{4.48}$$

4.2.3 *Lung Volumes Measured by Spirometry*

The *vital capacity* (VC) is the highest air volume displaced between the upper limit of inspiration and lower limit of expiration. The *functional residual capacity* (FRC) is the air volume left in the lungs at the end of a spontaneaous expiration. It corresponds to nearly 40% VC. The respiratory muscles are then completely relaxed. At FRC, the collapsing tendency of the lungs is balanced by the expanding tendency of the chest wall.

The *inspiratory capacity* is the maximal quantity of air (up to the maximal inspiratory level) that can be inhaled starting from a spontaneaous expiration (i.e., at functional residual capacity).

$$\text{IC} = V_T + \text{IRV} = \text{VC} - \text{ERV}. \tag{4.49}$$

The *inspiratory reserve volume* (IRV) is the maximal quantity of air (up to the maximal inspiratory level) that can be inhaled starting from the end of a spontaneaous inspiration (i.e., at FRC + V_T).

$$\text{IRV} = \text{VC} - (V_T + \text{ERV}). \tag{4.50}$$

Table 4.12 Standard lung volume (l) in average-size human genders

	Woman	Man
IRV	2.0	3.3
TV	0.5	0.5
ERV	0.7	1.0
RV	1.1	1.2
TLC	4.3	6.0

Table 4.13 Differences in main anatomy and physiology data of the ventilation apparatus between newborn and adult

Anatomy	Newborn	Adult
Body mass (kg)	3.5	70
Surface area (m^2)	0.21	1.90
Lung mass (g)	50	800
Tracheal caliber (mm)	8	20
Bronchiole caliber (mm)	0.1	0.2
Airway number ($\times 10^6$)	1.5	14
Alveolus number ($\times 10^6$)	24	296
Alveolar size (μ)	50-100	150-300
Cumulated alveolar surface area (m^2)	4	80
Physiology		
Vital capacity (ml/kg)	33	52
Functional residual capacity (ml/kg)	30	34
Dead space (ml/kg)	2.2	2.2
Tidal volume (ml/kg)	6	7
Rest repiratory frequency (Hz)	0.66	0.33
Alveolar ventilation (ml/kg/s)	1.6–2.5	1
Oxygen consumption at rest (ml/kg/s)	0.1	0.05

The *expiratory reserve volume* (ERV) is the maximal amount of air that can be expired (down to the maximal expiratory level) after a quiet breathing, starting at FRC.

$$\mathrm{ERV} = \mathrm{FRC} - \mathrm{RV}. \tag{4.51}$$

The *residual volume* (RV) is the volume remaining in lungs after a maximal expiration. The *total lung capacity* (TLC) is the lung volume after a maximal inspiration.

$$\mathrm{TLC} = \mathrm{RV} + \mathrm{ERV} + \mathrm{V_T} + \mathrm{IRV} = \mathrm{RV} + \mathrm{VC}. \tag{4.52}$$

The vital capacity is then the difference between TLC and RV:

$$\mathrm{VC} = \mathrm{ERV} + \mathrm{V_T} + \mathrm{IRV}. \tag{4.53}$$

The lung volumes depend on the subject size, sexe gender (Table 4.12), and age, particularly between children and adult (Table 4.13).

4.2.4 Dead Spaces

The dead space (V_D) is the portion of the tidal volume that does not participate in gas exchange. The *anatomical dead space* is the volume of the conducting airways that do not possess alveoli. It is equal to about 150 ml on the average in humans. Therefore, about 30% of inhaled air is wasted. The air indeed fills the anatomic dead space at the end of each inspiration and is exhaled unchanged.

The *physiological dead space* includes in addition to the anatomic dead space the contribution of well-ventilated poorly perfused alveoli. Because atmospheric p_{CO_2} is practically equal to zero, all the content in expired carbon dioxide can be assumed to come from functional alveoli and not from the dead space. Using the *Bohr equation* and measuring p_{CO_2} in expired air, the physiologic dead space can be computed.

The volume of a gas X equals its fractional concentration multiplied by the volume:

$$V_{E_X} = f_{E_X} V_T = f_{I_X} V_D + f_{A_X}(V_T - V_D). \tag{4.54}$$

Therefore,

$$V_D = V_T \frac{f_{E_X} - f_{A_X}}{f_{I_X} - f_{A_X}} = V_T \frac{p_{E_X} - p_{A_X}}{p_{I_X} - p_{A_X}} \tag{4.55}$$

The physiological dead space can be measured from another form of the Bohr equation:

$$V_D = V_T \frac{p_{A_{CO_2}} - p_{E_{CO_2}}}{p_{A_{CO_2}}}, \tag{4.56}$$

assuming that $p_{A_{CO_2}} = p_{a_{CO_2}}$. The formula is derived from the equation

$$V_D/V_E = (f_{A_{CO_2}} - f_{E_{CO_2}})/(f_{A_{CO_2}} - f_{D_{CO_2}}),$$

supposing that there is no mixture between gas in the dead space and alveolar gas ($f_{D_{CO_2}} \sim 0$), with $f_{CO_2} \propto p_{CO_2}$ ($V_T = V_E$).

In healthy individuals, the anatomic and physiologic dead spaces are roughly equivalent. However, in diseases characterized by lung regions with poorly perfused alveoli, the physiologic dead space may be considerably larger than the anatomic dead space. Hence, the physiologic dead space is clinically a more relevant index than is anatomic dead space. The *alveolar dead space* is the difference between the anatomic dead space and the physiologic dead space, being the part of gases that mixes with alveolar gases, but does not participate in gas exchange in well-ventilated nonperfused alveoli.

The value of the dead space varies with subject height $(H)^{10}$ and volume history.[11]

4.3 Mode of Ventilation

For a constant tidal volume, the usual breathing mode is characterized by a short inspiration followed by long expiration. The expiratory-to-inspiratory time ratio varies, but usually is equal to 1.4 in adults. Therefore, for a breathing frequency of 0.25 Hz (i.e., 15 cycles/mn, $T = 4$ s), inspiratory (t_I) and expiratory times (t_E) equal 1.67 and 2.33, respectively. For a breathing period of 5 s (12 cycles/mn, i.e., 0.20 Hz), t_I and t_I are equal to 2.1 and 2.9, respectively.

The expiratory-to-inspiratory time ratio varies during aging. In average, it is equal to about 1.1 in newborns, 1.3 in 7–24-month infants, and 1.46 in 2–3-year children. It then decreases slowly in older children. Critical thresholds are 1.4 in newborns, 1.5 in infants, and 1.7 in children and adults.

Breathing frequency, tidal volume, and inspiratory-to-expiratory time ratio are respiratory quantities that modulate *respiratory sinus arrhythmia* (RSA). Respiratory sinus arrhythmia corresponds to a variation in cardiac frequency that naturally occurs during a breathing cycle. Heart rate increases during inspiration and decreases during expiration. Cardiac frequency is controlled by centers in the medulla oblongata. Among these centers, the nucleus ambiguus increases parasympathetic nervous system input to the heart via the vagus nerve, hence decreasing sinusal node firing and heart rate. During expiration, cells of the nucleus ambiguus are activated and heart rate slows down. On the other hand, inspiration triggers inhibitory signals to the nucleus ambiguus. Inspiratory vagal blockade is sensitive to the steepness of inspiration. Electrocardiographic trace exhibits subtle changes in the R–R interval synchronized with respiration. The R–R interval is shortened during inspiration and prolonged during expiration.

Efficiency of pulmonary gas exchange is improved by RSA that aims at matching alveolar ventilation and perfusion during each respiratory cycle. Furthermore,

---

[10]For subject heights of 140 and 180 cm, V_D is respectively equal to ~0.35 and ~0.34 VC, or ~0.3 and ~0.24 TLC.

$$V_D = 2.1652\,H - 213$$

in a population of healthy adults, or

$$V_D = 7.585\,H^{2.363}.10^{-4}$$

in a population that includes children [13].

[11]The dead space measured after several deep expirations that mobilize ERV is lower than that after several deep inspirations that mobilize IRV. The existence of apnea modifies the value of the dead space in the first 5 seconds.

respiratory sinus arrhythmia does not fully control the heart vagal command, so that the respiratory modulation of cardiac vagal outflow does not represent the entire cardiac vagal tone.

Expiratory time is associated with inspiratory variables via mechanical, chemical, and neuronal mechanisms. Inspiratory timing influences expiratory timing independently of volume change and chemical drive, but expiratory timing does not significantly change inspiratory timing of the subsequent breath, although it has a small influence on tidal volume [14]. Expiratory time is regulated during speech and nonmetabolic respiratory functions.

4.4 Mechanics of Ventilation

The ventilation mechanics deals with forces leading to displacements of the lungs and of the chest wall, hence, to air flow and induced resistances. The quantities relevant to the ventilation mechanics are usually measured in static or quasistatic conditions. However, even during spontaneous breathing, the respiratory system does not behave as in static conditions. Respiratory muscle contraction generates an uneven distribution of pressure on the chest wall.

The magnitude of forces generated by activated respiratory muscles depends on the muscle properties (force–length and force–velocity characteristics). These forces are converted into pressures on the thoracic hollow structures that react according to their rheology. The applied pressure is used to: (1) mobilize the viscoelastic respiratory system; (2) change the lung volume and overcome the resistance of the respiratory system during air flow; and (3) accelerate the gas (the latter term is often neglected, except in high frequency ventilation):

$$p = q\mathrm{R} + V/\mathrm{C} + \dot{q}\mathrm{L}. \tag{4.57}$$

Air only flows in the entire airway tract when there is a difference between the pressures at the airway opening (mouth or nose) and in the alveoli, the opposite closed end of the airways. The inspiration is induced by a pressure higher at the airway opening that then becomes the airway entry. Conversely, the expiration is generated when the alveolar pressure exceeds the pressure at the airway opening (alveoli act as airway inlets). Therefore, the flow entirely reverses between the 2 main phases of the respiratory cycle, with strong inertial effects at the beginning and the end of each stage.

The greater the pressure difference between the entry and the exit of the respiratory tract, the faster the flow. The airway resistance, the ratio of driving pressure to the air flow rate, depends on the flow pattern, on the inhaled gas properties, and on the respiratory cycle phase (on the lung volume, hence on the airway bore).

The pressures are commonly given with respect to atmospheric pressure. This reference pressure is set to zero. The alveolar pressure determines whether air flows

Fig. 4.1 Monoaveolar model of the mechanics of the ventilation

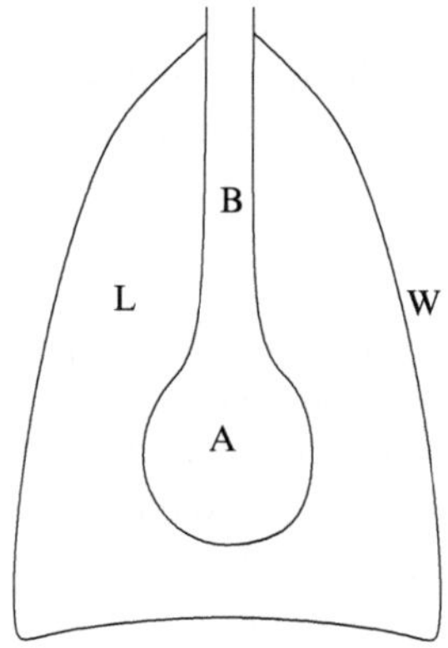

into or out of the air sacs, according to its negative (subatmospheric) or positive value. At end-inspiration or end-expiration, the alveolar pressure becomes equal to the atmospheric pressure (zero), and the air flow temporarily stops.

4.4.1 Monoalveolar Model

The monoaveolar model has been widely used in the mechanics of ventilation (Fig. 4.1). The lung–thorax system is represented by a mechanical system with one degree of freedom. Its motion is described by a single variable, the volume. The monoaveolar model includes 3 main components: (1) the chest wall with its musculature (W); (2) lung parenchyma (L); and (3) air conduits that are merged in a single bronchus (B) and a single alveolus (A).

The primary physical variables are (1) volume; (2) pressure set; (3) surface tension at the air–liquid interface of the alveolar wall; and (4) flow variables, resistance (R), compliance (C), and inertance (L), used in lumped parameter models of the air flow.

The involved pressures are the alveolar (p_A), intrabronchial (p_{ib}), end (p_m: mouth pressure owing to the mouthpiece of the measurement device), pleural[12] (p_{pl}) that is assumed to represent the intrathoracic (peribronchial) pressure, and perithoracic pressure (p_{atm}).[13]

These pressures allow the definition of a set of pressure difference, the *pressure drop* between 2 sections of the airpath as well as several *transmural pressures*.

[12]The pleural pressure within the pleural space is the pressure surrounding the lung. As a result of gravity in an upright subject (anatomical position), the pleural pressure at the lung base is greater (less subatmospheric) than at the lung apex.

[13]The atmospheric pressure, used as a reference, is the ambient air pressure, that averages 101.08 kPa (760 mm Hg) at sea level.

The *bronchial pressure difference* corresponds to the pressure difference between the bronchial ends (i.e. between the airway opening p_{ao} and the alveolus p_A) used for gas convection:

$$\Delta p = |p_{ao} - p_A| = R_G q + L_G q. \tag{4.58}$$

The pressure drop in the respiratory conduits $\Delta p = p_{v,aw} + p_{i,aw}$ is due to the frictional dissipation $p_{v,aw}$ and reversible variations in kinetic energy associated with geometry singularities ($p_{i,aw} \propto \rho U_q^2$, U_q: cross-sectional average velocity).

The *transpulmonary pressure*, or the so-called lung elastic recoil pressure[14] (p_L, usually measured in static conditions), is the pressure difference applied to the set constituted by alveolus walls (which represent the major part of the lung parenchyma) and alveolar gas. It is associated with the approximative behavior of the lung component:

$$p_L = p_A - p_{pl} = V_L/C_L + R_L q. \tag{4.59}$$

The transpulmonary pressure sets the lung size. The higher the transpulmonary pressure, the larger the lung.

The transpulmonary pressure is also supposed to be the sum of 3 components of the lung parenchyma (subscript L) reaction (elastic [$p_{el,L}$], viscous [$p_{v,L}$], and inertial [$p_{i,L}$] components):

$$p_L = p_{el,L} p_{v,L} + p_{i,L}. \tag{4.60}$$

The *transthoracic pressure* (p_{CW}) is associated with the approximative behavior of the chest wall component:

$$p_{CW} = p_{pl} - p_{atm} = V/C_{CW} + R_{CW} q. \tag{4.61}$$

The transthoracic pressure is also named the recoil pressure of the chest wall (subscript CW). When the chest wall is relaxed, the transthoracic pressure determines its size. When the transthoracic pressure equals zero, roughly at 75% of total lung capacity, the chest wall is at its unstressed volume.[15] When the pleural pressure is subatmospheric (with the chest wall muscles in relaxation state), the transthoracic pressure is negative, and the chest wall is pulled inward. Conversely, when the pleural pressure rises above the atmospheric pressure (with the chest wall muscles in relaxation state), $p_{CW} > 0$, and the chest wall bows out.

---

[14] In respiratory physiology, the recoil pressure is the difference in pressure between 2 sides of the deformable hollow structure (i.e., the transmural pressure). The transpulmonary pressure is also called the lung distending pressure.

[15] The unstressed chest volume refers to a zero transmural pressure at the chest wall, neglecting any prestress.

The transthoracic pressure corresponds to the pressure exerted by the respiratory muscles on the rib cage. It thus depends on the contraction state of the muscles. It is supposed to be the sum of an elastic $[p_{el,CW}]$, a viscous $[p_{v,CW}]$, and an inertial $[p_{i,CW}]$ component of the chest wall:

$$p_{CW} \sim p_{pl} - p_{musc} = p_{el,CW} + p_{v,CW} + p_{i,CW} \tag{4.62}$$

The *transthoracopulmonary pressure*, also designated as the recoil pressure of the respiratory system, is then given by:

$$p_{aA} = \Delta p_{tot} = |p_A - p_{atm}|. \tag{4.63}$$

The intrathoracic respiratory conduits are subjected to a transmural pressure. The *transbronchial pressure* is equal to:

$$p_B = p_{ib} - p_{pl} = \Delta p - p_L, \tag{4.64}$$

by adding and removing p_A.

Two classes of motor outputs exist that drive pump and resistance muscles. Muscles of the mouth and tongue or nose, glottis, and throat associated with upper airways as well as smooth muscles of the trachea and bronchi modulate airway resistance.

During breathing, the central nervous system receives sensory information and sends motor outputs. Breathing amplitude and rhythm adapt into an optimized pattern of thoracoabdominal movement appropriate to the metabolic demand. Chemosensors that are located primarily in carotid bodies and brainstem (pre-Bötzinger complex in the ventrolateral medulla and ventral zones of the medulla oblongata such as the retrotrapezoid nucleus) monitor blood gas concentrations.

4.4.2 *Quiet Breathing*

Rest breathing appears to be almost oscillatory (zero-mean sinusoidal flow), with the tidal volume amplitude and respiratory period. The tidal volume is mostly the sum of the dead space and alveolar volume. A given ventilation level can be achieved by different combinations of breathing pattern, varying the respiratory frequency and tidal volume.

The inspiratory time (t_I) is commonly shorter than the expiratory time (t_E). Moreover, volume changes are asymmetrical between inspiration and expiration.

When expiration ends and before inspiration begins, the respiratory muscles are at rest, the transpulmonary (+0.5 kPa) and transthoracic (−0.5 kPa) pressure are equal but opposite. The lung volume is equal to the functional residual capacity. The pleural pressure is subatmospheric (Table 4.14). The internal pressures along the tracheobronchial tree and in alveoli equal the atmospheric pressure. Air does not flow.

Table 4.14 Ventilation variable ranges during quiet breathing

Quantity	Range
Volume (V_T, ml)	[0,500]
Peak flow rate ($q \equiv \dot{V}$, ml/s)	[0,500]
Alveolar pressure (p_A, kPa)	[−0.1,0.1]
Pleural pressure (p_{pl}, kPa)	[−0.85,−0.5]

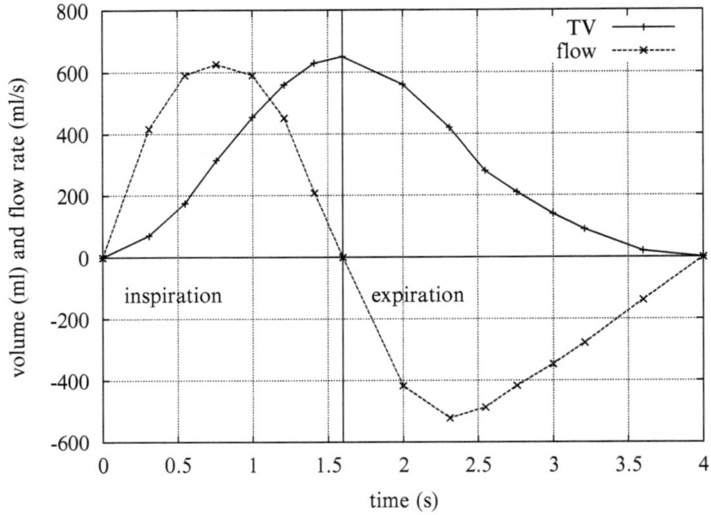

Fig. 4.2 Changes in volume and flow rate during rest breathing in adult male human subjects

During inspiration, the diaphragm (a pump muscle) and other inspiratory muscles, such as intercostals, contract. The contraction of the diaphragm, a dome-like muscular sheet that separates the thorax from the abdomen, increases the thoracic cage height; that of intercostal muscles raises the ribs and augments the cross section of the thorax.

The lungs expand with the thoracic cage. Because the diaphragm is covered by the inferior surface of the parietal pleura, when it contracts it pulls the pleura with it. The pleural pressure hence falls. Consequently, due to compliant properties of the alveolus wall, alveolar pressure becomes subatmospheric. The transpulmonary pressure increases. A pressure difference is generated between the airway opening and alveoli. Air flows into the lungs (Fig. 4.2).

The inspiration ends when a balance occurs between the inspiratory muscle pressure and the recoil pressure of the respiratory system (the transthoracopulmonary pressure). At the end of inspiration and before expiration begins, the alveolar pressure equals the atmospheric pressure (the transthoracopulmonary and driving pressure equal zero) and the pressure difference for flow in airways disappears.

The inspiratory muscles relax, causing an abrupt elevation in pleural pressure. The rise in pleural pressure causes the alveolar pressure to augment by the same amount, the lung volume, hence, the transpulmonary pressure being constant.

Table 4.15 Ventilation variables during quiet breathing. The tidal volume multiplied by the respiratory frequency gives the ventilation. The part of the tidal volume that participates in the gas exchange in the lungs (350 ml) multiplied by the respiratory frequency gives the alveolar ventilation

Respiratory variable	Usual value
Tidal volume (V_T, ml)	500
Rest ventilation ($\dot{V}$, ml/s)	100–125
Respiratory frequency (f_R, Hz)	0.25
Anatomical dead space (V_D, ml)	150
Alveolar volume (V_A, ml)	3000
Alveolar ventilation ($\dot{V}_A$, ml/s)	70–90
Pulmonary capillary volume (ml)	70
Pulmonary blood flow (ml/s)	83

During the following expiration, the increasing alveolar pressure establishes a positive pressure gradient from the alveolus to mouth, leading to air flow. The diaphragm passively relaxes. Thorax and lung volume decrease as air flows out, causing a reduction in transpulmonary pressure.

Expiration during quiet breathing is mainly a passive phenomenon. The lung passively returns to its preinspiratory volume. The respiratory muscles are relaxed. The lung and chest wall return to the equilibrium volume between these 2 structures, the functional residual capacity. In addition, during expiration, neural mechanisms prolong expiratory time above the passive duration.

The first portion of exhaled air at the airway opening corresponds to the air trapped in the anatomical dead space during the previous inspiration. The first portion of air entering in the alveoli is alveolar gas exhaled during the previous expiration and kept in the anatomical dead space. The neighboring environment thus collects a mixed expired air and the alveolus a mixed inspired air. The typical values of the ventilation parameters are given in Table 4.15.

The inspiration starting from FRC, with a gravity-related apicobasal difference in pleural pressure of about 0.75 kPa (lung apex: $p_{pl} = -1$ kPa, lung base: $p_{pl} = -0.25$ kPa), apical alveoli expands at a larger volume than the basal ones due to the difference in external pressure. However, in the range of lung volume during rest breathing, accommodation for volume increment remains similar in both apical and basal alveoli.

The combination of postinspiratory activity of the inspiratory muscles and the laryngeal control of expiratory flow explains that expiratory volumes are higher than in paralysed conditions.

During inspiration, laryngeal abductory muscles (posterior cricoarytenoid) widen the laryngeal lumen. During expiration, laryngeal adductory muscles (thyroarytenoid) narrow the laryngeal lumen. This muscle activity is controlled by the recurrent and superior laryngeal nerves.

During quiet breathing, two-thirds of the thoracic volume displacement is due to the diaphragm and one-third to the bony chest wall in both the sitting and standing position. In the lying position, the diaphragm contributes to three-quarters of the rest inspiration.

During passive inflation of the respiratory system, such as ventilation by a positive-pressure ventilator, both abdominal and pleural pressures increase. When the transdiaphragmatic pressure is assumed to be equal to zero, the changes in both cavity pressures are similar. When respiratory muscles are active, especially when the diaphragm contracts, the abdominal pressure increases.

4.4.3 *Effort Breathing*

Exercises disturb the body functioning, especially the ventilation and blood circulation. Exercise tests impose a muscular work in order to assess the body's reactions and its adaptation capabilities.

As any unsteady process, constant effort is characterized by 2 stages, an initial transient phase followed by a steady period, during which the oxygen consumption is relatively stable. Effort adaptation to adjust supplies to body needs is not immediate (oxygen debt).[16]

When effort stops, the oxygen consumption progressively returns to its rest level during a recovery step. During the recovery phase, the small oxygen stock is reconstituted, anaerobic metabolisms are replaced by aerobic processes, the tissue metabolism remains elevated due to temperature increase, and the oxygen supply to respiratory muscles remains augmented owing to their still increased activity.

The body adapts to effort by: (1) increasing muscular blood flow (up to 100 ml/100 g/mn); (2) augmenting oxygen captation in the alveolar septa (the mixed venous blood content in oxygen is reduced and the alveolocapillary difference in oxygen level rises); (3) rising ventilation, with an increase in both respiratory frequency and tidal volume, from the effort onset (ventilation hitching); and (4) elevating cardiac output (increase in stroke volume, cardiac frequency, blood pressure, and coronary flow).

During effort, the maximal voluntary ventilation can increase over 20-fold. Oxygen consumption can rise from 4.2 up to $\sim$52 ml/s (i.e. more than 12 times). Physiological responses include an increase in cardiac output, in ventilation, and in extraction of oxygen from the blood.

Owing to the apicobasal difference in pleural pressure especially during deep breathing, apical alveoli are more stretched than basal ones (Table 4.16). Hence, if inspiration starts from FRC, most of the incoming air goes preferentially to more compliant basal alveoli owing to the curvilinearity of the lung pressure–volume relationship (Sect. 4.4.4). When inspiration starts from RV, the air goes initially to

---

[16]The ratio of the oxygen debt to oxygen consumption allows distinguishing pulmonary disorders from cardiopathies, whether its value is lower than 40% (in favor of the former) or greater than this threshold (in favor of the latter).

Table 4.16 Pleural pressure (kPa), also transpulmonary pressure magnitude in noflow condition, at lung apex and base at residual functional capacity and at residual volume

	Apex	Base
FRC	−1.00	−0.25
RV	−0.40	+0.35

apical alveoli subjected to lower pleural pressure (−0.4 kPa). Basal lung regions undergo a positive pleural pressure (0.35 kPa) that prevents air passage through narrowed small airways.

4.4.4 Elastic Properties of the Lung and Chest Wall

Mechanical properties of the respiratory system are assessed in static conditions using more or less heavy experimental equipment and a set of assumptions that, most often, does not suit the dynamic conditions. Nevertheless, some mechanical variables can be measured during rest, or deep or forced breathing, using transducers with appropriate frequency response (Chap. 5).

4.4.4.1 Compliance

In mechanics and rheology, the *linear elastic modulus* (E [Pa]), also called tensile and Young modulus, describes the tensile elasticity, that is, the tendency of an object to lengthen along an axis when a traction [T] is applied along this axis. It is defined as the ratio of tensile stress (traction force per unit surface area [A]) to dimensionless tensile strain ($\Delta L/L_0$; L_0: initial length):

$$E = (T/A_0)/(\Delta L/L_0). \tag{4.65}$$

The Hooke law that describes the *stiffness* (k_c) of a purely elastic (ideal) spring, can be derived from the tensile stress–strain relation:

$$k_c = \Delta L/T. \tag{4.66}$$

The dimensional linear elastic coefficient also derived from the tension–length relation corresponds to $(T/A_0)/\Delta L$. The inverse of the *linear elastic coefficient* and *modulus* are the *distensibility coefficient* ($\Delta L/T$) and *modulus* ($(\Delta L/L_0)/T$).

The *rigidity modulus* (G), or *shear modulus*, describes the tendency of an object to shear (angular deformation; simple rotation of amount θ in the case of simple shear) under a torque (T). It is defined as the ratio of shear stress to dimensionless shear strain ($\Delta L/\ell_0$; ΔL: displacement; ℓ_0: initial transverse length):

$$G = (T/A)/(\Delta L/\ell_0) = (T/A)/\tan\theta. \tag{4.67}$$

The *bulk modulus* (B [Pa]) describes volumetric elasticity, that is, the tendency of an object to deform in all directions when it is uniformly loaded in all directions. It measures the material resistance to uniform traction or compression. It is defined as the ratio of the pressure change (volumetric stress) to the resulting relative volume change (volumetric strain):

$$B = \Delta p/(\Delta V/V_{\mathrm{ref}}) \quad \rightarrow \quad dp/(dV/V_{\mathrm{ref}}). \tag{4.68}$$

The material *compressibility* (K_m) is the inverse of the bulk modulus:

$$\mathrm{K}_m = (\partial V/V_{\mathrm{ref}})/\partial p. \tag{4.69}$$

For a fluid, the magnitude of the compressibility depends strongly on whether the process is adiabatic or isothermal. Isothermal and adiabatic compressibility are defined as:

$$\mathrm{K_T} = -\frac{1}{V}\left(\frac{\partial V}{\partial p}\right)_{\mathrm{T}}; \quad \mathrm{K_s} = -\frac{1}{V}\left(\frac{\partial V}{\partial p}\right)_{\mathrm{s}}, \tag{4.70}$$

where s is the entropy. The isothermal and adiabatic compressibility are related by the heat capacity ratio (γ):

$$\frac{\mathrm{K_T}}{\mathrm{K_S}} = \gamma. \tag{4.71}$$

When a material is strengthened or compressed in one direction, it usually retracts or expands in the other 2 directions (perpendicular to the direction of applied force). The Poisson ratio is the negative ratio of transverse to axial strain.

For a homogeneous isotropic material, simple relations exist among the elastic modulus, shear modulus, and bulk modulus, using the Poisson ratio (ν_{P}):

$$E = 2\,G(1 + \nu_{\mathrm{P}}) = 3\,B(1 - 2\nu_{\mathrm{P}}). \tag{4.72}$$

The *compliance* ($\mathtt{C}$) and its inverse, *elastance* ($\mathtt{E} = \mathtt{C}^{-1}$) refers to the distensibility of the respiratory structure. The compliance is given by the slope of the transmural pressure–volume curve for the lung or the relaxed chest wall. On the other hand, the elastance corresponds to the slope of the volume–transmural pressure curve.

Table 4.17 Quantities that measure volumic deformation of tube and hollow organ

Variable	Formula	Synonyms
Compliance	dV/dp	Capacitance, absolute distensibility
Specific compliance	$(dV/V)/dp$	Relative distensibility, compressibility
Elastance	dp/dV	
Relative elastance	$dp/(dV/V)$	Bulk modulus, Frank elasticity modulus

The transmural pressure is the difference between the internal and external pressure, that is, the difference in pressure exerted on the wetted surface minus that applied on the abluminal face of the wall of the considered respiratory structure (alveolus septum and airway and chest wall).

In the physiology of breathing and blood circulation, the compliance of a conduit (airway or blood vessel) describes the ease of stretching its wall. The greater the compliance, the higher is its stretchability.

Two ways exist to draw the relation between the transmural pressure and the volume of a pipe or hollow organ such as the lung whether the transmural pressure or the volume is taken as abscissa (horizontal coordinate), that is, the p–V or V–P relation are plotted. The slope at any point (or local curve gradient $[dy/dx]$) of the p–V or V–P (abscissa $[x]$–ordinate $[y]$) curve gives the local elastance or compliance, respectively.

In respiratory physiology, the compliance is usually defined as the ratio between the volume change and the variation of pressure across the structure (Table 4.17):

$$\mathrm{C} = \Delta V/\Delta p. \tag{4.73}$$

The compliance is measured during apneas at various lung volume (pointwise pressure–volume relationships).

In fact, several physiological quantities are used. The vascular compliance or *capacitance* is defined as $\Delta V/\Delta p$ [117]. Relative distensibility is defined as the ratio of compliance to volume [117]: $(\Delta V/V)/\Delta p$.

The lung compliance is the extent to which the lung expands for each increase in transpulmonary pressure (difference between the alveolar pressure and pleural pressure) at a steady state (enough time being allowed to reach the equilibrium, i.e., a nonphysiological procedure); the slope $\Delta V/\Delta p$ can be computed from the plotted p–V relation [117].

The pulmonary elastance is defined as $\Delta p/\Delta V$, which can be calculated from the V–p relation, the pressure being measured at the mouth and in the distal segment (downstream third part) of the esophagus, the latter being assumed to correspond to the pleural pressure [16]. In other words, the elastance is defined as the ratio between a pressure change to the volume variation caused by the pressure change ($\Delta p/\Delta V$) [133]. The relative volumic elastance, or Frank elasticity modulus, is given by $\Delta p/(\Delta V/V)$ [133].

The compliance is not only called capacitance, but also absolute volumic distensibility. The relative volumic distensibility $((\Delta V/V)/\Delta p)$ is also called the specific compliance [133]. Therefore, the specific compliance used by physiologists corresponds to the bulk modulus ($B = \Delta p/(\Delta V/V)$) used by mechanicians and rheologists.

In summary, the elasticity modulus and the *relative volumic elastance* represent a ratio between stress and dimensionless linear and volumic strain, respectively. On the other hand, the *specific compliance*, is the compliance relative to the lung volume ($\mathsf{C} = (\Delta V/V)/\Delta p$), also named relative distensibility, and corresponds to a ratio of strain to stress. The lung volume usually used for normalization is the functional residual capacity ($\mathsf{C} = (\Delta V/FRC)/\Delta p$). The relative *elastance* (in fact the bulk modulus) is the reciprocal of the compliance. It measures the variation in pressure achieved per unit volume change.

The lung compliance depends on: (1) the lung size; (2) inflation level and volume range over which measurements are carried out; (3) lung volume history; (4) airway permeability; (5) loading frequency; and (6) distensibility of air spaces. The higher the lung distensibility, the greater the compliance.

The "dynamic" compliance (compliance measured during very brief static condition [$\dot{V} = 0$] at end-inflation without interrupting rest breathing) is determined during spontaneous breathing, at small frequency to limit the influence of inertance, measuring the transpulmonary pressure, the volume and the flow rate at the airway opening. The "dynamic" compliance depends on the ventilation frequency. The higher the frequency, the smaller the compliance is.

The lung and chest wall are coupled. The chest wall tends toward expansion and lungs tend toward retraction. An equilibrium is established between the 2 anatomical structures during steady-state (nonphysiological) experiments. The rest volume of the respiratory system (lungs plus chest wall) is greater than the rest volume of the lung and lower than the rest volume of the chest wall. The lung being more distensible than the rib cage, the rest volume of the respiratory system is closer to the rest volume of the chest wall.

The compliance of the respiratory system ($\mathsf{C}_{\mathrm{CW,L}}$) is given by

$$\frac{1}{\mathsf{C}_{\mathrm{CW,L}}} = \frac{\Delta(p_{\mathrm{CW}} + p_{\mathrm{L}}}{\Delta V} = \frac{1}{\mathsf{C}_{\mathrm{CW}}} + \frac{1}{\mathsf{C}_{\mathrm{L}}}. \tag{4.74}$$

In a normal healthy lung at relatively low volume, either a small negative external pressure or a small positive internal pressure needs to be applied to significantly fill the lung. Because lung compliance decreases with increasing and decreasing lung volume, more pressure must be applied to get the same increase in volume.

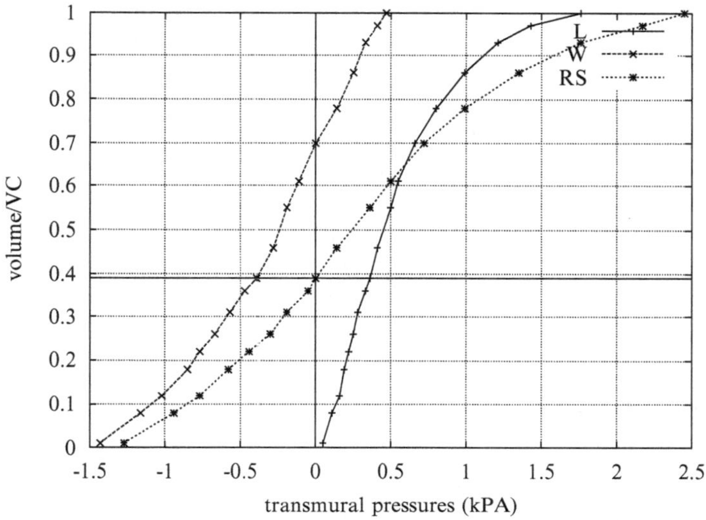

Fig. 4.3 Pressure–volume relationships of the respiratory system (RS; dotted line) and its components, the lung (L; solid line) and the chest wall (W; dashed line). At FRC (transmural pressure of the respiratory system equal to zero), the retraction forces of the lung balance the distension forces of the chest wall

Lung compliance changes in various diseases. In fibrosis, the lungs become stiffer. A greater transmural pressure is then necessary to maintain a moderate volume change. In emphysema, the lungs become too flexible. A small applied transmural pressure induces a large volume variation.

The relation between the airway caliber and the transmural pressure exhibits a sigmoidal curve. Several parameters affect airway compliance: (1) the smooth muscle tone; (2) connections between airway walls and adjoining structures; and (3) surface tension forces associated with liquid film on the respiratory epithelium surface (Vol. 5 – Chap. 13. Surfactant). In large airways, the dominant factor is the tone of the smooth muscle cells of the wall and the presence of cartilage. The more relaxed the parietal muscle is, the more distensible the respiratory conduit.

4.4.4.2 Pressure–Volume Relations

The elastic features of the chest wall and of the lung parenchyma are currently described by the relationships between the corresponding transmural pressure (transthoracic and transpulmonary pressures, respectively) and the volume (Fig. 4.3).

The contribution of the transpulmonary pressure and the transthoracic pressure to the transthoracopulmonary pressure varies throughout the vital capacity (Fig. 4.3).

At high lung volumes, both p_L and p_{CW} are positive. The lung parenchyma bears maximal distension forces and develops strong recoil forces. At low lung volumes, p_{CW} is negative and the chest wall tends to expand. At the resting volume of the respiratory system, the negative transthoracic pressure, in absolute value, is equal to the transpulmonary pressure. Lungs and chest are opposed each other. The former tends to collapse and the latter to expand. A negative pleural pressure hence results.

The pressure–volume relation of the lung parenchyma exhibits viscoelastic behavior, with the hysteresis between inflation and deflation curve parts. The nonlinearity is explained by the distensible elastic fibers and the stiffer collagen fibers that compose the matrix of the lung parenchyma.

4.4.5 Ventilatory Resistances

The total resistance of the respiratory system $\mathrm{R}_{CW,L}$ is computed from the difference between the total pressure ($p_{CW,L}$) and its elastic component ($p_{el,CWL}$) given by $p_{el,CWL} = V/\mathrm{C}_{CW,L}$:

$$\mathrm{R}_{CW,L} = (p_{CW,L} - p_{el,CWL})/\dot{V}. \tag{4.75}$$

The measurement is more accurate at maximal flow that usually occurs in the middle third of inflation.

The pulmonary resistance R_L is estimated during spontaneous breathing, from the difference between the pressure component $p_L \sim p_{ao} - p_{pl}$ and the elastic component $p_{el,L}$. The latter can be calculated once the compliance is measured at any lung volume during inspiration $p_{el,L} = V/\mathrm{C}_L$.

The same approach is used for the computation of the chest wall resistance (R_{CW}), from the transthoracic pressure (p_{CW}) and the corresponding elastic component ($p_{el,L}$).

4.4.6 Ventilatory Work

The mechanical work during a breathing cycle[17] can be estimated by measuring changes in the pleural pressure and lung volume: $W = pV$. The ventilatory work results from applied muscular forces and the potential energy stored in viscoelastic structures of lungs and chest wall.

The mechanical work necessary to inflate lungs from functional residual capacity to FRC+VT level and to deflate them to FRC is the area of the pressure–volume

[17]The ventilation work is accurately determined if applied to the whole breathing cycle and supposes that the pressure and volume are in phase.

loop and left from it. The left half of the pressure–volume loop area corresponds to elastic work that is recovered during expiration. The area left to the pressure–volume loop corresponds to dissipated energy with heat production.

The ventilatory work is used to overcome the strain resistance, which depends on lung volume, and flow resistance, which depends on the flow rate, then on the breathing frequency. The total inspiratory work is composed of several components. A first component is required to overcome the elastic resistance of the lung. A second component represents the work associated with the flow frictional resistance. For a given tidal volume, the work depends also on the inspiratory flow rate. During expiration, the work to overcome expiratory resistance is the area between the expiration pressure–volume curve and the elastic static pressure–volume curve of the lung. When the work is smaller than the stored quantity at the end of the inspiration, expiration can be an entirely passive process. Otherwise, expiratory muscular work is required.

For a given ventilation, a high-frequency small-volume breathing reduces the elastic work of breathing, but increases the nonelastic component. Conversely, a deep and slow breathing pattern reduces the frictional work, but increases the elastic work.

4.5 Ventilation–Perfusion Ratio

The primary function of the cardiovascular system is the supply of O_2 to and removal of CO_2 from the cells of the body. Respiratory gas exchange and transport can be limited by the diffusion alveolocapillary barrier. Inside lungs, the respiratory and cardiovascular systems combine their activities, such that locally alveolar gas and pulmonary blood volume on the one hand and breathing and cardiac frequency yield adequate regional alveolar ventilation and perfusion rate.

However, in normal lungs, neither ventilation nor perfusion are evenly distributed throughout the lung. Lung mass and shape determine the alveolar size and stress distribution, and, consequently, the ventilation distribution within the lungs. Apicobasal differences in compliance and resistance play a crucial role regarding the flow rates at given lung volumes. The distribution of ventilation across the lung is related to the position of each area on the compliance curve at FRC. Because the base state corresponds to a more favorable part of the compliance curve than the apices, the bases gain more volume change for similar applied pressure change and thus receive a greater ventilation.

The composition of the alveolar gas depends on 4 factors: (1) ventilation, (2) lung perfusion, (3) composition of inspired gas, and (4) composition of mixed venous blood. Early works develop a three-compartment model of pulmonary gas exchange composed of a first ideal compartment, a second with unperfused alveolus, and a third with unventilated alveolus [17, 18].

The respiratory exchange ratio R_R is calculated using the equation

$$R_R = \frac{\dot{V}_{CO_2}}{\dot{V}_{O_2}} = \frac{\dot{V}_E F_{E_{CO_2}}}{\dot{V}_E(F_{I_{O_2}} - F_{E_{O_2}})} = \frac{p_{E_{CO_2}}}{(p_{I_{O_2}} - p_{E_{O_2}})}. \quad (4.76)$$

The basic equation of the ratio between the alveolar ventilation ($\dot{V}_A$) and the alveolar perfusion q appears simple:

$$\frac{\dot{V}_A}{q} = 8.63\, R_R \frac{(c_{a_{O_2}} - c_{v_{O_2}})}{p_{A_{CO_2}}}, \quad (4.77)$$

where $c_{a_{O_2}}$ and $c_{v_{O_2}}$ are the oxygen concentrations in effluent and affluent blood, and $p_{A_{CO_2}}$ the alveolar partial pressure in carbon dioxide. However, the solution depends on the nonlinear oxygen and carbon dioxide dissociation curves. Consequently, the study of ventilation–perfusion relationships strongly relies on graphical analysis, especially the oxygen–carbon dioxide diagram [17].

The perfusion distribution throughout the lung depends on gravity. In the upright (anatomical) position, the perfusion pressure at the lung base is equal to the mean pulmonary artery pressure (~2 kPa) in addition to the hydrostatic pressure between the main pulmonary artery and lung base (~1.5 kPa). At the apices, the perfusion pressure is small, the hydrostatic pressure difference between the main pulmonary artery and lung apex being subtracted from the mean pulmonary artery pressure.

Ventilation and perfusion are relatively matched, with the lung bases receiving more of both than the apices. The inequality between bases and apices is indeed less marked for ventilation than for perfusion. It results an efficient oxygenation of blood passing through the lungs.

Using isotope analysis and arbitrarily dividing the lung in 9 thick slices from lung apex to base, the ventilation–perfusion ratio was shown to vary from 3.3 at the lung apex to 0.63 at the lung base, with a mean of 0.8 [19, 20].

Disturbance in ventilation and/or perfusion distribution leads to mismatch between the 2 processes (Fig. 4.4; Tables 4.18 and 4.19). When the ventilation–perfusion ratio either increases or decreases, the alveolar gas composition changes. The oxygen level in arterial blood is then reduced (hypoxemia). When ventilation occurs in a lung region of low ventilation–perfusion ratio, the hypoxemia can be corrected by restoring the alveolar oxygen delivery to a level sufficient to oxygenate blood appropriately.

At extreme ventilation–perfusion mismatch, a lung region being not perfused (infinite ventilation–perfusion ratio) corresponds to the *alveolar dead space*. In contrast, a lung region that is not ventilated after airway blockage leads to a *shunt* (zero ventilation–perfusion ratio). The well-ventilated lung regions cannot compensate for the shunt area because hemoglobin is almost fully saturated.

In supine position (i.e., dorsal decubitus), the regional blood flow increases from 1.2 ml/mn/cm^3 in ventral to 3.5 ml/mn/cm^3 in dorsal lung regions, whereas the craniocaudal gradient is small [22]. In prone position (i.e., ventral decubitus), regional alveolar ventilation matches regional blood flow, except at the dorsal lung

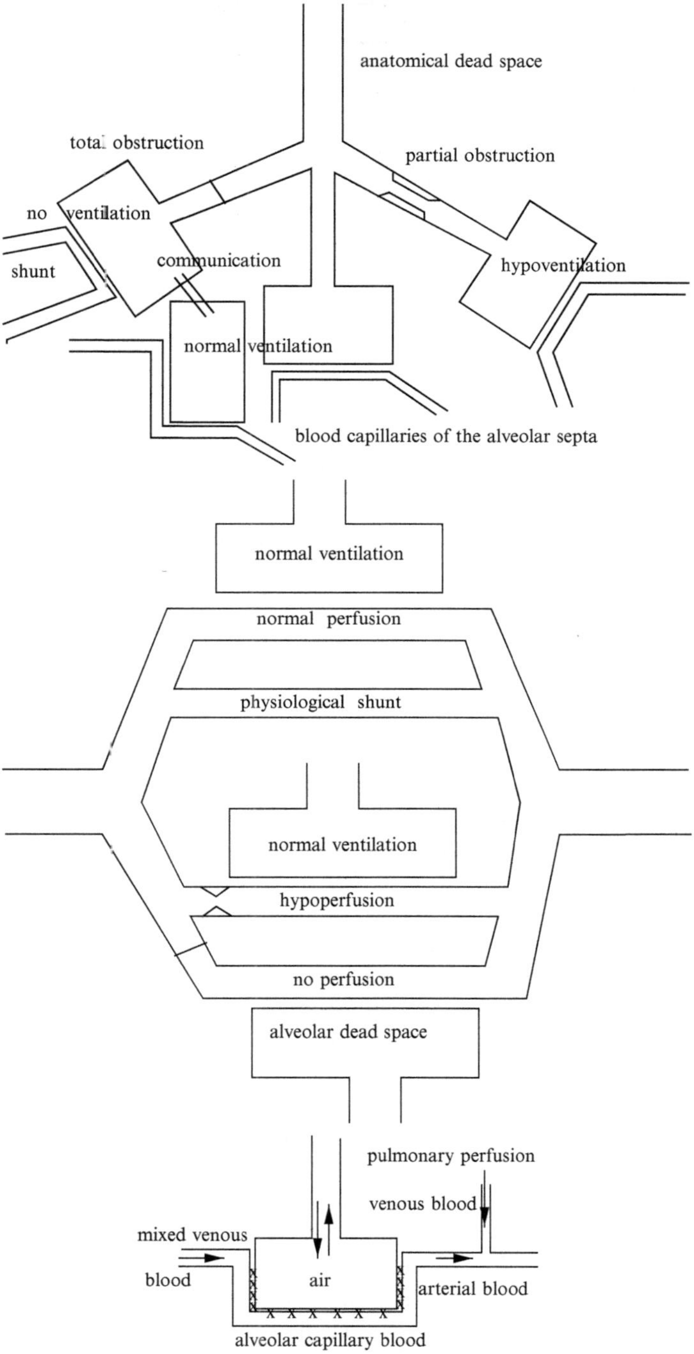

Fig. 4.4 Ventilation and perfusion of lungs. Normal and impaired distribution of the ventilation (**top**), perfusion (**middle**), and disturbances of the alveolocapillary transfer (**bottom**)

Table 4.18 Effects of various factors on alveoloarterial difference of oxygen partial pressure (p_{O_2}; Source: [21]; subscript$_{\bar{A}}$: mixed alveolar gas; subscript$_{\bar{a}}$: mixed arterial blood; +, ++, and + + +: slight, moderate, and large consequence)

	$p_{\bar{A}_{O_2}} - p_{\bar{a}_{O_2}}$	$p_{\bar{a}_{CO_2}} - p_{\bar{A}_{CO_2}}$
Venous admixture (shunt)	++	+
Transfer limitation	+	+
Overventilated or underperfused alveoli	++	+ + +
Underventilated or overperfused alveoli	++	~0

Table 4.19 Change in arterial gas partial pressure with falling alveolar ventilation assuming normal gas distribution and diffusion (respiratory quotient $R_R = 0.8$; Source: [21])

$\dot{V}_A$	p_{O_2} (mmHg)	Oxygen saturation (%)	p_{CO_2} (mmHg)	pH $[HCO]_3^-]$cst.
5	100	95.7	40	7.40
4	82	93.7	50	7.32
3	68	87.0	75	7.20
2	30	40.0	105	7.05

thoracic border, where a reduction in ventilation occasionally happens in normal subjects. Local vascular transit time is correlated to regional blood flow.

During exercise, the respiratory and cardiovascular systems coordinate their activity with that of the motor system, but all these physiological systems do not synergize their rhythms. The nervous system coupling ensures coordination among circulatory, respiratory, and motor centers.

Ordinary walking requires a 20-fold increase in O_2 consumption and CO_2 production by muscles with respect to the resting level. During steady-state exercise, ventilation increases in proportion to CO_2 production to regulate arterial CO_2 partial pressure (Pa_{CO_2}). The relationship between CO_2 output and O_2 input remains identical in unsteady exercise with respect to steady experiments [23]. However, the ventilation kinetics is quicker than O_2 uptake. Exercise-resulting hyperpnea is coupled with the pulmonary gas exchange rate via a signal proportional to CO_2 production in both steady and unsteady states of moderate exercise.

Therefore, the mechanism that controls the body's respiration during exercise aims at preventing rising Pa_{CO_2} when CO_2 production heigthens. The respiratory control system seems to respond to factors temporally associated with the pulmonary gas exchange rate, regardless of the motor activity. During exercise, among relevant information, the rate at which CO_2 is exchanged by the lungs appears to be the most appropriate [24].

4.6 Gas Transport

Gases are transported between the ambient air and tissues using both body's ventilation and blood circulation. Oxygen is needed by cells for energy fabrication. Carbon dioxide is removed with wastes from the cells.

Although gases dissolve in blood, neither O_2 nor CO_2 are primarily transported in the dissolved form. Oxygen is almost entirely carried by hemoglobin.[18] Carbon dioxide is carried predominantly in the form of bicarbonate ions (HCO_3^-).

Several types of disorders perturb gas transport from the ambient air to body tissues: (1) ventilation disturbances (hypoventilation, regulation impairment, dead space increase) that reduce the alveolar oxygen content; (2) alveolar gas exchange disturbances (perturbed ventilation–perfusion relation, increased tissular diffusion resistance); (3) blood circulation disturbances (heart failure, right-to-left shunt, vascular occlusion).

4.6.1 Inhaled Air Distribution

The distribution of inspired air depends on: (1) the regional convection-based ventilation described by the Navier–Stokes equations in large bronchi and Stokes equations in bronchioles; and (2) diffusion process in the multinary gas mixture within the last generations of bronchioles and alveolar ducts and sacs described by the Stefan–Maxwell equations.

The collateral ventilation is aimed at homogenizing the alveolar air composition in neighboring air spaces. However, it only has a slight effect. A stratification associated with a concentration gradient can occur due to a partial mixture in the respiratory terminal unit between inhaled gas and alveolar gas.

The regional ventilation depends on: (1) the local averaged airway path length and the local airway resistance; (2) the stress according to the location; and (3) the local rheological properties of the lung parenchyma. The air travel length in the intrathoracic airways down to the alveoli vary from $\sim$25 to $\sim$45 cm. Gas can reach proximal alveoli after an inhaled volume of $\mathcal{O}(10)$ ml and distal alveoli after an inhaled volume of $\mathcal{O}(100)$ ml.

Radioactive xenon administration shows that the lung base is more ventilated than the lung apex at FRC, and conversely at RV. All lung regions do not have the same inflation status with respect to the bulk motion, due to differences in applied pressures resulting from gravity. At TLC, all lung regions are maximally expanded, but the maximal volume can differ between the lung regions. The regional VC is larger in the lower lung regions than in upper ones. Similarly, the regional

[18]Hemoglobin is a protein that contains 4 subunits. Each subunit possesses a ferrous (Fe^{2+}) ion within a heme group. Up to 4 oxygen molecules can bind reversibly to each Hb molecule, one to each Fe^{2+} site.

Table 4.20 Partial pressure (kPa) of oxygen and carbon dioxide in serial compartments between atmospheric air and body tissues. In atmospheric air at a pressure of 101 kPa, the partial pressures of oxygen (fraction 20.9%) and carbon dioxide (fraction 0.04%) are 21.2 and 0.04 kPa, respectively

gas	atmosphere	airway	alveolus	artery	vein	cell	exhalation
O_2	19–21	20	13.3	12–13.3	4.6–6	<0.7	2.1
CO_2	0.04	0.04	5.3	5.3	6.1	>6	4.3

TV is greater in the lung top zones than in the bottom ones. The regional FRC (rFRC) also decreases from the lung apex to the lung base. At the lung base, rFRC corresponds to ~2/3 of the regional TLC (rTLC), and at the lung apex, to ~1/3 rTLC.

4.6.2 Gas Content in the Serial Compartments

The magnitude of gas partial pressure in the various compartments in series is determined by the level of ventilation and blood circulation (Table 4.20). From a partial pressure in the atmosphere of 21 kPa, the partial pressure of oxygen falls before reaching the arterial blood in 3 successive compartments: (1) upper airways, where the inspired air is humidified, reducing p_{O_2} to about 19.7 kPa; (2) alveoli that incorporate carbon dioxide, reducing p_{O_2} to about 14.4 kPa; and (3) a small physiological shunt that reduces p_{O_2} to approximately 13.3 kPa.

The mean alveolar oxygen partial pressure during the respiratory cycle is calculated using the simplified *alveolar gas equation*:

$$p_{\mathrm{A}_{O_2}} = (p_{\mathrm{I}_{O_2}} - p\mathrm{A}_{CO_2})/R_{\mathrm{R}}, \tag{4.78}$$

where $0.7 < R_{\mathrm{R}} < 0.95$, $p_{\mathrm{I}_{O_2}}$ the oxygen partial pressure in the inspired gas, and $p_{\mathrm{A}_{CO_2}}$ the alveolar carbon dioxide partial pressure.

4.6.3 Alveolar Ventilation

The alveolar gas composition results from the intermittent inflow of fresh air and outflow of alveolar gas (Fig. 4.5) and the permanent uptake of O_2 and release of CO_2 by the blood flow inside the alveolar septa. The alveolar ventilation can be estimated by the simple formula

$$\dot{V}_{\mathrm{A}} = (\mathrm{V_T} - \mathrm{V_D})\, f_{\mathrm{R}}. \tag{4.79}$$

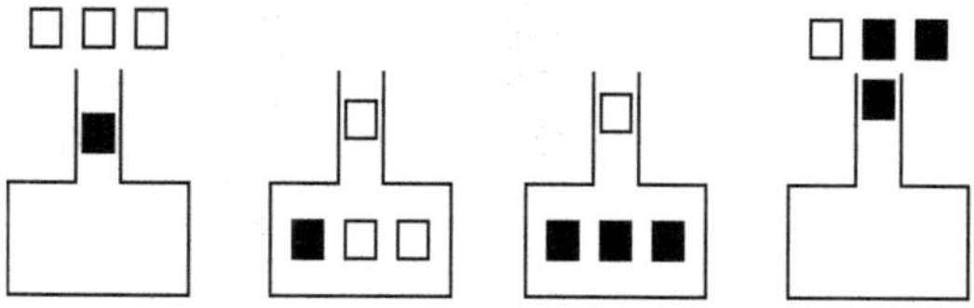

Fig. 4.5 Schematic explanation of the alveolar ventilation using a one-compartment model of the respiratory tract (assuming a uniform ventilation between the lung territories) at 4 instants of the respiratory cycle. Each square represents a gas volume of about 150 ml, empty square ambient air, filled square alveolar gas

Table 4.21 Transit time (ms) of lung capillary blood in the alveolar septum at rest and during exercise

Rest	Exercise
750	300

The breathing pattern (V_T, f_R) strongly influences the alveolar ventilation. The greater the tidal volume is, the more efficient the alveolar ventilation (the higher the V_T/V_D) ratio. The time taken by the pulmonary capillary blood to travel inside alveolar walls enables equilibrium between blood and alveolar gases (Table 4.21).

A slight difference in oxygen content between alveolar and systemic arterial values arises because of (1) a physiological ventilation–perfusion mismatch and (2) venous blood from lung tissues empties into the pulmonary veins, thus diluting the oxygen-rich blood that goes to the left heart. The alveoloarterial difference increases with age (in healthy young adults, <1.3 kPa).

A large change in alveolar ventilation has different consequences in $p_{A_{O_2}}$ whether the alveolar ventilation is high, inducing small variations in $p_{A_{O_2}}$, or low, causing great variations in $p_{A_{O_2}}$.

There are 2 kinds of increase in alveolar ventilation. An increase in alveolar ventilation relative to its reference value is called *hyperpnea*. An increase in alveolar ventilation relative to the corresponding metabolic level is termed *hyperventilation*. The latter generates a drop in $p_{A_{CO_2}}$. Isocapnia maintenance means that the level of alveolar ventilation remains proportional to the metabolic state ($\dot{V}_A/\dot{V}_{CO_2}$ = constant (normoventilation state irrespective of the absolute level of alveolar ventilation).

Alveolar Gas Equation

The exchange of oxygen and carbon dioxide with the environment depend on the pulmonary ventilation:

$$\dot{V}_{O_2} = \dot{V}_E(c_{I_{O_2}} - c_{E_{O_2}}), \tag{4.80}$$

$$\dot{V}_{CO_2} = \dot{V}_E(c_{I_{CO_2}} - c_{E_{CO_2}}), \tag{4.81}$$

where $c_{I_\bullet}$ and $c_{I_\bullet}$ are the mean inspired and expired concentrations of the gas component (O_2 or CO_2). Because the respiratory exchange ratio ($\dot{V}_{CO_2}/\dot{V}_{O_2}$) ranges from 0.7 to 1; inspired ($\dot{V}_I$) and expired ($\dot{V}_E$) ventilations differ. $\dot{V}_E$ is an important parameter according to ventilation regulation. However, with respect to gas exchange and adaptation to metabolic needs, the functional parameter of interest becomes the alveolar ventilation ($\dot{V}_A$):

$$\dot{V}_{O_2} = \dot{V}_A(c_{I_{O_2}} - c_{A_{O_2}}),$$
$$\dot{V}_{CO_2} = \dot{V}_A(c_{A_{CO_2}} - c_{I_{CO_2}}).$$

Using the partial pressure of O_2 ($p_{A_{O_2}}$) and CO_2 ($p_{A_{CO_2}}$) rather than the gas component concentrations, the alveolar gas equations become, neglecting the small difference between inspired and expired ventilation:

$$p_{A_{O_2}} = p_{I_{O_2}} - (\dot{V}_{O_2}/\dot{V}_A)p_{\text{atm}},$$
$$p_{A_{CO_2}} = p_{I_{CO_2}} + (\dot{V}_{CO_2}/\dot{V}_A)p_{\text{atm}}.$$

$p_{A_{CO_2}}$ can be measured by the end-expiratory p_{CO_2}. $p_{A_{CO_2}}$ is close to the arterial partial pressure ($p_{a_{CO_2}}$), except in the case of large shunts or major ventilation–perfusion inequalities.

4.6.4 Gas Transport and Mixing

The alveolar air does not have an uniform composition in the different pulmonary territories. These heterogeneities in composition of the gas mixture can be due to a heterogeneous distribution of inhaled air in these territories associated with convection and/or disturbances in diffusion in the distal region of the tracheobronchial tree.

A gas concentration gradient along the respiratory conduit axis that represents the last bronchiolar generations and the alveolar sacs, persists before the direction change of the respiratory flow (stratification). The diffusion indeed requires a finite duration for concentration equilibrium that is necessarily limited by the period of the respiratory cycle.

Table 4.22 Alveolocapillary membrane and various media crossed by gases between alveolar air and pulmonary capillary blood. Pulmonary alveoli are lined with a thin coating of pulmonary surfactant that adapts the surface tension to the alveolar volume and stabilizes alveoli. Changes in surfactant concentration and surface tension occur during respiration. They prevent alveolar collapse during deflation (exhalation), as it thickens and lowers the surface tension (down to 24.10^{-3} N/m), and rupture during inflation (inhalation), as it thins and elevates the surface tension. The basement membrane (thickness 100–400 nm) is made of the reticular and basal lamina, the latter being subdivided into the lamina lucida and lamina densa (thickness 30–80 nm). Each cell possesses a basement membrane with a specific thickness and constitution

Surfactant	~100 nm
Alveolar epithelium	
Pneumocyte-1	3–5 μm
Pneumocyte-2	8–12 μm
Macrophage	~20 μm
Fibroblast	10–15 μm
Basement membrane	10 nm (usually 100–200 nm)
Alveolar interstitium	$0<h<\mathcal{O}[10\,\mu m])$
Alveolar and capillary basement membrane can fuse	
Basement membrane	10 nm (usually 100–200 nm)
Capillary endothelium	1–8 μm
Plasma	$\mathcal{O}[100\,nm])$
Red blood capsule	Diameter 6.2–8.2 μm, outer ring thickness 2–2.5 μm, central depression thickness 0.8–1 μm
Thinnest septum	200–800 nm

4.6.5 Alveolocapillary Transfer

Pulmonary gas exchanges rely on gas transport across the thin tissue barrier that separates alveolar space from plasma and erythrocytes of surrounding capillaries (Table 4.22).

In a fluid, diffusion of a gas is a Brownian motion (i.e., a continuous random movement). In a steady regime, when the gas distribution is heterogeneous, molecules move according to a preferential direction from high to small concentration region until homogeneity is achieved.

The driving force is the partial pressure (p_X) of gas species X in the mixture between compartments separated by a virtual barrier ($\Delta p_X/\Delta L$), or, at the infinitesimal length scale by pressure gradient ($d\,p_X/d\,x_i$). The gas flow rate by diffusion across the fictitious barrier per mol per mm is:

$$\dot{M}_X = \mathcal{T}_L G(p_{X_1} - p_{X_2}), \tag{4.82}$$

where $\mathcal{T}_L$ is the transfer capacity of the alveolocapillary barrier for the gas species of interest (Table 4.23) that depends on material permeability ($\mathcal{P}$), surface area (A), and thickness h of the barrier. In particular,

Table 4.23 Physical analogies ($\mathcal{T}_{L_{O_2}} = 30$ ml/mn/mmHg)

Electricity	Hydrodynamics	Membrane gas transfer
$u = \mathrm{R}i$	$\Delta p = \mathrm{R}q$	$\Delta p_X = \mathrm{R}_{mb_X} J_X$
$i = \mathrm{G}u$	$q = \mathrm{G}\Delta p$	$J_X = \mathcal{T}_{L_X} \Delta p_X$

Table 4.24 Oxygen diffusivity and solubility in various media (Source: [25])

Medium	Diffusivity $\mathcal{D}_{O_2}$ (cm^2/s)	Solubility H_{O_2} (nmol/ml/mmHg)	κ_{O_2} (nmol/cm/s/mmHg)
Air (20°C)	0.2	55	11
Water (37°C)	33×10^{-6}	1.4	46×10^{-6}
Connective tissue (37°C)	10×10^{-6}	1.4	14×10^{-6}

$$\dot{V}_{O_2} = \mathcal{T}_{L_{O_2}} (p_{A_{O_2}} - p_{\bar{v}_{O_2}}). \tag{4.83}$$

In the alveolocapillary barrier,

$$\dot{V}_{O_2} = \mathcal{P}_K \frac{A}{h} (p_{A_{O_2}} - p_{b_{O_2}}) \quad = \mathcal{D}_{O_2} \frac{A}{h} (c_{A_{O_2}} - c_{b_{O_2}}), \tag{4.84}$$

where the gas conductance in the tissue $\mathrm{G}_{tis} = \mathcal{P}_{K_{tis}} A_{tis} / h_{tis}$; $\mathcal{P}_K$ is the Krogh permeability:

$$\mathcal{P}_K = \mathcal{D}_{O_2} \mathrm{H}_{O_2}, \tag{4.85}$$

$\mathcal{D}_{O_2}$ being the oxygen diffusivity and H_{O_2} the oxygen Henry parameter (oxygen solubility) in the material of interest (Table 4.24):

$$\mathrm{H}_{O_2} = \frac{\Delta c_{O_2}}{\Delta p_{O_2}}, \tag{4.86}$$

where p is the partial pressure and c the concentration of the dissolved gas in the liquid.

Convective transport of oxygen dissolved in blood and bound to a carrier depends on its concentration and flow rate (q) of blood supposed to be that of the carrier:

$$\dot{M}_{O_2} = q c_{O_2} = q \mathrm{H}_{O_2} p_{O_2}, \tag{4.87}$$

as the oxygen concentration is proportional to the product of its capacitance and partial pressure.

Any increase in blood–gas barrier thickness alters the exchange efficiency. In healthy alveoli, the harmonic mean thickness of the blood–air barrier is 770 nm. Oxygen traverses this barrier in less than 1 ms to saturate red blood cells in tens of milliseconds (i.e., before they exit the gas exchange region; 750 ms in humans) [26].

4.6.6 Pulmonary Gas Transfer Capacity

The *transfer capacity* ($\mathcal{T}_L$ [ml/mn/mmHg or ml/s/kPa]) of the alveolocapillary barrier for a gas is the gas volume per unit time and unit pressure gradient that crosses the barrier and goes to the blood or travels in the inverse direction.

The gas flux across the alveolocapillary membrane depends on: (1) membrane surface area (A_{mb}); (2) membrane thickness (h_{mb}); (3) membrane structure; (4) effective transport coefficient ($\mathcal{T}_L$); (5) alveolocapillary pressure gradient (Δp); (6) capillary septal density; and (7) blood flow rate (q). The travel length across the alveolar septum varies between 0.5 and 2.5 μm.

Oxygen uptake by pulmonary capillaries equals the sum of O_2 transfer through 2 biological tissues, the alveolocapillary membrane and plasma, and reaction rate between O_2 and hemoglobin.

The overall *resistance to gas transfer* depends then on 3 factors: (1) *alveolocapillary membrane factor* (i.e., the rate of gas transfer through the alveolar septum); (2) *plasma factor* (i.e., the rate at which the dissolved gas leaves the alveolar surface and enter in the flowing plasma); and (3) *cell factor* (i.e., the binding rate to hemoglobin).

The alveolocapillary membrane factor depends on: (1) the alveolocapillary partial pressure gradient (Δp); (2) the solubility and molecular mass of the gas; and (3) the thickness, surface area, and composition of the alveolocapillary membrane:

$$J_{ACM} = \mathcal{T}_{ACM} \Delta p_{A-c}. \quad (4.88)$$

The conductance of the alveolocapillary membrane (G_{ACM}) is given by:

$$\mathsf{G}_{ACM} = \mathcal{P}_{ACM} \frac{(A_A + A_{cap})/2}{h_{eff_{ACM}}}. \quad (4.89)$$

The plasma factor depends on: (1) capacitance (C); (2) local blood flow (q); and (3) the arteriovenous partial pressure gradient (Δp):

$$J_{pl} = \mathsf{C} q \Delta p_{a-v}. \quad (4.90)$$

The plasma conductance is given by

$$\mathsf{G}_{pl} = \mathcal{P}_{pl} \frac{A_{pl}}{h_{pl}}. \quad (4.91)$$

The cell factor depends on: (1) gas (CO or O_2)–hemoglobin binding rate (B_{RBC}); (2) pulmonary capillary blood volume (V_{cap}); and (3) the mixed venous–capillary partial pressure gradient (Δp):

$$J_{RBC} = \mathsf{B}_{RBC} V_{cap} \Delta p. \tag{4.92}$$

The RBC conductance is given by

$$\mathsf{G}_{RBC} = \mathsf{B}_{RBC} V_{cap} \times \frac{H_{mea}}{H_{ref}}. \tag{4.93}$$

The overall *transfer capacity* of the lung ($\mathcal{T}_L$ [mol/s/Pa])[19] is composed of 2 elements: (1) the transfer capacity of the pulmonary membrane ($\mathcal{T}_{mb}$) and (2) the transfer capacity of pulmonary capillary O_2-carrying capsule ($\mathcal{T}_{RBC}$). The transfer capacity of the pulmonary membrane is constituted of 2 terms, (1) the transfer capacity of the alveolocapillary membrane ($\mathcal{T}_{ACM}$) and (2) the transfer capacity of plasma ($\mathcal{T}_{pl}$):

$$\mathsf{R}_{\mathcal{T}_{L_{O_2}}} = \frac{1}{\mathcal{T}_{L_{O_2}}} = \frac{1}{\mathcal{T}_{ACM_{O_2}}} + \frac{1}{\mathcal{T}_{pl_{O_2}}} + \frac{1}{\mathcal{T}_{RBC_{O_2}}}. \tag{4.94}$$

The term linked to RBC compartment can be written as

$$\frac{1}{\mathcal{T}_{RBC_{O_2}}} = \frac{1}{\mathsf{B}_{RBC_{O_2}} V_{cap}}, \tag{4.95}$$

where $\mathsf{B}_{RBC_{O_2}}$ is the oxygen–hemoglobin binding rate (ml/mn/mmHg) per ml of pulmonary capillary blood for a reference hematocrit, that is, the gas amount (ml) taken up by red blood capsules in 1 ml of blood per minute for a gradient of 1 mmHg partial pressure of dissolved gas in the plasma and in the interior liquid of the red blood capsule, and V_{cap} the blood volume (80–100 ml at rest) in pulmonary capillaries exposed to the alveolar gas, the product $\mathsf{B}_{RBC_{O_2}} V_{cap}$ being the transfer capacity of blood.

Therefore [27, 139]:[20]

$$\frac{1}{T_L} = \frac{1}{T_{mb}} + \frac{1}{\mathsf{B}_{RBC} V_c}. \tag{4.96}$$

---

[19]The lung transfer capacity of carbon monoxide equals about 20 mmol/mn/mmHg (i.e., 6.7 mmol/mn/kPa).

[20]In other words, $\mathcal{T}_{mb} \propto (p_{A_{CO}} - p_{b_{CO}})^{-1}$ is the conductance from alveolar gas up to the plasma (subscript b), thus including both the alveolocapillary membrane and plasma conductances, $\upsilon_{RBC} V_c \propto (p_{b_{CO}} - p_{RBC_{CO}})^{-1}$ the oxygen-dependent conductance from erythrocyte (subscript RBC) plasma membrane to hemoglobin.

Table 4.25 Blood pressures (kPa) in the pulmonary circulation

Right ventricle	∼0–3.3 (2.3–4)
Pulmonary artery	min: 0.7–1.7, max: 2.3–4, mean: 1.3–2.7
Capillary bed inlet	1.6
Capillary bed outlet	1.1
Pulmonary vein	min: 0.5–0.7
Left atrium	0.5

The Roughton–Forster equation splits the alveolocapillary gas diffusion taking into account the gas kinetics with plasma and hemoglobin in solution [139].

At rest, the transit time of the capillary blood equals about 750 ms. Because the equilibrium between the alveoalr and blood phases is reached in 300–350 ms, there is no difference in gas level between the alveolus and the capillary downstream segment within the alveolar septa, when the transfer capacity is normal.

The alveoloarterial differences in gas partial pressure is determined by the membrane tranfer rate and mainly by the distribution of the ventilation–perfusion ratios. The gas transfer is optimal in lung regions characterized by a matching between ventilation and perfusion.

Imaging of gas transfer can rely on solubility and chemical shift of ^{129}Xe atoms [26].[21] This technique is sensitive to changes in blood–gas barrier thickness of approximately 5 μm.

4.6.7 Pulmonary Circulation

The pulmonary circulation supply the lung, carrying the entire cardiac output, with the pulmonary part of the systemic circulation, for alimentation of the lung parenchyma, mainly the airways, by a part of the cardiac output. The former is driven by a relatively low pressure difference between the right cardiac pump and the left atrium (Table 4.25 and 4.26).

The *pulmonary vascular resistances* (PVR) are much lower than the systemic vascular resistances[22] (SVR). Moreover, pulmonary vascular resistances must

[21]Tracer ^{129}Xe is particularly suited for gas transfer imaging. Xenon is soluble in the alveolocapillary barrier and erythrocytes. Inhaled ^{129}Xe enters alveoli, crosses the alveolocapillary barrier to reach erythrocytes in pulmonary capillaries, in which its concentration equilibrates that alveoli. Agent ^{129}Xe resonates at 3 distinct frequencies in the 3 compartments of interest (alveolar space, alveolocapillary barrier, and erythrocyte). Atom ^{129}Xe resonates in the alveolocapillary barrier (alveolar epithelium, tiny interstitium, and capillary endothelium; thickness 1–7.5 μm) as well as plasma at 197 ppm (4.64 kHz at 2 T) and erythrocytes at 211 ppm. Magnetic resonance signals of ^{129}Xe can be strongly enhanced by hyperpolarization.

[22]Assuming a pressure difference in the pulmonary circulation of 1.3 kPa and a flow rate of 80 ml/s, the pulmonary vascular resistances are equal to about 20 Pa/ml/s.

Table 4.26 Arterial pulmonary pressures (kPa) for a blood flow rate of 85 ml/s according to subject's age (Source: [29])

Age	Pressure (kPa)
20	Systolic: 2.1–3.1
	Diastolic: 0.5–1.1
	Mean: 1.2–2.0
	Capillary:ï¿½0.4–1.1
60	Systolic: 3.0–3.9
	Diastolic: 1.1–1.6
	Mean: 1.9–2.7
	Capillary:ï¿½1.1–1.7

Table 4.27 Effects of ventilation and perfusion on regional gas partial pressure in healthy resting seated upright men, lungs being subdivided into 9 horizontal slices (Source: [20])

	$\dot{V}/q$	Volume (%)	p_{CO_2} (mmHg)	p_{CO_2} (mmHg)
Top slice	3.3	7	132	28
Bottom slice	0.6	13	89	42

Table 4.28 Xe^{133} distribution of ventilation and perfusion (Source: [20])

Region	$\dot{V}$ (l/mn)	q (l/mn)	$\dot{V}/q$
Upper	0.95	0.69	1.38
Middle	1.68	1.78	0.94
Lower	2.36	3.52	0.67

remain low in order to avoid right heart failure, as the right heart is a low-pressure pump. When the blood pressure rises in the pulmonary arteries, capillary recruitment in lung regions where they are more or less collapsed at physiological pressures and blood vessel distensibility allow avoiding a strong increase in pulmonary vascular resistances.

Because the lung height is about 30 cm,[23] gravity exerts a significant role in blood flow distribution among the upper and lower lung regions (Tables 4.27 and 4.28) [20]. The blood flow rate decreases from the lung base to the lung apex. The lung is usually divided into 3 zones owing to a higher hydrostatic blood pressure component at the lung base than at the apex. Region 1 is characterized by arterial and venous pressures lower than the alveolar one that hinders strong perfusion due to vessel collapse. In region 2, the arterial pressure is greater than the alveolar pressure and the venous pressure is lower than the alveolar pressure. Blood flow is governed by the local transmural pressures, the arterioalveolar and venoalveolar pressures. In region 3, both the arterial and venous pressures exceed alveolar pressures. Blood flow is determined by the arteriovenous difference.

The pulmonary blood flow is mainly regulated in response to local alveolar hypoxia. A decrease in $p_{A_{O_2}} < 9.3$ kPa generates a vasoconstriction, in order to

[23]The lung height corresponds to the anatomical situation (i.e. to the standing position). A hydrostatic pressure difference of $\sim$30 cm $H_20 \equiv \sim$3 kPa results. In the prone position, the lung height with respect to the usual space reference frame is 12 to 15 cm.

Table 4.29 Factors that influence pulmonary vascular resistances

Angiotensin, histamine, $PGf_{2\alpha}$	PVR increase
Acetylcholine, bradykinin, $PGe_1 E1$, PGi_2	PVR decrease
Alveolar hypoxia	PVR increase
Acidemia	PVR increase

Table 4.30 Oxygen values in arterial and mixed venous bloods

	Arterial blood	Mixed venous blood
Partial pressure (kPa)	13.3	5.3
Hemoglobin saturation (%)	98	75
Content (%)	20	15

adjust the regional perfusion to the ventilation, reducing the irrigation of hypoxic territories.

The pulmonary resistances are altered by many factors (Table 4.29). Elevated interstitial pressure by squeezing the blood vessels as well as blood viscosity by rising friction increase the pulmonary vascular resistance. Increased pulmonary blood volume (reduced intrathoracic pressure) decreases PVR via the associated expansion of pulmonary blood vessels.

4.6.8 Blood O_2 Affinity Curve

Although gases dissolve in blood, the dissolved form is not a main transport source. Oxygen is almost entirely carried by hemoglobin[24] (Hb; Table 4.30), and CO_2 is carried predominantly in the form of bicarbonate ions[25] (Table 4.32). The hemoglobin concentration probably is a compromise between the advantages of a high O_2 concentration in the blood and the heart load, a higher hematocrit raising the blood viscosity.

The combination of oxygen with hemoglobin in Hb solutions obtained by lysed erythrocytes has been shown in the 1920's to be very quick, of order 1 ms [30, 31]. With intact erythrocytes, it is of order 10 ms. Later, a faster measurement technique was introduced, based on the photovoltaic cell and galvanometer [32].

[24]Each gram of Hb at 100% saturation can carry 1.34 ml of O_2, whereas each 100 ml of blood dissolves 3 μ l of O_2 per 13.3 kPa of oxygen partial pressure.

[25]A part of CO_2 dissolved in the plasma combines with water forming carbonic acid. This reaction is rapidly catalyzed by carbonic anhydrase. The carbonic acid then dissociates into bicarbonate and hydrogen ions. Bicarbonate ions diffuse out of the erythrocyte, being replaced by chloride ions. When the blood reaches the lung, the opposite reactions take place, and CO_2 is released into the alveolus.

Hemoglobin

Like any globin, hemoglobin is formed by 4 polypeptidic chains, the configuration of which leaves a pocket to accomodate heme. Heme is then easily accessible by oxygen. Iron is close to heme. Once oxygen is fixed to iron, the quaternary structure of the molecule changes with a small displacement and rotation of a part of the polypeptidic chains.

The 4 hemes bind oxygen successively rather than simultaneously. They hence interact. Oxygenation of the first hemes induces a configuration change and increases affinity for oxygen of polypeptidic chains with free hemes. When hemoglobin starts to release oxygen in the tissues, the process quickly rises because the freeing of one oxygen molecule facilitates the discharge of others.

The stimulators of glycolysis, like increments in intracellular pH, enhance the concentration in (2,3)-diphosphoglyceric acid (DPG).[26] The latter inhibits new formation of diphosphoglyceric acids. Binding of DPG with hemoglobin reduces free DPG and stimulates DPG synthesis. Carbon dioxide and DPG compete for the same binding sites on hemoglobin (terminal amin group of each polypeptidic chains).

Association–Dissociation Curve

Hemoglobin displays a sigmoidal oxygen association-dissociation curve. Oxygen is taken up more slowly in both the foot and shoulder curve parts at low and high oxygen partial pressures, and more quickly in the steep central part (Fig. 4.6). The initial relatively flat part of the curve occurs because the binding of the first oxygen molecule causes a small structural change to hemoglobin and facilitates the binding of subsequent oxygen molecules. The terminal relatively flat part of the curve means that a fall in $p_{\mathrm{aO_2}}$ has little effect on hemoglobin saturation and oxygen content until the steep part of the curve is reached ($p_{\mathrm{aO_2}}$ ~8 kPa). Once $p_{\mathrm{aO_2}}$ reaches this threshold, an additional decrease in $p_{\mathrm{aO_2}}$ results in a strong fall in hemoglobin saturation.

At normal $p_{\mathrm{aO_2}}$ of about 13.3 kPa, hemoglobin is nearly fully saturated with O_2 (97% saturation; Fig. 4.6). At normal $p_{\mathrm{aO_2}}$, about 20 ml of O_2[27] exist per 100 ml of blood (~1 l in a blood volume of 5 l), in addition to 0.23 ml per liter and per kPa of dissolved O_2 (0.3 ml/l/mmHg; i.e., 15.3 ml in a blood volume of 5 l).

[26] (2,3)-Diphosphoglyceric acid is formed from (1,3)-diphosphoglyceric acid using a mutase and degraded by a (2,3)DPG phosphatase.

[27] A hemoglobin concentration of 15 g/100 ml carries 1.34 × 15=20 ml of oxygen per 100 ml of blood. With a stroke volume of 80 ml/s, the amount of oxygen newly transported in the circulation is 16 ml. Hemoglobin in venous blood being about 75% saturated, 4 ml of oxygen is used at rest per second.

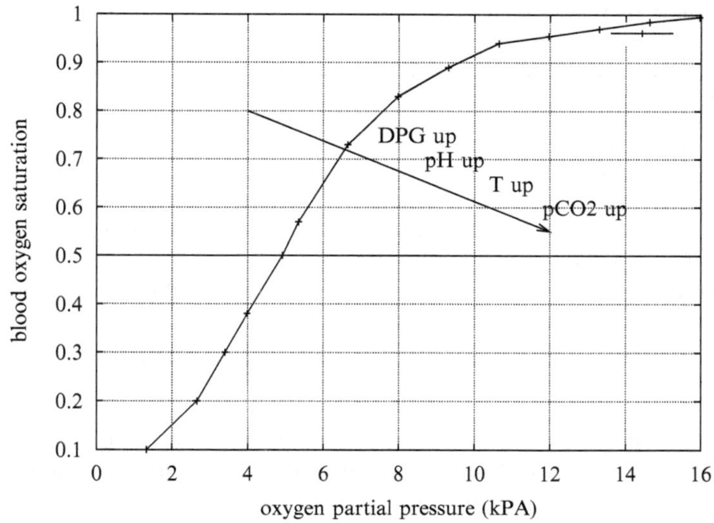

Fig. 4.6 Relationship between the oxygen blood saturation and partial pressure. Various factors (increase in temperature (T), acidosis (pH), and (2,3)-diphosphoglyceric acid concentration shift the curve rightward

The affinity of hemoglobin for O_2 is influenced by a number of factors including the blood pH (Fig. 4.6). With acidosis (high $[H^+]$), the hemoglobin dissociation curve shifts to the right, and the hemoglobin affinity for O_2 is decreased (Bohr effect). Increased temperature also decreases the affinity of hemoglobin for oxygen. The curve is shifted rightward when (2,3)-diphosphoglyceric acid level increases.[28] Conversely, alkalosis, decaying DPG concentration, and falling temperature move the Hb–O2 dissociation curve to the left.

The relative affinity of hemoglobin for oxygen can be expressed by $p50$, that is, p_{O_2} at which Hb is 50% saturated. This quantity is an index of the position of the Hb–O2 dissociation curve. It increases with decreasing affinity, and conversely. The decrease in affinity of hemoglobin for oxygen is useful in body tissues that yield a slightly acidic environment, where oxygen must be unloaded from the blood. The increase in affinity of hemoglobin for oxygen is needed to improve oxygen loading in the pulmonary capillary, a site that is slightly alkaline.

Hypoxia

Hypoxia defines a failure in the supply or use of oxygen. Tissues differ in their ability to withstand hypoxia or even anoxia. The brain and heart are the most sensitive organs.

[28] Diphosphoglyceric acid is a glycolytic product in erythrocytes that modulates hemoglobin affinity for oxygen. Diphosphoglyceric acid synthesis is increased by the blood alkalinity.

Table 4.31 Hypoxemia types

	$p_{A_{O_2}}$	$p_{a_{O_2}}$	$c_{a_{O_2}}$	$c_{\bar{v}_{O_2}}$
Alveolar hypoxia	Small	Small	Small	Small
Alveolocapillary membrane impairment	Normal	Small	Small	Small
Anemia	Normal	Normal	Small	Small
Hypoperfusion	Normal	Normal	Normal	Small
Right-to-left shunt	Normal	Small	Small	Small
Toxicity	Normal	Normal	Normal	High
Ventilation–perfusion mismatch	Normal	Small	Small	Small

Four kinds of hypoxia exist according to the cause (Table 4.31). Alveolar hypoxia is characterized by an inadequate oxygen supply to the arterial blood. It can be caused by hypoventilation or improper alveolocapillary transfer (e.g., dysfunctional alveolocapillary membrane and shunts). In severe anemia, the oxygen content falls in proportion to reduction in hemoglobin concentration. Circulatory disorders, even though the oxygen content of arterial blood is suitable, are responsible for inadequate oxygen delivery to the tissues. Tissue poisoning, although oxygen delivery to the tissues is normal, hinders normal use of oxygen by the cells.

4.6.9 CO_2 Blood Transport

Carbon dioxide employs 3 major kinds of shuttles: (1) dissolved CO_2, (2) carbamino compounds, and (3) bicarbonates (Table 4.32, Fig. 4.7). The quantity of dissolved CO_2 and $[H_2CO_3]$[29] is proportional to the partial pressure of CO_2.

Carbamino compounds represents 25–30% of the CO_2 in the body, a part combined to hemoglobin being conveyed by the blood. About 1 mEq of reduced hemoglobin gives 0.34 mmol of carbaminated CO_2. The CO_2 concentration in the venous blood is ~2.08 mmol/l. The formation of 1 mmol of carbaminated CO_2 releases 1.5 mmol of H^+. Whereas hemoglobin oxygenation transforms hemoglobin in strong acid and allows H^+ captation from the solution, oxyhemoglobin reduction leads to a weak acid and H^+ removal. The presence of CO_2 and the reduction of oxyhemoglobin cause 2 antagonist phenomena. The first is associated with carbaminated CO_2 and release of H^+ from Hb, the second with substraction of H^+ from the solution to Hb, Hb becoming a weak acid. For pH in the range [6.9,7.5], the latter process is dominant.

[29] $CO_2 + H_2O \rightleftharpoons H_2CO_3 \rightleftharpoons H^+ + HCO_3^-$.

Table 4.32 CO_2 Transport types (%). Carbon dioxide is carried in the blood in 3 major types: a small dissolved quantity, attached to hemoglobin as a carbamino compound, and, mainly, as bicarbonate

	Arterial blood	Arteriovenous difference
Dissolved	5	10
Carbamino	5	30
Bicarbonates	90	60

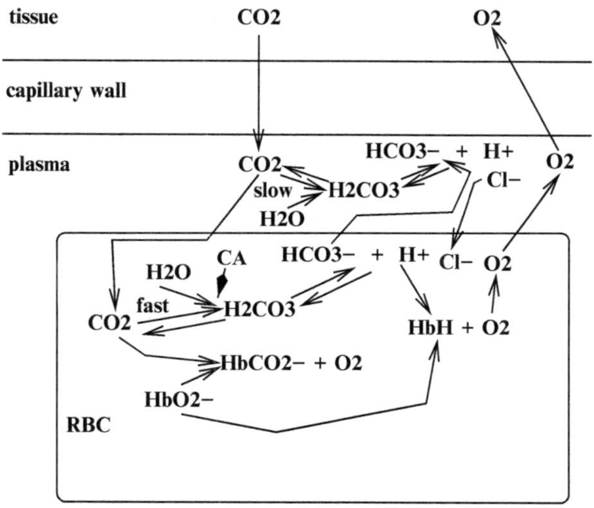

Fig. 4.7 Captation of CO_2 and O_2 release in capillaries. Synthesis of bicarbonates and carbamates. The formation of bicarbonates in the plasma is slow, whereas it is fast in the erythrocyte (RBC), catalyzed by carbonic anhydrase (AC)

Table 4.33 Transport of CO_2 in arterial (a) and venous (v) blood at rest when Ht = 45% (Source: [33]

	Blood	Plasma	Erythrocyte			
	a	v	a	v	a	v
H_2CO_3 (mmol/l)	1.15	1.35	1.2	1.4	1.1	1.3
HCO_3^- (mmol/l)	19	20.6	24	26	13	14.2
$HbCO_2$ (mmol/l)	0.4	0.6			0.8	1.2
total CO_2 (mmol/l)	20.6	22.6	25.2	27.4	14.9	16.7
p_{CO_2} (kPa)	5.3 6.3	5.3 6.3	5.3 6.3			

The CO_2 transport is thus related to the blood pH status:

$$pH = pK + \ln\left(\frac{[HCO_3^-]}{[\text{dissolved } CO_2] + [H_2CO_3]}\right) = pK + \ln\left(\frac{[HCO_3^-]}{\kappa p_{CO_2}}\right), \quad (4.97)$$

where $\kappa = 0.521$ in the plasma at body temperature and $pK = \ln K$ with K given by the mass action law:

$$K = \frac{[H^+][HCO_3^-]}{[\text{dissolved } CO_2] + [H_2CO_3]}.$$

The small blood quantity of H_2CO_3 is neglected without inducing significant errors.

4.7 Maintenance of pH Homeostasis—Hydrogen Ion Control

The acid–base state[30] of the living body must be regulated for a normal functioning. At neutral pH (Table 4.34), most catabolites are ionized, an efficient mechanism for maintaining the catabolites within the cell.

The acid–base status can be graphically assessed. The *Peters diagram* plots the relationship between p_{aCO_2} and c_{aCO_2} using logarithmic scales. Lines of isopH run at 45-degree angle in the diagram. Lines of CO_2 dissociation are also drawn, one for fully oxygenated blood, the other for upper limits of normal c_{CO_2}.

Respiratory (pH >7.45, $p_{aCO_2} < 4.7$ kPa) and nonrespiratory (pH >7.45, normal p_{aCO_2}, increased c_{aCO_2}) *alkalosis* and respiratory (nearly normal pH, $p_{aCO_2} > 6$ kPa, increased c_{aCO_2}) or nonrespiratory (reduced pH, normal p_{aCO_2}) *acidosis* can be represented on the chart.

The acid–base balance depends on various buffers that are acid–base conjugates such as proteic buffer (hemoglobin), phosphate buffer, and bicarbonate buffer:

$$H_2CO_3 \underset{k_2}{\overset{k_1}{\rightleftarrows}} HCO_3^- + H^+, \tag{4.98}$$

Table 4.34 Physiological ranges of quantities of the acid-base state at sea level

pH	7.35–7.45
HCO_3^-	23–28 mmol/l
p_{aCO_2}	4.7–5.6 kPa
H^+	40 ± 2 nEq/l

[30] Acids are H^+ donor, bases are H^+ acceptors. The activity of H^+ ions in a solution determines the acidity. In the blood, the activity coefficient ∼1. Therefore, pH = $-\ln[H^+]$, as in an infinitely dilute solution. An acid (AH) that transfers its H^+ can be transformed into its conjugated base anion (A^-) and H^+ cation, and conversely according to $[H^+]$ value. Once one H^+ has been released, any acid can become a base buffer able to bind free H^+. The mass action law relates the concentration of the 3 reactants: $[H^+] = k[AH]/[A^-]$. pK is the point at which the concentrations of the 2 conjugate substances are equal.

$$k_1 = \frac{[H^+][HCO_3^-]}{[H_2CO_3]}, \tag{4.99}$$

where the concentration $[H_2CO_3]$ is proportional to the concentration of dissolved carbon dioxide.

The equilibrium of a chemical acid–base reaction is given by:

$$HA \rightleftharpoons A^- + H^+, \tag{4.100}$$

where HA is a generic acid that dissociates into the conjugate base of the acid (A^-) and the hydrogen ion (H^+), or proton, which, in the case of an aqueous solution, exists as the hydronium ion (H_3O^+), in other words, a solvated proton formed by the protonation of water.

The chemical species HA, A^-, and H^+ are in equilibrium when their concentrations remain constant when a steady state is reached. The *acid dissociation constant* (K_{ac}; mol/l) is a measure of the level of an acid in solution. It equals the product of the equilibrium concentrations of the deprotonated form ($[A^-]$) and of the hydrogen ion ($[H^+]$) divided by the equilibrium concentration of the acid ([HA]):

$$K_{ac} = \frac{[A^-][H^+]}{[HA]}. \tag{4.101}$$

Its value is given using a logarithmic scale. The logarithmic constant (pK_{ac}) is then given by:

$$pK_{ac} = -\log_{10} K_{ac}. \tag{4.102}$$

The larger the pK_{ac} value, the smaller the extent of dissociation at any given pH. A weak acid has a pK_{ac} value that ranges from -2 to 12 in water. A strong acids ($pK_{ac} < -2$) is almost completely dissociated in aqueous solution.

> *The effectiveness of bicarbonates and phosphates to preserve neutrality in protoplasm is extraordinarily great. The bicarbonates constitute the first reserve, so to speak, in neutralizing acid. They are effective in a far greater degree than the salts of any acids in equal concentration could be because of the regulation of carbonic acid concentration by diffusion and excretion.*
>
> *(Henderson L.J.) [35]*

The Henderson–Hasselbalch equation yields pH from the negative log of the acid dissociation constant (pK_{ac}) in biochemical systems. It estimates pH of a buffer solution and gives the equilibrium pH in acid–base reactions:

$$pH = pK_{ac} + \log_{10}\left(\frac{[A^-]}{[HA]}\right). \tag{4.103}$$

Table 4.35 Changes in temperature, pH, and carbon dioxide content during exercise in 3 body tissues

	T(°C)	pH	p_{CO_2} (kPa)
Lung	37	7.40	5.3
Muscle	41	7.35	6.4
Skin	25	7.60	2.9

The Henderson–Hasselbalch equation can be applied to blood constituents, especially those of the bicarbonate buffer:

$$\mathrm{pH} = \mathrm{p}pK_{H_2CO_3} + \log_{10}\left(\frac{[HCO_3^-]}{[H_2CO_3]}\right), \tag{4.104}$$

or:

$$\mathrm{pH} = 6.1 + \log_{10}\left(\frac{[HCO_3^-]}{0.03\, p_{CO_2}}\right). \tag{4.105}$$

During hyperventilation, p_{CO_2} falls and pH augments (respiratory alkalosis). On the other hand, an uneven ventilation-to-perfusion ratio increases p_{CO_2}, hence pH lowers (respiratory acidosis).

Three important, temperature-dependent buffers in the intra- and extracellular media include proteins, phosphates, and the carbon dioxide–bicarbonate couple. When the temperature varies, intra- and extracellular pH change. The temperature evolves especially during exercise (Table 4.35).

The arterial partial pressure of CO_2 is a very important factor of the chemical regulation of the ventilation. Abrupt changes in p_{CO_2} of 130 Pa increase $[H^+]$ in the plasma of 0.77 nmol/l and in the cerebrospinal liquid of 4–5 nmol/l, this liquid being only slightly buffered. The solubility of CO_2 being high in water and lipids, any change in blood p_{CO_2} quickly induces variations of p_{CO_2} in interstitial fluids and cells, as well as pH changes. Any increase in $[H^+]$ stimulates the respiratory centers.

4.8 Control of the Respiration

Any biological system is affected by supply fluxes, assigned loads, and disturbances. The controlling process receives signals from peripheral and possibly central sensors. A comparator compares sensor signals (feedback) to command cues associated with the working setting. The comparator then delivers control signals to the effectors that correct the process (Table 4.36).

Ventilation and blood circulation mutually adjust and adapt to environmental cues. The regulation of the ventilatory system is based on feedback loops. The respiratory neurons send impulses to the respiratory muscles. Information sensed by specific receptors in the lungs and chest wall is conveyed to the respiratory

Table 4.36 Nervous control of breathing—Efferents (ACh: acetylcholine)

Parasympathetic	Dorsal motor nucleus of the vagus nerve Preganglionic fibers Hilar plexus (Ach neurotransmitter) Postganglionic fibers Innervation of bronchi (mainly) and blood vessels
Sympathetic	Preganglionic fibers Cervical ganglia (Ach neurotransmitter) T1–T4: middle and inferior cervical ganglion Postganglionic fibers Innervation of bronchi and blood vessels (mainly)

Table 4.37 Nervous control of breathing—Afferents (CSF: cerebrospinal fluid)

Site	Sensors
Upper airways	Mechano and chemoreceptors
Tracheobronchial tree	Irritant receptors (between airway epitheliocytes; rapidly adapting) (cause bronchoconstriction and hyperpnea)
Lung parenchyma	Pulmonary stretch receptors (airway smooth myocytes; slowly adapting) (reduction of respiratory frequency) Juxtacapillary (J) receptors (alveolar septa; unmyelinated C-fiber endings) (cause rapid, shallow breathing Bronchial C fibers (bronchial circulation)
Respiratory muscles, tendons, and joints	Proprioceptors (proprioceptive receptors)
Blood circulation	Chemoreceptors (sensitive to p_{O_2}, p_{CO_2}, and pH) Arterial baroreceptors
Medulla	Central chemoreceptors (cerebrospinal fluid) (sensitive to p_{CO_2}, $HCO^-_{3_{CSF}}$, and pH)

nervous centers (Table 4.37). Humoral stimuli, such as blood gases and acidity, also influence neuronal control of the respiratory motor. However, they act with a longer time scale.[31]

The ventilation control system includes the central nervous centers, their afferents and efferents, the sensors and the target organs, mainly the respiratory muscles.

[31]Increase in H^+ concentration and in arterial CO_2 partial pressure, decrease in arterial O_2 partial pressure stimulates the ventilation.

Table 4.38 Central regulation of breathing

Compartment	Centers
Pons	Pneumotaxic center (nucleus parabrachialis)
	Apneustic center
Medulla	Dorsal respiratory group (nucleus of the tractus solitarius)
	Ventral respiratory group nucleus ambiguus + nucleus retroambigualis
Spinal cord	Corticospinal tract

The regulation of ventilation: (1) ensures a periodic quasiregular ventilatory mode; (2) adjusts the tidal volume and respiration frequency such that the ventilation can respond to body needs; (3) controls the ventilatory mode that must be compatible with other body activities using the same muscles (e.g., speech and posture).

The central regulation of breathing comprises (Table 4.38): (1) voluntary control (cerebral cortex); (2) autonomous control (brainstem); and (3) integration of breathing activity (spinal cord).

Hypoxia transforms neuronal mechanisms of normal breathing (eupnea) generated by the central respiratory clock, that is, sequential activation of respiratory muscles innervated by cranial and spinal nerves (e.g., vagus, hypoglossal, and phrenic nerves) during inspiration and expiration, to a rudimentary pattern of inspiratory motions (gasping), owing to an oxygen-sensitive, burst-promoting, persistent sodium current within the pre-Bötzinger complex [36].[32]

Dysfunctions of the neural control of breathing include sudden infant death syndrome, sleep apnea, such as that resulting from neurodegenerative diseases (amyotrophic lateral sclerosis, multiple system atrophy, Parkinson's disease, etc.), and several genetic disorders, such as Rett syndrome[33] and congenital central hypoventilation syndrome (a.k.a. primary alveolar hypoventilation and Ondine's curse).[34]

[32]In eupnea, neurons discharge action potentials during the phrenic activity and inactivity. In gasping, the ramping inspiratory pattern is converted into a synchronous, decrementing inspiratory pattern with minimal expiratory activity. Gasping is characterized by an early onset of discharge in the vagal and hypoglossal nerves relative to phrenic nerve and loss of both postinspiratory vagal and preinspiratory hypoglossal discharges. A persistent sodium current that raises during hypoxia supports gasping, but is not required for the eupneic rhythm in the intact pontomedullary network of both neonatal and juvenile rats.

[33]The Rett syndrome results from mutations of the nuclear protein methyl CpG-binding protein MeCP2 that may repress gene expression, as it is able to bind methylated cytosine (C) on a peculiar DNA sequence (CpG).

[34]Ondine's curse is a type of central sleep apnea. In addition to respiratory arrests during sleep, Ondine's curse is associated with partial agenesis of the enteric nervous system, dysphagia

Table 4.39 Respiratory centers of the brainstem. The dorsal respiratory group produces every 5 s a 2-s ramp that receives neural input from chemical stimuli causes diaphragm contraction and inspiration

Group of respiratory neurons	Effect
Apneustic center	Inspiration prolongation
Pneumotaxic center	Inspiration termination
Dorsal respiratory group	Quiet inspiration
Ventral respiratory group	Forced inspiration and active expiration (exercise)

4.8.1 Respiration Nervous Centers

The central autonomous nervous centers include: (1) the respiratory centers[35] of the brainstem and (2) the integrative control centers for breathing in the spinal cord. The set of main central nervous centers is used for involuntary control of breathing.

Since the work of Lumsden (1920), the central operating system for the ventilation is divided into 3 centers: (1) pneumotaxic, (2) apneustic, and (3) inspiratory centers, the latter being subdivided into the expiratory and gasping centers (Table 4.39).

The *pneumotaxic center* corresponds to the *nucleus parabrachialis* in the *dorsolateral rostral pons*. It modulates, and possibly drives, the rhythmic respiratory neurons in the medulla. It controls the breathing pattern by changing the moment at which inspiration is interrupted that is determined by the inflation reflex operating via the vagus nerve as well as arterial oxygen and carbon dioxide content. It is also influenced by mechanoreceptors of upper airways.

The *apneustic center* localizes to the lower pons, in the formatio reticularis, close to the medulla oblongata. It increases inspiratory activity.

Several pools of respiratory neurons localize to the brainstem, either dorsally or ventrally (Fig. 4.8). The *dorsal respiratory group* (DRG) in the *nucleus of the tractus solitarius* (NTS) is composed of inspiratory neurons. Its inputs are transmitted by cranial IX and X afferent nerves. Its outputs are sent to the *ventral respiratory group* (VRG) and *spinal respiratory motoneurons*. Inspiratory motoneurons of the dorsal respiratory group are the first stimulated.

The *ventral respiratory group* with the *nucleus ambiguus* (NA) and *nucleus retroambigualis* (NRA) contain both inspiratory and expiratory neurons. The nucleus ambiguus consists of motoneurons of cranial nerves that innervate the ipsilateral accessory muscles of respiration. The nucleus retroambigualis controls motoneurons of ins- and expiratory intercostals and probably abdominal expiratory

(swallowing difficulty), etc. It is characterized by a complete loss of CO_2 sensitivity. Congenital central hypoventilation syndrome can result from mutations of the Phox2b gene.

[35] A nervous center refers to a pool of neurons.

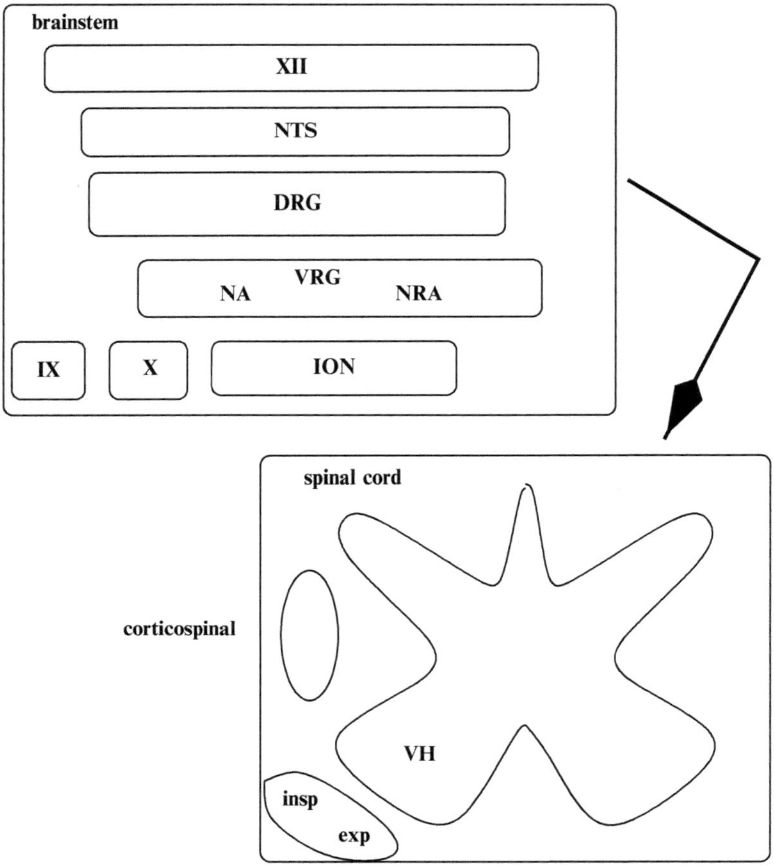

Fig. 4.8 Respiration nervous centers (NTS: nucleus of the tractus solitarius, DRG: dorsal respiratory group, VRG: ventral respiratory group, NA: nucleus ambiguus, NRA: nucleus retroambigualis, ION: inferior olivar nucleus, IX: glossopharingeal nucleus, X: vagus nucleus, XII: hypoglossus nucleus, VH: ventral horn

muscles, with contro- and ipsilateral projections. Delayed activation of inspiratory motoneurons of the ventral respiratory group requires a stronger stimulation.

About 60% of inspiratory motoneurons and about one-third of expiratory motoneurons have bulbospinal connections. *Spinal respiratory neurons* are not only relay stations, but also integrative centers. The central control of breathing projects in the dorsolateral columns of the spinal cord.

The automatic command of the rhythmic pattern and ventilatory reflexes (cough, hiccough, and yawning) travels through the ventral and lateral columns of the spinal cord.

In the spinal cord, nervous fibers implicated in the voluntary regulation of breathing localize to the *tractus corticospinalis*. Rhythmic impulses of NRA neurons travel along anterolateral paths to intercostals and abdominal motoneurons of the thoracic and lumbar column.

4.8.2 *Respiratory Clock, a Synchronous Bilateral Command*

The control of biological rhythms incorporates positive feedback that coordinates individual interacting elements to prime a collective temporal pattern and negative feedback that temporarily halts neuron clustering activated by the positive feedback. Positive and negative feedback alternate.

Breathing is regulated by a respiratory rhythm-generating network located in both sides of the brainstem that produces a synchronous bilateral command. Signals are transmitted to cranial and spinal populations of motoneurons that innervate cranial, thoracic, and abdominal respiratory muscles.

During normal cycles, expiratory effort occurs just before and again immediately after inspiration. In cycles without inspiration, expiratory activity is a single burst that lacks the inspiratory-related pause [37]. Preinspiratory neurons are actively inhibited during inspiration and then typically rebound to discharge again during the postinspiratory, early-expiratory phase. Lesions and transections are used to separate the regions that elicit inspiratory and expiratory activity.

4.8.2.1 Breathing Oscillators

The respiratory motor command directs sequential breathing phases (inspiratory, postinspiratory, and expiratory). Central respiratory circuits localize to both sides of the central nervous system and thus rely on commissural networks that project onto respiratory relay structures in the brainstem. Therefore, the robustness of the central control of breathing depends on excitatory commissural relays that function for both sensory integration and motor command. A system of 2 respiratory rhythm generators that responds differently to various sensory or modulatory inputs, such as lung, muscle, and joint mechanoreceptors, lung and airway nociceptors, and chemoreceptors, may yield robustness and rapid adjustmemt to changes in metabolic demand.

Neuron clusters implicated in the respiratory control lodge in the lateral rhombencephalon. More precisely, the *ventral respiratory column* (VRC) in the brainstem includes a set of ventilatory neuron clusters. From its cranial to caudal end, the ventral respiratory column comprises, at least in rats [37]: (1) the *caudal ventral respiratory group* (cVRG) that contains excitatory expiratory premotor neurons; (2) *rostral ventral respiratory group* (rVRG), in which reside excitatory inspiratory premotor neurons; (3) *pre-Bötzinger complex* (preBotC) with rhythm-generating interneurons [39]; (4) Bötzinger complex (BotC) with expiratory inhibitory interneurons; (5) retrotrapezoid nucleus (RTN)[36] and parafacial

[36]The retrotrapezoid nucleus corresponds to a small region that is located ventrally to the facial (VII) motor nucleus. It is defined by neurons that express PHox2b transcription factor. Mutations of the PHOX2B gene cause *congenital central hypoventilation syndrome* [38]. They cannot sustain robust breathing during sleep. This syndrome is also characterized by an insensitivity

respiratory group (pFRG)[37] with interneurons that may generate expiratory rhythms. Both caudal and rostral ventral respiratory groups are situated nearby the lateral reticular nucleus, pre-Bötzinger and Bötzinger complexes close to the nucleus ambiguus, and retrotrapezoid nucleus and parafacial respiratory group adjacent to the facial nucleus.

At least in fetal and newborn rodents, the respiratory clock in the brainstem includes 2 distinct, coupled oscillators, each one arising from bilateral, interconnected neuronal populations that are part of the continuous ventral respiratory column [37, 38]: (1) pacemaker neurons of a small region of the ventrolateral medulla oblongata near the nucleus ambiguus—the pre-Bötzinger complex—on the one hand and (2) a cluster of rhythmically active neurons among rostral respiratory networks concentrated in the retrotrapezoid nucleus and a related rhythmogenic area, the parafacial respiratory group. The retrotrapezoid nucleus serves as a CO_2 and H^+ sensor. The rhythms of these oscillators as well as activity of motor nerves are synchronous.

The pre-Bötzinger complex and the retrotrapezoid nucleus–parafacial respiratory group (RTN–pFRG) triggers inspiratory and expiratory activity, respectively [37]. The preBotC-induced inspiratory breathing pattern dominates at rest, whereas the retrotrapezoid nucleus–parafacial respiratory group may only have a CO_2-dependent activity to expiratory muscles. During exercise, when O_2 consumption and CO_2 production rise, the activity of RTN–pFRG neurons become progressively rhythmic [38].

Genesis of Breathing Oscillators

The anatomical structure of the central nervous system, hence the formation of functional circuits results from transcriptional programs. In particular, these programs coordinate anteroposterior (dorsoventral) patterning of the neural tube.

Different groups of interneurons and motor neurons differentiate according to homeodomain transcription factors in progenitor cells at given loci [40]. In mammals, commissural interneurons regulate the function of neural networks.[38]

---

to CO_2 stimulation of breathing. Hence, the retrotrapezoid nucleus serves as a site for central chemoreception.

[37]The parafacial respiratory group is also a region ventral to the facial motor nucleus. Some neurons of the retrotrapezoid nucleus and parafacial respiratory group project caudally to brainstem premotoneurons that excite spinal expiratory motoneurons.

[38]For example, ventral V0 or V3 classes of interneurons that originate from progenitor cells P0 and P3, respectively, express the transcription factors Developing brain homeobox gene product DBx1 and NKx2-2 for left–right alternation and robustness of the left–right patterning of locomotor activity [40].

As the performance of any neural command that relies on many neuronal populations requires training to coordinate activities of all the participants, mammals start breathing in utero [38]. Moreover, episodic fetal breathing is required for proper development of lungs and respiratory muscles.

Embryonic Parafacial Oscillator

In mouse embryos, the embryonic parafacial oscillator (epF), that is, the forerunner of the retrotrapezoid nucleus–parafacial respiratory group and the pre-Bötzinger complex arise independently at distinct anterior–posterior positions close to the facial branchiomotor nucleus and nucleus ambiguus, respectively [40].[39] In rodents, these 2 oscillators in the brainstem generate the respiratory rhythm. They are activated at about two-thirds of the gestational period. They become active sequentially. The embryonic parafacial oscillator ensures the onset and maintenance of episodic fetal breathing movements and contributes to the development of rhythmicity in the pre-Bötzinger complex. PreBötzinger complex neurons begin to oscillate 1 day later in synchrony with the embryonic parafacial oscillator [41]. Nonetheless, although the embryonic parafacial oscillator contributes to the establishment of the rhythmic activity in the pre-Bötzinger complex, it is dispensable for the development of the pre-Bötzinger complex.

The formation of the parafacial respiratory group requires the hindbrain transcription factor Early growth response EGR2 [41].[40] Transcription factor EGR2 controls the formation of hindbrain rhombomeric segments 3 and 5.[41]

In addition, survival of glutamatergic neurons in the retrotrapezoid nucleus–parafacial respiratory group necessitates another hindbrain transcription factor, Paired-like homeobox gene product PHox2b [41]. Transcription factor Phox2b is specifically expressed in neurons involved in reflex circuits that control digestive, cardiovascular, and respiratory functions.

---

[39] The embryonic parafacial oscillator corresponds to a population of neurons that flanks and partly caps the lateral aspect of the facial (VII) motor nucleus and extends about 200 μm caudal to VII nucleus.

[40] A.k.a. zinc finger protein Krox20.

[41] In vertebrate embryos, a rhombomere is a transient segment of the developing neural tube within the hindbrain region. Rhombomeres contribute to the genesis of the rhombencephalon (hindbrain) into the pons (metencephalon), cerebellum (metencephalon), and medulla oblongata (myelencephalon). Adult rhombencephalic regions map to their embryonic rhombomeres: rhombomeres 2 and 3 to basilar pontine nuclei, intertrigeminal nucleus, Kölliker–Fuse nucleus, motor nucleus of the trigeminal nerve, lateral parabrachial region, and part of superior cerebellar peduncle; rhombomeres 4 and 5 to the superior olive, facial nucleus, A5 noradrenergic neurons, and ventrolateral pons; rhombomeres 6 to 8 to the facial nucleus, RTN–pFRG clusters, BotC and preBotC complexes, nucleus ambiguus, rostral and caudal ventral respiratory groups, and lateral reticular nucleus.

PHox2b+ glutamatergic neurons derive from EGR2+ progenitors [41]. About 70% of these neurons synthesize type-1 tachykinin (neurokinin) receptor (NK_1 or NK1R). The rhythm generation in the embryonic parafacial oscillator clusters are independent of glutamatergic synaptic transmission and opioid modulation [41]. However, glutamatergic synaptic transmission mediated by AMPA-type[42] and kainate-type glutamate receptors is necessary for synchronization of bilateral parafacial regions across the midplane. Moreover, recurrent glutamatergic synaptic inputs are necessary for establishment of the rhythmic inspiratory command in preBotC neurons.

The rhythm generation in the embryonic parafacial oscillator clusters relies on a persistent Na^+ current ($i_{Na,p}$) as well as gap junctions. Furthermore, the embryonic parafacial oscillator rhythm is modulated by the hyperpolarization-activated cation (both K^+ and Na^+) current (i_h) through hyperpolarization-activated and cyclic nucleotide-gated channels (HCN).[43] Preinspiratory depolarization later arises in epF cells and transform them into pFRG neurons.

Pre-Bötzinger Oscillator

The genesis of glutamatergic interneurons of the pre-Bötzinger complex is controlled by the transcription factor Developing brain homeobox gene product DBx1 [40]. DBx1+ progenitors of the caudal hindbrain are fated to become confined sources of rhythmic neural activity with synchronicity across the midline.

Moreover, the navigation receptor for axon guidance Roundabout homolog Robo3 is necessary for axonal midline crossing to synchronize left and right pre-Bötzinger groups [40]. Robo3 Receptor could be a transcriptional target of DBx1 factor.

PreBötzinger Complex

The pre-Bötzinger complex constitutes a major commissural hub of central respiratory circuits. A large fraction of the pre-Bötzinger complex cells is actually endowed with a commissural identity. Under normal conditions, the dominant respiratory rhythm generator of the brainstem localizes to the opiate-sensitive pre-Bötzinger complex [37].[44] However, the pre-Bötzinger complex is not sufficient to ensure robust breathing, at least during the perinatal period.

---

[42]AMPA: α-amino 3-hydroxy 5-methyl 4-isoxasole propionic acid.

[43]HCN channels (HCN1–HCN4) are widespread in the heart and central nervous system (Vol. 3 – Chap. 2. Membrane Ion Carriers).

[44]Opiate-induced depression of the pre-Bötzinger complex promptly causes an unusual breathing pattern, quantal slowing, and a complete, prolonged, and fatal apnea.

The pre-Bötzinger complex is heterogeneous. It contains several populations of different neurons, such as glutamatergic and glycinergic neurons. Among glutamatergic neurons, some produce NK_1 receptor, others somatostatin (Sst). About 28% of preBotC, Sst+, glutamatergic neurons express neurokinin receptor NK_1; nearly 41% of preBotC NK_1+ neurons produce somatostatin [39].

The pre-Bötzinger complex is constituted mostly of glutamatergic interneurons that send inspiratory bursts to trigger inspiratory muscle activity, especially that of the diaphragm. Therefore, the breathing rhythm imposed by these neurons of the pre-Bötzinger complex oscillator relies on: (1) ionotropic glutamate receptors of AMPA and kainate receptor families as well as (2) type-2 vesicular glutamate transporter VGluT2, or SLC17a6.

These glutamatergic interneurons can synthesize the transcription factor Even-skipped homeobox gene product homolog Evx1,[45] type-1 tachykinin (neurokinin) receptor, and somatostatin (Evx1+, NK_1+, and Sst+ interneurons) [40].

Glycinergic neurons that contribute to about 20% of the pre-Bötzinger complex population can have a pacemaker function. They then may complement the population of DBx1-derived glutamatergic neurons.

These inspiratory neurons project across the midline and connect the 2 halves of the pre-Bötzinger complex to ensure a bilateral synchronous behavior for efficient breathing. These neurons slowly recover from postinspiratory hyperpolarization [37]. The most excitable neurons recover at mid-expiration and begin to spike before the next inspiration and recruit less excitable neurons.

Two populations of pacemaker neurons exist in the pre-Bötzinger complex. The voltage-dependent bursting of pacemaker neurons depends on a persistent Na^+ current. The proportion of PreBötzinger complex inspiratory neurons that operate as a voltage-dependent pacemaker ranges from 5 and 25% ($i_{Na,p}$-mediated pacemaker neurons) [37]. On the other hand, bursting of voltage-independent neurons of the pre-Bötzinger complex depends on Ca^{2+} and/or Ca^{2+}-activated, nonspecific, and voltage-insensitive cation current ($i_{Ca,ns}$-mediated bursting neurons). Both $i_{Na,p}$- and ($i_{Ca,ns}$)-mediated pacemaker neurons amplify the synaptic depolarization to provoke high-frequency spiking that causes the full inspiratory signal.

At rest (i.e., in the absence of active expiration), only the pre-Bötzinger complex may be rhythmic. Both gabaergic and glycinergic neurons, although dispensable, participate in the control of evolution of membrane potential of pre-Bötzinger complex neurons and formation of the normal respiratory pattern. Metabotropic glutamate receptors activate periodically during inspiration and cause $i_{Na,p}$ and $i_{Ca,ns}$ currents. The transcription factor musculoaponeurotic fibrosarcoma oncogene product homolog MAfb contributes to the breathing pattern.

[45] Even-skipped homeobox gene product homolog Evx1 acts as a repressor during embryogenesis.

Retrotrapezoid Nucleus–Parafacial Respiratory Group

Neurons of the retrotrapezoid nucleus–parafacial respiratory group (RTN–pFRG) discharge before the inspiratory burst and modulate respiratory rhythm [37]. They can indeed increase the frequency of the brainstem respiratory clock via coupling and entrainment of the pre-Bötzinger complex that controls the motor output.

Neurons of the RTN–pFRG cluster activate several hundred milliseconds before pre-Bötzinger complex neurons. They interact with pre-Bötzinger complex neurons to create a coupled oscillator system that regulates the ventilatory rhythm.[46] In certain circumstances, in the absence of activity of the pre-Bötzinger complex, the RTN–pFRG can generate ventilation by active expiration and passive inspiration.

4.8.3 Respiratory Nerves

The respiratory muscles are connected to the spinal cord by afferents, γ afferent fibers, and efferents, α- and γ efferent fibers, to myofibers and muscle spindles, respectively (Table 4.40). γ Afferent fibers have their cell bodies in the dorsal root ganglion and synapse with α-motoneurons in the spinal cord. The respiratory motoneuron output adapts to the needs via various ventilatory modes.

4.8.3.1 Afferences

Afferent nerves from the respiratory muscles project to the respiratory centers (supraspinal reflexes), or act via spinal reflexes. They are based on the proprioceptive control of muscle activation by muscle spindles and Golgi tendon organs.

The 2 types of *proprioceptive receptors* of respiratory muscles give the activity state of the muscle. They have opposite effects, either facilitating the contraction via α-motoneurons, or inhibiting it, when the muscle is stretched. Neuromuscular spindles stimulate muscle contraction, particularly of intercostal muscles that are rich in muscle spindles. The tendon receptors prevent contraction of muscles especially the diaphragm that is rich in Golgi tendon organs. The proprioceptive receptors send their information through γ-afferents via dorsal roots. The diaphragm and intercostals mutually influence via nervous and mechanical (via thoracic cage motions) interactions.

---

[46]Transections at the caudal end of the facial (VII) motor nucleus that disconnect the rostral portion of the RTN–pFRG from the medulla oblongata abolish expiratory motor activity without affecting inspiratory rhythm [37]. Transection between RTN–pFRG and preBotC completely abolishes active expiratory motor activity with only a modest effect on inspiratory pattern [38].

Table 4.40 Respiratory muscles and their motor innervation. The diaphragm, descent of which increases the volume of the thoracic cavity, is the most important muscle of respiration

Muscle	Nerve
Main inspiratory muscles	Phrenic nerve (C3–C5)
Diaphragm	
External intercostals	Intercostal nerves (T1–T11)
Parasternal intercostals	
Sternocostalis	Intercostal nerves (T1–T6)
Accessory inspiratory muscles	Hypoglossal nerve
Genioglossus	
(enlarges upper airway segment vulnerable to collapse)	Cervical nerve (C2–C3)
Sternocleidomastoid	
Scalene	Cervical spinal nerve (C4–C6)
Pectoralis	Medial and lateral pectoral nerve (C5–T1)
Posterior cricoarytenoids	Recurrent laryngeal branch (vagus nerve)
(open true vocal folds, enbling breathing)	Intercostal nerves (T1–T11)
(deep) Expiratory muscles	
Internal intercostals	
Rectus abdominis	Thoracoabdominal nerves (T7–L1)
External abdominal oblique	Thoracoabdominal and subcostal nerves (T7–L1)
Internal abdominal oblique	Iliohypogastric and -inguinal nerves (T12–L1)
Transverse abdominal	Lower intercostal nerves (T7–T11), iliohypogastric and ilioinguinal nerves

Supraspinal reflexes consist of an inhibition of phrenic nerve activity by intercostal nerve stimulation. This reflex inhibits not only bulbospinal inspiratory neurons, but also motoneurons of laryngeal abductory muscles.

The vagi are mixed (sensory and motor) bilateral nerves. Their sensory neurons are located in the nodus and petrosus ganglions. The afferent fibers innervate various thoracic and abdominal organs (Fig. 4.9). Vagus nerve damage lowers the breathing frequency and augments the tidal volume.

The Hering–Breuer inflation reflex corresponds to breathing inhibition during maintained lung inflation after an end-inspiration occlusion at the artificial airway opening. Such maneuver induces a delayed next inspiratory effort. The larger the lung volume, the longer the delay. Any blockage of vagus signaling eliminates the response delay. The opposite maneuver maintains the lung deflated at a volume between FRC and RV level. Consequently, the phrenic activity increases. When vagus nerves are cut, lung deflation has no effect on the breathing timing. Therefore, vagal afferents act on inspiration according to the lung volume. When the occlusion happens at high lung volume, the inspiration is inhibited.

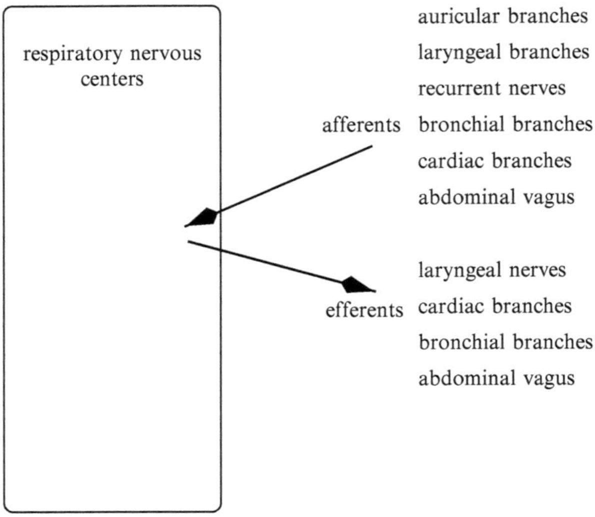

Fig. 4.9 Afferent and efferent nerves of the respiratory system

When the lung volume is maintained at a low level, the inspiratory stimulation increases.

The nose, mouth, and pharynx send afferent innervation mostly by the trigeminal nerve. In addition, ventilatory reflexes, either excitatory or inhibitory, can have cardiac or visceral origins.

4.8.3.2 Efferences

The larynx is innervated by branches of the vagus nerve, the superior laryngeal and recurrent (inferior laryngeal) nerves. The lung parenchyma is innervated by both types of autonomic nervous system, sympathetic and parasympathetic (Fig. 4.10).

The main and accessory respiratory muscles of the thoracic and abdominal walls are innervated by cranial nerves (trigeminal, facial, glossopharyngeal, vagus, and hypoglossal nerves). Cervical nerves (C1–C8) are spinal nerves that originate above their corresponding cervical vertebrae, except the C8 nerve that exits below the C7 vertebra. The phrenic nerve originates mainly from the C4 nerve as well as C3 and C5 nerves (Table 4.41). Thoracic nerves (T1–T12) are spinal nerves that emerge below thoracic vertebrae. Intercostal nerves are anterior branches of T1 to T11 nerves. They are connected with the adjoining ganglion of the sympathetic trunk by rami communicans. Lumbar nerves (L1–L5) are spinal nerves that emerge below the lumbar vertebrae. The iliohypogastric nerve is constituted by the superior branch of the anterior ramus of L1 nerve T12 (subcostal nerve). The inferior branch is the ilioinguinal nerve.

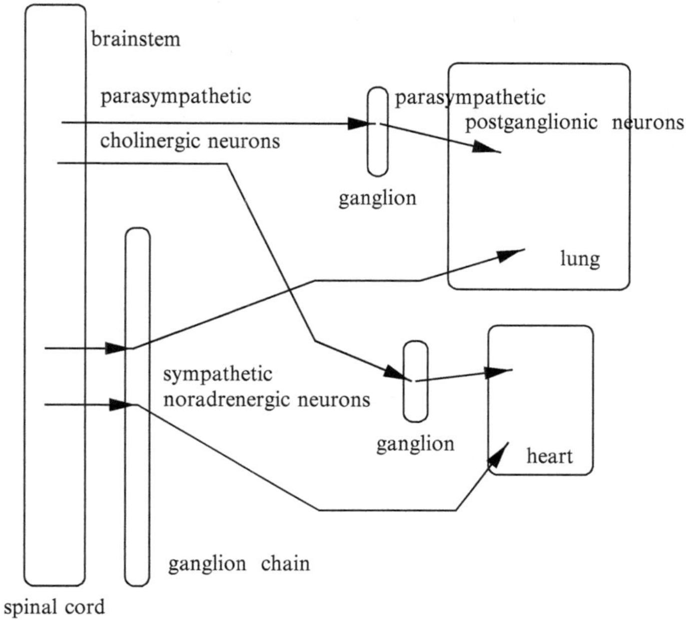

Fig. 4.10 Sympathetic and parasympathetic innervation of coupled lungs and heart. The sympathetic innervation uses nodes in the paravertebral ganglion chain, especially the stellate and middle cervical ganglion. The parasympathetic innervation synapses in plexi in the neighborhood of organs

Table 4.41 Innervation of respiratory muscles

Respiratory muscle	Nerve
Inspiratory muscles	
Diaphragm	Phrenicus (C3–C5)
Intercostalis externus	Intercostalis (T1–T11)
Sternocostalis	Intercostalis (T1–T6)
Sternocleidomastoideus	Cervicalis XII (C2–C4)
Scalenus	Cervicalis spinalis (C3–C8)
Pectoralis	Pectoralis (C5–T1)
Expiratory muscles	
Intercostalis internus	Intercostalis (T1–T11)
Rectus abdominis	Ilioinguinalis (T7–L1)
Obiquus abdominis	Iliohypogastrus (T7–L1)

4.8.4 Ventilation Receptors

Receptors involved in the ventilation regulation include chemo- and mechanoreceptors (stretch receptors or baroreceptors), irritant receptors, and J receptors. In particular, the larynx contains chemo-, mechano-, osmo-, and thermoreceptors.

Chemoreceptors are either centrally or peripherally located. Central chemoreceptors are sensible to the variations of p_{aCO_2} and pH of the cerebrospinal fluid. Major central chemoreceptors localize within 0.2 mm of the anterior surface of the medulla. Peripheral chemoreceptors of the aortic arch and carotid sinus that have mainly vascular and respiratory effects, respectively, are essentially sensitive to p_{aCO_2} variations. They are innervated by the *Hering's nerve*, a branch of the glossopharyngeal nerve (or cranial nerve IX) to the carotid sinus and carotid body. They receive a very high blood supply that renders their environment close to the arterial blood, hence with small arteriovenous differences in p_{CO_2}, p_{O_2}, and H^+ level. They respond with time constants of a few seconds.

Stretch receptors of the muscular layer of airway walls respond to changes in intratissular tension. They are connected to large myelinated fibers of the vagus nerves. They are responsible of the Hering–Breuer reflex (apnea at the end of inspiration). Pulmonary slowly adapting stretch (mechano[chemo]) receptors cause a bronchodilation and small increase in cardiac frequency. Certain stretch receptors are activated by lung expansion when the lung volume exceeds FRC (phasic activity), whereas others are also active at expiration end (tonic activity).

Rapidly adapting stretch (mechano[chemo]) receptors, or irritant receptors, between epitheliocytes of large airway epithelia generate a bronchoconstriction as well as cough and tachypnea. They respond rapidly to stretch. In addition, many noxious agents (dust, cigarette smoke, certain inhaled pollutant gas) can activate irritant receptors. Their activity is inversely proportional to the lung compliance, involving the tissue tension. They are innervated by small myelinated fibers of vagus nerves.

Juxtacapillary receptors (J receptors) belong to the alveolar wall close to capillaries. They are innervated by unmyelinated fibers of vagus nerves. They respond to pulmonary edema and emboli, pneumonia, and barotrauma that cause a decrease in oxygenation. They heighten the respiratory frequency and reduce the tidal volume and cardiac frequency, hence provoking a rapid superficial breathing.

The autonomic sympathic system includes adrenal glands and sympathic nerves. The sympathic nerves innervate airways from sympathic ganglions, where acetylcholine is the neuromediator. Bronchial adrenergic receptors are mainly of β type (bronchodilation), in which the neurotransmittor is noradrenaline. Noradrenaline generates the formation of series-E prostaglandins that, like noradrenaline, exerts a negative feedback. α Adrenergic receptors also exist.

4.9 Oxygen Sensing

The maintenance of respiratory and cardiovascular functions ensures an optimal oxygen delivery to the body's cells, hence adequate ATP production. Impaired O_2 supply affects cellular redox homeostasis.

During hypoxia, reflexes initiated by arterial chemosensors that monitor arterial blood O_2 level such as those of the carotid body as well as catecholamine secretion by chromaffin cells of the adrenal medulla stimulated by hypoxia participate in this maintenance.

Various signaling mediators are implicated in the response of the carotid body and adrenal medulla to hypoxia, such as K^+ channels as well as Ca^{2+}, biogenic amines, and gas messengers.

Hypoxia-inducible factors HIF1 and HIF2 adapt the transcriptional response to hypoxia (Vol. 4 – Chaps. 10. Other Major Signaling Mediators and 11. Signaling Pathways). Heterodimers HIF1 and HIF2 consist of an O_2-regulated HIF1α or HIF2α subunit and a constitutively expressed HIF1β subunit. Subunit HIF1α is ubiquitous, but HIF2α production is restricted to certain cell types such as those of the carotid body and adrenal medulla.

In the carotid body, HIF1 and HIF2 support synthesis of NADPH oxidase NOx2, a prooxidant enzyme, and superoxide dismutase SOD2, an antioxidant enzyme, respectively, thereby priming expression of genes, products of which have opposite functions in the regulation of the redox state. The redox state relies on feedforward and feedback between ROS, HIF, and redox regulatory enzymes. Redox-dependent mechanisms contribute to the reciprocal regulation of HIFα subunit isoforms.

Balanced activity of hypoxia-inducible factors HIF1 and HIF2 controls oxygen sensing by the carotid body and adrenal medulla [34]. Mutual antagonism between HIF1α and HIF2α actually determines the redox state, hence the set point for O_2 sensing in these 2 regulatory organs.

In Hif2$\alpha^{+/-}$ mice, partial HIF2α deficiency increases HIF1α and NOx2 concentrations, causing an oxidized intracellular redox state and exaggerated hypoxic sensitivity [34]. Conversely, in Hif1$\alpha^{+/-}$ mice, partial HIF1α deficiency raises HIF2α and SOD2 concentrations, reducing intracellular redox state and blunting oxygen sensing, hence impairing carotid body and ventilatory responses to chronic hypoxia.

Partial loss of HIF1α or HIF2α raises expression of the other isoform in the carotid body and adrenal medulla. These abnormalities are not observed in Hif1$\alpha^{+/-}$, Hif2$\alpha^{+/-}$ mice. Mutual functional antagonism between HIF1 and HIF2 isoforms determines the proper redox set point in the carotid body and adrenal medulla.

4.10 Immune Defense of the Ventilatory Apparatus

The respiratory tract has a huge epithelial surface (in human adults, estimated surface area $70\,m^2$) that is continuously exposed to the atmospheric air with its pollutants and allergenic and pathogenic agents. Particle deposition in the respiratory conduits is currently supposed to be done by several mechanisms: sedimentation, impaction, and interception.

Table 4.42 First-line defense molecules produced by airway epitheliocytes (Source: [9])

Product	Effect
Mucins	Binding to pathogens
Surfactant protein-A	Opsonins for pathogen clearance
Surfactant protein-C	Binding to pathogens
Surfactant protein-D	Opsonins for pathogen clearance
Complement and cleavage products	Promotion of phagocytosis, Bridging of innate and adaptive immunity
Antimicrobial peptides	Direct antipathogen activity, Activation of adaptive immunity

Cough aims at eliminating foreign materials in the thoracic airways. It includes a deep inspiration followed by a forced expiration, at first against closure of the glottis, during which airway pressure and expiratory velocity can reach very high values (Mach number close to 1). Sneezing is another defense mechanism aimed at cleaning the upper airways. The pharynx partially collapses at the beginning of a forced expiration.

Three principal elements are involved in the defense processes: (1) mucociliary clearance, (2) macrophage activity, and (3) immune reactions. Innate mechanisms are important components of the defense system, especially before the development of the adaptive immune system. In addition to lung macrophages, airway epitheliocytes, airway mucosal dendritic cells, memory helper and regulatory T lymphocytes, natural killer cells and mastocytes participate in the maintenance of the local immune defense. Especially in highly vascularized regions, the lung vasculature contains a large pool of marginated T lymphocytes.

The respiratory system is involved in the defense of its mucosal surface, especially via the secretory immune system that releases in particular immunoglobulin A (IgA), in addition to other types of immunoglobulins (IgE, IgG, and IgM).

Airway epitheliocytes synthesize defense molecules against pathogens and and their products (Table 4.42). Surfactant proteins-A and -D belong to the collectin family.[47] Surfactant proteins SPa and SPd (Vol. 5 – Chap. 13. Surfactant) are associated with the innate immune system, as they are able to recognize a broad spectrum of pathogens.

Antimicrobial peptides include defensins, cathelicidins, histatins, lysozyme, lactoferrin, secretory leukocyte peptidase inhibitor (SLPI),[48] which protects

[47] Collectins are soluble pattern recognition receptors of the superfamily of collagen-containing C-type lectins.

[48] A.k.a. antileukoproteinase (ALk1 or ALP).

epithelia from serine proteases, elafin,[49] palate, lung, and nasal epithelium clone protein (PLUNC),[50] and endogenous antibiotic Bactericidal and permeability-increasing protein (BPI).

Pattern-recognition receptors such as Toll-like receptors (Vol. 3 – Chap. 11. Receptors of the Immune System) are major actors of the local immune surveillance. Toll-like receptors primarily coat sentinel dendritic cells of airway mucosa. Toll-like receptors foster the generation of effector memory responses against antigens, as airway mucosal dendritic cells transport antigens to regional lymph nodes.

4.10.1 Organization of the Immune System of the Ventilatory Apparatus

The ventilatory immune system must discriminate among usual, harmless, air-immersed antigens and much rarer, pathogen-associated antigens to minimally damage the respiratory epithelium. It must maintain the defense against eventual incoming pathogens at different compartments of the respiratory tract (mainly conducting airways and lung parenchyma). On the other hand, the immune system must avoid inappropriate responses to inhaled immunogenic components of plant and animal origin that could lead to chronic airway inflammation.

In thick-walled conducting airways, the air–wall interface lies on a mucosa. The lung parenchyma comprises mainly thin-walled alveoli. The pulmonary alveolus wall is a component of the lung parenchyma. However, the lung parenchyma corresponds to the collection of bronchovasculonervous sheaths and septa between lung acini, lobules, and lobes, in addition to the set of alveolar walls. Different populations of immune cells reside in these 2 compartments of the respiratory tract. They capture of inhaled antigens for subsequent recognition by T lymphocytes in draining lymph nodes [9].

4.10.1.1 Immunity in Conducting Airways

The immune system in conducting airways (i.e., the proximal compartment of the respiratory tract) relies on not only usual innate and adaptive immune responses, but also the mucociliary clearance.

[49] A.k.a. elastase-specific inhibitor (ESI), peptidase inhibitor-3 (PI3), and skin-derived antileukoproteinase (SkALP).

[50] A.k.a. secretory protein in the upper respiratory tract (SPURT).

Mucociliary Clearance

The respiratory epithelium encompasses ciliated cells and secretory goblet cells that yield the mucociliary clearance of inhaled antigens. Mucus (Vol. 5 – Chap. 12. Airway Surface Liquid and Respiratory Mucus) is made of exsudated plasma and secretion products from the respiratory mucosa. It is moved by ciliary beats with a frequency that can reach 50 Hz. The speed of the mucus motion ranges from 7 μm/s in bronchioles to 0.2 mm/s in the trachea. The deposited material is eliminated in about one hour.

Immunocytes

In addition, the airway mucosa contains pools of myeloid (predominantly) and plasmacytoid dendritic cells (Vol. 5 – Chap. 3. Blood Cells) as well as macrophages. The lamina propria also contains mastocytes, plasmocytes that mainly produce IgA, and some scattered B lymphocytes [9]. Relatively high amounts of T lymphocytes also lodge in the airway mucosa, both within the epithelium and lamina propria.

Recruitment of lymphoid cells to the respiratory tract requires chemokine recognition and adhesion molecules. IgA-secreting cells that home to the bronchial mucosa and T lymphocytes possess $\alpha_4\beta_1$ integrins that target vascular cell-adhesion molecule VCAM1 on vascular endothelial cells in the bronchial mucosa [9]. In addition, T lymphocytes have $\alpha_L\beta_2$ integrins that connect to intercellular adhesion molecules ICAM1, also constitutively expressed on vessel walls in the bronchial mucosa. IgA-secreting cells that move into the bronchial mucosa also express chemokine receptor CCR10 for locally produced CCL28 chemokine.

Dendritic Cells

Airway mucosa-resident dendritic cells are specialized for antigen uptake and processing within the epithelium as well as in near-wall lumen owing to protrusions into the airway lumen, but lack efficient antigen presentation [9]. However, in some circumstances such as atopic (extrinsic) asthma, lung dendritic cells can become transiently potent antigen-presenting cells. Lung dendritic cells are characterized by a high turnover rate in normal conditions and rapid CCR1- and CCR5-dependent recruitment of dendritic cells and their precursors during inflammation.

Resident pulmonary dendritic cells highly synthesize class-2 MHC molecules and C-type lectin CLec13b (or CD205), and, at a low level, CD8 coreceptors, TNFRSF5, and costimulatory or inhibitory T-lymphocyte antigens CD80 and CD86.

Several distinct subpopulations of pulmonary dendritic cells have been identified. Plasmacytoid dendritic cells have a distinct pattern of TLR expression w.r.t. myeloid dendritic cells, but they have poor antigen presentation activity, unlike myeloid dendritic cells. They highly produce interferon-α in response to microbial infections.

Luminal antigens are taken up by intraepithelial dendritic cells in the bronchial wall. Antigen-endowed dendritic cells upregulate chemokine receptor CCR7 and migrate through lymphatics to reach regional draining lymph nodes, where they present antigens to T lymphocytes [9]. Activated T lymphocytes proliferate and migrate through efferent lymphatics and then blood vessels. Afterward, effector T lymphocytes exit into the bronchial mucosa through postcapillary venules in the lamina propria.

Macrophages

Macrophages closely adhere to airway epitheliocytes. They secrete transforming growth factor-β that fosters the expression of $\alpha_V\beta_6$-integrins on airway epitheliocytes [9].

T Lymphocytes

Most CD8+ T lymphocytes reside in the epithelium, whereas CD4+ T lymphocytes localize preferentially in the lamina propria [9]. Most T lymphocytes in the tracheobronchial mucosa are in close contact with dendritic cells.

Helper T Cells

The helper T-cell (T_H) subpopulation of lymphocytes is decomposed into several categories. T_H progenitors can differentiate into effector and memory T_H cells. Effector T_H cells secrete activators of other types of leukocytes as well as T_H cells. Memory T_H cells retain the affinity for the originally activating antigen.

Type-1 (T_{H1}) and -2 (T_{H2}) helper T cells interact mainly with macrophages and B lymphocytes, respectively. T_{H17} cells produce large amounts of interleukins IL17, IL22, and IL23 [9]. They are stimulated by a combination of interleukins IL1β and IL6. Interleukin-17 mobilizes neutrophils.

Regulatory T Cells

FoxP3+, CD4+, CD25+ (IL2Rα+) regulatory T cells (T_{Reg}) control T-cell responses, mainly via interleukin-10 and transforming growth factor-β, thereby protecting against inflammatory sequelae of airway infections, especially after viral bronchiolitis. Regulatory T cells can also attenuate airway hyperresponsiveness after aeroallergen inhalation [9].

Natural Killer Cells

Natural killer (NK) cells control host responses to virus, in particular influenza and respiratory syncytial viruses, as well as bacteria such as Bordetella pertussis and Mycobacterium tuberculosis [9]. Invariant natural killer (iNKT) cells differ from NK cells, as they express T-cell coreceptor CD3 and invariant T-cell-receptor α chain [9]. They respond to glycolipid antigens presented by CD1d glycoproteins. They then produce large quantities of interferon-γ and interleukin-4. They are involved in defense against a wide range of bacteria, viruses, and parasites. Activated iNKT cells by glycolipid antigens may prime airway hyperresponsiveness.

Mastocytes

Mastocytes localize mainly near mucosal surfaces exposed to the environment such as airway mucosa. They promote innate immunity, as they activate Toll-like receptors as well as receptors for complement and other inflammatory peptides [9].

Mastocytes are activated by multiple agents and release, according to the stimulus, various products, such as tumor-necrosis factor, antimicrobial cathelicidins (such as neutrophils, macrophages, and epitheliocytes), and other mediators (e.g., histamine, prostaglandins, and free radicals), as well as growth factors, cytokines such as thymic stromal lymphopoietin for monocytes and dendritic cells, and chemokines for inflammatory leukocytes.

Mastocytes can behave as phagocytic cells. They can recruit neutrophils, stimulate IgE production in B lymphocytes, promote proliferation and migration of T lymphocytes, as well as mediate tolerance by regulatory T lymphocytes.

4.10.1.2 Bronchus-Associated Lymphoid Tissue

The common mucosal immune system includes lymphocyte travel between mucosal surfaces and lymphoid tissues. The bronchus-associated lymphoid tissue (BALT; Vol. 5 – Chap. 4. Lymphatic System) comprises: (1) bronchial lymphoid nodules, especially at airway bifurcations, with a high number of IgA-secreting precursor lymphocytes; (2) aggregates of lymphoid and nonlymphoid cells within the connective tissue; and (3) lymphocytes, mainly helper T cells, in airway walls and alveolar lining layer.

Bronchus-associated lymphoid tissue can be usually observed in children. It mostly contains isolated lymphoid follicles. Lymphocyte homing to bronchial-associated lymphoid tissue relies on interaction between VCAM1 and $\alpha_4\beta_1$ integrins. Specialized endotheliocytes of lung capillaries facilitate transit of lymphocytes from blood into bronchus-associated lymphoid tissue. Extravasation (Vol. 5 – Chaps. 9. Vascular Endothelium and 11. Tissue Development, Repair, and Remodeling) is mediated by plasmalemmal addressins and lymphocyte homing receptors. Antigen-specific antibody-producing cells are recruited to the lungs after antigen stimulation.

4.10.1.3 Immunity in Lung Acini

Bronchioles extend to alveolar ducts that drive air into alveoli. Immunocytes localize to the walls of the last generations of airways and alveoli. In normal conditions, the leukocyte population in alveolar walls and spaces is dominated by alveolar macrophages (>90% of the total cell population), in addition to dendritic cells and T lymphocytes [9]. The lung parenchyma contains scattered macrophages, dendritic cells, and T and B lymphocytes, as well as mastocytes, but no plasmocytes.

Alveolar Macrophages

Alveolar macrophages are mature lung-resident cells. They can uptake particles and possibly digest them using lysosomes. Phagocytosis permits sequestration of antigen from the immune system to shield from the development of specific immune responses [9]. In addition, alveolar macrophages also suppress the triggering of adaptive immunity using nitric oxide, interleukin-10, prostaglandins, and transforming growth factor-β, as well as regulate dendritic-cell migration. However, alveolar macrophages also produce small amounts of proinflammatory cytokines.

Inert materials are incorporated in alveolar macrophages and conveyed to the bronchi in order to be eliminated by the mucociliary clearance in about one day. Other particles will be moved to the interstitial tissue, either directly transported or conveyed by alveolar macrophages, in order to be removed by the lymph circulation toward lymph nodes and blood circulation. Such a purification takes about 3 to 10 days. However, some particles migrate from the interstitium to the septa of subpleural pulmonary lobules. Particle elimination is then very slow (about 3 months). Certain particles can induce inflammation (Vol. 5 – Chap. 11. Tissue Development, Repair, and Remodeling).

Alveolar macrophages secrete TGFβ that supports the expression of $\alpha_V\beta_6$-integrins on alveolar epitheliocytes [9]. Ligation of Toll-like receptors on alveolar macrophages causes their detachment from alveolar epitheliocytes.

The turnover of alveolar macrophage populations is normally extremely slow w.r.t. to neighboring dendritic cells. In normal conditions, alveolar macrophage renewal results from the proliferation of local precursor cells. During inflammation, macrophage renewal essentially arises from migrating monocytes using CCL2 chemokine and CCR2 receptor. In addition, monocytes can operate as antigen-presenting cells before their maturation into alveolar macrophages [9].

Dendritic Cells

As in bronchial walls, inhaled antigens can be taken up by pulmonary dendritic cells that reside in alveolus septa. These cells then migrate through afferent lymphatics to draining lymph nodes to present antigens to T lymphocytes. Activated T lymphocytes proliferate and migrate through efferent lymphatics and blood vessels to reach pulmonary capillaries.

T Lymphocytes

T Lymphocytes enter the lung parenchyma owing to $\alpha_L\beta_2$ integrin and selectin-P ligand (P-selectin glycoprotein ligand PSGL1) on the one hand as well as chemokine receptors CCR4, CCR5, and CXCR6 for chemokines CCL5, CCL17, and CCL22, as well as CXCL16, respectively [9]. Their retention relies on downregulation of $\alpha_L\beta_2$ integrin and upregulation of $\alpha_1\beta_1$ integrin. In addition, recruitment of invariant natural killer T cells to the lung parenchyma, including alveolar spaces, depends on CCR4 that tethers to CCL17 and CCL22 chemokines [9].

4.11 Metabolic Functions of the Lung

The lungs are not only involved in the gas transport from the ambient air down to the blood circulation and conversely, but also have other functions. They are loci of synthesis and/or metabolism of various substances. Lungs also act as a buffer for inhaled and circulating xenobiotics, such as pollutants and certain drugs.

4.11.1 Detoxificafion Enzymes

Chemical detoxificafion aims at eliminating or excreting pollutants and drugs out of the body. In particular, drug metabolism modifies drugs via chemical reactions to terminate drug action and promotes removal upon transformation that renders drugs less reabsorbable (less lipid-soluble). Transformation enzymes encompass oxidases, reductases, and hydrolases. Conjugation reactions add to drugs or their metabolites a chemical group, such as methyl, acetyl, sulfate, amino acid, or glucuronic acid (catalyzed by microsomal glucuronyltransferase) and generate usually inactive products.

The major site is the liver, but other organs such as lungs can be involved. In the cell, drug processing occurs in the cytosol (e.g., alcohol deshydrogenase) as well as in the endoplasmic reticulum (microsomes) and mitochondria (e.g., monoamine oxidase and acetyltransferase). In the intestinal lumen and mucosa, drugs can undergo the action of glucuronidase and azoreductase from the gut microbiota as

Table 4.43 Reactions carried out by mixed-function oxidase (Source: [42])

Reaction	Substrate
Alcohol oxidation	Ethanol (drinking alcohol)
Aliphatic hydroxylation	Pentobarbital (sedative)
Aromatic hydroxylation	Lidocaine (common local anesthetic drug)
	Phenacetin (analgesic)
^{N}-Dealkylation	Morphine, diazepam (anxiolytic, anticonvulsant, and skeletal muscle relaxant)
^{O}-Dealkylation	Codeine (analgesic, anxiolytic, antitussive, antihypertensive)
^{S}-Dealkylation	6-Methylthiopurine (purine synthesis inhibitor)
Dehalogenation	Halothane (anesthetic)
Epoxidation	Benzo(a)pyrene (pollutant)
^{N}-Oxidation	Methylpyridine (pollutant)
(amine oxidation)	Paracetamol (analgesic)
^{S}-Oxidation (sulfoxidation)	Chlorpromazine (antidopaminergic, -adrenergic, -serotonergic, -cholinergic, and -histaminergic)
Oxidative deamination	Amphetamine (dopamine and norepinephrine potentiator)
Phosphothionate oxidation	Parathion (insecticide and acaricide)

well as mucosal monoamine oxidase and sulfatase. In the plasma, drug-processing enzymes include catechol ^{O}methyltransferase (COMT), esterases, and amidases.

Chemical detoxificafion relies, at least partly, on cytochrome-P450 reductase,[51] a membrane-bound enzyme required for electron transfer to cytochrome-P450 in the endoplasmic reticulum of cells from a FAD- and FMN-containing enzyme NADPH:cytochrome-P450 reductase.

Mixed-function oxidase (MFO) is a microsomal (endoplasmic reticulum) enzymatic complex that oxidizes toxic compounds to render them more susceptible to processing and excretion. Oxidative and reductive metabolism is carried out by hepatic and extrahepatic (lung, kidney, adrenal, gastrointestinal tract, and skin) mixed-function oxidases.

Mixed-function oxidase catabolizes many compounds, such as fatty acids, steroids, alkanes, polycyclic carcinogens, drugs, and environmental pollutants. It indeed performs many types of reactions using molecular oxygen and Mg^{2+} ions (Table 4.43). It is composed of: (1) cytochrome-P450 associated with NAD^{+} and FAD); (2) cytochrome-B5, a ubiquitous, microsomal, membrane-bound electron transport hemoprotein; and (3) the flavoprotein NADPH–cytochrome-P450 reductase (i.e., cytochrome reductase associated with NADPH coenzyme).

[51] A.k.a. NADPH:ferrihemoprotein oxidoreductase, NADPH:hemoprotein oxidoreductase, NADPH:P450 oxidoreductase, and P450 reductase.

Table 4.44 Blood substance catabolism rate through the pulmonary circulation (PG: prostaglandin)

Subtance	Removal rate (%)
Adenosine monophosphate	~100
Adenosine triphosphate	~100
Adrenaline	~0
Angiotensin-1	100 (conversion in angiotensin-2)
Bradykinin	≤80
Histamine	~0
Leukotrienes	~100
Noradrenaline	≤30
PGe_2, $PGf_{2\alpha}$	~100
PGa_2, PGi_2	~0
Serotonin	~100

Flavin adenine dinucleotide (FAD)[52] is a redox cofactor. In its fully oxidized form (quinone form), it accepts 2 electrons and 2 protons to become $FADH_2$ (hydroquinone form) that can be oxidized to the semireduced FADH form (semiquinone) . Flavin mononucleotide (FMN), or riboflavin 5′-phosphate, another flavoprotein, is produced from riboflavin (vitamin-B2) by riboflavin kinase. It functions as the prosthetic group of various oxidoreductases, undergoing a reversible interconversion of the oxidized (FMN), semiquinone (FMNH), and reduced ($FMNH_2$) forms.

Mixed-function oxidases catalyze a reaction in which each of the 2 atoms of molecular oxygen, an electron acceptor, is used for a different function. Agent FAD serves as an electron acceptor for NADPH cofactor. On the other hand, FMN is a electron carrier from NADPH to cytochrome-P450 during the mixed-function catalysis.

4.11.2 Circulating Substance Fate

Lungs are able to process or degrade and remove several circulating vasoactive substances, such as angiotensin 1, bradykinin, prostaglandins E and F (PGe_1 and $PGf_{2\alpha}$), serotonin (or 5-hydroxitryptamin), and noradrenaline, in particular owing to the large vascular surface within the lung parenchyma (Table 4.44).

Serotonin and noradrenaline are degraded by monoamine oxidase (MAO). Ninety percent of circulating serotonin as well as 40% of circulating noradrenaline can be inactivated during a single passage.

Angiotensin 1 is transformed into angiotensin 2 by the convertase of vascular endotheliocytes. Bradykinin is inactivated by bradykininase at the surface of

[52]The flavin group is bound to ribitol, a sugar alcohol, by a carbon–nitrogen bond, not a glycosidic bond. Hence, riboflavin is not a nucleotide (the name flavin adenine dinucleotide is a misnomer).

endotheliocytes. Ninety-five percent of bradykinin activity is lost in a single passage. Ninety to 95% of targeted prostaglandins are inactivated after endocytosis and transformation.

Glucose is metabolized in lungs. Lipolysis occurs in the capillary endothelium in lungs. Fatty acids are oxidized in lungs; the produced energy is used for lipogenesis. Phospholipids are synthetized by type-2 alveolar pneumocytes to form the surfactant. Prostaglandins, in particular $PGf_{2\alpha}$, are produced by lungs. Synthesized proteins are used for the extracellular matrix.

Histamine is manufactured by mastocytes of the respiratory tract. Its release is primed by α-adrenergic stimulation and precluded by β-adrenergic stimulation.

4.11.3 Respiratory Cell Production

4.11.3.1 Bronchoconstrictors

Pulmonary cells such as mastocytes are able to release various substances, such as histamine, eosinophil (ECFA)[53] and neutrophil (NCFA)[54] chemotactic factor of anaphylaxis, slow-reacting substance of anaphylaxis (SRSA),[55] platelet-activating factor, and so one. All these substances engender a bronchoconstriction. They act directly on parietal smooth muscle cells or via nerve excitation. These molecules also modify the capillary permeability and cause inflammation.

4.11.3.2 Epithelial Integrity and Mucociliary Clearance

Most airways are lined by an epithelium made up of ciliated and secretory cells with a wetted (luminal) surface and undifferentiated basal progenitor cells aimed at regenerating the respiratory epithelium after damage. This selective polarized barrier produces an liquid–air interface and fulfills the mucociliary clearance and

---

[53]Eosinophil chemotactic factor of anaphylaxis includes 2 small-molecular-weight tetrapeptides (hence not classified as cytokines). Tetrapeptides (Ala–Gly–Ser–Glu [AGSE] and Val–Gly–Ser–Glu [VGSE]) of eosinophil chemotactic factor of anaphylaxis are released from mastocytes and basophils [43]. They serves as mediators of immediate hypersensitivity (anaphylaxis). They attract eosinophils.

[54]Both ECFA and NCFA are chemoattractants for eosinophils and neutrophils during the late stage of anaphylaxis. Leukotriene-B_4 constitutes a major constituent of neutrophil chemotactic factor of anaphylaxis on release of IgE and IgG by lung tissue.

[55]Slow-reacting substance of anaphylaxis is a mixture of leukotrienes LTc_4, LTd_4, and LTe_4 secreted by mastocytes and basophils [44]. Human pulmonary slow-reacting substance of anaphylaxis has indeed been identified as the peptidolipid 5-hydroxy 6-cysteinyl glycinyl (7,9,11,14)-eicosatetraenoic acid (Cys-Gly SRSA), a metabolite of arachidonic acid in leukocytes [45]. It is a potent bronchoconstrictor originally decribed as an active nonhistamine molecule synthesized and released by antigen stimuli, especially anaphylaxis, from lung tissue and human leukocytes. This mediator of immediate hypersensitivity is secreted with other substances such as histamine, but more slowly [46].

innate immunity against pathogens and allergens. The lateral part of the plasma membrane of respiratory epitheliocytes contain specialized junctions that enable cell adhesion and selective transcellular passage of ions and molecules as well as immunocytes, thereby regulating inflammation and struggling against infection.

Transcription factors of the GRHL (grainyhead-like) category coordinate multiple cellular processes required for transcellular transport and lipid metabolism, cell morphogenesis, adhesion, and motility, as well as epithelial morphogenesis, differentiation, remodeling, and repair. In the embryonic mouse lung, GrHL1 to GrHL3 have distinct spatiotemporal expression patterns in the epithelium; GrHL2 operates in airway branching morphogenesis [47].

In human bronchial epitheliocytes and their undifferentiated progenitors, GrHL2 is a transcriptional activator or repressor, directly or not, of several hundreds of genes, thereby regulating the ciliated and secretory epithelium [47]. Among GrHL2-controlled genes, several (e.g., MPZL2,[56] nectin-4,[57] RHBDL2,[58] VAV1,[59] and Znf750)[60] intervene in epithelial barrier integrity and repair. It regulates intercellular adhesion, in particular apical junctions, in coordination with other factors regulated by GrHL2, to ensure junction assembly and cell polarity at the post-transcriptional or post-translational level. It may operate in alternative splicing of transcripts via epithelial splicing regulatory proteins ESRP1 and ESRP2, transfer of membrane components via Rab GTPases, and dynamic interaction of membrane proteins with the cytoskeleton via RhoGEF and RhoGAP regulators.

4.11.3.3 Surfactant

Production of surfactant proteins (Vol. 5 – Chap. 13. Surfactant) relies on kinases of the Hippo pathway (Vol. 4 – Chap. 10. Other Major Signaling Mediators), that is, STK3 and STK4 (Vol. 4 – Chap. 5. Cytosolic Protein Ser/Thr Kinases) [51]. They coordinate activity of transcription factors involved in the synthesis of surfactant proteins by type-2 pneumocytes. Both STK3 and STK4 kinases phosphorylate (activate) Forkhead box FoxA2 transcription factor that regulates pneumocyte maturation and surfactant protein expression.

---

[56]The MPZL2 gene encodes myelin protein zero-like protein-2, also called epithelial V-like antigen EVA1. It is a member of the immunoglobulin superfamily expressed in embryonic epithelia. In particular, it is produced in the thymic epithelium in the early stage of thymus organogenesis. It is then downregulated during the thymocyte developmental progression [48]. This immunoglobulin-like adhesion molecule regulates integrity of the blood–cerebrospinal fluid barrier, where occurs primarily the immune surveillance of the brain. It is involved in adhesion of CD4+ T lymphocytes on choroid plexus epitheliocytes [49].

[57]A.k.a. PVRL4 gene that encodes poliovirus receptor-related protein-4. Nectins and Nectin-like molecules are Ca^{2+}-independent cell adhesion molecules.

[58]The RHBDL2 gene encodes the proenzyme of intramembrane rhomboid, veinlet-like serine peptidase-2 (RhbdL2).

[59]The VAV1 gene encodes the Vav1 guanine nucleotide-exchange factor.

[60]The ZNF750 gene encodes zinc finger protein ZNF750 expressed in the lung, skin, thymus, prostate, and placenta [50].

Chapter 5
Medical Images and Physiological Signals

Si l'activité scientifique expérimente, il faut raisonner; si elle raisonne, il faut expérimenter
[If the scientific activity experiments, one must reason; if it reasons, one must experiment].

(Bachelard [1884–1962]) [1]

Many visualization techniques are available to explore the cardiovascular system from usual ultrasound echography and velocimetry, multislice spiral computed tomography particularly for cardiac imaging, and magnetic resonance imaging for blood flow assessment, to magnetocardiography, diode laser, and optical coherence tomography. Functional magnetic resonance imaging is based on increased blood flow gushing in target regions that are responding to imposed stimuli. Ultrasound scattering from Rayleigh fractal aggregates is proposed for imaging flow dynamics of deformable or hardened red cell clusters in dense suspension.

5.1 Imaging and Geometry Modeling

Computerized medical imaging provides subject-dependent 3D geometry of any body organ, in particular the heart and large blood vessels. Medical images taken of the human body are mainly displayed in 3 main planes: (1) *coronal* planes that divide the body into front and back regions; (2) *sagittal* planes that separate the body into left and right parts; and (3) *transverse* planes perpendicular to the body axis that split the body into upper and lower domains.

Numerical simulations are performed in computational domains based on imaging data after 3D reconstruction. Input data for the surface reconstruction of the target compartment of the physiological apparatus usually come either from X-ray computed tomography [2–4], magnetic resonance images [5–8], or 3D ultrasound images [9] (Table 5.1).

M. Thiriet, *Anatomy and Physiology of the Circulatory and Ventilatory Systems*,
BBMCVS 6, DOI 10.1007/978-1-4614-9469-0_5,

Table 5.1 Principle of medical imaging techniques

Radiography	X-rays absorption
Nuclear magnetic resonance imaging	Proton environment
US echography	Tissue physical (mechanical) properties
Nuclear imaging	Radioactive isotope concentration

Table 5.2 Relative radiopacity (opacity of target/opacity of water) of various biological materials. The gray intensity is related to the average intensity in the voxel. Usually, human tissues range from −100 to +100 HU

Medium	Haunsfield unit	Gray level
Air	−1,000	Black
Adipose tissue	−95	Dark gray
Bone	+1,000	Clear gray
Water	0	White

5.1.1 *Imaging Techniques*

Standard thoracic X-ray images correspond to the posteroanterior projection and additionally to the lateral projection, hence allowing us to see the retrosternal and retrocardiac spaces. These images are taken in both maximal inspiration and maximal expiration.

5.1.1.1 Computed Tomography

Computed tomography (CT) uses special X-ray equipment to obtain cross-sectional pictures of regions inside the body with gray-level scaling. To enhance vascular anatomy, an intravenous injection of a radiopaque solution can be made prior to or during the scan (computed angiography [CA]).

The Hounsfield unit scale in each voxel expresses the attenuation coefficient coded in 4096 gray levels. It is calibrated upon the attenuation coefficient for water and air, with water reading 0 HU and air −1000 HU (Table 5.2).

The circular rotating gantry contains both the X-ray source and detectors. Emergent X-ray beams are received on a scintillation counter and electronic impulses are recorded and processed. Tomograms are then reconstructed.

In *spiral CT* (SCT, or helical CT), X-rays have a helical path because data acquisition is combined with continuous motion of the acquisition system. The scanner rotates continuously as the patient's couch continuously glides. Scanning is short enough that it can be practiced during a single inspiratory apnea, thereby reducing artifacts due to breathing. Spiral CT images entire anatomical regions in a 20- to 30-s breath hold. Data can be reconstructed to get 3D organ displays and virtual endoscopy.

Table 5.3 Main features (size [mm] and time) of computed tomography

	SCT	MSSCT	EBCT
Slice thickness	10	1.5→0.6	8
Between-slice thickness		30–50% overlapping	Pair gap of 4, contiguous dual slicing
Scanning acquisition time (s)	≥ 160	20	8–12
Slice acquisition time (ms)			100

Table 5.4 Spatial resolution of X-ray sanners

CT	Δx, Δy (mm)	Δz (mm)	Matrix	Slice number
MSSCT function mode	0.5	0.6	512 × 512	4
MSSCT volume mode	0.5	0.6	512 × 512	$\mathcal{O}(100)$
SS-EBCT	0.7	1.5–3	512 × 512	~100
MS-EBCT	1	8	360 × 360	2–8

Multislice spiral CT (MSSCT) scanners can acquire many slices in a single rotation. The rapid development in CT technology from 16-slice MSSCT with already high spatial resolution and reduced examination time to 320-slice acquisition lowering scanning time from 250 to 87 ms as well as X-ray exposure.

Precontrast scanning (low-dose acquisition) is advisable when bleeding is suspected; it can evaluate wall calcifications, intramural hematomas, and dissections.

Contrast scanning is triggered when the bolus reaches the region of interest (e.g., ROI intensity measurement > 180 HU), the injection duration matching the acquisition time.

Special protocols can: (1) evaluate the vessel caliber; (2) exhibit a mural thrombus on the aneurysmal wall and demonstrate the arterial supply of the aneurysm as well as initial flap, and false and true channels of arterial dissection; and (3) help to determine the therapy choice between surgery and endovascular procedures (Vol. 7 – Chap. 4. Treatments of Cardiovascular Diseases).

In *electron beam CT* (EBCT), an electron beam is generated and focused on a circular array of tungsten X-ray anodes. Emitted X-rays, which can be triggered by ECG, are collimated and detected as in conventional computed tomography. Electron beam CT, which has no mechanical motion of the X-ray source and a stationary detector array, allows very quick scanning in 50 to 100 ms. The high temporal resolution is associated with a limited spatial resolution.

The scan speed allows getting a complete image set of the heart during a single breath hold, or any moving and deformable ducts [12]. Electron beam CT is useful to evaluate right and left ventricular muscle mass, chamber volumes, and systolic and diastolic ejection fractions. It can also measure calcium deposits in the coronary arteries. Features of these main CTs are summarized in Tables 5.3, 5.4, and 5.5.

Fractional flow reserve (FFR) can be evaluated from computed tomography and computational fluid dynamics during adenosine-induced hyperemia in the presence of a stenosis [10]. The severity of a stenosis can also be assessed by attenuation of the concentration of a traveling contrast bolus through epicardial coronary

Table 5.5 Temporal resolution of X-ray scanners

CT	Slice scan time (ms)	Temporal image number
MSSCT function mode	300	60
MSSCT volume mode	300	4
SS-EBCT	50	1
MS-EBCT	30	6–40

arteries [11]. The transluminal attenuation gradient is defined as the regression coefficient between attenuation (HU) per 10-mm length of the coronary artery from the ostium. The corrected coronary opacification is determined from intracoronary attenuation values before and after the stenosis normalized to the attenuation values in the descending aorta in the same CT slice. However, these quantities have a limited value for the functional evaluation of stenosis by computed tomography.

Cardiovascular interventional imaging associated with electrocardiogram and hemodynamic monitoring and possibly electrophysiological mapping is used during closure of atrial or ventricular septal defects, electrophysiological firing site ablation, pulmonary balloon valvuloplasty, valve placement, and dilatation of aortic coarctation. Unlike X-ray fluoroscopy, real-time MRI can be used for navigation guidance during radiation-free catheterization.

5.1.1.2 Magnetic Resonance Imaging

Bref l'art poétique de la physique se fait avec des nombres, avec des groupes, avec des spins, etc.
[Brief, the poetic art of physics is done with numbers, sets, and spins, etc.]

(Bachelard) [13]

Nuclear magnetic resonance imaging (MRI) uses a strong, constant magnetic field and radiofrequency waves[1] to excite hydrogen nuclei in selected regions of the body [14]. The precessing nuclei in turn generate a magnetic field oscillating at radiofrequency. The emerging signals have frequency, amplitude, and phase components that are processed to construct images of the human body. Magnetic field gradients are introduced to determine the spatial location of reemitted microwaves. Frequency and timing characteristics of the excitation pulse are modified to image particular types of molecules and motions. Each slice usually represents a thickness

[1]The wavelength is reduced in the human body according to the permittivity and conductivity of crossed tissues.

of 2 to 10 mm. Each pixel represents from 1 to 5 mm. *MR angiography* (MRA) focuses on blood vessels without using any contrast material, although contrast agents may be given to provide better MR images [15].

Various sequence types can be employed. Rephasing consists of returning out-of-phase magnetic moments back into phase coherence either by rapidly reversing a magnetic gradient (field echo) or by applying a 180° radio frequency pulse (spin echo). The steady-state free precession corresponds to any field echo or gradient echo sequence in which a nonzero steady state develops for both transverse and longitudinal components of magnetization.

In spin echo, a radio frequency pulse echo 90°–180°–180° train is repeated several times within the same slice with a given repetition time (TR). The time between the 90° pulse and the output signal time after the first and second 180° pulse is the first and second time echo (TE).

Weighing procedures lead to T1-weighted MRI (TR <600 ms; TE <40 ms) and T2-weighted MRI (TR >1,500 ms; T E>60 ms), in addition to proton or spin density MRI (TR >1,500 ms; TE <40 ms).

In gradient echo imaging, or field echo imaging, the echo produced by reversing the direction of the magnetic field gradient cancels out the position-dependent phase shifts accumulated due to the gradient. The initial flip angle is smaller than 90°. The signal is refocused by gradient reversal (reversing whichever side of the magnet is of higher [lower] magnetic field strength).

Functional MR (FMR) can display the myocardium function [16]. High-performance magnetic resonance imaging now allows high resolution and quick image acquisition, which can provide particularly good delineation of organs and thin-walled vessels.

Magnetic resonance imaging can be used to assess regional tissue perfusion, viability, scarring, and deformation. Perfusion and scarring assessment rely on the first passage and late uptake of gadolinium contrast, respectively. Perfusion is estimated from the upslope of the signal intensity in the myocardium when a gadolinium-based contrast agent is injected. On the other hand, the delayed increase in signal intensity is caused by the retarded washout of injected gadolinium-based contrast agent from the expanded extracellular space of the myocardial scar. Late gadolinium enhancement enables investigation of myocardial fibrosis.

Phase-locking MRI is used to record the vasculature deformation by imaging the selected region of the cardiovascular system at different phases of the cardiac cycle.

Imaging of soft tissue dynamics uses *MR tagging* that allows for encoding of a grid of signal voids on cardiac MR images produced with various techniques [17–21].

Diffusion tensor magnetic resonance imaging (DTMRI) is particularly used to estimate the myofiber orientation in the myocardium. Effective diffusion tensor ($\mathbf{D}_{\mathrm{eff}}$) MRI noninvasively evaluates the structure of biological tissues by measuring water diffusion[2] using magnetic field gradient and diffusion- and nondiffusion weighted images [22]. The water Brownian motion in a medium characterized by ordered rod-like elements has a preferential path. Consequently, the probable molecule location is in an ellipsoid displacement domain rather than a sphere profile when the medium is isotropic. The effective diffusion tensor is computed from the measured apparent diffusion tensor once the eigenvalues[3] have been determined using 6 different directions.

The regional cardiac deformation can be estimated from 3D images to derive the wall displacement field, and in association with a biomechanical model, compute the strain field. The viability of the myocardium when any coronary artery becomes occluded can then be assessed. The wall strain field provides a better analysis than the endocardial motion and allows updating of the remodeling [27]. Using MRI, a shape-tracking approach has been proposed [28]. After segmentation of the left ventricle inner and outer walls in each slice of the initialization cycle phase, contours are propagated in the corresponding slice of the different acquisition instants, and after checking, assembled into endo- and epicardial surfaces. Each small patch of the original surface is mapped to a plausible window of the ventricle deformed surface at a given time, and the patch of the deformed surface having the most similar shape is selected. This method is questionable. The motion of the heart wall can

---

[2]Diffusion of water within a tissue excited by a magnetic field gradient causes MRI signal attenuation. The eigenvectors and eigenvalues of the voxel-averaged diffusion tensor specify the principal directions and rates of water diffusion in each voxel of the tissue image. The eigenvector corresponding to the maximum eigenvalue of the diffusion tensor points in the direction of maximum rate of diffusion assumed to be the direction of the axis of a cylindrical fiber. The orientation of the eigenvectors can be defined by inclination and transverse angles. The inclination angle of the myofiber is the angle between: (1) the intersection line of the image plane and the plane parallel to the epicardial tangent plane at the corresponding azimuthal position (tangent plane); and (2) the projection of the eigenvector onto the tangent plane. The transverse angle is the angle between: (1) the intersection line, and (2) the projection of the eigenvector onto the image plane. The correlation has been checked by experiments performed in an excised portion of the right ventricle by comparison of DTMRI and histology myofiber angles [23, 24], but DTMRI time and space resolutions, especially in vivo, were too large to get an accurate map of myofiber angles. More recently, space resolution of 310 to 390 μm in the slice plane with a slice thickness of 0.8 to 1 mm has been obtained in isolated dog hearts [25]. These authors extract 2 local angles, the myofiber main axis angle and the cross-sheet (from endocardium to epicardium) angle. The myofiber angle is defined by the angle between the local circumferential tangent vector of the reconstructed mesh of the heart wall and projection of the primary eigenvector of the voxel-related water diffusion tensor onto the epicardial tangent plane. The cross-sheet angle is determined by the radial vector and the projection of the tertiary eigenvector of the diffusion tensor, which is parallel to the cardiac sheet normal, onto the plane defined by the radial and circumferential vectors. Images can be obtained using a slice-selection fast spin-echo diffusion-weighted technique coupled with gradient recalled acquisition in the steady-state (GRASS) imaging mode to define epicardial and endocardial surfaces [26].

[3]The diffusivities along the 3 principal axes of the ellipsoid.

be tracked by phase-contrast cineMRI, like blood movement. Heart cyclic motion can be also measured at grid points from tagged MRI [29]. MR tagging modifies the magnetization of selected targets (tags) within the heart wall. The wall motion between the tagging and image acquisition is captured by tag displacement. Spatial modulation of magnetization is used to create tagged plane.

With a relatively large acquisition window, MRI is not perfectly suited to image the thin mobile heart valves located in rapidly moving regions during the cardiac cycle.[4] Valve position tracking and one-dimensional motion-compensated transvalvar velocimetry have been developed using Comb[5] tagging [30]. Coronary arteries undergo a complex displacement due to both respiratory and cardiac motion. Variability in heartbeat frequency is not only associated with chaotic heart behavior, but also with the subject's psychological response to an unusual environment. The variability in respiratory and cardiac cycle durations within and between subjects hinders movement prediction. However, real-time low-resolution tracking of coronary grooves and surrounding fats has been proposed in specific orientations, assessing minimal motion acquisition windows and vessel locations for high-resolution imaging (the so-called image-based navigators). Proposed Cartesian frames are defined by the long axis of the left ventricle and the atrioventricular groove.

Another new technique is magnetic particle imaging (MPI) [31]. Using magnetic resonance imaging, contrast agents that incorporate strongly magnetic particles can be introduced into the body to highlight specific anatomical structures (blood vessels) or serve as markers for nanoscale processes. An external time-constant magnetic field (selection field) is applied with a space-dependent strength. It vanishes in the center of the field of view, the field-free point (FFP), and increases in magnitude toward the edges, where the magnetization reaches saturation. When the particles in a magnetic field are further excited by an additional oscillating radiofrequency field (modulation field), saturated magnetic particles remain in the same state, whereas unsaturated ones, in the FFP area, reply to the modulation field with oscillating magnetization. The latter thus induce a signal in the detector, which can be assigned to the poorly magnetized regions. The FFP position can be changed through the sample, for example, by moving the object within the coil assembly, to generate a tomographic image. The resulting map can give the spatial distribution of the magnetic particles.

During interventional cardiovascular MRI, the guide wire is compatible and the tip and shaft of the catheter are visualized by markers integrated into the catheter. Active markers consist of small resonant wire loops that respond to radiofrequency pulses. Navigation of balloon-tipped catheters is tracked by real-time MRI, as the balloon is inflated with gadolinium [32].

---

[4]The displacement of the aortic valve has been estimated to be equal to 15–20 mm.

[5]Comb excitation enables simultaneous tagging in multiple parallel planes during breathhold.

5.1.1.3 Magnetocardiography

Magnetocardiography (MCG) aims at measuring and recording the weak local time-varying magnetic field associated with the cardiac electrical activity using an extremely sensitive equipment, such as the superconducting quantum interference device (SQUID) [33, 34] or room temperature cardiomagnetometers based on optically detected magnetic resonance (ODMR) [35]. The optically pumped magnetometer combines laser spectroscopy and magnetic resonance in a vapor of paramagnetic atoms (Cs) using a frequency-stabilized diode-laser as light source. The main difficulty of this noninvasive technique is the measurement of the magnetic field of the heart, the intensity of which on the chest is lower than 100 pT (i.e., 6 orders of magnitude smaller than that of the Earth [$\sim$50 μT]).

The objective of magnetocardiography is to give a more complete information on the electrophysiological activity with sufficient signal-to-noise ratio and spatiotemporal resolution than the electrocardiogram.

Once a map of the magnetic field is obtained over the chest, using an array of multiple SQUID sensors placed at fixed positions (e.g., vectormagnetometer with more than 300 SQUID sensors [36]) and adequate mathematical processing that take into account the conductivity and permittivity of the torso, the source of the activity can be localized.

A sequence of magnetic field maps depicts the dynamics of the depolarization wave in a given cardiac region at a specified period during the cardiac cycle. Consecutive maps can be evenly spaced with a time interval of 2 ms.

5.1.1.4 Ultrasonography and Echography

Ultrasound imaging (USI) involves ultrasound propagation through biological tissues, where ultrasounds are partly reflected at each acoustical interface on the US beam path. The echoing waves are then interpreted to create anatomical images.

A ultrasound wave is produced by a piezoelectric transducer. The probe emits ultrasound waves 10% of time and received reflected waves remaining 90% of time. An excited quartz crystal generates ultrasound waves at high frequency (up to 20 MHz). A gel with a sound conductivity similar to that of biological tissues is interposed between the probe and skin surface. Mechanical perturbation in the target elastic medium results from rarefaction and compression of medium particles.

Most acoustic energy is lost from acoustic absorption during sound propagation in media of given physical properties such as viscosity. Acoustic scattering is an additional factor of acoustic energy drop. The attenuation coefficient is the log relative energy intensity loss per traveled cm (dB/cm).

The sound wave is partially reflected at interfaces with density change. The return sound wave vibrates the transducer. Vibrations are transformed into electrical pulses that travel to the ultrasonic scanner where they are processed to yield images. Axial spatial resolution is related to the wave frequency. Ultrasound emitted at very high frequencies has a lower penetration into tissues.

Table 5.6 Attenuation coefficient of ultrasounds

Tissue	Attenuation coefficient (dB/cm/MHz)
Adipose tissue	0.6
Blood	0.2
Liver	0.9
Muscle	1.8

In biological soft tissues, the ultrasound attenuation coefficient is roughly proportional to the ultrasound frequency for the frequency range used in medical imaging. The attenuation coefficient divided by the frequency (dB/cm/MHz) is nearly constant in a given tissue (Table 5.6).

The quality of echographic images depends on the: (1) axial (in the US propagation direction) and transverse resolution, (2) ultrasonic attenuation, and (3) echo dynamics.

Because the resulting images are associated with the interaction between US waves and tissues, the collected information can reveal the mechanical properties of tissues through which ultrasounds travel (compressibility and density).

The ultrasound attenuation, by diffusion and absorption, can also provide additional data on the tissue heterogeneity level at the wavelength scale and intercellular cohesion, respectively.

The ultrasound transducer functions as both a stereo loudspeaker to generate streams of high-frequency sound waves and a microphone to receive echoing waves back from the internal structures and contours of organs. Because of the freely maneuverable probe, spatial sampling of the produced data is both inhomogeneous and unpredictable. Mechanical arms can be fixed to the probe and its location and orientation can be measured. Landmarks can be attached to the probe and cameras can track it. However, once the position and orientation of the probe are known, the data are still noisy.

M-mode echocardiography records amplitude and rate of motion (hence M) of a moving structure in real-time by repeatedly measuring the distance of the structure from the transducer at given moments.

High-resolution *bidimensional (B)-mode echography*, also called echotomography and ultrasonic tomography, aims at imaging predetermined cross sections of the body. It is the tool of choice for measuring artery caliber, as it is noninvasive, portable, and reproducible and has a low cost. The main challenge with B-mode imaging is to identify clear vascular boundaries.

Doppler echography, also termed Doppler ultrasonography and fluximetry (Sect. 5.2.2.1), employs the Doppler effect to assess motions of blood flow as well as other types of body tissues toward or away from the probe as well as the motion velocity.

Doppler echocardiography combines echocardiogram that images the heart using high-frequency ultrasound waves and Doppler echography that determines the direction and magnitude of blood flow within the heart.

Table 5.7 Echocardiographic right atrial and ventricular dimensions in adults (Source: [37])

Dimension	Value (mm)
Right atrium	
Length	<53
Width	<44
Right ventricle	
Apical 4-chamber view	
Basal width (basal one-third of RV cavity)	<42
Midcavity width (middle third of RV cavity at the papillary muscle level)	<35
Apicobasal length (from the middle of the tricuspid valve to RV apex)	<86
Parasternal view	
Parasternal short axis (just proximal to pulmonary valve)	27
Parasternal long axis	33
Right ventricular functional measurements	
Apical 4-chamber view	
Tricuspid annular plane systolic excursion	>16

Duplex ultrasonography combines simultaneous acquisition of real-time B-mode diameter and pulsed-wave Doppler velocity signals. However, the same transducer detect signals for both the Doppler frequency shift and arterial caliber that have antagonist incidences for optimal data acquisition. B-mode echoes have a greater intensity when the ultrasound beam is perpendicular to the local vessel axis (not accurately known), whereas pulsed-wave Doppler signals require incidence less than 60 degrees w.r.t. the local streamwise direction. The Doppler beam windowing enables to measure Doppler frequency shifts separately with a given incidence angle between the Doppler beam and the estimated local vessel orientation, whereas it maintains an optimal B-mode imaging.

Three-dimensional transthoracic and transesophageal echocardiographies are used for cardiac valve explorations. The right ventricular morphology is generally adequately described by transthoracic echocardiography (Table 5.7) [37]. However, additional imaging methods can be required (transoesophageal echocardiography, magnetic resonance imaging, or multidetector computed tomography). Three-dimensional echocardiography can visualize 3 leaflets and commissures of the tricuspid valve in a single image.

Intravascular ultrasonography (IVUS) is a catheter-based technique that provides real-time high-resolution images of both the lumen and arterial wall of a vascular segment. Axial and transverse resolutions are 80 to 100 μm and 200 to 250 μm, respectively. Thirty images per second can be obtained. IVUS is used to detect atherosclerotic plaques (Vol. 7 – Chap. 2. Vascular Diseases), which give a much better estimate of stenosis degree than angiography. The angiographic evaluation indeed depends on the imaging incidence angle with respect to the stenosed artery. Moreover, echogenicity of atherosclerotic plaques is correlated with their composition. However, conventional IVUS is unable to identify vulnerable plaques. Integrated backscatter IVUS provides a color-coded tissue map of plaque components (lipidic, fibrous, and calcified zones),[6] hence enabling tissue characterization. IVUS is also used in interventional cardiology to control stent placement and assess stent restenosis. Tissue characterization is expected to identify patients at high risk of restenosis, embolization, and myonecrosis after stenting.

Color flow Doppler imaging aims at visualizing direction and velocity of flowing red blood capsules using reflexed waves on RBC membranes. Direction and velocity are coded as colors and shades, respectively (red: blood coming to the probe; blue: blood flowing away from the probe).

Power Doppler imaging is sensitive to blood flow, but does not provide velocity or directional information. It is based on the amplitude of echoes received from moving cells, but not on frequency shifts.

In addition to Doppler echography, *tissue Doppler echography* (TDUS), or echodopplervelocimetry, is applied for tissue motion estimation using appropriate signal processing. The inherent limitation of the measurements is an inability to estimate more than one velocity component. However, combined with conventional US echography the tissue Doppler ultrasonography can estimate cardiac wall motion.

Intravascular ultrasound palpography assesses the mechanical properties of the vessel wall using the deformation caused by intraluminal pressure. Regions of higher strains are found in fatty than in fibrous plaques [38].

Ultrasound *elastography* is used for tissue characterization. Tissue motion (displacement and deformation) can be inferred from speckle tracking, the spatial and temporal speckle pattern being expressed by the optical flow equations.

Ultrasound ECG-triggered elastography of the beating heart provides real-time strain data[7] at selected phases of the cardiac cycle. Periodic myocardial thickening associated with normal heart function as well as tissue ischemia or infarction can be detected [39].

Myocardial strain can be measured using either tissue Doppler ultrasonography or 2D speckle tracking echocardiography (STE), each method having advantages and disadvantages. However, measurements of strain and strain rate (strain temporal derivative) have inherent limitations due to assessment of at most 2 dimensions at a given time, compromise between spatial resolution and noise, reverberation artefacts, and drop-out [40].

Ultrasound-based *electromechanical wave imaging* is based on acquisition of radiofrequency frames at very high frame rates (390–520 Hz) to observe the onset of small, localized, transient strains.

---

[6]IVUS-based virtual histology results from signal processing by autoregressive spectral analysis of radiofrequency ultrasound backscatter signals to assess plaque composition that is not complex. However, necrotic cores that surround calcified zones can be artefacts. Wavelet analysis of radiofrequency US signals represents an alternative modality.

[7]Displacement of the tissue between 2 images can be used to assess the bulk rheology of a region of interest of the explored tissue. Elastographic scanning maps strain magnitude (image brightness) and sign (color hue associated with compression or distention, for instance).

Perfusion measurement by ultrasounds relies on the uptake of an ultrasound contrast microbubbles in the myocardium. The local contractile reserve is assessed by dobutamine (a positive inotropic stimulator) stress echocardiography, the local wall motion reflecting the local contractile reserve.

Contrast echocardiography can be useful for the checkup of acute myocardial infarction in patients suffering from chest pain without obvious ECG signs. This echocardiographic technique requires the injection of a contrast agent bolus.

Contrast-enhanced ultrasonography (CEU) aims at exploring angiogenesis and progenitor cell engraftment into neovessels [41]. This technique is based on the ultrasonic detection of microbubbles (2–5 μm) with various shell (lipid or polymer) and gas (nitrogen and octafluoropropane) composition. Some microbubbles have a binding affinity for α_V-integrins of the endothelial surface of arterioles and capillaries in angiogenic vessels. Others target α_5-integrins and vascular cell adhesion molecule VCAM1 on activated neutrophils and monocytes to assess inflammatory contribution to vasculo- and arteriogenesis in response to ischemia. Certain microbubbles connect to vascular endothelial growth factor receptor VEGFR2 to evaluate the development or regression of a tumor vasculature.

Echocardiographic particle image velocimetry tracks the position of acoustic signatures of contrast microbubbles, thereby assessing intracavital vortex and transvalvular jet.

Left atrial function can be estimated by 2D echocardiography combined to Doppler analysis of the transmitral and pulmonary vein flow and tissue Doppler assessment of left atrial myocardial velocities. Two-dimensional speckle-tracking echocardiography can detect early abnormalities of the left atrial mechanics [42]. Three-dimensional echocardiography can be used for the analysis of left atrial shape and volume changes. In addition, 3D speckle-tracking echocardiography enables the spatial analysis of left atrial strains.

Echocardiography enables to determine the indication of left ventricular assist device implantation for patients with advanced heart failure as well as to assess postimplantation state and serial long-term follow-up to manage postimplant care [43]. Four types of echocardiographic evaluations comprise preoperative transthoracic (TTE), perioperative transoesophageal (TEE), postoperative (TEE/TTE), and serial (TTE) echocardiography.

Ultrasound guidance in interventional medicine is the cheapest and easiest procedure. However, 2 main drawbacks, border and mirror-image shadows, limit its use.

5.1.1.5 Nuclear Medicine Imaging

In nuclear medicine imaging (NMI), or scintigraphy, radioactive tracers, which have a short lifetime, attached to selected substances, are administred into the patient. Tagging molecules seek specific sites. The distribution within the body of the radioactive isotope provides information on irrigation and the chemicophysical functions of the explored organ.

Nuclear imaging is most often used to assess tissue perfusion from the early tissue uptake of a perfusion tracer as well as its viability, in addition to wall motion and eventual fibrosis.

The patient is placed in a detector array and the radiation emitted from the body is measured by a scintillation γ-camera containing 60 to 70 photomultipliers each shielded by lead and equipped with a pinhole collimator. Scintigraphy produces images with low resolution and a high amount of noise, due to necessary low-radiation doses.

The 2 most common types of NMI are *single photon emission computed tomography* (SPECT) and *positron emission tomography* (PET). The SPECT method uses photon-emiting radiotracers, whereas PET utilizes radiotracers that produce positron–electron pairs. Different tracers have distinct uptake mechanisms and radiation exposure. Data fusion with CT or MRI images allows coupling of images of the physiological activity to images of the anatomy.

Both SPECT and PET techniques enable estimation of regional perfusion defects. Whereas SPECT yields qualitative imaging, yielding the relative perfusion compared with the region with the highest tracer uptake, PET with its higher spatial resolution assesses a quantification of the local blood flow.

Cardiac PET combines tomographic imaging with radionuclide tracers of blood flow metabolism and tracer kinetics for quantifying regional myocardial blood flow, substrate fluxes, and biochemical reaction rates. PET assesses the regional blood volume and flow on the one hand, and local O_2 extraction rate and consumption. Cardiac SPECT studies myocardial perfusion with agents, such as thallium-201 and technetium tracers, at rest and during testing.

5.1.1.6 Optical Coherence Tomography

High-resolution optical coherence tomography (OCT) is based on light interferometry. When a light signal is subdivided in 2 parts running over a path of a given length, interference signals occur that contains information, such as the structure and composition of crossed media.

Traveling light interferometry for tissue morphology commonly uses near-infrared light emitted from low-coherence, broad-bandwidth light sources[8] to capture 10 to 20μm-resolution, three-dimensional images from optical scattering within short-depth biological tissues. Near-infrared light has shorter wavelength and higher frequency than ultrasound beam (Table 5.8).

Like intravascular ultrasonography, OCT is based on emission of waves and detection of backscattered signals, but with at least 5- to 10-fold greater resolution.[9] Moreover, OCT employed for vascular lumen measurements exhibits a higher

[8]Broad bandwidth lights can be generated from superluminescent diodes or femtosecond laser pulses.

[9]IVUS technique has a 100 μm axial resolution.

Table 5.8 Comparative technical features of the 3 main imaging modalities used in interventional cardiology: optical coherence tomography (OCT), intravascular ultrasound (IVUS), and computed tomography (CT; Source: [44]; NA: not applicable)

Feature	OCT	IVUS	CT
Wave type	Light	US	X-ray
Frequency (THz)	192	$20–45\times10^{-6}$	$3\times10^{3}–3\times10^{7}$
Wavelength (μm)	1.3	35–80	$10^{-5}–10^{-2}$
Axial resolution (μm)	10–20	100–200	60–140
Lateral resolution (μm)	20–90	200–300	NA
Rotation speed (Hz)	16–160	30	NA
Pull-back speed (mm/s)	1–20	0.5–1.0	NA
Tissue penetration depth (mm)	1–3	10	200–450
Field of view (mm)	7–11	15	NA

accuracy and reproducibility than IVUS imaging. A bolus of crystalloid solution injected through the guiding catheter can serve as a contrast agent. The main drawback is the requirement of blood flushing from the explored lumen.

Optical coherence tomography enables visualization of local microvasculature. It is safely and efficiency used to assess the size of luminal areas. It detects microstructures in the vessel wall near the lumen due to its limited tissue penetration (<1.5 mm).

Optical coherence tomography is associated with time (tdOCT) or frequency (fdOCT) domain analysis. The frequency domain optical coherence tomography has an improved lateral resolution w.r.t. tdOCT (but same axial resolution) and maximal field of view (11 mm), but a limited depth (0.5–2.0 mm). The catheter employs a single-mode optical fiber compatible with a conventional angioplasty guide wire.

Optical coherence tomography is used to characterize vulnerable atheromatous plaques (Vol. 7 – Chap. 2. Vascular Diseases). In atherosclerosis, this imaging technique can directly measure the fibrous cap thickness and possibly detect plaques with a risk of a rupture with greater accuracy than intravascular ultrasonography. This method can detect ruptured plaques with a thrombus attached to the site of rupture of the fibrous cap.

In addition to IVUS-guided stent implantation, optical coherence tomography evaluates stent deployment quality as well as healing after stenting (Vol. 7 – Chap. 4. Treatments of Cardiovascular Diseases). Two features indicate a strut: a highly reflective spot and/or an associated shadow behind the strut [45]. The presence of a single criterion enables strut identification.

Optical coherence tomography is also used to evaluate clinical outcomes of drug-eluting stents. Biolimus-eluting stent (BES) is based on biodegradable polymers, whereas sirolimus-eluting stent (SES) uses a durable polymer. Durable polymer surface coatings may be one of the causes for incomplete endothelialization. Follow-up OCT indeed reveals that most drug-eluting stents, including those with a biodegradable polymer, are partly covered with a thin neointima because of a delayed healing and poor endothelialization. Incomplete stent strut endothelialization is associated with the long-term risk of stent thrombosis.

Optical coherence tomography identifies underexpansion, malapposition, uneven stent strut distribution, and small intrastent thrombotic formations. An incomplete stent apposition, or malapposition, means that a stent strut is separated from the vessel wall by a distance greater than the strut thickness for bare metal stents or greater than the sum of the thickness of the strut plus polymer for drug eluting stents. It favors stent thrombosis. A stent overlap corresponds to 2 superimposed stent struts.

An intrastent tissular protrusion results from tissue prolapse between stent struts. A thrombus is identified as an intraluminal mass discontinuous with the wetted surface of the vessel wall or as a highly backscattered luminal protrusion in continuity with the vessel wall and resulting in signal-free shadowing. Dissections that are frequent at the stent edges are defined by their width and longitudinal and circumferential extension. In late in-stent restenosis, the strong reflective power of the stent struts allows their detection through a thick tissue layer.

5.1.1.7 Structured Light Imaging

Fast (up to 667 frames/s) three-dimensional structured light imaging (in-plane spatial resolution 87 μm; depth resolution 10 μm) combined to a deformable surface-matching, motion-tracking algorithm (mean tracking error 0.37±0.10 mm; maximal error 1.23 mm) has been developed to study epicardial motion of an unloaded rabbit heart [46]. Unlike 2D speckle tracking, structured light imaging accounts for out-of-plane motion and surface curvature.

5.1.1.8 Confocal Laser Endomicroscopy

The thinnest available bronchoscope cannot image the respiratory tract beyond bronchioles, the caliber of which is smaller than 3 mm. However, a 1.4-mm miniprobe can be introduced into the bronchoscope working channel. Cellvizio is an endomicroscopic system developed by Mauna Kea Technologies (Paris, France) that enables real-time microscopic imaging as well as biopsies in a minimally invasive way. Anatomical conduits such as the tracheobronchial tree can be explored from proximal airways down to distal bronchioles, from which alveoli can be imaged.

Fibred confocal fluorescence microscopy (9 frames/s; field of view 600 μm; lateral resolution 3.5 μm; and focus depth 0–50 μm) allows imaging of pulmonary acinar content as demonstrated by measurements of calibers of pulmonary lobular microvessels (~90 μm) and of openings of alveoli into alveolar sacs (278±53 μm), in addition to acinar elastic fiber[10] thickness (10±2.7 μm) [47]. Alveolar mouths are smaller in the right upper lobe and right medial basal segment. Alveolar macrophages are detected in smokers, but not in nonsmokers, their amount being correlated with the number of cigarettes smoked per day.

[10]Elastin serves as the axial backbone of alveolar ducts and alveolar entrances. It also resides in the sheath of extraalveolar microvessels.

5.1.2 3D Reconstruction

Imaging devices provide noninvasively accurate and very large datasets of discrete information on explored organs. However, output data are usually not suitable for archiving and data processing, as well as for representing 3D geometry; polygonal models are largely preferred. This requirement for piecewise approximations of the domain boundaries is reinforced by numerical applications. Most of the current reconstruction algorithms convert the initial sampled data into surface triangulations having the same degree of complexity as the original data (i.e., a number of triangles on the order 10^5 to 10^6). Hence, to be easily manageable, the complexity of such polygonal models needs to be simplified drastically. Surface simplification algorithms are aimed at finding a compromise between the minimal number of triangles and the preservation of the geometric accuracy of the surface model. In addition, specific requirements can be imposed on resulting meshes, for instance, element shape and size for numerical simulations. Moreover, smoothing is required to limit computational flow errors [48].

The common technique to create a mesh from imaging data is segmenting[11] and faceting. This 2-step method consists first of segmenting the selected organ in collected images, and then using the segmentation surface to create the facetization. The automatic mesh generator must be able to cope with such a surface, which is frequently full of gaps, overlaps, and other imperfections.

Various algorithms have been proposed to reconstruct a polygonal model (a piecewise linear approximation) depending on the nature of the sampled data (series of slices, range images, point clouds, etc.). Three classes can be defined.

Slice connection algorithms work for a series of planar parallel sections of the target vessel. At first, a closed contour is extracted in each slice, then contours are connected to each other between each pair of adjacent slices [49, 50].

Marching-cube approaches attempt to extract an implicit surface from a 3D range image based on a "voxelhood" analysis.

Delaunay tetrahedralization algorithms first generate 3D triangulation over a point cloud and then extract a bounded surface triangulation from this set of tetrahedra using suitable topological and geometrical criteria [51].

5.1.2.1 Level Set Methods

Level set methods are numerical techniques designed to track the evolution of fronts [52]. Level set methods exploit a strong link between moving fronts and equations from computational fluid equations [53]. This technique, based on high-order upwind formulations, is stable and accurate, and preserves monotonicity. Furthermore, it handles problems in which separating fronts develop, the existence

[11] Vessel bore and wall smoothness depend on the threshold.

of sharp corners and cusps, and topology changes. Level set methods are designed for problems in which the front can move forward in some places and backward in others. The solution starts at an initial position and evolves in time (initial value formulation). The level set method tracks the motion of the front by embedding the front as the zero level set of the signed distance function. The motion of the front is matched with the zero level set of the level set function, and the resulting initial value partial differential equation for the evolution of the level set function resembles a Hamilton–Jacobi equation. In this setting, curvatures and normals may be easily evaluated; topological changes occur in a natural manner. This equation is solved using entropy-satisfying schemes borrowed from the numerical solution of hyperbolic conservation laws. The interface between a vessel and the surrounding tissues can be detected, the border being defined by the intensity gradient.

5.1.2.2 Marching Cubes

The marching cube algorithm is used in volume rendering to reconstruct an isosurface from a 3D field of values [54]. The basic principle behind the marching cube algorithm is to subdivide the space into a series of cubes. In the framework of medical image processing, the matrix of cubes or cells is defined by the set of voxel barycenters. The imaged region is represented as a field of values through which the surface to be determined is defined by a threshold, which is provided by a previous step of the image processing. The first step is to calculate the corner values. The mean intensity values of the voxels are assigned to the corresponding barycenters. The algorithm then instructs to "march" through each of the cubes, testing the corner points and replacing the cube with an appropriate set of polygons, most often triangles. This step is done by inserting vertices at the cell edges using linear interpolation; each vertex is positioned according to the ratio between the selected threshold and the values of the neighboring corners. The result formed by joining the vertices with facets is a piecewise surface that approximates the isosurface. The isosurface can be defined by the set of intersection points between the voxel mesh and a 3D implicit function, the value of which is given from image thresholding. This operation is easy in the case of high-quality acquired images.

5.1.2.3 Slice Connections

In some circumstances, the surface is reconstructed from a point set, which defines planar parallel contours of vessels. A triangulated surface is drawn between each pair of consecutive sections, ensuring that each point in a contour is connected to its closest point in the next contour. To fit up a surface on a set of contours amounts to constructing a volume enclosed by these contours. The global volume is considered as the union of independent pieces resulting from pair treatment. Consider a set of input points that define the vessel contours obtained by the level set method. After a cubic-spline fit of each contour associated with smoothing, a new node set can be defined by equally spaced points along the vessel contour. Then 2 successive slices are projected orthogonally to the local axis in a same plane and a 2D constrained Delaunay triangulation is built. The surface triangles are finally extracted by elevation of the 2 planes. The projection direction based on the set of intrinsic axes of the vessels gives a better slice-pair treatment than using a projection direction normal to the slice planes, as is done for general purposes. Furthermore, entry and exit sections are rotated to be normal to the local axis.

5.1.2.4 Deformable Models

Contour shape can be known a priori, as a closed contour with a regular surface of a given region of the image. It can then be considered as an elastic contour in equilibrium under a set of forces. Among the strategies used to create a computational mesh from imaging data, there is a direct generation procedure, which starts from an average template mesh for the organ of interest and performs an elastic deformation of the mesh onto the image set [55].

5.1.2.5 Finite Octree Method

The vascular segment of interest is placed in a cube that is subdivided into octants of length scale determined by the mesh size to suit the possible caliber changes of the explored vascular segment [56]. The octants containing the vessel walls are trimmed to match the wall surface with a given tolerance level. Smoothing can then be performed. The octants within the vessel are subdivided into tetrahedra.

5.1.2.6 Implicit Surfaces

Smooth implicit functions for 3DR have been associated with spectral/hp high-order elements for blood flow computations [57]. The vessel edges are first detected, segmented, and smoothed using B-spline interpolation in each image of the slice set. The resulting contours are fitted by an implicit function defined as a linear combination of radial basis functions associated with the contour nodes ($\sim$30 points) and a set of interior constraint points along the normal direction to the spline at the corresponding nodes. An isosurface extraction leads to vessel surface triangulation using an implicit surface polygonizer. The vessel surface is then smoothed using a boundary representation of its edges and surfaces by spline curves and surfaces, preserving the main curvatures. Bicubic spline patches, which are interactively defined and projected onto the implicit surface, serve as the initial element for meshing [58]. Mesh size and shape optimization are determined by the eigenvalues of the Hessian matrix of the implicit function.

5.1.2.7 Snakes

Snakes, or active closed contours defined within an image domain, can be used in image processing, particularly for image segmentation [59]. Snakes move under the influence of internal forces coming from the curve and external forces computed from the image data. The functional to be minimized is analogous to the deformation energy of an elastic material subjected to a loading. The properties of the deformable snakes are specified by a potential associated with a contraction–expansion term by analogy to a mechanical thin heterogeneous membrane. The internal and external forces are defined so that the snake will conform to an object boundary (image intensity gradient) within an image. The initial contour can be a small circle centered on a selected point. The iterative deformation of the initial curve in the force field can be done by convolving gradient images to Gaussian-type functions and modeling the deformable curve by splines. The coefficient number of these splines rises during the iterations to decrease the curve energy progressively. An additional force has been proposed to deform the snakes to avoid tracking spurious isolated edge points [60]. This method has been successfully applied to heart ventricule extraction.[12] External force, like gradient vector flow (GVF), can be computed as a diffusion of the gradient vectors of a gray-level or binary edge map derived from the image [61]. Superposition of a simplicial grid over the image domain and using this grid to iteratively reparameterize the deforming snakes model, the model is able to flow into complex shapes, even shapes with significant protrusions or branches, and dynamically change the topology [62].

5.1.2.8 Axis-Based Method

Bioconduit modeling can be based on vessel-axis determination. Once the set of axes is known, a "response function" is computed for each vessel slice from a vector rotating around the axis point and the intensity gradient. The voxel positions for which the response function is maximum give the vessel contour in the investigated slices when the axis is correctly determined. Precisely approximated contours a posteriori confirm that the vessel axis was reasonably well estimated. Several techniques may be used for axis determination. The vessel contours can be detected by derivative operators.[13] The eigenvalues of the Hessian matrix $\nabla^2 i$ of the intensity i can be computed; the eigenvalue that is the nearest to zero estimates the axis location [63].

---

[12]The snakes are dilated by external and internal forces using a finite element method to solve the minimization problem, only taking into account the suitable edge points that have been extracted by an edge detector.

[13]The contour is then defined as the location of the maxima of the gradient of the image intensity in the gradient direction. Vascular modeling relies on the assumption of a Gaussian intensity profile with a first derivative maximal at the vessel wall and a second derivative maximal on the vessel axis, the vessel being supposed to have a circular cross section. When the vessel axis is determined and local radii are estimated, the vessel wall is reconstructed.

Reconstruction of the intricate vessel tree in any organ can use a marking procedure to detect all the connected components of the lumen of the vessel network and set up a mark for each of them serving as starting subset for high-order reconstruction. A suitable morphological filter is then needed. A morphological filter based on selective marking and depth constrained connection cost, which labels the vessel by binarization of the difference between the original image and connection cost image, can be proposed [64]. An energy-based aggregation model is applied to the marking set for tree 3DR with respect to voxel values. The marking set progressively grows by state change of boundary voxels, according to local minimization of an energy functional (Markov process-like method). The energy functional is composed of 3 propagation potentials $\mathcal{E} = \mathcal{U}_\mathrm{r} + \mathcal{U}_\mathrm{l} + \mathcal{U}_\mathrm{c}$ ($\mathcal{U}_\mathrm{r}$: potential associated with pipe topography, $\mathcal{U}_\mathrm{l}$: potential based on similar density for all ducts, $\mathcal{U}_\mathrm{c}$: limiting growth potential for bounded growth within lumen limits) [64]. The state is determined with respect to the states of 26-connected adjoining voxels $y \in \mathcal{V}_{26}(x)$, and their gray levels $F(y)$. The smoothing is adaptative according to vessel caliber based on adaptation of Gaussian kernel to size of labeled vessels. Vessel axis computation is based on geodesic front propagation (GFP) with respect to a source point, the axis being defined by the set of centroids of successive fronts. The determination of the axis tree uses GFP combined with a 3D distance map associated with vessel wall geometry, which allows space partitioning. This method provides robust branching point detection with axis hierarchy preservation [65].

5.1.2.9 Limitations

Despite being efficient and robust, 3D reconstruction generally suffers several drawbacks. Discrete data may be very noisy: points can be off the surface. The accuracy of scanning and sensing devices leads to unnecessary dense datasets, with a density that is not related to the local geometric complexity, and consequently to very large polygonal models.

Reconstruction algorithms often introduce artifacts in the polygonal approximation, such as "staircase" effects. The element shape quality does not always fit the requirement in numerical simulations. To overcome these problems, a 2-step integrated approach consists of first generating a simplified geometric surface mesh, possibly preceded by a surface smoothing stage, and then constructing a computational surface mesh by taking into account shape and size requirements for the mesh elements.

5.1.3 Meshing

Two main issues in mesh studies applied to numerical analysis are the surface mesh quality, which must not significantly affect the problem solution, and the algorithms associated with volumic meshes that are automatically generated [66–69].

The second work class focuses on mesh adaptation and adaptativity, including mesh refinement/coarsening, edge refinement and swapping, and node displacement, based either on metrics and error estimations independently of equation types [70, 71] or minimization of the hierarchical estimator norm [72–75].

The objective is to construct surface meshes that match strong requirements related to the accuracy of the surface approximation (geometry) and boundary conditioning as well as the element shape and size quality (computation). Patches such as B-splines and NURBS that match the facetization can be meshed. However, facetization can be directly processed. Determined corners and ridges define patches that are triangulated by an advancing front technique [76]. In any case, the first stage consists of simplifying the initial dense surface mesh to produce a geometric surface mesh. First, the initial reference mesh must be simplified to remove redundant elements while preserving the accuracy of the geometric approximation of the underlying surface. A simplification procedure based on the *Hausdorff distance* can thus be used. This algorithm involves vertex deletion, edge flipping, and node smoothing local mesh modifications. Often, especially for surface triangulations supplied by marching-cube algorithms, a smoothing stage based on a bi-Laplacian operator is required to remove the "staircase" artifacts [77]. This stage yields to a geometric surface mesh that is a good approximation of the surface geometry and contains far fewer nodes than the initial reference mesh [78].

Surface discretization obtained from the 3D reconstruction needs further treatment to be suitable for numerical simulations. Boundary conditions must be set sufficiently far from the exploration volume, otherwise they affect the flow within the fluid domain. Moreover, any geometry change along a vessel (bends, branching segment, lumen narrowing or enlargment, wall cavity, taper, etc.) induces flow disturbances over a given length both upstream and downstream from the causal segment. Consequently, short straight ducts in the direction of the local axis can be connected to every vessel end. Furthermore, vessel-end sections must be cross sections because of stress-free boundary conditions usually applied at outlets. The blood vessels are continuously curved; curvature induces transverse pressure gradient in any bend cross section as well as in upstream and downstream cross sections of possible straight pipe over a given length, which depends on the values of the flow governing parameters [79]. In addition, axial pressure gradient is exhibited in vessel sections that are nonperpendicular to the vessel axis.[14]

Numerical simulations are the final objective, so element shapes and sizes must be controlled as they usually affect the accuracy of the numerical results.[15] Therefore, an *anisotropic geometric metric map* based on the local principal directions and radii of curvatures is constructed in tangent planes related to mesh vertices.

[14] In straight pipes, the axial pressure difference, which varies either nonlinearly in the entry length or linearly when the flow is fully developed, is exhibited in any duct section that is not normal to the centerline.

[15] The surface element size depends on the local surface curvature. The stronger the curvature, the smaller the size.

This metric map prescribes element sizes proportional to the local curvature of the surface [80]. The metric map can also be combined with a computational metric map (e.g., supplied by an a posteriori error estimate) and eventually modified to account for a desired mesh gradation. Then, a surface mesh generation algorithm is governed by the metric map and based on local topological and geometrical mesh modifications. This approach can be easily extended to *mesh adaptation* as it already involves mixing geometric and computational metrics to govern the mesh generation stage.

Dynamic meshing is aimed at meshing an organ during its displacement, like the cardiac pump, without computing the whole mesh at each iteration. Three main steps are required: (1) node displacement, (2) mesh coarsening with removal of the collapsed element, determined from a triangle degradation criterion, and (3) mesh enrichment after a mesh smoothing, especially in high-error regions.

Heat and wave propagation can be efficiently computed using mesh adaptivity and anisotropic mesh adaption to the propagation front, avoiding inaccurate predictions due to numerical diffusion, especially for free-surface problems [81, 82]. Accurate predictions of the interface require a refined mesh in the vicinity of the interface (Fig. 5.1).

5.2 Hemodynamics Signals

5.2.1 Volume and Pressure

In back decubitus, blood vessels are assumed to be approximately situated at the same height, which is supposed to be given by that of the right atrium. The reference point of the intravascular pressure is located at the level of the tricuspid valve [7]. The mean pressure at this point does not significantly vary with body position (variations <0.15 kPa). In the lying position, the gravity can be neglected, whereas in the upright position the hydrostatic component varies linearly with height H ($-\rho g H$, $H < 0$ and $H > 0$ under and over the reference point, respectively). The hydrostatic pressure then rises or decays with the vertical distance from the reference point, whether the vascular site is located under or over the reference point. The generating pressure takes blood pump action into account.

The central arterial pressure refers to the aortic root pressure in opposition to peripheral pressure (e.g., brachial pressure). The peripheral arterial pressure differs from the central aortic pressures, especially because of amplification of the pulse pressure wave. This amplification results from arterial wall stiffening during aging and diseases. It also depends on various factors, such as cardiac frequency, subject's height and gender, as well as eventual drugs.

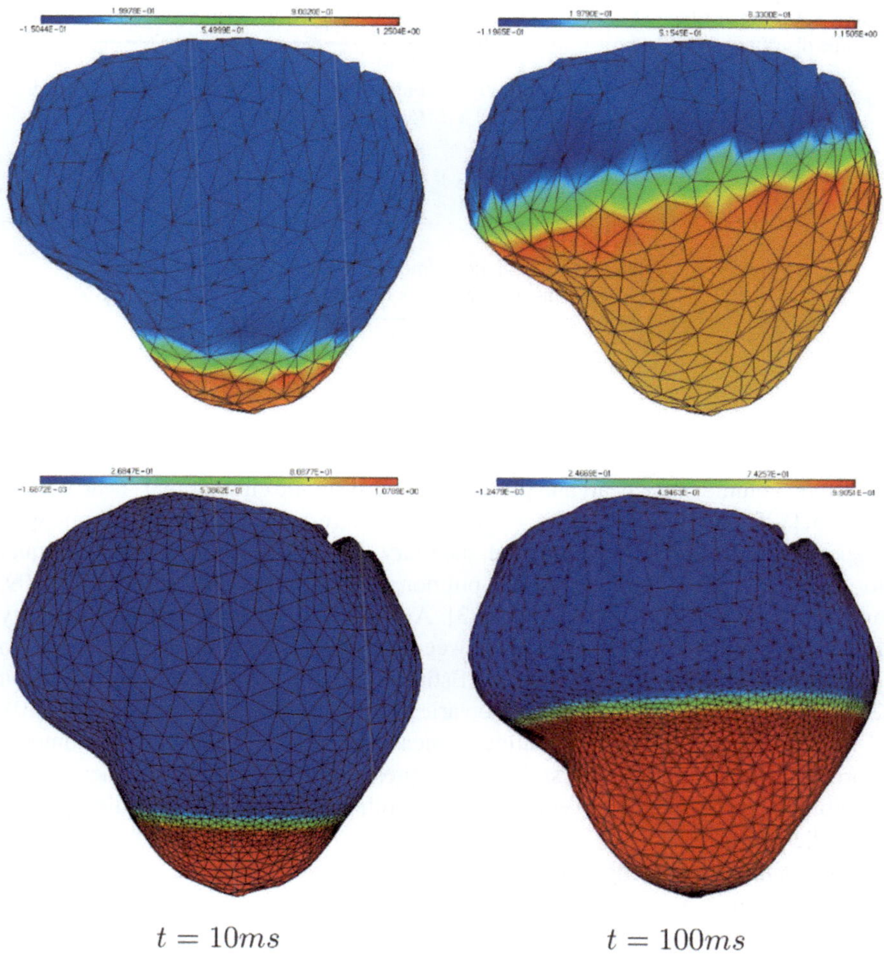

Fig. 5.1 Mesh adaptivity for numerical simulation of the propagation of the action potential in the heart wall, based on Fitzhugh–Nagumo equations (from Y. Belhamadia)

Time variations of pressure and of volume of heart cavities are estimated from catheter-based measurements.[16] Pressure–volume loops throughout the complete cardiac cycle are then plotted. Alternatively, chamber volume changes can be assessed using echocardiography, radionuclide imaging, or tagged MRI.

---

[16]Quickly varying pressures must be measured by transducers, which convert pressures into electrical signals, able to accurately sense high frequencies. Sensitivity, linear output for the whole pressure range, suitable frequency response, and spatial resolution are the main features of transducers. In particular, the transducer size must be smaller than the distance over which exist spatial pressure variations.

Table 5.9 Estimated blood volume distribution (%) in the circulation compartments (between-author variability)

Compartment	[118]	[119]
Heart	18	7
Pulmonary circulation	12	9
Systemic circulation	70	84
Arteries	15	15
Capillaries	5	5
Veins	50	64

Table 5.10 Circulation time (s) estimated from indicator bolus measurements (Source: [83])

Thoracic aorta to femoral artery	2–4
Cubital vein to carotid sinus	12–33
Right heart to aorta	6–9

Blood volume in the heart and pulmonary and systemic circulation is nearly equal to 0.05, 0.25, and 0.70 with respect to total volume, respectively. Blood volume with respect to total volume in the arterial, capillary, and venous compartments is equal to about 0.30, 0.05, and 0.65 in the pulmonary circulation and about 0.17, 0.09, and 0.74 in the systemic circulation [83]. An important between-author variability in the distribution of blood volume between the main compartments of circulation is found in the literature (Table 5.9). Estimates of the circulation times between 2 compartments of the circulation also varies among the literature data (Table 5.10).

The arterial pressure is routinely measured using a sphygmomanometer. Korotkoff sounds indicate systolic blood pressure; they might be generated by instabilities of coupling between blood and collapsed artery [84] and disappear at diastolic pressure.

Impedance plethysmography is another noninvasive technique to measure the blood pressure via vessel volume changes, and using several measuring sites, to estimate the wave speed [85]. Artery bore changes are small and produce low impedance variation. The recorded signal is then very sensitive to noise and must be adequately filtered. Moreover, the wave speed depends on the chosen reference point, electrode distribution, and artery compartment (proximal/distal). It is also affected by the subject's age and position. In normal subjects, the wave speed is higher in a standing than supine position.

Other pressure measurement techniques include *tonometry* [86] and arterial *photoplethysmography*, which is based on the optical determination of blood volume changes and pulsations in superficial arteries (in fingers) [87].

In tonometry, the resulting pulse wave is calibrated to conventional brachial pressure measurements. Various mathematical processing tools estimate the central aortic pressures. The central augmentation index is a measure of the timing and magnitude of pressure wave reflections from the peripheral circulation and their superimposition on the incident pressure wave. Computation of the central augmentation index depends on accurate recording of high frequency harmonics and precise identification of an inflection point on the upstroke of the central aortic pressure wave. However, the central augmentation index does not have any independent clinical utility w.r.t. arterial pressure [88]. The central arterial pressure does not have a significantly better predictive ability than that of peripheral arterial pressure [89]. Although the predictive role of the augmentation index is independent of that of the peripheral pressure, its ability to discriminate, calibrate, and reclassify risks cannot be assessed.

Photoplethysmography uses the reflection or transmission of infrared light. Artery volume variations modulate light intensity recorded by the photodetector. To be valid, the method requires several measurements made over long periods. The reliability of the Finapress technique has been studied [90–92]. Vessel caliber can be simultaneously measured by US echography. Blood pressure and wall radial velocity simultaneous measurements can evaluate the elastic modulus of the vessel wall.

The *Korteweg–de Vries equations* (KdV equation: $g_t + \kappa_1 g_x + \kappa_2 g_{xxx} = 0$) have traveling solutions,[17] the solitons,[18] the shape and speed of which are preserved after interactions. Two- and three-solitons[19] are peculiar solutions of the KdV equation. Their analytical expression can be found in [94]. Soliton-based signal processing, coupled with a windkessel model, can be used to compute the blood pressure wave in large proximal arteries, even at the left ventricle outlet, the pressure being distally measured (at the finger), using 2- and 3-solitons [95].

Heart pressures are invasively measured using catheter-mounted microtransducers with suitable frequency response. Catheters are introduced in peripheral veins or arteries for pressure measurements in right or left heart, respectively. Pressure time variations and gradients (dp/dt) have been assessed using MR measurements of velocity and acceleration within the ascending aorta [96].

A new signal-processing technique is based on scattering-based analysis of pulse-shaped signals theory [97]. This inverse scattering technique consists of solving the spectral problem associated with a 1D Schrödinger operator perturbed by a potential depending upon the signal to analyze, and optimized in order to approximate this signal. Some functions associated with the Schrödinger equation

---

[17]The nonlinear soliton equation was developed by D.J. Korteweg and G. de Vries at the end of the nineteenth century. The nonlinear term balances the dispersion. The existence and uniqueness of solution to the Cauchy problem for the nonlinear Korteweg–de Vries equation and local controllability around the origin given by the nonlinear term has been proved [93]. Signal processing using progressive wave speed analysis is more appropriate than employing frequency analysis suitable for stationary waves.

[18]The soliton is a wave that propagates without dispersion. Solitons interact without losing their identity, keeping shape and amplitude. An n-soliton solution refers to n components of different amplitude that interact. The propagation speed is proportional to wave amplitude. The higher the amplitude, the faster the propagation. Soliton solutions are used to model fast dynamics of pressure wave propagation and an associated windkessel model to take slow dynamics into account.

[19]The second wave of the 2-soliton model is associated with the dicrotic wave due to the aortic valve closure. The 3-soliton model is used to fit bifid pressure waves. A bifid curve exhibits an incisure in the ascending systolic part near the peak value rather than an usual monotonic soaring aspect.

(the squared Jost solutions) are analogous to sine and cosine in the Fourier analysis of signals. Low- and high-frequency components of the standard Fourier analysis are replaced by low- and high-velocity components using an interpretation in term of traveling waves (i.e., N-solitons). The method was applied to physiological signals, in particular arterial blood pressure [97].

Nonlinear spectral analysis of arterial blood pressure waveforms uses a reduced model of the input–output behavior of an arterial compartment, including the short systolic phase, during which wave phenomena are predominant.

5.2.2 *Flow Rate and Velocity*

Flow rates have been previously estimated according to the Fick principle or dilution methods, using injected tracers. Calibration and positioning of catheter flowmeters are difficult. Remote flowmetry is now used. Vascular ultrasound imaging (US angiography) is used to monitor the blood flow and evaluate possible flow blockages. The quality of velocity measurements, whatever the technique, depends on the resolution: (1) spatial resolution (size of the sample volume), (2) temporal resolution (quickly varying flow), and (3) amplitude resolution (signal-to-noise ratio).

5.2.2.1 Doppler Ultrasound Velocimetry

The Doppler effect is a change in the frequency of a wave, resulting in the case of a reflected wave, from the motion of the reflector. The Doppler shift frequency f_D (the difference between transmitted and received frequencies) depends upon the transmitted frequency f, blood velocity v, and Doppler angle θ, which is the angle of incidence between the US beam and the estimated flow direction (the local vessel axis; Doppler equation):

$$f_D = 2\,f\,v\,\cos\theta/c, \tag{5.1}$$

where c is the speed of sound. Local blood velocity is calculated using the Doppler equation:

$$v = f_D\,c/(2\,f\,\cos\theta). \tag{5.2}$$

The angle θ is evaluated by the sonographer by aligning an indicator on the duplex image along the longitudinal axis of the vessel. If the US beam is perpendicular to the blood stream direction, there is no Doppler shift. The angle also should be less than 60 degrees, because the cosine function has a steeper curve above this angle and errors in angle correction are therefore magnified.

The Doppler ultrasonography (DUS) detects the Doppler effect on circulating blood cells. Several depictions of blood flow are used in medical Doppler imaging:

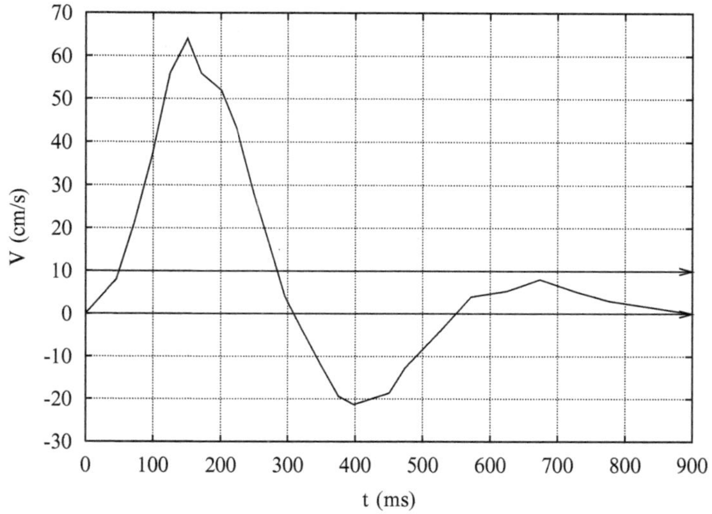

Fig. 5.2 Ensemble-averaged (25 cycles) cross-sectional averaged velocity $\langle V_q \rangle$ in the femoral artery of a healthy volunteer measured by US Doppler technique. The ensemble averaging assumes a constant cycle period (here 900 ms, $f_c \sim 1.1$ Hz), neglecting chaotic heart behavior. In pulsatile flows, the boundary layer moves forward and backward, especially in certain proximal arteries, such as the femoral artery. Positive velocities during the systolic ejection are followed by negative velocities during the diastole before the final recovery stage with a very small velocity amplitude. The time-mean cross-sectional averaged velocity is then small ($\overline{V}_q \sim 10$ cm/s)

(1) color Doppler (CDUS), for a global description of blood flow, an estimation of the mean velocity in the investigated vessel region, and display of bulk motion using a graphical color map; (2) pulsed Doppler (PDUS), for detailed analysis at a selected site (velocity distribution within the measurement volume; Fig. 5.2); (3) and power Doppler, which determines the amplitude of Doppler signals rather than frequency shift. Duplex scanning pulsed US velocimetry is used for real-time imaging guidance. Multichannel pulsed US Doppler velocimetry with phase lock loop frequency tracking of the Doppler signal gives good agreement with LDV [98]. Phase-shift averaging reduces physiological variability. Phase errors are minimized by a cross-correlation between the velocity and its ensemble average $\langle v(t) \rangle = (1/N) \sum_{k=0}^{N-1} v_k(t \pm \Delta t_w/2 + kt)$ (Δt_w: time window, Fig. 5.2).

Duplex or color flow Doppler ultrasound is used for blood velocimetry. Every ultrasound transmission from the probe is directed forward and contains several pulses to measure Doppler shift. *Intravascular Doppler ultrasonography* (IVDUS) is used to assess flow pertubations induced by arterial lesions, to provide continuous monitoring of flow velocity throughout angioplasties, and so one. However, IVUS probes operate in a high-resolution B-mode, emitting at right angles from the transducer tip only 1 or 2 pulses per transmission. Real-time imaging can be done with up to 30 conventional IVUS frames. Differences in the position of blood cells between sequential images are computed to assess the local magnitude of blood flow, without any quantification.

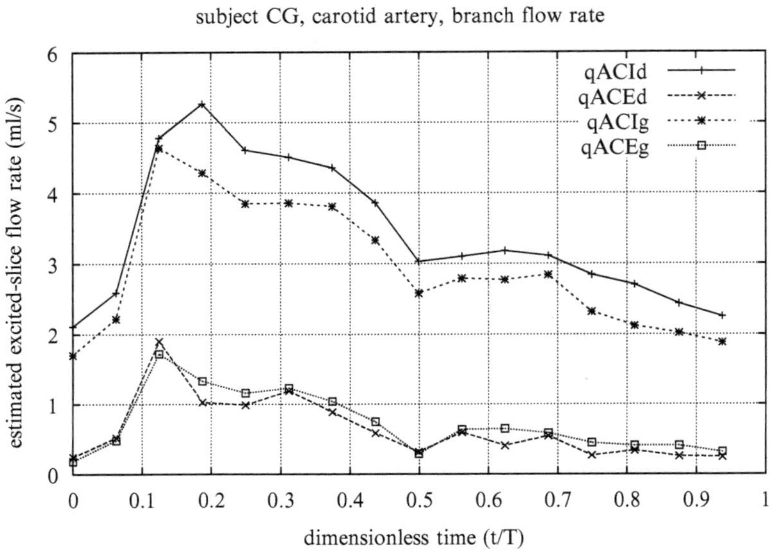

Fig. 5.3 Variations of the flow rates in the external (E) and internal (I) left (g) and right (d) carotid arteries measured by MRI in a healthy volunteer

5.2.2.2 Nuclear Magnetic Resonance Velocimetry

Magnetic resonance velocimetry (MRV) is directly capable of 3D blood flow velocity measurements across the whole selected vessel section [99]. MRV allows quantification in deep vessels that cannot be explored by DUS techniques. Inhaling hyperpolarized helium, MRV can be performed on respiratory flows [100, 101].

Among several operation modes, phase-contrast magnetic resonance velocimetry (PCMRV) is used to analyze and quantify blood flow [102, 103]. This technique employs the signal from blood-conveyed protons stimulated by specially designed magnetic field gradients. The output phase shift is proportional to flow velocity. In contrast, the net phase shift of the nuclei in stationary tissue is equal to zero. At multiple instants during the cardiac cycle, a phase image and corresponding magnitude image are constructed. The magnitude images depict the anatomy at the specified location. On phase images, the intensity of each voxel within the vessel lumen corresponds to the velocity of blood flow at that location (Fig. 5.3, Table 5.11). Dynamic MRA analyzes transient image features, such as retrograde flow [104]. Bipolar flow-encoding gradients can be used in a three-dimensional MRI procedure to visualize small vessels having relatively slow flow [105]. Moreover, multiple station acquisition with possible wave velocity calculation can be done using comb excitation and Fourier velocity encoding [106]. Multiple components of velocity and acceleration by Fourier phase encoding can be simultaneously measured with a few encoding steps and efficient velocity-to-noise ratio [107]. Among the influence parameters, the number of flow-encoding steps, signal samples

Table 5.11 Coefficients of the Fourier series of cross-sectional velocity (cm/s) in femoral and internal carotid arteries:
$U_q(t) = A_0 + \sum_{1}^{12} \Big(A_k \cos(\omega k t) + B_k \sin(\omega k t) \Big)$

	Femoral	Carotid		Femoral	Carotid
A_0	2.95	33.29	B_1	−7.29	9.52
A_1	15.41	−5.27	B_2	5.72	0.72
A_2	29.40	−5.10	B_3	29.30	0.29
A_3	22.87	−2.81	B_4	42.25	−3.25
A_4	0.10	−2.15	B_5	35.66	−0.95
A_5	−36.97	−0.12	B_6	−13.21	−1.17
A_6	−45.89	0.79	B_7	−32.07	−0.50
A_7	−8.91	0.27	B_8	−16.09	−0.41
A_8	10.67	0.29	B_9	−4.71	−0.18
A_9	11.10	0.06	B_{10}	2.18	−0.17
A_{10}	10.31	0.12	B_{11}	8.11	−0.20
A_{11}	4.97	0.07	B_{12}	7.30	−0.16
A_{12}	−1.66	0.09			

for averaging, and dimensions of the flow field, maximizing the number of steps is the most efficient way of improving the precision of measurements, keeping a reasonable acquisition time [108].

5.3 Measurement of Heart Electric and Magnetic Properties

Basic bioelectric source models include the single dipole with variable orientation and magnitude, either with a fixed location or moving, and multiple dipoles, fixed in space, each representing an anatomical region of the heart. The cardiac generator has been modeled in first works at the end of the nineteenth century as a dipole. The human body is considered as a resistive, piecewise homogeneous (isotropic or not) conductor. The thoracic conductor can incorporate the following components: heart wall, with pericardium and fat, high-conductivity intracardiac blood,[20] low-conductivity lung parenchyma,[21] intermediate-conductivity thoracic muscle layer, and nonconducting bones,[22] as well as other organs such as large vessels. The electric generator corresponds to the heart wall.

The propagating depolarization of cardiomyocytes can be modeled as a double layer. A double layer at the activation surface can be approximated by a single

[20]Blood resistivity depends on blood flow [109] and hematocrit [110].

[21]The resistivity of the lungs is approximately 20 times that of the blood.

[22]The resistivity of bones increases about a 100-fold with respect to blood.

resultant dipole. At the wavefront, a lumped negative and positive point source constitute a dipole in the direction of propagation. A double layer, with a positive side pointing to the recording electrode, produces a positive signal.

Unlike depolarization, repolarization is not a propagating phenomenon but rather a propagating-like process. Any cardiomyocyte repolarizes at a certain time after its depolarization, independently of the repolarization of the adjoining cells.

5.3.1 Electrocardiogram

Body surface electrodes record the electrical activity generated by traveling action potentials (combination of all electrical signals) to provide an electrocardiogram (ECG). ECG thereby depends on the electric behavior of tissues between the thoracic skin and heart wall. Owing to very narrow between-cell space and the existence of many gap junctions, the cardiomyocytes can be represented as clusters of myofibers receiving at different times the action potential according to their situation with respect to the electrochemical wave propagation, with preferential propagation along the myofiber axis.

The ECG signal relies on several observations and assumptions. Nodal cells and cardiomyocytes are characterized by 2 electrophysiological states: polarization (rest) and depolarization followed by repolarization (variable duration according to the site, up to several hundred ms) during the genesis of the action potential in the pacemaker (normally the sinoatrial node) and its propagation in the cardiac wall. The propagating activation front can be defined by its resultant vector. The propagation speed of the activation front is assessed in the main regions of the conduction tissue and the myocardium. In the standard 12-lead ECG system the source is a dipole in a fixed location and the volume conductor is either infinite homogeneous or spherical homogeneous. The thorax is supposed to be homogeneous. The atria and ventricles are activated during separate phases of the cardiac cycle, the physiological diastole and systole, respectively. Their sequential activation hence produces different signals, at least for the depolarization. The size, the location, and the orientation of the heart have a limited variability between human subjects.

However, many quantities affect the measurements. The resistivities of the intracardiac blood, of the cardiac wall, and of the lungs are about 1.6, 5.6, and 10 to 20 Wm, respectively. The stronger conducting intracardiac blood mass leads to an increased sensitivity to radial dipole elements (decreased sensitivity to tangential dipoles) with respect to the homogeneous model (Brody effect) [111]. However, the Brody effect is reduced when other heterogeneities are included. Blood mass, lung resistivity, and position of lungs and heart change during cardiac and respiratory cycles. These changes affect electrocardiology [112].

Table 5.12 Signals of ECG leads (standard (L1, L2, L3) and augmented (aVR, aVL, aVF) limb leads, precordial leads V_i, $i = 1, \ldots, 6$) (u: voltage, subscript F: standard left leg lead, subscript L: standard left arm lead, subscript R: standard right arm lead, $u_s = (u_F + u_L + u_R)/3$

Lead	Voltage
L1	$u_R - u_L$
L2	$u_R - u_F$
L3	$u_L - u_F$
aVR	$3(u_R - u_s)/2$
aVL	$3(u_L - u_s)/2$
aVF	$3(u_F - u_s)/2$
V_i	$u_{V_i} - u_s$

5.3.1.1 ECG Leads

The *Einthoven standard limb leads* explore cardiac electrical activity in the coronal plane. The standard bipolar leads at the right arm (R), left arm (L), and left leg (F) are supposed to constitute the vertices of an equilateral triangle, the so-called Einthoven triangle [113–115]. Signal L1 is positive in the direction R to L, L2 in the direction R to F, and L3 in the direction L to F (Table 5.12). A simple model assumes that the cardiac dipole is located at the center of a homogeneous sphere representing the thorax, hence at the center of the equilateral triangle. Hence, voltages measured by the 3 limb leads are proportional to projections of the electric heart vector on the sides of the lead triangle.

After the sinoatrial node depolarization, the action potential spreads in the atrial walls. The projections of the resultant vector of the atrial electric activity on each of the 3 Einthoven limb leads is positive. Depolarization reaches the atrioventricular node. Propagation through the atrioventricular junction slows down, allowing complete ventricular filling. A delay in activation progression is thus observed. Once activation has reached the ventricles, propagation proceeds along the Purkinje fibers to the inner walls of the ventricles. Activation wavefronts proceed from endocardium to epicardium and mainly from apex to base. Ventricular depolarization starts from the left side of the interventricular septum. Therefore, the resultant dipole from this septal activation points to the right, causing a negative signal in left-to-right arm and left foot-to-right arm leads. As depolarization on both sides of the septum, then apex occurs, the resultant vector points to the apex. Depolarization propagates through the wall of the right ventricle to reach the epicardial surface of the right ventricle free wall, whereas it continues to move in the thicker wall of the left ventricle. The resultant vector points leftward with a maximal magnitude. The amplitude then decreases until the whole ventricular myocardium is depolarized. Once the ventricle basis reached, there is a second delay.

The action potentials of epicardial myocytes have shorter duration than those of the endocardial myocytes. The repolarization appears to proceed from epicardium to endocardium. Recovery and activation dipoles are thus in the same direction. Because repolarization is more disperse, the signal has a much smaller amplitude and longer duration than those of depolarization.

Three *Goldberger augmented limb leads*—aVR, aVL, and aVF—are currently added to the standard limb leads [116]. Goldberger unipolar leads use the same electrodes; each one is a positive pole and the 2 others negative. Lead aVR is oriented northwest, aVL northeast, and aVF south.

Precordial leads—V1 to V6—are located on the chest wall. Leads V1 and V2 are located in the fourth intercostal space on the right and left side of the sternum. Lead V4 localizes to the left fifth intercostal space at the midclavicular line. Lead V3 lodges halfway between V2 and V4 leads. Leads V5 and V6 are at the same horizontal level as V4 but on the left anterior axillary line and at the midaxillary line, respectively.

Commonly, ECGs are carried out with a 12-lead technique. Leads R, L, F, aVR, aVL, aVF are derived from the same 3 measurement points. Limb leads R, L, F more or less reflect the frontal components. Precordial leads target the transverse components. In the standard 12-lead ECG, the source is supposed to be a dipole in a fixed location and the volume conductor homogeneous and either infinite or spherical homogeneous. However, the thorax is heterogeneous. Tissue resistivity change between the skin and the cavital blood is, at least partially, responsible for the Brody effect, with an increased and decreased sensitivity to radial and tangential components, respectively. Lung resistivity, position, and shape change during the respiratory cycle, and hence affect the electric heart vector. Heart motion during the cardiac cycle also influences ECG [117].

Endocavital exploration uses catheter-based sensors. The travel timing of the action potential is thus much more precise. In particular, conduction in the atrioventricular node can be differentiated from propagation in the trunk of the His bundle.

5.3.1.2 ECG Trace

Normal ECG consists of 3 basic features, a P wave, QRS complex, and T wave, corresponding to the atrial and ventricular depolarization and ventricular repolarization, respectively. The shape of the different waves, which depends on electrode location, and the time intervals between them (PQ, or PR in the absence of Q wave, QT, ST, as well as QRS duration) are analyzed for medical checkup (Table 5.13). Any alteration in action potential transmission and cardiac frequency is revealed by the ECG trace.

Normal P wave amplitude and duration are about 100 μV (<250 μV) and 85 to 100 ms. PR interval ranges 120 to 200 ms. Normal QRS complex amplitude and

Table 5.13 Depolarization and repolarization of the atria and the ventricles produce the usual ECG sequence P–QRS–T. Amplitude of ECG waves (μV). Duration of ECG waves and intervals (ms)

P wave	<110
PR spacing	120–200
Q wave	<40
QR wave	<30 (V1–V2), <50 (V5–V6)
S wave	<40
QRS wave	<100
QT interval	350–450

duration are about 1 mV and 800 to 900 ms. Q wave normally has an amplitude lower than 10 to 20 μV and a duration shorter than 50 ms. All complexes are normally quasi-evenly spaced with a rate of 60 to 100 per minute. T waves follow QRS-complex after about 200 ms.

QT interval duration depends on heart rate as well as heart rate history over at least several minutes. T waves are not altered by ECG leads and are more reproducible than the QT interval. Beat-to-beat variability of T-wave descriptors could better describe the autonomic ventricular regulation than QT-interval indices, as beat-to-beat QT interval variability is not related to cardiac sympathetic regulation of ventricular function [118].

5.3.1.3 Mathematical Electrocardiography

Mathematical electrocardiography is aimed at simulating the electrical activity of the heart using the so-called bidomain model (Sect. 3.1.11.5; Vol. 5 – Chap. 5. Cardiomyocytes and. 8 – Chap. 7. Numerical Simulations). The bidomain model involves transmembrane and extracellular electrical potentials, intracellular and extracellular conductivities, as well as a simplified model of ion tranport across the cardiomyocyte membrane and in the extracellular medium and gating variables. An additional equation governs the electrical potential in the torso. The thoracic cage is commonly assumed to contain the heart, lungs, ribs, and remaining tissues [119]. Adequate boundary conditions are defined at each interface.

One goal of mathematical electrocardiography is to compute, at least roughly, the location and size of myocardium infarctions. A three-dimensional reconstruction of the thorax and its main components is used with possible adaptative mesh to assess electrical activity in several selected points of the torso. A model of the cardiomyofiber structure is incorporated because conductivity in the heart wall is anisotropic. Another objective of numerical simulations is to optimize the position of the probes of an implanted pacemaker, which detects and corrects conduction defects of the action potential.

5.3.2 *Vectorelectrocardiography*

The measurement and display of the electric heart vector is called vectorcardiography (VCG), or vectorelectrocardiography [120]. The 3 components of the electric heart vector are measured with respect to a selected Cartesian frame, which is defined by coordinate axes corresponding either to body axes (usually) or cardiac axes. Although the VCG information is similar to that of the ECG, VCG allows better analysis of the activation front. Seven electrodes are used, one serving as a reference.

5.3.3 *Magnetocardiography*

Sarcolemmal depolarization–repolarization cycles produce temporal changes in the magnetic field around the heart. The magnetocardiography (MCG) noninvasively records the magnetic field (a vector field), whereas ECG estimates the electric potential field (a scalar field).[23] Both methods can be combined to give a better measurement of the electrical activity of the myocardium, hence a better diagnosis [121, 122].

The magnetocardiogram corresponds to the first detected biomagnetic signal [123]. The local magnetic fields created by the small electrical currents[24] in the heart wall can be detected using superconducting quantum interference device sensors (SQUID) without any contact with the skin. An array of multiple SQUID sensors is placed at selected positions over the torso. However, MCG technology is more complicated than that of the ECG and requires expensive equipment.

5.3.4 *Impedance Plethysmography*

Impedance plethysmography is aimed at determining changes in body tissue volumes by measuring tissular frequency-dependent electric impedance at the body surface. Impedance plethysmography measures tissue impedance. Impedance cardiography has been proposed for the noninvasive estimation of stroke volume,

[23]Electric and magnetic leads are different. The signal-to-noise ratio for the electrical and magnetic recordings are affected by different factors. Although the electrical resistivity of the lung parenchyma is relatively high, the magnetic permeability of body tissues resembles that of a free space, allowing easy recordings from the posterior face of the thorax.

[24]Magnetocardiographic signals have been computed using a model with a source represented by an uniform double layer and with a heterogeneous, multicompartmental model of the thorax, the geometry of which is derived from magnetic resonance imaging [124]. Computed and measured magnetic signals were in good agreement. The magnetocardiogram and electrocardiogram have a common basis.

using a constant current fed to the thorax by an electrode pair with a frequency range of 20 to 100 kHz and measuring the resulting thoracic voltage via separate electrode pairs [125]. Impedance cardiography permits noninvasive, continuous, reproducible assessment of stroke volume and ejection time and volume, and their modifications [126]. Changes in electrical impedance of the chest are indeed related to blood displacements in the thorax during the cardiac cycle (mainly by blood ejection from heart ventricles): $\varDelta V = \varrho(L\mathrm{Z}_0)^2\varDelta\mathrm{Z}$, where ϱ is the blood resistivity (usually assumed to be constant; 130–160 Ω.cm, L the distance between electrodes, and Z_0 the basic impedance of the body's region bound by electrodes. The impedance cardiography signal depends on the size of the heart and elastic arteries, and blood resistivity and volume. It fluctuates due to breathing and the body's motion.

Impedance plethysmography has also been used for the detection of thromboses in leg veins. However, modeling addressed for impedance cardiography relies on crude assumptions, more or less neglecting changes in blood volume in the thoracic vessels and organs, and above all flow-dependent changes in blood conductivity.

5.3.4.1 Magnetic Susceptibility Plethysmography

Magnetic susceptibility plethysmography is aimed at measuring blood volume changes in the thorax. The motions of the heart and other thoracic tissues and organs during the cardiac cycle induce variations in magnetic flux when a strong magnetic field is applied to the thorax. Magnetic susceptibility plethysmography does not currently have any clinical applications.

5.4 Upper Airway Patency and Resistance

Nasal patency is a term used to describe the cross-sectional surface area available for airflow, but does not mean airflow resistance. Computed tomography and nuclear magnetic resonance imaging provide a good assessment of air passage permeability.

Volumetry is based on closure of the anterior and posterior ends of the nose. Closure of the posterior end of the nasal cavity is achieved by pressurization to elevate the soft palate against the posterior pharyngeal wall. The closed chamber is then connected to equipment for pressure or volume recordings.

Rhinomanometry enables assessing nasal resistance via measurements of transnasal pressure and air flow during quiet breathing in one cavity. Active anterior rhinomanometry is the most common rhinomanometric method.

Acoustic rhinometry estimates surface areas of nasal cross sections. A pulse wave or a sound wave with given amplitude, frequency, and harmonic content is transmitted along the nose, and is then reflected back, affected by changes in the cross-sectional area of the nasal cavity. Signal processing enables the display of an area–distance graph.

Rhinostereometry records changes in the thickness of the nasal mucosa using a surgical microscope firmly fixed by a personalized splint. This technique allows the study of nasal reactivity to locally applied chemical agents such as histamine.

5.5 Lung Function Testing

The exploration of the lung mechanics does not provide the airway flow distribution, air flow velocity field and derived wall shear stress, and the local static and dynamical pressures. These flow parameters depend on the airway anatomy, which is obtained at a given moment in a given subject by three-dimensional medical imaging. This additional exploration can be performed in the clinical routine. Following a three-dimensional reconstruction and meshing of the visible airways, the flow can be computed using a Navier–Stokes solver. Thus, the characteristics of airflow can be added to provide a better evaluation of airway dysfunctions, especially before or after a suitable surgical therapy or to estimate the efficiency of a medical treatment.

5.5.1 Breathing Mechanics—Main Measuring Devices and Procedures

The exploration of the ventilation mainly provides values of lung volumes and flow rates, thereby quantifying the functional consequences of respiratory diseases. However, the measured data of breathing mechanics are restricted to few bulk parameters.

The pressure p_i and the flow rate q at the airway opening are measured using a mouthpiece connected to a rigid duct with a calibrated pressure transducer and a calibrated *pneumotachograph*. Lung volume V changes are measured by a *spirograph* (spirometry).

The esophageal pressure can be estimated by means of a latex balloon placed in the mid or lower segment of the esophagus [127]. It is supposed to represent the intrathoracic pressure which is assumed to be uniformly distributed. The intrathoracic pressure is assumed to be equal to the pleural pressure (p_{pl}).

The alveolar pressure (p_{A}) is evaluated by means of a *body plethysmograph*, a rigid airtight box in which the subject is seated, breathing via an opening across the box wall [128, 129].

5.5.1.1 Pressure Transducers

Various types of pressure transducers exist. A common one is made of 2 chambers separated by a diaphragm. The first chamber, at ambient pressure, serves as a reference chamber. An air-filled or liquid-filled tube connects the measurement site to the second chamber of the transducer. The pressure difference between the 2 chambers distorts the diaphragm, which is part of a Wheatstone bridge circuit, and therefore translates the pressure difference into an electrical signal.

5.5.1.2 Flowmeters

The air flow rate is usually measured by a pneumotachograph connected to the airways via a suitable connector (face mask, mouthpiece, endotracheal tube, etc.). The pneumotachograph acts as a linear resistor for a specified range of flow rates. The pressure difference measured with a differential pressure transducer between inflow and outflow ports of the pneumotachograph is proportional to airflow. The flow resistance (<0.2 kPa) consists either in a set of numerous in-parallel tubes (Fleisch pneumotachograph) or in a fine-mesh metal gauze (Lilly pneumotachograph). The associated pressure transducer must have a suitable frequency response. The flow rate signal can be intregrated in order to provide the respiratory volumes

5.5.1.3 Propagation Speed of Sound Waves

Sound waves are used to assess the cross-sectional surface area of upper airways [130, 131]. The propagating sound waves undergo dispersion and attenuation. However, experiments show that for a frequency greater than 3 Hz, the airway wall behaves like a rigid material for sound waves and does not significantly affect them.

Several factors alter the propagation of a sound wave, the airway geometry, gas composition, humidity level,[25] temperature, and gas convection.

The flow rate q and the transit time t are recorded and the derivative $\partial q/\partial t|_{q=0}$ is computed. When the wave speed c is known, the area can be assessed by $A = t/(c(\partial q/\partial t|_{q=0}))$ in a tube of uniform cross section, assuming a potential flow and the absence of interaction with the tube wall. However, the airway cross-sectional area varies with the distance from the source–receptor device. Furthermore, the walls of the upper airways are deformable. The upper airways are supposed to be composed of a set of segments, each segment being characterized by averaged properties (lumped parameter model).

5.5.1.4 Respiratory Impedance

The respiratory impedance Z can be estimated using small-amplitude, high-frequency oscillations, the so-called *forced oscillations*, which are superimposed on the quiet breathing. They are produced by a loudspeaker connected to the

---

[25] The propagation speed of sound waves at 310 K is equal to 354, 357, and 352 m/s in partially humidified air, in air saturated with water vapor, and in exhaled air, respectively.

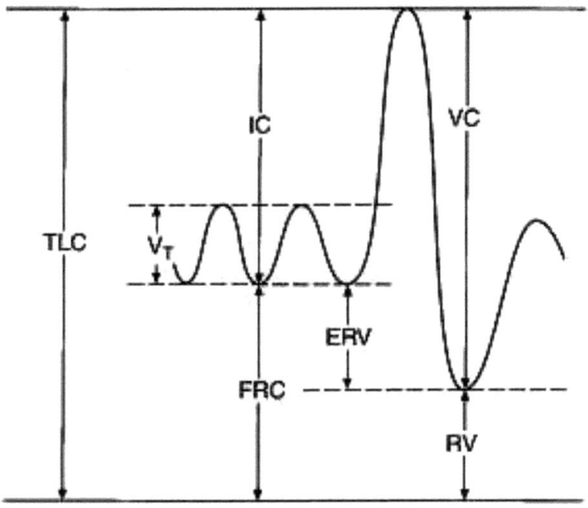

Fig. 5.4 Lung volumes measured by spirometry

mouthpiece via a low-impedance rigid tube, which contains a flowmeter and pressure measurement sites. The signal can either be sinusoidal or contain several harmonics. The analysis is based on lumped parameter models.

The respiratory impedance is subdivided into the resistance `R`, the real part, and the reactance `X`, the imaginary part. The reactance is the sum of compliance `C` and the inertance `L`. When the esophageal pressure is recorded, the impedance of lungs and of chest wall can be separated.

5.5.1.5 Spirography

Spirometry is a standard method aimed at measuring and monitoring exhaled and inhaled lung volumes during rest and deep breathing (TV, IRV, ERV, VC; Fig. 5.4). Spirometry is also used to measure forced expiration volumes and flow rates. Respiratory volume measurements require the cooperation of the examined subjects. Moreover, oxygen consumption ($\dot{V}_{O2}$) can be measured using a spirometer with separate inspired and expired conduits. The latter is equipped with a CO_2 scrubber. The reduction of the spirometer volume thus reflects the subject oxygen consumption. In addition, provocation tests can explore the airway reactivity to inhaled drugs.

Spirometry is not able to provide absolute lung volumes. Other techniques are required to measure the residual volume, such as gas dilution tests and body plethysmography. This relatively old measurement technique has been replaced by flow rate recording and integration.

5.5.1.6 Magnetic Field-Based Techniques

Chest wall displacements can be measured by magnetometry via coils placed in opposite spots of the chest, one coil generating a magnetic field and the other one recording it. Induction plethysmography uses 2 coils, one around the chest, and another around the abdomen. Breathing motions change the inductance of each coil proportionally to the change in volume of the thorax and abdomen, and the sum of the 2 signals is proportional to the variation in respiratory volume.

5.5.1.7 Body Plethysmography

The subject sits inside an airtight box, the body plethysmograph, of known volume. He breathes through a mouthpiece connected to a shutter. The pressure is monitored in the box (p_{box}) and at the mouth (p_{ao}).

The lungs are assumed to behave as a one-compartment model. In the case of airway obstruction, 2-compartment models have been used, assuming linearity and a sinusoidal input.

During breathing, a shutter closes the breathing tube, usually at the end of a normal exhalation (FRC). The absence of flow leads to $p_{ao} = p_A$. The subject makes respiratory efforts against the closed shutter (panting). The chest volume expands. The lung air decompresses. Meanwhile, the body plethysmograph volume slightly decreases and the box pressure increases.

Using the Boyle–Mariotte law at constant temperature (Sect. 4.2.1), the initial pressure and volume of the body plethysmograph being known, the pressure at the end of a chest expansion being measured, the box volume can be calculated during the respiratory effort. The difference between this volume and the initial volume of the box gives the volume change of the chest.

5.5.1.8 Gas Dilution Test

Gas dilution has been aimed at determining lung volumes that cannot be measured by spirometry (functional residual capacity, residual volume, and total lung capacity). The subject is connected to a spirometer containing a known concentration of inert and insoluble gas, other than nitrogen, such as helium.

After several minutes of breathing, the helium concentrations in the spirometer and lung become the same. The law of mass conservation states that the total amount of helium before and after remains constant. Therefore, the fractional concentration times the volume before is equal to the fractional concentration times the volume after: $f_1 V_1 = f_2(V_1 + V_2)$ (V_1: spirometer volume, V_2: lung volume). This equation allows the calculation of the lung volume.

5.5.1.9 Esophageal Balloon

The esophageal balloon is used in order to assess the pleural pressure. It consists of a small polyethylene tube with an open end covered by a thin-walled flexible balloon, connected to a pressure transducer. It is passed via the nose into the lower third of the esophagus. The balloon is slightly inflated with a small amount of air so that the balloon pressure reflects the esophageal pressure.

The compliant esophagus (quasi-closed at rest) is located in the mediastinum with the lungs. When the esophagus is not working, it is assumed that its transmural pressure equals zero (the internal pressure equals the external pressure) and that it is related to the pleural pressure. Changes in intraesophageal pressure are assumed to mimic variations in pleural pressure. The esophageal balloon technique assesses changes in pleural pressure, rather than in the pleural pressure values. In particular, it does not take into account the apicobasal difference in pleural pressure.

5.5.1.10 Tracer Gas Bolus

Transport can be analyzed injecting a bolus (V_{inj}) of tracer gas. Ideal administration would be a Dirac-like bolus injection in the inhaled air via a suitable connector. The real injection mimicks rather a step function of short duration. During transport, the concentration front spreads out. The response curve can be analyzed using the transit (or residence) time distribution via its moments:

$$\mathcal{M}_0 = \int_0^\infty c\, dt \equiv V_{inj}/q \quad (T),$$
$$\mathcal{M}_1 = \int_0^\infty ct\, dt, \quad (T^2),$$
$$\mathcal{M}_2 = \int_0^\infty ct^2\, dt, \quad (T^2), \tag{5.3}$$

where c is the relative tracer concentration (ratio between tracer gas volume and transporter gas volume). The mean transit time cumulated over N generations of respiratory conduits $\bar{t} = \mathcal{M}_1/\mathcal{M}_0$, and the variance of the cumulated transit times (the square of time intervals separating arrival of diverse bolus parts) $\sigma_t^2 = \mathcal{M}_2/\mathcal{M}_0 - (\mathcal{M}_1/\mathcal{M}_0)^2$. The variance is proportional to the degree of axial mixing $\sigma_t^2 \propto \mathcal{D}_{\text{app}}$.[26]

[26] $\sigma_t^2 \propto \overline{\mathcal{D}}_{\text{app}}\bar{t}/V_q^2$, where $\mathcal{D}_{\text{app}}/\mathcal{D} = \kappa \text{Pe}_T^2$ (Pe_T: Péclet number in the trachea) varies according to literature data both at inspiration (1.1–1.5) and expiration (0.4–0.5) [132, 133].

Table 5.14 Carbon monoxide transfer capacity. Estimated values of the bulk transfer factor and the membrane–plasma transfer factor (ml/mn/mmHg; Source: [141])

	Rest	Exercise
$\mathcal{T}_{L_{CO}}$	~25	~40
$\mathcal{T}_{M_{CO}}$	~46	~60

5.5.2 *Breathing Function—Blood Gas Measurements*

Arterial blood gas measurements are aimed at determining the blood concentration of oxygen, carbon dioxide, and bicarbonate, as well as pH. Small (80–200 μl) arterial blood samples are removed from any easily accessible artery (either radial, brachial, or femoral artery) into a special heparinized syringe, eliminating any gas bubble that disturbs the measurements. Blood gas and pH can be quickly and automatically measured using microelectrodes.

Oxygen saturation of hemoglobin is measured by spectrophotometry. Spectrophotometry also enables the measurement of relative blood concentrations of oxy-, carboxy-, met-, and reduced hemoglobin. Oxygen saturation can also be calculated using the Hill nomogram and knowing p_{O_2} and pH. Transcutaneous pulse oximetry allows a quick checking of the oxygen saturation; a beam of red and infrared light crosses a pulsating vascular bed, the ratio of red (absorbed by oxygenated hemoglobin) to infrared (absorbed by deoxygenated hemoglobin) transmission varying with the oxygen saturation of blood.

Oxygen and carbon dioxide content can be measured by the Van Slyke method. In addition, O_2 content can also be assessed using a galvanic oxygen cell [134] or a polarographic method.

5.5.3 *Carbon Monoxide Transfer Capacity*

The carbon monoxide[27] transfer capacity was proposed by M. Krogh and A. Krogh at the beginning of twentieth century [135–137]. The rest carbon monoxide transfer capacity ($\mathcal{T}_{L_{CO}}$) equals about 39.10^{-4} ml/s/Pa (or ~31 ml/mn/mmHg), which is close to values found with modern techniques (Table 5.14). It is greater in men than women and in adults than children. M. Krogh showed a 20–40% increase in $\mathcal{T}_{L_{CO}}$ during exercise (56 ml/min/mmHg), because of improved ratios between the inspired gas flow and the alveolar volume and between the membrane transfer capacity and the capillary volume.

[27] The association speed of carbon monoxide on hemoglobin is slower than that of oxygen, but its dissociation rate is a thousand times slower than its association rate. Carbon monoxide is thus said to have a high affinity for hemoglobin.

The lung transfer capacity is assessed using 2 different techniques based on inhalation of a gas mixture with carbon monoxide (CO). The single breath method introduced by M. and A. Krogh was later modified. W. Fowler added helium as an inert reference gas to the inspired mixture [138]. Further refinements lead to a reliable measurement.

In the breath-holding method, after a deep expiration, the subject inhales up to its TLC (single breath) a gas mixture with less than 0.3% of CO and a small quantity of He (also with 21% of O_2 and the remaining of N_2), followed by 10-s apnea. Steady-state CO uptake techniques have also been proposed. In the steady-state method, the subject inspires a gas mixture with 0.1% of CO (quiet breathing). After 3 mn, the expiratory gas is collected during 2 mn and analyzed. This method is more inaccurate than the single breath technique when the inspired gas is unevenly distributed between the lung territories.

The transfer capacity $\mathcal{T}_{L_{CO}}$ depends on $\dot{V}_{CO}$ and on $p_{A_{CO}}$. The latter cannot be accurately assessed when significant disturbances of the ventilation distribution exist. Moreover, the single CO breath method overestimates the alveolocapillary transfer of rest ventilation, breath-holding favoring the penetration of CO in not well-ventilated areas. Furthermore, the results depend on the lung volume. Empirical relations between $\mathcal{T}_{L_{CO}}$ and hemoglobin concentration ([Hb]) have been proposed in the literature [139, 140].

5.5.4 Clearance by Respiratory Epithelium

The single method to assess in vivo the permeability of the respiratory epithelium is based on measurements of the uptake rate of tracers by the airway lumen and alveoli [142]. The radiolabeled tracer is administered as an aerosol.[28] The uptake rate is usually expressed by the clearance or decrease rate of the pulmonary tracer content per minute.[29]

The thickness of the airway surface film is calculated from the known permeability coefficient and measured clearance. The measurement of the permeability coefficient of a tracer through a tissue is based on knowledge of the luminal

[28] The usual tracer is radiolabeled diethylenetriaminepentaacetic acid (DTPA).

[29] The clearance $(dQ/dt)/cV$ (dimension: T^{-1}) is the flux of tracer (dQ/dt, Q: substance quantity) divided by the luminal tracer content, the latter being the product of tracer luminal concentration c by volume V of airway surface film, itself the product of the airway surface area (tracer uptake area) by the liquid film thickness. The uptake rate can also be evaluated from the decay rate constant, assuming a monoexponential relationship with time, although the clearance can follow a biexponential evolution. The clearance of DTPA from human lungs ranges 0.59–1.56.10^{-2}/s, with a mean of about 10^{-2}/s, varying with the aerosol size and the working group [142].

Table 5.15 Respiratory parameters in a 25-yr-old man (height 170 cm; weight 70 kg), in a 25-yr-old woman (height 170 cm; weight of 67 kg), and in a 6-yr-old child (height 115 cm; weight 20 kg)

Index	Child	Woman	Man
VC (l, BTPS)	1.52	4.54	5.04
RV (l, BTPS)	0.40	1.22	1.35
RV/VC (%)	26	27	27
FEV_1/VC (%)	90	80	80
DV (ml, BTPS)		104	156
$\dot{V}_{O_2}$ (l/mn)	0.23	0.28	0.34

concentration. Consequently, the permeability coefficient is assessed in vitro. The clearance component by mucociliary transport is only known for albumin. Furthermore, the permeability of the airway mucosa can depend on the airway size.

5.5.5 *Respiratory Indices*

The measured volumes obtained from the spirogram include the tidal volume, vital capacity, and functional residual capacity. The residual volume and total lung capacity are indirectly assessed. Certain respiratory indices are given in Table 5.15 for an adult young woman and man as well as a young child.

Common expressions of gas volume are computed for given temperature, pressure, and humidity: (1) at ambient temperature and pressure, saturated with water vapor (ATPS); (2) at ambient temperature and pressure, dry (ATPD); (3) at body temperature (37°C, or 310 K) and ambient pressure, saturated with water vapor (47 mmHg, or 6.2 kPa; BTPS); and (4) at standard temperature (0°C, or 273 K) and pressure (760 mmHg, or 101.3 kPa), dry (STPD). Normal values depend on age, sex, height, weight, and ethnicity.

5.5.5.1 Forced Expiration

Forced expiration maneuver is a maximal exhalation following a maximal inspiration, starting from the total lung capacity (TLC). The spirometry tracings obtained during forced expiration tests allow the separation of obstructive disorders from restrictive states. The reproducibility of forced expiration curves is usually good in motivated subjects strongly supported by the medical team.

During a forced expiration test, the forced expiratory volume in the first second (FEV_1) is calculated, as well as the ratio between the forced expiratory volume in the first second and the forced vital capacity (FEV_1/FVC). Both indices are used in diagnosing obstructive and restrictive disorders. Other quantities are used, such as the flow rates at 50 and 25% of FVC.

In a normal forced expiration curve, FEV_1 is usually about 80% of FVC. In an obstructive disorder, FVC is reduced, as well as FEV_1, but at a greater extent, which leads to a decay in the FEV_1/FVC ratio. A restrictive ventilatory defect

Table 5.16 Normal ranges of values of arterial blood gases

p_{O_2}	12–13 kPa (90–100 mmHg)
O_2 content	15–23%
S_{aO_2}	94–100%
p_{CO_2}	4.5–6.2 kPa (34–46 mmHg)
total CO_2	23–27 mmol/l
HCO_3^-	22–26 mmol/l
H_2CO_3	1.02–1.38 mmol/l
pH	7.35–7.45
Base excess	−2–2 mmol/l

is obviously characterized by a reduction in total lung capacity, as well as any measurable lung volumes (total lung capacity, functional residual capacity, etc.). In a restrictive impairment, FVC and FEV_1 decrease by the same amount, so that FEV_1/FVC ratio is normal.

Although the usual analysis of the forced expiration is used to detect airway obstruction, standard indices are insensitive indicators of slight airway dysfunction. Impairments of the ventilatory function, validated by a FEV_1 decrease, indeed occur when alterations of the bronchial walls are already significant in many pulmonary compartments. Early manifestations of airway diseases can be assumed to be associated with heterogeneities of the lung emptying. Two tests have then been proposed to detect diseases at an early stage: the multibreath nitrogen washout test and forced expiration. Not only the usual forced expiratory indices, but also the time domain analysis of the spirogram are calculated.

The forced expiration can indeed be defined as the cumulative distribution of transit times of exhalation at the mouth of small fractions ΔV of FVC. The moment analysis of the distribution of the transit times has been performed in a male population,[30] the impairment of the ventilatory function being assessed by the smoking history. However, the reproducibility of the transit time moments was found to be less good than for the other forced expiratory indices (unpublished communication).[31] In addition, the transit time indices slightly increase with age and weakly depend on the subject height. Unfortunately, they are much less related to tobacco consumption than the conventional quantities.

[30]The population comprises 114 volunteers, 27 to 58-yr old, with 15 nonsmokers, and 43 ex-smokers. Smoking habits have been defined according to the amount (1 cigar being equivalent to 5 cigarettes, 1 cigarillo to 2 cigarettes, 1 g of tabacco to 1 cigarette), to the quality (with or without smoke inhalation), ex-smokers being subdivided into 2 groups whether they have stopped more than one month or more than one year. Slight smokers were defined by less than 5 cigarettes per day over any period of time or more than 5 cigarettes per day for less than one year. Smokers include 2 categories, according to the existence of respiratory symptoms.

[31]Transit time moments are computed using the entire maneuver. They then depend on the effort performed at the end of the test. Forced expirations terminate more or less prematurely, at unrepeatable times in a given subject. A better repeatability is obtained using truncated spirograms. The values then strongly depend on the arbitrary chosen threshold.

5.5.6 *Blood Gas*

Typical normal ranges of arterial blood gases are given in Table 5.16. Acidosis or alkalosis can be found whether pH <7.35 or >7.45. Oxygen and carbon dioxide partial pressure and bicarbonate values indicate whether the cause is respiratory or metabolic, with or without any compensation.

Conclusion

The fox knows many things; the hedgehog one big thing.
(Archilochos of Paros [c.−680–c.−645])

Human physiology describes mechanisms that operate in the body in normal (healthy) conditions. Pathology copes with diseases. Pathophysiology details causes whereby any abnormal condition with its structural and functional changes develops and progresses.

Any substance, such as reactive oxygen and nitrogen species, that serve as a nutrient, signaling mediator, or cofactor at a physiological concentration can become toxic in excessive amount. Inappropriate activation or inhibition of signaling pathways contributes to the pathogenesis of illnesses.

For example, the *renin–angiotensin axis*, an endo-, para-, auto-, and intracrine regulatory pathway, participates in the control of sodium homeostasis and arterial pressure. Its dysregulation contributes to the pathogenesis of hypertension and renal injury via inappropriate activation of widespread AT_1 receptors [1].

Briefly, the renal aspartyl protease renin is released primarily from juxtaglomerular cells on afferent arterioles (but also other cell types).[1] The secreted active form of renin results from cleavage of the N-terminus of prorenin. Prorenin is also secreted into the blood circulation, where it can be converted to renin upon proteolysis by trypsin and taken up by cells. In addition, the (pro)renin receptor can initiate an intracellular signaling to activate ERK1, ERK2, and P38MAPK

[1]The juxtaglomerular apparatus consists of renin-producing cells in the media of afferent glomerular arterioles and extraglomerular mesangial cells between the glomerulus and distal tubule, cells of which form the macula densa. Reduced renal arterial perfusion or sodium supply to the macula densa, increased renal β1-adrenergic nerve stimulation, circulating catecholamines and prostaglandins cause secretion of renin. On the other hand, angiotensin-2, atrial natriuretic peptide, and vasopressin inhibit renin release.

M. Thiriet, *Anatomy and Physiology of the Circulatory and Ventilatory Systems*, BBMCVS 6, DOI 10.1007/978-1-4614-9469-0,

kinases [1]. In the heart and kidney, the renin receptor binds to both renin and prorenin, hence increasing the catalytic efficiency of angiotensin-1 processing from angiotensinogen.

Angiotensinogen primarily synthesized and liberated by hepatocytes into the blood circulation is cleaved by renin into angiotensin-1. The latter is processed to angiotensin-2 by circulating and endothelium-linked dipeptidyl carboxypeptidases, angiotensin-1-converting enzyme (ACE). The latter localize also to brush border membranes of proximal tubular cells. Several angiotensinases and peptidases catabolize angiotensin-2.

In addition, chymase produced by mastocytes can convert angiotensin-1 to angiotensin-2. In the human heart, chymase is synthesized and stored in endothelial and mesenchymal cells. It is secreted into the interstitium. Cathepsins and kallikreins can process angiotensinogen into angiotensin-1 and -2 [1]. Another carboxypeptidase, ACE2, which abounds in renal endotheliocytes and proximal tubule epitheliocytes, removes a single amino acid from the C-terminus of angiotensin-1 to form angiotensin$_{(1--9)}$. Enzyme ACE2 also converts angiotensin-2 to angiotensin$_{(1--7)}$. Transmembrane protein TMem27, or collectin, a ACE2 homolog without the dipeptidyl carboxypeptidase domain, localizes to the luminal surface and in the cytoplasm of collecting duct cells.

Angiotensin-2 is the most potent product of the renin–angiotensin axis, although other angiotensin family peptides, such as angiotensin-3 and -4 as well as angiotensin$_{(1--7)}$, are active. Angiotensin-2 and -3 preferentially bind to AT_1 and AT_2 receptor, respectively. In the kidney, AT_1 receptor resides in the proximal tubule and thick ascending limb of the loop of Henle as well as glomerulus, vasa recta, and arcuate arteries, in addition to juxtaglomerular cells [1].

Angiotensin-2 not only operates as a hormone, but also as a paracrine factor. It activates multiple intracellular signaling cascades, thereby functioning in cell growth and proliferation, differentiation, migration, and apoptosis, as well as inflammation.

Angiotensin-2 binds to its cognate receptors and, subsequently, provokes contraction and hypertrophy of vascular smooth myocytes, enhances myocardial contractility, stimulates aldosterone production and secretion as well as release of pituitary hormones and catecholamines from the adrenal medulla and sympathetic nerve endings, activates the sympathetic nervous system, regulates sodium transport by intestinal and renal epitheliocytes (in particular promoting sodium and water conservation via its effects on renal hemodynamics and tubular reabsorption), and stimulates thirst and salt appetite [2].

In the kidney as well as adrenal glands, angiotensin-2 concentration is much greater than the amount delivered by arterial blood. Renal angiotensin-2 results from angiotensinogen that is supplied as well as locally produced by proximal tubule cells. In rats at least, angiotensin convertase abounds in the proximal and distal tubules and collecting ducts as well as renal endotheliocytes.

Angiotensin-2 is one of the most powerful sodium-retaining hormones. Furthermore, it regulates *aldosterone* that stimulates sodium reabsorption via mineralocorticoid receptors in the connecting and cortical segments of the collecting tubule.

The circumventricular organs, median preoptic nucleus, and medium surrounding the anteroventral third ventricle in the lamina terminalis (AV3V), which are connected to the hypothalamus, limbic system, and brainstem, yield the neuroanatomic substratum for cardiovascular control as well as thirst and sodium appetite [2]. The cerebral AV3V region is endowed with angiotensinergic nerve endings and AT_1 receptors. The nucleus tractus solitarius and other structures in the brainstem that constitute a inhibitory axis for blood-volume control, as they respond to baro- and voloreceptor send noradrenergic and other projections to the AV3V region. The subfornical organ, organum vasculosum of the lamina terminalis, and area postrema contain angiotensin receptors that enable circulating angiotensin-2 to function in hypovolemic thirst and sodium appetite and blood pressure control.

In specific regions of the kidney, angiotensin-2 concentration is selectively regulated, that is, in a manner distinct from the circulating angiotensin-2 level [1]. The medulla is characterized by high angiotensin-2 level (stronger than cortical content) and the large density of its receptors. Within the cortex, angiotensin-2 is distributed in the interstitial and tubular fluid as well as within the cell. Its interstitial concentration is much higher than that in plasma. In the cell, it can be detected in endosomes, from which it can be recycled and secreted .

Angiotensin-2 synthesized in the kidney influences renal hemodynamics. It also induces hypertrophy of proximal tubular epitheliocytes. Angiotensin-2 is involved in the regulation of the glomerular filtration, as both AT_1 and AT_2 receptors lodge on mesangial cells. In addition, it elicits constriction of both afferent and efferent arterioles (both pre- and postglomerular arterioles) [1]. It raises the proximal tubular reabsorption rate and subsequently lowers fluid volume delivery in the distal nephron. Angiotensin-2 can also damage podocytes. Moreover, it modulates the sensitivity of vascular and mesangial elements of the tubuloglomerular feedback that links reabsorption in tubules and glomerular filtration rate. Changes in the tubular fluid solute concentration in the terminal segment of the loop of Henle are signaled to afferent arterioles and mesangium to contract or dilate to stabilize the filtration.

In fact, angiotensin-2 influences proximal tubule sodium transport according to its dose. At physiological concentrations (pmol/l), intrarenally produced angiotensin-2 stimulates sodium reabsorption, which is coupled with bicarbonate reabsorption, upon activation of apical Na^+–H^+ exchanger, basolateral Na^+–HCO_3^- symporter, and basolateral Na^+–K^+ ATPase as well as insertion of H^+ ATPase into the apical plasma membrane using SRC family kinases as well as the MAPK module and inhibiting adenylate cyclase. At higher levels (μ mol/l), it prevents transfer [1].

Moreover, at physiological concentrations (1 nmol/l), it increases albumin endocytosis in the proximal tubule via luminal AT_2 receptor and protein kinase-B [1].

In the distal tubule, angiotensin-2 also increases sodium and bicarbonate reabsorption [1]. It stimulates Na^+–H^+ exchanger in both up- and downstream segments upon activation of AT_1 receptor and vacuolar H^+ ATPase in the downstream distal segment. In both the proximal and distal tubules, angiotensin-2 raises H^+ secretion and Na^+ absorption, as Na^+ serves as a counterion for H^+ secretion.

In the medullary collecting duct, angiotensin-2 increases the production of the B1 subunit of H^+ ATPase, at least in rodents. It also stimulates activity of the apical epithelial sodium channel. The inner medullary collecting duct is responsible for the final concentration of urine. Basolateral angiotensin-2 elevates vasopressin-stimulated urea permeability and causes phosphorylation of the urea transporter, thereby favoring water reabsorption.

Angiotensin-2 acts on both the renal microcirculation and nephron epithelial transport. Arterial hypertension is a major cardiovascular risk factor. In hypertension, autoregulation can be blunted upon chronic elevation of the angiotensin-2 level. Intrarenal angiotensin-2 can inappropriately regulate salt and water transfer in the nephron. Angiotensin-2 may induce oxidative stress in both proximal and distal tubules. An increased secreted amount of angiotensinogen by proximal tubular epitheliocytes spills over into the distal nephron, thereby heightening the effect of angiotensin-2 on the distal tubule. Sustained stimulation of the intrarenal renin–angiotensin axis augments arterial pressure and causes renal injury and fibrosis. Angiotensin-2 hypertensinogenic action is mostly attributed to the AT_1 receptor [1].

Aldosterone is a steroid hormone synthesized by the outer zona glomerulosa of the adrenal gland. The storage capacity for aldosterone within producing cells is very limited. The sustained production of aldosterone requires persistent Ca^{2+} entry through voltage-gated Ca_V1 and Ca_V3 channels upon inhibition of K^+ channels and Na^+–K^+ ATPase that causes membrane depolarization. Its synthesis by aldosterone synthase (CyP11b2) is mainly controlled by angiotensin-2 and extracellular potassium, in addition to angiotensin-3 and adrenocorticotropic hormone (ACTH). Aldosterone acts via the mineralocorticoid receptor, or nuclear NR3c2 receptor. Angiotensin-2 provokes Ca^{2+} influx from intracellular store. Calcium–calmodulin activates calcium–calmodulin-dependent protein kinases CamK1 and CAmK4 that excite transcription factors, such as NR4a1, NR4a2, ATF1, and CREB, which, in turn, stimulate promoter of the CYP11B2 gene [3].

Aldosterone participates in the regulation of blood pressure, mainly by acting on the distal tubule and collecting duct of the nephron, where it increases reabsorption of sodium and water and promotes secretion of potassium, as it activates basolateral Na^+–K^+ ATPase and apical epithelial Na^+ channel. Choride ion is also reabsorbed, but K^+ and H^+ are secreted into the tubular lumen, the latter in intercalated cells of the cortical collecting tubule.

When its production is dysregulated, aldosterone contributes to the development of cardiovascular and renal diseases. In particular, elevated aldosterone concentration can predict an adverse outcome of arterial hypertension.

Inter- and multidisciplinary approaches combine methodologies used in physiology, histology, molecular biology, genetics, and bioinformatics to understand causes, analyze relevant signaling mediators, and unravel pathophysiological mechanisms involved in diseases.

References

Introduction

1. Bachelard G (1938 [1969]) La psychanalyse du feu [The Psychoanalysis of Fire]. Gallimard, Paris

Chap. 1. Anatomy of the Cardiovascular System

1. Echtler K, Stark K, Lorenz M, Kerstan S, Walch A, Jennen L, Rudelius M, Seidl S, Kremmer E, Emambokus NR, von Bruehl ML, Frampton J, Isermann B, Genzel-Boroviczény O, Schreiber C, Mehilli J, Kastrati A, Schwaiger M, Shivdasani RA, Massberg S (2010) Platelets contribute to postnatal occlusion of the ductus arteriosus. Nature – Medicine 16:75–82
2. Hong Z, Kutty S, Toth PT, Marsboom G, Hammel JM, Chamberlain C, Ryan JJ, Zhang HJ, Sharp WW, Morrow E, Trivedi K, Weir EK, Archer SL (2013) Role of dynamin-related protein 1 (drp1)-mediated mitochondrial fission in oxygen sensing and constriction of the ductus arteriosus. Circulation Research 112:802–815
3. Gray H (1995) Gray's anatomy: anatomy descriptive and surgical. Barnes and Noble, New York
4. Rouvière H, Delmas A (2002) Anatomie humaine descriptive, topographique et fonctionnelle [Descriptive, Topographical, and Functional Human Anatomy], Vols. I–IV. Masson, Paris
5. Iacobellis G, Corradi D, Sharma AM (2005) Epicardial adipose tissue: anatomic, biomolecular and clinical relationships with the heart. Nature – Clinical Practice Cardiovascular Medicine 2:536–543.
6. Badano LP, Agricola E, Perez de Isla L, Gianfagna P, Zamorano JL (2009) Evaluation of the tricuspid valve morphology and function by transthoracic real-time three-dimensional echocardiography. European Journal of Echocardiography 10:477–484
7. Kozlowski D, Owerczuk A, Piwko G, Kozlowska M, Bigus K, Grzybiak M (2002) The topography of the subthebesian fossa in relation to neighbouring structures within the right atrium. Folia Morphologica 62:65–70
8. Stradins P, Lacis R, Ozolanta I, Purina B, Ose V, Feldmane L, Kasyanov V (2004) Comparison of biomechanical and structural properties between human aortic and pulmonary valve. European Journal of Cardio-thoracic Surgery 26:634–639
9. Keith A, Flack M (1907) The form and nature of the muscular connections between the primary divisions of the vertebrate heart. Journal of Anatomy and Physiology 41:172–189

M. Thiriet, *Anatomy and Physiology of the Circulatory and Ventilatory Systems*, BBMCVS 6, DOI 10.1007/978-1-4614-9469-0,

10. Anderson RH, Razavi R, Taylor AM (2004) Cardiac anatomy revisited. Journal of Anatomy 205:159–177
11. Anderson RH, Webb S, Brown NA (1999) Clinical anatomy of the atrial septum with reference to its developmental components. Clinical Anatomy 12:362–374
12. Staszewsky L, Latini R (2013) What is the atrium trying to tell us? European Heart Journal 34:255-257
13. Gupta S, Matulevicius SA, Ayers CR, Berry JD, Patel PC, Markham DW, Levine BD, Chin KM, de Lemos JA, Peshock RM, Drazner MH (2013) Left atrial maximal volume and left atrial emptying fraction as predictors of cardiovascular events in community-based or population studies. European Heart Journal 34:278–285
14. Swanson WM, Clark RE (1974) Dimensions and geometric relationships of the human aortic valve as a function of pressure. Circulation Research 35:871–882
15. Grande KJ, Kunzelman KS, Cochran RP, David TE, Verrier ED (1993) Porcine aortic leaflet arrangement may contribute to clinical xenograft failure. ASAIO Journal 39:918–922
16. Bäck M, Gasser TC, Michel JB, Caligiuri G (2013) Biomechanical factors in the biology of aortic wall and aortic valve diseases. Cardiovascular Research 99:232–241
17. Valsalva AM (1740) Opera. Venice
18. Bellhouse BJ (1969) Velocity and pressure distributions in the aortic valve. Journal of Fluid Mechanics 37:587–600
19. Jatene MB, Monteiro R, Guimaraes MH, Veronezi SC, Koike MK, Jatene FB, Jatene AD (1999) Aortic valve assessment. Anatomical study of 100 healthy human hearts. Arquivos Brasileiros de Cardiologia 73:81-86
20. Tank PW, Gross Anatomy, Department of Neurobiology and Developmental Sciences, University of Arkansas for Medical Sciences, http://anatomy.uams.edu/anatomyhtml/heart2.html
21. Jiamsripong P, Honda T, Reuss CS, Hurst RT, Chaliki HP, Grill DE, Schneck SL, Tyler R, Khandheria BK, Lester SJ (2007) Three methods for evaluation of left atrial volume. European Journal of Echocardiography 9:351–355
22. Herregods MC, De Paep G, Bijnens B, Bogaert JG, Rademakers FE, Bosmans HT, Bellon EP, Marchal GJ, Baert AL, Van de Werf F, De Geest H (1994) Determination of left ventricular volume by two-dimensional echocardiography: comparison with magnetic resonance imaging. European Heart Journal 15:1070–1073
23. Armour JA (2008) Potential clinical relevance of the 'little brain' on the mammalian heart. Experimental Physiology 93:165–176
24. Gray AL, Johnson TA, Ardell JL, Massari VJ (2004) Parasympathetic control of the heart. II. A novel interganglionic intrinsic cardiac circuit mediates neural control of heart rate. Journal of Applied Physiology 96:2273–2278
25. Waldmann M, Thompson GW, Kember GC, Ardell JL, Armour JA (2006) Stochastic behavior of atrial and ventricular intrinsic cardiac neurons. Journal of Applied Physiology 101:413–419
26. Sharma V (2009) Deterministic chaos and fractal complexity in the dynamics of cardiovascular behavior: perspectives on a new frontier. Open Cardiovascular Medicine Journal 3:110–123
27. Avnir D, Biham O, Lidnar D, Malcai O (1998) Is the geometry of nature fractal? Science 279:39–40
28. Kleiber M (1947) Body size and metabolic rate. Physiological Reviews 27:511–541
29. West GB, Brown JH, Enquist BJ (1997) A general model for the origin of allometric scaling laws in biology. Science 276:122–126
30. Sapoval B, Gobron T, Margolina A (1991) Vibrations of fractal drums. Physical Review Letters 67:2974–2977
31. D'Arcy Thompson W (1917) On Growth and Form. Cambridge University Press, Cambridge, UK
32. Van Vliet P, Wu SM, Zaffran S, Pucéat M (2012) Early cardiac development: a view from stem cells to embryos. Cardiovascular Research 96:352–362
33. Senyo SE, Steinhauser ML, Pizzimenti CL, Yang VK, Cai L, Wang M, Wu TD, Guerquin-Kern JL, Lechene CP, Lee RT (2013) Mammalian heart renewal by pre-existing cardiomyocytes. Nature 493:433–436

34. Mercola M (2012) Cardiovascular biology: A boost for heart regeneration. Nature 492: 360–362
35. Eulalio A, Mano M, Dal Ferro M, Zentilin L, Sinagra G, Zacchigna S, Giacca M (2012) Functional screening identifies miRNAs inducing cardiac regeneration. Nature 492:376–381
36. Chien KR, Domian IJ, Parker KK (2008) Cardiogenesis and the complex biology of regenerative cardiovascular medicine. Science 322:1494–1497
37. Smart N, Riley PR (2009) Derivation of epicardium-derived progenitor cells (EPDCs) from adult epicardium. Current Protocols in Stem Cell Biology 2:unit2c.2
38. Zhou B, Ma Q, Rajagopal S, Wu SM, Domian I, Rivera-Feliciano J, Jiang D, von Gise A, Ikeda S, Chien KR, Pu WT (2008) Epicardial progenitors contribute to the cardiomyocyte lineage in the developing heart. Nature 454:109–113
39. Wagner KD, Wagner N, Schedl A (2003) The complex life of WT1. Journal of Cell Science 116:1653–1658
40. Wagner KD, Wagner N, Bondke A, Nafz B, Flemming B, Theres H, Scholz H (2002) The Wilms' tumor suppressor Wt1 is expressed in the coronary vasculature after myocardial infarction. FASEB Journal 16:1117–1119
41. Koninckx R, Daniëls A, Windmolders S, Mees U, Macianskiene R, Mubagwa K, Steels P, Jamaer L, Dubois J, Robic B, Hendrikx M, Rummens JL, Hensen K (2013) The cardiac atrial appendage stem cell: a new and promising candidate for myocardial repair. Cardiovascular Research 97:413–423
42. Nam YJ, Song K, Luo X, Daniel E, Lambeth K, West K, Hill JA, DiMaio JM, Baker LA, Bassel-Duby R, Olson EN (2013) Reprogramming of human fibroblasts toward a cardiac fate. Proceedings of the National Academy of Sciences of the United States of America 110:5588–5593
43. Mulligan-Kehoe MJ (2010) The vasa vasorum in diseased and nondiseased arteries. American Journal of Physiology – Heart and Circulatory Physiology 298:H295–H305
44. Chambers R, Zweifach BW (1994) Topography and function of the mesenteric capillary circulation. American Journal of Anatomy 75:175–205
45. Lee JS (2000) Biomechanics of the microcirculation, an integrative and therapeutic perspective. Annals of Biomedical Engineering 28:1–13
46. Hilgers RHP, Schiffers PMH, Aartsen WM, Fazzi GE, Smits JFM, De Mey JGR (2004) Tissue angiotensin-converting enzyme in imposed and physiological flow-related arterial remodeling in mice. Arteriosclerosis, Thrombosis, and Vascular Biology 24:892–897
47. Batra S, Rakusan K (1992) Capillary length, tortuosity, and spacing in rat myocardium during cardiac cycle. American Journal of Physiology – Heart and Circulatory Physiology 263:H1369–H1376
48. Hudlicka O, Tyler KR (1984) The effect of long-term high-frequency stimulation on capillary density and fibre types in rabbit fast muscles. Journal of Physiology 353:435–445
49. Krenz GS, Lin J, Dawson CA, Linehan JH (1994) Impact of parallel heterogeneity on a continuum model of the pulmonary arterial tree. Journal of Applied Physiology 77:660–670
50. Marxen M, Henkelman RM (2003) Branching tree model with fractal vascular resistance explains fractal perfusion heterogeneity. American Journal of Physiology – Heart and Circulatory Physiology 284:H1848–H1857
51. Liew G, Mitchell P, Rochtchina E, Wong TY, Hsu W, Lee ML, Wainwright A, Wang JJ (2003) Fractal analysis of retinal microvasculature and coronary heart disease mortality. European Heart Journal 32:422–429
52. Schelin AB, Károlyi G, de Moura AP, Booth NA, Grebogi C (2010) Fractal structures in stenoses and aneurysms in blood vessels. Philosophical Transactions of the Royal Society – London – A Mathematical, Physical, and Engineering sciences 368:5605–5617
53. Bassingthwaighte JB, King RB, Roger SA (1989) Fractal nature of regional myocardial blood flow heterogeneity. Circulation Research 65:578–590
54. Karch R, Neumann F, Podesser BK, Neumann M, Szawlowski P, Schreiner W (2003) Fractal properties of perfusion heterogeneity in optimized arterial trees: a model study. Journal of General Physiology 122:307–321

55. Kurz H, Sandau K (1998) Allometric scaling in biology. Science 281:751
56. Kurz H, Wilting J, Sandau K, Christ B (1998) Automated evaluation of angiogenic effects mediated by VEGF and PlGF homo- and heterodimers. Microvascular Research 55:92–102
57. Mori D, Yamaguchi T (2002) Computational fluid dynamics modeling and analysis of the effect of 3-D distortion of the human aortic arch. Computer Methods in Biomechanics and Biomedical Engineering 5:249–260
58. Ruddy JM, Jones JA, Spinale FG, Ikonomidis JS (2008) Regional heterogeneity within the aorta: relevance to aneurysm disease. Journal of Thoracic and Cardiovascular Surgery 136:1123–1130
59. Cebral JR (2005) www. scs. gmu. edu/~jcebral/
60. Bouthillier A, van Loveren HR, Keller JT (1996) Segments of the internal carotid artery: a new classification. Neurosurgery 38:425–432
61. Schaller B (2004) Physiology of cerebral venous blood flow: from experimental data in animals to normal function in humans. Brain Research Reviews 46:243–260
62. Harmon JV, Edwards WD (1987) Venous valves in subclavian and internal jugular veins. Frequency, position, and structure in 100 autopsy cases. American Journal of Cardiovascular Pathology 1:51–54
63. Henry JL, Calaresu FR (1974) Excitatory and inhibitory inputs from medullary nuclei projecting to spinal cardioacceleratory neurons in the cat. Experimental Brain Research 20:485–504
64. Hildebrandt JR (1974) Central connections of aortic depressor and carotid sinus nerves. Experimental Neurology 45:590–605
65. Gebber GL, Taylor DG, Weaver LC (1973) Electrophysiological studies on organization of central vasopressor pathways. American Journal of Physiology 224:470–481
66. Snyder DW, Gebber GL (1973) Relationships between medullary depressor region and central vasopressor pathways. American Journal of Physiology 225:1129–1137
67. Weaver LC, Gebber GL (1974) Electrophysiological analysis of neural events accompanying active dilatation. American Journal of Physiology 226:84–89
68. Abboud FM (2010) In search of autonomic balance: the good, the bad, and the ugly. American Journal of Physiology — Regulatory, Integrative and Comparative Physiology 298:R1449–R1467
69. Klabunde RE (2011) Cardiovascular Physiology Concepts, 2nd edition, Wolters Kluwer – Lippincott Williams and Wilkins, Philadelphia, Pennsylvania
70. Prabhakar NR, Peng YJ (2004) Peripheral chemoreceptors in health and disease. Journal of Applied Physiology 96:359–366
71. Schultz HD, Li YL (2007) Carotid body function in heart failure. Respiratory Physiology and Neurobiology 157:171–185
72. Eyzaguirre C (2007) Electric synapses in the carotid body–nerve complex. Respiratory Physiology and Neurobiology 157:116–122
73. De Castro F (1928) Sur la structure et l'innervation du sinus carotidien de l'homme et des mammifères. Nouveaux faits sur l'innervation et la fonction du glomus caroticum. [On the structure and innervation of carotid sinus in humans and mammals. New facts on innervation and function of the glomus caroticum.] Travaux du Laboratoire de Recherches en Biologie 25:331–380
74. Kummer W, Gibbins IL, Heym C (1989) Peptidergic innervation of arterial chemoreceptors. Archives of Histology and Cytology 52:361–364
75. Rey S, Del Rio R, Alcayaga J, Iturriaga R (2006) Endothelins in the cat petrosal ganglion and carotid body: effects and immunolocalization. Brain Research 1069:154–158
76. Prabhakar NR (2000) Oxygen sensing by the carotid body chemoreceptors. Journal of Applied Physiology 88:2287–2295
77. Schultz HD, Li YL, Ding Y (2007) Arterial chemoreceptors and sympathetic nerve activity: implications for hypertension and heart failure. Hypertension 50:6–13
78. Molenda O (1975) Morphology and topography of the carotid body and carotid sinus in sheep. Polskie Archiwum Weterynaryjne 18:343–364

79. Sadik AH, Al-Shaikhly AK, Khamas WA (1993) Anatomic location of the carotid body and carotid sinus in sheep and goats. Small Ruminant Research 12:371–377
80. Peng YJ, Nanduri J, Raghuraman G, Souvannakitti D, Gadalla MM, Kumar GK, Snyder SH, Prabhakar NR (2010) H_2S mediates O_2 sensing in the carotid body. Proceedings of the National Academy of Sciences of the United States of America 107:10719–10724
81. Doan TN, Stephans K, Ramirez AN, Glazebrook PA, Andresen MC, Kunze DL (2004) Differential distribution and function of hyperpolarization-activated channels in sensory neurons and mechanosensitive fibers. Journal of Neuroscience 24:3335–3343

Chap.2. Anatomy of the Ventilatory Apparatus

1. Ochs M, Nyengaard JR, Jung A, Knudsen L, Voigt M, Wahlers T, Richter J, Gundersen HJ (2004) The number of alveoli in the human lung. American Journal of Respiratory and Critical Care Medicine 169:120–124
2. Horsfield K (1978) Morphometry of the small pulmonary arteries in man. Circulation Research 42:593–597
3. Glenny RW (2011) Emergence of matched airway and vascular trees from fractal rules. Journal of Applied Physiology 110:1119–1129
4. Fetita C, Mancini S, Perchet D, Prêteux F, Thiriet M, Vial L (2005) An image-based computational model of oscillatory flow in the proximal part of tracheobronchial trees. Computer Methods in Biomechanics and Biomedical Engineering 8:279–293
5. Christophe JJ, Ishikawa T, Matsuki N, Imai Y, Takase K, Thiriet M, Yamaguchi T (2010) Patient-specific morphological and blood flow analysis of pulmonary artery in the case of severe deformations of the lung due to pneumothorax. Journal of Biomechanical Science and Engineering 5:485–498
6. Robinson RJ, Russo J, Doolittle RL (2009) 3D airway reconstruction using visible human data set and human casts with comparison to morphometric data. Anatomical Record – Advances in Integrative Anatomy and Evolutionary Biology 292:1028–1044
7. Guyton AC (1985) Anatomy and Physiology, Saunders College Publishing, New York
8. Golde AR (2006) Rhinoplastic techniques for the nasal valve for the patient with sleep apnea. Operative Techniques in Otolaryngology-Head and Neck Surgery 17:242–251
9. Haight JS, Cole P (1983) The site and function of the nasal valve. Laryngoscope 93:49–55
10. Martínez MM, Román LJ, C Villalobos (2004) Biofluid dynamics of the ear–nose–throat system. Congress on Biofluid Dynamics of Human Body Systems at University of Puerto Rico, Mayagëz
11. Bachmann W, Legler U (1972) Studies on the structure and function of the anterior section of the nose by means of luminal impressions. Acta Oto-Laryngologica 73:433–442
12. Ingels KJ, Meeuwsen F, van Strien HL, Graamans K, Huizing EH (1990) Ciliary beat frequency and the nasal cycle, European Archives of Otorhinolaryngology 248:123–126
13. Sappey MPC (1867–1874) Traité d'anatomie descriptive (Treatise of Descriptive Anatomy), Paris; and Sappey MPC (1879) Atlas d'Anatomie Descriptive (Atlas of Descriptive Anatomy), Paris (see also "The Larynx – Gray's Anatomy of the Human Body" Yahoo – Education. education.yahoo.com/reference/gray/subjects/subject/236)
14. Begis D, Delpuech C, Le Tallec P, Loth L, Thiriet M, Vidrascu M (1988) A finite-element model of tracheal collapse. Journal of Applied Physiology 64:1359–1368
15. Thiriet M (1994) Etude des écoulements dans les voies aériennes proximales et les artères de gros calibre. Mémoire pour l'habilitation à diriger des recherches, UFR de physique, Université Paris VII. (Study of flows in proximal airways and large arteries. Thesis for Accreditation to Supervise Research, Paris Diderot University)

16. Thiriet M, Maarek JM, Chartrand DA, Delpuech C, Davis L, Hatzfeld C, Chang HK (1989) Transverse images of the human thoracic trachea during forced expiration. Journal of Applied Physiology 67:1032–1040
17. Rhodin J, Dalhamn T (1956) Electron microscopy of the tracheal ciliated mucosa in rat. Zeitschrift für Zellforschung und mikroskopische Anatomie 44:345–412
18. Reinhardt JM, Ding K, Cao K, Christensen GE, Hoffman EA, Bodas SV (2008) Registration-based estimates of local lung tissue expansion compared to xenon CT measures of specific ventilation. Medical Image Analysis 12:752–763
19. Ukil S, Reinhardt JM (2009) Anatomy-guided lung lobe segmentation in X-ray CT images. IEEE Transactions on Medical Imaging 28:202–214
20. Miserocchi G (1997) Physiology and pathophysiology of pleural fluid turnover. European Respiratory Journal 10:219–225
21. Vasilescu DM, Gao Z, Saha PK, Yin L, Wang G, Haefeli-Bleuer B, Ochs M, Weibel ER, Hoffman EA (2012) Assessment of morphometry of pulmonary acini in mouse lungs by nondestructive imaging using multiscale microcomputed tomography. Proceedings of the National Academy of Sciences of the United States of America 109:17105–17110
22. Bucher U, Reid L (1961) Development of the intrasegmental bronchial tree: the pattern of branching and development of cartilage at various stages of intra-uterine life. Thorax 16:207–218
23. Sera T, Uesugi K, Yagi N (2005) Localized morphometric deformations of small airways and alveoli in intact mouse lungs under quasi-static inflation. Respiratory Physiology and Neurobiology 147:51–63
24. Weibel ER (1963) Morphometry of the human lung, Academic Press, New York
25. Choi J, Tawhai MH, Hoffman EA, Lin CL (2009) On intra- and intersubject variabilities of airflow in the human lungs. Physics of Fluids 21:101901
26. Lin CL, Tawhai MH, McLennan G, Hoffman EA (2007) Characteristics of the turbulent laryngeal jet and its effect on airflow in the human intra-thoracic airways. Respiratory Physiology and Neurobiology 157:295–309
27. Mandelbrot BB (1982) The Fractal Geometry of Nature. Henry Holt and Company (Macmillan), New York
28. Weibel ER, Gomez DM (1962) Architecture of the human lung. Use of quantitative methods establishes fundamental relations between size and number of lung structures. Science 137:577–585
29. Nelson TR, West BJ, Goldberger AL (1990) The fractal lung: universal and species-related scaling patterns. Experientia (Cellular and Molecular Life Sciences) 46:251–254
30. Mauroy B, Filoche M, Weibel ER, Sapoval B (2004) An optimal bronchial tree may be dangerous. Nature 427:633-636
31. West BJ, Bhargava V, Goldberger AL (1986) Beyond the principle of similitude: renormalization in the bronchial tree. Journal of Applied Physiology 60:1089–1097
32. Imre A (1999) Ideas in theoretical biology – Comment about the fractality of the lung. Acta Biotheoretica 47:79–81
33. Kitaoka H, Takaki R (1999) Fractal analysis of the human fetal lung development. Forma 14:205–212
34. Hou C, Gheorghiu S, Coppens MO, Huxley VH, Pfeifer P (2005) Gas diffusion through the fractal landscape of the lung: how deep does oxygen enter the alveolar system? (p. 17–30) In: Losa GA, Merlini D, Nonnenmacher TF, Weibel ER (Eds). Fractals in Biology and Medicine, Vol. IV, Birkhäuser, Basel
35. Hou C, Gheorghiu S, Huxley VH, Pfeifer P (2010) Reverse engineering of oxygen transport in the lung: adaptation to changing demands and resources through space-filling networks. PLoS Computational Biology 6:e1000902
36. Basset F, Poirier J, Le Crom M, Turiaf J (1971) Ultrastructural study of the human bronchiolar epithelium. Zeitschrift für Zellforschung und mikroskopische Anatomie 116:425–442
37. Clara M (1937) Zur Histologie des Bronchialepithels Zeitschrift für mikroskopisch-anatomische Forschung 41:321–347

38. ten Have-Opbroek AA, Otto-Verberne CJ, Dubbeldam JA, Dykman JH (1991) The proximal border of the human respiratory unit, as shown by scanning and transmission electron microscopy and light microscopical cytochemistry. Anatomical Record 229:339–354
39. Metzger RJ, Krasnow MA (1999) Genetic control of branching morphogenesis. Science 284:1635–1639
40. Metzger RJ, Klein OD, Martin GR, Krasnow MA (2008) The branching programme of mouse lung development. Nature 453:745–750
41. Hilfer SR (1996) Morphogenesis of the lung: control of embryonic and fetal branching. Annual Review of Physiology 58:93–113
42. Jeffery PK (1998) The development of large and small airways. American Journal of Respiratory and Critical Care Medicine 157:S174–S180
43. Hirashima T, Iwasa Y, Morishita Y (2009) Mechanisms for split localization of Fgf10 expression in early lung development. Developmental Dynamics 238:2813–2822
44. Thurlbeck WM (1975) Postnatal growth and development of the lung. American Review of Respiratory Diseases 111:803–844
45. Hislop A, Muir DCF, Jacobsen M, Simon G, Reid L (1972) Postnatal growth and function of the pre-acinar airways. Thorax 27:265–274
46. Hislop AA, Haworth SG (1989) Airway size and structure in the normal fetal and infant lung and the effect of premature delivery and artificial ventilation. American Review of Respiratory Diseases 140:1717–1726
47. Horsefield K, Cordon WI, Kemp W, Phillips S (1987) Growth of bronchial tree in man. Thorax 42:383–388
48. Masters JR (1976) Epithelial-mesenchymal interaction during lung development: the effect of mesenchymal mass. Developmental Biology 51:98–108
49. Mollard R, Dziadek M (1998) A correlation between epithelial proliferation rates, basement membrane component localization patterns, and morphogenetic potential in the embryonic mouse lung. American Journal of Respiratory Cell and Molecular Biology 19:71–82
50. Lu P, Werb Z (2008) Patterning mechanisms of branched organs. Science 322:1506–1509
51. Bellusci S, Grindley J, Emoto H, Itoh N, Hogan BL (1997) Fibroblast growth factor 10 (FGF10) and branching morphogenesis in the embryonic mouse lung. Development 124:4867–4878
52. Zhou S, Degan S, Potts EN, Foster WM, Sunday ME (2009) NPAS3 is a trachealess homolog critical for lung development and homeostasis. Proceedings of the National Academy of Sciences of the United States of America 106:11691–11696
53. Serra R, Pelton RW, Moses HL (1994) TGF β1 inhibits branching morphogenesis and N-myc expression in lung bud organ cultures. Development 120:2153–2161
54. Mahlapuu M, Enerbäck S, Carlsson P (2001) Haploinsufficiency of the forkhead gene Foxf1, a target for sonic hedgehog signaling, causes lung and foregut malformations. Development 128:2397–2406
55. Li Y, Zhang H, Choi SC, Litingtung Y, Chiang C (2004) Sonic hedgehog signaling regulates Gli3 processing, mesenchymal proliferation, and differentiation during mouse lung organogenesis. Developmental Biology 270:214–231
56. Motoyama J, Liu J, Mo R, Ding Q, Post M, Hui CC (1998) Essential function of Gli2 and Gli3 in the formation of lung, trachea and oesophagus. Nature – Genetics 20:54–57
57. Kicheva A, Cohen M, Briscoe J (2012) Developmental pattern formation: insights from physics and biology. Science 338:210–212
58. Mollard R, Dziadek M (1997) Homeobox genes from clusters A and B demonstrate characteristics of temporal colinearity and differential restrictions in spatial expression domains in the branching mouse lung. International Journal of Developmental Biology 41:655–666
59. Gjorevski N, Nelson CM (2010) The mechanics of development: models and methods for tissue morphogenesis. Birth Defects Research. Part C, Embryo Today 90:193–202
60. Schittny JC, Miserocchi G, Sparrow MP (2000) Spontaneous peristaltic airway contractions propel lung liquid through the bronchial tree of intact and fetal lung explants. American Journal of Respiratory Cell and Molecular Biology 23:11–18

61. Hislop A, Fairweather DV, Blackwell RJ, Howard S (1984) The effect of amniocentesis and drainage of amniotic fluid on lung development in Macaca fascicularis. British Journal of Obstetrics and Gynaecology 91:835–842
62. Perlman M, Williams J, Hirsch M (1976) Neonatal pulmonary hypoplasia after prolonged leakage of amniotic fluid. Archives of Disease in Childhood 51:349–353
63. Moore KA, Polte T, Huang S, Shi B, Alsberg E, Sunday ME, Ingber DE (2005) Control of basement membrane remodeling and epithelial branching morphogenesis in embryonic lung by Rho and cytoskeletal tension. Developmental Dynamics 232:268–281
64. Turing AM (1952) The chemical basis of morphogenesis. Philosophical Transactions of the Royal Society of London. Series B, Biological Sciences 237:37–72
65. Gierer A, Meinhardt H (1972) A theory of biological pattern formation. Kybernetik 12:30–39
66. Murray JD (2003) On the mechanochemical theory of biological pattern formation with application to vasculogenesis. Comptes Rendues Biologies 326:239–252
67. Elliott FM, Reid L (1965) Some new facts about the pulmonary artery and its branching pattern. Clinical Radiology 16:193–198
68. Pump KK (1972) Distribution of bronchial arteries in human lung. Chest 62:447–451
69. Hislop A, Reid L (1972) Intrapulmonary arterial development during fetal life: branching pattern and structure. Journal of Anatomy 113:35–48
70. Hislop A, Reid L (1973) Fetal and childhood development of the intrapulmonary veins in man: branching pattern and structure. Thorax 28:313–319
71. West JB (2008) Respiratory Physiology: The Essentials, Lippincott Williams and Wilkins, Baltimore, MD
72. Plank L, James J, Wagenvoort CA (1980) Caliber and elastin content of the pulmonary trunk. Archives of Pathology and Laboratory Medicine 104:238–241
73. Rhoades R, Bell DR (2009) Medical Physiology: Principles for Clinical Medicine, 3rd ed., Wolters Kluwer – Lippincott Williams and Wilkins, Philadelphia, Pennsylvania
74. Sant'Ambrogio G (1987) Nervous receptors of the tracheobronchial tree. Annual Review of Physiology 49:611–627
75. Kang S, Jang JH, Price MP, Gautam M, Benson CJ, Gong H, Welsh MJ, Brennan TJ (2012) Simultaneous disruption of mouse ASIC1a, ASIC2 and ASIC3 genes enhances cutaneous mechanosensitivity. PLoS ONE 7:e35225
76. Manzke T (2005) Expression and function of serotonin receptor isoforms in the respiratory system. PhD Thesis, Göttingen
77. Heistad DD, Abboud FM, Mark AL, Schmid PG (1974) Interaction of baroreceptor and chemoreceptor reflexes. Modulation of the chemoreceptor reflex by changes in baroreceptor activity. Journal of Clinical Investigation 53:1226–1236
78. Miura M, Reis DJ (1971) The paramedian reticular nucleus: a site of inhibitory interaction between projections from fastigial nucleus and carotid sinus nerve acting on blood pressure. Journal of Physiology 216:441–460
79. Miura M, Reis DJ (1972) The role of the solitary and paramedian reticular nuclei in mediating cardiovascular reflex responses from carotid baro- and chemoreceptors. Journal of Physiology 223:525–548
80. Van De Borne P, Mezzetti S, Montano N, Narkiewicz K, Degaute JP, Somers VK (2000) Hyperventilation alters arterial baroreflex control of heart rate and muscle sympathetic nerve activity. American Journal of Physiology – Heart and Circulatory Physiology 279:H536–H541
81. Somers VK, Mark AL, Zavala DC, Abboud FM (1989) Influence of ventilation and hypocapnia on sympathetic nerve responses to hypoxia in normal humans. Journal of Applied Physiology 67:2095–2100
82. Narkiewicz K, van de Borne P, Montano N, Hering D, Kara T, Somers VK (2006) Sympathetic neural outflow and chemoreflex sensitivity are related to spontaneous breathing rate in normal men. Hypertension 47:51–55
83. Schocken Roth (1977) Reduced β-adrenoceptor concentrations in ageing man. Nature 267:856–858

84. Barnes P, Jacobs M, Roberts JM (1984) Glucocorticoids preferentially increase fetal alveolar β-adrenoreceptors: autoradiographic evidence. Pediatric Research 18:1191–1194
85. Schell DN, Durham D, Murphree SS, Muntz KH, Shaul PW (1992) Ontogeny of β-adrenergic receptors in pulmonary arterial smooth muscle, bronchial smooth muscle and alveolar lining cells in the rat. American Journal of Respiratory Cell and Molecular Biology 7:317–324
86. Pulerà N, Bernard P, Carrara M, Bencini C, Pacifici GM (1988) Muscarinic cholinergic receptors in lung of developing rats. Developmental Pharmacology and Therapeutics 11:142–146
87. Sheppard MN, Marangos PJ, Bloom SR, Polak JM (1984) Neuron specific enolase: a marker for the early development of nerves and endocrine cells in the human lung. Life Sciences 34:264–271
88. Hislop AA, Wharton J, Allen KM, Polak JM, Haworth SG (1990) Immunohistochemical localization of peptide-containing nerves in the airways of normal young children. American Journal of Respiratory Cell and Molecular Biology 3:191–198
89. Sheppard MN, Polak JM, Allen JM, Bloom SR (1984) Neuropeptide tyrosine (NPY): a newly discovered peptide is present in the mammalian respiratory tract. Thorax 39:326–330
90. Abdel-Samad D, Perreault C, Ahmarani L, Avedanian L, Bkaily G, Magder S, D'Orléans-Juste P, Jacques D (2012) Differences in neuropeptide Y-induced secretion of endothelin-1 in left and right human endocardial endothelial cells. Neuropeptides pii: S0143-4179(12)00099-6
91. Allen KM, Wharton J, Polak JM, Haworth SG (1989) A study of nerves containing peptides in the pulmonary vasculature of healthy infants and children and of those with pulmonary hypertension. British Heart Journal 62:353–360
92. Sharma RK, Addis BJ, Jeffery PK (1995) The distribution and density of airway vasoactive intestinal polypeptide (VIP) binding sites in cystic fibrosis and asthma. Pulmonary Pharmacology 8:91–96
93. Klabunde RE (2004) Cardiovascular Physiology Concepts. Lippincott Williams and Wilkins, Philadelphia, Pennsylvania (cvphysiology.com)
94. Kraske S, Cunningham JT, Hajduczok G, Chapleau MW, Abboud FM, Wachtel RE (1998) Mechanosensitive ion channels in putative aortic baroreceptor neurons. American Journal of Physiology – Heart and Circulatory Physiology 275:1497–1501
95. Krauhs JM (1979) Structure of rat aortic baroreceptors and their relationship to connective tissue. Journal of Neurocytology 8:401–414

Chap. 3. Physiology of the Cardiovascular Apparatus

1. Bestel J, Clément F, Sorine M (2001) A biomechanical model of muscle contraction (p. 1159–1161). In Niessen WJ, Viergever MA (eds) Medical Image Computing and Computer-Assisted Intervention (MICCAI'01), Lecture Notes in Computer Science (LNCS), vol. 2208, Springer
2. Krejci P, Sainte-Marie J, Sorine M, Urquiza JM (2006) Solutions to muscle fiber equations and their long time behaviour. Nonlinear Analysis: Real World Applications 7:535558
3. Sainte-Marie J, Chapelle D, Cimrman R, Sorine M (2006) Modeling and estimation of the cardiac electromechanical activity. Computers and Structures 84:1743–1759
4. Fernández MA, Gerbeau JF, Grandmont C (2007) A projection semi-implicit scheme for the coupling of an elastic structure with an incompressible fluid. International Journal for Numerical Methods in Engineering 69:794821
5. Burman E, Fernández MA (2007) Stabilized explicit coupling for fluid–structure interaction using Nitsche's method. Comptes Rendus de l'Académie des sciences, Paris, Ser. I (Mathematics) 345:467–472

6. Prassl AJ, Kickinger F, Ahammer H, Grau V, Schneider JE, Hofer E, Vigmond EJ, Trayanova NA, Plank G (2009) Automatically generated, anatomically accurate meshes for cardiac electrophysiology problems. IEEE Transactions on Biomedical Engineering 56:1318–1330
7. Diniz dos Santos N, Gerbeau JF, Bourgat JF (2008) A partitioned fluid–structure algorithm for elastic thin valves with contact. Computer Methods in Applied Mechanics and Engineering 197:1750–1761
8. Deng W, Bukiya AN, Rodríguez-Menchaca AA, Zhang Z, Baumgarten CM, Logothetis DE, Levitan I, Rosenhouse-Dantsker A (2012) Hypercholesterolemia induces up-regulation of KACh cardiac currents via a mechanism independent of phosphatidylinositol 4,5-bisphosphate and $G\beta\gamma$. Journal of Biological Chemistry 287:4925–4935
9. Moireau P, Chapelle D, Le Tallec P (2008) Joint state and parameter estimation for distributed mechanical systems. Computer Methods in Applied Mechanics and Engineering, 197:659677
10. Chapelle D, Moireau P, Le Tallec P (2009) Robust filtering for joint state-parameter estimation in distributed mechanical systems. Discrete and Continuous Dynamical Systems, Series A 23:65–84
11. Poon CS, Merrill CK (1997) Decrease of cardiac chaos in congestive heart failure, Nature 389:492–495
12. Pironet A, Dauby PC, Paeme S, Kosta S, Chase JG, Desaive T (2013) Simulation of left atrial function using a multi-scale model of the cardiovascular system. PLoS One 8:e65146
13. Penney D (2003) Cardiac cycle, www.coheadquarters.com/PennLibr/MyPhysiology/
14. Al-Rubaiee M, Gangula PR, Millis RM, Walker RK, Umoh NA, Cousins VM, Jeffress MA, Haddad GE (2013) Inotropic and lusitropic effects of calcitonin gene-related peptide in the heart. American Journal of Physiology – Heart and Circulatory Physiology 304:H1525–H1537
15. Robinson TF, Factor SM, Sonnenblick EH (1986) The heart as a suction pump. Scientific American 6:62-69
16. Pagel PS, Kehl F, Gare M, Hettrick DA, Kersten JR, Warltier DC (2003) Mechanical function of the left atrium: new insights based on analysis of pressure-volume relations and Doppler echocardiography. Anesthesiology 98:975–994
17. Malik ZA, Kott KS, Poe AJ, Kuo T, Chen L, Ferrara KW, Knowlton AA (2013) Cardiac myocyte exosomes: stability, HSP60, and proteomics. American Journal of Physiology – Heart and Circulatory Physiology 304:H954–H965
18. Zhang P, Su J, Mende U (2012) Cross talk between cardiac myocytes and fibroblasts: from multiscale investigative approaches to mechanisms and functional consequences. American Journal of Physiology – Heart and Circulatory Physiology 303:H1385–H1396
19. Sipido KR, Cheng H (2013) T-tubules and ryanodine receptor microdomains: on the road to translation. Cardiovascular Research 98:159–161
20. Zhang H, Gomez AM, Wang X, Yan Y, Zheng M, Cheng H (2013) ROS regulation of microdomain Ca^{2+} signalling at the dyads. Cardiovascular Research 98:248–258
21. Kohl T, Lehnart SE (2013) Imaging T-tubules: dynamic membrane structures for deep functions. Cardiovascular Research 98:162–164
22. Guo A, Zhang C, Wei S, Chen B, Song LS (2013) Emerging mechanisms of T-tubule remodelling in heart failure. Cardiovascular Research 98:204–215
23. Shaw RM, Colecraft HM (2013) L-type calcium channel targeting and local signalling in cardiac myocytes. Cardiovascular Research 98:177–186
24. Scriven DR, Asghari P, Moore ED (2013) Microarchitecture of the dyad. Cardiovascular Research 98:169–176
25. Tanskanen AJ, Greenstein JL, Chen A, Sun SX, Winslow RL (2007) Protein geometry and placement in the cardiac dyad influence macroscopic properties of calcium-induced calcium release. Biophysical Journal 92:3379–3396
26. Maier SK, Westenbroek RE, Schenkman KA, Feigl EO, Scheuer T, Catterall WA (2002) An unexpected role for brain-type sodium channels in coupling of cell surface depolarization to contraction in the heart. Proceedings of the National Academy of Sciences of the United States of America 99:4073–4078

27. Zobel C, Cho HC, Nguyen TT, Pekhletski R, Diaz RJ, Wilson GJ, Backx PH (2003) Molecular dissection of the inward rectifier potassium current (IK1) in rabbit cardiomyocytes: evidence for heteromeric co-assembly of Kir2.1 and Kir2.2. Journal of Physiology 550:365–372
28. Liu GX, Derst C, Schlichthörl G, Heinen S, Seebohm G, Brüggemann A, Kummer W, Veh RW, Daut J, Preisig-Müller R (2001) Comparison of cloned Kir2 channels with native inward rectifier K^+ channels from guinea-pig cardiomyocytes. Journal of Physiology 532:115–126
29. Gorelik J, Wright PT, Lyon AR, Harding SE (2013) Spatial control of the Ŏ3b2AR system in heart failure: the transverse tubule and beyond. Cardiovascular Research 98:216–224
30. Wu CY, Jia Z, Wang W, Ballou LM, Jiang YP, Chen B, Mathias RT, Cohen IS, Song LS, Entcheva E, Lin RZ (2011) PI3Ks maintain the structural integrity of T-tubules in cardiac myocytes. PLoS One 6:e24404
31. Chopra N, Knollmann BC (2013) Triadin regulates cardiac muscle couplon structure and microdomain Ca^{2+} signalling: a path towards ventricular arrhythmias. Cardiovascular Research 98:187–191
32. Kohlhaas M, Maack C (2013) Calcium release microdomains and mitochondria. Cardiovascular Research 98: 259–268
33. Plovanich M, Bogorad RL, Sancak Y, Kamer KJ, Strittmatter L, Li AA, Girgis HS, Kuchimanchi S, De Groot J, Speciner L, Taneja N, Oshea J, Koteliansky V, Mootha VK (2013) MICU2, a paralog of MICU1, resides within the mitochondrial uniporter complex to regulate calcium handling. PLoS One 8:e55785
34. Mallilankaraman K, Cárdenas C, Doonan PJ, Chandramoorthy HC, Irrinki KM, Golenár T, Csordás G, Madireddi P, Yang J, Müller M, Miller R, Kolesar JE, Molgó J, Kaufman B, Hajnóczky G, Foskett JK, Madesh M (2012) MCUR1 is an essential component of mitochondrial Ca^{2+} uptake that regulates cellular metabolism. Nature – Cell Biology 14:1336–1343
35. Domenech RJ, Sánchez G, Donoso P, Parra V, Macho P (2003) Effect of tachycardia on myocardial sarcoplasmic reticulum and Ca^{2+} dynamics: a mechanism for preconditioning? Journal of Molecular and Cellular Cardiology 35:1429–1437
36. Sánchez G, Pedrozo Z, Domenech RJ, Hidalgo C, Donoso PJ (2005) Tachycardia increases NADPH oxidase activity and RyR2 S-glutathionylation in ventricular muscle. Journal of Molecular and Cellular Cardiology 39:982–991
37. Beigi F, Gonzalez DR, Minhas KM, Sun QA, Foster MW, Khan SA, Treuer AV, Dulce RA, Harrison RW, Saraiva RM, Premer C, Schulman IH, Stamler JS, Hare JM (2012) Dynamic denitrosylation via S-nitrosoglutathione reductase regulates cardiovascular function. Proceedings of the National Academy of Sciences of the United States of America 109:4314–4319
38. Lopaschuk GD, Ussher JR, Folmes CD, Jaswal JS, Stanley WC (2010) Myocardial fatty acid metabolism in health and disease. Physiological Reviews 90:207–258
39. Kahn BB, Alquier T, Carling D, Hardie DG (2005) AMP-activated protein kinase: ancient energy gauge provides clues to modern understanding of metabolism. Cell Metabolism 1: 15–25
40. Pawlikowska P, Orzechowski A (2007) Role of transmembrane GTPases in mitochondrial morphology and activity [Article in Polish]. Postepy Biochemii 53:53–59
41. Legros F, Lombès A, Frachon P, Rojo M (2002) Mitochondrial fusion in human cells is efficient, requires the inner membrane potential, and is mediated by mitofusins. Molecular Biology of the Cell 13:4343–4354
42. Shi Y, Dierckx A, Wanrooij PH, Wanrooij S, Larsson NG, Wilhelmsson LM, Falkenberg M, Gustafsson CM (2012) Mammalian transcription factor A is a core component of the mitochondrial transcription machinery. Proceedings of the National Academy of Sciences of the United States of America 109:16510–16515
43. Metodiev MD, Lesko N, Park CB, Cámara Y, Shi Y, Wibom R, Hultenby K, Gustafsson CM, Larsson NG (2009) Methylation of 12S rRNA is necessary for in vivo stability of the small subunit of the mammalian mitochondrial ribosome. Cell Metabolism 9:386–397

44. Spåhr H, Habermann B, Gustafsson CM, Larsson NG, Hallberg BM (2012) Structure of the human MTERF4-NSUN4 protein complex that regulates mitochondrial ribosome biogenesis. Proceedings of the National Academy of Sciences of the United States of America 109:15253–15258
45. Beck H, Flynn K, Lindenberg KS, Schwarz H, Bradke F, Di Giovanni S, Knöll B (2012) Serum Response Factor (SRF)-cofilin-actin signaling axis modulates mitochondrial dynamics. Proceedings of the National Academy of Sciences of the United States of America 109:E2523-E2532
46. Murphy MP (2012) Modulating mitochondrial intracellular location as a redox signal. Science Signaling 5:pe39
47. Carrer M, Liu N, Grueter CE, Williams AH, Frisard MI, Hulver MW, Bassel-Duby R, Olson EN (2012) Control of mitochondrial metabolism and systemic energy homeostasis by microRNAs 378 and 378. Proceedings of the National Academy of Sciences of the United States of America 109:15330–15335
48. Yu E, Mercer J, Bennett M (2012) Mitochondria in vascular disease. Cardiovascular Research 95:173–182
49. O'Rourke B (2007) Mitochondrial ion channels. Annual Review of Physiology 69:19–49
50. O'Rourke B, Cortassa S, Aon MA (2005) Mitochondrial ion channels: gatekeepers of life and death. Physiology 20:303–315
51. Zaobornyj T, Ghafourifar P (2012) Strategic localization of heart mitochondrial NOS: a review of the evidence. American Journal of Physiology – Heart and Circulatory Physiology 303:H1283–H1293
52. Virkki LV, Forster IC, Biber J, Murer H (2005) Substrate interactions in the human type IIa sodium-phosphate cotransporter (NaP_i-IIa). American Journal of Physiology – Renal Physiology 288:F969–F981
53. Aprille JR (2003) Mechanism and regulation of the mitochondrial ATP-Mg/P_i carrier. Journal of Bioenergetics and Biomembranes 25:473–481
54. Rich PR (2003) The molecular machinery of Keilin's respiratory chain. Biochemical Society Transactions 31:1095–1105
55. Covian R, Balaban RS (2012) Cardiac mitochondrial matrix and respiratory complex protein phosphorylation. American Journal of Physiology – Heart and Circulatory Physiology 303:H940–H966
56. Illingworth J (Faculty of Biological Sciences, University of Leeds) Bioenergetics, www.bmb.leeds.ac.uk/illingworth/oxphos/index.htm
57. Lu G, Sun H, Korge P, Koehler CM, Weiss JN, Wang Y (2009) Functional characterization of a mitochondrial Ser/Thr protein phosphatase in cell death regulation. Methods in Enzymology 457:255–273
58. Castro L, Demicheli V, Tórtora V, Radi R (2011) Mitochondrial protein tyrosine nitration. Free Radical Research 45:37–52
59. Wirstam M, Blomberg MRA, Siegbahn PEM (1999) Reaction mechanism of compound I formation in heme peroxidases: a density functional theory study. Journal of the American Chemical Society 121:10178–10185
60. Hurd TR, Costa NJ, Dahm CC, Beer SM, Brown SE, Filipovska A, Murphy MP (2005) Glutathionylation of mitochondrial proteins. Antioxidants and Redox Signaling 7:999–1010
61. Murphy E, Kohr M, Sun J, Nguyen T, Steenbergen C (2012) S-nitrosylation: a radical way to protect the heart. Journal of Molecular and Cellular Cardiology 52:568–577
62. Tarze A, Deniaud A, Le Bras M, Maillier E, Molle D, Larochette N, Zamzami N, Jan G, Kroemer G, Brenner C (2007) GAPDH, a novel regulator of the pro-apoptotic mitochondrial membrane permeabilization. Oncogene 26:2606–2620
63. Sirover MA (2005) New nuclear functions of the glycolytic protein, glyceraldehyde-3-phosphate dehydrogenase, in mammalian cells. Journal of Cellular Biochemistry 95:45–52
64. Jandu SK, Webb AK, Pak A, Sevinc B, Nyhan D, Belkin AM, Flavahan NA, Berkowitz DE, Santhanam L (2011) Nitric oxide regulates tissue transglutaminase localization and function in the vasculature. Amino Acids 0939-4451:1-9

65. Gundemir S, Johnson GVW (2009) Intracellular localization and conformational state of transglutaminase 2: implications for cell death. PLoS One 4:e6123
66. Kim Y, Park J, Kim S, Song S, Kwon SK, Lee SH, Kitada T, Kim JM, Chung J (2008) PINK1 controls mitochondrial localization of Parkin through direct phosphorylation. Biochemical and Biophysical Research Communications 377:975–980
67. Junn E, Jang WH, Zhao X, Jeong BS, Mouradian MM (2009) Mitochondrial localization of DJ-1 leads to enhanced neuroprotection. Journal of Neuroscience Research 87:123–129
68. Smirnova E, Griparic L, Shurland DL, van der Bliek AM (2001) Dynamin-related protein Drp1 is required for mitochondrial division in mammalian cells. Molecular Biology of the Cell 12:2245–2256
69. Schroeder MA, Ali MA, Hulikova A, Supuran CT, Clarke K, Vaughan-Jones RD, Tyler DJ, Swietach P (2013) Extramitochondrial domain rich in carbonic anhydrase activity improves myocardial energetics. Proceedings of the National Academy of Sciences of the United States of America 110:E958–E967
70. Des Rosiers C, Labarthe F, Lloyd SG, Chatham JC (2011) Cardiac anaplerosis in health and disease: food for thought. Cardiovascular Research 90:210–219
71. Cotter DG, Schugar RC, Crawford PA (2013) Ketone body metabolism and cardiovascular disease. American Journal of Physiology – Heart and Circulatory Physiology 304:H1060–H1076
72. Robinson AM, Williamson DH (1978) Utilization of D-3-hydroxy[3-14C]butyrate for lipogenesis in vivo in lactating rat mammary gland. Biochemical Journal 176:635–638
73. Wise A, Foord SM, Fraser NJ, Barnes AA, Elshourbagy N, Eilert M, Ignar DM, Murdock PR, Steplewski K, Green A, Brown AJ, Dowell SJ, Szekeres PG, Hassall DG, Marshall FH, Wilson S, Pike NB (2003) Molecular identification of high and low affinity receptors for nicotinic acid. Journal of Biological Chemistry 278:9869–9874
74. Tunaru S, Kero J, Schaub A, Wufka C, Blaukat A, Pfeffer K, Offermanns S (2003) PUMA-G and HM74 are receptors for nicotinic acid and mediate its anti-lipolytic effect. Nature – Medicine 9:352–355
75. Jeninga EH, Bugge A, Nielsen R, Kersten S, Hamers N, Dani C, Wabitsch M, Berger R, Stunnenberg HG, Mandrup S, Kalkhoven E (2009) Peroxisome proliferator-activated receptor γ regulates expression of the anti-lipolytic G-protein-coupled receptor 81 (GPR81/Gpr81). Journal of Biological Chemistry 284:26385–26393
76. Irukayama-Tomobe Y, Tanaka H, Yokomizo T, Hashidate-Yoshida T, Yanagisawa M, Sakurai T (2009) Aromatic D-amino acids act as chemoattractant factors for human leukocytes through a G protein-coupled receptor, GPR109B. Proceedings of the National Academy of Sciences of the United States of America 106:3930–3934
77. Yaniv Y, Spurgeon HA, Ziman BD, Lyashkov AE, Lakatta EG (2013) Mechanisms that match ATP supply to demand in cardiac pacemaker cells during high ATP demand. American Journal of Physiology – Heart and Circulatory Physiology 304:H1428–H1438
78. Traaseth N, Elfering S, Solien J, Haynes V, Giulivi C (2004) Role of calcium signaling in the activation of mitochondrial nitric oxide synthase and citric acid cycle. Biochimica et Biophysica Acta 1658:64–71
79. Takahashi E, Asano K (2002) Mitochondrial respiratory control can compensate for intracellular O_2 gradients in cardiomyocytes at low PO2. American Journal of Physiology – Heart and Circulatory Physiology 283:H871–H878
80. Timmer SAJ, Knaapen P (2013) Coronary microvascular function, myocardial metabolism, and energetics in hypertrophic cardiomyopathy: insights from positron emission tomography. European Heart Journal – Cardiovascular Imaging 14:95–101
81. Chauhan VS, Tuvia S, Buhusi M, Bennett V, Grant AO (2000) Abnormal cardiac Na^+ channel properties and QT heart rate adaptation in neonatal ankyrin$_B$ knockout mice. Circulation Research 86:441–447
82. Mohler PJ, Schott JJ, Gramolini AO, Dilly KW, Guatimosim S, duBell WH, Song LS, Haurogne K, Kyndt F, Ali ME, Rogers TB, Lederer WJ, Escande D, Le Marec H, Bennett V (2003) Ankyrin-B mutation causes type 4 long-QT cardiac arrhythmia and sudden cardiac death. Nature 421:634–639

83. Bhasin N, Cunha SR, Mudannayake M, Gigena MS, Rogers TB, Mohler PJ (2007) Molecular basis for PP2A regulatory subunit B56α targeting in cardiomyocytes. American Journal of Physiology – Heart and Circulatory Physiology 293:H109–H119
84. Marx SO, Kurokawa J, Reiken S, Motoike H, D'Armiento J, Marks AR, Kass RS (2002) Requirement of a macromolecular signaling complex for β adrenergic receptor modulation of the KCNQ1–KCNE1 potassium channel. Science 295:496–499
85. Wehrens XHT, Lehnart SE, Marks AR (2005) Intracellular calcium release channels and cardiac disease. Annual Review of Physiology 67:69–98
86. Roberts BN, Yang PC, Behrens SB, Moreno JD, Clancy CE (2012) Computational approaches to understand cardiac electrophysiology and arrhythmias. American Journal of Physiology – Heart and Circulatory Physiology 303:H766–H783
87. Ha CH, Kim JY, Zhao J, Wang W, Jhun BS, Wong C, Jin ZG (2010) PKA phosphorylates histone deacetylase 5 and prevents its nuclear export, leading to the inhibition of gene transcription and cardiomyocyte hypertrophy. Proceedings of the National Academy of Sciences of the United States of America 107:15467–15472
88. Hallaq H, Yang Z, Viswanathan PC, Fukuda K, Shen W, Wang DW, Wells KS, Zhou J, Yi J, Murray KT (2006) Quantitation of protein kinase A-mediated trafficking of cardiac sodium channels in living cells. Cardiovascular Research 72:250–261
89. Rook MB, Evers MM, Vos MA, Bierhuizen MF (2012) Biology of cardiac sodium channel $Na_V1.5$ expression. Cardiovascular Research 93:12–23
90. Nichols CB, Rossow CF, Navedo MF, Westenbroek RE, Catterall WA, Santana LF, McKnight GS (2010) Sympathetic stimulation of adult cardiomyocytes requires association of AKAP5 with a subpopulation of L-type calcium channels. Circulation Research 107:747–756
91. Taylor SS, Ilouz R, Zhang P, Kornev AP (2012) Assembly of allosteric macromolecular switches: lessons from PKA. Nature Reviews – Molecular Cell Biology 13:646–658
92. Aguiar CJ, Andrade VL, Gomes ER, Alves MN, Ladeira MS, Pinheiro AC, Gomes DA, Almeida AP, Goes AM, Resende RR, Guatimosim S, Leite MF (2009) Succinate modulates Ca^{2+} transient and cardiomyocyte viability through PKA-dependent pathway. Cell Calcium 47:37–46
93. Houser SR (2009) Ca^{2+} signaling domains responsible for cardiac hypertrophy and arrhythmias. Circulation Research 104:413–415
94. Chiang CS, Huang CH, Chieng H, Chang YT, Chang D, Chen JJ, Chen YC, Chen YH, Shin HS, Campbell KP, Chen CC (2009) The $Ca_V3.2$ T-type Ca^{2+} channel is required for pressure overload-induced cardiac hypertrophy in mice. Circulation Research 104:522–530
95. Terentyev D, Belevych AE, Terentyeva R, Martin MM, Malana GE, Kuhn DE, Abdellatif M, Feldman DS, Elton TS, Györke S (2009) miR-1 overexpression enhances Ca^{2+} release and promotes cardiac arrhythmogenesis by targeting PP2A regulatory subunit B56alpha and causing CaMKII-dependent hyperphosphorylation of RyR2. Circulation Research 104: 514–521
96. Carusi A, Burrage K, Rodríguez B (2012) Bridging experiments, models and simulations: an integrative approach to validation in computational cardiac electrophysiology. American Journal of Physiology – Heart and Circulatory Physiology 303:H144–H155
97. Quinn TA, Kohl P (2013) Combining wet and dry research: experience with model development for cardiac mechano-electric structure-function studies. Cardiovascular Research 97:601–611
98. Kraeutler MJ, Soltis AR, Saucerman JJ (2010) Modeling cardiac β-adrenergic signaling with normalized-Hill differential equations: comparison with a biochemical model. BMC Systems Biology 4:157
99. Grandi E, Puglisi JL, Wagner S, Maier LS, Severi S, Bers DM (2007) Simulation of Ca-calmodulin-dependent protein kinase II on rabbit ventricular myocyte ion currents and action potentials. Biophysical Journal 93:3835–3847
100. Hashambhoy YL, Greenstein JL, Winslow RL (2010) Role of CaMKII in RyR leak, EC coupling and action potential duration: a computational model. Journal of Molecular and Cellular Cardiology 49:617–624

101. Saucerman JJ, Zhang J, Martin JC, Peng LX, Stenbit AE, Tsien RY, McCulloch AD (2006) Systems analysis of PKA-mediated phosphorylation gradients in live cardiac myocytes. Proceedings of the National Academy of Sciences of the United States of America 103:12923–12928
102. Iancu RV, Jones SW, Harvey RD (2007) Compartmentation of cAMP signaling in cardiac myocytes: a computational study. Biophysical Journal 92:3317–3331
103. Himeno Y, Sarai N, Matsuoka S, Noma A (2008) Ionic mechanisms underlying the positive chronotropy induced by beta1-adrenergic stimulation in guinea pig sinoatrial node cells: A simulation study. Journal of Physiological Sciences 58:53–65
104. Soltis AR, Saucerman JJ (2010) Synergy between CaMKII substrates and β-adrenergic signaling in regulation of cardiac myocyte Ca^{2+} handling. Biophysical Journal 99:2038–2047
105. Hodgkin AL and Huxley AF (1952) A quantitative description of membrane current and its application to conduction and excitation in nerve. Journal of Physiology 117:500–544
106. ten Tusscher KH, Noble D, Noble PJ, Panfilov AV (2004) A model for human ventricular tissue. American Journal of Physiology – Heart and Circulatory Physiology 286:H1573–H1589
107. Bernus O, Wilders R, Zemlin CW, Verschelde H, Panfilov AV (2002) A computationally efficient electrophysiological model of human ventricular cells. American Journal of Physiology – Heart and Circulatory Physiology 282:H2296–H2308
108. Beeler GW, Reuter H (1977) Reconstruction of the action potential of ventricular myocardial fibres. Journal of Physiology 268:177-210
109. Mitchell CC, Schaeffer DG (2003) A two-current model for the dynamics of cardiac membrane. Bulletin of Mathematical Biology 65:767–793
110. Aliev RR, Panfilov AV (1996) A simple two-variable model of cardiac excitation. Chaos Solitons Fractals 7:293–301
111. Fenton F, Karma A (1998) Vortex dynamics in three-dimensional continuous myocardium with fiber rotation: Filament instability and fibrillation. Chaos 8:20-47
112. Luo CH, Rudy Y (1991) A model of the ventricular cardiac action-potential: depolarization, repolarization, and their interaction. Circulation Research 68:1501–1526
113. Djabella K, Sorine M. (2006) A reduced differential model for cardiac action potentials. SIAM Conference on the Life Sciences, Raleigh, USA
114. Tolkacheva EG, Schaeffer DG, Gauthier DJ, Mitchell CC (2002) Analysis of the Fenton-Karma model through an approximation by a one-dimensional map. Chaos 12:1034–1042
115. Boulakia M, Cazeau S, Fernández MA, Gerbeau JF, Zemzemi N (2010) Mathematical modeling of electrocardiograms: a numerical study. Annals of Biomedical Engineering 38:1071–1097
116. Jiao Q, Bai Y, Akaike T, Takeshima H, Ishikawa Y, Minamisawa S (2009) Sarcalumenin is essential for maintaining cardiac function during endurance exercise training. American Journal of Physiology – Heart and Circulatory Physiology 297:H576–H582
117. Guyton AC, Hall JE (2006) Textbook of medical physiology (7th Edition) Elsevier – Saunders, Philadelphia, Pennsylvania
118. Milnor WR (1982) Haemodynamics. Williams and Wilkins, Baltimore, MD
119. Silbernagl S, Despopoulos A (2001) Atlas de poche de physiologie [Pocket Atlas of Physiology]. Flammarion, Paris
120. Davies JI, Struthers AD (2003) Pulse wave analysis and pulse wave velocity: a critical review of their strengths and weaknesses. Journal of Hypertension 21:463–472
121. Meaney, E, Alva F, Moguel R, Meaney A, Alva J, Webel R (2000) Formula and nomogram for the sphygmomanometric calculation of the mean arterial pressure. Heart 84:64
122. Learoyd BM, Taylor MG (1996) Alterations with age in the viscoelastic properties of human arterial walls. Circulation Research 18:278–292
123. Mills CJ, Gabe IT, Gault JH, Mason DT, Ross J Jr, Braunwald E, Shillingford JP (1970) Pressure-flow relationships and vascular impedance in man. Cardiovascular Research 4: 405–417

124. Anliker M et al (1977) Non-invasive measurement of blood flow, In: Hwang NHC, Normann NA (eds) Cardiovascular flow dynamics and measurements. University Park Press, Baltimore
125. Bergfeld GR, Forrester T (1992) Release of ATP from human erythrocytes in response to a brief period of hypoxia and hypercapnia. Cardiovascular Research 26:40–47
126. McCullough WT, Collins DM, Ellsworth ML (1997) Arteriolar responses to extracellular ATP in striated muscle. American Journal of Physiology – Heart and Circulatory Physiology 272:H1886–H1891
127. Arciero JC, Carlson BE, Secomb TW (2008) Theoretical model of metabolic blood flow regulation: roles of ATP release by red blood cells and conducted responses. American Journal of Physiology – Heart and Circulatory Physiology 295:H1562–H1571
128. Kalmanson D, Veyrat C (1978) Clinical aspects of venous return: a velocimetric approach to a new system dynamics concept. In: Baan J, Noordegraaf A, Raines J (eds) Cardiovascular System Dynamics. MIT Press, Cambridge
129. Moreno AH (1978) Dynamics of pressure in the central veins. In: Baan J, Noordegraaf A, Raines J (eds) Cardiovascular System Dynamics. MIT Press, Cambridge
130. Hoffman JI, Spaan JA (1990) Pressure-flow relations in coronary circulation. Physiological Reviews 70:331–390
131. Shukla P, Sun C, O'Rourke ST (2012) Melatonin inhibits nitric oxide signaling by increasing PDE5 phosphorylation in coronary arteries. American Journal of Physiology – Heart and Circulatory Physiology 303:H1418–H1425
132. Scaramucci J (1695) De motu cordis, theorema sexton. Theoremata familiaria viros eruditos consulentia de variis physico medicis lucubrationibus iucta leges mecanicas. Urbino, Italy: Apud Joannem Baptistam Bustum, 70–81
133. Lenègre J, Blondeau M; Bourdarias JP, Gerbaux A, Himbert J, Maurice P (1973) Cœur et Circulation [Heart and Circulation]. In Vallery-Radot P, Hamburger J, Lhermitte F (eds) Pathologie Médicale. [Medical Pathology] (Vol.3), Flammarion Médecine Sciences, Paris
134. Mori H, Tanaka E, Hyodo K, Mohammed MU, Sekka T, Ito K, Shinozaki Y, Tanaka A, Nakazawa H, Abe S, Handa S, Kubota M, Tanioka K, Umetani K, Ando M (1999) Synchrotron microangiography reveals configurational changes and to-and-fro flow in intramyocardial vessels. American Journal of Physiology. Heart Circulation Physiology 276:H429–H437
135. Carlson BE, Arciero JC, Secomb TW (2008) Theoretical model of blood flow autoregulation: roles of myogenic, shear-dependent, and metabolic responses. American Journal of Physiology – Heart and Circulatory Physiology 295:H1572–H1579
136. Olsson RA (1981) Local factors regulating cardiac and skeletal muscle blood flow. Annual Review of Physiology 43:385–395
137. Momen A, Mascarenhas V, Gahremanpour A, Gao Z, Moradkhan R, Kunselman A, Boehmer JP, Sinoway LI, Leuenberger UA (2009) Coronary blood flow responses to physiological stress in humans. American Journal of Physiology – Heart and Circulatory Physiology 296:H854–H861
138. Caesar K, Offenhauser N, Lauritzen M (2008) Gamma-aminobutyric acid modulates local brain oxygen consumption and blood flow in rat cerebellar cortex. Journal of Cerebral Blood Flow and Metabolism 28:906–915
139. Stefanovic B, Hutchinson E, Yakovleva V, Schram V, Russell JT, Belluscio L, Koretsky AP, Silva AC (2008) Functional reactivity of cerebral capillaries. Journal of Cerebral Blood Flow and Metabolism 28:961–972
140. Sirotin YB, Das A (2009) Anticipatory haemodynamic signals in sensory cortex not predicted by local neuronal activity. Nature 457:475–479
141. van Beek AHEA, Claassen JAHR, Rikkert MGMO, Jansen RWMM (2008) Cerebral autoregulation: an overview of current concepts and methodology with special focus on the elderly. Journal of Cerebral Blood Flow and Metabolism 28:1071–1085

142. Tzeng YC, Ainslie PN, Cooke WH, Peebles KC, Willie CK, Macrae BA, Smirl JD, Horsman HM, Rickards CA (2012) Assessment of cerebral autoregulation: the quandary of quantification. American Journal of Physiology – Heart and Circulatory Physiology 303:H658–H671
143. Fonck E, Feigl GG, Fasel J, Sage D, Unser M, Rüfenacht DA, Stergiopulos N (2009) Effect of aging on elastin functionality in human cerebral arteries. Stroke 40:2552–2556
144. West JB (1974) Respiratory Physiology. Williams and Wilkins, Baltimore, MD
145. Wagner WW, Latham LP (1975) Pulmonary capillary recruitment during airway hypoxia in the dog. Journal of Applied Physiology 39:900–905
146. Hanson WL, Emhardt JD, Bartek JP, Latham LP, Checkley LL, Capen RL, Wagner WW (1989) Site of recruitment in the pulmonary microcirculation. Journal of Applied Physiology 66:2079–2083
147. Dawson CA, Rickaby DA, Linehan JH (1986) Location and mechanisms of pulmonary vascular volume changes. Journal of Applied Physiology 60:402–409
148. Barman SA, Taylor AE (1990) Effect of pulmonary venous pressure elevation on vascular resistance and compliance. American Journal of Physiology – Heart and Circulatory Physiology 258:H1164–H1170
149. Ryan JW, Ryan US, Schultz DR, Whitaker C, Chung A (1975) Subcellular localization of pulmonary antiotensin-converting enzyme (kininase II). Biochemical Journal 146:497–499
150. Curry FRE, Adamson RH (2010) Vascular permeability modulation at the cell, microvessel, or whole organ level: towards closing gaps in our knowledge. Cardiovascular Research 87:218–229
151. Perktold K, Prosi M, Zunino P (2009) Mathematical models of mass transfer in the vascular walls (Chap. 7). In Formaggia L, Quarteroni A, Veneziani A (eds.) Cardiovascular Mathematics: Modeling and Simulation of the Circulatory System, Springer, Milano
152. Shen Q, Rigor RR, Pivetti CD, Wu MH, Yuan SY (2010) Myosin light chain kinase in microvascular endothelial barrier function. Cardiovascular Research 87:272–280
153. Curry FRE, Noll T (2010) Spotlight on microvascular permeability. Cardiovascular Research 87:195–197
154. VanTeeffelen JWGE, Brands J, Vink H (2010) Agonist-induced impairment of glycocalyx exclusion properties: contribution to coronary effects of adenosine. Cardiovascular Research 87:311–319
155. Spindler V, Schlegell N, Waschkel J (2010) Role of GTPases in control of microvascular permeability. Cardiovascular Research 87:243–253
156. Durán WN, Breslin JW, Sánchez FA (2010) The NO cascade, eNOS location, and microvascular permeability. Cardiovascular Research 87:254–261
157. Bates DO (2010) Vascular endothelial growth factors and vascular permeability. Cardiovascular Research 87:262–271
158. Zhou X, He P (2010) Endothelial [Ca2+]i and caveolin-1 antagonistically regulate eNOS activity and microvessel permeability in rat venules. Cardiovascular Research 87:340–347
159. Rabiet M-J, Plantier J-L, Rival Y, Genoux Y, Lampugnani MG, Dejana E (1996) Thrombin-induced increase in endothelial permeability is associated with changes in cell-to-cell junction organization. Arteriosclerosis, Thrombosis, and Vascular Biology 16:488–496
160. Sun C, Wu MH, Guo M, Day ML, Lee ES, Yuan SY (2010) ADAM15 regulates endothelial permeability and neutrophil migration via Src/ERK1/2 signalling. Cardiovascular Research 87:348–355
161. He P (2010) Leucocyte/endothelium interactions and microvessel permeability: coupled or uncoupled? Cardiovascular Research 87:281–290
162. Ngok SP, Geyer R, Liu M, Kourtidis A, Agrawal S, Wu C, Seerapu HR, Lewis-Tuffin LJ, Moodie KL, Huveldt D, Marx R, Baraban JM, Storz P, Horowitz A, Anastasiadis PZ (2012) VEGF and Angiopoietin-1 exert opposing effects on cell junctions by regulating the Rho GEF Syx. Journal of Cell Biology 199:1103–1115
163. Michel CC, Curry FE (1999) Microvascular permeability. Physiological Reviews 79:703–761

164. Weinbaum S, Curry FE (1995) Modelling the structural pathways for transcapillary exchange. Symposia of the Society for Experimental Biology 49:323–345
165. Agre P, Brown D, Nielsen S (1995) Aquaporin water channels: unanswered questions and unresolved controversies. Current Opinion in Cell Biology 7:472–483
166. Tarbell JM, Demaio L, Zaw MM (1999) Effect of pressure on hydraulic conductivity of endothelial monolayers: role of endothelial cleft shear stress. Journal of Applied Physiology 87:261–268
167. Chen SC, Liu KM, Wagner RC (1998) Three-dimensional analysis of vacuoles and surface invaginations of capillary endothelia in the eel rete mirabile. Anatomical Record 252:546–553
168. Tarbell JM (2010) Shear stress and the endothelial transport barrier. Cardiovascular Research 87:320–330
169. Reed RK, Rubin K (2010) Transcapillary exchange: role and importance of the interstitial fluid pressure and the extracellular matrix. Cardiovascular Research 87:211–217
170. Levick JR, Michel CC (2010) Microvascular fluid exchange and the revised Starling principle. Cardiovascular Research 87:198–210
171. Minami Y, Kasukawa T, Kakazu Y, Iigo M, Sugimoto M, Ikeda S, Yasui A, van der Horst GT, Soga T, Ueda HR (2009) Measurement of internal body time by blood metabolomics. Proceedings of the National Academy of Sciences of the United States of America 106:9890–9895
172. Curtis AM, Cheng Y, Kapoor S, Reilly D, Price TS, FitzGerald GA (2007) Circadian variation of blood pressure and the vascular response to asynchronous stress. Proceedings of the National Academy of Sciences of the United States of America 104:3450–3455
173. Fuller PM, Lu J, Saper CB (2008) Differential rescue of light- and food-entrainable circadian rhythms. Science 320:1074–1077
174. Davidson AJ, London B, Block GD, Menaker M (2005) Cardiovascular tissues contain independent circadian clocks. Clinical and Experimental Hypertension 27:307–311
175. McNamara P, Seo SP, Rudic RD, Sehgal A, Chakravarti D, FitzGerald GA (2001) Regulation of CLOCK and MOP4 by nuclear hormone receptors in the vasculature: a humoral mechanism to reset a peripheral clock. Cell 105:877–889
176. Damiola F, Le Minh N, Preitner N, Kornmann B, Fleury-Olela F, Schibler U (2000) Restricted feeding uncouples circadian oscillators in peripheral tissues from the central pacemaker in the suprachiasmatic nucleus. Genes and Development 14:2950–2961
177. Balsalobre A, Brown SA, Marcacci L, Tronche F, Kellendonk C, Reichardt HM, Schutz G, Schibler U (2000) Resetting of circadian time in peripheral tissues by glucocorticoid signaling. Science 289:2344–2347
178. Lemmer B (1992) Cardiovascular chronobiology and chronopharmacology. In: Touitou Y, Haus E (eds) Biologic Rhythms in Clinical and Laboratory Medicine, 418–427. Springer-Verlag, Berlin.
179. Bray MS, Young ME (2008) Diurnal variations in myocardial metabolism. Cardiovascular Research 79:228–237
180. Martino TA, Oudit GY, Herzenberg AM, Tata N, Koletar MM, Kabir GM, Belsham DD, Backx PH, Ralph MR, Sole MJ (2008) Circadian rhythm disorganization produces profound cardiovascular and renal disease in hamsters. American Journal of Physiology – Regulatory, Integrative and Comparative Physiology 294:R1675–1683
181. Ivanov PC, Hu K, Hilton MF, Shea SA, Stanley HE (2007) Endogenous circadian rhythm in human motor activity uncoupled from circadian influences on cardiac dynamics. Proceedings of the National Academy of Sciences of the United States of America 104:20702–20707
182. Pennes HH (1948) Analysis of tissue and arterial blood temperatures in the resting human forearm. Journal of Applied Physiology 1:93–122
183. Valvano JW, Bioheat transfer. users.ece.utexas.edu/∼valvano/research/jwv.pdf
184. Werner J, Brinck H (2001) A three-dimensional vascular model and its application to the determination of the spatial variations in the arterial, venous, and tissue temperature distribution, In: Leondes C (ed) Biofluid Methods in Vascular and Pulmonary Systems. CRC Press, Boca Raton, FL

185. Wissler EH (1998) Pennes' 1948 paper revisited. Journal of Applied Physiology 85:35–41
186. Arkin H, Xu LX, Holmes KR (1994) Recent developments in modeling heat transfer in blood perfused tissues. IEEE Transactions on Biomedical Engineering 41:97–107
187. Tedgui A, Lévy B (1994) Biologie de la paroi artérielle [Biology of the Arterial Wall]. Masson, Paris
188. Dongaonkar RM, Nguyen TL, Quick CM, Hardy J, Laine GA, Wilson E, Stewart RH (2013) Adaptation of mesenteric lymphatic vessels to prolonged changes in transmural pressure. American Journal of Physiology – Heart and Circulatory Physiology 305:H203–H210
189. Bayliss W (1902) On the local reactions of the arterial wall to changes of internal pressure. Journal of Physiology 28:220–231
190. Davis MJ (2012) Perspective: physiological role(s) of the vascular myogenic response. Microcirculation 19:99-114
191. Lidington D, Schubert R, Bolz SS (2013) Capitalizing on diversity: an integrative approach towards the multiplicity of cellular mechanisms underlying myogenic responsiveness. Cardiovascular Research 97:404–412
192. Storch U, Schnitzler MM, Gudermann T (2012) G protein-mediated stretch reception. American Journal of Physiology – Heart and Circulatory Physiology 302:H1241–H1249
193. Pluznick JL, Protzko RJ, Gevorgyan H, Peterlin Z, Sipos A, Han J, Brunet I, Wan LX, Rey F, Wang T, Firestein SJ, Yanagisawa M, Gordon JI, Eichmann A, Peti-Peterdi J, Caplan MJ (2013) Olfactory receptor responding to gut microbiota-derived signals plays a role in renin secretion and blood pressure regulation. Proceedings of the National Academy of Sciences of the United States of America 110:4410–4415
194. Inoue R, Jian Z, Kawarabayashi Y (2009) Mechanosensitive TRP channels in cardiovascular pathophysiology. Pharmacology and Therapeutics 123:371–385
195. Yin J, Kuebler WM (2010) Mechanotransduction by TRP channels: general concepts and specific role in the vasculature. Cell Biochemistry and Biophysics 56:1–18
196. Baumgarten CM (2007) Origin of Mechanotransduction: Stretch-Activated Ion Channels (Chap. 2) In Weckstrom M, Tavi P (eds.) Cardiac Mechanotransduction. Landes Bioscience, Austin, Texas, and Springer, New York
197. Coste B, Xiao B, Santos JS, Syeda R, Grandl J, Spencer KS, Kim SE, Schmidt M, Mathur J, Dubin AE, Montal M, Patapoutian A (2012) Piezo proteins are pore-forming subunits of mechanically activated channels. Nature 483:176–181
198. Hill MA, Trippe KM, Li QX, Meininger GA (1992) Arteriolar arcades and pressure distribution in cremaster muscle microcirculation. Microvascular Research 44:117–124
199. Chlopicki S, Nilsson H, Mulvany MJ (2001) Initial and sustained phases of myogenic response of rat mesenteric small arteries. American Journal of Physiology – Heart and Circulatory Physiology 281:H2176–H2183
200. Zou H, Ratz PH, Hill MA (2000) Temporal aspects of Ca^{2+} and myosin phosphorylation during myogenic and norepinephrine-induced arteriolar constriction. Journal of Vascular Research 37:556–567
201. Bolz SS, Vogel L, Sollinger D, Derwand R, Boer C, Pitson SM, Spiegel S, and Pohl U (2003) Sphingosine kinase modulates microvascular tone and myogenic responses through activation of RhoA/Rho kinase. Circulation 108:342–347
202. Yeon DS, Kim JS, Ahn DS, Kwon SC, Kang BS, Morgan KG, Lee YH (2002) Role of protein kinase C- or RhoA-induced Ca^{2+} sensitization in stretch-induced myogenic tone. Cardiovascular Research 53:431–438
203. Frisbee JC, Roman RJ, Krishna UM, Falck JR, Lombard JH (2001) 20-HETE modulates myogenic response of skeletal muscle resistance arteries from hypertensive Dahl-SS rats. American Journal of Physiology – Heart and Circulatory Physiology 280:H1066–H1074
204. Obara K, Koide M, Nakayama K (2002) 20-Hydroxyeicosatetraenoic acid potentiates stretch-induced contraction of canine basilar artery via PKC alpha-mediated inhibition of KCa channel. British Journal of Pharmacology 137:1362–1370
205. Martinez-Lemus LA, Wu X, Wilson E, Hill MA, Davis GE, Davis MJ, Meininger GA (2003) Integrins as unique receptors for vascular control. Journal of Vascular Research 40:211–233

206. Sun Z, Martinez-Lemus LA, Trache A, Trzeciakowski JP, Davis GE, Pohl U, Meininger GA (2005) Mechanical properties of the interaction between fibronectin and $\alpha_5\beta_1$-integrin on vascular smooth muscle cells studied using atomic force microscopy. American Journal of Physiology – Heart and Circulatory Physiology 289:H2526–H2535
207. Martinez-Lemus LA, Crow T, Davis MJ, Meininger GA (2005) $\alpha_V\beta_3$- and $\alpha_5\beta_1$-integrin blockade inhibits myogenic constriction of skeletal muscle resistance arterioles. American Journal of Physiology – Heart and Circulatory Physiology 289:H322–H329
208. Hong Z, Sun Z, Li Z, Mesquitta WT, Trzeciakowski JP, Meininger GA (2012) Coordination of fibronectin adhesion with contraction and relaxation in microvascular smooth muscle. Cardiovascular Research 96:73–80
209. Schmid-Schönbein GW (2012) The integrin–cortex complex under control of GPCRs. Cardiovascular Research 96:7–8
210. Nelson MT, Cheng H, Rubart M, Santana LF, Bonev AD, Knot HJ, Lederer WJ (1995) Relaxation of arterial smooth muscle by calcium sparks. Science 270:633–637
211. Bagher P, Beleznai T, Kansui Y, Mitchell R, Garland CJ, Dora KA (2012) Low intravascular pressure activates endothelial cell TRPV4 channels, local Ca^{2+} events, and IKCa channels, reducing arteriolar tone. Proceedings of the National Academy of Sciences of the United States of America 109:18174–18179
212. Harder DR, Roman RJ, Gebremedhin D, Birks EK, Lange AR (1998) A common pathway for regulation of nutritive blood flow to the brain: arterial muscle membrane potential and cytochrome P450 metabolites. Acta Physiologica Scandinavica 164:527–532
213. Fischell TA, Bausback KN, McDonald TV (1990) Evidence for altered epicardial coronary artery autoregulation as a cause of distal coronary vasoconstriction after successful percutaneous transluminal coronary angioplasty. Journal of Clinical Investigation 86:575–584
214. Davis MJ, Hill MA (1999) Signaling mechanisms underlying the vascular myogenic response. Physiological Reviews 79:387–423
215. Mederos y Schnitzler M, Storch U, Meibers S, Nurwakagari P, Breit A, Essin K, Gollasch M, Gudermann T (2008) Gq-coupled receptors as mechanosensors mediating myogenic vasoconstriction. EMBO Journal 27:3092–3103
216. Zou Y, Akazawa H, Qin Y, Sano M, Takano H, Minamino T, Makita N, Iwanaga K, Zhu W, Kudoh S, Toko H, Tamura K, Kihara M, Nagai T, Fukamizu A, Umemura S, Iiri T, Fujita T, Komuro I (2004) Mechanical stress activates angiotensin II type 1 receptor without the involvement of angiotensin II. Nature – Cell Biology 6:499–506
217. Hofmann T, Obukhov AG, Schaefer M, Harteneck C, Gudermann T, Schultz G (1999) Direct activation of human TRPC6 and TRPC3 channels by diacylglycerol. Nature 397:259–263
218. Nilius B, Owsianik G, Voets T, Peters JA (2007) Transient receptor potential cation channels in disease. Physiological Reviews 87:165–217
219. Sriram K, Salazar Vázquez BY, Tsai AG, Cabrales P, Intaglietta M, Tartakovsky DM (2012) Autoregulation and mechanotransduction control the arteriolar response to small changes in hematocrit. American Journal of Physiology – Heart and Circulatory Physiology 303:H1096–H1106
220. Chu C, Thai K, Park KW, Wang P, Makwana O, Lovett DH, Simpson PC, Baker AJ (2013) Intraventricular and interventricular cellular heterogeneity of inotropic responses to $\alpha 1$-adrenergic stimulation. American Journal of Physiology – Heart and Circulatory Physiology 304:H946–H953
221. Guyenet PG (2006) The sympathetic control of blood pressure. Nature Reviews – Neuroscience 7:335–346
222. Pyke KE, Poitras V, Tschakovsky ME (2008) Brachial artery flow-mediated dilation during handgrip exercise: evidence for endothelial transduction of the mean shear stimulus. American Journal of Physiology – Heart and Circulatory Physiology 294:H2669–H2679
223. Green DJ, Bilsborough W, Naylor LH, Reed C, Wright J, O'Driscoll G, Walsh JH (2005) Comparison of forearm blood flow responses to incremental handgrip and cycle ergometer exercise: relative contribution of nitric oxide. Journal of Physiology 562:617–628

224. Guo ZL, Tjen-A-Looi SC, Fu LW, Longhurst JC (2009) Nitric oxide in rostral ventrolateral medulla regulates cardiac-sympathetic reflexes: role of synthase isoforms. American Journal of Physiology – Heart and Circulatory Physiology 297:H1478–H1486
225. Izumi H, Karita K (1994) The parasympathetic vasodilator fibers in the trigeminal portion of the distal lingual nerve in the cat tongue. American Journal of Physiology 266:R1517–R1522
226. Oakley AE, Clifton DK, Steiner RA (2009) Kisspeptin signaling in the brain. Endocrine Reviews 30:713–743
227. Rometo AM, Rance NE (2008) Changes in prodynorphin gene expression and neuronal morphology in the hypothalamus of postmenopausal women. Journal of Neuroendocrinology 20:1376–1381
228. Mittelman-Smith MA, Williams H, Krajewski-Hall SJ, McMullen NT, Rance NE (2012) Role for kisspeptin/neurokinin B/dynorphin (KNDy) neurons in cutaneous vasodilatation and the estrogen modulation of body temperature. Proceedings of the National Academy of Sciences of the United States of America 109:19846–19851
229. Ieda M, Kanazawa H, Kimura K, Hattori F, Ieda Y, Taniguchi M, Lee JK, Matsumura K, Tomita Y, Miyoshi S, Shimoda K, Makino S, Sano M, Kodama I, Ogawa S, Fukuda K (2007) Sema3a maintains normal heart rhythm through sympathetic innervation patterning. Nature – Medicine 13:604–612
230. Fu LW, Guo ZL, Longhurst JC (2012) Ionotropic glutamate receptors in the external lateral parabrachial nucleus participate in processing cardiac sympathoexcitatory reflexes. American Journal of Physiology – Heart and Circulatory Physiology 302:H1444–H1453
231. Koba S, Gao Z, Xing J, Sinoway LI, Li J (2006) Sympathetic responses to exercise in myocardial infarction rats: a role of central command. American Journal of Physiology – Heart and Circulatory Physiology 291, H2735-H2742
232. Sin PY, Galletly DC, Tzeng YC (2010) Influence of breathing frequency on the pattern of respiratory sinus arrhythmia and blood pressure: old questions revisited. American Journal of Physiology – Heart and Circulatory Physiology 298:H1588–H1599
233. Blain G, Meste O, Blain A, Bermon S (2009) Time-frequency analysis of heart rate variability reveals cardiolocomotor coupling during dynamic cycling exercise in humans. American Journal of Physiology-Heart and Circulatory Physiology 296:H1651-H1659
234. Aletti F, Bassani T, Lucini D, Pagani M, Baselli G (2009) Multivariate decomposition of arterial blood pressure variability for the assessment of arterial control of circulation. IEEE Transactions on Biomedical Engineering 56:1781–1790
235. Kamiya A, Kawada T, Mizuno M, Shimizu S, Sugimachi M (2010) Parallel resetting of arterial baroreflex control of renal and cardiac sympathetic nerve activities during upright tilt in rabbits. American Journal of Physiology – Heart and Circulatory Physiology 298:H1966–H1975
236. Tigerstedt R, Bergman P (1898) Niere und Kreislauf [Kidney and circulation]. Skandinavisches Archiv für Physiologie 8:223–271
237. James PF, Grupp IL, Grupp G, Woo AL, Askew GR, Croyle ML, Walsh RA, Lingrel JB (1999) Identification of a specific role for the Na,K-ATPase $\alpha 2$ isoform as a regulator of calcium in the heart. Molecular Cell 3:555–563
238. Blaustein MP, Leenen FH, Chen L, Golovina VA, Hamlyn JM, Pallone TL, Van Huysse JW, Zhang J, Wier WG (2012) How NaCl raises blood pressure: a new paradigm for the pathogenesis of salt-dependent hypertension. American Journal of Physiology – Heart and Circulatory Physiology 302:H1031–H1049
239. Gao J, Wymore RS, Wang Y, Gaudette GR, Krukenkamp IB, Cohen IS, Mathias RT (2002) Isoform-specific stimulation of cardiac Na/K pumps by nanomolar concentrations of glycosides. Journal of General Physiology 119:297–312
240. Crowley SD, Gurley SB, Herrera MJ, Ruiz P, Griffiths R, Kumar AP, Kim HS, Smithies O, Le TH, Coffman TM (2006) Angiotensin II causes hypertension and cardiac hypertrophy through its receptors in the kidney. Proceedings of the National Academy of Sciences of the United States of America 103:17985–17990

241. Machnik A, Neuhofer W, Jantsch J, Dahlmann A, Tammela T, Machura K, Park JK, Beck FX, Müller DN, Derer W, Goss J, Ziomber A, Dietsch P, Wagner H, van Rooijen N, Kurtz A, Hilgers KF, Alitalo K, Eckardt KU, Luft FC, Kerjaschki D, Titze J (2009) Macrophages regulate salt-dependent volume and blood pressure by a vascular endothelial growth factor-C-dependent buffering mechanism. Nature – Medicine 15:545–552
242. Smith OA, Astley CA, Spelman FA, Golanov EV, Bowden DM, Chesney MA, Chalyan V (2000) Cardiovascular responses in anticipation of changes in posture and locomotion. Brain Research Bulletin 53:69–76
243. Taylor JA, Eckberg DL (1996) Fundamental relations between short-term RR interval and arterial pressure oscillations in humans. Circulation 93:1527–1532
244. Zhang R, Khoo MSC, Wu Y, Yang Y, Grueter CE, Ni G, Price EE, Thiel W, Guatimosim S, Song LS, Madu EC, Shah AN, Vishnivetskaya TA, Atkinson JB, Gurevich VV, Salama G, Lederer WJ, Colbran RJ, Anderson ME (2005) Calmodulin kinase II inhibition protects against structural heart disease. Nature – Medicine 11:409–417
245. Goyal RK (1989) Muscarinic receptor subtypes: physiology and clinical implications. New England Journal of Medicine 321:1022–1029
246. Lymperopoulos A, Rengo G, Funakoshi H, Eckhart AE, Koch WJ (2007) Adrenal GRK2 upregulation mediates sympathetic overdrive in heart failure. Nature – Medicine 13:315–323
247. Monti A, Médigue C, Sorine M (2002) Short-term modelling of the controlled cardiovascular system. ESAIM: Proceedings 12:115–128
248. Hammer PE, Saul JP (2005) Resonance in a mathematical model of baroreflex control: arterial blood pressure waves accompanying postural stress. American Journal of Physiology – Regulatory, Integrative and Comparative Physiology 288:R1637–R1648.
249. Degtyarenko AM, Kaufman MP (2005) MLR-induced inhibition of barosensory cells in the NTS. American Journal of Physiology – Heart and Circulatory Physiology 289:H2575–H2584
250. Ulrich-Lail YM, Herman JP (2009) Neural regulation of endocrine and autonomic stress responses. Nature Reviews – Neuroscience 10:397–409
251. Joëls M, Baram TZ (2009) The neuro-symphony of stress. Nature Reviews – Neuroscience 10:459–466
252. Watts SW, Morrison SF, Davis RP, Barman SM (2012) Serotonin and blood pressure regulation. Pharmacological Reviews 64:359–388
253. Ramage AG, Villalón CM (2008) 5-hydroxytryptamine and cardiovascular regulation. Trends in Pharmacological Sciences 29:472–481
254. Kurbel S, Dodig K, Radic R (2002) The osmotic gradient in kidney medulla: a retold story. Advances in Physiology Education 26:278–281
255. Lombes M, Farman N, Oblin ME, Baulieu EE, Bonvalet JP, Erlanger BF, Gasc J (1990) Immunohistochemical localization of renal mineralocorticoid receptor by using an anti-idiotypic antibody that is an internal image of aldosterone. Proceedings of the National Academy of Sciences of the United States of America 87:1086–1088
256. Sasano H, Fukushima K, Sasaki I, Matsuno S, Nagura H, Krozowski ZS (1992) Immunolocalization of mineralocorticoid receptor in human kidney, pancreas, salivary, mammary and sweat glands: a light and electron microscopic immunohistochemical study. Journal of Endocrinology 132:305–310
257. Cantin M, Genest J (1986) Le cœur est une glande endocrine [The heart is an endocrine gland]. Pour la Science 102:43–49
258. Kondoh G, Tojo H, Nakatani N, N Komazawa, Murata C, Yamagata K, Maeda Y, Kinoshita T, Okabe M, Taguchi R, Takeda J (2005) Angiotensin-converting enzyme is a GPI-anchored protein releasing factor crucial for fertilization. Nature – Medicine 11:160–166
259. Lalioti MD, Zhang J, Volkman HM, Kahle KT, Hoffmann KE, Toka HR, Nelson-Williams C, Ellison DE, Flavell R, Booth CJ, Lu Y, Geller DS, Lifton RP (2006) Wnk4 controls blood pressure and potassium homeostasis via regulation of mass and activity of the distal convoluted tubule. Nature – Genetics 38:1124–1132

260. Coffman TM (2006) A WNK in the kidney controls blood pressure. Nature – Genetics 38:1105–1106
261. Jentsch TJ, Hübner CA, Fuhrmann JC (2004) Ion channels: Function unravelled by dysfunction. Nature – Cell Biology 6:1039–1047 (2004)
262. Oberleithner H, Riethmüller C, Schillers H, MacGregor GH, de Wardener HE, Hausberg M (2007) Plasma sodium stiffens vascular endothelium and reduces nitric oxide release. Proceedings of the National Academy of Sciences of the United States of America 104:16281–16286
263. de Kloet AD, Krause EG, Scott KA, Foster MT, Herman JP, Sakai RR, Seeley RJ, Woods SC (2011) Central angiotensin II has catabolic action at white and brown adipose tissue. American Journal of Physiology – Endocrinology and Metabolism 301:E1081–E1091
264. Yoshida T, Semprun-Prieto L, Wainford RD, Sukhanov S, Kapusta DR, Delafontaine P (2012) Angiotensin II reduces food intake by altering orexigenic neuropeptide expression in the mouse hypothalamus. Endocrinology 153:1411–1420
265. Grobe JL, Grobe CL, Beltz TG, Westphal SG, Morgan DA, Xu D, de Lange WJ, Li H, Sakai K, Thedens DR, Cassis LA, Rahmouni K, Mark AL, Johnson AK, Sigmund CD (2010) The brain renin-angiotensin system controls divergent efferent mechanisms to regulate fluid and energy balance. Cell Metabolism 12:431–442
266. Hilzendeger AM, Morgan DA, Brooks L, Dellsperger D, Liu X, Grobe JL, Rahmouni K, Sigmund CD, Mark AL (2012) A brain leptin–renin–angiotensin system interaction in the regulation of sympathetic nerve activity. American Journal of Physiology – Heart and Circulatory Physiology 303:H197–H206
267. Bełtowski J (2012) Leptin and the regulation of endothelial function in physiological and pathological conditions. Clinical and Experimental Pharmacology and Physiology 39: 168–178
268. Benkhoff S, Loot AE, Pierson I, Sturza A, Kohlstedt K, Fleming I, Shimokawa H, Grisk O, Brandes RP, Schröder K (2012) Leptin potentiates endothelium-dependent relaxation by inducing endothelial expression of neuronal NO synthase. Arteriosclerosis, Thrombosis, and Vascular Biology 32:1605–1612
269. Abboud FM, Floras JS, Aylward PE, Guo GB, Gupta BN, Schmid PG. Role of vasopressin in cardiovascular and blood pressure regulation. Blood Vessels 27:106–115
270. Koshimizu TA, Nasa Y, Tanoue A, Oikawa R, Kawahara Y, Kiyono Y, Adachi T, Tanaka T, Kuwaki T, Mori T, Takeo S, Okamura H, Tsujimoto G (2006) V1a vasopressin receptors maintain normal blood pressure by regulating circulating blood volume and baroreflex sensitivity. Proceedings of the National Academy of Sciences of the United States of America 103:7807–7812
271. Chassin C, Hornef MW, Bens M, Lotz M, Goujon JM, Vimont S, Arlet G, Hertig A, Rondeau E, Vandewalle A (2007) Hormonal control of the renal immune response and antibacterial host defense by arginine vasopressin. Journal of Experimental Medicine 204:2837–2852
272. Knepper MA, Star RA (2008) Vasopressin: friend or foe? Nature – Medicine 14:14–16
273. de Bold AJ (1985) Atrial natriuretic factor: a hormone produced by the heart. Science 230:767–770
274. Tsujita Y, Muraski J, Shiraishi I, Kato T, Kajstura J, Anversa P, Sussman MA (2006) Nuclear targeting of Akt antagonizes aspects of cardiomyocyte hypertrophy. Proceedings of the National Academy of Sciences of the United States of America 103:11946-11951
275. Suga SI, Itoh H, Komatsu Y, Ishida H, Igaki T, Yamashita J, Doi K, Chun TH, Yoshimasa T, Tanaka I, Nakao K (1998) Regulation of endothelial production of C-type natriuretic peptide by interaction between endothelial cells and macrophages. Endocrinology 139:1920–1926
276. Vesely BA, Eichelbaum EJ, Alli AA, Sun Y, Gower WR, Vesely DL (2006) Urodilatin and four cardiac hormones decrease human renal carcinoma cell numbers. European Journal of Clinical Investigation 36:810–819
277. Callaghan B, Hunne B, Hirayama H, Sartor DM, Nguyen TV, Abogadie FC, Ferens D, McIntyre P, Ban K, Baell J, Furness JB, Brock JA (2012) Sites of action of ghrelin receptor ligands in cardiovascular control. American Journal of Physiology – Heart and Circulatory Physiology 303:H1011–H1021

278. Hökfelt T, Lundberg JM, Schultzberg M, Johansson O, Skirboll L, Anggård A, Fredholm B, Hamberger B, Pernow B, Rehfeld J, Goldstein M (1980) Cellular localization of peptides in neural structures. Proceedings of the Royal Society of London. Series B, Biological Sciences 210:63–77
279. Sartor DM, Verberne AJ (2008) Abdominal vagal signalling: a novel role for cholecystokinin in circulatory control? Brain Research Reviews 59:140–154
280. Mahapatra NR (2008) Catestatin is a novel endogenous peptide that regulates cardiac function and blood pressure. Cardiovascular Research 80:330–338

Chap. 4. Physiology of the Ventilation

1. Bernard C (1966) Introduction à l'étude de médecine expérimentale [Introduction to the Study of Experimental Medicine]. Garnier – Flammarion, Paris
2. Dejours P (1981) Principles of Comparative Respiratory Physiology, Elsevier and North-Holland Biomedical Press, Amsterdam and New York
3. Atkins PW, de Paula J (2002) Physical Chemistry (7th Edition). Freeman WH, New York
4. Thiriet M, Douguet D, Bonnet JC, Canonne C, Hatzfeld C (1979) Influence du mélange He–O_2 sur la mixique dans les bronchopneumopathies obstructives chroniques [Influence of a He–O_2 mixture on gas mixing in chronic obstructive lung diseases]. Bulletin Européen de Physiopathologie Respiratoire 15:1053–1068
5. Cantor CR, Schimmel PR (1980) Biophysical Chemistry, Part 2: Techniques for the Study of Biological Structure and Function. Freeman WH, New York
6. Anonymous authors (2000-2013) Diffusion Time Calculator (www.physiologyweb.com)
7. Keener JP (2013) Mathematical Biology (www.math.utah.edu/ keener/classes/math5120)
8. Mircea D, Panaitescu M (2013) Green Pack Online: Environmental Components – Air composition. The Regional Environmental Center for Central and Eastern Europe www.greenpackonline.org/english/environmental-components.php?id=01-01
9. Holt PG, Strickland DH, Wikström ME, Jahnse FL (2008) Regulation of immunological homeostasis in the respiratory tract. Nature Reviews – Immunology 8:142-152
10. Salathe M (2007) Regulation of mammalian ciliary beating. Annual Review of Physiology 69:401–422
11. Ingber DE (2006) Cellular mechanotransduction: putting all the pieces together again. FASEB Journal 20:811–827
12. Wright JL, Thurlbeck WM (2005) Quantitative Anatomy of the Lung. In: Churg AM, Myers JL, Tazelaar HD, Wright JL (eds), Thurlbeck's Pathology of the Lung, third edition, Thieme, New York
13. Hart MC, Orzalesi MM, Cook CD (1963) Relation between anatomic respiratory dead space and body size and lung volume. Journal of Applied Physiology 18:519–522
14. Rafferty GF, Gardner WN (1996) Control of the respiratory cycle in conscious humans. Journal of Applied Physiology 81:1744–1753
15. Rafferty GF, Gardner WN (1995) Interaction between expiratory time and inspiration in conscious humans. Biological Psychology 41:96–97
16. Israël-Asselain R, Pocidalo JJ (1971) Respiration et maladies respiratoires. [Respiration and respiratory diseases]. In Vallery-Radot P, Hamburger J, Lhermitte F (eds.) Pathologie Médicale. [Medical Pathology] (Vol.2), Flammarion Médecine Sciences, Paris
17. Fenn WO, Rahn H, and Otis AB (1946) A theoretical study of the composition of the alveolar air at altitude. American Journal of Physiology 146:637-653
18. Riley RL, Cournand A (1949) "Ideal" alveolar air and the analysis of ventilation-perfusion relationships in the lungs. Journal of Applied Physiology 1:825–847

19. West JB, Dollery CT (1960) Distribution of blood flow and ventilation-perfusion ratio in the lung, measured with radioactive carbon dioxide. Journal of Applied Physiology 15:405–410
20. West JB (1970) Ventilation/Blood Flow and Gas Exchange. Blackwell, Oxford
21. Bates DV, Christie RV (1964) Respiratory Function in Disease, WB Saunders, Philadelphia
22. Brudin LH, Rhodes CG, Valind SO, Jones T, Hughes JM (1994) Interrelationships between regional blood flow, blood volume, and ventilation in supine humans. Journal of Applied Physiology 76:1205–1210
23. Wasserman DH, Whipp BJ (1983) Coupling of ventilation to pulmonary gas exchange during nonsteady-state work in men. Journal of Applied Physiology 54:587-593
24. Haouzi P (2006) Theories on the nature of the coupling between ventilation and gas exchange during exercise. Respiratory Physiology and Neurobiology 151:267–279
25. Altman PL, Dittmer DS [eds.] (1971) Biological Handbooks: Respiration and Circulation, Federation of American Societies for Experimental Biology, Bethesda, MD
26. Driehuys B, Cofer GP, Pollaro J, Mackel JB, Hedlund LW, Johnson GA (2006) Imaging alveolar-capillary gas transfer using hyperpolarized 129Xe MRI. Proceedings of the National Academy of Sciences of the United States of America 103:18278–18283
27. Kruhoffer P (1954) Lung diffusion coefficient for CO in normal human subjects by means of C14O. Acta Physiollogica Scandinavica 32:106–123.
28. Roughton FJW, Forster RE (1957) Relative importance of diffusion and chemical reaction in determining rate of exchange of gases in the human lung. Journal of Applied Physiology 11:290–302
29. Tartullier M, Ritz B, Ferrini M (1982) Physiologie de la Circulation Pulmonaire [Pulmonary Circulation Physiology]. In Denolin H (ed.) Physio-Pathologie Cardio-Pulmonaire [Cardiopulmonary Pathophysiology], SIMEP, Villeurbanne, France
30. Hartridge H, Roughton FJW (1923) Measurement of the rates of oxidation and reduction of haemoglobin. Nature 111:325–326
31. Hartridge H, Roughton FJW (1923) Method of measuring the velocity of very rapid chemical reactions. Proceedings of the Royal Society London Series A 104:376–394
32. Millikan GAA (1933) A simple photoelectric calorimeter. Journal of Physiology, London 79:158–179
33. Even P (1983) La respiration (p. 1087–1358). In Physiologie humaine, Meyer P (ed), Flammarion Médecine–Sciences, Paris
34. Yuan G, Peng YJ, Reddy VD, Makarenko VV, Nanduri J, Khan SA, Garcia JA, Kumar GK, Semenza GL, Prabhakar NR (2013) Mutual antagonism between hypoxia-inducible factors 1α and 2α regulates oxygen sensing and cardio-respiratory homeostasis. Proceedings of the National Academy of Sciences of the United States of America 110:E1788–E1796
35. Henderson LJ (1908) Concerning the relationship between the strength of acids and their capacity to preserve neutrality. American Journal of Physiology 21:173–179
36. Paton JFR, Abdala APL, Koizumi H, Smith JC, St-John WM (2006) Respiratory rhythm generation during gasping depends on persistent sodium current. Nature – Neuroscience 9:311–313
37. Feldman JL, Del Negro CA (2006) Looking for inspiration: new perspectives on respiratory rhythm. Nature Reviews – Neuroscience 7:232–241
38. Feldman JL, Kam K, Janczewski WA (2009) Practice makes perfect, even for breathing. Nature – Neuroscience 12:961–963
39. Tan W, Janczewski WA, Yang P, Shao XS, Callaway EM, Feldman JL (2008) Silencing preBötzinger complex somatostatin-expressing neurons induces persistent apnea in awake rat. Nature – Neuroscience 11:538–540
40. Bouvier J, Thoby-Brisson M, Renier N, Dubreuil V, Ericson J, Champagnat J, Pierani A, Chédotal A, Fortin G (2010) Hindbrain interneurons and axon guidance signaling critical for breathing. Nature – Neuroscience 13:1066–1074
41. Thoby-Brisson M, Karlén M, Wu N, Charnay P, Champagnat J, Fortin G (2009) Genetic identification of an embryonic parafacial oscillator coupling to the preBötzinger complex. Nature – Neuroscience 12:1028–1035

42. Gibson GG, Skett P (2001) Introduction to Drug Metabolism, Nelson Thornes Publishers, Cheltenham, UK
43. Capron M, Capron A, Goetzl EJ, Austen KF (1981) Tetrapeptides of the eosinophil chemotactic factor of anaphylaxis (ECF-A) enhance eosinophil Fc receptor. Nature 289:71–73
44. Sirois P, Borgeat P (1982) Leukotrienes: a new approach to the biochemistry of hypersensitivity. Survey of Immunologic Research l:279–285
45. Morris HR, Taylor GW, Jones CM, Scully N, Piper PJ, Tippins JR, Samhoun MN (1981) Structure elucidation and biosynthesis of slow reacting substances and slow reacting substance of anaphylaxis from guinea pig and human lungs. Progress in Lipid Research 20:719–725
46. Piper PJ (1978) Slow reacting substance of anaphylaxis. Annals of the Royal College of Surgeons of England 60:201–204
47. Gao X, Vockley CM, Pauli F, Newberry KM, Xue Y, Randell SH, Reddy TE, Hogan BL (2013) Evidence for multiple roles for grainyheadlike 2 in the establishment and maintenance of human mucociliary airway epithelium. Proceedings of the National Academy of Sciences of the United States of America 110:9356–9361
48. Guttinger M, Sutti F, Panigada M, Porcellini S, Merati B, Mariani M, Teesalu T, Consalez GG, Grassi F (1998) Epithelial V-like antigen (EVA), a novel member of the immunoglobulin superfamily, expressed in embryonic epithelia with a potential role as homotypic adhesion molecule in thymus histogenesis. Journal of Cell Biology 141:1061–1071
49. Wojcik E, Carrithers LM, Carrithers MD (2011) Characterization of epithelial V-like antigen in human choroid plexus epithelial cells: potential role in CNS immune surveillance. Neuroscience Letters 495:115–120
50. Yang CF, Hwu WL, Yang LC, Chung WH, Chien YH, Hung CF, Chen HC, Tsai PJ, Fann CS, Liao F, Chen YT (2008) A promoter sequence variant of ZNF750 is linked with familial psoriasis. Journal of Investigative Dermatology 128:1662–1668
51. Chung C, Kim T, Kim M, Kim M, Song H, Kim TS, Seo E, Lee SH, Kim H, Kim SK, Yoo G, Lee DH, Hwang DS, Kinashi T, Kim JM, Lim DS (2013) Hippo-Foxa2 signaling pathway plays a role in peripheral lung maturation and surfactant homeostasis. Proceedings of the National Academy of Sciences of the United States of America 110:7732–7737

Chap. 5. Medical Images and Physiological Signals

1. Bachelard G (1934) Le nouvel esprit scientifique [The New Scientific Spirit]. Presses Universitaires de France, Paris
2. Cebral JR, Löhner R (2001) From medical images to anatomically accurate finite element grids. International Journal for Numerical Methods in Engineering 51:985–1008
3. Thiriet M, Brugières P, Bittoun J, Gaston A (2001) Computational flow models in cerebral congenital aneurisms I: Steady flow. Revue Mécanique et Industries 2:107–118
4. Salmon S, Thiriet M, Gerbeau J-F (2003) Medical image-based computational model of pulsatile flow in saccular aneurisms. Mathematical Modelling and Numerical Analysis 37:663–679
5. Milner JS, Moore JA, Rutt BK, Steinman DA (1998) Hemodynamics of human carotid artery bifurcations: computational studies with models reconstructed from magnetic resonance imaging of normal subjects. Journal of Vascular Surgery 1998:143–156
6. Moore JA, Rutt BK, Karlik SJ, Yin K, Ethier CR (1999) Computational blood flow modeling based on in vivo measurements. Annals of Biomedical Engineering 27:627–640
7. Ladak HM, Milner JS, Steinman, DA (2000) Rapid 3D segmentation of the carotid bifurcation from serial MR images. Journal of Biomechanical Engineering 122:96–99
8. Papaharilaou Y, Doorly DJ, Sherwin SJ, Peiró J, Griffith C, Chesire N, Zervas V, Anderson J, Sanghera B, Watkins N, Caro CG (2002) Combined MRI and computational fluid dynamics detailed investigation of flow in idealised and realistic arterial bypass graft models. Biorheology 39:525–532

9. Gill JD, Ladak HM, Steinman DA, Fenster A (2000) Accuracy and variability assessment of a semiautomatic technique for segmentation of the carotid arteries from 3D ultrasound images. Medical Physics 27:1333–1342
10. de Feyter PJ (2012) CT functional imaging using intracoronary gradient analysis: an indispensable boost for CT coronary angiography. European Heart Journal – Cardiovasc Imaging 13:971–972
11. Choi JH, Koo BK, Yoon YE, Min JK, Song YB, Hahn JY, Choi SH, Gwon HC, Choe YH (2012) Diagnostic performance of intracoronary gradient-based methods by coronary computed tomography angiography for the evaluation of physiologically significant coronary artery stenoses: a validation study with fractional flow reserve. European Heart Journal – Cardiovasc Imaging 13:1001–1007
12. Thiriet M, Maarek JM, Chartrand DA, Delpuech C, Davis L, Hatzfeld C, Chang HK (1989) Transverse images of the human thoracic trachea during forced expiration. Journal of Applied Physiology 67:1032–1040
13. Bachelard G (1940) La philosophie du non: essai d'une philosophie du nouvel esprit scientifique [The Philosophy of No: a Philosophy of the New Scientific Mind]. Presses Universitaires de France, Paris
14. Bittoun J (1998) Basic principles of magnetic resonance imaging. In: Cerdan S, Haase A, Terrier F (eds.) Spectroscopy and Clinical MRI. Springer, New York
15. Singer JR (1959) Blood flow rates by nuclear magnetic resonance measurements. Science 130:1652–1653
16. McCready VR, Leach M, Ell PJ (1987) Functional Studies Using NMR. Springer, New York
17. Zerhouni EA, Parish DM, Rogers WJ, Yang A, Shapiro EP (1988) Human heart: tagging with MR imaging – a method for noninvasive assessment of myocardial motion. Radiology 169:59–63
18. Axel L, Dougherty L (1989) Heart wall motion: improved method of spatial modulation of magnetization for MR imaging. Radiology, 172:349–350
19. Mosher TJ, Smith MB (1990) A DANTE tagging sequence for the evaluation of translational sample motion. Magnetic Resonance in Medicine 15:334–339
20. Fischer SE, McKinnon GC, Maier SE, Boesiger P (1993) Improved myocardial tagging contrast. Magnetic Resonance in Medicine 30:191–200
21. McVeigh ER (1996) MRI of myocardial function: motion tracking techniques. Magnetic Resonance Imaging 14:137-150
22. Basser PJ, Mattiello J, LeBihan D (1994) MR diffusion tensor spectroscopy and imaging. Biophysical Journal 66:259–267
23. Hsu EW, Muzikant AL, Matulevicius SA, Penland RC, Henriquez CS (1998) Magnetic resonance myocardial fiber-orientation mapping with direct histological correlation. American Journal of Physiology – Heart and Circulatory Physiology 274:H1627–H1634
24. Scollan DF, Holmes A, Winslow R, Forder J (1998) Histological validation of reconstructed myocardial microstructure obtained from diffusion tensor magnetic resonance imaging. American Journal of Physiology – Heart and Circulatory Physiology 275 44:H2308–H2318
25. Helm PA, Tseng HJ, Younes L, McVeigh ER, Winslow RL (2005) Ex vivo 3D diffusion tensor imaging and quantification of cardiac laminar structure. Magnetic Resonance in Medicine 54:850-859
26. Winslow RL, Scollan DF, Holmes A, Yung CK, Zhang J, Jafri MS (2000) Electrophysiological modeling of cardiac ventricular function: from cell to organ. Annual Review of Biomedical Engineering 2:119–155
27. Papademetris X, Sinusas AJ, Dione DP, Constable RT, Duncan JS (2002) Estimation of 3-D left ventricular deformation from medical images using biomechanical models. IEEE Transactions on Medical Imaging 21:786–800
28. Sinusas AJ, Papademetris X, Constable RT, Dione DP, Slade MD, Shi P, Duncan JS (2001) Quantification of 3-D regional myocardial deformation: shape-based analysis of magnetic resonance images. American Journal of Physiology – Heart and Circulatory Physiology 281:698-714

29. Declerck J, Ayache N, McVeigh ER (1999) Use of a 4D planispheric transformation for the tracking and the analysis of LV motion with tagged MR images. In: Chen C-T, Clough AV (eds.) Medical Imaging 1999: Physiology and Function from Multidimensional Images. SPIE, Bellingham
30. Kozerke S, Scheidegger MB, Pedersen EM, Boesiger P (1999) Heart motion adapted cine phase-contrast flow measurements through the aortic valve. Magnetic Resonance in Medicine 42:970–978
31. Gleich B, Weizenecker J (2005) Tomographic imaging using the nonlinear response of magnetic particles. Nature 435:1214–1217
32. Ratnayaka K, Faranesh AZ, Hansen MS, Stine AM, Halabi M, Barbash IM, Schenke WH, Wright VJ, Grant LP, Kellman P, Kocaturk O, Lederman RJ (2013) Real-time MRI-guided right heart catheterization in adults using passive catheters. European Heart Journal 34: 380–389
33. Zimmerman JE, Theine P, Harding JT (1970) Design and operation of stable rf-biased superconducting point-contact quantum devices, and a note on the properties of perfectly clean metal contacts. Journal of Applied Physics 41:1572–1580
34. Cohen D, Edelsack EA, Zimmerman JE (1970) Magnetocardiograms taken inside a shielded room with a superconducting point-contact magnetometer. Applied Physics Letters 16: 278–280
35. Groeger S, Bison G, Knowles PE, Wynands R, Weis A (2006) Laser-pumped cesium magnetometers for high-resolution medical and fundamental research. Sensors and Actuators – A:Physical 129:1–5
36. Koch H (2004) Recent advances in magnetocardiography. Journal of Electrocardiology 37:117-122
37. Valsangiacomo Buechel ER, Mertens LL (2012) Imaging the right heart: the use of integrated multimodality imaging. European Heart Journal 33:949–960
38. Schaar JA, de Korte CL, Mastik F, Baldewsing R, Regar E, de Feyter P, Slager CJ, van der Steen AF, Serruys PW (2003) Intravascular palpography for high-risk vulnerable plaque assessment. Herz 28:488–495
39. Kanai H, Hasegawa H, Chubachi N, Koiwa Y, Tanaka M (1997) Noninvasive evaluation of local myocardial thickening and its color-coded imaging. IEEE Transactions on Ultrasonics, Ferroelectrics, and Frequency Control 44:752–768
40. Argyle RA, Ray SG (2009) Stress and strain: double trouble or useful tool? European Journal of Echocardiography 10:716–722
41. Leong-Poi H (2009) Molecular imaging using contrast-enhanced ultrasound: evaluation of angiogenesis and cell therapy. Cardiovascular Research 84:190–200
42. Todaro MC, Choudhuri I, Belohlavek M, Jahangir A, Carerj S, Oreto L, Khandheria BK (2012) New echocardiographic techniques for evaluation of left atrial mechanics. European Heart Journal – Cardiovascular Imaging 13:973–984
43. Ammar KA, Umland MM, Kramer C, Sulemanjee N, Jan MF, Khandheria BK, Seward JB, Paterick TE (2012) The ABCs of left ventricular assist device echocardiography: a systematic approach. European Heart Journal – Cardiovascular Imaging 13:885–899
44. Gutiérrez-Chicol JL, Alegría-Barrero E, Teijeiro-Mestre R, Chan PH, Tsujioka H, de Silva R, Viceconte N, Lindsay A, Patterson T, Foin N, Akasaka T, di Mario C (2012) Optical coherence tomography: from research to practice. European Heart Journal – Cardiovascular Imaging 13:370–384
45. Prati F, Guagliumi G, Mintz GS, Costa M, Regar E, Akasaka T, Barlis P, Tearney GJ, Jang IK, Arbustini E, Bezerra HG, Ozaki Y, Bruining N, Dudek D, Radu M, Erglis A, Motreff P, Alfonso F, Toutouzas K, Gonzalo N, Tamburino C, Adriaenssens T, Pinto F, Serruys PW, Di Mario C; for the Expert's OCT Review Document (2012) Expert review document part 2: methodology, terminology and clinical applications of optical coherence tomography for the assessment of interventional procedures. European Heart Journal 33:2513–2520.

46. Laughner JI, Zhang S, Li H, Shao CC, Efimov IR (2012) Mapping cardiac surface mechanics with structured light imaging. American Journal of Physiology – Heart and Circulatory Physiology 303:H712–H720
47. Thiberville L, Salaün M, Lachkar S, Dominique S, Moreno-Swirc S, Vever-Bizet C, Bourg-Heckly G (2009) Human in vivo fluorescence microimaging of the alveolar ducts and sacs during bronchoscopy. European Respiratory Journal 33:974–985
48. Moore JA, Steinman DA, Ethier CR (1998) Computational blood flow modelling: errors associated with reconstructing finite element models from magnetic resonance images. Journal of Biomechanics 31:179–184
49. Boissonnat J-D (1988) Shape reconstruction from planar cross-sections. Computer Vision, Graphics, and Image Processing 44:1–29
50. Boissonnat J-D, Chaine R, Frey P, Malandain G, Salmon S, Saltel E, Thiriet M (2005) From arteriographies to computational flow in saccular aneurisms: the INRIA experience. Medical Image Analysis 9:133–143
51. Boissonnat, J-D, Cazals F (2002) Smooth surface reconstruction via natural neighbour interpolation of distance functions. Computational Geometry 22:185–203
52. Osher S, Sethian JA (1988) Fronts propagating with curvature-dependent speed: algorithms based on Hamilton–Jacobi formulations. Journal of Computational Physics 79:12–49
53. Sethian, JA (1996) Level Set Methods: Evolving Interfaces in Geometry, Fluid Mechanics, Computer Vision, and Materials Science. Cambridge University Press, Cambridge
54. Lorensen, WE, Cline HE (1987) Marching cubes: a high resolution 3D surface construction algorithm. Computer Graphics 21:163–169
55. Delingette H, Hébert M, Ikeuchi K (1992) Shape representation and image segmentation using deformable surfaces. Image and Vision Computing 10:132–144
56. Taylor CA, Hughes TJR, Zarins CK (1998) Finite element modeling of blood flow in arteries. Computer Methods in Applied Mechanics and Engineering 158:155–196
57. Peiró J, Giordana S, Griffith C, Sherwin SJ (2002) High-order algorithms for vascular flow modelling. International Journal for Numerical Methods in Fluids 40:137–151
58. Sherwin SJ, Peiró J (2002) Mesh generation in curvilinear domains using high-order elements. International Journal for Numerical Methods in Engineering 53:207–223
59. Giachetti A, Tuveri M, Zanetti G (2003) Reconstruction and web distribution of measurable arterial models. Medical Image Analysis 7:79–93
60. Cohen LD (1991) On active contour models and balloons. Computer Vision, Graphics, and Image Processing 53:211–218
61. Xu C, Prince JL (1998) Snakes, shapes and gradient vector flow. IEEE Transactions on Image Processing 7:359–369
62. McInerney T, Terzopoulos D (1995) Topologically adaptable snakes. In: Proceedings of the fifth international conference on computer vision. IEEE
63. Krissian K, Malandain G, Ayache N (1998) Model based multiscale detection and reconstruction of 3D vessels. INRIA Research Report RR-3442
64. Fetita C, Prêteux F, Beigelman-Aubry C, Grenier P (2004) Pulmonary airways: 3D reconstruction from multi-slice CT and clinical investigation, IEEE Transactions on Medical Imaging 23:1353–1364
65. Perchet D, Fetita CI, Vial L, Prêteux F, Sbiêrlea-Apiou G, Thiriet M (2004) Virtual investigation of pulmonary airways in volumetric computed tomography, Computer Animation and Virtual Worlds 15:361–376
66. George P-L, Hecht H, Saltel E (1990) Fully automatic mesh generator for 3D domains of any shape. Impact of Computing in Science and Engineering 2:187–218
67. George P-L (1990) Génération automatique de maillages [Automatic Mesh Generation]. Masson, Paris
68. George P-L, Borouchaki H (1997) Triangulation de Delaunay et maillage [Delaunay Triangulation and Mesh]. Hermès, Paris
69. Frey PJ, George P-L (1999) Maillages [Meshes]. Hermès, Paris

70. George P-L, Hecht F (1999), Nonisotropic grid. In: Thompson JF, Soni BK, Weatherill NP (eds.) Handbook of Grid Generation. CRC Press, Boca Raton, FL
71. Mohammadi B, George P-L, Hecht F, Saltel, E (2000) 3D Mesh adaptation by metric control for CFD. Revue européenne des éléments finis 9:439–449
72. Habashi WG, Dompierre J, Bourgault Y, Fortin M, Vallet M-G (1998) Certifiable computational fluid mechanics through mesh optimization. AIAA Journal 36:703–711
73. Fortin, M (2000) Anisotropic mesh adaptation through hierarchical error estimators. In: Minev, P, Yanping L (eds.) Scientific Computing and Applications. Nova Science, Commack, NY
74. Dompierre J, Vellet M-G, Bourgault Y, Fortin M, Habashi WG (2002) Anisotropic mesh adaptation: towards user-independent, mesh-independent and solver-independent CFD III: Unstructured meshes. International Journal for Numerical Methods in Fluids 39:675–702
75. Fortin A, Bertrand F, Fortin M, Boulanger-Nadeau PE, El Maliki A, Najeh N (2004) Adaptive remeshing strategy for shear-thinning fluid flow simulations. Computers and Chemical Engineering 28:2363–2375
76. Cebral JR, Lohner R (2001) From medical images to anatomically accurate finite element grids. International Journal for Numerical Methods in Engineering 51:985–1008
77. Taubin G (1995) Curve and surface smoothing without shrinkage. In: Proceedings of the Fifth International Conference on Computer Vision. IEEE
78. Frey PJ, Borouchaki H (1998) Geometric surface mesh optimization. Computing and Visualization in Science 1:113–121
79. Thiriet M, Graham JMR, Issa RI (1992) A pulsatile developing flow in a bend. Journal de Physique III 2:995–1013
80. Frey PJ, Borouchaki H (2003) Surface meshing using a geometric error estimate. International Journal for Numerical Methods in Engineering 58:227–245
81. Belhamadia Y, Fortin A, Chamberland E (2004) Anisotropic mesh adaptation for the solution of the Stefan problem. Journal of Computational Physics 194:233–255
82. Belhamadia Y, Fortin A, Chamberland E (2004) Three-dimensional anisotropic mesh adaptation for phase change problems, Journal of Computational Physics 201:753–770
83. McGraw-Hill Encyclopedia of Science and Technology (1960) McGraw-Hill, New York
84. Conrad WA, McQueen DM, Yellin EL (1980) Steady pressure flow relations in compressed arteries: Possible origin of Korotkoff sounds. Medical and Biological Engineering and Computing 18:419–426
85. Risacher F (1995) Étude de la propagation de l'onde de pouls par pléthysmographie d'impédance électrique [Study of the propagation of pulse waves by electric impedance plethysmography]. PhD Thesis, University Claude Bernard, Lyon
86. O'Rourke MF, Avolio AP, Kelly RP (1992) The Arterial Pulse. Lea & Febiger, Baltimore, MD
87. Penaz J (1992) Criteria for set point estimation in the volume clamp method of blood pressure measurement. Physiological Research 41:5–10
88. Williams B, Lacy PS (2010) Central haemodynamics and clinical outcomes: going beyond brachial blood pressure? European Heart Journal 31:1819–1822
89. Vlachopoulos C, Aznaouridis K, O'Rourke MF, Safar ME, Baou K, Stefanadis C (2010) Prediction of cardiovascular events and all-cause mortality with central haemodynamics: a systematic review and meta-analysis. European Heart Journal 31:1865–1871
90. Omboni S, Parati G, Frattola A, Mutti E, Di Rienzo M, Castiglioni P, Mancia G (1993) Spectral and sequence analysis of finger blood pressure variability. Comparison with analysis of intra-arterial recordings. Hypertension 22:26–33
91. Novak V, Novak P, Schondorf R (1994) Accuracy of beat-to-beat noninvasive measurement of finger arterial pressure using the Finapres: a spectral analysis approach. Journal of Clinical Monitoring 10:118–126
92. Imholz BP, Wieling W, van Montfrans GA, Wesseling KH (1998) Fifteen years experience with finger arterial pressure monitoring: assessment of the technology. Cardiovascular Research 38:605–616

93. Coron J-M, Crépeau E (2003) Exact boundary controllability of a nonlinear KdV equation with critical lengths. INRIA Research Report RR-5000
94. Whitman GB (1999) Linear and Nonlinear Waves. Wiley-Interscience, New York
95. Crépeau E, Sorine M (2005) personal communication
96. Tasu J-P, Mousseaux E, Colin P, Slama MS, Jolivet O, Bittoun J (2002) Estimation of left ventricle performance through temporal pressure variations measured by MR velocity and acceleration mappings. Journal of Magnetic Resonance Imaging 16:246–252
97. Laleg TM, Crépeau E, Papelier Y, Sorine M (2007) Arterial blood pressure analysis based on scattering transform. 29th Annual International Conference of the IEEE Engineering in Medicine and Biology Society (IEEE EMBC), Lyon, France
98. Kitney RI, Giddens DP (1983) Analysis of blood velocity waveforms by phase shift averaging and autoregressive spectral estimation. Journal of Biomechanical Engineering 105:398–401
99. Thiriet M, Cybulski G, Darrow RD, Doorly DJ, Dumoulin C, Tarnawski M, Caro CG (1997) Apports et limitations de la vélocimétrie par résonance magnétique nucléaire en biomécanique. Mesures dans un embranchement plan symétrique [Contributions and limitations of the nuclear magnetic resonance velocimetry in biomechanics. Measures in a two plane symmetrical bifurcation]. Journal de Physique III 7:771–787
100. Durand E, Guillot G, Darrasse L, Tastevin G, Nacher PJ, Vignaud A, Vattolo D, Bittoun J (2002) CPMG measurements and ultrafast imaging in human lungs with hyperpolarized helium-3 at low field (0.1 T). Magnetic Resonance in Medicine 47:75–81
101. de Rochefort L., Vial L., Fodil R., Maître X., Louis B., Isabey D., Caillibotte G., Thiriet M., Bittoun J., Durand E., Sbirlea-Apiou G. (2007) In vitro validation of CFD simulation in human proximal airways reconstructed from medical images with hyperpolarized helium-3 MRI phase contrast velocimetry. Journal of Applied Physiology: Respiratory, Environmental and Exercise Physiology 102:2012–2023
102. Dumoulin CL, Hart HR Jr (1986) Magnetic resonance angiography. Radiology 161 717–720
103. Dumoulin CL, Souza SP, Hart HR (1987) Rapid scan magnetic resonance angiography. Magnetic Resonance in Medicine 5:238–245
104. Dumoulin CL, Souza SP, Walker MF, Yoshitome E (1988) Time-resolved magnetic resonance angiography. Magnetic Resonance in Medicine 6:275–286
105. Dumoulin CL, Souza SP, Walker MF, Wagle W (1989) Three-dimensional phase contrast angiography. Magnetic Resonance in Medicine 9:139–149
106. Dumoulin CL, Doorly DJ, Caro CG (1993) Quantitative measurement of velocity at multiple positions using comb excitation and Fourier velocity encoding. Magnetic Resonance in Medicine 29:44–52
107. Bittoun J, Jolivet O, Herment A, Itti E, Durand E, Mousseaux E, Tasu J-P (2000) Multidimensional MR mapping of multiple components of velocity and acceleration by Fourier phase encoding with a small number of encoding steps. Magnetic Resonance in Medicine 44:723–730
108. Durand E, Jolivet O, Itti E, Tasu J-P, Bittoun J (2001) Precision of magnetic resonance velocity and acceleration measurements: theoreticals issues and phantom experiments. Magnetic Resonance in Medicine 13:445–451
109. Liebman FM, Pearl J, Bagnol S (1962) The electrical conductance properties of blood in motion. Physics in Medicine and Biology 7:177–194
110. Geddes LA, Sadler C (1973) The specific resistance of blood at body temperature. Medical and Biological Engineering 11:336–339
111. Brody DA (1956) A theoretical analysis of intracavitary blood mass influence on the heart-lead relationship. Circulation Research 4:731–738
112. Gulrajani RM, Roberge FA, Mailloux GE (1989) The forward problem of electrocardiography. In Comprehensive Electrocardiology: Theory and Practice in Health and Disease. Macfarlane PW, Lawrie TDV (eds.), 237–288, Pergamon Press, New York
113. Einthoven W (1895) Uber die Form des menschlichen Electrokardiograms [On the shape of the electrocardiogram in men]. Pflügers Archiv fur die gesamte Physiologie des Menschen und der Tiere 60:101-123

114. Einthoven W (1908) Weiteres über das Elektrokardiogram [More on the electrocardiogram]. Pflügers Archiv fur die gesamte Physiologie des Menschen und der Tiere 122:517–548
115. Einthoven W, Fahr G, de Waart A (1913) Uber die Richtung und die Manifeste Grösse der Potentialschwankungen im mennschlichen Herzen und über den Einfluss der Herzlage auf die Form des Elektrokardiogramms [On the direction and manifest size of the variations of potential in the human heart and on the influence of the heart position in the shape of the electrocardiogram]. Pflügers Archiv fur die gesamte Physiologie des Menschen und der Tiere 150:275-315
116. Goldberger E (1942) The aVL, aVR, and aVF leads. A simplification of standard lead electrocardiography. American Heart Journal 24:378–396
117. Einthoven W, Fahr G, de Waart A (1950) On the direction and manifest size of the variations of potential in the human heart and on the influence of the position of the heart on the form of the electrocardiogram. American Heart Journal 40:163–211
118. Baumert M, Lambert GW, Dawood T, Lambert EA, Esler MD, McGrane M, Barton D, Nalivaiko E (2008) QT interval variability and cardiac norepinephrine spillover in patients with depression and panic disorder. American Journal of Physiology – Heart and Circulatory Physiology 295:H962–H968
119. Boulakia M, Fernández MA, Gerbeau JF, Zenzemi N (2007) Numerical simulation of ECG. Functional Imaging and Modeling of the Heart: FIMH07, Springer, NY
120. Frank E (1956) An accurate, clinically practical system for spatial vectorcardiography. Circulation 13: 737–749
121. Nousiainen J, Oja S, Malmivuo J (1994) Normal vector magnetocardiogram. I. Correlation with the normal vector ECG. Journal of Electrocardiology 27:221–231
122. Nousiainen J, Oja S, Malmivuo J (1994) Normal vector magnetocardiogram. II. Effect of constitutional variables. Journal of Electrocardiology 27:233–241
123. Baule GM, McFee R (1963) Detection of the magnetic field of the heart. American Heart Journal 66:95–96
124. van Oosterom A, Oostendorp TF, Huiskamp GJ, ter Brake HJ (1990) The magnetocardiogram as derived from electrocardiographic data. Circulation Research 67:1503–1509
125. Penney BC (1986) Theory and cardiac applications of electrical impedance measurements. CRC Critical Reviews in Biomedical Engineering 13:227-281
126. Cybulski G (2005) Dynamic impedance cardiography – the system and its applications. Polish Journal of Medical Physics and Engineering 11:127–209
127. Mead J, Whittenberger JL (1953) Physical properties of human lungs measured during spontaneous respiration. Journal of Applied Physiology 5:779–796
128. DuBois AB, Botelho SY, Comroe JH (1956) A new method for measuring airway resistance in man using a body plethysmograph: values in normal subjects and in patients with respiratory disease. Journal of Clinical Investigation 35:327–335
129. Jaeger MJ, Otis AB (1964) Measurement of airway resistance with a volume displacement body plethysmograph. Journal of Applied Physiology 19:813–819
130. Rice DA (1980) Sound speed in the upper airways. Journal of Applied Physiology 49: 326–336
131. Rice DA (1983) Sound speed in pulmonary parenchyma. Journal of Applied Physiology 54:304–308
132. Scherer PW, Shendalman LH, Greene NM, Bouhuys A (1975) Measurement of axial diffusivities in a model of the bronchial airways. Journal of Applied Physiology 38:719–723
133. Ultman JS, Blatman HS (1977) Longitudinal mixing in pulmonary airways. Analysis of inert gas dispersion in symmetric tube network models. Respiration physiology 30:349–367
134. Grubb BR, Mills CD (1981) Blood oxygen content in microliter samples using an easy-to-build galvanic oxygen cell. Journal of Applied Physiology 50:456-464
135. Krogh A, Krogh M (1909) Rate of diffusion into lungs of man. Skandinavisches Archiv für Physiologie 23:236–247
136. Krogh A (1909) On the mechanism of gas exchange in the lungs. Skandinavisches Archiv für Physiologie 23:248–278

137. Krogh M (1915) The diffusion of gases through the lungs of man. Journal of Physiology, London 49:271–296
138. Forster RE, Fowler WS, Bates DV (1954) The absorption of carbon monoxide by the lungs during breathholding. Journal of Clinical Investigation 33:1135–1145
139. Roughton FJW, Forster RE (1957) Relative importance of diffusion and chemical reaction in determining rate of exchange of gases in the human lung. Journal of Applied Physiology 11:290–302
140. Burrows B, Kasik JE, Niden AH, Barclay WR (1961) Clinical usefulness of the single-breath pulmonucy diffusing capacity test. American Review of Respiratory Diseases 84:789–806
141. Hughes JMB, Bates DV (2003) Historical review: the carbon monoxide diffusing capacity and its membrane and red cell components. Respiratory Physiology and Neurobiology 138:115–142
142. Widdicombe J (1997) Airway and alveolar permeability and surface liquid thickness: theory. Journal of Applied Physiology 82:3–12

Conclusion

1. Kobori H, Nangaku M, Navar LG, Nishiyama A (2007) The intrarenal renin-angiotensin system: from physiology to the pathobiology of hypertension and kidney disease. Pharmacological Reviews 59:251–287
2. Fitzsimons JT (1998) Angiotensin, thirst, and sodium appetite. Physiological Reviews 78:583–686
3. Beuschlein F (2013) Regulation of aldosterone secretion: from physiology to disease. European Journal of Endocrinology 168:R85–R93

Appendices

1. Hoffmann R, Valencia A (2004) A gene network for navigating the literature. Nature – Genetics 36:664 (Information Hyperlinked over Proteins. www.ihop-net.org)
2. BioGRID: General Repository for Interaction Datasets; database of physical and genetic interactions for model organisms (www.thebiogrid.org)
3. GeneCards human gene database. Crown Human Genome Center, Department of Molecular Genetics, the Weizmann Institute of Science (www.genecards.org)
4. Universal Protein Resource (UniProt) Consortium (European Bioinformatics Institute, Swiss Institute of Bioinformatics, and Protein Information Resource. www.uniprot.org)

Notation Rules: Aliases and Symbols

A given molecule usually possesses many aliases. Conversely, a given alias commonly refers to various types of molecules [1–4].

Aliases that designate different types of molecules as well as those that do not carry an obvious meaning should be eliminated; they are thus not used in the present text.

For example, P35 is an alias for annexin-A1, brain syntaxin-1A, ficolin-2, interleukin-12A, the cyclin-H assembly factor Ménage à trois homolog-1, regulatory subunit-1 of cyclin-dependent kinase CDK5, and uroplakin-3B, among others. It is substituted by AnxA1, Stx1a, Fcn2, IL12a, MAT1, $CDK5_{r1}$, and UPk3b aliases, respectively.

Protein P39 corresponds to the subunit D1 of the lysosomal V-type H^+ ATPase (ATP6v0d1), Jun transcription factor, a component of Activator protein AP1, and regulatory subunit-2 of cyclin-dependent kinase CDK5 ($CDK5_{r2}$).

Extracellular signal-regulated protein kinases ERK1 and ERK2, members of the mitogen-activated protein kinase (MAPK) module (last tier), are also abbreviated P44 and P42 (also P40 and P41). However, both P42 and P44 correspond to the 26S protease regulatory AAA ATPase subunit (PSMC6). Alias P42 is also utilized for cyclin-dependent kinase CDK20, cyclin-dependent kinase-like protein CDKL1, and 43-kDa NuP43 nucleoporin. Alias P44 can also refer to interferon-induced protein IFI44 (or microtubule-associated protein MTAP44) and androgen receptor cofactor P44 (a.k.a. methylosome protein MeP50 and WD repeat-containing protein WDR77).

The numbering of mitogen-activated protein kinase (MAPK) isoforms that are categorized into 3 families (ERK, JNK, and P38) is neither straighforward nor founded on unicity (ERK2 is also called MAPK1 and MAPK2 and MAPK15 refers to both ERK7 and ERK8). In the present text, stress-activated members of the P38 family (P38α–P38δ)[1] are designated as P38MAPKs to avoid confusion

[1]Protein P38α is also known as MAPK14, cytokine suppressive anti-inflammatory drug (CSAID)-binding protein CSBP, CSBP1, or CSBP2, and stress-activated protein kinase SAPK2a; P38β

M. Thiriet, *Anatomy and Physiology of the Circulatory and Ventilatory Systems*, BBMCVS 6, DOI 10.1007/978-1-4614-9469-0,

with other molecules, the alias of which is also P38. Alias P38 indeed stands for: (1) extracellular signal-regulated kinase ERK3 and ERK6; (2) adaptor CRK (chicken tumor virus regulator of kinase, or v-crk sarcoma virus CT10 oncogene homolog); (3) growth factor receptor-binding protein GRB2-related adaptor protein GRAP2 (a.k.a. GRID, GADS, GRB2L, GRF40, GRPL, and Mona); (4) ubiquitin ligase RING finger protein RNF19a, or dorfin; (5) 38-kDa DNA polymerase-δ-interacting protein Polδ IP2 (a.k.a. polymerase [DNA-directed] PDIP38 and PolD4); (6) activator of 90-kDa heat shock protein ATPase homolog AHSA1; and (7) aminoacyl tRNA synthase complex-interacting multifunctional protein AIMP2, or tRNA synthase complex component JTV1 [1].

Abbreviation, Acronym, Alias, Alphabetism, Initialism, and Portmanteau

Aliases include all written variants, i.e., any abbreviation[2] such as acronyms. An *acronym* corresponds to a word made from the initial letters or syllables of nouns that are pronounceable as a word. Acronyms are generally written with all letters in uppercase. Yet, some acronyms are treated as words and written in lowercase (e.g., laser [originally LASER] is an acronym for light amplification by stimulated emission of radiation, sonar [originally SONAR] for sound navigation and ranging). A substance's name can derive from its chemical name (e.g., amphetamine: α-methylphenethylamine).

Acronyms can give rise to molecule names by adding a scientific suffix such as "-in", a common ending of molecule nouns (e.g., sirtuin, a portmanteau, that comes from the alias SIRT, which stands for silent information regulator-2 [two]). Other scientific prefixes and suffixes can be frequently detected throughout the present text. Their meaning is given in the appendix List of Currently Used Prefixes and Suffixes, particularly for readers from Asia. Many prefixes are used to specify position, configuration and behavior, quantity, direction and motion, structure, timing, frequency, and speed.

A *portmanteau* is a word that combines initials and some inner letters of at least 2 words (e.g., calmodulin stands for calcium modulated protein; caspase for cysteine-dependent aspartate-specific protease; chanzyme for ion channel

---

as MAPK11 and SAPK2b; P38γ as MAPK12, ERK6, and SAPK3; and P38δ as MAPK13 and SAPK4.

[2]In general, abbreviations exclude the initials of short function words, such as "and", "or", "of", or "to". However, they are sometimes included in acronyms to make them pronounceable (e.g., radar [originally RADAR] for radio detection and ranging). These letters are often written in lower case. In addition, both cardinal (size, molecular weight, etc.) and ordinal (isoform discovery order) numbers in names are represented by digits.

and enzyme; chemokine for chemoattractant cytokine;[3] emilin for elastin microfibril interfacer; endorphins and endomorphins for endogenous morphines; ephrin for erythropoietin-producing hepatocyte (EPH) receptor kinase interactor; granzyme for granule enzyme; moesin for membrane-organizing extension spike protein; porin for pore-forming protein; restin for Reed-Steinberg cell-expressed intermediate filament-associated protein, an alias for cytoplasmic linker protein CLiP1 (or CLiP170); serpin for serine protease inhibitor; siglec for sialic acid-binding Ig-like lectin; transceptor for transporter-related receptor; and Prompt for promoter upstream transcript).[4]

Initialisms are abbreviations that are formed from initial letters of a single long noun or several nouns and, instead of being pronounced like an ordinary word, are read letter-by-letter (e.g., DNA stands deoxyribonucleic acid).

Some abbreviations can give rise to alphabetisms that are written as new words (e.g., Rho-associated, coiled-coil-containing protein kinase [RoCK] that is also called Rho kinase). In biochemistry, multiple-letter abbreviations can also be formed from a single word that can be long (e.g., Cam stands for calmodulin, which is itself a portmanteau word, Trx for thioredoxin, etc.) as well as short (e.g., Ttn for titin, etc.). In addition, single-letter symbols of amino acids are often used to define a molecule alias (e.g., tyrosine can be abbreviated as Tyr or Y, hence SYK stands for spleen tyrosine kinase).

Aliases use, in general, capital letters and can include hyphens and dots. Yet, as a given protein can represent a proto-oncogene[5] encoded by a gene that can give rise to an oncogene (tumor promoter) after gain- or loss-of-function mutations,[6] the same acronym represents 3 different entities.[7]

---

[3]Cytokines are peptidic, proteic, or glycoproteic regulators that are secreted by cells of the immune system. These immunomodulating agents serve as auto- or paracrine signals.

[4]The uppercase initial P in Prompt is used to avoid confusion with command-line interpreter prompt or prompt book to direct precise timing of actions on the theater stage.

[5]In 1911, P. Rous isolated a virus that was capable of generating tumors of connective tissue (sarcomas) in chicken. Proteins were afterward identified, the activity of which, when uncontrolled, can provoke cancer, hence the name oncogene given to genes that encode these proteins. Most of these proteins are enzymes, more precisely kinases. The first oncogene was isolated from the avian Rous virus by D. Stéhelin and called Src (from sarcoma). This investigator demonstrated that the abnormal functioning of the Src protein resulted from mutation of a normal gene, or proto-oncogene, which is involved in cell division.

[6]Loss-of-function mutations cause complete or partial loss of function of gene products that operate as tumor suppressors, whereas gain-of-function mutations generate gene products with new or abnormal function that can then act as oncogenes. Typical tumor-inducing agents are enzymes, mostly regulatory kinases and small guanosine triphosphatases, that favor proliferation of cells, which normally need to be activated to exert their activities. Once their genes are mutated, these enzymes become constitutively active. Other oncogenes include growth factors (a.k.a. mitogens) and transcription factors. Mutations can also disturb signaling axis regulation, thereby raising protein expression. Last, but not least, chromosomal translocation can also provoke the expression of a constitutively active hybrid protein.

[7]Like Latin-derived shortened expressions—as well as foreign words—that are currently written in italics, genes can be italicized. However, this usage is not required in scientific textbooks published by Springer. Italic characters are then used to highlight words within a text to target them easily.

In addition, a given abbreviation can designate distinct molecules without necessarily erroneous consequence in a given context (e.g., PAR: poly^{ADP}ribose or protease-activated receptor and GCK: germinal center kinases or glucokinase; in the latter case, the glucokinase abbreviation should be written as GcK or, better, GK).

Molecule Aliases and Adopted Notation Rules

Numerous aliases that designate a single molecule can result from the fact that molecules have been discovered independently several times with possibly updated functions.

Some biochemists uppercase the name of a given molecule, whereas others lowercase (e.g., cell division cycle guanosine triphosphatase of the Rho family CDC42 or Cdc42, adaptor growth factor receptor-bound protein GRB2 or Grb2, chicken tumor virus regulator of kinase CRK or Crk, guanine nucleotide-exchange factor Son of sevenless SOS or Sos, etc.).

Acronyms are then not always entirely capitalized. Printing style of aliases should not only avoid confusion, but also help one in remembering alias meaning.

In the present textbook, choice of lower- and uppercase letters in molecule aliases is dictated by the following criteria.

(1) An uppercase letter is used for initials of words that constitute molecule nouns (e.g., receptor tyrosine kinase RTK). An alias of any compound takes into account added atoms or molecules (e.g., PI: phosphoinositide and PIP: phosphoinositide phosphate) as well as their number (e.g., PIP_2: phosphatidylinositol bisphosphate, DAG: diacylglycerol, and PDE: [cyclic nucleotide] phosphodiesterases).

(2) A lowercase letter is used when a single letter denotes a subfamily or an isoform when it is preceded by a capital letter (e.g., PTPRe: protein tyrosine phosphatase receptor-like type-E). Nevertheless, an uppercase letter is used in an alias after a single or several lowercase letters to distinguish the isoform type (e.g., RhoA isoform and DNA-repair protein RecA for recombination protein-A), but OSM stands for oncostatin-M, not osmole Osm[8] to optimize molecule identification.

These criteria enable the use of differently written aliases with the same sequence of letters for distinct molecules (e.g., CLIP for corticotropin-like intermediate peptide, CLiP: cytoplasmic CAP-Gly domain-containing linker protein, and iCliP: intramembrane-cleaving protease).

Proteins are currently romanized (ordinary print), but with a capital initial. Nevertheless, names (not aliases) of chemical species are entirely lowercased like in most (if not all) scientific articles, except to avoid confusion with a usual word (e.g., hedgehog animal vs. Hedgehog protein and raptor [bird of prey] vs. Raptor molecule).

[8]Osmole: the amount of osmotically active particles that exerts an osmotic pressure of 1 atm when dissolved in 22.4 l of solvent at 0°C.

As the exception proves the rule, current aliases, such as PKA and PLA that designate protein kinase-A and phospholipase-A, respectively, have been kept. Preceded by only 2 uppercase letters, a lowercase letter that should be used to specify an isoform can bring confusion with acronyms of other protein types (e.g., phospholamban alias PLb).

Nouns (e.g., hormone-like fibroblast growth factor [hFGF] and urokinase-type plasminogen activator [uPA]) or adjectives (e.g., intracellular FGF isoform [iFGF]) that categorize a subtype of a given molecule correspond to a lowercase letter to emphasize the molecule species. Hence, an uppercase letter with a commonly used hyphen (e.g., I[R]-SMAD that stands for inhibitory [receptor-regulated] SMAD; V-ATPase for vacuolar adenosine triphosphatase; MT1-MMP for membrane type-1 matrix metalloproteinase; and T[V]-SNARE for target [vesicle-associated] soluble ^{N}ethylmaleimide-sensitive factor-attachment protein receptor) is then replaced by a lowercase letter (e.g., i[r]SMAD, vATPase, mt1MMP, and t[v]SNARE), as is usual for RNA subtypes (mRNA, rRNA, snRNA, and tRNA for messenger, ribosomal, small nuclear, and transfer RNA, respectively). Similarly, membrane-bound and secreted forms of receptors and coreceptors that can derive from alternative mRNA splicing are defined by a lowercase letter (e.g., sFGFR for secreted extracellular FGFR form and sFRP for soluble Frizzled-related protein), as well as eukaryotic translation elongation (eEF) and initiation (eIF) factors.

(3) Although `l`, `r`, and `t` can stand for molecule-like, -related, and -type, respectively, when a chemical is related to another one, in general, uppercase letters are used for the sake of homogenity and to clearly distinguish between the letter L and numeral 1 (e.g., KLF: Krüppel-like factor, CTK: C-terminal Src kinase (CSK)-type kinase, and SLA: Src-like adaptor).

(4) An uppercase letter is most often used for initials of adjectives contained in the molecule name (e.g., AIP: actin-interacting protein; BAX: BCL2-associated X protein; HIF: hypoxia-inducible factor; KHC: kinesin heavy chain; LAB: linker of activated B lymphocytes; MAPK: mitogen-activated protein kinase; and SNAP: soluble N-ethylmaleimide-sensitive factor-attachment protein).

(5) Lowercase letters are used when alias letters do not correspond to initials (e.g., Fox—not fox— [forkhead box]), except for portmanteau words that are entirely written in minuscules (e.g., gadkin: γ1-`ad`aptin and `kin`esin `in`teractor).

This rule applies, whether alias letters correspond to successive noun letters (e.g., Par: partitioning defective protein and Pax: paxillin, as well as BrK: breast tumor kinase and ChK: checkpoint kinase, whereas CHK denotes C-terminal Src kinase [CSK]-homologous kinase) or not (e.g., Fz: Frizzled and HhIP: Hedgehog-interacting protein),[9] except for composite chemical species (e.g., DAG: diacylglyc-

[9]The Hedgehog gene was originally identified in the fruit fly Drosophila melanogaster. It encodes a protein involved in the determination of segmental polarity and intercellular signaling during morphogenesis. Homologous gene and protein exist in various vertebrate species. The name of the mammal hedgehog comes from hecg and hegge (dense row of shrubs or low trees), as it resides in hedgerows, and hogg and hogge, due to its pig-like, long projecting nose (snout). The word Hedgehog hence is considered as a seamless whole.

erol). However, some current usages have been kept for short aliases of chemical species name (e.g., Rho for Ras homolog rather than RHo).

In any case, molecule (super)family (class) aliases as well as those of their members are written in capital letters, such as the IGSF (IGSF*i*: member *i*; immunoglobulin), KIF (KIF*i*; kinesin), SLC (SLC*i*; solute carrier), TNFSF (TNFSF*i*; tumor-necrosis factor), and TNFRSF (TNFRSF*i*; tumor-necrosis factor receptor) superfamily.

Gene names are also written with majuscules when the corresponding protein name contains at least one minuscule, otherwise only the gene name initial is written with an uppercase letter that is then followed by lowercase letters.

To highlight its function, substrate aliases (e.g., ARF GTPases) contained in a molecule alias are partly written with lowercase letters (e.g., ArfRP, ArfGEF, ArfGAP stand for ARF-related protein, ARF guanine nucleotide-exchange factor, and ARF GTPase-activating protein, respectively).

Last, but not least, heavy and pedantic designation of protein isoforms based on roman numerals has been avoided and replaced by usual arabic numerals (e.g., angiotensin-2 rather than angiotensin-II), except for coagulation (or clotting) factors. Moreover, the character I can mean either letter I or number 1 without obvious discrimination at first glance (e.g., GAPI that stands for Ras GTPase-activating protein GAP1, but can be used to designate a growth-associated protein inhibitor).

Unnecessary hyphenation in aliases of substances (between an upper case letter, which can define the molecule function, and the chemical alias, or between it and assigned isotype number) has been avoided. In any case, the Notation section serves not only to define aliases, but also, in some instances, as disambiguation pages.

A space rather than hyphen is used in: (1) structural components at the picoscale (e.g., P loop), nanoscale (e.g., G protein [G standing for guanine nucleotide-binding]), microscale (e.g., H zone, M line, A band, I band, and Z disc of the sarcomere and T tubule of the cardiomyocyte); (2) process stages (e.g., M phase of the cell division cycle); and (3) cell types (e.g., B and T lymphocytes). When these terms are used as adjectives, a hypen is then employed (e.g., P-loop Cys–X_5–Arg (CX_5R) motif, G-protein-coupled receptor, Z-disc ligand, M-phase enzyme, and T-cell activation).

In terms incorporating a Greek letter, similarly, a space is used in: (1) structural components (e.g., α and β chains and subunits); (2) cellular organelles (e.g., α granule); and (3) cell types (e.g., pancreatic β cell). On the other hand, terms are hyphenated when they refer to (1) structural shape (e.g., α-helix and α (β)-sheet) and (2) molecule subtype (e.g., α-actinin, β-glycan, and γ-secretase).

Symbols for Physical Variables

Unlike substances aliases, symbols for physical quantities are most often represented by a single letter of the Latin or Greek alphabet (i: current; J: flux; L: length; m: mass; p: pressure; P: power; T: temperature; t: time; u: displacement;

v: velocity; x: space; λ: wavelength; μ: dynamic viscosity; ρ: mass density; etc.). These symbols are specified using sub- and superscripts (c_p and c_v: heat capacity at constant pressure and volume, respectively; $\mathcal{D}_T$: thermal diffusivity; G_h: hydraulic conductivity; G_T: thermal conductivity; α_k: kinetic energy coefficient; α_m: momentum coefficient; etc.).

A physical quantity associated with a given point in space at a given time can be: (1) a scalar uniquely defined by its magnitude; (2) a vector characterized by a magnitude, a support, and a direction represented by an oriented line segment defined by a unit vector; and (3) a tensor specified by a magnitude and a few directions. To ensure a straightforward meaning of symbols used for scalar, vectorial, and tensorial quantities, boldface upper- (**T**) and lowercase (**v**) letters are used to denote a tensor and a vector, respectively, whereas both roman (plain, upright)-style upper- and lowercase letters designate a scalar.

The n dash is used rather than the hyphen to distinguish a double-barreled name from cases for which 2 different researcher's names as well as their derived adjectives (e.g., Newtonian) are joined up to define equations (e.g., Kedem–Katchalsky, Navier–Stokes, and Stefan–Maxwell equations); laws (e.g., Boyle–Mariotte law); chemical reactions (e.g., Michaelis–Menten enzyme kinetics); model types (e.g., Mitchell–Schaeffer model); effects (e.g., Fahraeus–Lindqvist effect); and numerical procedures (e.g., arbitrary Lagrangian–Eulerian formulation, Chorin–Temam projection scheme, and Dirichlet–Neumann domain decomposition algorithm).

List of Currently Used Prefixes and Suffixes

Prefixes (localization)

"ab-" (Latin) and "apo-" (Greek: απ ο): away from or off (abluminal: endothelial edge opposite to wetted surface; apolipoproteins: lipid carriers that cause egress [also ingress] from cells; aponeurosis (απ ονευρωσις; νευρ ον: sinew, tendon) muscle sheath that limits radial motion and enhances axial contraction; and apoptosis: separation ["-ptosis": fall (πτωσ ις): as leaves fall away from a tree], a type of programmed cell death)

"acr-" (variant "acro-" [ακρ ος]): top or apex

"ad-" (adfecto: to reach; adfio: to blow toward; adfluo: to flow toward): toward (ad- becomes "ac-" before c, k, or q; "af-" before f [afferent]; "ag-" before g [agglutination]; "al-" before l; "ap-" before p [approximation]; "as-" before s; and "at-" before t)

"cis-", "juxta-", and "para-" (παρα): near, beside, or alongside

"contra-": opposite side; "ipsi-" (ipse): same side; "latero-": side;

"ecto-" (εκτ ος), "exo-" (εξ ο), and "extra-": outside, outer, external, or beyond (exogenous chemicals produced by an external source, or xenobiotics ["xeno-": foreigner])

"endo-" (ενδ ον) and "intra-": inside (endogenous substances synthesized by the body's cells; endomembranes at organelle surfaces within the cell)

"ep-" (variant "eph-", or "epi-" [επι]): upon (epigenetics refers to the inheritance ("-genetic": ability to procreate [γεννητικ ος]) of variations in gene expression beyond ("epi-": on, upon, above, close to, beside, near, toward, against, among, beyond, and also) change in the DNA sequence.

"front-" and "pre-": anterior or in front of

"post-": behind

"infra-" and "sub-": under or below

"super-" and "supra-": above

"inter-": between or among

"peri-" (περι): around

"tele-" (τελε): remote

"trans-": across

Prefixes (composition)

"an-" and "aniso-" (ανισ ος): unequal, uneven, heterogeneous

"iso-" (ισ ος): equal, alike (isomer [μερ ος: part, portion]

"mono-" (μ ον ος) and "uni-" (unicus): single

"oligo-" (ολιγ ος): few, little, small

"multi-" (multus), "pluri-" (plus, plures), and "poly-" (π ολυς): many, much

"ultra-": in excess.

Prefixes (quantity)

"demi-" (dimidius) and "hemi-" (ημι): half

"sesqui-": one and a half (half more)

M. Thiriet, *Anatomy and Physiology of the Circulatory and Ventilatory Systems*, BBMCVS 6, DOI 10.1007/978-1-4614-9469-0,

"di-" or "dis-" (δυ ο; δις) as well as "bi-" or "bis-": 2, twice
"tri" (τρεις, τρι-; tres, tria): 3
"tetra-" (τετρα), "quadri-" (variant: "quadr-" and "quadru-"): 4
"penta-" (πεντας; pentas), "quinqu-", and "quint-": 5
"hexa-" (εξ) and "sexa-": 6
"hepta-" (επτα): 7
"octa-" (οκτα): 8
"nona-" (εννεα): 9 (ninth part)
"deca-" (δεκα): 10
"quadra-" (quadragenarius): 40 (elements)
"quinqua-" (quinquagenarius): 50
"sexa-" (sexagenarius [sex: 6]: 60
"septua-" (septuagenarius [septem: 7]): 70
"nona-" (nonagenarius): 90

Prefixes (motion and direction)

"af-": toward the center (single master object); e.g., nerve and vascular afferents (ferre: to carry) to brain and heart, respectively, rather than toward any slave, supplied tissue from the set of the body's organs; also affector, i.e., chemical messenger that brings a signal to the cell considered as the object of interest, this exploration focus being virtually excised from the organism with its central command system, except received signals
"ef-" (effero: to take away): from the center (efferent; effector, i.e., chemical transmitter recruited by the previous mediator of a signaling cascade at a given locus to possibly translocate to another subcellular compartment)
"antero-" (anterior): before, in front of, facing, or forward
"retro-": behind or backward
"tropo-" (τρ οπ ος): duct direction; (tropa: rotation; celestial revolution); e.g., tropomyosin (μυς, musculus: muscle; μυ ο-: refers to muscle [μυ οτρωτ ος: injured at a muscle])

Prefixes (structure and size)

"macro-" (μακρ ος): large, long, or big
"mega-" (μεγας): great, large
"meso-" (μεσ ος): middle
"micro-" (μικρ ος): small
"nano-" (ναν ος): dwarf, tiny
"homo-" (ομ ο-): same (ομ ολ ογ ος: agreeing, corroborating; variant: "homeo-" [homeostasis])
"hetero-" (ετερ ο-): other

Prefixes (timing)

"ana-" (ανα): culminating (anaphase of the cell division cycle), up, above (αν οδ ος: a way up, anode [positive electrode; οδ ος: way, path, road, track])
"ante-": before
"circa-": approximately, around (circadian: approximately one day)
"infra-": below, shorter (infradian: rhythm with lower frequency than that of circadian rhythm, not smaller period)
"inter-": among, between, during
"meta-" (μετα): after, beyond, behind, later; in the middle of (metaphase of the cell division cycle); as well as connected to, but with a change of state (metabolism) and about (metadata)
"post-": after
"pre-": earlier
"pro-" (πρ ο): preceding, first, before (prophase of the cell division cycle)
"telo-" (τελος): end, completion
"ultra-": beyond, longer (ultradian: period smaller than that of 24–28-hour cycle, i.e., frequency greater than that of the circadian rhythm)

Prefixes (functioning modality)

"auto-" (αυτ ος): same, self
"brady-" (βραδυς): slow (decelerate)
"tachy-" (ταχ ος): rapid (accelerate)
"amphi-" (αμφι): both (amphiphilic substances are both hydrophilic and lipophilic; amphisomes are generated by both autophagosomes and endosomes)
"ana-" : upward (anabolism) or against (anaphylaxis)

"cata-" (κατα): downward (catabolism, cathode [negative electrode; οδ ος; way, path, road, track])
"anti-" (αντι): against
"pro-": favoring
"co-" (coaccedo: add itself to): together
"contra-": adverse, against, beside, next to, opposite
"de-": remove, reduce, separation after association (Latin de; e.g., deoxy-)
"dys-" (δυς): abnormal (δυσαης): ill-blowing)
"equi-" (æque): equal or alike
"hem-" or "hemat-" (αιμα: blood): related to blood
"hypo-" (υπ ο): under, beneath, and low
"hyper-" (υπερ): above, beyond, and large
"per-": through (e.g., percutaneous) and during (e.g., peroperative)
"pseudo-" (ψευδ ο): pretended, false
"re-"; again

Scientific suffixes

"-ase": enzyme (synthase, lipase, etc.)
"-ate": salt of a base
"-cyte" (κυτ ος): cell (erythro- [ερυθρ ος: red], leuko- [λευκ ος: light, bright, clear, white], thrombo- [θρ ομβ ος: lump, clot], adipo- [adeps: fat; adipalis, adipatus, adipeus, adipinus: fatty], fibro- [fibra: fiber, filament], myo- [μυς: muscle, mouse, mussel], myocardiocyte [κραδια: heart; cardiacus: related to heart, stomach; to have heart trouble, stomach trouble], etc.);
"-crine" (κρινω): to decide, to separate, and to secrete (e.g., endocrine regulator) (ευκρινεω: keep in order)
"-elle": small (organelle in a cell [like an organ in a body])
"-ium", "-ion", "-isk", and "-iscus": little ("-ium": tissue interface and envelope, such as endothelium and pericardium)
"-phil" (φιλια): attracted (αφιλια: want of friends)
"-phob" (φ οβια): repulsed (υδρ οφ οβια, hydrophobia [Latin]: horror of water)
"-phore" (φερω): carrier (αμφερω: to bring up)
"-yl" denotes a radical (molecules with unpaired electrons)
"-ploid" (πλ οω): double, fold (diploid, twofold; διπλ οω: to double; διαπλ οω: unfold)
"-emia": in relation to flow (ανεμια: flatulence; ευηνεμια: fair wind), particularly blood condition
"-genesis" (γενεσις): cause, generation, life source, origin, productive force
"-iasis": for diseased condition
"-itis": inflammation
"-lemma" (λεμμα: skin): sheath
"-ole" and "-ule": small (arteriole and venule; variant "-ula" [blastula] and "-ulum")
"-plasma" (πλασμα): anything molded (plasma: creature generated from silt of earth)
"-plasia" (πλασια): formation, molding
"-podium" (π οδ ος: foot; podium [Latin]: small knoll, small protuberance): protrusion
"-poiesis" (π οιεω): production
"-soma" (σωμα): body
"-sclerosis" (σκλημα): hardness, induration
"-stasis" (στασις): stabilization (απ οκαταστασις: restoration; ανυπ οστασις: migration)
"-stomosis" (στ ομα: mouth): equipped with an outlet
"-taxy/tactic" (ταχ υ: rapid; τακτικ ος: to maneuver): related to motion (also prefix, i.e., ταχυκινησις: quick motion; ταχυνω: to accelerate; and ταχυπν οια: short breath; not [δια]ταξις: disposition, arrangement)
"-trophy/trophic" (τρ οφις: well fed): related to growth
"-oma": tumor of
"-pathy" (παθ ος, παθεια): disease of
"-tomy" (τ ομια) and "-ectomy": surgical removal (απλ οτ ομια: simple incision; φαûrhουγγ οτ ομια: laryngotomy)

List of Aliases and Primary Symbols

A

$\mathcal{A}$: Avogadro number
$\mathcal{A}(p)$: area–pressure relation
A: Almansi strain tensor
A: cross-sectional area
A: actin-binding site
A: surface area-to-volume ratio
a: acceleration
a: major semiaxis
AA: arachidonic acid
AAA: ATPase associated with diverse cellular activities
AAA: abdominal aortic aneurysm
AAAP: aneurysm-associated antigenic protein
AAK: adaptin-associated kinase
AAS: acute aortic syndrome
AATK: apoptosis-associated tyrosine kinase
ABC: ATP-binding cassette transporter (transfer ATPase)
AbI: Abelson kinase interactor
Abl: Abelson leukemia viral proto-oncogene product (NRTK)
ABLIM: actin-binding LIM domain-containing protein
ABP: actin-binding protein
ABR: active breakpoint cluster region (BCR)-related gene product (GEF and GAP)
AC: atrial contraction
ACAA: acetylCoA acyltransferase
ACAP: ArfGAP with coiled-coil, ankyrin repeat, PH domains
ACase: adenylate cyclase
ACi: adenylate cyclase isoform i
ACAT: acylCoA–cholesterol acyltransferase
ACC: acetyl coenzyme-A carboxylase
ACD: adrenocortical dysplasia homolog
ACE: angiotensin-converting enzyme
ACh: acetylcholine
ACK: activated CDC42-associated kinase
ACL: ATP–citrate lyase
Aco: aconitase
ACP1: acid phosphatase-1, soluble (lmwPTP)
ACTH: adrenocorticotropic hormone
^{F}actin: filamentous actin
(Cav–actin: caveolin-associated ^{F}actin)
^{G}actin: monomeric globular actin
AcvR: activin receptor (TGFβ receptor superfamily)
Ad: adrenaline
ADAM: a disintegrin and metallopeptidase (adamalysin)
ADAMTS: a disintegrin and metallopeptidase with thrombospondin
ADAP: adhesion and degranulation-promoting adaptor protein
ADAP: ArfGAP with dual PH domains
ADCF: adipocyte-derived constricting factor
ADF: actin-depolymerizing factor (cofilin-related destrin)
ADH: antidiuretic hormone (vasopressin)
ADMA: asymmetric dimethylarginine
ADP: adenosine diphosphate
ADRF: adipocyte-derived relaxing factor
aDuSP: atypical dual specificity phosphatase
AE: anion exchanger
AEA: N-arachidonoyl ethanolamine (anandamide)
AF: atrial fibrillation
AFAP: ArfGAP with phosphoinositide-binding and PH domains
aFGF: acidic fibroblast growth factor (FGF1)

M. Thiriet, *Anatomy and Physiology of the Circulatory and Ventilatory Systems*, BBMCVS 6, DOI 10.1007/978-1-4614-9469-0,

AGAP: ArfGAP with GTPAse, ankyrin repeat, and PH domains
AGF: autocrine growth factor
AGFG: ArfGAP with FG repeats
Ago: Argonaute protein
AGS: activator of G-protein signaling
AHI: apnea–hypopnea index
AHR: airway hyperresponsiveness
AHR: aryl hydrocarbon receptor
AIF: apoptosis-inducing factor
AIP: actin-interacting protein
AIRe: autoimmune regulator
AIx: augmentation index
AKAP: A-kinase (PKA)-anchoring protein
ALE: arbitrary Eulerian Lagrangian
ALIX: apoptosis-linked gene-2-interacting protein-X
ALK: anaplastic lymphoma kinase
ALK*i*: type-*i* activin receptor-like kinase (TGFβ receptor superfamily)
ALOx5: arachidonate 5-lipoxygenase
ALOx5AP: arachidonate 5-lipoxygenase activation protein
ALP: actinin-associated LIM protein (PDLIM3)
alsin: amyotrophic lateral sclerosis protein (portmanteau)
ALX: adaptor in lymphocytes of unknown function X
AMAP: A multidomain ArfGAP protein
AMBRA: activating molecule in beclin-1-regulated autophagy protein
AMHR: anti-Müllerian hormone receptor (TGFβ receptor superfamily)
AMIS: apical membrane initiation site (lumenogenesis)
AMP: adenosine monophosphate
AMPAR: α-amino 3-hydroxy 5-methyl 4-isoxazole propionic acid receptor
AMPK: AMP-activated protein kinase
AMSH: associated molecule with SH3 domain (deubiquitinase)
AmyR: amylin receptor
Ang (AngPt): angiopoietin
AngL: angiopoietin-like molecule
Ank: ankyrin
ANP: atrial natriuretic peptide
ANPR (NP_1): atrial natriuretic peptide receptor (guanylate cyclase)
ANS: autonomic nervous system
ANT: adenine nucleotide transporter
Anx: annexin
AOC: amine oxidase copper-containing protein
AoV: aortic valve
AP: (clathrin-associated) adaptor proteic complex
AP: Activator protein (transcription factor)
AP: activating enhancer-binding protein
AP_4A: diadenosine tetraphosphate
APAF: apoptotic peptidase-activating factor
APAP: ArfGAP with PIX- and paxillin-binding domains
APC: antigen-presenting cell
APC: adenomatous polyposis coli protein (Ub ligase)
APC/C: anaphase-promoting complex (or cyclosome; Ub ligase)
APH: anterior pharynx defective phenotype homolog
aPKC: atypical protein kinase-C
APl: action potential
Apn: adiponectin
Apo: apolipoprotein
ApoER: apolipoprotein-E receptor
APPL: adaptor containing phosphoTyr interaction, PH domain, and Leu zipper
APS: adaptor with a PH and SH2 domain
Aqp: aquaporin
AR: adrenergic receptor (adrenoceptor)
AR: androgen receptor (nuclear receptor NR3c4; transcription factor)
AR: area ratio
ARAP: ArfGAP with RhoGAP, ankyrin repeat, PH domains
ARDS: acute respiratory distress syndrome
ARE: activin-response element
ARE: androgen response element
ARE: antioxidant response element
Areg: amphiregulin (EGF superfamily member)
ARF: ADP-ribosylation factor
ArfRP: ARF-related protein
ARFTS: CKI2A-locus alternate reading frame tumor suppressor (ARF or $p14^{ARF}$)
ARH: autosomal recessive hypercholesterolemia adaptor (low-density lipoprotein receptor adaptor)
ARH: aplysia Ras-related homolog
ArhGEF: RhoGEF
ARL: ADP-ribosylation factor-like protein
ARNO: ARF nucleotide site opener
ARNT: aryl hydrocarbon nuclear receptor translocator
ARP: absolute refractory period
ARP: actin-related protein
ARPP: cAMP-regulated phosphoprotein

Arr: arrestin
ART: arrestin-related transport adaptor (α-arrestin)
ART: adpribosyltransferase
Artn: artemin
ARVCF: armadillo repeat gene deleted in velocardiofacial syndrome
ARVD: arrythmogenic right ventricular dystrophy
AS: Akt (PKB) substrate
ASAP: artery-specific antigenic protein
ASAP: ArfGAP with SH3, ankyrin repeat, PH domains
ASIC: acid-sensing ion channel
ASK: apoptosis signal-regulating kinase
aSMC: airway smooth muscle cell
ASO: arteriosclerosis obliterans
ASP: actin-severing protein
AT: antithrombin
ATAA: ascending thoracic aortic aneurysm
ATF: activating transcription factor
AtG: autophagy-related gene product
ATGL: adipose triacylglycerol lipase
ATMK: ataxia telangiectasia mutated kinase
ATn: angiotensin
ATng: angiotensinogen
AtOx: antioxidant protein (metallochaperone)
ATP: adenosine triphosphate
ATPase: adenosine triphosphatase
ATR ($AT_{1/2}$): angiotensin receptor
ATRK: ataxia telangiectasia and Rad3-related kinase
AVN: atrioventricular node
AVV: atrioventricular valves
AW: analysis window

B

B: Biot–Finger strain tensor
B: bulk modulus
$\mathcal{B}$: bilinear form
B: binding rate
b: minor semiaxis
b: body force
$\hat{\mathbf{b}}$: unit binormal
b: birth rate
B lymphocyte (B cell): bone marrow lymphocyte
BACE: β-amyloid precursor protein-converting enzyme
BAD: BCL2 antagonist of cell death
BAF: barrier-to-autointegration factor
BAG: BCL2-associated athanogene (chaperone regulator)
BAI: brain-specific angiogenesis inhibitor (adhesion GPCR)
BAIAP: brain-specific angiogenesis inhibitor-1-associated protein (insulin receptor substrate)
BAK: BCL2-antagonist killer
(i)BALT: (inducible) bronchus-associated lymphoid tissue
BAMBI: BMP and activin membrane-bound inhibitor homolog
BAnk: B-cell scaffold with ankyrin repeats
Barkor: beclin-1-associated autophagy-related key regulator
BAT: brown adipose tissue
BATF: basic leucine zipper ATF-like transcription factor (B-cell-activating transcription factor)
BAX: BCL2-associated X protein
BBB: blood–brain barrier
BBS: Bardet–Biedl syndrome protein
BBSome: BBS coat complex (transport of membrane proteins into cilium)
BC: boundary condition
bCAM: basal cell adhesion molecule (Lutheran blood group glycoprotein)
BCAP: B-cell adaptor for phosphatidylinositol 3-kinase
BCAR: Breast cancer antiestrogen resistance docking protein
BCL: B-cell lymphoma (leukemia) protein
BCLxL: B-cell lymphoma extralarge protein
BCR: B-cell receptor
BCR: breakpoint cluster region protein (GAP and GEF)
Bdk: bradykinin
BDNF: brain-derived neurotrophic factor
Be: Bejan number
Becn, beclin: BCL2-interacting protein
BEM: boundary element method
BES: biolimus-eluting stent
Best: bestrophin
bFGF: basic fibroblast growth factor (FGF2)
BFUe: burst-forming unit-erythroid
BFUmeg: burst-forming unit-megakaryocyte
BGP: bone γ-carboxyglutamate acid (Gla)-containing protein (osteocalcin)
BGT: betaine–GABA transporter
BH_4: tetrahydrobiopterin (enzyme cofactor)
BHR: bronchial hyperresponsiveness
BID: BH3-interacting domain death agonist
BIG: brefeldin-A-inhibited GEFs for ARFs

BIK: BCL2-interacting killer
BIM: BH3-containing protein BCL2-like 11 (BCL2L11)
BK: high-conductance, Ca^{2+}-activated, voltage-gated K^+ channel
BLK: B-lymphoid tyrosine kinase
Blm: Bloom syndrome, RecQ DNA helicase-like protein
BLnk: B-cell linker protein
BLOC: biogenesis of lysosome-related organelles
BM: basement membrane
BMAL: brain and muscle ARNT-like protein (gene Bmal)
BMAT: bone marrow adipose tissue
BMF: BCL2 modifying factor
BMP: bone morphogenetic protein (TGFβ superfamily)
BMPR: bone morphogenetic protein receptor
BNIP: BCL2/adenovirus E1B 19-kDa protein-interacting protein
BNP: B-type natriuretic peptide
BMS: bare-metal stent
BMX: bone marrow Tyr kinase gene in chromosome-X product
Bo: Boltzmann constant
BOC: brother of CDO
BOK: BCL2-related ovarian killer
BORG: binder of Rho GTPase
Br: Brinkman number
BRAG: brefeldin-resistant ArfGEF
BrCa: breast cancer-associated (susceptibility) protein (tumor suppressor; DNA-damage repair; a.k.a. FancD1)
BrD: bromodomain-containing protein
BrK: breast tumor kinase
BrSK: brain-selective kinase
BSCB: blood–spinal cord barrier
BSP: bone sialoprotein
BSEP: bile salt export pump
BTF: basic transcription factor
BTK: Bruton Tyr kinase
BUB: budding uninhibited by benzimidazoles

C

C: stress tensor
C: compliance, capacitance
C: heat capacity
C: chronotropy
Cx: type-x chemokine C (γ)
C_D: drag coefficient
C_f: friction coefficient
C_L: lift coefficient
C_p: pressure coefficient
C_{VM}: van Mises stress
c: stress vector
c_τ: shear
c_w: wall shear stress
c_X: concentration of species X
$c(p)$: wave speed
c_p: isobar heat capacity
c_v: isochor heat capacity
C1P: ceramide 1-phosphate
C-terminus: carboxy (carboxyl group COOH) terminus
C/EBP: CCAAT/enhancer-binding protein
CA: computed angiography
CAi: carbonic anhydrase isoform i
Ca: calcium
Ca_V: voltage-gated Ca^{2+} channel
Ca_V1.x: L-type high-voltage-gated Ca^{2+} channel
Ca_V2.x: P/Q/R-type Ca^{2+} channel
Ca_V3.x: T-type low-voltage-gated Ca^{2+} channel
CAAT: cationic amino acid transporter
CABG: coronary artery bypass grafting
Cables: CDK5 and Abl enzyme substrate
CACT: carnitine–acetylcarnitine transferase
CAK: CDK-activating kinase (pseudokinase)
Cam: calmodulin (calcium-modulated protein)
CamK: calmodulin-dependent kinase
cAMP: cyclic adenosine monophosphate
CAN: cardiovascular autonomic neuropathy
CAP: adenylate cyclase-associated protein
CAP: carboxyalkylpyrrole protein adduct
CAP: chromosome-associated protein (BrD4)
CAPN: calpain gene
CaPON: C-terminal PDZ ligand of NOS1 (NOS1AP)
CAR: constitutive androstane receptor (NR1i3)
CaR: calcium-sensing receptor
CARD: caspase activation and recruitment domain-containing protein
CARMA: CARD and membrane-associated guanylate kinase-like (MAGuK) domain-containing protein
CARP: cell division cycle and apoptosis regulatory protein
CAS: cellular apoptosis susceptibility protein
CAS: CRK-associated substrate (or P130CAS and BCAR1)
CAs: cadherin-associated protein
CASK: calcium–calmodulin-dependent serine kinase (pseudokinase)

CASL: CRK-associated substrate-related protein (CAS2)
CASP: cytohesin-associated scaffold protein
caspase: cysteine-dependent aspartate-specific peptidase
CAT: carnitine acetyltransferase
Cav: caveolin
CBF: coronary blood flow
CBF: core-binding factor
CBL: Casitas B-lineage lymphoma adaptor and Ub ligase
CBLb: CBL-related adaptor
CBP: cap-binding protein
CBP: CREB-binding protein
CBP: C-terminal Src kinase-binding protein
CBS: cystathionine β-synthase (H_2S production)
CCD: cortical collecting duct
CCDC: coiled-coil domain-containing protein
CCE: capacitative Ca^{2+} entry channel (SOC channel)
CCHS: congenital central hypoventilation syndrome
CCICR: calcium channel-induced Ca^{2+} release
CCK4: colon carcinoma kinase-4 (PTK7)
CCL: chemokine CC-motif ligand
CCN: CyR61, CTGF, and NOv (CCN1–CCN3) family
Ccn: cyclin
Ccnx–CDK*i*: type-x cyclin–type-*i* cyclin-dependent kinase dimer
CCPg: cell cycle progression protein
CCS: copper chaperone for superoxide dismutase
CCT: chaperonin containing T-complex protein
CCx: type-x chemokine CC (β)
CCR: chemokine CC motif receptor
CD: cluster determinant protein (cluster of differentiation)
CDase: ceramidase
CDC: cell division cycle protein
cDC: classical dendritic cell
CDH: CDC20 homolog
Cdh: cadherin
CDK: cyclin-dependent kinase
Cdm: caldesmon
CDO: cell adhesion molecule-related/ downregulated by oncogenes
CE (CsE): cholesteryl esters
CEC: circulating endothelial cell
CELSR: cadherin, EGF-like, LAG-like, and seven-pass receptor
CenP: centromere protein
CEP: carboxyethylpyrrole
CeP: centrosomal protein
CEPC: circulating endothelial progenitor cell
Cer: ceramide
CerK: ceramide kinase
CerT: ceramide transfer protein
CETP: cholesterol ester transfer protein
CFD: computational fluid dynamics
CFLAR: caspase-8 and FADD-like apoptosis regulator
CFTR: cystic fibrosis transmembrane conductance regulator
CFU: colony-forming unit
CFUb: CFU-basophil (basophil-committed stem cells)
CFUc: CFU in culture (granulocyte precursors, i.e., CFUgm)
CFUe: CFU-erythroid
CFUeo: CFU-eosinophil
CFUg: CFU-granulocyte
CFUgm: CFU-granulocyte–macrophage
CFUgemm: CFU-granulocyte–erythroid–macrophage–megakaryocyte
CFUm: CFU-macrophage
CFUmeg: CFU-megakaryocyte
CFUs: CFU-spleen (pluripotent stem cells)
CG: chromogranin
cGK: cGMP-dependent protein kinase (protein kinase-G)
cGMP: cyclic guanosine monophosphate
CGN: cis-Golgi network
CGRP: calcitonin gene-related peptide
chanzyme: ion channel and enzyme
chemokine: chemoattractant cytokine
CHIP: C-terminus heat shock cognate-70-interacting protein
ChK: checkpoint kinase
CHK: CSK homologous kinase
Chn: chimerin (GAP)
CHOP: CCAAT/enhancer-binding protein homologous protein
CHREBP: carbohydrate-responsive element-binding protein
ChT: choline transporter
CI: cardiac index
CICR: calcium-induced calcium release
Cin: chronophin
CIP: CDC42-interacting protein
CIP2a: cancerous inhibitor of protein phosphatase-2A
CIPC: CLOCK-interacting protein, circadian
CIS: cytokine-inducible SH2-containing protein

CITED: CBP/P300-interacting transactivator with glutamic (E) and aspartic acid (D)-rich C-terminus-containing protein
CK: creatine kinase
CK: casein kinase
CKI: cyclin-dependent kinase inhibitor
CLAsP: CLiP-associated protein (microtubule binder)
ClASP: clathrin-associated sorting protein
CLC: cardiotrophin-like cytokine
ClC: voltage-gated chloride channel
ClCa: calcium-activated chloride channel
ClIC: chloride intracellular channel
CLINT: clathrin-interacting protein located in the trans-Golgi network
CLIP: corticotropin-like intermediate peptide
CLiP: cytoplasmic CAP-Gly domain-containing linker protein
iCliP: intramembrane-cleaving peptidase (that clips)
CLK: CDC-like kinase
ClNS: Cl^- channel nucleotide-sensitive
CLOCK: circadian locomotor output cycles kaput
CLP: common lymphoid progenitor
CLS: ciliary localization signal
Cmi: chylomicron
CMLP: common myeloid–lymphoid progenitor
CMP: common myeloid progenitor
CMC: cardiomyocyte
Col: collagen
CoLec: collectin
ColF: collagen fiber
CORM: carbon monoxide (CO)-releasing molecule
CNG: cyclic nucleotide-gated channel
CnK: connector enhancer of kinase suppressor of Ras
CNS: central nervous system
CNT: connecting tubule
CNTi: concentrative nucleoside transporter (SLC28ai)
CNTF: ciliary neurotrophic factor
CntnAP: contactin-associated protein
CO: cardiac output
CoA: coenzyme-A
CoBl: Cordon-bleu homolog (actin nucleator)
COLD: chronic obstructive lung disease
COOL: cloned out of library (RhoGEF6/7)
coSMAD: common (mediator) SMAD (SMAD4)
COx: cyclooxygenase (prostaglandin endoperoxide synthase)
COx17: cytochrome-C oxidase copper chaperone
CoP: coat protein
CoP: constitutive photomorphogenic protein (Ub ligase)
COPD: chronic obstructive pulmonary disease
COUPTF: chicken ovalbumin upstream promoter transcription factor (NR2f1/2)
CP4H: collagen prolyl 4-hydroxylase
CPC: chromosomal passenger complex
CpG: cytidine^P–guanosine oligodeoxynucleotide (motif)
cPKC: conventional protein kinase-C
CPT: carnitine palmitoyltransferase
Cpx: complexin
CR: complement component receptor
Cr: creatine
cRABP: cellular retinoic acid-binding protein
cRBP: cellular retinol-binding protein
CRAC: Ca^{2+} release-activated Ca^{2+} channel
CRACR: CRAC regulator
Crb: Crumbs homolog polarity complex
CRE: cAMP-responsive element
CREB: cAMP-responsive element-binding protein
CRF: corticotropin-releasing factor (family)
CRH: corticotropin-releasing hormone
CRIB: CDC42/Rac interactive-binding protein
CRIK: citron Rho-interacting, Ser/Thr kinase (STK21)
CRK: chicken tumor virus CT10 regulator of kinase
CRKL: CRK avian sarcoma virus CT10 homolog-like
CRL4: cullin-4A RING ubiquitin ligase
CRLR: calcitonin receptor-like receptor
CRP: C-reactive protein
Crt: calreticulin
CRTC: CREB-regulated transcription coactivator
CRU: Ca^{2+} release unit (couplon or dyad)
Cry: cryptochrome
CS: coronary sinus
CS: citrate synthase
Cs: cholesterol
CSBP: cytokine-suppressive anti-inflammatory drug-binding protein
CSE: cystathionine γ-lyase (H_2S production)
CSF: cerebrospinal fluid
CSF: colony-stimulating factor
CSF1: macrophage colony-stimulating factor (mCSF)

CSF2: granulocyte–macrophage colony-stimulating factors (gmCSF and sargramostim)
CSF3: granulocyte colony-stimulating factors (gCSF and filgrastim)
CSK: C-terminal Src kinase
Csk: cytoskeleton
Csq: calsequestrin
CSS: candidate sphingomyelin synthase
CT: cardiotrophin
CT: computed tomography
CTBP: C-terminal-binding protein
CTen: C-terminal tensin-like protein
CTF: C-terminal fragment
CTGF: connective tissue growth factor
CTL: cytotoxic T lymphocyte
CTLA: cytotoxic T-lymphocyte-associated protein
Ctnn: catenin
CTr: copper transporter
CtR: calcitonin receptor
CTRC: CREB-regulated transcription coactivator
Cul: cullin
CUT: cryptic unstable transcript
CVI: chronic venous insufficiency
CVLM: caudal ventrolateral medulla
CVP: central venous pressure
CVS: cardiovascular system
Cx: connexin
CXCLi: type-i CXC (C–X–C motif; α) chemokine ligand
CXCRi: type-i CXC (C–X–C motif; α) chemokine receptor
$CX_3CL$$i$: type-$i$ CX_3C (δ) chemokine ligand
$CX_3CR$$i$: type-$i$ CX_3C (δ) chemokine receptor
cyCK: cytosolic creatine kinase
Cyld: cylindromatosis tumor suppressor protein (deubiquitinase USPL2)
CyP: member of the cytochrome-P450 superfamily
C3G: Crk SH3-binding GEF

D

D: dromotropy
D: vessel distensibility
$\mathcal{D}$: diffusion coefficient
$\mathcal{D}_T$: thermal diffusivity
D: deformation rate tensor
d: displacement vector
D: flexural rigidity
D: demobilization function (from proliferation to quiescence)
D_{RBC}: damage rate
D_f: fractal dimension
d: death, decay, degradation rate
d: distance
d: duration
d_h: hydraulic diameter
Dab: Disabled homolog
DAD: delayed afterdepolarization
DAG: diacylglycerol
DAPC: dystrophin-associated protein complex
DAPK: death-associated protein kinase
DARC: Duffy antigen receptor for chemokine
DAT: dopamine active transporter
DAX: dosage-sensitive sex reversal, adrenal hypoplasia critical region on chromosome X (NR0b1)
DBC: deleted in breast cancer protein
DBF: dumbbell formation kinase (in Saccharomyces cerevisiae; e.g., DBF2)
DBP: albumin D-element binding protein (PAR/b–ZIP family)
DC: dendritic cell
DCA: directional coronary atherectomy
DCAF: DDB1- and Cul4-associated factor
DCC: deleted in colorectal carcinoma (netrin receptor)
DCT: distal convoluted tubule
Dctn: dynactin
DDAH: dimethylarginine dimethylaminohydrolase
DDB: damage-specific DNA-binding protein
DDEF: development and differentiation-enhancing factor (ArfGAP)
DDR: discoidin domain receptor
De: Dean number
Deb: Deborah number
DEC: differentially expressed in chondrocytes (DEC1 and DEC2 are a.k.a bHLHe40 and bHLHe41, bHLHb2 and bHLHb3, or HRT2 and HRT1)
DEC: deleted in esophageal cancer
DEG: delayed-early gene
deoxyHb: deoxyhemoglobin (deoxygenated hemoglobin)
DES: drug-eluting stent
DETC: dendritic epidermal $\gamma\delta$ T cell
DGAT: diacylglycerol acyltransferase
DH: Dbl homology
DHA: docosahexaenoic acid
DHEA: dehydroepiandosterone
DHET: dihydroxyeicosatrienoic acid
DHh: desert Hedgehog

Dia: Diaphanous
DICOM: digital imaging and communication for medicine
DICR: depolarization-induced Ca^{2+} release
DISC: death-inducing signaling complex
Dkk: Dickkopf
DLg: Disc large homolog
DLL: Delta-like (Notch) ligand
DLx: distal-less homeobox protein
DM: double minute
DMM: DNA methylation modulator
DMPK: myotonic dystrophy-associated protein kinase
DMT: divalent metal transporter
DN1: double-negative-1 cell
DN2: double-negative-2 cell
DN3: double-negative-3 cell
DNA: deoxyribonucleic acid
DNAPK: DNA-dependent protein kinase
DoC2: double C2-like domain-containing protein
DOCK: dedicator of cytokinesis (GEF)
DOK: downstream of Tyr kinase docking protein
DOR: δ-opioid receptor
DPG: diphosphoglyceric acid
DPLD: diffuse parenchymal lung disease
DRAM: damage-regulated modulator of autophagy
DRF: Diaphanous-related formin (for GTPase-triggered actin rearrangement)
DRG: dorsal root ganglion
Drl: Derailed
Dsc: desmocollin
Dsg: desmoglein
Dsh: Disheveled (Wnt-signaling mediator)
DSK: dual-specificity kinase
dsRNA: double-stranded RNA
Dst: dystonin
DUb: deubiquitinase
DuOx: dual oxidase
DUS: Doppler ultrasound
DuSP: dual-specificity phosphatase
DV: dead space volume
Dvl: Disheveled (cytoplasmic phosphoprotein; other alias Dsh)
DVT: deep-vein thrombosis
dynactin: dynein activator
DYRK: dual-specificity Tyr (Y) phosphorylation-regulated kinase

E

$\mathbf{E}$: strain tensor
$\mathbf{E}$: electric field
E: elastance modulus
$\mathtt{E}$: elastance
$\mathcal{E}$: energy
$\{\hat{\mathbf{e}}_i\}_{i=1}^3$: basis
$\mathbf{e}$: strain vector
e: specific free energy
E-box: enhancer box sequence of DNA
E2: ubiquitin conjugase
E3: ubiquitin ligase
E_1: estrone (a single hydroxyl group in its molecule)
E_2: estradiol (2 hydroxyl groups), or 17β-estradiol
E_3: estriol (3 hydroxyl groups)
EAAT: excitatory amino acid (glutamate–aspartate) transporter
EAD: early afterdepolarization
EAR: V-ErbA-related nuclear receptor (NR2f6)
EB: end-binding protein
EBCT: electron beam CT
EBF: early B-cell factor
EC: endothelial cell
Ec: Eckert number
ECA: external carotid artery
ECF: extracellular fluid
ECG: electrocardiogram
ECM: extracellular matrix
ED1L: EGF-like repeat- and discoidin-1-like domain-containing protein
EDGR: endothelial differentiation gene receptor
EDHF: endothelial-derived hyperpolarizing factor
EDIL: EGF-like repeats and discoidin-1 (I)-like domain-containing protein
EDV: end-diastolic volume
EEA: early endosomal antigen
eEF: eukaryotic translation elongation factor
EEL: external elastic lamina
EET: epoxyeicosatrienoic acid
EFA6: exchange factor for ARF6 (ArfGEF)
EF-Tu: elongation factor Tu
EGF: epidermal growth factor
EGFL: EGF-like domain-containing protein

EGFR: epidermal growth factor receptor
EGR: early growth response transcription factor
EHD: C-terminal EGFR substrate-15 homology domain-containing protein
EHHADH: enoylCoA hydratase/3-hydroxyacylCoA dehydrogenase
eIF: eukaryotic translation initiation factor
EL: endothelial lipase
ELAM: endothelial–leukocyte adhesion molecules
ELCA: excimer laser coronary angioplasty
ELk: ETS-like transcription factor (ternary complex factor [TCF] subfamily)
ElMo: engulfment and cell motility adaptor
Eln: elastin
ElnF: elastin fiber
ELP: early lymphoid progenitor
EMI: early mitotic inhibitor
EMR: EGF-like module-containing, mucin-like, hormone receptor-like protein
EMT: epithelial–mesenchymal transition
ENA–VASP: Enabled homolog and vasoactive (vasodilator)-stimulated phosphoprotein family
ENaC: epithelial Na^+ channel
EnaH: Enabled homolog
endo-siRNA: endogenous small interfering RNA
ENPP: ectonucleotide pyrophosphatase–phosphodiesterase
Ens: endosulfine
ENT: equilibrative nucleoside transporter
ENTPD: ectonucleoside triphosphate diphosphohydrolase
EPAC: exchange protein activated by cAMP
EPAS: endothelial PAS domain protein
EPC: endothelial progenitor cell
EPCR: endothelial protein-C receptor
EPDC: epicardial-derived cell
Epgn: epigen (EGF superfamily member)
EPH: erythropoietin-producing hepatocyte receptor kinase or pseudokinase (EPHa10 and EPHb6)
ephrin: EPH receptor interactor
EPo: erythropoietin
EPS: epidermal growth factor receptor pathway substrate
ER: endoplasmic reticulum
ERx: type-x estrogen receptor (NR3a1/2)
eRas: embryonic stem cell-expressed Ras (or hRas2)
ErbB: erythroblastoma viral gene product-B (HER)
ERE: estrogen response element (DNA sequence)
Ereg: epiregulin (EGF superfamily member)
eRF: eukaryotic release factor
ERGIC: endoplasmic reticulum–Golgi intermediate compartment
ERK: extracellular signal-regulated protein kinase
ERK1/2: usually refers to ERK1 and ERK2
ERM: ezrin–radixin–moesin
ERMES: endoplasmic reticulum–mitochondrion encounter structure
ERP: effective refractory period
ERR: estrogen-related receptor (NR3b1–NR3b3)
ESCRT: endosomal sorting complex required for transport
ESL: E-selectin ligand
ESRP: epithelial splicing regulatory protein
ESV: end-systolic volume
ET: endothelin
ETC: electron transport chain
ETP: early thymocyte progenitor
ETR ($ET_{A/B}$): endothelin receptor
ETS: E-twenty six (transcription factor; erythroblastosis virus E26 proto-oncogene product homolog)
ETV: ETS-related translocation variant
EVAR: endovascular aneurysm repair
Exo: exocyst subunit
Ext: exostosin (glycosyltransferase)

F

$\mathbf{F}$: transformation gradient tensor
F: function fraction of proliferating cells
F: erythrocytic rouleau fragmentation rate
$\mathbf{f}$: surface force
$\hat{\mathbf{f}}$: fiber direction unit vector
f: binding frequency
f_C: cardiac frequency
f_R: breathing frequency
f: friction shape factor
f_v: head loss per unit length
f_X: molar fraction of gas component X
FA: fatty acid
FABP: fatty acid-binding protein
FABP: filamentous actin-binding protein
FACAP: ^{F}actin complex-associated protein
FACoA: fatty acylCoA

FACS: fatty acylCoA synthase
FAD: flavine adenine dinucleotide
FADD: Fas receptor-associated death domain
FAK: focal adhesion kinase
Fanc: Fanconi anemia protein
FAN: Fanconi anemia-associated nuclease
FAPP: phosphatidylinositol four-phosphate adaptor protein
Fas: death receptor (TNFRSF6a)
FasL: death ligand (TNFSF6)
FAST: Forkhead activin signal transducer
FATP: fatty acid transport protein (SLC27a)
FB: fibroblast
Fbln (Fibl): fibulin
Fbn: fibrillin
FBS: F-box, Sec7 protein (ArfGEF)
FBx: F-box only protein (ArfGEF)
FC: fibrocyte
FCHO: FCH domain only protein
Fcα R: Fc receptor of IgA
Fcγ R: Fc receptor of IgG
Fcϵ R: Fc receptor of IgE
FCP: TF2F-associating C-terminal domain phosphatase
FDM: finite difference method
FEM: finite element method
FERM: four point-1, ezrin–radixin–moesin domain
FeR: FeS-related Tyr kinase
FeS: feline sarcoma kinase
FFA: free fatty acid
FGF: fibroblast growth factor
FGFR: fibroblast growth factor receptor
FGR: viral feline Gardner–Rasheed sarcoma oncogene homolog kinase
FHL: four-and-a-half LIM-only protein
FHoD: formin homology domain-containing protein (FmnL)
FIH: factor inhibiting HIF1α (asparaginyl hydroxylase)
FIP: family of Rab11-interacting protein
FIP: focal adhesion kinase family-interacting protein
FIT: Fat-inducing transcript
FKBP: FK506-binding protein
FlIP: flice-inhibitory protein
FLK: fetal liver kinase
fMLP: ^{N}formyl methionyl-leucyl-phenylalanine
FN: fibronectin
Fn: fibrin
Fng: fibrinogen
Fos: Finkel Biskis Jinkins murine osteosarcoma virus sarcoma proto-oncogene product
Fox: forkhead box transcription factor
Fpn: ferroportin
FR: flow ratio
FRK: Fyn-related kinase
FrmD: FERM domain-containing adaptor
FRNK: FAK-related nonkinase
FRS: fibroblast growth factor receptor substrate
FSH: follicle-stimulating hormone
FSI: fluid–structure interaction
Fum: fumarase
FVM: finite volume method
FXR: farnesoid X receptor (NR1h4)
Fz: Frizzled (Wnt GPCR)

G

$\mathbf{G}$: Green–Lagrange strain tensor
G: shear modulus
G': storage modulus
G'': loss modulus
$\mathcal{G}$: Gibbs function
G: conductance
$\mathbf{G}$: conductivity tensor
G_{p}: pressure gradient
$\mathcal{G}_{\mathrm{b}}$: perfusion conductivity
$\mathcal{G}_{\mathrm{e}}$: electrical conductivity
$\mathcal{G}_{\mathrm{h}}$: hydraulic conductivity
$\mathcal{G}_{\mathrm{T}}$: thermal conductivity
$\mathbf{g}$: gravity acceleration
g: physical quantity
g: gravity
g: detachment frequency
g: free enthalpy
G protein: guanine nucleotide-binding protein (Gαβγ trimer)
Gα: α subunit (signaling mediator) of G protein
$G\alpha_i$ (Gi): inhibitory Gα subunit
$G\alpha_s$ (Gs): stimulatory Gα subunit
$G\alpha_t$ (Gt): transducin, Gα subunit of rhodopsin
Gs_{XL}: extralarge Gs protein
$G\alpha_{i/o}$ (Gi/o): Gα subunit class
$G\alpha_{q/11}$ (Gq/11): Gα subunit class
$G\alpha_{12/13}$ (G12/13): Gα subunit class
Gβγ: dimeric subunit (signaling effector) of G protein
G_{gust}: gustducin, G protein α subunit (Gi/o) of taste receptor
G_{olf}: G protein α subunit (Gs) of olfactory receptor
GAB: GRB2-associated binder
GABA: γ-aminobutyric acid

GABA$_A$: GABA ionotropic receptor (Cl^- channel)
GABA$_B$: GABA metabotropic receptor (GPCR)
GABARAP: GABA$_A$ receptor-associated protein
GaBP: globular actin-binding protein
GADD: growth arrest and DNA-damage-induced protein
gadkin: γ1-adaptin and kinesin interactor
GAG: glycosaminoglycan
GAK: cyclin G-associated kinase
Gal: galanin
GAP: GTPase-activating protein
GAPDH: glyceraldehyde 3-phosphate dehydrogenase
GARP: Golgi body-associated retrograde protein complex
GAS: growth arrest-specific noncoding, single-stranded RNA
GAT: γ-aminobutyric acid transporter
GATA: DNA sequence GATA-binding transcription factor
GBF: Golgi body-associated brefeldin-A-resistant guanine nucleotide-exchange factor
GBP: guanylate-binding protein
GCAP: guanylate cyclase-activating protein
GCC: Golgi coiled-coil domain-containing protein
GCK: germinal center kinase
GCKR: GCK-related kinase
GCNF: germ cell nuclear factor (NR6a1)
GCN2: general control non-derepressible-2 (pseudokinase)
GCS: glutamylcysteine synthase
gCSF: granulocyte colony-stimulating factor (CSF3)
GCV: great cardiac vein
GD: disialoganglioside
GDP: guanosine diphosphate
GDF: growth differentiation factor
GDF: (Rab)GDI displacement (dissociation) factor
GDI: guanine nucleotide-dissociation inhibitor
GDNF: glial cell line-derived neurotrophic factor
GEF: guanine nucleotide (GDP-to-GTP)-exchange factor
GF: growth factor
GFAP: glial fibrillary acidic protein (intermediate filament)
GFL: GDNF family of ligands
GFP: geodesic front propagation
GFR: growth factor receptor
GFRαi: type-i GDNF family receptor-α
GGA: Golgi body-localized γ-adaptin ear-containing Arf-binding protein
Ggust: (G protein) Gα subunit gustducin
GH: growth hormone
GHR: growth hormone receptor
GHRH: growth hormone-releasing hormone
GIP: GPCR-interacting protein
GIRK: G$\beta\gamma$-regulated inwardly rectifying K^+ channel
GIT: GPCR kinase-interacting protein
GKAP: G-kinase-anchoring protein
GKAP: glucokinase-associated phosphatase (DuSP12)
GKAP: guanylate kinase-associated protein
GLK: GCK-like kinase
GluK: ionotropic glutamate receptor (kainate type)
GluN: ionotropic glutamate receptor (NMDA type)
GluR: ionotropic glutamate receptor (AMPA type)
GluT: glucose transporter
GlyCAM: glycosylation-dependent cell adhesion molecule
GlyR: glycine receptor (channel)
GlyT: glycine transporter
GM: monosialoganglioside
gmCSF: granulocyte–monocyte colony-stimulating factor (CSF2)
GMP: granulocyte–monocyte progenitor
GMP: guanosine monophosphate
GnRH: gonadotropin-releasing hormone
GP: glycoprotein
Gpc: glypican
GPI: glycosylphosphatidylinositol anchor
gpiAP: GPI-anchored protein
GPCR: G-protein-coupled receptor
GPx: glutathione peroxidase
GQ: quadrisialoganglioside
GR: glucocorticoid receptor (NR3c1)
Gr: Graetz number
GRAP: GRB2-related adaptor protein (or GAds)
GRB: growth factor receptor-bound protein
GRC: growth factor-regulated, Ca^{2+}-permeable, cation channel (TRPV2)
GRE: glucocorticoid response element (DNA sequence)
GRHL: grainyhead-like transcription factor
GRK: G-protein-coupled receptor kinase
GRP: G-protein-coupled receptor phosphatase
G^{SH} (GSH): reduced form of glutathione

G^{SS} (GSSG): oxidized form of glutathione (glutathione disulfide)
GSK: glycogen synthase kinase
GSR: glutathione disulfide reductase
GsS: glutathione synthase
GST: glutathione ^{S}transferase
GT: trisialoganglioside
GTF: general transcription factor
GTP: guanosine triphosphate
GTPase: guanosine triphosphatase
GuCy: guanylate cyclase (CyG)
GWAS: genome-wide association study

H

H: height
$\mathcal{H}$: history function
H: dissipation
H: Henry parameter (solubility)
h: head loss
h: thickness
h: specific enthalpy
h_m: mass transfer coefficient
h_T: heat transfer coefficient
HA: hyaluronic acid
HAD: haloacid dehalogenase
HADH: hydroxyacylCoA dehydrogenase
HAP: huntingtin-associated protein
HAT: histone acetyltransferase
HAAT: heterodimeric amino acid transporter
HAND: heart and neural crest derivatives expressed protein
Hb: hemoglobin
Hb^{SNO}: ^{S}nitrosohemoglobin
HBEGF: heparin-binding EGF-like growth factor
HCK: hematopoietic cell kinase
HCLS: hematopoietic lineage cell-specific Lyn substrate protein
HCN: hyperpolarization-activated, cyclic nucleotide-gated K^+ channel
HCNP: hippocampal cholinergic neurostimulatory peptide
HCT: helical CT
HDAC: histone deacetylase complex
HDL: high-density lipoprotein
HDL^{Cs}: HDL–cholesterol
HDL^{CsE}: HDL–cholesteryl ester
HDM: human double minute (Ub ligase)
HEET: hydroxyepoxyeicosatrienoic acid
hemin: heme oxygenase-1 inducer
HERG: human ether-a-go-go related gene
HER: human epidermal growth factor receptor (HER3: pseudokinase)
HES: Hairy enhancer of split
HETE: hydroxyeicosatetraenoic acid
HETEE: HETE ethanolamide
HEV: high endothelial venule
HGF: hepatocyte growth factor
HGFA: hepatocyte growth factor activator (serine peptidase)
HGFR: hepatocyte growth factor receptor
HGNET: high-grade neuroendocrine tumor
HGS: HGF-regulated Tyr kinase substrate (HRS)
HhIP: Hedgehog-interacting protein
HIF: hypoxia-inducible factor
HIP: huntingtin-interacting protein
HIP1R: HIP1-related protein
His: histamine
Hjv: hemojuvelin
HK: hexokinase
HL: hepatic lipase
HMG: high-mobility group protein
HMGB: high-mobility group box protein
HMGCL: HMGCoA lyase
HMGCoA: 3-hydroxymethylglutarylCoA;
HMGCoAR (HMGCR): hydroxymethylglutaryl coenzyme-A reductase
HMGCS: HMGCoA synthase
HMT: histone methyl transferase
HMWK: high-molecular-weight kininogen
HNF: hepatocyte nuclear factor (NR2a1/2)
HNP: human neutrophil peptide
hnRNP: heterogeneous nuclear ribonucleoprotein
HODE: hydroxyoctadecadienoic acid
HOP: HSP70–HSP90 complex-organizing protein
HoPS: homotypic fusion and vacuole protein sorting complex
HotAIR: HOX antisense intergenic RNA (large intergenic noncoding RNA)
HOx: heme oxygenase
Hox: homeobox DNA sequence (encodes homeodomain-containing morphogens)
HpCa: hippocalcin
HPETE: hydroperoxyeicosatetraenoic acid
HPETEE: HPETE ethanolamide
HPK: hematopoietic progenitor kinase (MAP4K)
hpRNA: long hairpin RNA
hRas: Harvey Ras
HRE: hormone response element (DNA sequence)

HRM: hypoxia-regulated microRNA
hRNP: heterogeneous ribonucleoprotein
HRS: hepatocyte growth factor-regulated Tyr kinase substrate
HRT: Hairy and enhancer of Split-related transcription factor
HS: heparan sulfate
HSC: hematopoietic stem cell
HSC: heat shock cognate
HSER: heat stable enterotoxin receptor (guanylate cyclase-2C)
HSP: heat shock protein (chaperone)
HSPG: heparan sulfate proteoglycan
Ht: hematocrit
HTR: high temperature requirement endopeptidase
HUNK: hormonally upregulated Neu-associated kinase

I

I: identity tensor
I: inotropy
i: current
IAP: inhibitor of apoptosis protein
IBABP: intestinal bile acid-binding protein
IC: isovolumetric contraction
ICA: internal carotid artery
ICAM: intercellular adhesion molecule (IgCAM member)
ICDH: isocitrate dehydrogenase
IgCAM: immunoglobulin-like cell adhesion molecule
ICF: intracellular fluid
ICliP: intramembrane-cleaving peptidase
ID: inhibitor of DNA binding
IDL: intermediate-density lipoprotein
IDmiR: immediately downregulated microRNA
IDOL: inducible degrader of LDL receptor (Ub ligase)
IEG: immediate-early gene
IEL: internal elastic lamina
IEL: intraepithelial lymphocyte
IfIH: interferon-induced with helicase-C domain-containing protein
Ifn: interferon
IfnAR: interferon-α/β/ω receptor
IFT: intraflagellar transport complex
Ig: immunoglobulin
IGF: insulin-like growth factor
IGFBP: IGF-binding protein
IgHC: immunoglobulin heavy chain
IgLC: immunoglobulin light chain
iGluR: ionotropic glutamate receptor
IH: intimal hyperplasia
IHh: indian Hedgehog
IK: intermediate-conductance Ca^{2+}-activated K^+ channel
Iκ B: inhibitor of NFκ B
IKK: Iκ B kinase
IL: interleukin
iLBP: intracellular lipid-binding protein
ILC: innate lymphoid cell
ILD: interstitial lung disease
ILK: integrin-linked (pseudo)kinase
ILKAP: integrin-linked kinase-associated Ser/Thr phosphatase-2C
IMH: intramural hematoma
IMM: inner mitochondrial membrane
IMP: Impedes mitogenic signal propagation
INADl: inactivation no after-potential D-protein
InCenP: inner centromere protein
InF: inverted formin
InsIG: insulin-induced gene product (ER anchor)
InsL: insulin-like peptide
InsR (IR): insulin receptor
InsRR: insulin receptor-related receptor
IP: inositol phosphate
IP_3: inositol (1,4,5)-trisphosphate
IP_3R: IP_3 receptor (IP_3-sensitive Ca^{2+}-release channel)
IP_4: inositol (1,3,4,5)-tetrakisphosphate
IP_5: inositol pentakisphosphate
IP_6: inositol hexakisphosphate
IPCEF: interaction protein for cytohesin exchange factor
IPOD: (perivacuolar) insoluble protein deposit
IPP: inositol polyphosphate phosphatase
IPP: ILK–PINCH–parvin complex
iPSC: induced pluripotent stem cell
IQGAP: IQ motif-containing GTPase-activating protein (IQ: first 2 amino acids of the motif: isoleucine [I; commonly] and glutamine [Q; invariably]).
IR: isovolumetric relaxation

IRAK: IL1 receptor-associated kinase (IRAK2: pseudokinase)
IRE: irreversible electroporation
IRES: internal ribosome entry site
IRF: interferon-regulatory protein (transcription factor)
IRFF: interferon-regulatory factor family
IRP: iron regulatory protein
IRS: insulin receptor substrate
ISA: intracranial saccular aneurysm
ISG: interferon-stimulated gene product
iSMAD: inhibitory SMAD (SMAD6 or SMAD7)
ITAM: immunoreceptor tyrosine-based activation motif
Itch: Itchy homolog (Ub ligase)
Itg: integrin
ITIM: immunoreceptor tyrosine-based inhibitory motif
ITK: interleukin-2-inducible T-cell kinase
ITPK: inositol trisphosphate kinase
IVC: inferior vena cava
IVP: initial value problem
IVUS: intravascular ultrasound

J

J: flux
J_m: cell surface current density
JAM: junctional adhesion molecule
JaK: Janus (pseudo)kinase
JIP: JNK-interacting protein (MAPK8IP1 and -2)
JMy: junction-mediating and regulatory protein
JNK: Jun N-terminal kinase (MAPK8–MAPK10)
JNKBP: JNK-binding protein
JNKK: JNK kinase
JP: junctophilin
JSAP: JNK/SAPK-associated protein
jSR: junctional sarcoplasmic reticulum
Jun: avian sarcoma virus-17 proto-oncogene product (Japanese juunana: seventeen [17]; TF)
JUNQ: juxtanuclear quality-control compartment

K

$\mathbf{K}$: conductivity tensor
K: bending stiffness
K: reflection coefficient
K_d: dissociation constant (index of ligand–target affinity: ([L][T])/[C]; [L], [T], [C]: molar concentrations of the ligand, target, and created complex, respectively)
K_d: dissociation constant
K_M: Michaelis constant (chemical reaction kinetics)
K_m: material compressibility
K_R: resistance coefficient
k: cross-section ellipticity
k_ATP: myosin ATPase rate
k_B: Boltzmann constant (1.38×10^{-23} J/K)
k_c: spring stiffness
k_i: kinetic coefficient
k_m: mass transfer coefficient
k_P: Planck constant
KaP: karyopherin
$\mathrm{K_{ATP}}$: ATP-sensitive K^+ channel
$\mathrm{K_{Ca}}$1.x: BK channel
$\mathrm{K_{Ca}}$2/3/4.x: SK channel
$\mathrm{K_{Ca}}$5.x: IK channel
$\mathrm{K_{IR}}$: inwardly rectifying K^+ channel
$\mathrm{K_V}$: voltage-gated K^+ channel
KAP: kinesin (KIF)-associated protein
Kap: karyopherin
KAT: lysine (K) acetyltransferase
KCC: K^+–Cl^- cotransporter
KChAP: K^+ channel-associated protein
KChIP: $\mathrm{K_V}$ channel-interacting protein
KDELR: KDEL (Lys–Asp–Glu–Leu) endoplasmic reticulum retention receptor
KDR: kinase insert domain receptor
KGDH: ketoglutarate dehydrogenase
KHC: kinesin heavy chain
KIF: kinesin family
KIR: killer cell immunoglobulin-like receptor
KIT: cellular kinase in tyrosine (SCFR)
Kk: kallikrein
KLC: kinesin light chain
KLF: Krüppel-like factor
KLR: killer cell lectin-like receptor
Kn: Knudsen number
KOR: κ-opioid receptor

kRas: Kirsten Ras
Krt: keratin
KSR: kinase suppressor of Ras (adaptor; pseudokinase)

L

L: velocity gradient tensor
L: inertance
L: length
L_e: entry length
LA: left atrium
LAB: linker of activated B lymphocyte
LAd: LCK-associated adaptor
LAMTOR: late endosomal and lysosomal adaptor, MAPK and TOR activator
LANP: long-acting natriuretic peptide
LAP: leucine-rich repeat and PDZ domain-containing protein (4-member family)
LAP: latency-associated peptide (4 isoforms LAP1–LAP4)
LAP: nuclear lamina-associated polypeptide
LAR: leukocyte common antigen-related receptor (PTPRF)
LAT: linker of activated T lymphocytes
LaTS: large tumor suppressor
LAX: linker of activated X cells (both B and T cells)
LBR: lamin-B receptor
LCA: left coronary artery
LCAD: long-chain acylCoA dehydrogenase
LCAT: lysolecithin cholesterol acyltransferase
LCC: left coronary cusp
lcFA: long-chain fatty acid (10–16 carbon atoms)
LCK: leukocyte-specific cytosolic (nonreceptor) Tyr kinase
LCP: lymphocyte cytosolic protein (adaptor SLP76)
LDH: lactate dehydrogenase
LDL: low-density lipoprotein
LDL^{Cs}: LDL–cholesterol
LDLR: low-density lipoprotein receptor
LDV: laser Doppler velocimetry
Le: entry length
LEF: lymphoid enhancer-binding transcription factor
LGalS: lectin, galactoside-binding, soluble cell adhesion molecule
LGIC: ligand-gated ion channel
LGL: lethal giant larva protein
LH: luteinizing hormone
LIF: leukemia-inhibitory factor
LIFR: leukemia-inhibitory factor receptor
LIMA: LIM domain and actin-binding protein
LIME: LCK-interacting molecule
LIMK: Lin1, Isl1, and Mec3 motif-containing kinase
LIMS: LIM and senescent cell antigen-like-containing domain protein
LiNC: linker of nucleoskeleton and cytoskeleton
lincRNA: large intergenic noncoding RNA
LipC: hepatic lipase
LipD: lipoprotein lipase
LipE: hormone-sensitive lipase
LipG: endothelial lipase
LipH: lipase-H
liprin: LAR PTP-interacting protein
LIR: leukocyte immunoglobulin-like receptor
LIS: lissencephaly protein
LKb: liver kinase-B
LKLF: lung Krüppel-like factor
LLTC: large latent TGFβ complex
LMan: lectin, mannose-binding
LMO: LIM domain-only-7 protein
Lmod: leiomodin (actin nucleator)
LMPP: lymphoid-primed multipotent progenitor
LMR: laser myocardial revascularization
Lam: laminin
LOx: lipoxygenase
LOxLDLR: lectin-type oxidized low-density lipoprotein receptor
LP: lipoprotein
LPA: lysophosphatidic acid
LPLase: lipoprotein lipase
lpDC: lamina propria dendritic cell
Lphn: latrophilin (adhesion GPCR)
LPL: lysophospholipid
LPLase: lysophospholipase
LPP: lipid phosphate phosphatase
LPR: lipid phosphatase-related protein
LPS: lipopolysaccharide
LQTS: long-QT syndrome
LRAT: lecithin–retinol acyltransferase
LRH: liver receptor homolog (NR5a2)
LRO: lysosome-related organelle
LRP: LDL receptor-related protein
LRRTM: leucine-rich repeat-containing transmembrane protein
LSK: Lin−, SCA1+, KIT+ cell
LST: lethal with Sec-thirteen
LSV: long saphenous vein
LT (Lkt): leukotriene

LTBP: latent TGFβ-binding protein
LTCC: L-type Ca^{2+} channel (Ca_V1)
LTK: leukocyte tyrosine kinase
LUbAC: linear ubiquitin chain assembly complex
LV: left ventricle
LVAD: left ventricular assist device
LX: lipoxin
LXR: liver X receptor (NR1h2/3)
LyVE: lymphatic vessel endothelial hyaluronan receptor

M

M: molar mass
M: metabolic rate
$\mathcal{M}$: moment
m: mass
Ma: Mach number
MACF: microtubule–actin crosslinking factor
mAChR: acetylcholine muscarinic receptor (metabotropic; GPCR)
MaCoA: malonylCoA
MAD: mothers against decapentaplegic homolog
MAD: mitotic arrest-deficient protein
MAdCAM: mucosal vascular addressin cell adhesion molecule
MAF: musculoaponeurotic fibrosarcoma oncogene homolog (TF)
MAGI: membrane-associated guanylate kinase-related protein with inverted domain organization
MAGP: microfibril-associated glycoprotein
MAGuK: membrane-associated guanylate kinase
MAIT: mucosal-associated invariant T lymphocyte
MALT: mucosa-associated lymphoid tissue
MALT1: mucosa-associated lymphoid tissue lymphoma translocation peptidase
MAO: monoamine oxidase
MAP: microtubule-associated protein
MAP1LC3: microtubule-associated protein-1 light chain-3 (LC3)
mAP: mean arterial pressure
MAPK: mitogen-activated protein kinase
MAP2K: MAPK kinase
MAP3K: MA2KP kinase
MAP3K7IP: MAP3K7-interacting protein
MAPKAPK: MAPK-activated protein kinase
MARCKS: myristoylated alanine-rich C kinase substrate
MaRCo: macrophage receptor with collagenous structure (ScaRa2)
MARK: microtubule affinity-regulating kinase
MASTL: microtubule-associated Ser/Thr kinase-like protein
MAT: ménage à trois
MATK: megakaryocyte-associated Tyr kinase
MAVS: mitochondrial antiviral signaling protein
MAX: MyC-associated factor-X (bHLHd4–bHLHd8)
MBP: myosin-binding protein
MBP: myeloid–B-cell progenitor
MBTPSi: membrane-bound transcription factor peptidase site i
MCAD: medium [C4 to C12]-chain acylCoA dehydrogenase
MCAK: mitotic centromere-associated kinesin
MCAM: melanoma cell adhesion molecule
MCC: monocarboxylate carrier
MCD: medullary collecting duct
MCD: malonylCoA decarboxylase
mcFA: medium-chain fatty acid (6–12 carbon atoms)
MCL1: BCL2-related myeloid cell leukemia sequence protein-1
MCLC: stretch-gated Mid1-related chloride channel
MCM: minichromosome maintenance protein
MCP: monocyte chemoattractant protein
mCSF: macrophage colony-stimulating factor (CSF1)
MCT: monocarboxylate–proton cotransporter
mDC: myeloid dendritic cell
MDH: malate dehydrogenase
MDM: mitochondrial distribution and morphology protein
MDR: multiple drug resistance (ABC transporter)
MEF: myocyte enhancer factor
megCSF: megakaryocyte colony-stimulating factor
MEJ: myoendothelial junction
MELK: maternal embryonic leucine zipper kinase
MEP: megakaryocyte erythroid progenitor
MEP: myeloid–erythroid progenitor
MET: mesenchymal–epithelial transition factor (proto-oncogene; HGFR)
MFO: mixed-function oxidase
metHb: methemoglobin
MGIC: mechanogated ion channel

mGluR: metabotropic glutamate receptor
MGP: matrix γ-carboxyglutamate acid (Gla)-containing protein
MHC: major histocompatibility complex
MHC: myosin heavy chain
MyHC or MYH: myosin heavy chain gene
Mid: midline
MIF: macrophage migration-inhibitory factor
MinK: misshapen-like kinase
miR: microRNA
miRNP: microribonucleoprotein
MiRP: MinK-related peptide
MIRR: multichain immune-recognition receptor
MIS: Müllerian-inhibiting substance
MIS: mini-invasive surgery
MIS: mitochondrial intermembrane space
MIST: mastocyte immunoreceptor signal transducer
MIT: mini-invasive therapy
MiV: mitral valve
MIZ: Myc-interacting zinc finger protein
MJD: Machado–Joseph disease protein domain-containing peptidase (DUb)
MKL: megakaryoblastic leukemia-1 fusion coactivator
MKnK: MAPK-interacting protein Ser/Thr kinase (MnK)
MKP: mitogen-activated protein kinase phosphatase
MLC: myosin light chain
MLCK: myosin light chain kinase
MLCP: myosin light chain phosphatase
MLK: mixed lineage kinase
MLKL: mixed lineage kinase-like pseudokinase
MLL: mixed lineage [myeloid–lymphoid] leukemia factor
MLLT: mixed lineage leukemia translocated protein
MLP: muscle LIM protein
mmCK: myofibrillar creatine kinase
MME: membrane metalloendopeptidase
MMM: maintenance of mitochondrial morphology protein
MMP: matrix metallopeptidase
MO: mouse protein
Mo: monocyte
MOMP: mitochondrial outer membrane permeabilization
MOR: μ-opioid receptor
MP: MAPK partner
MPF: mitosis (maturation)-promoting factor (CcnB–CDK1 complex)
MPG: ^{N}methylpurine (^{N}methyladenine)-DNA glycosylase
MPO: median preoptic nucleus
MPo: myeloperoxidase
MP_P: membrane protein, palmitoylated
MPP: multipotent progenitor
MR: mineralocorticoid receptor (NR3c2)
mRas: muscle Ras (or rRas3)
MRCK: myotonic dystrophy kinase-related CDC42-binding kinase
MRI: (nuclear) magnetic resonance imaging
mRNA: messenger RNA
mRNP: messenger ribonucleoprotein
MRTF: myocardin-related transcription factor
MSC: mesenchymal stem cell
MSH: melanocyte-stimulating hormone
MSIC: mechanosensitive ion channel
MSSCT: multislice spiral CT
MST: mammalian sterile-twenty-like kinase
MSt1R: macrophage-stimulating factor-1 receptor (RON)
MT: metallothionein
mtCK: mitochondrial creatine kinase
mtCU: mitochondrial calcium uniporter (Ca^{2+} uptake protein)
mtTE: mitochondrial thioesterase
mtETC: mitochondrial electron transport chain
mtTerF: mitochondrial transcription termination factor
MTM: myotubularin (myotubular myopathy-associated gene product)
mtMMP: membrane-type MMP (mt*i*MMP: type-*i* mtMMP)
MTMR: myotubularin-related phosphatase
MTOC: microtubule organizing center
MTP: myeloid–T-cell progenitor
MTP: microsomal triglyceride transfer protein
MUFA: monounsaturated fatty acid
MuRF: muscle-specific RING finger (Ub ligase)
MuSK: muscle-specific kinase
MVB: multivesicular body
MVE: multivesicular endosome (MVB)
MVO2: myocardial oxygen consumption
MWSS: maximal wall shear stress
MyB: myeloblastosis viral oncogene homolog (TF)
MyBPc: myosin-binding protein-C
MyC: myelocytomatosis viral oncogene homolog (TF)
MyD88: myeloid differentiation primary response gene product-88
MyHC: myosin heavy chain
MyLC or MYL: myosin light chain gene

MyPT: myosin phosphatase targeting subunit
MyT: myelin transcription factor

N

N: sarcomere number
$\hat{\mathbf{n}}$: unit normal vector
n: mole number
n: PAM density with elongation x
$\mathtt{n}$: myosin head density
$\mathcal{N}_A$: Avogadro number
N-terminus: amino (amine group NH_2) terminus
NAADP: nicotinic acid adenine dinucleotide phosphate
nAChR: acetylcholine nicotinic receptor (ionotropic; LGIC)
NAD: nicotine adenine dinucleotide
NADPH: reduced form of nicotinamide adenine dinucleotide phosphate
NAd: noradrenaline
NAF: nutrient-deprivation autophagy factor
NALT: nasal-associated lymphoid tissue
NAmPT: nicotinamide phosphoribosyltransferase
Nanog: ever young (Gaelic)
NAP: NCK-associated protein (NCKAP)
NAT: nucleobase–ascorbate transporter
NAT1: noradrenaline transporter
Na_V: voltage-gated Na^+ channel
NBC: Na^+–HCO_3^- cotransporters
NCC: noncoronary cusp
NCC: Na^+–Cl^- cotransporter
Ncdn: neurochondrin
NCK: noncatalytic region of Tyr kinase adaptor
NCoA: nuclear receptor coactivator
NCoR: nuclear receptor corepressor
NCR: natural cytotoxicity-triggering receptor
ncRNA: noncoding RNA
NCS: neuronal calcium sensor
NCKX: Na^+–Ca^{2+}–K^+ exchanger
NCLX: Na^+–Ca^{2+}–Li^+ exchanger
NCX: Na^+–Ca^{2+} exchanger
NDCBE: Na^+-dependent Cl^-–HCO_3^- exchanger
NecL: nectin-like molecule
NEDD: neural precursor cell expressed, developmentally downregulated
NDFIP: NEDD4 family-interacting protein
NeK: never in mitosis gene-A (NIMA)-related kinase
NES: nuclear export signal
NESK: NIK-like embryo-specific kinase
nesprin: nuclear envelope spectrin repeat protein
NET: neuroendocrine tumor
NeuroD: neurogenic differentiation protein
NF: neurofilament protein (intermediate filament)
NF: neurofibromin (RasGAP)
NFAT: nuclear factor of activated T cells
NFe2: erythroid-derived nuclear factor-2
NFH: neurofilament, heavy polypeptide
NFκ B: nuclear factor κ light chain enhancer of activated B cells
NFL: neurofilament, light polypeptide
NFM: neurofilament, medium polypeptide
NGAL: neutrophil gelatinase-associated lipocalin
NGF: nerve growth factor
Ngn: neogenin (netrin receptor)
NHA: Na^+–H^+ antiporter
NHE: Na^+–H^+ exchanger
NHERF: NHE regulatory factor
NHR: nuclear hormone receptor
NIc: nucleoporin-interacting protein
NIK: NFκ B-inducing kinase
NIK: NCK-interacting kinase
NIP: neointimal proliferation
NK: natural killer cell
NKCC: Na^+–K^+–$2Cl^-$ cotransporter
NKG: NK receptor group
NKT: natural killer T cell
NKx2: NK2 transcription factor-related homeobox protein
NLR: NOD-like receptor (nucleotide-binding oligomerization domain, Leu-rich repeat-containing)
NLS: nuclear localization signal
NMDAR: ^{N}methyl ^{D}aspartate receptor
NmU: neuromedin-U
NMVOC: nonmethane volatile organic compound
NO: nitric oxide (nitrogen monoxide)
NO_x: nitrogen oxides
NOD: nucleotide-binding oligomerization domain
NonO: non-POU domain-containing octamer-binding protein
NOR: neuron-derived orphan receptor (NR4a3)
NOS: nitric oxide synthase
NOS1: neuronal NOS
NOS1AP: NOS1 adaptor protein
NOS2: inducible NOS
NOS3: endothelial NOS
NOx: NAD(P)H oxidase

Noxa: damage (Latin)
NPAS: neuronal PAS domain-containing transcription factor
NPC: nuclear-pore complex
NPC: Niemann–Pick disease type-C protein
NPC1L: Niemann–Pick protein-C1-like
nPKC: novel protein kinase-C
NPY: neuropeptide-Y
NR: nuclear receptor
NRAP: nebulin-related actinin-binding protein
nRas: neuroblastoma Ras
NRBP: nuclear receptor-binding protein
NREM: nonrapid eye movement sleep
NRF: nuclear factor erythroid-derived-2 (NF-E2)-related factor
NRF1: nuclear respiratory factor-1
Nrg: neuregulin (EGF superfamily member)
Nrgn: neuroligin
Nrp: neuropilin (VEGF-binding molecule; VEGFR coreceptor)
NRPTP: nonreceptor protein Tyr phosphatase
NRSTK: nonreceptor Ser/Thr kinase
NRTK: nonreceptor protein Tyr kinase
NRx: nucleoredoxin
Nrxn: neurexin
NSC: nonselective cation channel
NSCLC: non-small-cell lung cancer
NSF: ^{N}ethylmaleimide-sensitive factor
NSLTP: nonspecific lipid-transfer protein
NST: nucleus of the solitary tract
NT: neurotrophin
NT5E: ecto-5′-nucleotidase
NTCP: sodium–taurocholate cotransporter polypeptide
NTF: N-terminal fragment
NTP: nucleoside triphosphate
NTPase: nucleoside triphosphate hydrolase superfamily member
NTRK: neurotrophic tyrosine receptor kinase (TRK)
NTRKR: neurotrophic protein Tyr receptor kinase-related protein ($ROR_{(RTK)}$)
NTS: nucleus tractus solitarius
Nu: Nusselt number
NuAK: nuclear AMPK-related kinase
NuP: nucleoporin (nuclear-pore complex protein)
NuRD: nucleosome remodeling and histone deacetylase
NuRR: nuclear receptor-related factor (NR4a2)
nWASP: neuronal WASP

O

^{O}Glc^{N}Ac: β^{N}acetyl ^{D}glucosamine
OCRL: oculocerebrorenal syndrome of Lowe phosphatase
OCT: optical coherence tomography
Oct: octamer-binding transcription factor
ODE: ordinary differential equation
OGA: ^{O}Glc^{N}Acase (β^{N}acetylglucosaminidase)
OHS: obesity hypoventilation syndrome
OI: osteogenesis imperfecta
OMCD: outer medullary collecting duct
OMM: outer mitochondrial membrane
ORC: origin recognition complex
ORF: open reading frame
ORP: OSBP-related protein
OSA: obstructive sleep apnea
OSHAS: obstructive sleep hypopnea–apnea syndrome
OSBP: oxysterol-binding protein
OSI: oscillatory shear index
OSM: oncostatin-M
OSMR: oncostatin-M receptor
OSR (OxSR): oxidative stress-responsive kinase
OTK: off-track (pseudo)kinase
OTU: ovarian tumor superfamily peptidase (deubiquitinase)
OTUB: otubain (Ub thioesterase of the OTU superfamily)
OVLT: organum vasculosum lamina terminalis
oxyHb: oxyhemoglobin (oxygenated hemoglobin)

P

$\mathcal{P}$: permeability
P: power
P: cell division rate
p: production rate
p: pressure
P_i: inorganic phosphate (free phosphate ion)
p_X: partial pressure of gas component X
PA: phosphatidic acid
PAAT: proton–amino acid transporter
PACS: phosphofurin acidic cluster sorting protein
PAF: platelet-activating factor

PAFAH: platelet-activating factor acetylhydrolase
PAG: phosphoprotein associated with glycosphingolipid-enriched microdomains
PAH: polycyclic aromatic hydrocarbon
PAH: pulmonary arterial hypertension
PAI: plasminogen activator inhibitor
PAK: P21 (CKI1a)-activated kinase
PALR: promoter-associated long RNA
PALS: protein associated with Lin-7
PAMP: pathogen-associated molecular pattern
PAMP: proadrenomedullin peptide
PAR: $poly^{ADP}ribose$
PAR: promoter-associated, noncoding RNA
PARi: type-i peptidase-activated receptor
Par: partitioning defective protein
PARG: $poly^{ADP}ribosyl$ glycosidase
PARP: $poly^{ADP}ribose$ polymerase
PASR: promoter-associated short RNA
PAT: pulse amplitude tonometry
PATJ: protein (PALS1) associated to tight junctions
PAU: penetrating atherosclerotic ulcer
Pax: paxillin
Pax$i$: paired box protein-i (transcription regulator)
PBC: pre-Bötzinger complex (ventilation frequency)
PBIP: Polo box-interacting protein
PC: phosphatidylcholine
PC: polycystin
PC: protein-C
PCMRV: phase-contrast MR velocimetry
PCr: phosphocreatine
PCT: proximal convoluted tubule
PCTP: phosphatidylcholine-transfer protein
PD: pharmacodynamics
pDC: plasmacytoid dendritic cell
PdCD: programmed cell death protein
PdCD6IP: PdCD-6-interacting protein
PdCD1Lg: programmed cell death-1 ligand
PDE: phosphodiesterase
PDE: partial differential equation
PDGF: platelet-derived growth factor
PDGFR: platelet-derived growth factor receptor
PDHK: pyruvate dehydrogenase kinase
PDHP: pyruvate dehydrogenase phosphatase
PDI: protein disulfide isomerase
PDK: phosphoinositide-dependent kinase
PDP: pyruvate dehydrogenase phosphatase
Pe: Péclet number
PE: phosphatidylethanolamine
PE: pulmonary embolism
PEBP: phosphatidylethanolamine-binding protein
PECAM: platelet–endothelial cell adhesion molecule
PEDF: pigment epithelium-derived factor (serpin-F1)
PEEP: positive end-expiratory pressure mechanical ventilation
PEn2: presenilin enhancer-2
PEO: proepicardial organ
Per: Period homolog
PERK: protein kinase-like endoplasmic reticulum kinase
PERP: P53 apoptosis effector related to peripheral myelin protein PMP22
PET: positron emission tomography
Pex: peroxin
PF: platelet factor
PFK: phosphofructokinase
pFRG: parafacial respiratory group
PG: prostaglandin
PGC: PPARγ coactivator
pGC: particulate guanylate cyclase
PGEA: prostaglandin ethanolamide
PGF: paracrine growth factor
PGG: prostaglandin glycerol ester
PGi2: prostacyclin
PGP: permeability glycoprotein
PGx: type-x (D, E, F, H, I) prostaglandin
PGxS: type-x prostaglandin synthase
PH: pleckstrin homology domain
PHD: prolyl hydroxylase
PhK: phosphorylase kinase
PHLPP: PH domain and Leu-rich repeat protein phosphatase
PI: phosphoinositide (phosphorylated phosphatidylinositol)
PI(4)P: phosphatidylinositol 4-phosphate
PI(i)PiK: phosphatidylinositol i-phosphate i-kinase
PI(i,j)P_2: phosphatidylinositol (i,j)-bisphosphate (PIP_2)
PI(3,4,5)P_3: phosphatidylinositol (3,4,5)-trisphosphate (PIP_3)
PI3K: phosphatidylinositol 3-kinase
PI3KAP: PI3K adaptor protein
PIiK: phosphatidylinositol i-kinase
PIAS: protein inhibitor of activated STAT (SUMo ligase)
PIC: preinitiation complex
PICK: protein that interacts with C-kinase
PIDD: P53-induced protein with a death domain

PIKE: phosphoinositide 3-kinase enhancer (GTPase; ArfGAP)
PIKK: phosphatidylinositol 3-kinase-related kinase (pseudokinase)
PIM: provirus insertion of Molony murine leukemia virus gene product
PIN: peptidyl prolyl isomerase interacting with NIMA
PINCH: particularly interesting new Cys–His protein (or LIMS1)
PInK: PTen-induced kinase
PIP: phosphoinositide monophosphate
PIP*i* K: phosphatidylinositol phosphate *i*-kinase
PIP_2: phosphatidylinositol bisphosphate
PIP_3: phosphatidylinositol triphosphate
PIPP: proline-rich inositol polyphosphate 5-phosphatase
PIR: paired immunoglobulin-like receptor
piRNA: P-element-induced wimpy testis-interacting (PIWI) RNA
PIRT: phosphoinositide-interacting regulator of TRP channels
PITP: phosphatidylinositol-transfer protein
Pitx: pituitary (or paired-like) homeobox transcription factor
PIV: particle image velocimetry
PIX: P21-activated kinase (PAK)-interacting exchange factor (RhoGEF6/7)
PK: pharmacokinetics
PK: protein kinase
PKA: protein kinase-A
PKB: protein kinase-B
PKC: protein kinase-C
PKD: protein kinase-D
PKG: protein kinase-G
PKL: paxillin kinase linker
PKM: pyruvate kinase muscle isozyme
PKMYT (MYT): membrane-associated Tyr/Thr protein kinase
PKN: protein kinase novel
Pkp: plakophilin
PL: phospholipase
PLA2: phospholipase-A2
PLC: phospholipase-C
PLD: phospholipase-D
PLb: phospholamban
PLd: phospholipid
PlGF: placental growth factor
PLK: Polo-like kinase
PLTP: phospholipid transfer protein
PLV: (coronary) posterolateral vein
Plxn: plexin
PM: particulate matter (particulates or particle pollution)
PMCA: plasma membrane Ca^{2+} ATPase
PML: promyelocytic leukemia protein
PMR: percutaneous (laser) myocardial revascularization
PMRT: protein arginine methyltransferase
Pn: plasmin
Png: plasminogen
PNS: peripheral nervous system
PoG: proteoglycan
PoM: pore membrane protein
Pon: paraoxonase
POP: persistent organic pollutant
POPx: partner of PIX
POSH: scaffold plenty of SH3 domains
POT: Protection of telomeres (single-stranded telomeric DNA-binding protein)
PP: protein phosphatase
PP3: protein phosphatase-3 (PP2b or calcineurin)
PPAR: peroxisome proliferator-activated receptor (NR1c1–3)
PPG: photoplethysmography
PPId: peptidyl prolyl isomerase-D
PPIP: monopyrophosphorylated inositol phosphate
$(PP)_2IP$: bispyrophosphorylated inositol phosphate
PPK: PIP kinase
PPM: protein phosphatase (magnesium-dependent)
PPR: pathogen-recognition receptor
PPRE: PPAR response element (DNA sequence)
PPTC: protein phosphatase T-cell activation (TAPP2c)
PR: progesterone receptor (NR3c3)
Pr: Prandtl number
PRC: protein regulator of cytokinesis
PRC: Polycomb repressive complex
pre-cDC: preclassical dendritic cell
pre-miR: precursor microRNA
preBotC: pre-Bötzinger complex
preKk: prekallikrein
PREx: PIP_3-dependent Rac exchanger (RacGEF)
PRG: plasticity-related gene product
PRH: prolactin-releasing hormone
pri-miR: primary microRNA
PRL: phosphatase of regenerating liver
Prl: prolactin
PrlR: prolactin receptor
PRMT: protein arginine (R) ^{N}methyltransferase

Prompt: promoter upstream transcript
Protor: protein observed with Rictor
PROX: prospero homeobox gene
Prox: PROX gene product (transcription factor)
PrP: processing protein
PRPK: P53-related protein kinase
PRR: pattern recognition receptor
PRR: prorenin and renin receptor
PRx: peroxiredoxin
PS: presenilin
PS: protein-S
PSC: pluripotent stem cell
PSD: postsynaptic density adaptor
PsD: postsynaptic density
PSEF: pseudo-strain energy function
PSer: phosphatidylserine
PSGL: P-selectin glycoprotein ligand
PSKh: protein serine kinase-H
Psm: proteasome subunit
PSTPIP: Pro–Ser–Thr phosphatase-interacting protein
PTA: plasma thromboplastin antecedent
PTES: paclitaxel-eluting stent
Ptc: Patched receptor (Hedgehog signaling)
PTCA: percutaneous transluminal coronary angioplasty
PtcH: Patched Hedgehog receptor
PTCRA: PTC rotational burr atherectomy
PtdCho (PC): phosphatidylcholine
PtdEtn (PE): phosphatidylethanolamine
PtdSer (PS): phosphatidylserine
PtdIns (PI): phosphatidylinositol
PTen: phosphatase and tensin homolog deleted on chromosome ten (phosphatidylinositol 3-phosphatase)
PTFE: polytetrafluoroethylene
PTH: parathyroid hormone
PTHRP: parathyroid hormone-related protein
PTK: protein Tyr kinase
PTK7: pseudokinase (RTK)
PTP: (mitochondrial) permeability transition pore
PTP: protein Tyr phosphatase
PTPn*i*: protein Tyr phosphatase nonreceptor type *i*
PTPR: protein Tyr phosphatase receptor
PTRF: RNA polymerase-1 and transcript release factor
PUFA: polyunsaturated fatty acid
PUMA: P53-upregulated modulator of apoptosis
PuV: pulmonary valve
PVF: PDGF- and VEGF-related factor
PVNH: paraventricular nucleus of hypothalamus
PVR: pulmonary vascular resistance
PWS: pulse wave speed
Px: pannexin
PXR: pregnane X receptor (NR1i2)
PYK: proline-rich tyrosine kinase
P2X: purinergic ligand-gated channel
P53AIP: P53-regulated apoptosis-inducing protein
$P75^{NTR}$: pan-neurotrophin receptor (TNFRSF16)

Q

$\mathtt{Q}$: material quantity
Q_e: electric current density
Q_T: thermal energy (heat)
q_T: transfer rate of thermal energy (power)
q_{met}: metabolic heat source
q: flow rate

R

R: resistance
$\mathcal{R}$: local reaction term
R_h: hydraulic radius
R_g: gas constant
R_R: respiratory quotient
R: recruitment function (from quiescence to proliferation)
r: cell renewal rate
$\mathtt{r}$: electrical resistivity
r: radial coordinate
RA: right atrium
RAAS: renin–angiotensin–aldosterone system
Rab: Ras from brain
Rab11FIP: Rab11 family-interacting protein
Rac: Ras-related C3-botulinum toxin substrate
RACC: receptor-activated cation channel
RACK: receptor for activated C-kinase
RAD: recombination protein-A (RecA)-homolog DNA-repair protein
Rad: radiation sensitivity protein
Rag: Ras-related GTP-binding protein
Ral: Ras-related protein
RAlBP: retinaldehyde-binding protein
RalGDS: Ral guanine nucleotide-dissociation stimulator

RAMP: (calcitonin receptor-like) receptor activity-modifying protein
Ran: Ras-related nuclear protein
RAP: receptor-associated protein
Rap: Ras-proximate (Ras-related) protein
Raptor: regulatory-associated protein of TOR
RAR: retinoic acid receptor (NR1b2/3)
Ras: rat sarcoma viral oncogene homolog (small GTPase)
RasA: Ras p21 protein activator
rasiRNA: repeat-associated small interfering RNA (PIWI)
RASR: rapidly adapting stretch receptor
RASSF: Ras interaction/interference protein RIN1, afadin, and Ras association domain-containing protein family member
RB: retinoblastoma protein
RBC: red blood capsule (cell, or erythrocyte [without nucleus])
RBP: retinoid-binding protein
RC: ryanodine calcium channel (RyR)
RCA: right coronary artery
RCan: regulator of calcineurin
RCC: right coronary cusp
RCC: regulator of chromosome condensation
Re: Reynolds number
REDD: regulated in development and DNA-damage response gene product
Rel: reticuloendotheliosis proto-oncogene product (TF; member of NFκ B)
REM: rapid eye movement sleep
REP: Rab escort protein
ReR: renin receptor (PRR)
restin: Reed–Steinberg cell-expressed intermediate filament-associated protein (CLiP1)
ReT: rearranged during transfection (receptor Tyr kinase)
RevRE: reverse (Rev)-ErbA (NR1d1/2) response element (DNA sequence)
RFA: radiofrequency ablation
RGL: Ral guanine nucleotide-dissociation stimulator-like protein (GEF)
RGS: regulator of G-protein signaling
RHEB: Ras homolog enriched in brain
RHI: reactive hyperemia index
RHS: equation right-hand side
Rho: Ras homologous
RIAM: $Rap1^{GTP}$-interacting adaptor molecule
RIBP: RLK- and ITK-binding protein
RICH: RhoGAP interacting with CIP4 homolog
RICK: receptor for inactive C-kinase
Rictor: rapamycin-insensitive companion of TOR
RIF: Rho in filopodium
RIN: Ras-like protein expressed in neurons (GTPase)
RIn: Ras and Rab interactor (RabGEF)
RIP: regulated intramembrane proteolysis
RIPK: receptor-interacting protein kinase
RIRR: ROS-induced ROS release
RISC: RNA-induced silencing complex
RIT: Ras-like protein expressed in many tissues
RKIP: Raf kinase inhibitor protein
RlBP: retinaldehyde-binding protein
RLC: RISC-loading complex
RLK: resting lymphocyte kinase (TXK)
RNA: ribonucleic acid
RNABP: RNA-binding protein
RNase: ribonuclease
RnBP: renin-binding protein
RNF2: RING finger protein-2 (Ub ligase)
RNP: ribonucleoprotein
RNS: reactive nitrogen species
Robo: roundabout
ROC: receptor-operated channel
RoCK: Rho-associated, coiled-coil-containing protein kinase
ROI: region of interest
ROMK: renal outer medullary potassium channel
ROR: RAR-related orphan receptor (NR1f1–NR1f3)
$ROR_{(RTK)}$: receptor Tyr kinase-like orphan receptor
ROS: reactive oxygen species
Ros: ros UR2 sarcoma virus proto-oncogene product (RTK)
RPIP: Rap2-interacting protein
RPS6: ribosomal protein S6
RPTP: receptor protein Tyr phosphatase
rRas: related Ras
rRNA: ribosomal RNA
RSA: respiratory sinus arrhythmia
RSE: rapid systolic ejection
RSK: P90 ribosomal S6 kinase (^{P90}RSK)
RSKL: ribosomal protein S6 kinase-like (pseudokinase)
rSMAD: receptor-regulated SMAD (SMAD1–SMAD3, SMAD5, and SMAD9)
RSMCS: robot-supported medical and surgical system
RSpo: R-spondin
RSTK: receptor Ser/Thr kinase
RTK: receptor protein Tyr kinase

RTN: retrotrapezoid nucleus
Rubicon: RUN domain and Cys-rich domain-containing, beclin-1-interacting protein
Runx: Runt-related transcription factor
RV: right ventricle
RVF: rapid ventricular filling
RVLM: rostral ventrolateral medulla
RVMM: rostral ventromedial medulla
RXR: retinoid X receptor (NR2b1–NR2b3)
RYK: receptor-like (or related to receptor) protein Tyr (Y) kinase (pseudokinase)
RyR: ryanodine receptor (ryanodine-sensitive Ca^{2+}-release channel)

S

S: Cauchy–Green deformation tensor
S: hemoglobin saturation of a given gas species (%)
s: entropy
s: sarcomere length
s: sieving coefficient
s: evolution speed
s: solubility
SAA: serum amyloid A
$SAC_{Cl(K)}$: stretch-activated Cl^- (K^+)-selective channel
SAc: suppressor of actin domain-containing 5-phosphatase
sAC: soluble adenylate cyclase
$SACC_{NS}$: stretch-activated cation nonselective channel
SACM1L: suppressor of actin mutation-1-like
SAH: subarachnoid hemorrhage
SAIC: stretch-activated ion channel
SAN: sinoatrial node
SAP: SLAM-associated protein
SAP: stress-activated protein
SAPi: synapse-associated protein i
SAPK: stress-activated protein kinase (MAPK)
SAR: secretion-associated and Ras-related protein
SASR: slowly adapting stretch receptor
SBE: SMAD-binding element
SBF: SET-binding factor
Sc: Schmidt number
SCA: stem cell antigen
SCAMP: secretory carrier membrane protein
SCAP: SREBP cleavage-activating protein (SREBP escort)
SCAR: suppressor of cAMP receptor (WAVe)
ScaR: scavenger receptor
SCF: SKP1–Cul1–F-box Ub-ligase complex
SCF: stem cell factor
scFA: short-chain fatty acid (4–8 carbon atoms)
SCFR: stem cell factor receptor (KIT)
Scgb: secretoglobin
SCLC: small-cell lung cancer
scLC: squamous-cell lung cancer (NSCLC subtype)
SCN: suprachiasmatic nucleus
SCO: synthesis of cytochrome-C oxidase
SCoAS: succinylCoA synthase
SCOT: succinylCoA:3-oxoacidCoA transferase
SCP (CTDSP): small C-terminal domain (CTD)-containing phosphatase
Scp: stresscopin (urocortin-3)
Scrib: Scribble polarity protein
SDH: succinate dehydrogenase
Sdc: syndecan
SDF: stromal cell-derived factor
SDPR: serum deprivation protein response
SE: systolic ejection
SEF: strain-energy function
SEF: similar expression to FGF genes (inhibitor of RTK signaling)
SEK: SAPK/ERK kinase
Sema: semaphorin (Sema, Ig, transmembrane, and short cytoplasmic domain)
SERCA: sarco(endo)plasmic reticulum calcium ATPase
serpin: serine peptidase inhibitor
SerT: serotonin transporter
SES: sirolimus-eluting stent
SF: steroidogenic factor (NR5a1)
SFK: SRC family kinase
SFO: subfornical organ
SFPQ: splicing factor proline and glutamine-rich
sFRP: secreted Frizzled-related protein
SftP (SP): surfactant protein
sGC: soluble guanylate cyclase
SGK: serum- and glucocorticoid-regulated kinase
SGlT: Na^+–glucose cotransporter (SLC5a)
Sgo: shugoshin (Japanese: guardian spirit)
SH: Src homology domain
Sh: Sherwood number
SH3P: Src homology-3 domain-containing adaptor protein
SHAnk: SH3 and multiple ankyrin repeat domain-containing protein
SHAX: SNF7 (VSP32) homolog associated with ALIX

SHB: Src homology-2 domain-containing adaptor
SHC: Src-homologous and collagen-like substrate
SHC: Src homology-2 domain-containing transforming protein
SHh: sonic Hedgehog
SHIP: SH-containing inositol phosphatase
SHP: SH-containing protein Tyr phosphatase (PTPn6/11)
SHP: small heterodimer partner (NR0b2)
shRNA: small (short) hairpin RNA
SIAH: Seven in absentia homolog (Ub ligase)
siglec: sialic acid-binding Ig-like lectin
SIK: salt-inducible kinase
SIn: stress-activated protein kinase-interacting protein
SIP: steroid receptor coactivator-interacting protein
siRNA: small interfering RNA
SiRP: signal-regulatory protein
SIRT: sirtuin (silent information regulator-2 [two]; histone deacetylase)
SIT: SHP2-interacting transmembrane adaptor
SK: small conductance Ca^{2+}-activated K^+ channel
SK*i*: sphingosine kinase-*i*
SKIP: sphingosine kinase-1-interacting protein
SKIP: skeletal muscle and kidney-enriched inositol phosphatase
SKP: S-phase kinase-associated protein
SLA: Src-like adaptor
SLAM: signaling lymphocytic activation molecule
SLAMF: SLAM family member
SLAP: Src-like adaptor protein
SLC: solute carrier superclass member
SLCO: solute carrier organic anion class transporter
SLK: Ste20-like kinase
Sln: sarcolipin
SLPI: secretory leukocyte peptidase inhibitor
SLTC: small latent TGFβ complex
SM: sphingomyelin
SMA: smooth muscle actin
SMAD: small (son of, similar to) mothers against decapentaplegia homolog
SMAP: Small ArfGAP protein, stromal membrane-associated GTPase-activating protein
SMase: sphingomyelinase
SMC: smooth muscle cell
Smo: Smoothened
SMPD: sphingomyelin phosphodiesterase
SMRT: silencing mediator of retinoic acid and thyroid hormone receptor (NCoR2)
SMS: sphingomyelin synthase
SMURF: SMAD ubiquitination regulatory factor
SNAAT: sodium-coupled neutral amino acid transporter
SNAP: soluble ^{N}ethylmaleimide-sensitive factor-attachment protein
SnAP: synaptosomal-associated protein
SNARE: SNAP receptor
SNF7: sucrose nonfermenting (VPS32)
SNIP: SMAD nuclear-interacting protein
snoRNA: small nucleolar RNA
snoRNP: small nucleolar ribonucleoprotein
SNP: single-nucleotide polymorphism
snRNA: small nuclear RNA
snRNP: small nuclear ribonucleoprotein
SNx: sorting nexin
SO_x: sulfur oxides
SOC: store-operated Ca^{2+} channel
SOCE: store-operated Ca^{2+} entry
SOCS: suppressor of cytokine signaling protein
SOD: superoxide dismutase
SorbS: sorbin and SH3 domain-containing adaptor
SOS: Son of sevenless (GEF)
Sost: sclerostin
SostDC: sclerostin domain-containing protein
SOX: sex-determining region Y (SRY)-box gene
Sox: SOX gene product (transcription factor)
SP1: specificity protein (transcription factor)
SPARC: secreted protein acidic and rich in cysteine
SPC: sphingosylphosphorylcholine
SPCA: secretory pathway Ca^{2+} ATPase
SPECT: single-photon emission CT
Sph: sphingosine
SphK: sphingosine kinase
SPI: spleen focus-forming virus (SFFV) proviral integration proto-oncogene product (transcription factor)
SPInt: serine peptidase inhibitor
SPN: supernormal period
SPP: sphingosine phosphate phosphatase
SpRED: Sprouty-related protein with an EVH1 domain
SPURT: secretory protein in upper respiratory tract
SQTS: short-QT syndrome
SR: sarcoplasmic reticulum

SR: Arg/Ser domain-containing protein (alternative splicing)
SRA: steroid receptor RNA activator
SRC: steroid receptor coactivator
Src: sarcoma-associated (Schmidt–Ruppin A2 viral oncogene homolog) kinase
SREBP: sterol regulatory element-binding protein
SRF: serum response factor
SRM/SMRS: Src-related kinase lacking regulatory and myristylation sites
SRP: stresscopin-related peptide (urocortin-2)
SRPK: splicing factor RS domain-containing protein kinase
SRY: sex-determining region Y
SSAC: shear stress-activated channel
SSE: slow systolic ejection
Ssh: slingshot homolog phosphatase
SSI: STAT-induced STAT inhibitor
SSR: sympathetic skin response
ssRNA: single-stranded RNA
Sst: somatostatin
SSV: short saphenous vein
St: Strouhal number
STAM: signal-transducing adaptor molecule
STAMBP: STAM-binding protein (Ub isopeptidase)
StAR: steroidogenic acute regulatory protein
StART: StAR-related lipid transfer protein
STAT: signal transducer and activator of transduction
STEAP: six transmembrane epithelial antigen of the prostate
STICK: substrate that interacts with C-kinase
StIM: stromal interaction molecule
STK: protein Ser/Thr kinase
STK1: stem cell protein Tyr kinase receptor
STLK: Ser/Thr kinase-like (pseudo)kinase
Sto: Stokes number
StRAd: Ste20-related adaptor
STRAP: Ser/Thr kinase receptor-associated protein
StRAP: stress-responsive activator of P300
Stx: syntaxin ($SNARE^Q$)
SUMo: small ubiquitin-related modifier
SUn: Sad1 and Unc84 homology protein
SUR: sulfonylurea receptor
SUT: stable unannotated transcript
SV: stroke volume
SVC: superior vena cava
SVCT: sodium-dependent vitamin-C transporter
SVF: slow ventricular filling
SVP: synaptic vesicle precursor
SVR: systemic vascular resistance
SVR: surface area-to-volume ratio
SW: stroke work
SwAP70: 70-kDa switch-associated protein (RacGEF)
SYK: spleen tyrosine kinase
Synj: synaptojanin
Syp: synaptophysin
Syt: synaptotagmin
S1P: sphingosine 1-phosphate
S6K: P70 ribosomal S6 kinase (^{P70}RSK)

T

T: extrastress tensor
T: transition rate from a cell cycle phase to the next
T: temperature
T_3: triiodothyronine
T_4: thyroxin
$\mathcal{T}$: transport parameter
$\mathcal{T}_L$: transfer capacity of the alveolocapillary membrane for gas species
T_s: surface tension
T lymphocyte (T cell): thymic lymphocyte
T_C: cytotoxic T lymphocyte (CD8+ effector T cell; CTL)
T_{C1}: type-1 cytotoxic T lymphocyte
T_{C2}: type-2 cytotoxic T lymphocyte
T_{CM}: central memory T lymphocyte
T_{Conv}: conventional T lymphocyte
T_{Eff}: effector T lymphocyte
T_{EM}: effector memory T lymphocyte
T_{FH}: follicular helper T lymphocyte
T_H: helper T lymphocyte (CD4+ effector T cell)
T_{Hi}: type-i helper T lymphocyte ($i = 1/2/9/17/22$)
T_{H3}: TGFβ-secreting T_{Reg} lymphocyte
T_L: lung transfer capacity (alveolocapillary membrane)
T_{R1}: type-1, IL10-secreting, regulatory T lymphocyte
T_{Reg}: regulatory T lymphocyte
aT_{Reg}: CD45RA−, FoxP3^{hi}, activated T_{Reg} cell
iT_{Reg}: inducible T_{Reg} lymphocyte
nT_{Reg}: naturally occurring (natural) T_{Reg} lymphocyte
rT_{Reg}: CD45RA+, FoxP3^{low}, resting T_{Reg} cell
$\hat{\mathbf{t}}$: unit tangent
t: time

Tβ Ri: type-i TGFβ receptor
TAA: thoracic aortic aneurysm
TAB: TAK1-binding protein
TACE: tumor-necrosis factor-α-converting enzyme (ADAM17)
TACE: transarterial chemoembolization
TAF: TBP-associated factor
TAK: TGFβ-activated kinase (MAP3K7)
TALK: TWIK-related alkaline pH-activated K^+ channel
TANK: TRAF family member-associated NFκ B activator
TASK: TWIK-related acid-sensitive K^+ channel
TASR: terminus-associated short RNA
TAP: transporter associated with antigen processing (ABC transporter)
TATN: transverse and axial tubular network
Taz: taffazin
TBC1D: Tre2 (or USP6), BUB2, CDC16 domain-containg RabGAP
TBCK: tubulin-binding cofactor kinase (pseudokinase)
TBK: TANK-binding kinase
TBP: TATA box-binding protein (subclass-4F transcription factor)
TBx: T-box transcription factor
TC: thrombocyte (platelet)
TCA: tricarboxylic acid cycle
TCC: tricarboxylate carrier
TCF: T-cell factor
TCF: ternary complex factor
TcFi: type-i transcription factor
TCP: T-complex protein
TCR: T-cell receptor
TEA: transluminal extraction atherectomy
TEC: Tyr kinase expressed in hepatocellular carcinoma
TEF: thyrotroph embryonic factor (PAR/b–ZIP family)
TEK: Tyr endothelial kinase
TEM: transendothelial migration
Ten: tenascin
TF: transcription factor
Tf: transferrin
TFM: traction force microscopy
TFPI: tissue factor pathway inhibitor
TfR: transferrin receptor
TG (TAG): triglyceride (triacylglycerol)
TGm: transglutaminase
TGF: transforming growth factor
TGFBR: TGFβ receptor gene
TGFβ RAP: TGFβ receptor-associated protein
TGN: trans-Golgi network
THET: trihydroxyeicosatrienoic acid
THIK: tandem pore-domain halothane-inhibited K^+ channel
THR: thyroid hormone receptor (NR1a1/2)
TIAM: T-lymphoma invasion and metastasis-inducing protein (RacGEF)
TICE: transintestinal cholesterol efflux
TIE: Tyr kinase with Ig and EGF homology domains (angiopoietin receptor)
TIEG: TGFβ-inducible early gene product
TIF: transcription intermediary factor (kinase and Ub. ligase)
TIGAR: TP53-inducible glycolysis and apoptosis regulator
TIM: T-cell immunoglobulin and mucin domain-containing protein
Tim: timeless homolog
TIMM: translocase of inner mitochondrial membrane
TIMP: tissue inhibitor of metallopeptidase
TIRAP: Toll–IL1R domain-containing adaptor protein
tiRNA: transcription initiation RNA
TJ: tight junction
TKR: Tyr kinase receptor
TLC: total lung capacity
TLR: Toll-like receptor
TLT: TREM-like transcript
TLX: tailless receptor (NR2e1)
TM: thrombomodulin
TMi: transmembrane segment-i of membrane protein
TMC: twisting magnetocytometry
TMePAI: transmembrane prostate androgen-induced protein
TMy: tropomyosin
Tnn (TN): troponin
Tn: thrombin
TNF: tumor-necrosis factor
TNFα IP: tumor-necrosis factor-α-induced protein
TNFR: tumor-necrosis factor receptor
TNFRSF: tumor-necrosis factor receptor superfamily member
TNFSF: tumor-necrosis factor superfamily member
TNK: Tyr kinase inhitor of NFκ B
Tns: tensin
TOR: target of rapamycin
TORC: target of rapamycin complex
TORC: transducer of regulated CREB activity (a.k.a. CRTC)
TP: thromboxane-A2 Gq/11-coupled receptor
TP53I: tumor protein P53-inducible protein

tPA: tissue plasminogen activator
TPo: thrombopoietin
TpM: tropomyosin
TPPP: tubulin polymerization-promoting protein
TPST: tyrosylprotein sulftotransferase
TR: testicular receptor (NR2c1/2)
TRAAK: TWIK-related arachidonic acid-stimulated K^+ channel
TRADD: tumor-necrosis factor receptor-associated death domain adaptor
TRAF: tumor-necrosis factor receptor-associated factor
TRAM: TRIF-related adaptor molecule
transceptor: transporter-related receptor
TRAP: TNF receptor-associated protein (HSP75)
TraPP: transport protein particle
TRAT: T-cell receptor-associated transmembrane adaptor
Trb: Tribbles homolog (pseudokinase)
TRE: TPA-response element (AP1/CREB-binding site on promoters)
TRE: trapped in endoderm
TREK: TWIK-related K^+ channel
TREM: triggering receptor expressed on myeloid cells
TRESK: TWIK-related spinal cord K^+ channel
TRF: TBP-related factor
TRF: double-stranded telomeric DNA-binding repeat-binding factor
TRH: thyrotropin-releasing hormone
TRIF: Toll–IL1R domain-containing adaptor inducing Ifnβ
TRIM: T-cell receptor-interacting molecule
TRIP: TGFβ receptor-interacting protein (eIF3S2)
TRK: tropomyosin receptor kinase (NTRK)
tRNA: transfer RNA
TRP: transient receptor potential channel
TRPA: ankyrin-like transient receptor potential channel
TRPC: canonical transient receptor potential channel
TRPM: melastatin-related transient receptor potential channel
TRPML: mucolipin-related transient receptor potential channel
TRPN: no mechanoreceptor potential C
TRPP: polycystin-related transient receptor potential channel
TRPV: vanilloid transient receptor potential channel
TrrAP: transactivation (transformation)/transcription domain-associated protein (pseudokinase)
TrV: tricuspid valve
TRx: thioredoxin
TRxIP: thioredoxin-interacting protein
TSC: tuberous sclerosis complex
TSH: thyroid-stimulating hormone
TSLP: thymic stromal lymphopoietin
Tsp: thrombospondin
Tspan: tetraspanin
TsPO: translocator protein of the outer mitochondrial membrane
tSNARE: target SNARE
tsRNA: tRNA-derived small RNA
tssaRNA: transcription start site-associated RNA
TTbK: Tau-tubulin kinase
TTG: tissue transglutaminase
TTK: dual-specificity Thr/Tyr kinase
Ttn: titin (pseudokinase)
TUT: terminal uridine transferase
TWIK: tandem of P domains in a weak inwardly rectifying K^+ channel
TxA2: thromboxane-A2 (thromboxane)
TxB2: thromboxane-B2 (thromboxane metabolite)
TXK: Tyr kinase mutated in X-linked agammaglobulinemia
TxaS: thromboxane-A synthase
TyK: tyrosine kinase
T_3: tri-iodothyronine
T_4: thyroxine
$^+$TP: plus-end-tracking proteins

U

U: right stretch tensor
u: displacement vector
u: electrochemical command, electrical potential
u: specific internal energy
Ub: ubiquitin
UbC: ubiquitin conjugase
UbE2: E2 ubiquitin conjugase
UbE3: E3 ubiquitin ligase
UbL: ubiquitin-like protein
UCH: ubiquitin C-terminal hydrolase (DUb)
Ucn: urocortin
UCP: uncoupling protein
UDP: uridine diphosphate
^{UDP}glucose: UDP–glucose

UK: urokinase
ULK: uncoordinated-51-like kinase (pseudokinase)
Unc: uncoordinated receptor
uPA: urokinase-type plasminogen activator (urokinase)
uPAR: uPA receptor
uPARAP: uPAR-associated protein (CLec13e)
UPR: unfolded protein response
UPS: ubiquitin–proteasome system
UP4A: uridine adenosine tetraphosphate
Uro: urodilatin
US: ultrasound
USC: unipotential stem cell
USF: upstream stimulatory factor
USI: ultrasound imaging
USP: ubiquitin-specific peptidase (deubiquitinase)
UTP: uridine triphosphate
UTR: untranslated region
UVRAG: ultraviolet wave resistance-associated gene product

V

V: left stretch tensor
V: volume
V_c: pulmonary capillary blood volume in alveolar walls
V_q: cross-sectional average velocity
V_s: specific volume
v: velocity vector
v: recovery variable
v_{X}: volume of gas component X
V1(2)R: type-1(2) vomeronasal receptor
$V_{1A/1B/2}$: type-1A/1B/2 arginine vasopressin receptor
VAAC: volume-activated anion channel
$VAC_{Cl(K)}$: volume-activated Cl^- (K^+)-selective channel
VACamKL: vesicle-associated CamK-like (pseudokinase)
$VACC_{NS}$: volume-activated cation nonselective channel
VAChT: vesicular acetylcholine transporter
VAIC: volume-activated ion channel
VAMP: vesicle-associated membrane protein (synaptobrevin)
VanGL: Van Gogh (Strabismus)-like protein
VAP: VAMP-associated protein
VASP: vasoactive stimulatory phosphoprotein
VAT: vesicular amine transporter
vATPase: vesicular-type H^+ ATPase
VAV: ventriculoarterial valve
Vav: GEF named from Hebrew sixth letter
VC: vital capacity
VCAM: vascular cell adhesion molecule
VCt: vasoconstriction
VDAC: voltage-dependent anion channel (porin)
VDACL: plasmalemmal, volume- and voltage-dependent, ATP-conductive, large-conductance, anion channel
VDCC: voltage-dependent calcium channel
VDP: vesicle docking protein
VDt: vasodilation
VEGF: vascular endothelial growth factor
VEGFR: vascular endothelial growth factor receptor
VF: ventricular fibrillation
VF: ventricular filling
VGAT: vesicular GABA transporter
VGC: voltage-gated channel
VgL: Vestigial-like protein
VGluT: vesicular glutamate transporter
VHL: von Hippel–Lindau Ub ligase
VIP: vasoactive intestinal peptide
VLCAD: very-long-chain acylCoA dehydrogenase
vlcFA: very-long-chain fatty acids (17–26 carbon atoms)
VLDL: very-low-density lipoprotein
VLDLR: very-low-density lipoprotein receptor
VMAT: vesicular monoamine transporter
VN: vitronectin
VOC: volatile organic compound
VOM: vein of Marshall
VPO: vascular peroxidase
VPS: vacuolar protein sorting-associated kinase
VR: venous return
VRAC: volume-regulated anion channel
VRC: ventral respiratory column
VRK: vaccinia-related kinase
VS: vasostatin
vSMC: vascular smooth myocyte
vSNARE: vesicular SNAP receptor (SNARE)
VSOR: volume-sensitive outwardly rectifying anion channel
VSP: voltage-sensing phosphatase
VVO: vesiculovacuolar organelle
vWF: von Willebrand factor

W

W: vorticity tensor
$\mathcal{W}$: strain energy density
W: work, deformation energy
w: weight
w: grid velocity
WASH: WASP and SCAR homolog
WASP: Wiskott–Aldrich syndrome protein
WAT: white adipose tissue
WAVe: WASP-family verprolin homolog
WBC: white blood cell
WDR: WD repeat-containing protein
Wee: small (Scottish)
WHAMM: WASP homolog associated with actin, membranes, and microtubules
WIP: WASP-interacting protein
WIPF: WASP-interacting protein family protein
WIPI: WD repeat domain-containing phosphoinositide-interacting protein
WNK: with no K (Lys) kinase (Lys-deficient kinase)
Wnt: wingless-type
WPWS: Wolff–Parkinson–White syndrome
WNRRTK: Wnt and neurotrophin receptor-related receptor Tyr kinase ($ROR_{(RTK)}$)
WSB: WD-repeat and SOCS box-containing protein (Ub ligase)
WSS: wall shear stress
WSSTG: WSS transverse gradient
WWTR: WW domain-containing transcription regulator

X

$\mathcal{X}$: trajectory
X: reactance
X: Lagrangian position vector
x: position vector
$\{x, y, z\}$: Cartesian coordinates
XBE: X-factor-binding element
XBP: X-box-binding protein (transcription factor)
XIAP: X-linked inhibitor of apoptosis protein (Ub ligase)
XOx: xanthine oxidase

Y

Y: admittance coefficient
YAP: Yes-associated protein
YBP: Y-box-binding protein (transcription factor)
YY: yin yang (transcriptional repressor)

Z

Z: impedance
ZAP70: 70-kDa TCRζ chain-associated protein
ZBTB: zinc finger and BTB (Broad complex, Tramtrack, and bric-à-brac) domain-containing transcription factor
ZnF: zinc finger protein
ZO: zonula occludens

Miscellaneous

1D: one-dimensional
2D: two-dimensional
2-5A: 5′-triphosphorylated, (2′,5′)-phosphodiester-linked oligoadenylate
2AG: 2-arachidonyl glycerol
3D: three-dimensional
3DR: three-dimensional reconstruction
3BP2: Abl Src homology-3 domain-binding adaptor
4eBP1: inhibitory eIF4e-binding protein
5HT: 5-hydroxytryptamine (serotonin)
7TMR: 7-transmembrane receptor (GPCR)

Complementary Lists of Notations

Greek Symbols

α: volumic fraction
α: convergence/divergence angle
α: attenuation coefficient
α_{k}: kinetic energy coefficient
α_{m}: momentum coefficient
β: inclination angle
β_g: gas g solubility
$\{\beta_i\}_1^2$: myocyte parameters
β_T: coefficient of thermal expansion
Γ: domain boundary
Γ_{L}: local reflection coefficient
Γ_{G}: global reflection coefficient
γ: heat capacity ratio
γ: activation factor
γ_{G}: amplitude ratio (modulation rate) of G
$\dot{\gamma}$: shear rate
δ: boundary layer thickness
ϵ_T: emissivity (thermal energy radiation)
ϵ_e: electric permittivity
ϵ: strain
ε: dimensionless small quantity
ζ: singular head loss coefficient
ζ: transmural coordinate
$\{\zeta_j\}_1^3$: local coordinate
η: azimuthal spheroidal coordinate
θ: circumferential polar coordinate
θ: $(\hat{\mathbf{e}}_x, \hat{\mathbf{t}})$ angle
κ: wall curvature
κ_{c}: curvature ratio
κ_{d}: drag reflection coefficient
κ_{f}: frictional sieving coefficient
κ_{h}: hindrance coefficient
κ_{o}: osmotic reflection coefficient
κ_{r}: reflection coefficient
κ_{s}: size ratio
$\{\kappa_k\}_{k=1}^9$: tube law coefficients
κ_{e}: correction factor
Λ: head loss coefficient
λ_{L}: Lamé coefficient
λ: stretch ratio
λ: wavelength
λ_{A}: area ratio
λ_{a}: acceleration ratio
λ_{L}: length ratio
λ_{Ld}: length-to-diameter ratio
λ_{p}: molecule radius-to-pore radius ratio
λ_{q}: flow rate ratio
λ_{t}: time ratio
λ_{v}: velocity ratio
μ: dynamic viscosity
μ_{L}: Lamé coefficient
ν: kinematic viscosity
ν_{P}: Poisson ratio
Π: osmotic pressure
ρ: mass density
τ: time constant
Φ: potential
$\phi(t)$: creep function
φ: phase
ffl: Lagrangian label
χ_i: molar fraction of species i
χ_i: wetted perimeter
$\psi(t)$: relaxation function
Ψ: porosity
ω: angular frequency
Ω: computational domain

Dual Notations

Bφ: basophil
Eφ: eosinophil

M. Thiriet, *Anatomy and Physiology of the Circulatory and Ventilatory Systems*,
BBMCVS 6, DOI 10.1007/978-1-4614-9469-0,

Lφ: lymphocyte
Mφ: macrophage
aaMφ: alternatively activated macrophage
caMφ: classically activated macrophage
Nφ: neutrophil
Σc: sympathetic
pΣc: parasympathetic

Subscripts

$_A$: alveolar, atrial
$_{\bar{A}}$: mixed alveolar
$_{ACM}$: alveolocapillary membrane
$_{Ao}$: aortic
$_a$: arterial
$_{\bar{a}}$: mixed arterial
$_{ac}$: acid
$_{ao}$: airway opening (mouth or nose)
$_{app}$: apparent
$_{atm}$: atmospheric
$_{aw}$: airway
$_b$: blood
$_{\mathbf{c}}$: contractile
$_c$: center
$_{\mathrm{c}}$: point-contact
$_{cap}$: capillary
$_{co}$: core (flow)
$_{CW}$: chest wall
$_{cy}$: cytosolic
$_D$: Darcy (filtration)
$_D$: dead space (airway)
$_{dias}$: diastolic
$_{dyn}$: dynamic
$_E$: expiration, Eulerian
$_{ECF}$: extracellular fluid
$_e$: external
$_{\mathtt{e}}$: extremum
$_{eff}$: effective
$_{se}$: systolic ejection
$_f$: fluid
$_{fast}$: fast (inward current)
$_g$: grid
$_H$: heart
$_h$: heat
$_I$: inspiration
$_i$: internal
$_{ib}$: intrabronchial
$_{in}$: (ionic) influx
$_{inc}$: incremental
$_{int}$: interstitial
$_{ion}$: sum of transmembrane ionic currents
$_L$: Lagrangian
$_L$: lung
$_l$: limit
$_\ell$: line-contact
$_M$: macroscopic
$_m$: mass (e.g., q_m mass flow rate)
$_{\mathrm{m}}$: mean
$_{\mathrm{m}}$: muscle, mouth
$_{max}$: maximum
$_{mb}$: membrane
$_{md}$: mesodiastolic
$_{mea}$: measured
$_{met}$: metabolic
$_{min}$: minimum
$_{ms}$: mesosystolic
$_{mt}$: mitochondrial
$_{musc}$: muscular
$_\mu$: microscopic
$_{out}$: (ionic) outflux
$_P$: pulmonary
$_{\mathbf{p}}$: parallel
$_p$: particle
$_{pd}$: protodiastolic
$_{ps}$: protosystolic
$_{pa}$: pulmonary arterial
$_{pl}$: plasma, pleural
$_{pv}$: pulmonary veinous
$_q$: quasi-ovalization
$_r$: radial
$_{ref}$: reference
$_{rel}$: relative
$_S$: systemic
$_s$: solute
$_{\mathbf{s}}$: serial
$_{syst}$: systolic
$_{sa}$: systemic arterial
$_{si}$: sink
$_{slow}$: slow (inward current)
$_{so}$: source
$_{stim}$: external stimulus
$_{sv}$: systemic veinous
$_T$: tidal (breathing)
$_t$: turbulence
$_t$: stream division
$_t$: time derivative of order 1
$_{tt}$: time derivative of order 2
$_{td}$: telediastolic
$_{tis}$: tissue
$_{tot}$: total
$_{ts}$: telesystolic
$_{ung}$: ungated
$_V$: ventricular
$_v$: systemic venous blood
$_{\bar{v}}$: pulmonary (mixed) venous blood
$_w$: wall

$_w$: water (solvent)
$_\Gamma$: boundary
$_\theta$: azimuthal
$_+$: positive command
$_-$: negative command
$_*$: at interface
$_0$: reference state ($\cdot_0$: unstressed or low shear rate)
$_\infty$: high shear rate

Superscripts

^{a}: active state
^{e}: elastic
^{f}: fluid
^{h}: hypertensive
^{n}: normotensive
^{p}: passive state
^{P}: power
^{s}: solid
^{T}: transpose
^{v}: viscoelastic
*: scale
*: complex variable ($z^* = \Im m\, z + \imath \Re e\, z$)
$\cdot'$: first component of complex elastic and shear moduli
$\cdot''$: second component of complex elastic and shear moduli
$^\natural$: static, stationary, steady variable

Mathematical Notations

T: bold face capital letter means tensor
v: bold face lower case letter means vector
S, s: upper or lower case letter means scalar
$\Delta\bullet$: difference
$\delta\bullet$: increment
$d \bullet /dt$: time gradient
∂_t: first-order time partial derivative
∂_{tt}: second-order time partial derivative
∂_i: first-order space partial derivative with respect to spatial coordinate x_i
∇: gradient operator
$\nabla\mathbf{u}$: displacement gradient tensor
$\nabla\mathbf{v}$: velocity gradient tensor
$\nabla\cdot$: divergence operator
∇^2: Laplace operator
$|\ |_+$: positive part
$|\ |_-$: negative part
$\dot{\bullet}$: time derivative
$\bar{\bullet}$: time mean
$\breve{\bullet}$: space averaged
$\check{\bullet}$: conduit generation averaged
$\langle\bullet\rangle$: ensemble averaged
$\tilde{\bullet}$: dimensionless
$\bullet^+$: normalized ($\in [0, 1]$)
$\hat{\bullet}$: peak value
$\bullet_\sim$: modulation amplitude
det($\bullet$): determinant
cof($\bullet$): cofactor
tr($\bullet$): trace

Cranial Nerves

I: olfactory nerve (sensory)
II: optic nerve (sensory)
III: oculomotor nerve (mainly motor)
IV: trochlear nerve (mainly motor)
V: trigeminal nerve (sensory and motor)
VI: abducens nerve (mainly motor)
VII: facial nerve (sensory and motor)
VIII: vestibulocochlear (auditory-vestibular) nerve (mainly sensory)
IX: glossopharyngeal nerve (sensory and motor)
X: vagus nerve (sensory and motor)
XI: cranial accessory nerve (mainly motor)
XII: hypoglossal nerve (mainly motor)

Chemical Notations

[X]: concentration of X species
X (x): upper and lower case letters correspond to gene and corresponding protein or conversely (i.e., Fes, FES, and fes designate protein, a proto-oncogene product that acts as a kinase, and corresponding gene and oncogene product, respectively)
$^\bullet$: radical (unpaired electron[s])
$^{\Delta NT}$: truncated form without the N-terminal domain
$^{\Delta CT}$: truncated form without the C-terminal domain
^{C}X: cardiac-specific isoform of X molecule
$^{D(L)}$X: D (L)-stereoisomer of amino acids and carbohydrates (chirality prefixes for dextro- [dexter: right] and levorotation [lævus: left]), i.e., dextro(levo)rotatory enantiomer
^{G}X: globular form of X molecule

$^{F(G)}$actin: polymeric, filamentous (monomeric, globular) actin

$_C$X, $_{CT}$X: carboxy (carboxyl group COOH [C])-terminal cleaved part of X molecule

$_c$X: cytosolic molecule

$_{L,Ac}$X: lysosomal, acidic X molecule (e.g., sphingomyelinase)

$_m$X: membrane-bound molecule

$_N$X, $_{NT}$X: amino (amine group NH_2 [N])-terminal cleaved part of X molecule

$_S$X: secreted form of X molecule

$_{S,Ac}$X: secreted, acidic molecule X (e.g., sphingomyelinase)

$_t$X: truncated isoform

X_i: type-i isoform of the receptor of ligand X (i: integer)

XRi: receptor isoform i of ligand X (i: integer)

X+: molecule X expressed (X-positive)

X^+: cation; also intermediate product X of oxidation (loss of electron) from a reductant (or reducer) by an oxidant (electron acceptor that removes electrons from a reductant)

X−: molecule X absent (X-negative)

X^-: anion; also intermediate product X of reduction (gain of electron) from an oxidant (or oxidizer) by a reductant (electron donor that transfers electrons to an oxidant)

X^A: activator form of molecule X

X^a: active form of molecule X

X^{ECD}: soluble fragment correponding to the ectodomain of molecule X after extracellular proteolytic cleavage and shedding (possible extracellular messenger or sequestrator)

$X^{(ER)}$: endoplasmic reticulum type of molecule X

small GTPase$^{GTP(GDP)}$: active (inactive) form of small (monomeric), regulatory guanosine triphosphatase

$X^{GTP(GDP)}$: GTP (GDP)-loaded molecule X

X^{high}: molecule X produced at high levels

X^{ICD}: soluble fragment correponding to intracellular domain of molecule X after intracellular proteolytic cleavage (possible messenger and/or transcription factor; e.g., Notch^{ICD}: intracellular Notch fragment)

X^{low}: molecule X produced at low levels

X^{MT}: mitochondrial type of molecule X

X^{PM}: plasmalemmal type of molecule X

X^R: repressor form of molecule X

X^S: soluble form

X_{alt}: alternative splice variant

X_{FL}: full-length protein X

$X_{h(l,m)MW}$: high (low, mid)-molecular-weight isotype

$X_{L(S)}$: long (short) isoform (splice variants)

X_c: catalytic subunit

X_i: number of molecule or atom (i: integer, often 2 or 3)

$(X_1–X_2)_i$: oligomer made of i complexes constituted of molecules X_1 and X_2 (e.g., histones)

a, c, nX: atypical, conventional, novel molecule X (e.g., PKC)

al, ac, nX: alkaline, acidic, neutral molecule X (e.g., sphingomyelinase)

asX: alternatively spliced molecule X (e.g., asTF)

cX: cellular, cytosolic, constitutive (e.g., cNOS), or cyclic (e.g., cAMP and cGMP) X molecule

caX: cardiomyocyte isoform (e.g., caMLCK)

dX: deoxyX

eX: endothelial isoform (e.g., eNOS and eMLCK)

hX: human form (ortholog); heart type (e.g., hFABP); hormone-like isoform (FGF)

iX: inhibitory mediator (e.g., iSMAD) or intracellular (e.g., iFGF) or inducible (e.g., iNOS) isoform

kX: renal type (kidney) X molecule

ksX: kidney-specific isoform of X molecule

lX: lysosomal X molecule

lpX: lipoprotein-associated X molecule (e.g., lpPLA2)

mX: mammalian species or membrane-associated X molecule (e.g., mTGFβ)

mtX: mitochondrial type of X molecule

nX: neutral X; neuronal type (e.g., nWASP)

oxX: oxidized X molecule (e.g., oxLDL)

plX: plasmalemmal type of X molecule

rX: receptor-associated mediator or receptor-like enzyme; also regulatory type of molecular species (e.g., rSMAD)

skX: skeletal myocyte isoform (e.g., skMLCK)

smcX: smooth muscle cell isoform (e.g., smcMLCK)

tX: target type of X (e.g., tSNARE); tissue type (e.g., tPA)

tmX: transmembrane type of X

vX: vesicle-associated (e.g., vSNARE) or vacuolar (e.g., vATPase) type of X

GPx: glycoprotein (X: molecule abbreviation or assigned numeral)

Xx: (x: single letter) splice variants

X1: human form (ortholog)
Xi: isoform type i (paralog or splice variant; i: integer)
Xi/j: (i,j: integers) refers to either both isoforms (i.e., Xi and Xj, such as ERK1/2) or heterodimer (i.e., Xi–Xj, such as ARP2/3)
X1/X2: molecular homologs or commonly used aliases (e.g., contactin-1/F3)
PI(i)P, PI(i,j)P_2, PI(i,j,k)P_3: i,j,k (integers): position(s) of phosphorylated OH groups of the inositol ring of phosphatidylinositol mono-, bis-, and trisphosphates

Post-Translational Modification

X_A: acetylated molecule X
acX: acetylated molecule X (e.g., acLDL)
X^M: methylated molecule X
X_M: myristoylated molecule X
X_P: palmitoylated molecule X
X^P: phosphorylated molecule X
pAA: phosphorylated amino acid (pSer, pThr, and pTyr)
X^{SNO}: ^{S}nitrosylated molecule X
X^{SSG}: ^{S}glutathionylated molecule X
X^U: ubiquitinated protein X

Amino Acids

Ala (A): alanine
Arg (R): arginine
Asn (N): asparagine
Asp (D): aspartic acid
Asp^{COO^-}: aspartate
CysH (C): cysteine
Cys: cystine
Gln (Q): glutamine
Glu (E): glutamic acid
Glu^{COO^-}: glutamate
Gly (G): glycine
His (H): histidine
Iso, Ile (I): isoleucine
Leu (L): leucine
Lys (K): lysine
Met (M): methionine
Orn: ornithine (not encoded by DNA, but use in the urea cycle)
Phe (F): phenylalanine
Pro (P): proline
Ser (S): serine
Thr (T): threonine
Trp (W): tryptophan
Tyr (Y): tyrosine
Val (V): valine

Ions

Asp^-: aspartate (carboxylate anion of aspartic acid)
ADP^{3-}: ADP anion
ATP^{4-}: ATP anion
Ca^{2+}: calcium cation
Cl^-: chloride anion
Co^{2+}: cobalt cation
Cu^+: copper monovalent cation
Cu^{2+}: copper divalent cation
Fe^{2+}: ferrous iron cation
Fe^{3+}: ferric iron cation
Glu^-: glutamate (carboxylate anion of glutamic acid)
H^+: hydrogen cation (proton)
H_3O^+: hydronium (oxonium or hydroxonium) cation
HCO_3^-: bicarbonate anion
HPO_4^{2-}: hydrogen phosphate divalent anion (inorganic phosphate species)
$H_2PO_4^-$: dihydrogen phosphate monovalent ion (inorganic phosphate species)
HS^-: hydrosulfide anion
K^+: potassium cation
Mg^{2+}: magnesium cation
$^{Mg}ATP^{2-}$: ATP anion
Mn^{2+}: manganese cation
Na^+: sodium cation
Ni^{2+}: nickel cation (common oxidation state)
OH^-: hydroxide anion
PO_4^{3-}: phosphate anion (inorganic phosphate species)
S^{2-}: sulfide anion;
SO_3^{2-}: sulfite anion
SO_4^{2-}: sulfate anion
Zn^{2+}: zinc cation (common oxidation state)

Atmospheric Pollutants

CH_4: methane
HNO_2: nitrous acid
HNO_3: nitric acid

H_2SO_4: sulfuric acid
NO_x: nitrogen oxides
NO_2: nitrogen dioxide
O_3: ozone
PM_{10}: inhalable coarse particulate matter (2.5 < size < 10 μ m)
$PM_{2.5}$: fine particulate matter (0.1 < size ≤ 2.5 μ m)
$PM_{0.1}$: ultrafine particulate matter (aerodynamic diameter ≤ 0.1 μ m)
SO_x: sulfur oxides
SO_2: sulfur dioxide

Inhaled and Signaling Gas

CO: carbon monoxide (or carbonic oxide; signaling gas and pollutant [air level ∼ 0.1 ppm])
CO_2: carbon dioxide (cell waste)
H_2S: hydrogen sulfide (signaling gas) [air level ∼ 0.0001 ppm])
He: helium (inert monatomic gas)
N_2: nitrogen (inert diatomic gas)
NH_3: ammonia (trihydrogen nitride; trace quantities in air)
NO: nitric oxide (or nitrogen monoxide; signaling gas and pollutant [air level ∼ 0.1 ppm])
O_2: oxygen (cell energy producer)

Nitric Oxide Derivatives

HNO: protonated nitroxyl anion
$NO^\bullet$: free radical form
NO^+: nitrosyl or nitrosonium cation
NO^-: nitroxyl or hyponitrite anion (inodilator)
NO_2^-: nitrite anion
NO_3^-: nitrate anion

Reactive Oxygen and Nitrogen Species

$CO_3^{\bullet -}$: carbonate radical
H_2O_2: hydrogen peroxide
HOCl: hypochlorous acid
N_2O_3: dinitrogen trioxide
$NO_2^\bullet$: nitrogen dioxide
$^1O_2$: singlet oxygen
O_2^-: superoxide ($O_2^{\bullet -}$)
$O{=}C(O^\bullet)O^-$: carbonate radical
$OH^\bullet$: hydroxyl radical (hydroxide ion neutral form)
$ONOO^-$: peroxynitrite
$RO^\bullet$: alkoxyl
$RO_2^\bullet$: peroxyl

Moieties (R denotes an organic group)

R: alkyl group (with only carbon and hydrogen atoms linked exclusively by single bonds)
R–CH_3: methyl group (with 3 forms: methanide anion [CH_3^-], methylium cation [CH_3^+], and methyl radical [$CH_3^\bullet$])
R–CHO: aldehyde group
R–CN: nitrile group
R–CO: acyl group
R–CO–R: carbonyl group
R–COO^-: carboxylate group
R–COOH: carboxyl group
R–NC: isonitrile group
R–NCO: isocyanate group
R–NH_2: amine group
R–NO: nitroso group
R–NO_2: nitro group
R–O: alkoxy group
R=O: oxo group
R–OCN: cyanate group
R–OH: hydroxyl group
R–ONO: nitrosooxy group
R–ONO_2: nitrate group
R–OO–R: peroxy group
R–OOH: hydroperoxy group
R–S–R: sulfide group
R–SH: thiol (or sulfhydryl) moiety
R–SN: sulfenyl-amide moiety
R–SNO: nitrosothiol (or thionitrite) moiety
R–SO: sulfinyl
R–SO–R: sulfoxide group
R–SO_2: sulfonyl group
R–SO_2H: sulfinic acid (sulfinyl moiety)
R–SO_2N: sulfonyl-amide moiety
R–SO_3H: sulfonic acid (sulfonyl moiety)
R–SOH: sulfenic acid (sulfenyl moiety)
R–SON: sulfinyl-amide moiety
R–SS–R: disulfide group

Lung Function Testing

ERV: expiratory reserve volume
f_C: cardiac frequency
f_R: breathing frequency
FEF_f: forced expiratory flow at a fraction (f [%]) of forced expiration (FEF_{25}, FEF_{50}, and FEF_{75})
FEV_1: volume expired at the end of the first second of forced expiration
FEV_τ: forced expiratory volume at time τ (fraction of a second over which maximally fast exhaled volume is measured) starting from full inspiration
FRC: functional residual capacity (lung volume at end of rest expiration)
IC: inspiratory capacity (IRV + V_T)
IRV: inspiratory reserve volume
MBC: maximum breathing capacity (per mn of effort)
MVV: maximal voluntary ventilation (volume of air breathed in a specified period during repetitive maximal exercise)
PEF: peak expiratory flow
R_R: respiratory quotient ($\dot{V}_{CO_2}/\dot{V}_{O_2}$)
RV: residual volume
TLC: total lung capacity
V_A: alveolar gas volume
V_D: dead space volume
V_L: lung volume
V_T: tidal volume
$\dot{V}$: total ventilation (air volume exhaled par mn)
$\dot{V}_A$: alveolar ventilation ($f_R(V_T - V_D)$)
$\dot{V}_{O_2}$: oxygen consumption
$\dot{V}_{CO_2}$: carbon dioxide production
VC: vital capacity (air volume quietly expelled from full inspiration)

Time Units

s: second
mn: minute
h: hour
d: day
wk: week
mo: month
yr: year

SI-Based and Non-SI Units of Quantity

mmHg: millimeter of mercury (133.322 Pa [~ 0.1333 kPa])
mmol, nmol, μ mol: milli-, nano-, micromoles
(amount of a chemical species, one mole containing about $6.02214078 \times 10^{23}$ molecules)
mosm: milliosmole
(osm: number of moles of a osmotically active chemical compound)
kDa: kiloDalton
(Da: atomic or molecular mass unit)
ppm: parts per million
l: liter

Temperature and Pressure Conditions

ATPS: ambient temperature and pressure, saturated with water at body temperature, i.e., at 37°C, $p_{H_2O} = 6.27$ kPa (47 mmHg)
BTPS: body temperature and ambient pressure, saturated with water
STPD: standard temperature (0°C) and pressure (101 kPa [760 mmHg]), dry air

$$V_{BTPS} = V_{ATPS} \times \frac{273 + 37}{273 + T} \times \frac{p - p_{H_2O}}{p - 47}$$

$$V_{STPD} = V_{ATPS} \times \frac{273}{273 + T} \times \frac{p - p_{H_2O}}{760}$$

Index

A
α-actinin, 301
α-adrenergic receptor, 149, 227, 255, 259, 313, 329, 439
acetoacetate, 219
acetoacetylCoA synthase, 221
acetylcholine, 58, 144, 153, 254, 272, 330, 428
acetylCoA, 219, 220, 228
acetylCoA carboxylase, 190, 221
acetylCoA thiolase, 221
ACh muscarinic receptor, 149, 236, 330
ACh nicotinic receptor, 331
acid-sensing channel, 70
acidosis, 65, 66, 69, 412
acoustic rhinometry, 475
actin, 198, 266
action potential, 16, 19, 153, 163, 166, 172, 228, 331
acylCoA dehydrogenase, 192
adamlysin, 275
addressin, 435
adenine nucleotide translocase, 193, 209
adenosine, 187, 254, 270, 274, 308
adenylate cyclase, 178, 225, 232, 236, 329, 330
adherens junction, 172, 266
adipocyte, 222, 347
adipokine, 195
adipose tissue, 5, 347
adipose triglyceride lipase, 190
ADP, 182, 186, 353
adrenaline, 58, 329, 334, 352
adrenal cortex, 350
adrenal medulla, 58, 312, 320, 331, 334, 429
afterload, 160, 166, 168, 291
aging, 160, 170, 246
airway peristalsis, 126
airway reflex defense, 141
airway smooth myocyte, 126, 149
AKAP, 178, 229, 232
albumin, 188
aldehyde dehydrogenase, 34
aldosterone, 314, 325, 345, 350
alkalosis, 412
allergy, 274
alveolar duct, 74, 100, 101, 108
alveolar gas equation, 398
alveolar macrophage, 435
alveolar pressure, 384
alveolar sac, 73, 100, 101, 105, 108
alveolocapillary membrane, 40, 96, 109, 115, 116, 359
alveolus, 73, 96, 100, 101, 104, 105, 109, 115, 383, 385, 386
amiloride-sensitive cation channel, 69, 70
amniotic liquid, 127
AMP, 187
AMPA-type glutamate receptor, 318, 422, 423
amphiphysin, 179
AMPK, 190, 191, 194
amygdala, 333
anaplerosis, 187, 218, 224, 225
anastomosis, 39
angiopoietin, 277
angiopoietin-like protein, 190
angiotensin, 66, 67, 69, 134, 149, 254, 259, 261, 308, 314, 325, 327, 336, 340, 345, 438
angiotensin-converting enzyme, 345
angiotensinogen, 345
angiotensin receptor, 308
ankyrin, 229
Anrep effect, 166
antilipolytic GPCR, 223
antiporter, 199, 207

M. Thiriet, *Anatomy and Physiology of the Circulatory and Ventilatory Systems*, BBMCVS 6, DOI 10.1007/978-1-4614-9469-0,

aorta, 13, 36, 45, 49, 67, 94, 152, 287
aortic body, 64, 331
aortic dissection, 443
aortic valve, 7, 12, 15, 163
apneustic center, 137, 138, 417
aquaporin, 279
arachidonic acid, 307
arcuate nucleus, 317, 347
area ratio, 36, 37, 104, 112, 136
arterial pressure, 245, 246, 287, 313, 321, 464
ASIC channel, 300
astrocyte, 307
ATP, 58, 186, 249, 250, 308, 316, 325, 331, 353
ATP synthase, 182
atrioventricular node, 10, 16
atrium, 5, 8
augmentation index, 246, 464
autonomic nervous system, 141
autoregulation, 254, 257, 293
Avogadro constant, 355
axial tension, 3
azimuthal tension, 3
azygos vein, 48, 94

B

β-adrenergic receptor, 149, 178, 184, 185, 189, 225, 227, 230, 233, 236, 255, 329, 439
β-hydroxybuty, 219
β-hydroxybutyrate dehydrogenase, 219, 221
B-mode echography, 449
Bötzinger complex, 419
Bachman bundle, 10
baroreceptor, 67, 146, 152, 314, 323, 332, 428
baroreceptor reflex, 292, 300, 325
basement membrane, 121, 125, 127, 265
basilar trunk, 39
basilic vein, 56
bathmotropy, 313
Bayliss effect, 293
bidomain model, 242, 243
blood–brain barrier, 256
BMAL factor, 286
BMP, 31
body plethysmograph, 476
Bohr effect, 366, 409
Bohr equation, 378
Boltzmann constant, 358, 373
boundary layer, 4
Bowditch–Treppe effect, 166
Boyle–Mariotte law, 374, 376, 479
brachiocephalic trunk, 48
brachiocephalic vein, 54, 94
bradykinin, 144, 261, 272, 438
bradypnea, 366
brain, 295, 347
brainstem, 333, 334
branching angle, 36, 37, 94, 110, 112, 130, 136
branching ratio, 42
breathing frequency, 379
Brody effect, 470
bronchial artery, 50, 99
bronchial pressure difference, 382
bronchiole, 73, 106
bronchioloalveolar communication, 116
bronchoconstriction, 428, 439
bronchodilation, 366
bronchus, 383
bronchus-associated lymphoid tissu, 434
Brownian motion, 278
brush cell, 93
bulk modulus, 284, 388

C

C-fiber receptor, 141
C1 cell, 153
Ca^{2+}–H^+ antiporter, 175, 218
Ca_V channel, 66, 69, 176, 228, 230, 235, 308
cadherin, 268, 278
calcitonin gene-related peptide, 151, 166
calcium, 3, 173, 182, 202, 232, 250, 370
calcium blink, 178
calcium puff, 307
calcium pulsar, 307
calcium release unit, 177
calcium spark, 178, 183, 184, 306
calcium sparklet, 178, 307
caliber ratio, 42, 120
calmodulin, 266
calsequestrin, 244
CamK kinase, 184, 225, 230, 233, 235, 236, 272, 330
cAMP, 184, 190, 236, 271, 275
capillary, 41, 100, 245, 255, 259, 260
capillary density, 42
capillary length, 42
capillary recruitment, 261
carbohydrate oxidation, 187
carbonate radical, 211
carbonic anhydrase, 217
carbon dioxide, 217
carbon dioxide production, 369
carbon monoxide, 69
cardiac accelerating center, 20, 313
cardiac efficiency, 166
cardiac frequency, 21, 159, 167, 227, 287, 288, 291, 319, 321, 329, 379

cardiac index, 160
cardiac inhibitory center, 20, 313
cardiac output, 159, 287, 291, 313
cardiac reserve, 160
cardiac valve, 6
cardiomyocyte, 299
cardioprotection, 218
carnitine–acylcarnitine translocase, 190
carnitine palmitoyltransferase, 188, 190, 210, 219
carotid artery, 52
carotid artery, 39, 48, 51, 64, 82
carotid body, 64, 154, 429
carotid body (glomus), 40, 62, 63, 65, 69, 331
carotid sinus, 51, 63, 67, 146, 152, 331
carotid sinus nerve, 60, 146, 152, 154
cartilage, 92
cataplerosis, 218
catecholamine, 66, 222, 286, 329, 331, 340, 429
catestatin, 352
caudal ventral respiratory group, 419
caveola, 180, 236
caveolin, 179, 272, 274
CDC42 GTPase, 270
CDK kinase, 3
celiac artery, 50
central nervous system, 137, 347
cephalic vein, 56
cerebral artery, 39, 307
cerebral drainage, 53
cerebrospinal fluid, 325, 428
cervical nerve, 426
cGMP, 67, 272, 275
chaos, 160
Charles law, 374
chemokine receptor, 432, 433
chemoreceptor, 64, 139, 146, 314, 331, 427
chemoreceptor reflex, 66, 300
chemotactic GPCR, 223
cholecystokinin, 351
chordae tendineae, 7, 11
chromogranin, 352
chronotropy, 21, 313
chylomicron, 188
ciliary beat frequency, 370, 371
ciliated cell, 93
cilium, 370
cilium caliber, 370
cilium length, 370
circadian rhythm, 246, 286, 321
circumventricular organs, 314
citrate synthase, 221
citric acid cycle, 196
Cl^-–HCO_3^- exchanger, 69
Clara cell, 113
claudin, 265
CLOCK factor, 286
CoA transferase (SCOT), 221
cofilin, 198, 267
collagen, 93, 100, 125
complexity, 35
compliance, 246, 388
compressibility, 388
computed tomography, 442
concha, 81
confluence of sinuses, 53
confocal laser endomicroscopy, 455
contrast-enhanced ultrasonography, 452
convection, 262, 367
coronary artery, 13, 18, 226, 250, 251, 308, 443, 446, 447
coronary reserve, 251
coronary ridge, 9
coronary sinus, 9
cortactin, 267
corticosteroid, 286
corticotropin-releasing hormone, 153, 337
corticotropin (ACTH), 326
cortisol, 337
cotransporter, 199
couplon, 177
cranial nerve, 426
creatine, 186, 355
creatine kinase, 186
Crumbs, 278
Cryptochrome repressor, 286
cyclooxygenase, 183
cytochrome-C oxidase, 212, 226
cytochrome P450 epoxygenase, 299

D

Dalton law, 356
DBx factor, 420
DBx factor, 422
dead space, 102, 378, 385
DEG–ENAC family channel, 70
dendritic cell, 430–432, 436
denitrosylation, 185
depressor response, 146
diacylglycerol, 276, 302, 308
diad, 176, 180, 183, 202
Diaphanous, 278
diaphragm, 89, 368, 384
diastole, 161, 163, 166, 168, 252, 319
diastolic pressure time interval, 246
dietary salt, 325

diffusion, 262, 357, 367
diffusional permeability, 262, 281
diffusion tensor MRI, 445
diffusivity, 357
dilution method, 375
distensibility modulus, 387
diversity, 35
dopamine, 352
Doppler echography, 449, 466
Doppler equation, 466
dorsal nucleus of vagus nerve, 61, 335
dorsal respiratory group, 417
dorsal root ganglia, 21
dromotropy, 21, 313
dual-specificity phosphatase, 207
ductus arteriosus, 2
duplex ultrasonography, 449
dynamin, 3
dynein, 118
dynorphin, 317

E

Early growth response factor (EGR), 421
echocardiographic particle image velocimetry, 452
echodopplervelocimetry, 451
ectopic calcification, 443
EET, 303, 307, 308
elastance, 388, 390
elastic modulus, 3, 387
elastin, 93, 100, 455
electron beam CT, 443
electron transport chain, 182, 195, 196, 202, 203, 218
ENaC channel, 70, 325
end-diastolic volume, 160
end-systolic volume, 160
endoplasmic reticulum, 173
endothelin, 66, 67, 340
endotheliocyte, 261, 299, 331, 438
endothelium, 261, 277, 326
enoylCoA hydratase, 192
enthalpy, 375
entropy, 372, 388
epithelial–mesenchymal transition, 126
epithelium, 81, 83, 87, 93, 113–115
ERK, 272, 273, 275, 301
esophageal balloon, 476
estrogen receptor (NR3a), 317
ETS variant factor (ETV), 32
eupnea, 366
Eustachian ridge, 9, 10, 16
Evx factor, 423
exercise, 163, 244, 249, 255, 312, 316, 330, 333, 355, 366, 369, 386, 396, 420
exosome, 171
expiratory-to-inspiratory time ratio, 379
expiratory center, 137
expiratory reserve volume, 377

F

facial nucleus, 420
Fahraeus effect, 267
fastigial nucleus, 146
fastigial pressor response, 146
fatty acid, 188, 353
fatty acid-binding protein, 189
fatty acid anion, 193
fatty acid oxidation, 186, 187, 219, 225
fatty acid transport protein, 189
fatty acylCoA synthase, 188, 191
feedback, 67, 124, 138, 189, 192, 225, 286, 308, 331, 337, 352, 419, 428
femoral vein, 57
Fenn effect, 227
FEV_1/FVC ratio, 483
FGF, 33, 122
fibroblast, 98, 100
fibronectin, 125
Fick law, 278, 357
filtration coefficient, 278
flow division ratio, 36
flux, 357
foramen ovale, 9
forced oscillations, 477
FoxA factor, 440
FoxF factor, 123
fractal, 27, 109
Frank–Starling effect, 165
functional hyperemia, 256
functional residual capacity, 102, 143, 376, 385

G

GABA, 153
$GABA_A$ channel, 146
gap junction, 171, 242, 422
GATA factor, 32, 35
Gay–Lussac law, 374
gene regulatory network, 124
ghrelin, 351
ghrelin receptor, 351
Gibbs free energy, 203
Gli transcription factor, 123
glomerular filtration, 340
glomus cell, 65, 70

glossopharyngeal nerve, 51, 62, 68
glucagon, 219, 222
glucocorticoid, 337
glucocorticoid receptor, 337
gluconeogenesis, 222
glucose, 218, 439
glutamate, 153
glutathiolation, 185
glutathione, 213, 214
glycerolphosphate acyltransferase, 189
glycine channel, 146
glycocalyx, 265, 267, 279
glycogen phosphorylase, 187
glycolysis, 222, 353
glycosaminoglycan, 267
glypican, 267
goblet cell, 87, 93
Graham law, 357, 359
grainyhead-like factor (GrHL), 440
great cerebral vein of Galen, 53
GRK kinase, 331
guanylate cyclase, 272, 273
G protein, 329, 331

H

H^+–Pi cotransporter, 199
H^+-gated ion channel, 70, 141
HAND factor, 32, 35
HCN channel, 70, 228, 422
heart, 295
heart failure, 69, 318, 331
heart valve, 161, 447, 450
heat capacity, 372
heat capacity ratio, 375, 388
heat shock protein, 171, 272
heat transfer coefficient, 291
Hedgehog, 123
helper T cell, 430, 433
hematocrit, 267
heme oxygenase, 3
hemoglobin, 394, 397, 405, 407, 408
Henderson–Hasselbach equation, 204, 413
Henry law, 359
heparan sulfate proteoglycan, 190, 268
hepatic artery, 50
Hering–Breuer reflex, 138, 365, 425, 428
HETE, 303, 307
hexosamine synthesis, 187
HIF, 429
hilum, 94
hippocampus, 333
histamine, 144, 254, 352, 439
histone deacetylase, 223
His bundle, 16
HMGCoA synthase, 221
homeobox (Hox), 125
Hooke law, 387
hormone-sensitive lipase, 190
hyaluronan, 267
hyaluronic acid, 267, 371
hydraulic conductivity, 262, 278, 280
hydrogen (proton), 202
hydrogen peroxide, 199, 203
hydrogen sulfide, 69
hydroxyacylCoA dehydrogenase, 192
hydroxyl radical, 199, 211
hypercapnia, 65, 66
hyperpnea, 366, 399
hypertension, 69, 312, 324
hyperventilation, 366, 399
hypopharynx, 83
hypophysis, 39, 337
hypopnea, 366
hypotension, 292
hypothalamo–pituitary–adrenocortical axis, 333
hypothalamus, 154, 286, 314, 325, 333, 334, 347
hypoxia, 65, 66, 69, 198, 259, 261, 409, 416, 429

I

ICAM, 432
IGF, 272
iliac artery, 50
immunoglobulin, 430
impedance plethysmography, 464
inferior petrosal sinus, 54
inferior sagittal sinus, 53
inflammation, 274
inositol trisphosphate, 276
inotropy, 21, 165, 168, 291, 313
inspiratory capacity, 376
inspiratory center, 137
inspiratory reserve volume, 376
insulin, 189, 195, 218, 219, 222
integrin, 125, 126, 284, 301, 304, 432, 433, 436
interalveolar pore, 116
intercalated disc, 242
intercellular cleft, 268
intercostal, 90, 368, 384
intercostal nerve, 426
internal energy, 372, 373
interstitium, 100
intestine, 285
intramural hematoma, 443

intravascular ultrasonography, 450
intravascular ultrasound palpography, 451
intricacy, 35
ionotropic glutamate receptor, 146
IP_3 receptor, 180, 229, 307
IQGAP (RasGAP), 271
irritant receptor, 139, 428
ischemic preconditioning, 185
isocitrate dehydrogenase, 182
isovolumetric contraction, 161
isovolumetric relaxation, 161

J
JNK, 207
jugular vein, 54
junctophilin, 179
juxtacapillary receptor, 139, 428
juxtaglomerular apparatus, 297, 342

K
K_{2P} channel, 300
K_{ACh} channel, 300
K_{ATP} channel, 300, 308
K_{Ca} (BK) channel, 69, 303, 306
K_{Ca} (IK) channel, 306
K_{Ca} channel, 67, 300, 303, 307
K_{IR} channel, 178, 305
K_V channel, 66, 69, 228, 230
kainate-type glutamate receptor, 422, 423
Kedem–Katchalsky equation, 281
ketoacid dehydrogenase, 210
ketoacylCoA thiolase, 192
ketogenesis, 219
ketoglutarate dehydrogenase, 182
ketolysis, 220
ketone, 219
ketone body oxidation, 220, 225
kidney, 295, 324
Kirchhoff law, 243
kisspeptin, 317
Kleiber law, 27
Korteweg–de Vries equation, 465

L
lamina propria, 93
laminin, 125
larynx, 84, 385, 426, 427
lateral reticular nucleus, 420
left atrium, 10, 36
left auricle, 17
left ventricle, 11, 36
length/diameter ratio, 110
length ratio, 42, 110
leptin, 347, 352
leukotriene, 254, 439
limbic ridge, 10
LIMK kinase, 267, 270
lipogenesis, 220
lipolysis, 219, 222, 439
lipoprotein lipase, 190
lipoxygenase, 183
liver, 39, 219
liver kinase-B (STK11), 195
lumbar nerve, 426
lung, 94, 426
lung parenchyma, 96
lusitropy, 21, 168
lymph, 259
lymphatic, 100, 259
lymphocyte, 430, 433
lymph node, 285

M
M-mode echography, 449
macrophage, 69, 99, 285, 327, 430, 432, 433
magnesium, 202
magnetic particle imaging, 447
magnetic resonance imaging, 444
magnetocardiography, 448, 474
main bronchus, 91, 94
malonylCoA, 190
malonylCoA decarboxylase, 190
mass concentration, 355
mass density, 355
mass fraction, 356
mastocyte, 352, 430, 432, 434, 439
Mayer wave, 323, 333
meatus, 81
mechanical strain, 3, 129
mechanical stress, 3, 126, 128
mechanotransduction, 70, 125, 183, 268, 293, 296
median preoptic nucleus, 314
mediastinum, 5, 91
medulla, 334, 428
melatonin, 251
mesenchymal stem cell, 126
mesotheliocyte, 98
mesothelium, 98
metabolic syndrome, 198
metabotropic glutamate receptor, 423
microcirculation, 40, 259
microRNA, 33, 198, 233
mineralocorticoid receptor, 337

mitochondrial Ca^{2+} uniporter, 175, 180, 181, 200, 201
mitochondrial permeability transition pore, 201
mitochondrion, 172, 219
mitofusin, 180, 195
mitral valve, 7, 161
mixed-function oxidase, 437
MLCK, 266, 275
MLCP, 266
mobility, 358
molality, 355
molarity, 356
molar concentration, 356
molar fraction, 356
molar mass, 355
molar volume, 355
mole, 355
monoamine oxidase, 183, 438
monodomain model, 242, 243
mouth, 83
mucociliary clearance, 430
mucus, 93, 432
multipotent stem cell, 29
multislice spiral CT, 442
muscle mechanoreflex, 318
muscle metaboreflex, 318
musculoaponeurotic fibrosarcoma homolog (MAF), 423
MyC transcription factor, 123
myocardial infarction, 288, 318
myocardin, 35
myocardin-related transcription factor (MRTF), 126
myocardium, 158, 225, 254, 330
myoendothelial junction, 306
myogenic response, 293

N

Na^{+}–Ca^{2+} exchanger, 175, 178, 181, 199, 201, 228, 229, 244
Na^{+}–H^{+} exchanger, 69, 181, 199
Na^{+}–K^{+} ATPase, 178, 229, 324
Na_V channel, 178, 230
NADPH oxidase, 69, 183, 185, 429
nasal (pseudo)valve, 78
nasal cycle, 81
nasal vestibule, 78
nasopharynx, 83
natriuretic peptide, 66, 254, 271, 276, 324, 349
natural killer cell, 430, 434
natural killer T lymphocyte, 434
Navier–Stokes equation, 397
NEDD Ub ligase, 325
nephron, 285, 340
Nernst equation, 203
neurokinin, 317
neuron, 108, 140, 299
neuropeptide Y, 150, 325
neuropilin, 317
neurotrophin RTK, 317
neutrophil, 275
NGF, 317
niacin, 223
niacin receptor, 223
nicotinic acid, 223
nidogen, 125
nitric oxide, 66, 149, 185, 201, 211, 214, 226, 249, 251, 255, 261, 267, 269, 271, 274, 306, 308, 316, 330, 340, 348, 369
nitric oxide synthase, 202, 316, 347
nitrogen dioxide, 211
nitrosylation, 185
NKx2-5 factor, 32
NKx factor, 420
NMDA-type glutamate receptor, 318
nociceptor, 139
Nodal, 31
nodal cell, 16, 158
nodose ganglion, 21, 61, 70, 300
noradrenaline, 58, 69, 329, 334, 345, 352, 428, 438
NOS, 69, 183, 271, 273
nose, 75
nostril, 78
Notch, 123
nuclear receptor, 286
nucleus ambiguus, 22, 61, 63, 335, 379, 417, 420
nucleus of the solitary tract, 60, 61, 68, 69, 137, 146, 152, 314, 335, 347, 348, 352, 417
nucleus retroambigualis, 417
number density, 355
number fraction, 356

O

occipital sinus, 53
occludin, 265
Oct factor, 31
odorant GPCR (chemosensor), 297
Ohm law, 243
olfactory cleft, 81
optical coherence tomography, 453
orexin, 153
oropharynx, 83
osmoreceptor, 314
osmotic pressure, 263

osmotic reflection coefficient, 262, 280
ouabain, 325
oxidative phosphorylation, 172, 187, 217, 225, 227, 354
oxidative stress, 199
oxygen, 3, 186, 202, 226
oxygen consumption, 369
oxytocin, 153

P
P2Y GPCR, 171
Péclet number, 282
pacemaker, 16, 228, 286, 420
pain receptor, 139
Paired-like homeobox (PHox), 419
Paired-like homeobox (PHox), 421
PAK kinase, 270
pannexin, 171
papillary muscle, 7, 11, 162
parabrachial nucleus, 314, 417
parafacial respiratory group, 419
paramedian medullary reticular group, 152
paranasal sinus, 78, 82
parasympathetic, 19, 58, 141, 312, 313, 330, 426
paraventricular nucleus, 67, 154, 314, 335
paravertebral ganglia, 21, 334
partial pressure, 356
pattern recognition receptor, 431
Pennes bioheat equation, 289
perforating vein, 56
peribronchium, 105
pericardium, 5
pericyte, 100, 265
Period repressor, 286
peroneal vein, 57
peroxidase, 211
peroxiredoxin, 213
peroxyl radical, 212
peroxynitrite, 199, 211, 212
Peters diagram, 412
PGC factor, 191, 198
pharynx, 83
PHLPP phosphatase, 207
phosphodiesterase, 178, 236, 251, 275
phosphofructokinase, 187, 193
phospholamban, 232, 244
phosphorylation, 207
photoplethysmography, 464
phrenic nerve, 426
PI3K, 179, 284
Piezo, 300
PInK kinase, 207
PKA, 178, 184, 207, 225, 230, 232, 233, 235, 236, 370
PKB, 207, 272
PKC, 207, 232, 267, 272, 273, 284, 302, 307, 370
PKD, 278
PKG, 251, 272, 370
PLA, 276, 299, 302
plasmocyte, 432
platelet-activating factor, 272
PLC, 273, 276, 302, 307, 308
PlekHg GEF, 278
plethysmography, 376
pleura, 97
pleural pressure, 384, 385
pneumocyte, 149, 439
pneumotachograph, 476
pneumotaxic center, 137, 138, 417
Poisson ratio, 388
polycystin, 300
popliteal vein, 57
pore, 279
portal drainage, 39, 57
positron emission tomography, 453
postload, 291
PP1, 211, 230, 233, 236
PP2, 211, 229, 230, 233, 236
PP3, 232, 233
PPAR factor, 223
PPAR response element (PPRE), 223
PPAR factor, 191
PPM1, 207, 211
pre-Bötzinger complex, 383, 416, 419, 420
prefrontal cortex, 333
pregnancy, 42
preload, 160, 168, 291
pressure, 3, 126
pressure drop, 382
prevertebral ganglia, 334
proprioceptive receptor, 424
proprioceptor, 139
prostacyclin, 285, 306, 307
prostaglandin, 144, 254, 271, 285, 307, 340, 428, 438
proteoglycan, 327
PTPMt1, 211
PTPn11/SHP2, 207
pulmonary vascular resistance, 405
pulmonary acinus, 73, 74, 101, 108, 111, 119
pulmonary artery, 94, 99
pulmonary lobule, 73, 96, 98, 99, 108, 131
pulmonary trunk, 11, 36
pulmonary valve, 7, 36
pulmonary vein, 7, 10, 17, 36

Purkinje fiber, 16, 179, 234
pyruvate, 193, 218
pyruvate carboxylase, 224
pyruvate dehydrogenase, 182, 209, 224

R

Rac GTPase, 185, 270, 271, 273, 275
RapGEF, 271
raphe, 312
rapidly adapting stretch receptor, 141, 144, 428
Rap GTPase, 271, 275
RBC, 249, 267
reactive nitrogen species, 182, 199, 211
reactive oxygen species, 63, 171, 172, 182, 199, 211, 225, 276, 277
reciprocal dysautonomia, 68
regulatory T cell, 430, 433
renal artery, 50, 308
renal glomerulus, 340
renin, 298, 342, 345
renin–angiotensin axis, 324, 327, 347
renin-angiotensin-aldosterone axis, 344
residual volume, 375, 377
resistance, 133, 152, 160, 245, 250, 251, 258–260, 291, 295, 343
respiratory bronchiole, 73, 101, 108
respiratory exchange ratio, 394
respiratory sinus arrhythmia, 319, 329, 379
reticular nuclei, 60, 146
retrotrapezoid nucleus, 383, 419
rhinomanometry, 475
rhinostereometry, 475
RhoGEF, 270, 271
Rho GTPase, 266, 270, 273, 275, 277
rib, 88
right atrium, 9, 36
right auricle, 9
right ventricle, 11, 36
rigidity modulus, 387
Robo, 422
RoCK kinase, 3, 266, 270, 275
rostral ventrolateral medulla, 312
rostral ventral respiratory group, 419
rostral ventrolateral medulla, 154, 314, 316
Roughton–Forster equation, 405
ryanodine receptor, 176, 185, 230, 231, 235, 244

S

saphenous sinus, 56
sarcalumenin, 244
sarcomere, 227
scavenger receptor, 188
SCFA GPCR, 298
Schrödinger equation, 465
scintigraphy, 452
selectin, 436
selectivity coefficient, 278
semaphorin, 317
SERCA pump, 185
serotonin, 144, 153, 254, 438
serum response factor (SRF), 198
shear, 3, 126
shear rate, 3
Sieder–Tate equation, 291
sigmoid sinus, 54
single photon emission CT, 453
sinoatrial node, 9, 16, 236
size ratio, 109
SLC transporter, 219, 221
slingshot, 198
slowly adapting stretch receptor, 141, 142, 428
smoking, 141, 217
smooth myocyte, 2
solubility, 359, 402, 414
solvent drag coefficient, 262, 281
somatostatin, 151, 423
Sox transcription factor, 31
specific heat capacity, 373
speckle tracking echography, 451
sphingosine 1-phosphate, 271, 301
sphingosine kinase, 301
spinal cord, 21, 314, 334, 418
spinal nerve, 426
spine, 88
spirometry, 83, 375, 476, 478, 483
Sprouty, 118, 123
Src kinase, 207, 209, 271, 273, 275, 278, 301, 302
Starling law, 280
Stefan–Maxwell equation, 358, 397
stem cell, 33
sternum, 88
steroid, 347
sterol synthesis, 220
STK3 kinase, 440
STK4 kinase, 440
Stokes–Einstein equation, 358
Stokes equation, 397
straight sinus, 53
stratification, 400
stretch, 3, 126
stretch receptor, 428
stroke, 288
stroke volume, 159, 165, 287, 291, 321
stroke work, 166

subclavian artery, 48, 94
subclavian vein, 54
subendocardial viability ratio, 246
subfornical organ, 347
substance P, 151, 153
succinate, 232
superior petrosal sinus, 54
superior sagittal sinus, 53
superoxide, 199, 203
superoxide dismutase, 213, 429
superoxide flash, 183
surface/volume ratio, 110
surface/volume ratio, 101, 109
surface tension, 381
surfactant, 439
sympathetic, 20, 58, 65, 100, 141, 152, 227, 254, 259, 291, 312, 313, 324, 329, 331, 345, 347, 426
sympatheticoadrenal axis, 352
sympatho–adrenomedullary axis, 333
sympathoexcitatory reflex, 314
sympathovagal imbalance, 69
symporter, 199, 207
syndecan, 267
systemic vascular resistance, 185, 246, 259, 313, 324, 345
systole, 161, 166, 168, 252
systolic ejection, 161
systolic pressure time interval, 246

T

T-box factor (TBx), 33, 35
tachycardia, 185
tachykinin receptor, 422, 423
tachypnea, 366
telethonin, 179
terminal bronchiole, 73, 102, 108
terminal crest, 9
terminal sulcus, 9
TGF, 123, 125, 433, 435
Thebesian ridge, 9, 10, 16
thermoreceptor, 139
thoracic cage, 87
thoracic duct, 48, 91, 94
thoracic nerve, 426
thromboxane, 307, 308
thymosin, 34
thyroid hormone, 320
TIAM GEF, 271
tibial vein, 57
tidal volume, 371, 379, 383
tight junction, 265, 277
time to reflection, 246
tissue-Doppler echography, 451
Toll-like receptor, 171, 431
tonometry, 464
tonotropy, 313
torque, 3
total lung capacity, 102, 377
trabeculae carnae, 7
trachea, 75, 84, 91, 92, 102, 383
tracheobronchial tree, 73, 75, 102, 367, 383, 400
transalveolar pressure, 367
transbronchial pressure, 383
transfer capacity, 368, 401, 403, 404
transitional bronchiole, 101
translocase (TOM/TIM), 209
transpulmonary pressure, 382, 384
transthoracic pressure, 382
transthoracopulmonary pressure, 383, 384
transverse sinus, 54
TREK1 channel, 300
triadin, 180
triangle of Koch, 10
tricarboxylic acid cycle, 172, 182, 187, 196, 218, 220, 224, 228, 353
tricuspid valve, 6, 10
triglyceride, 189, 219, 222
TRP channel, 273, 299, 307, 308, 326
trunk-to-branch diameter ratio, 110
tumoral cell, 228
turbinate, 81
T tubule, 172

U

ultrasonography, 448
ultrasound elastography, 451
uncoupling protein, 181, 192
uniporter, 200, 207
universal gas constant, 373
upper airways, 73, 75
urodilatin, 350

V

vagus nerve, 20, 58, 60, 61, 63, 68, 100, 141, 379, 425, 426, 428
Valsalva sinus, 7, 13
variability, 35
vascular permeability, 277
vascular smooth myocyte, 149, 298, 299
vasoactive intestinal peptide, 150, 351
vasoconstriction, 255, 258, 269, 295, 316, 324
vasodilation, 185, 255, 259, 269, 295, 316, 347, 366

vasomotor center, 60, 152, 313
vasopressin, 153, 286, 324, 337, 345, 347
Vav GEF, 271
VCAM, 432
VDAC channel/porin, 181, 209
VEGF, 272, 278, 327
VEGF receptor, 32, 273
vein, 37, 57
vena cava, 9, 16, 36
venous return, 166, 247
ventilation/perfusion ratio, 113, 367, 394
ventral respiratory group, 417, 419
ventricle, 5
ventricular filling, 161
vesicular glutamate transporter, 423
Vieussens valve, 9
vinculin, 301
vital capacity, 376
VLDL, 188
voloreceptor, 146, 152, 314, 332
volume-to-body surface area ratio, 11
volume fraction, 356

W

wall remodeling, 325
wall shear stress, 4
wall thickness/lumen caliber ratio, 41
Willis circle, 39
Willis circle, 39
WNK kinase, 345
Wnt, 32
WT1 transcription factor, 32

X

xanthine oxidase, 183

MIX
Papier aus verantwortungsvollen Quellen
Paper from responsible sources
FSC® C105338

If you have any concerns about our products,
you can contact us on
ProductSafety@springernature.com

In case Publisher is established outside the EU,
the EU authorized representative is:
Springer Nature Customer Service Center GmbH
Europaplatz 3, 69115 Heidelberg, Germany

Printed by Libri Plureos GmbH
in Hamburg, Germany